Tests of hypotheses

Hypotheses	Rejection Region of H_0
$H_0: \mu = \mu_0$ $H_1: \mu > \mu_0$	$\dfrac{\bar{x} - \mu_0}{\sigma/\sqrt{n}} \geq z(\alpha)$
$H_0: \mu = \mu_0$ $H_1: \mu > \mu_0$	$\dfrac{\bar{x} - \mu_0}{s/\sqrt{n}} \geq t(\alpha; n - 1)$
$H_0: \mu_1 = \mu_2$ $H_1: \mu_1 > \mu_2$	$\dfrac{\bar{x} - \bar{y}}{\sqrt{\sigma_1^2/n_1 + \sigma_2^2/n_2}} \geq z(\alpha)$
$H_0: \mu_1 = \mu_2$ $H_1: \mu_1 > \mu_2$	$\dfrac{\bar{x} - \bar{y}}{\sqrt{\dfrac{(n_1 - 1)s_x^2 + (n_2 - 1)s_y^2}{n_1 + n_2 - 2}\left(\dfrac{1}{n_1} + \dfrac{1}{n_2}\right)}} \geq t(\alpha; n_1 + n_2 - 2)$
$H_0: p = p_0$ $H_1: p > p_0$	$\dfrac{y/n - p_0}{\sqrt{\dfrac{p_0(1 - p_0)}{n}}} \geq z(\alpha)$
$H_0: p_1 = p_2$ $H_1: p_1 > p_2$	$\dfrac{y_1/n_1 - y_2/n_2}{\sqrt{\left(\dfrac{y_1 + y_2}{n_1 + n_2}\right)\left(1 - \dfrac{y_1 + y_2}{n_1 + n_2}\right)\left(\dfrac{1}{n_1} + \dfrac{1}{n_2}\right)}} \geq z(\alpha)$
$H_0: \beta_1 = \beta_{10}$ $H_1: \beta_1 > \beta_{10}$	$\dfrac{\hat{\beta}_1 - \beta_{10}}{\sqrt{\dfrac{\sum (y_i - \hat{y}_i)^2/(n - 2)}{\sum (x_i - \bar{x})^2}}} \geq t(\alpha; n - 2)$

Tolerance regions

With normal assumptions, $\bar{x} \pm ks$ covers at least $(100)p$ percent of the population with probability $1 - \alpha$.

APPLIED STATISTICS FOR ENGINEERS AND PHYSICAL SCIENTISTS

JOHANNES LEDOLTER
ROBERT V. HOGG

University of Iowa

Third Edition

Prentice Hall
is an imprint of

Upper Saddle River, NJ 07458

Library of Congress Cataloging-in-Publication Data

Ledolter, Johannes.
  Applied statistics for engineers and physical scientists / Johannes Ledolter, Robert V. Hogg. — 3rd ed.
    p. cm.
  Order of authors reversed on 2nd ed.
  ISBN-13: 978-0-13-601798-1
  ISBN-10:   0-13-601798-3
  1. Engineering — Statistical methods. I. Hogg, Robert V. II. Hogg, Robert V. Applied statistics for
engineers and physical scientists. III. Title.
  TA340.H64 2010
  620.0072'7 — dc22   2008029904

Editor-in-Chief: *Deirdre Lynch*
Senior Project Editor: *Chere Bemelmans*
Assistant Editor: *Christina Lepre*
Editorial Assistant: *Dana Jones*
Project Manager: *Robert S. Merenoff*
Associate Managing Editor: *Bayani Mendoza De Leon*
Senior Managing Editor: *Linda Mihatov Behrens*
Senior Operations Supervisor: *Diane Peirano*
Marketing Assistant: *Kathleen DeChavez*
Senior Art Director: *Jayne Conte*
Cover Designer: *Bruce Kenselaar*
AV Project Manager: *Thomas Benfatti*
Compositor: *TexTech International Pvt. Ltd*
Art Studio: *Laserwords*

Prentice Hall
is an imprint of

PEARSON

© 2010 by Pearson Education, Inc.
Pearson Prentice Hall
Pearson Education, Inc.
Upper Saddle River, NJ 07458

Printed in the United States of America

10 9 8 7 6 5 4 3 2 1

ISBN-13: 978-0-13-601798-1
ISBN-10:   0-13-601798-3

Pearson Education Ltd., *London*
Pearson Education Singapore, Pte. Ltd
Pearson Education Canada, Inc.
Pearson Education–Japan
Pearson Education Australia PTY, Limited
Pearson Education North Asia, Ltd., *Hong Kong*
Pearson Educación de Mexico, S.A. de C.V.
Pearson Education Malaysia, Pte. Ltd.
Pearson Education Upper Saddle River, New Jersey

To our spouses, Lea Vandervelde and Ann Hogg

CONTENTS

2 PROBABILITY MODELS AND DISCRETE DISTRIBUTIONS 87

3 CONTINUOUS PROBABILITY MODELS 153

4 STATISTICAL INFERENCE: SAMPLING DISTRIBUTIONS, CONFIDENCE INTERVALS, AND TESTS OF HYPOTHESES   217

5   STATISTICAL PROCESS CONTROL **293**

PREFACE

This is the third edition of our text on statistics for engineers and scientists. Our work started with the book *Engineering Statistics,* published by Macmillan in 1987. The second edition, *Applied Statistics for Engineers and Physical Scientists,* was published in 1992. Much has happened in engineering statistics since then, and we have updated the manuscript accordingly. What has not changed during the last 17 years, however, is the need for a text that covers both basic statistical theory and interesting applications. We believe that we have filled this gap.

Our audience comprises engineers and scientists at the undergraduate and graduate levels who want an introduction to applied statistics. We expect our students to have a calculus background. Our book is practical and illustrates how engineers and scientists use statistics to solve many of their problems. But the book also discusses the needed underlying theoretical concepts so that readers can build on the introductory material. One needs a strong foundation for lifelong learning, and a "cookbook" approach to statistics alone would not be sufficient. Our book is also well suited for self-study by engineers, scientists, quality-control professionals, management consultants, and other practitioners.

In teaching statistical methods, we have found that students learn best if they see the relevance of the material, learn clearly how to apply the tools, and are shown the underlying statistical concepts. People learn statistics by solving exercises, analyzing real data sets, and working on larger scale projects that they design and carry out. In every chapter, each section includes many exercises ranging from straightforward "drill-type" problems to more challenging problems that test tools and concepts. The projects at the end of the chapters are an important and unique part of the book. They apply the methods studied in the text to the solution of real-world problems. They go beyond simple exercises and are designed to challenge students. Several projects describe how statistical studies were carried out and give readers the opportunity to analyze and interpret the results; others are written so that students can develop their own studies and compare their approaches with what was actually done.

We believe that the projects are best assigned to groups or teams of students. Written term papers on one or the other of these projects and subsequent oral presentations of the results can also be used for student evaluation. We hope that you will find the projects useful. Add your own projects if you have some that work better—and let us know what works and what hasn't.

Each chapter ends with a section of notes titled "Additional Remarks." Including these sections allowed us to elaborate on basic concepts by sharing interesting historical comments. The remarks are meant to be informative as well as fun.

The development of excellent statistical computer software has made analyzing data as well as designing experiments easy. Much of the discussion in this book is tied to Minitab, although other packages, such as SAS, SPSS, and JMP, can also be used. These packages are spreadsheet based, with simple and intuitive pull-down menus for carrying out data analysis, and they can be used interchangeably in the course. We encourage students to try several programs; all it usually takes is to "copy and paste" the information into program-specific spreadsheets and work with similar pull-down menus. The R software environment for statistical computing and graphics provides another excellent, very general, and inexpensive (free!) platform for computation. It is somewhat different from the spreadsheet-based packages, as it requires the user to execute R-specific language instructions. But the language is easy to learn, and this should not cause difficulties for engineering students who are already used to writing programs in Matlab. Instructions on how to use Minitab and R are included in the instructor's manual and are available online. Files of all data sets used in the text and the exercises can be downloaded from the Prentice-Hall resource website and from J. Ledolter's website www.biz.uiowa.edu/faculty/jledolter/AppliedStatistics. Files are given in both Minitab and Text formats.

Answers to approximately one-third of the exercises are listed in Appendix B of the text. These answers help students check their work. We have tried to include at least one answer from each type of problem. A detailed instructor's manual with solutions to all exercises and most projects is also available.

The entire book, including the projects, can be used for a strong two-semester sequence in applied statistics. At the end of this sequence, students will be familiar with tools for summarizing and interpreting data, basic probability and probability distributions, sampling distributions, and basic statistical inference. In addition, they will know how to design experiments and analyze the resulting data and carry out regressions. Further, students will be familiar with statistical process control, including control charts, capability assessment, and sample inspection. Students will gain familiarity with a wide range of useful statistical tools and will understand why these tools work.

The book can also be used for one-semester courses, depending on the interests of the students and their instructor. Chapters 1 through 5 provide the basis for a solid introductory statistics course, which includes statistical quality control, but not the design of experiments and regression, except for the introduction to these topics in Chapter 1. More of design of experiments or regression can be covered if the following sections of Chapters 1 through 5 are omitted:

Chapter 2: Sections 2.4 and 2.7-1.

Chapter 3: Sections 3.3, 3.5, 3.6-1, and 3.7.

Chapter 4: Sections 4.4, 4.6-3, 4.6-4, and 4.7.

Chapter 5: Section 5.3.

Omitting these sections would free up time for an expanded coverage of the design of experiments (Chapters 6 and 7) and regression (Sections 8.1 through 8.3). In the later chapters, the following sections can be omitted:

Chapter 6: Sections 6.3 and Section 6.4.

Chapter 7: Section 7.2 and some of the more difficult material (on fractional factorials) in Section 7.4.

Chapter 8: Details on regression modeling in Sections 8.1 through 8.5, and Section 8.6.

We are indebted to many students and colleagues who have made numerous valuable suggestions and who have shared data and exercises with us. We would like to thank all the students who took our classes at the University of Iowa. We treasure the interactions we have had with our students and value all we have gained from them. We are also grateful to the participants at two conferences on statistical education held in Iowa City in the late 1980s and early 1990s that helped sharpen our focus. We gratefully acknowledge the help of three very special friends in industry who made unusual contributions to the first edition of this book: Ron Snee (formally with DuPont, but now a principal partner at Tunnell Consulting), Gerry Hahn (formally with General Electric and now a statistical consultant), and Bert Gunter (a statistical consultant at Genentech). People who greatly influenced our thinking and learning include George Box, Norman Draper, and the late Bill Hunter, all of whom taught courses on engineering statistics and statistical modeling when one of the authors (J.L.) was a graduate student at the University of Wisconsin–Madison. The very lively "Monday night beer seminars" in George Box's basement also had a profound impact, imparting the importance of well-designed studies for learning and providing a strategy for implementing statistical methods in real-world settings.

We would like to thank Milo Koretsky (Oregon State University), Melinda McCann (Oklahoma State University), James Simpson (Florida A&M University and the Florida State University), Kristofer Jennings (Purdue University), Robert Mnatsakanov (West Virginia University), Jayant Rajgopal (University of Pittsburgh), and Syed Kirmani (Northern Iowa University) for their helpful feedback on the third edition. We appreciate the comments of our editor, Deirdre Lynch, at Pearson Education, who gave us valuable feedback at the times we needed it most and who helped guide this project to its finish.

We thank Rob Merenoff at Pearson Education for his expert oversight of the production process and TexTech International for the careful typesetting of our manuscript.

We could not have completed this book without the encouragement of our families and closest friends. We will always be thankful for the patience and support we received from those closest to us.

Throughout the book we have tried to convey our passion for the subject and to share with readers our strongly-felt beliefs in the power of statistical methods and their practical value. We will have succeeded if our book encourages you to apply these tools to your problems. Please send us your feedback (johannes-ledolter@uiowa.edu).

Johanness Ledolter
Robert V. Hogg

1

COLLECTION AND ANALYSIS
OF INFORMATION

1.1 Introduction

Statistics deals with the *collection* and *analysis of data* to solve *real-world problems*. Statistics teaches us how to obtain relevant information in an efficient manner and how to best analyze the resulting data. Of course, all engineers and scientists engage in such activities, so all engineers and scientists deal with statistics. What makes the discipline of statistics particularly useful is that it can teach us how to solve problems in the presence of *uncertainty*. Statistical methods are designed to deal with variability; they help us make inferences about underlying processes on the basis of often imperfect and incomplete information.

1.1-1 Data Collection

Informative data are essential for solving problems successfully. The saying "garbage in, garbage out" applies to statistical analyses. If the data have not been collected appropriately and are not of high quality, even the best statistician will not be able to make much sense out of them.

Not all data are alike, and it is important to discuss the various ways in which data can be collected. It is useful to distinguish between *data collected in designed experiments* and *data arising from observational studies*.

Many data sets in engineering and the physical sciences are the result of *designed studies* that usually are carefully planned and executed. For example, an engineer studying the impact of pressure and temperature on the yield of a production process may manufacture several products under varying levels of pressure and temperature. He or she may select three different pressures and four different settings for temperature, and conduct one or several experiments at each of the $(3)(4) = 12$ different combinations of pressure and temperature. The experimenter will suspect that other factors have an impact on the yield, but initially may not know which ones. It could be the purity of the raw materials, environmental conditions in the plant during the manufacture of the product, and so on. To spread the effects of these uncontrolled factors evenly among all those studied, the investigator will probably want to randomize the times at which he or she performs the experimental runs. That way, the effects of unknown time trends will be minimized. One certainly would not want to run all experiments with the lowest temperature on one day and all with the highest temperature on another. If the process is sensitive to daily fluctuations in plant conditions, an observed difference in the results from the two days of experimentation may be due not to temperature, but to the different conditions in the plant. Also, a good experimenter will be careful when changing the two factors of interest, keeping other factors as uniform as possible. Moreover, the experimenter may carry out a number of replications of the experiment because averages of several responses taken at the same experimental conditions will vary less than an individual response. What is important is that the experimenter be actively involved in all aspects of obtaining the data, following certain well-established principles of good experimentation. These principles are reviewed in Section 1.6, and discussed more fully in Chapters 6 and 7 on the statistical design of experiments.

Observational data are somewhat different, as the investigator has little opportunity to affect the process that generates the data. Instead, the data are taken directly from that process. Observational data are often referred to as "happenstance" data, as they just happen to be available for analysis. A further problem with many observational data sets from economics and the social sciences is that they are "secondhand": They were collected by others who may have had different purposes in mind. The data may have been collected through a *census* (i.e., every single event is recorded) or through a *survey* (i.e., only certain portions of the events are recorded). Several things may have gone wrong during the data-gathering process, and the analyst who uses the data has no chance to correct these problems. A survey may not be representative of the population that one wants to study. Data definitions may not exactly match the factors that one wants to measure, and the data that were gathered may be poor proxies at best. Also, the quality of the data may be poor, because there may not have been enough time for careful data collection and processing. In addition, some data may be missing, and the data at hand may not be "rich" enough to separate the effects of competing factors and theories.

Consider the following illustration: Suppose that you want to explain success in college, as measured by the students' grade point average (GPA). Your admissions office provides the students' American College Test (ACT) or Scholastic Aptitude Test (SAT) scores on tests taken prior to admission, and you have survey data on the number of hours students studied per week. Does this information allow you to develop a good model for predicting success in college? Yes, to a certain degree. However, there are also significant problems. First, the GPA is quite a narrow definition of success in college. GPA figures are readily available, but one needs to discuss whether they constitute the information one really wants. Second, the range of ACT scores may not be wide enough to find a big impact of ACT scores on college GPA. Most good universities do not accept marginal students with low ACT scores. As a consequence, the range of ACT scores at your institution will be somewhat narrow, and over such a limited range, the impact of ACT score on GPA will be weak. Third, the study hours are self-reported, and students may have a tendency to make themselves look better or worse than they really are. Fourth, ACT scores and study hours tend to be related. Students with high ACT scores tend to have good study skills, and it is rare to find someone with a very high ACT score who does not study. Thus, the relationship between the two explanatory variables, ACT score and study hours, makes it difficult to separate their effects on college GPA.

If you are able to conduct your own *survey,* then you have a better chance to affect the quality of the data collected. You can make sure that the questions are clear and unambiguous. In addition, you can administer the survey in a way that leads to acceptable response rates. You can decide whether you want to distribute the survey personally or by mail or telephone, and you may offer certain incentives for people to complete the survey. Also, you can ensure that the results of your sample are representative of the population you want to study. The concept of a *random sample*—which gives each possible sample the same chance of being selected—is central to many statistical techniques and will be discussed in detail in a later section.

1.1-2 Types of Data

Measurements on dimensions such as the length of an egg, the weight of a quarter-pound hamburger, the length of a steel flat, and the tensile strength of a steel billet are what we call *continuous measurements*—measurements that can take on any value within certain intervals. For example, the weight of a quarter-pound hamburger patty probably is within 0.20 and 0.30 pound; we certainly would hope for a value that is close to the target, 0.25 pound. Due to rounding, we may ignore differences beyond the third decimal digit; that is, we would record 0.254 for any measurement within 0.0005 of 0.254. The weight of the contents of a 16-ounce cereal box is another example of a continuous measurement; the weight could be any value within a certain interval (say, between 15 and 17 ounces).

Categorical measurements represent the other commonly encountered type of data. For example, products may be divided into defective and good products, and we may code this information as 1 (defective) and 0 (good). Similarly, we may describe the quality of a steel surface as "smooth," "mildly scratched," or "rough."

A questionnaire of first-time car buyers measures customer satisfaction on (say) a five-point scale, from 1, representing extreme dissatisfaction, to 5, representing total satisfaction. We call the resulting data *ordered categorical,* because there is an order to the different categories. Similarly, we might collect data on family income from a survey question like the following: "1" represents income below (or equal to) $40,000, "2" income between $40,000 and $60,000 (inclusive), "3" income between $60,000 and $80,000, "4" income between $80,000 and $120,000, and "5" income exceeding $120,000. Again, income is an ordered categorical variable.

We may collect data on the type of engineer (1 for electrical, 2 for chemical, 3 for mechanical, 4 for civil/industrial, 5 for others) or data on ethnicity (1 for white, 2 for black, 3 for Hispanic, and 4 for others). These are examples of *unordered categorical* measurements, as there is no natural ordering among the categories. Clearly, we could have arranged the categories in other ways.

It is often the case that continuous measurements are taken, but that measurements are subsequently transformed into categorical data. For example, the target value for the gauge of a steel flat may be 0.25 inch, with lower and upper specification limits of 0.235 and 0.265 inch, respectively. A quality inspector checking the dimension of a certain steel flat records a gauge of 0.234. The steel flat is coded as defective, because its dimension is outside the specification limits. The item is then given a "1" for being defective. Keep in mind, however, that such an approach will throw away valuable information. A good item with gauge 0.236 (which is barely within the specification limits) is certainly worse than a good item that is right on the target value, 0.25. Similarly, a defective item (with gauge 0.234) is bad, but certainly better than a defective item with gauge 0.220. Categorizing continuous data leads to a loss of information. We don't want to be just within the specification limits; we want to be right on target! In quality applications, it is common to refer to the coded categorical data as *attribute data* (they characterize attributes of the items, good or bad); continuous data are also referred to as *measurement* data.

1.1-3 The Study of Variability

Variability (or dispersion) in measurements and processes is a fact of life. Virtually all processes vary. Take, for example, a few items from a production line, and measure a certain characteristic of these items. If your measurements have sufficient resolution, you will find that they vary. Or count the number of flaws on different bolts of fabric, and you will notice variability among the counts from bolt to bolt. Or measure the thickness of certain thin wafers, the diameter of knobs, the yields of chemical batch processes, or the percentage of defective items in successive lots of 10,000. In all these cases, your measurements will vary.

There are several reasons for this variability. It can come from the slightly different conditions under which each item is made, reflecting differences in raw materials, differences among machines or operators, differences among operating conditions caused by changes in such things as furnace temperature, humidity, production-floor temperature, and so on. Since this variability comes from the process, we refer to it as *process variability*.

Part of the variability among the recorded values is also *measurement variability*. For example, when you record the weight of an object, the measurement depends on the exact location of the item on the scale, whether the scale is on a level surface, how the scale is calibrated, and so on. When you measure the size of a very small particle, the measurement depends, of course, on the accuracy of the instrument, but also on the way you have set up the measurement and how you place the small object into the measurement apparatus. For example, in determining chemical substances in identical water samples, the results of a gas chromatography test do not always give the same concentrations. Similarly, the measured moisture content of a certain pigmented paste will not always be the same. Indeed, measurement itself can be thought of as a process and is subject to inputs and outputs, environmental change, and so on; hence, it is subject to variability. We refer to this variability as *measurement variability, analytical test variability, variability due to the measurement process,* or, simply, the *noise* that obscures the true signal of the process.

The simple example that follows illustrates these concepts and helps us to distinguish between process and measurement variability. Take a carton of 12 eggs from your refrigerator and measure the length of each egg. Even if you can determine the length of an egg exactly, you find variability among the 12 measurements. Differences among the 12 eggs arise because of process variability: The process of producing eggs is not totally uniform, because different hens are involved and the feed is not always the same.

In addition to process variability, measurement variability is present. Suppose that you measure the length of an egg by putting it "lengthwise" into a handheld gage. Because this is a rather crude way of measuring length, you find that repeated measurements on the same egg are not identical. Each recorded number contains a measurement error, as measurements depend on how the egg gets placed in the gage, how numbers are rounded, and so on. If you now look at the length measurements of your 12 eggs, you realize that the variation reflects a combination of process and measurement variability.

There is yet a third source of variability: *variability due to sampling*. In many instances, we do not measure every item, because that would be too time consuming and expensive. Moreover, in some cases it would be impossible to measure each item, as the measurement results in the destruction of the object. For example, the strength of a welding process is often determined by a pull test to destruction, and the lifetime of a light bulb is determined by measuring the time elapsed until it burns out. Instead of considering the set of all possible items (which, in statistics, is called the *population*), we select and measure a subset of the items (which, in statistics, is referred to as a *sample*). Consider, for example, the production of ballpoint pens produced in lots of 10,000. The number of defective pens will probably vary from lot to lot, due to minor changes in the manufacturing process. But let us suppose that there are exactly 200 defective pens in a certain lot of 10,000. Then, assuming that there are no measurement errors (i.e., errors that result from falsely classifying a perfectly good pen as defective or a defective pen as working), we could find that there are exactly 200 defective pens by testing *each* of the 10,000 pens. Obviously, this is not practical, so we take a sample. Let us say that we choose a sample of $n = 100$ pens from the total of $N = 10,000$ pens, and determine the sample fraction of defectives. Now, it could be that $0, 1, 2, \ldots, 50, \ldots$, or even 100 of the pens that we selected are defective. However, it would be highly unlikely that 50 percent, and even more unlikely that 100 percent, of the sampled items are defective if the entire lot contains only 2 percent defectives. We would expect to obtain about 2 percent in the sample also, but clearly there could be 3, or zero, or some other percent. It depends on the items that we have selected in our sample. This selection process, called *sampling,* introduces *sampling variability* into what we observe in the sample.

Say that we select the $n = 100$ items from the total of $N = 10,000$ items such that each subset of size 100 has the same "chance" of being selected; that is, each possible subset is "equally likely" to be selected. Then we can use the information from the sample to draw conclusions—that is, to make inferences—about the underlying population of those 10,000 pens. For example, if we find that one of the 100 pens selected is defective, our best "guess," or *estimate,* of the fraction of defectives in the population is 1 percent. Of course, there is uncertainty in this estimate due to sampling variability. But how much sampling variability is there? In this book, we show how "confidence bands" are developed around an estimate to express its uncertainty. These bands are so named because they reflect our "confidence" that the unknown fraction of defectives is within the specified bands. Similarly, we could use the information from the sample to confirm or refute certain theories, or hypotheses, about the fraction of defectives in the population. For example, if we find that 30 percent of the sample is defective, we probably would reject the claim that the process produces defectives at a rate of not more than 2 percent. But what if we find that 3 percent of the sampled pens are defective? Would that be enough evidence to reject the claim that the process allows for no more than 2 percent defectives? In this case, the evidence is not as clear. Obviously, we do not want to reject a lot as being defective if it actually is not defective; that would cost the company money. But we would not want to accept it either if the process has gone out of control and the population contains more than 2 percent defectives. Failure to reject a lot would lead to consumer complaints and loss of confidence in a product. To make a good decision, it is essential to study the sampling variability that we ordinarily can expect, and that topic is one of the areas that we consider in this book.

To make inferences from the sample to the population, we must select a sample that is *representative* of the population. Selecting the first 100 pens that come off the production line, or the last 100, or 100 from a particular machine may not lead to a representative sample because there may be special start-up problems or a particular machine may be more or less precise than others. Ideally, we should select a *random sample,* according to which each subset of size $n = 100$ has the same chance of being selected. Now, there are many possible ways of selecting n items from a population of N items; in fact, elementary algebra tells us that there are

$$\binom{N}{n} = \frac{N!}{n!(N - n)!}$$

ways, where "k factorial" is $k! = (k)(k - 1)(k - 2) \ldots (3)(2)(1)$, and $0! = 1$. For example, there are

$$\binom{4}{2} = \frac{4!}{2!2!} = \frac{4 \cdot 3 \cdot 2 \cdot 1}{2 \cdot 1 \cdot 2 \cdot 1} = 6$$

different ways of selecting two items from the four items a, b, c, and d, namely, (a, b), (a, c), (a, d), (b, c), (b, d), and (c, d). One way to obtain a random sample is to take six slips of paper, write the name of a different possible sample on each one, put the slips into a bowl, mix them, and select one slip at random. Say that it is the slip with (a, d). This means that we select items a and d for our random sample. We call the sample chosen in this way a *simple random sample,* as we give each possible subset of size $n = 2$ the same chance of being selected. Now, if N and also the sample size $n < N$ become large, this drawing mechanism becomes a very cumbersome procedure, since we would draw one of $\binom{N}{n}$ slips. Just take $N = 10{,}000$ and $n = 100$ as an illustration; there are many, many possibilities now. The easier, but entirely equivalent, approach of obtaining a random sample takes *repeated* samples of a *single* item from the N items until n items in all are selected. You could make up N slips of paper that identify the elements of the population, place them in a bowl, and select one of them—say, item 251. From the remaining $N - 1$ items, take another element—it could be number 9,010. Continue this procedure until you have selected $n = 100$ items. Since we do not return the sampled element to the population, we refer to this process as *sampling without replacement.*

Even though this procedure of obtaining a random sample is somewhat easier to carry out than the earlier method, in practice it is often impossible to take genuine random samples. For example, it is rare that we would pick 100 pens from each lot according to such an approach. We would probably pick those items "more or less" at random. However, it is important to realize the danger in "heuristic" or "quasi-random" sampling plans: We could end up with a sample that is not quite representative. Wanting to sample whenever it is convenient is understandable, but items that are conveniently sampled are not always representative.

Often, one must make practical trade-offs between taking a random sample and not interfering too much with the process. Also, there are instances in which a purely random sampling approach may in fact generate bad information. Take, for example, a process that involves the firing of ceramic elements in large ovens that can accommodate elements at many different positions. Suppose that every time we obtain a sample, we have to open the oven. However, opening the oven may change the temperature and affect the process. In a purely random sampling approach in which we continuously pick items from randomly selected positions, we may open the oven too often, destabilizing the process and, as a consequence, obtaining inferior information. It may be better to open the oven less frequently and sample not just from one position, but end up taking a few items at a time.

1.1-4 Distributions

There is another important concept that we want to introduce early on in this book, namely, that of a *distribution*. Later in the chapter, we consider the compressive strengths of a certain type of concrete block. We find that the compressive strength for this type of block varies from about 2,800 to 6,800 pounds per square inch (psi). The majority of the blocks, however, have strengths somewhere between 4,000 and 5,200. We find fewer blocks with compressive strengths between 2,800 and 4,000 psi and between 5,200 and 6,800 psi. In other words, observations "bunch up" toward the middle of the range from 2,800 to 6,800 psi. Statisticians search for curves like the one in Figure 1.1-1 to describe the variability (in other words, the distribution) among measurements. The figure depicts one possible distribution of the compressive strength of concrete blocks. On the horizontal axis, we record compressive strength x. On the vertical axis, we record the values of a nonnegative function $f(x)$. By selecting this curve carefully (we will discuss such functions in Chapter 3), we ensure that the percentage of the total area under the curve between two marks (representing numbers) on the x-axis reflects the percentage of blocks that have

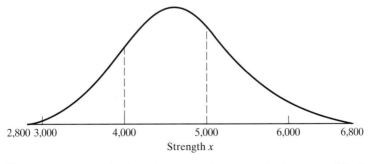

Figure 1.1-1 Distribution of compressive strength of concrete blocks

strengths between those two numbers. For illustration, roughly 50 percent of the blocks have strengths between 4,000 and 5,000 psi, only about 20 percent have strengths between 3,000 and 4,000 psi, and just a few percent have strengths larger than 6,000 or smaller than 3,000 psi. Here, these percentages have been estimated "by eye," but if the equation of the curve is known, the percentages can, of course, be determined by mathematical integration. Furthermore, the total area under such a curve must always be 100 percent, as all observations must be between 2,800 and 6,800 psi.

Such a curve tells us a great deal about the variation in the compressive strength of these concrete blocks. About which value do the observations vary? Here we might answer 4,600 psi, which is the value at which the curve assumes its highest point. We call 4,600 psi the *mode* of the distribution. How much variation is there? Of course, we see that the strengths vary between 2,800 and 6,800 psi. However, more importantly, the curve describes the percentage of the blocks that are between any two points. For example, about 60 percent are between 4,000 and 5,200 psi. Or, put another way, 60 percent of the observations are within 600 psi of the mode (4,600 psi).

Let us discuss another example to get a better intuitive understanding of variability. It happens quite often that whenever a rookie in a certain sport has had a particularly good first year, his or her performance slips during the second. How can we explain this phenomenon, commonly referred to as the "sophomore jinx"? We know that the "rookie of the year" in major league baseball does not win that title with an average or below-average year. His batting percentage that year is probably about where the *dot* is on the graph in Figure 1.1-2, which depicts the variability in his performance throughout his career. This implies that in the next year, assuming that there are no sudden changes (improvements or slumps) in his performance, he has only a 10 percent chance to improve on the first performance. Thus, his performance in the second year is likely to be poorer than that in the first, explaining the sophomore jinx. (Of course, our analysis ignores a learning effect.)

An illustration analogous to the sophomore jinx can be found in the movie industry. Let us ask movie fans how often they find sequels of hit movies worse than the originals. Although sequels often use the same casts and the same directors as in the original movies, frequently the answer to this question is "around 90 to 95 percent." The reason is that the performance in the original "hit" is usually far above the average of that combination of people, and as a consequence, it is extremely difficult to improve upon it. Thus, the sequel is frequently worse than the original.

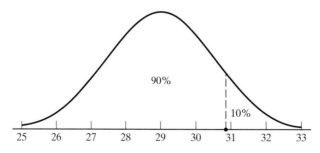

Figure 1.1-2 Theoretical distribution of yearly batting percentage of a rookie baseball player

1.1-5 Importance of Variability (or Lack Thereof) for Quality and Productivity Improvement

Managers and engineers must have a good understanding of variability, as it will help them improve their businesses and industries. But what do distributions and the sophomore jinx have to do with improving quality in business or industry? Actually, a lot! Baseball players are not the only ones who have their "ups and downs." So do processes.

Usually, it is the system—in other words, the process—that is responsible for the variability, and engineers can make major improvements to that. Suppose that, as in Figure 1.1-3(a), the variability of a process is not centered around the target value. Assume that in this example we are concerned about the bore size in a certain gear blank. Suppose that the target value is 10 centimeters, with lower and upper specification limits of 9.9 and 10.1 centimeters, respectively. Production outside the specification limits cannot be shipped and must be scrapped. Then our production process in Figure 1.1-3(a) is not very successful in meeting those requirements: The bore sizes of a relatively large proportion (about 15 percent) of our production are below the lower specification limit. We also notice that our production is off target. Changes in the process must be made to shift the distribution to the right. We notice that the production in Figure 1.1-3(b) is right on target and that most of the gear blanks produced satisfy the required specification limits.

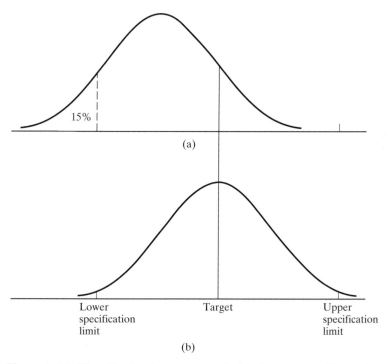

Figure 1.1-3 Two distributions compared with the target and lower/upper specification limits

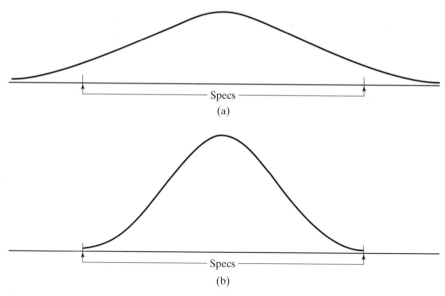

Figure 1.1-4 Two distributions compared with their specification limits

The example in Figure 1.1-3 illustrates how the shift of a distribution can bring a process close to a desired target. Such shifts require adjustments in levels, and these adjustments are often easy to carry out. However, they leave the extent of the variability unchanged. So, making such adjustments and shifting the distribution to the right or left does not always solve the problem.

Let us now consider a second example, shown in Figure 1.1-4(a). Here, the percentage of items outside the specification limits is much too large. Adjustments to reduce the variability must be made to bring most, if not all, items within specifications. But how should we make the appropriate adjustments? For that, you need to be a good detective. The tools you learn in this course will improve your chances of finding the root causes of the undesirable variability. By understanding the principles of good experimentation (i.e., good data collection) and by knowing how to summarize and interpret the information collected (i.e., good data analysis), you can determine whether certain factors matter. Figure 1.1-4(b) shows the results after a successful adjustment. The variability is reduced and the items are within the specification limits. Changes that decrease the variability of a process are often more expensive than those needed to change the level of the process.

We also want to point out that producing zero defects is often not good enough. Consider the two distributions in Figure 1.1-5. In either case, all items are within the specification limits, and yet in case (b), a larger percentage of items is closer to the target T. Manufacturers have learned that it is often not sufficient just to produce within specifications, and they know that each deviation from the target decreases the quality of the product. Efforts are directed toward a continual reduction of the variability and toward coming closer and closer to the target. The parts of the final product will simply fit better if each component is closer to its target value.

There is another danger about which we should warn you. Suppose the system is such that on a particular day the dimension of a manufactured part is on the high

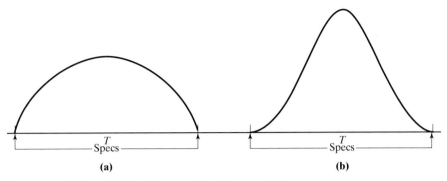

Figure 1.1-5 Two distributions resulting in zero defects

side, but is still consistent with the variability that is normally associated with that dimension. Nevertheless, a change is made to reduce the dimension. Having made this adjustment, we notice that on the next day the dimension is on the low side; so the following day, another adjustment is made. This is done each day, always hoping to bring the product closer to the target value. After two weeks of daily adjustments, suppose it is found that the variation in the output is worse than ever. What went wrong? Most likely, the original process was already on target and the observed fluctuations around the target value were part of the natural variation of the process. Our daily adjustments introduced additional variation and made matters worse. That is, in addition to the natural variation of the process around its target value, more variation was introduced by these daily adjustments. This is called "tampering with the system." Unless we see clearly how the process is to be improved, we should not tamper with it.

When we argue against tampering with the process, we are not against making changes for improvement. We, as statisticians, know that we must experiment in our search to find good operating conditions for our processes. However, our experimentation must be done in a systematic manner, a topic we will discuss when we consider the design of experiments.

Exercises 1.1

*1.1-1 Give several illustrative examples of studies that collect data through (a) a census, (b) a survey, and (c) a designed experiment.

*1.1-2 Give several illustrative examples of categorical variables (with ordered as well as unordered categories). Give examples of measurement variables.

1.1-3 Explain the difference between observational data and data that originate from designed experiments. Which type of information would you prefer, and why?

1.1-4 Explain "population" and "sample." What do we mean by a random sample, and why is random sampling so important? How many samples are possible if your population consists of 10 distinct elements and if you want to sample 3 of them? Discuss a strategy for obtaining a random sample of 3 elements.

*1.1-5 A plant produces widgets according to a certain specified dimension. You measure the dimensions of several widgets and you notice variation. Discuss possible sources of that variation. How would you display (i.e., communicate) this variability?

1.1-6   Discuss why 100 percent inspection of your production may not keep you in business. Discuss whether it is possible to "inspect quality into" a product. (See Section 1.2-2 for a detailed discussion.)

1.1-7   Use the random-number functions of your computer software. Select the elements of a random sample of size $n = 5$ from a population of $N = 300$ distinct elements, numbered 1 through 300.

Hint: Many books list tables of random digits, and not too long ago (1955) a book with a table with a million random digits (between 0 and 9) was published by the Rand Corporation. A part of this table (2,000 digits) is given as Table C.10 in Appendix C. With today's computer software, however, such tables are no longer needed. In Minitab, for example, you create a worksheet column C1 containing the elements of the population (here, the numbers $1, 2, \ldots, 300$). The command "Calc > Random Data > Sample from Columns" selects $n = 5$ numbers from among the numbers in column C1 at random and stores the selected elements in another column of the worksheet; here, you use sampling without replacement as you do not want to pick the same item more than once.

You can also achieve the sample selection with the numbers in Table C.10. Pick a random entry in this table—say, row 10 and column 3—and select nonoverlapping triples by reading across (or down, or up). The sequence 074, 941, 388, 763, 791, 976, 355, 840, 440, 110, 518, . . . then tells you to select items 74, 110, ... You ignore numbers that are outside the range 1 through 300 and that have been selected once before. The reason for considering nonoverlapping triples is that you need three placeholders to represent all elements of the population. For a population with $N = 76,000$ elements you work with strings of five numbers.

*1.1-8   The total workforce of your company (its population) consists of $N = 400$ employees: 100 managers and 300 hourly workers. You would like to determine (i.e., estimate) the population's average donation to charitable organizations. You base your estimate on the results of a sample of size $n = 4$. You draw this sample by using the following procedure: You select (at random) 1 manager from the 100 managers and 3 hourly workers from the 300 hourly workers. You ask the individuals selected about their donations, and you use the sample mean as an estimate of the average donation in the population.

*(a) Does this sampling procedure lead to a (simple) random sample from the population of 400 employees? Why or why not?

Hint: A random sample gives each possible sample the same chance of being selected. Is this the case here?

*(b) What is good about this particular sampling procedure?

Hint: Think about samples that are being ruled out by this stratified sampling approach and why that may be a good thing.

1.2 Measurements Collected over Time

1.2-1 Time-Sequence Plots

Often, measurements are collected sequentially, and time is an important factor that affects the measurements and contributes to their variation. In a time-sequence display, we plot the measurements against time showing the order in which measurements

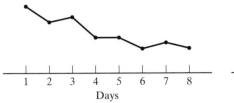 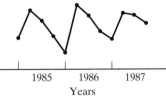

Figure 1.2-1 Two time sequences

are collected. To emphasize the time-series nature of the observations, it is common to display them on a graph as dots and to connect successive observations. Time-sequence plots can tell us about unusual observations, trends or runs in the data, cycles, periods of increased variability, and unusual time patterns, such as low measurements followed by high ones and high measurements followed by low ones.

The two plots in Figure 1.2-1 illustrate such trends and cycles. The first one shows trends that are perhaps due to tool wear; the second illustrates the cyclic demand for a seasonal product.

Trends and cycles in measurements are fairly easy to recognize, as long as we plot the measurements in a time sequence. Recognizing trends is important, because it shows that time is an important factor. A display of variation without reference to time (such as the curve in Figure 1.1-1) combines the variation that is due to time with the variation arising from factors affecting the process at each given point in time. The resulting variation may appear quite large. Additional information is provided in this case by the time-sequence plot, as it isolates the part of the variation that can be attributed to the time variable.

Let us consider a few examples that illustrate the importance of time-sequence plots. Project 1 at the end of this chapter describes a metal-cutting operation on a lathe. The objective of this application is to cut holes of a specified dimension into gear blanks that go into the engines of large tractors. The data in this exercise are successive deviations from the target bore size. In Figure 1.2-2, we plot the results of the first three experiments, keeping the cutting tool in a fixed position. The three data sets come from three different cutting tools.

The time-sequence plots from all three experiments show decreasing levels. Except for the last segment of the observations in experiment 3, the trends are roughly linear; also, the coefficients of the slope are fairly similar. Of course, one would not expect to get exactly the same slope in all three cases, as there may be variability among the different cutting tools.

Figure 1.2-3 is a time-sequence plot of annual Northern Hemisphere temperatures from 1600 to 1979. These data are taken from a paper by Moberg et al. published in *Nature* [Vol. 433, No. 7026, pp. 613–617, February 10, 2005] that reconstructs almost 2,000 years of Northern Hemisphere temperatures from low- and high-resolution proxy data. The data represent temperature differences from the 1961–1990 average of the Northern Hemisphere annual mean temperatures. The plot shows large multi-centennial variability, A.D. 1600 minimum temperatures that are about 0.7 degree below the average of 1961–90, and a steady rise in temperatures starting with the Industrial Revolution in the early 19th century.

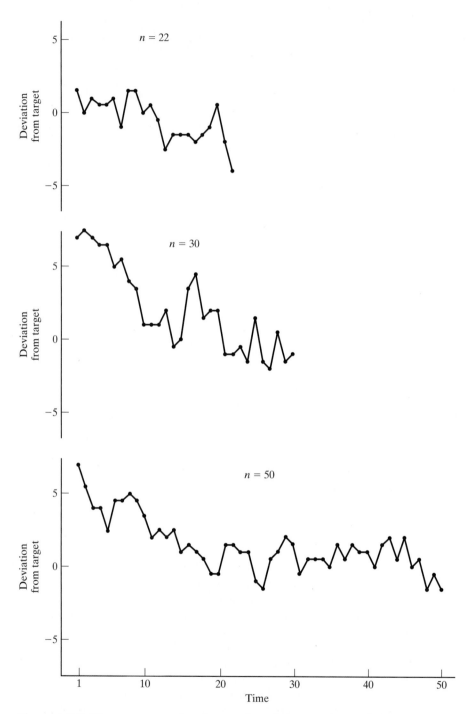

Figure 1.2-2 Time-sequence plots for metal-cutting example. Deviations from target are expressed as inches $\times 10^4$

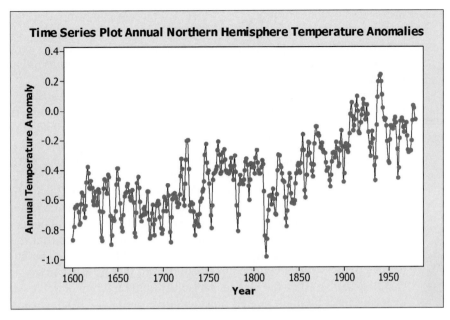

Figure 1.2-3 Time sequence plot of annual northern hemisphere temperature anomalies (1600 to 1997). This graph was created with Minitab

1.2-2 Control Charts: A Special Case of Time-Sequence Plots

We have emphasized the fact that processes vary. Too much variation in industrial processes, however, affects the quality of the product and, as a consequence, the productivity of a company. If a large proportion of manufactured items is outside the specification limits, and if products are not inspected before they are shipped to consumers, the producer will receive many complaints from unhappy customers and, if a warranty exists, an excessive number of warranty claims. Ultimately, customers will lose confidence in the product and will select a different and more reliable supplier.

However, a rigorous inspection program cannot inspect quality into a product. Suppose, for example, that products are 100 percent inspected and that products outside the specification limits are in fact identified. Because unacceptable products have to be scrapped, reworked, or sold as seconds, the company faces additional costs. If the producer wants to maintain profits, these costs have to be passed on to the consumer; thus, prices have to be raised. As a consequence, consumers will look for suppliers who succeed in reducing the variability of their products and can keep their processes within tight specification limits, because such suppliers can offer their products at a lower price. This simple illustration shows that the competitive position of a company depends to a large extent on its success in controlling variability in the production process. The upshot is that it is very important to keep a close watch on the stability of a process and to find ways of reducing process variability.

There exist very simple statistical techniques that can be used to *monitor* the variability of a process; they require taking periodic measurements on the process and displaying the results in a prominent place. This effectively communicates the most current state of the process. More will be said about monitoring process variability in later sections of this book, and in Chapter 5 we give an introduction to control charts. We think that it is important to introduce control charts in an introductory book on engineering statistics, as these charts represent an important, successful application of statistical methods.

Control charts are basically time-sequence plots. Let us illustrate the basic ideas behind control charts with two simple examples. Let us first assume that our plant produces ballpoint pens in lots of 10,000. Now, an easy check of whether the quality of our production has changed is to take representative samples of size 100 from each lot, determine the fraction of defectives, and enter the fractions from successive lots on a chart. A plot of fractions of defectives from successive lots will give a useful description of the overall quality of the production process. Such a time-sequence chart, displayed in a prominent place, will alert workers and management to changes (either slippage or improvement) in the process. Usually, control limits will be added to these charts; points above the upper control limit indicate that the process has deteriorated and that certain modifications should be made. The construction of control limits will be explained in Chapter 5; all that we want to say at this point is that they are based on the natural variability in the fraction of defectives that can be expected from taking repeated representative samples. A control chart that graphs the fraction of defectives among consecutive samples of size $n = 100$ is illustrated in Figure 1.2-4. In the past, the fraction of defectives varied around 2 percent. The last two samples, with six and seven defectives, indicate that there was an unexpected increase in the fraction of defectives.

In the preceding example, we measured the quality of a component on a pass–fail basis. We analyzed attribute data—that is, data which describe the presence

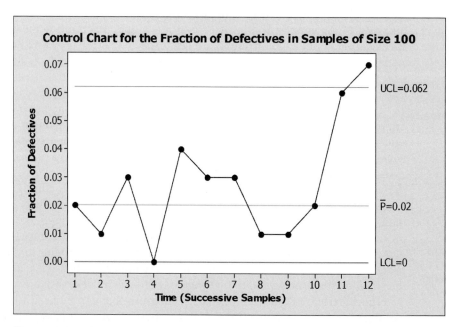

Figure 1.2-4 Control chart for the fraction of defectives from consecutive samples of size $n = 100$

or absence of a certain characteristic. Just as we can monitor the variability in the fraction of defectives from successive lots, we can monitor the diameter of knobs. We can sample a small number of knobs each hour or from each production shift, determine the diameters of the sampled knobs, and display the sample average (which is the sum of the observations divided by the number of observations) and the range (which is the difference between the largest and smallest measures in the sample) on control charts. (Details are given in Chapter 5.)

We pointed out that we cannot inspect quality into an item; improvements in quality can come only from changes in the process. Control charts, however, are very helpful in suggesting that changes should be made. Points outside the limits, as well as patterns in these charts, will lead to discussions and investigations aimed at finding the reason certain lots have more defectives than others. We can search for patterns in the charts and relate them to changes that have taken place in the production process. For example, why is it that there were many more defectives during the last several lots? Was it due to a bad batch of ink used in the pens? If the records show that these batches were produced with ink purchased from a new supplier, we might hypothesize that it is the switch to the new supplier that is responsible for the sudden increase. Or why is it that the number of defectives is drifting upward? Is it because the machines have not been serviced for a while? Checks of the service records and whether previous overhauling of machines has eliminated such drifts could provide some justification for this hypothesis. Also, if we find that one of the machines has not been serviced for a long time, it may be useful to observe that machine and check whether it could be the cause of the decrease in quality. Suppose that every tenth lot contains more defectives than the others. Then if we produce 10 lots a day and the problem lot is always the first one produced, the problem could be related to the start-up of the production line. If the number of defectives from the weekend's production is always higher than the number of defectives produced on weekdays, we could hypothesize that the inexperienced weekend staff needs more training. This discussion shows that information from control charts is very helpful because it leads to discussions and investigations that in turn may lead to an understanding of the causes of such deviations. Usually, the discussion leads to the formation of certain hypotheses about the factors that could have affected the response.

Exercises 1.2

*1.2-1 Consider the following listing of the $n = 52$ weekly sales of thermostat replacement parts (read across):

206	245	185	169	162	177	207	216	193	230	212	192	162
189	244	209	207	211	210	173	194	234	156	206	188	162
172	210	205	244	218	182	206	211	273	248	262	258	233
255	303	282	291	280	255	312	296	307	281	308	280	345

Construct a time-sequence plot and discuss your findings. What do you think will happen in the next month? The next two months?

1.2-2 Consider the monthly number of motor vehicle fatalities for Iowa shown in Project 2 at the end of this chapter. Construct and interpret the time-sequence plot of these observations.

The data in Project 2 go through 1979. Collect recent monthly or yearly data on traffic fatalities for Iowa (or the state you live in), and construct a time-sequence plot. What do you learn? Has the number of accidents increased (decreased)? Does an increase (decrease) in the number imply that it has become more dangerous (safer) to drive?

1.2-3   Look up the monthly unemployment rates of your state (or the United States as a whole) for the last five years. Construct a time-series plot of the information and interpret your findings. Do you notice trends? Do you notice seasonality, and what are the reasons for the seasonality? Note that the government reports raw (unadjusted) rates, as well as seasonally adjusted rates. What is the reason for adjusting the rates for seasonality?

1.2-4   You may have encountered control charts during your summer jobs or internships. If so, discuss.

1.3 Data Display and Summary

1.3-1 Summary and Display of Measurement Data

Let us now discuss how to summarize and display data that are obtained by measuring objects. For example, we may have measured several items, determining their weight, width, length, strength, diameter, and deflection.

Data sets are often large, so data must be organized, summarized, and displayed before any interpretation can be attempted. Graphical displays, such as plots and diagrams, are especially useful to uncover unknown features of the data. Pictures stimulate insight and force us to notice what we frequently do not expect to see.

Let us start with an example. Table 1.3-1 lists measurements of compressive strength (in units of 100 pounds per square inch, or psi) that were made on 90 concrete blocks. These measurements were taken to investigate the variability among blocks purchased from the traditional supplier. Lacking both a summary and a display, the table is not very informative.

To get more insight, we could construct a *dot diagram* of the observations. In a dot diagram, we mark the observations on the horizontal axis; for illustration, we have used the first 10 observations of Table 1.3-1. The dot diagram in Figure 1.3-1(a) gives a visual display of the variability among these 10 observations.

Table 1.3-1 Compressive Strength of Concrete Blocks (100 Pounds per Square Inch)									
49.2	53.9	50.0	44.5	42.2	42.3	32.3	31.3	60.9	47.5
43.5	37.9	41.1	57.6	40.2	45.3	51.7	52.3	45.7	53.4
51.0	45.7	45.9	50.0	32.5	67.2	55.1	59.6	48.6	50.3
45.1	46.8	47.4	38.3	41.5	44.0	62.2	62.9	56.3	35.8
38.3	33.5	48.5	47.4	49.6	41.3	55.2	52.1	34.3	31.6
38.2	46.0	47.0	41.2	39.8	48.4	49.2	32.8	47.9	43.3
49.3	54.5	54.1	44.5	46.2	44.4	45.1	41.5	43.4	39.1
39.1	41.6	43.1	43.7	48.8	37.2	33.6	28.7	33.8	37.4
43.5	44.2	53.0	45.1	51.9	50.6	48.5	39.0	47.3	48.8

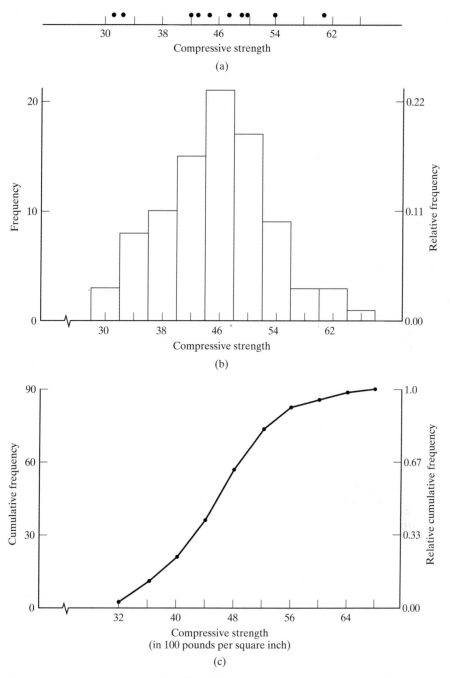

Figure 1.3-1 Graphical display of the compressive strength measurements in Table 1.3-1: (a) dot diagram of the first 10 observations; (b) histogram; (c) cumulative frequencies and a plot of the ogive

Table 1.3-2 Frequency Distribution

Interval	Midpoint	Tally	Frequency f_i	Relative Frequency f_i/n	Cumulative Frequency $fcum_i = f_1 + \cdots + f_i$	Relative Cumulative Frequency $fcum_i/n$
28.0–32.0	30.0	|||	3	0.033	3	0.033
32.0–36.0	34.0	ℋ |||	8	0.089	11	0.122
36.0–40.0	38.0	ℋ ℋ	10	0.111	21	0.233
40.0–44.0	42.0	ℋ ℋ ℋ	15	0.167	36	0.400
44.0–48.0	46.0	ℋ ℋ ℋ ℋ		21	0.233	57	0.633
48.0–52.0	50.0	ℋ ℋ ℋ ||	17	0.189	74	0.822
52.0–56.0	54.0	ℋ ||||	9	0.100	83	0.922
56.0–60.0	58.0	|||	3	0.033	86	0.955
60.0–64.0	62.0	|||	3	0.033	89	0.988
64.0–68.0	66.0			1	0.012	90	1.000
Sums			90	1.000		

For 90 observations, such a diagram would not be very informative, because we would have many points bunched together. With large data sets, it is better to construct a frequency distribution, as in Table 1.3-2, and to display the results in the form of a *histogram,* as in Figure 1.3-1(b).

To determine the frequency distribution, we first rank the n observations, say, $x_1, x_2, \ldots, x_n$, from the smallest value, x_{min}, to the largest, x_{max}. The range $R = x_{max} - x_{min}$ of the data tells us about the extent of their variability. In our example, $x_{min} = 28.7$, $x_{max} = 67.2$, and $R = 38.5$. Then we group the data into nonoverlapping intervals (cells or classes) that are usually of equal length (width). There are no universally applicable rules for determining the number of cells. Experience and experimentation with different lengths of the class intervals usually provide the best guide. However, it is generally desirable to have somewhere between 7 and 15 classes (cells). Of course, the number of classes will depend on the number of observations: If there are more observations, we can use more classes with shorter class widths (lengths). Let us assume, in our example, that we have 10 classes, each of width 4.0. The first class starts with 28.0 and goes up to (but not including) 32.0, the second from 32.0 to 36.0, ..., and the last from 64.0 to 68.0. The endpoints of each class are called the *class boundaries*; the middle is called the *midpoint.* Next, we count the number of values that belong to each class, and we make a frequency distribution table. We use the convention that an observation which falls exactly on a boundary is allocated to the class that has this boundary as the lower limit. Thus, each observation in Table 1.3-2 falls into one and only one class. (Note that statistical software packages may use different rules. Minitab, for example, allocates such an observation to the class that has this boundary as the upper limit.) Apart from the *(absolute) frequencies* f_i ($i = 1, 2, \ldots, k$, where k is the number of classes), we also list the *relative frequencies* f_i/n, the *cumulative frequencies*

$$fcum_i = f_1 + \cdots + f_i \ (i = 1, 2, \ldots, k),$$

and the relative cumulative frequencies

$$\frac{fcum_i}{n} = \frac{1}{n}(f_1 + \cdots + f_i).$$

Note that the absolute frequencies of the classes sum to n, and the relative frequencies sum to 1.

The frequency table can be better visualized if it is displayed graphically. The *histogram* in Figure 1.3-1(b) is a bar graph that associates (either the absolute or the relative) frequencies with the data intervals. The histogram shows how the values of our variable of interest are distributed. It condenses a set of data for easy visual comprehension of its general characteristics, such as typical values, spread, and shape. It also helps to detect unusual observations in a data set.

The histogram in Figure 1.3-1(b) is approximately *symmetric*, with a single peak, or hump, in the middle. The peak represents the most frequent class; in our example, it is the class from 44 to 48. The midpoint of this class, 46, is also called the *mode*; it stands for the most frequent value in the data set if we assume that the values in each class equal its midpoint.

However, as the histograms in Figure 1.3-2 show, symmetric distributions with a single hump in the middle are not the only possibilities. The first is a histogram of the CO emissions of 794 cars. (See Exercise 1.3-9 for the data.) The second is a histogram of the lengths of life of more than 1 million human beings. (See Exercise 1.3-10 for the data.) The third is a histogram of the thickness of the ears of 150 paint cans. (*Ears* are tabs used to secure the lids of large paint cans.) The data are listed in Exercise 1.3-8.

The histogram in Figure 1.3-2(a) has a single hump, but is *not* symmetric. We call such a nonsymmetric distribution *skewed*. The histogram in Figure 1.3-2(a) is said to be *skewed to the right*, because it has a long right tail of relatively large numbers. The histogram in Figure 1.3-2(b) is also skewed. Because it has a long left tail of relatively small numbers, we call it *skewed to the left*. Note that the frequency of death (failure) is relatively high in the first age interval, from 0 to 5 years. This is due to the fact that the infant mortality rate is somewhat higher than the mortality rate during the teen years. Histograms of the failure time of items for which there is an increasing failure rate (i.e., the older it gets, the more likely it is to fail) are skewed to the left. However, some items may also fail early in life, and the increasing failure rate may apply only to items that survived past a certain point. In these cases, we would observe a pattern such as that in Figure 1.3-2(b).

Figure 1.3-2(c) shows a histogram with two peaks (or two modes). We call such a histogram *bimodal*, compared with the *unimodal* histogram in Figure 1.3-2(a). The bimodal nature can be explained by the fact that the hopper, from which the paint cans were selected, collected cans from two different machines that produced ears of different thickness. (See Exercise 1.3-8.)

So far, we have considered absolute and relative frequencies. The cumulative frequencies in the last column of Table 1.3-2 show that 36 of 90 measurements, or 40 percent, are smaller than 44; similarly, 89 of 90 measurements, or 98.8 percent, are smaller than 64. In Figure 1.3-1(c), we plot the (relative) cumulative frequencies against the upper limits of the intervals. An *ogive* is a line graph that connects these points. The definition of cumulative frequencies implies that this ogive is nondecreasing and always between zero and 1, including those values.

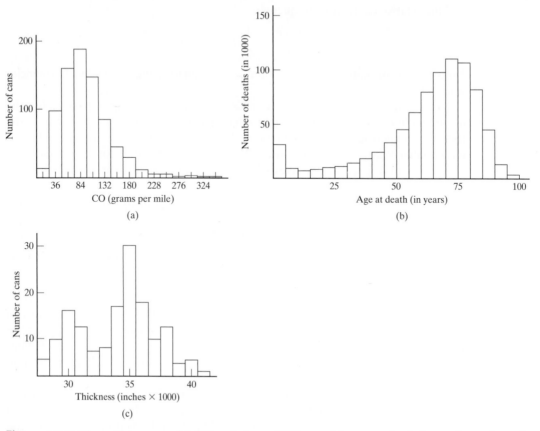

Figure 1.3-2 Three histograms: (a) CO emissions of 794 cars; (b) ages at death for a cohort of people; (c) thickness of the ears of 150 paint cans

A histogram condenses a set of data for easy *visual* comprehension of its general characteristics. Frequently, however, we also want to summarize the information *numerically* and obtain a few statistics that characterize the data set. In particular, we want measures of location and dispersion.

1.3-2 Measures of Location

For a given data set $x_1, x_2, \ldots, x_n$, the most familiar measure of location is the arithmetic average of the n observations. Because the data set usually corresponds to a sample from a larger population (in our example, the population could be the set of all concrete blocks that are in our inventory), we refer to this average as the *sample mean* and denote it by

$$\bar{x} = \frac{1}{n}(x_1 + x_2 + \cdots + x_n) = \frac{1}{n}\sum_{i=1}^{n} x_i.$$

Note that the deviations of the observations from the sample mean, $x_i - \bar{x}$, $i = 1, 2, \ldots, n$, always add to zero; that is,

$$\sum_{i=1}^{n}(x_i - \bar{x}) = \sum_{i=1}^{n}x_i - n\bar{x} = \sum_{i=1}^{n}x_i - \sum_{i=1}^{n}x_i = 0.$$

One can think of the sample mean as the fulcrum that keeps a weightless ruler on which each observation is represented by the same weight in perfect balance.

If we consider only the first 10 observations in Table 1.3-1, we find that

$$\bar{x} = \frac{454.1}{10} = 45.41.$$

For the complete set of 90 observations, the average is 45.51.

One problem with the mean as the center of a data set is that it is affected by the presence of a single observation that lies very far to one side of the other values. We call such an observation an *outlier*. For example, assume that the third observation in the first row of Table 1.3-1 gets recorded incorrectly as 5.0 instead of 50.0. This is clearly an outlying observation, as a simple dot diagram of the 10 observations would show. (See Figure 1.3-3.) The sample mean of the first 10 observations then becomes $\bar{x} = 40.91$, a considerable change from 45.41.

The *median* of a sample (*Md,* for short) is a measure of location that is much less sensitive than the mean is to a single observation. The median is the middle observation when observations are arranged in increasing order of magnitude and if n is an odd number. If n is even, there is no unique middle and the median is usually taken to be the average of the middle pair of numbers. The median has the property that one-half of the observations lie below it and one-half lie above it. (If n is odd, we can think of splitting the middle value.)

Take the first 10 observations in Table 1.3-1 and arrange them in increasing order:

31.3   32.2   42.2   42.3   44.5   47.5   49.2   50.0   53.9   60.9

Since n is even, we take the average of the fifth- and sixth-largest observations. We find that the median is $Md = (44.5 + 47.5)/2 = 46.0$. Note that the one outlier (5.0 instead of 50.0) in the sample of the first 10 observations affects the median just slightly: $Md = (42.3 + 44.5)/2 = 43.4$. Note also that if the outlier had been 500 (instead of 50.0), the median would not have changed at all. In general, changes in the median are usually much smaller than changes in the mean when outliers are introduced. The median for all 90 observations is the average of the 45th- and 46th-largest observations, or $(45.3 + 45.7)/2 = 45.5$.

We could have used the ogive in Figure 1.3-1(c) to determine the median, at least approximately. Since, by definition, half of the observations are below the median, we look for the intersection of the ogive and the relative cumulative frequency of 0.5. The x-coordinate of the point of intersection is our median. Of course,

Figure 1.3-3 Dot diagram of the first 10 observations in Table 1.3-1, with the third observation, 50.0, replaced by 5

this method is only approximate (but still very good), because we are using just the grouped data in Table 1.3-2, not the individual observations.

For symmetric distributions such as the one in Figure 1.3-1(b), the mean, the median, and the mode (which is the most frequent value, or, in the case of grouped data, the midpoint of the most frequent class) are about the same. For this particular data set of $n = 90$ observations, we have found that $\bar{x} = 45.51$, $Md = 45.5$, and Mode $= 46.0$. For skewed distributions, these three measures of location are no longer about equal. Take, for example, the CO data in Exercise 1.3-9, which were used to construct the skewed histogram in Figure 1.3-2(a); there, the mode is 84, and the median and the mean are, respectively, $Md = 87.9$ and $\bar{x} = 94.2$. For frequency distributions that are skewed to the right, it is usually true that Mode $<$ Median $<$ Mean. Similarly, for distributions that are skewed to the left, we usually find that Mode $>$ Median $>$ Mean. You can use the data in Exercise 1.3-10 and Figure 1.3-2(b) to illustrate these facts.

The median divides a data set into two parts. A finer partition can be obtained by dividing the data set into more than two parts. For example, we can order the n observations $x_1, x_2, \ldots, x_n$ from the smallest one $x_{(1)}$, to the second-smallest one $x_{(2)}, \ldots$, to the largest one $x_{(n)}$. The ordered observations are called *order statistics* $x_{(1)} \leq x_{(2)} \leq \ldots \leq x_{(n)}$; the numbers in parentheses are the *ranks* of the observations. The smallest observation $x_{(1)}$ represents the first $100/n$ percent of the distribution of the sample. If we take the midpercentage, $100/2n = 100[(1 - 0.5)/n]$, as the representative value, we say that smallest observation $x_{(1)}$ is the $100[(1 - 0.5)/n]$th *percentile* of the sample. The second-smallest observation $x_{(2)}$ represents the second $100/n$ percent of the distribution of the sample. The midpoint of this second $100/n$ percent region, $100[(2 - 0.5)/n]$, is the representative percentage; thus, we say that $x_{(2)}$ is the $100[(2 - 0.5)/n]$th percentile of the sample. The observation $x_{(i)}$ is the $100[(i - 0.5)/n]$th percentile; we also call it the *quantile of order* $(i - 0.5)/n$. For example, take $n = 10$. Here each order statistic represents $100/10 = 10$ percent of the distribution. The order statistics $x_{(1)}, x_{(2)}, \ldots, x_{(10)}$ are the respective 5th, 15th, . . ., 95th percentiles of the sample.

Frequently, we want to calculate a certain percentile from the data. Because the ith-order statistic is the $100P = 100[(i - 0.5)/n]$th percentile, it follows that the $(100P)$th percentile of the sample is the observation with rank $nP + 0.5$. If this is not an integer, we take the average of the two order statistics whose ranks are closest to that number. Take the 50th percentile, or the median, as an example; it is the observation with rank $(n/2) + 0.5 = (n + 1)/2$. If n is an odd number, we take the middle observation with rank $(n + 1)/2$. If n is even, we take the average of the two middle observations with ranks $(n/2)$ and $(n/2) + 1$. In a similar manner, to find the 25th percentile, which is also called the *lower (first) quartile* Q_1, we take the observation with rank $(n/4) + 0.5$. Again, if this is not an integer, but a fraction, we average the two order statistics whose ranks are closest to that number. Similarly, the 75th percentile, or *upper (third) quartile* Q_3, is the observation with rank $n(3/4) + 0.5$. With $n = 10$ observations, the 75th percentile is the eighth-largest observation. The 50th percentile (the median or second quartile Q_2) is the average of $x_{(5)}$ and $x_{(6)}$. For the 10 observations in the first row of Table 1.3-1, we find that $Q_1 = 42.2$, $Q_2 = 46.0$, and $Q_3 = 50.0$. With $n = 90$, Q_1 is the observation with rank 23, Q_3 the one with rank 68, and the median Q_2 is the average of $x_{(45)}$ and $x_{(46)}$.

Remark   Some authors and some computer programs use a slightly different definition for percentiles. One popular scheme is the following: Because the order statistics divide the data set into $n + 1$ segments, they define the ith-order statistics as the $100[i/(n + 1)]$th percentile. Accordingly, the $(100P)$th percentile is the observation with rank $(n + 1)P = nP + P$. If n is not too small and P is between 0.1 and 0.9, there is very little difference between the two definitions.

Example 1.3-1   A manufacturer claims that his fabric consists of 80 percent cotton. To check his claim, we take a small swatch from each bolt of fabric and determine its cotton content. The results of 25 such measurements are as follows:

77	81	76	76	79	79	80	77	89	77	78	85	80
75	79	88	81	78	82	80	76	83	81	85	79	

The 25 order statistics are

$$75 \leq 76 \leq 76 \leq 77 \leq 77 \leq 77 \leq 78 \leq 78 \leq 79 \leq 79 \leq 79 \leq 79 \leq 80 \leq$$
$$80 \leq 80 \leq 81 \leq 81 \leq 81 \leq 82 \leq 83 \leq 85 \leq 85 \leq 86 \leq 88 \leq 89$$

The fourth order statistic, $x_{(4)} = 77$, is the $100[(4 - 0.5)/25)] = 14$th percentile, and $x_{(19)} = 82$ is the $100[(19 - 0.5)/25] = 74$th percentile. The median is the order statistic with rank $25/2 + 0.5 = 13$; that is, $x_{(13)} = 80$. The first and third quartiles have ranks $25/4 + 0.5 = 6.75$ and $(25)3/4 + 0.5 = 19.25$, respectively. Thus, those respective quartiles are $(x_{(6)} + x_{(7)})/2 = (77 + 78)/2 = 77.5$ and $(x_{(19)} + x_{(20)})/2 = (82 + 83)/2 = 82.5$. Alternatively, one could interpolate by taking the weighted averages $(0.25)x_{(6)} + (0.75)x_{(7)} = 77.75$ and $(0.75)x_{(19)} + (0.25)x_{(20)} = 82.25$. The differences, however, will usually be quite small.   ■

There is yet one other measure of location: the *trimmed mean*. In calculating the 100α percent trimmed mean $\overline{x}_\alpha$, we ignore (or trim) the $n\alpha$ smallest and the $n\alpha$ largest observations and calculate the arithmetic average of the remaining $n(1 - 2\alpha)$ observations. For example, a 10 percent trimmed mean, calculated from a sample size $n = 10$, trims $(10)(0.1) = 1$ observation from each end of the ordered data set. Ignoring the smallest and largest observations of the first 10 observations in Table 1.3-1 (i.e., 31.3 and 60.9) leads to the 10 percent trimmed mean, $\overline{x}_{0.1} = 45.24$.

1.3-3 Measures of Variation

We have discussed the fact that variability or dispersion is an integral and natural part of all measurements. The sample mean, median, mode, and trimmed mean do not capture dispersion, as they are estimates of location.

The simplest measure of dispersion is the *range*, defined as the difference between the largest and the smallest observations. For example, the minimum of the first 10 observations in Table 1.3-1 is 31.3, and the maximum is 60.9; therefore, the range of those $n = 10$ observations is $R = 60.9 - 31.3 = 29.6$.

A drawback of this measure is that it is highly sensitive to outliers. Just one large (or small) observation can influence it a great deal. A more reliable measure of dispersion is the *interquartile range*, $IQR = Q_3 - Q_1$, the difference between the third and first quartiles. For our 10 observations, the interquartile range is $50.0 - 42.2 = 7.8$.

The units for the range and the interquartile range are the same as the units of the observations x. In our example, strength is measured in units of 100 psi. Thus, the interquartile range is 780 psi.

The most common measures of variability, however, are the *sample variance* and its square root, the *sample standard deviation*. The square $(x_i - \bar{x})^2$ of the distance $|x_i - \bar{x}|$ of an observation $x_i, i = 1, 2, \ldots, n$, from the overall average $\bar{x}$ provides some information about the variability. The *sample variance* is a special average of these n squared distances and is defined as

$$s^2 = \frac{1}{n-1}[(x_1 - \bar{x})^2 + (x_2 - \bar{x})^2 + \ldots + (x_n - \bar{x})^2] = \frac{1}{n-1}\sum_{i=1}^{n}(x_i - \bar{x})^2.$$

Notice that we just said that the sample variance is the average of n such components, but we have divided the sum by $n - 1$ instead of n. Why? A formal reason for this change will be given in a later chapter. For the time being, however, consider the following intuitive explanation: We noted that the n deviations $x_i - \bar{x}, i = 1, \ldots, n$, add up to zero. Thus, we really need only $(n - 1)$ of these deviations to calculate s^2. We can always obtain the last deviation from

$$(x_n - \bar{x}) = -\sum_{i=1}^{n-1}(x_i - \bar{x}).$$

Hence, it seems reasonable to divide by $n - 1$ instead of n. At any rate, using the divisor $(n - 1)$ instead of n makes almost no numerical difference, provided that the number of observations is reasonably large.

The following convenient shortcut equation for calculating the sample variance avoids working out the deviations from the mean (see Exercise 1.3-4):

$$s^2 = \frac{1}{n-1}\left[\sum_{i=1}^{n}x_i^2 - \frac{\left(\sum_{i=1}^{n}x_i\right)^2}{n}\right].$$

For the first 10 observations in Table 1.3-1, we find that $\sum x_i = 454.1$ and $\sum x_i^2 = 21,364.27$; thus, $s^2 = (1/9)[21,364.27 - (454.1)^2/10] = 82.62$. Similarly, we can calculate the sample variance of the 90 observations in Table 1.3-1: $s^2 = (1/89)[191,605.04 - (4095.6)^2/90] = 58.74$.

The variance $s^2 \geq 0$ measures the scatter of the observations around their mean. If $s^2 = 0$, there is no variation at all, because all n observations must be the same. However, there is a drawback to the variance as a measure of variation, as its unit is not the same as the unit of our observations. Since s^2 is an average of squares, its unit is the square of the unit of the measurements x; if the x's are measured in pounds, the unit of s^2 is (pounds)^2.

To return to the original unit of measurement, we take the square root of the sample variance. This gives a measure known as the *sample standard deviation*:

$$s = \sqrt{s^2} = \sqrt{\frac{1}{n-1}\sum_{i=1}^{n}(x_i - \bar{x})^2}.$$

For our 10 observations, the standard deviation is $s = \sqrt{82.62} = 9.1$. Its unit is the same as that of the observations—that is, 100 psi. Thus, the standard deviation is 910 psi. For all 90 observations, the standard deviation is $s = \sqrt{58.74} = 7.7$, or 770 psi.

So far, we have learned how to calculate sample variances and sample standard deviations. Calculation, of course, is important, but in today's computer age, when these summary statistics are calculated by machines, it is even more important to understand their *meaning*. Here we give an intuitive interpretation of the sample standard deviation. Looking at the preceding equation for s, we see that the distances from the observations to the sample mean average about one standard deviation. Or, to say this differently, observations, on average, are *approximately* one standard deviation from the mean. Here we wish to emphasize the word *approximately*, because the average actually involves the square, after which the square root is taken. In our example of $n = 10$ observations, we find that 7 of the 10 observations lie within one standard deviation, $s = 9.1$, on either side of the mean $\bar{x} = 45.4$. The other three observations, 31.3, 32.3, and 60.9, lie beyond these limits, so that all 10 distances from the mean average to *about* one standard deviation. In general, we find that usually over half of the observations will be within plus or minus one standard deviation from the mean. More than three-fourths of the observations, and often even much more than that, will lie within plus or minus two standard deviations of the mean, and almost all the observations will be within three standard deviations. In a subsequent chapter, we discuss this situation in more detail.

We often express the standard deviation as a percentage of the mean. The measure

$$\text{CV} = \frac{100s}{\bar{x}}$$

is called the *coefficient of variation*. Since s and $\bar{x}$ are expressed in the same units, we find that this measure does not depend on the unit of measurement. The coefficient of variation says that the observations lie, on the average, within approximately CV percent of the mean. In our example with 10 observations, we found that $\bar{x} = 45.41$ and $s = 9.1$. Thus, the observations lie, on average, within $(100)(9.1)/45.41 = 20$ percent of the mean.

A note on computation is now in order. Measures such as the mean and the variance are easy to calculate for small samples. However, in many realistic applications, we may be facing very large data sets. Sophisticated electronic recording devices may give you measurements every second, and it may be necessary to summarize 10,000 or more observations. Certainly, we would use a computer program in such circumstances. However, in writing such programs, statisticians have to worry about the efficient calculation of these statistics. Take, for example, the sample variance that we get by averaging the squares of 10,000 deviations. Since, typically, $\bar{x}$ is rounded off, we introduce this rounding error into 10,000 squares $(x_i - \bar{x})^2$, and the result will probably not be very satisfactory. The shortcut method for calculating the sample variance is considerably better, even though it can run into difficulties if the observations are large in magnitude, but vary only little. Here, we do not go into these issues. (If you want to learn more about them, consult books on statistical computing.)

Excellent computer software for displaying and summarizing data is available, and packages such as Minitab, SPSS, SAS, R, and Excel make it very easy to construct histograms and obtain summary statistics such as the mean, median, and standard deviation. Undoubtedly, you are using one or more of these programs in the course, and we expect you to carry out the analysis of the data sets in this book with those packages. Section 1.7-1 will tell you more about statistical software.

1.3-4 Exploratory Data Analysis: Stem-and-Leaf Displays and Box-and-Whisker Plots

Histograms are very valuable tools for displaying the variability of observations, and they help the analyst develop an understanding of a data set. However, a disadvantage of histograms is that individual data points cannot be identified because all data falling into an interval are indistinguishable. Consequently, statisticians have looked for other, more informative graphical descriptions of the data. John Tukey, in particular, has created several imaginative and very useful data displays. Because they are usually performed at the initial exploratory stage of the analysis, these displays are known as *exploratory data analysis* (EDA). Here, we introduce a few of the simpler ideas; for additional discussion consult J. W. Tukey, *Exploratory Data Analysis* (Reading, MA: Addison-Wesley, 1977) or P. Velleman and D. Hoaglin, *Applications, Basics, and Computing of Exploratory Data Analysis* (Boston: Duxbury Press, 1981).

Stem-and-Leaf Displays The easiest way to begin is with an illustration. Suppose that we have the following $n = 58$ test scores:

76	93	42	66	60	56	60	75	78	81
61	70	58	64	67	73	49	52	74	91
76	82	86	59	69	73	74	64	94	65
48	59	72	68	66	64	88	80	51	82
77	60	66	86	85	51	90	91	53	63
69	73	67	61	86	80	72	65		

We note that 42 and 94 are the smallest and largest scores, respectively. Of course, we could construct a histogram, together with the usual tabulation. But we can do much the same thing with a *stem-and-leaf display* and not lose the original values as we do with tally marks. In this data set, take the first number 76 and record it as follows: The 7 in the "tens" place is treated as the *stem,* and the 6 in the "units" place is the corresponding *leaf.* Note that the leaf 6 comes after the stem 7 in Table 1.3-3. The second number, 93, is represented by the leaf 3 after the stem 9; the third number, 42, by the leaf 2 and after the stem 4; the fourth number, 66, by the leaf

Table 1.3-3 Stem-and-Leaf Display

Stem	Leaf	Frequency
4	2 9 8	3
5	6 8 2 9 9 1 1 3	8
6	6 0 0 1 4 7 9 4 5 8 6 4 0 6 3 9 7 1 5	19
7	6 5 8 0 3 4 6 3 4 2 7 3 2	13
8	1 2 6 8 0 2 6 5 6 0	10
9	3 1 4 0 1	5

Table 1.3-4 Ordered Stem-and-Leaf Display

Stem	Leaf	Frequency
4	2 8 9	3
5	1 1 2 3 6 8 9 9	8
6	0 0 0 1 1 3 4 4 4 5 5 6 6 6 7 7 8 9 9	19
7	0 2 2 3 3 3 4 4 5 6 6 7 8	13
8	0 0 1 2 2 5 6 6 6 8	10
9	0 1 1 3 4	5

6 after the stem 6; the fifth number, 60, by the leaf 0 after the stem 6 (note that this is the second leaf on the stem 6); and so on. Continue and complete the table, carefully lining up the leaves vertically to give the same effect as a histogram. Of course, here the original numbers are not lost, as they are with tally marks.

Sometimes, to help us find percentiles and other characteristics of the data, we order the leaves according to magnitude, giving an *ordered stem-and-leaf display* as in Table 1.3-4. For illustration, since $n = 58$ is even, the median is the average of observations with rank 29 and 30; that is, $Md = (69 + 69)/2 = 69$. The first quartile is the observation with rank $(58)(0.25) + 0.5 = 15$, or $Q_1 = 61$. The third quartile is the observation with rank $(58)(0.75) + 0.5 = 44$, or $Q_3 = 80$. The 20th percentile, or the second *decile,* as it is often called, is the observation with rank $(58)(0.2) + 0.5 = 12.1 \cong 12$, which is 60. Note here that we violated our earlier rule and did not average the 12th- and 13th-order statistics to obtain the 20th percentile. Because there really is no unique rule among statisticians, we prefer this particular approximation, as 12.1 is about equal to 12. The 85th percentile is the $(58)(0.85) + 0.5 = 49.8 \cong $ 50th − largest observation, 86 (again noting that the rank 49.8 is about equal to 50).

The stem-and-leaf tables as given here are equivalent to a histogram with six classes. Suppose that we desire more classes. It is easy to increase the number of classes with the modification given in Table 1.3-5. Here, a stem with * has the leaves, 0, 1, 2, 3, 4 and a stem with • has the leaves 5, 6, 7, 8, 9.

Table 1.3-5 Ordered Stem-and-Leaf Display with More than Six Classes

Stem	Leaf	Frequency
4*	2	1
4•	8 9	2
5*	1 1 2 3	4
5•	6 8 9 9	4
6*	0 0 0 1 1 3 4 4 4	9
6•	5 5 6 6 6 7 7 8 9 9	10
7*	0 2 2 3 3 3 4 4	8
7•	5 6 6 7 8	5
8*	0 0 1 2 2	5
8•	5 6 6 6 8	5
9*	0 1 1 3 4	5

As a second illustration, consider $n = 44$ scores on a test that was based upon 40 points. The scores, ranging from 12 to 37, are as follows:

17	22	36	28	30	33	19	21	20	29
34	26	27	23	20	12	18	24	14	37
25	30	24	27	18	15	35	33	29	24
26	20	15	24	16	22	31	23	32	25
29	26	32	19						

A stem-and-leaf display produced after these data are ordered is given in Table 1.3-6. Here, the leaves are recorded in the following manner: 0, 1 after a stem with *; 2, 3 after a stem with t (t for twos and threes); 4, 5 after a stem with f (f for fours and fives); 6, 7 after a stem with s (s for sixes and sevens); and 8, 9 after a stem with •. The median is the observation with rank 22.5—that is, the average of the 22nd- and 23rd-largest scores, or $(24 + 25)/2 = 24.5$.

Imagination must be used to create stem-and-leaf displays. In some instances, modifications must be made that may lead to a small loss of information. As an illustration, consider the compressive strengths of $n = 90$ concrete blocks given in Table 1.3-1. First, we could round off our numbers a little, writing 49.2 as 49, 53.9 as 54, 43.5 as 44, 53.4 as 53, and so on. Then we could start the stem-and-leaf display with the 2• having possible leaves of 5, 6, 7, 8, 9; continue with 3* having leaves of 0, 1, 2, 3, 4; and end with stem 6• having possible leaves of 5, 6, 7, 8, 9. Such a scheme would provide nine classes.

If, however, the person describing the data does not want to lose information by rounding (although this is usually not serious), we could create stems with double leaves: 49.2 would be recorded as the leaf 92 after the stem 4•, 53.9 as the leaf 39 after the stem 5t, and so on. Thus, in the concrete-block illustration, stems would run from 2• to 6s, for a total of 20 classes. As an illustration, with the stem 4s, we have

4s: 75   68   74   74   60   70   79   62   73

representing the numbers 47.5, 46.8, 47.4, 47.4, 46.0, 47.0, 47.9, 46.2, and 47.3. Thus, the frequency of the class with stem 4s is 9. We ask you to complete this stem-and-leaf display in Exercise 1.3-11.

Table 1.3-6 Ordered Stem-and-Leaf Display for the Second Example

Stem	Leaf						Frequency
1t	2						1
1f	4	5	5				3
1s	6	7					2
1•	8	8	9	9			4
2*	0	0	0	1			4
2t	2	2	3	3			4
2f	4	4	4	4	5	5	6
2s	6	6	6	7	7		5
2•	8	9	9	9			4
3*	0	0	1				3
3t	2	2	3	3			4
3f	4	5					2
3s	6	7					2

Box-and-Whisker Displays In *box-and-whisker* displays (also called *box plots*), we depict the three quartiles, together with the two extremes of the data. The box in these displays, aligned either horizontally or vertically, encloses the interquartile range, with the left (or lower) line identifying the 25th percentile (lower quartile) and the right (or upper) line identifying the 75th percentile (upper quartile). The box contains the middle 50 percent of the data. The width of the box (i.e., the interquartile range IQR $= Q_3 - Q_1$) is a measure of the spread of the distribution. If the box is short, then the data are tightly packed around the middle. The line sectioning the box displays the 50th percentile (the median) and its relative position within the interquartile range. The whiskers at either end extend to the extreme values. In large data sets, with sample sizes of at least 50 or 100, the whiskers may extend only to the 10th and 90th percentiles or the 5th and 95th percentiles instead of the extreme values. Then, extreme values may also be displayed as unconnected dots. Different computer programs have various ways of calculating the ends of whiskers. John Tukey, who studied such displays in great detail, calculates upper and lower limits by adding 1.5 times the interquartile range to the upper quartile and subtracting 1.5 times the interquartile range from the lower quartile. Then the upper whisker extends to the highest data value within the upper limit, and the lower whisker extends to the lowest value within the lower limit. (See *Exploratory Data Analysis* (Reading, MA: Addison-Wesley, 1977).)

Figure 1.3-4 illustrates the box-and-whisker display of the 58 test scores in Table 1.3-4; recall that those five numbers (the minimum, first quartile, median, third quartile, and maximum), in that order, are 42, 61, 69, 80, and 94. The stem-and-leaf display in Table 1.3-4 shows that the distribution of these scores is fairly symmetric around the center value. We can also see from the box-and-whisker display in Figure 1.3-4 that the distribution is symmetric; the left and right whiskers are about equal in length, and the lengths of the left and right boxes around the median are about the same.

For the data in Table 1.3-6, the minimum, first quartile, median, third quartile, and maximum are 12, 20, 24.5, 29.5, and 37, respectively. The box-and-whisker display is depicted in Figure 1.3-5. Again, the distribution of these observations looks quite symmetric.

In contrast, many data sets are not symmetric. For example, consider the CO concentrations shown in Figure 1.3-2(a), the incomes of engineers, or losses from automobile accidents; these distributions are skewed to the right. Consider the $n = 794$ CO concentration in Exercise 1.3-9. There, the 5th, 25th, 50th, 75th, and 95th percentiles are given by 30.5, 61.0, 87.9, 117.8, and 180.2 grams/mile, respectively. The box-and-whisker display in Figure 1.3-6 shows that the right whisker is longer than

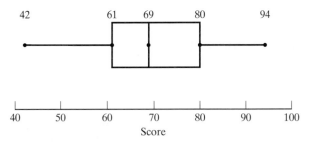

Figure 1.3-4 Box-and-whisker display for the 58 test scores

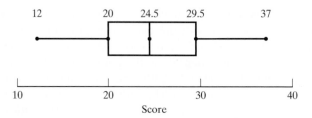

Figure 1.3-5 Box-and-whisker display of the data in Table 1.3-6

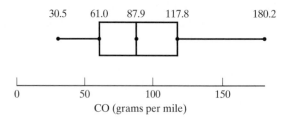

Figure 1.3-6 Box-and-whisker display for the CO emissions in Exercise 1.3-9

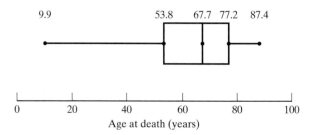

Figure 1.3-7 Box-and-whisker display of ages at death in Exercise 1.3-10

the left one, and the right box is larger than the left; we say that the distribution is skewed to the right.

The histogram of the number of deaths in various age groups in Figure 1.3-2(b) illustrates a distribution that is skewed to the left. From the data in Exercise 1.3-10, we calculate the 5th, 25th, 50th, 75th, and 95th percentiles as 9.9, 53.8, 67.7, 77.2, and 87.4 years, respectively. The box-and-whisker display in Figure 1.3-7 shows that the left whisker and the left box are larger than the right whisker and the right box. Thus, these data are skewed to the left.

1.3-5 Analysis of Categorical Data

Dot diagrams, histograms, box plots, and summary statistics such as averages and standard deviations are appropriate for describing continuous measurement data. But how should one proceed if the information is *categorical* in nature? Consider, for example, the numbers of engineering bachelor's-degree graduates in the years 2001 and 2002, classified according to the field of study and gender. The data are from the 2003 national survey of recent graduates conducted by the National

Table 1.3-7 Numbers of Engineering Graduates with Bachelor Degrees in 2001 and 2002

Engineering Field	Male	Female	Total
Aerospace/aeronautical/astronautical	2,600	600	3,200
Chemical	6,900	3,700	10,600
Civil/architectural	12,200	4,100	16,300
Electrical/computer	30,100	5,700	35,800
Industrial	4,200	2,300	6,500
Mechanical	21,700	3,100	24,800
Other	10,700	3,700	14,400
Total	88,400	23,200	111,600

Sciences Foundation. Table 1.3-7 is already a summary of the raw information, which most likely was organized into rows and columns of a spreadsheet. The first row, for the first person listed, probably had an identifier in the first column, a code for the field of study in the second column (say, 1 for aerospace, 2 for chemical, and so on, but clearly the coding is arbitrary), a code for gender in the third column (say, 1 for male and 2 for female), and information on other variables of interest (such as starting salary, cumulative grade point average, etc.) in subsequent columns.

Starting salary and GPA are continuous (measurement) variables, whereas "engineering field" and "gender" represent unordered categorical variables. The tools for summarizing and visualizing categorical data are tallies and tables. Let us first look at each categorical variable separately. Table 1.3-8 presents tallies for the field of study and for gender, including both absolute frequencies (counts) and relative frequencies (proportions).

Table 1.3-8 Tallies with Frequencies and Relative Frequencies, by Field of Study and Gender

Field of Study	Frequency	Relative Frequency
Aerospace	3,200	0.029
Chemical	10,600	0.095
Civil/architectural	16,300	0.146
Electrical/computer	35,800	0.321
Industrial	6,500	0.058
Mechanical	24,800	0.222
Other	14,400	0.129
Total	111,600	1.000

Gender	Frequency	Relative Frequency
male	88,400	0.792
female	23,200	0.208
Total	111,600	1.000

The information in tallies can be displayed graphically through *bar charts,* in which the heights of the bars are proportional to frequencies, or *pie charts,* in which the areas of the pie segments are proportional to frequencies. A bar chart for field of study and a pie chart for gender are shown in Figure 1.3-8. Electrical and mechanical engineering are the two most popular majors, representing about 55 percent of all bachelor's-degree graduates. In those years, engineering was—and it still is—a male-dominated undergraduate field, with only about 21 percent female graduates.

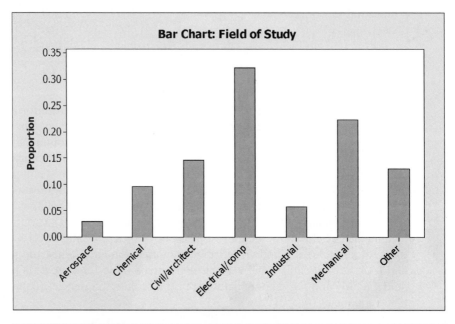

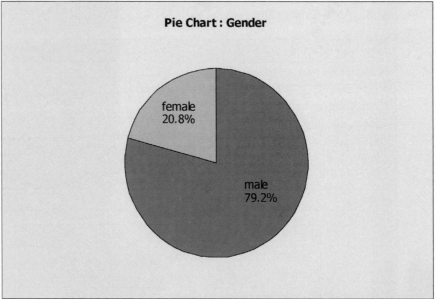

Figure 1.3-8 Bar and pie charts for a single categorical variable

Tallies and bar and pie charts display the information contained in a single categorical variable. The simultaneous information in two categorical variables can be displayed through summary tables like Table 1.3-7. This table classifies bachelor's-degree recipients simultaneously according to the field of study and gender. The frequencies in the margins of the table represent the summary counts for each of the two variables separately and were displayed in Tables 1.3-8 and Figure 1.3-8. Frequencies in a two-way table can also be expressed as percentages, either as a percentage of the grand total, as a percentage of the row total, or as a percentage of the column total. The appropriate standardization depends on the comparison one wishes to make. Here, our objective is to investigate possible gender differences in the choice of engineering major, and Table 1.3-9 expresses the frequencies as percentages of the column (gender) sums. Note that the entries in both columns sum to 1. Figure 1.3-9 represents the information in terms of a clustered bar chart, with the

Table 1.3-9 Distributions of Field of Study, Separated by Gender		
Engineering Field	Male	Female
Aerospace	0.029	0.026
Chemical	0.078	0.159
Civil/architectural	0.138	0.177
Electrical/computer	0.341	0.246
Industrial	0.048	0.099
Mechanical	0.245	0.134
Other	0.121	0.159
Total	1.000	1.000

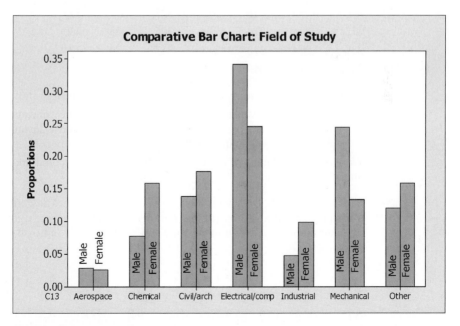

Figure 1.3-9 Clustered bar chart of the proportions of engineering graduates in various fields of study

proportions for males and females clustered within each field of study. The table and the figure show gender differences: Chemical, civil, and industrial engineering are more popular among women, while electrical and mechanical engineering are more popular among men.

Besides displaying the frequencies in two-way tables, we can add information from other variables, which can represent continuous measurement data such as the GPA or the starting salary. For example, for each of the cells in the field of study and gender cross-classification, we can add summary statistics on the variable that denotes the salary.

Exercises 1.3

*1.3-1 The tensile strengths (in pounds per square inch) of eight synthetic fibers are as follows:

12   15   18   16   15   14   16   17

(a) Construct a dot diagram.

*(b) Calculate the sample mean, variance, and standard deviation.

*(c) Calculate the quartiles and the interquartile range.

*1.3-2 A manufacturer of metal alloys is concerned about customer complaints regarding the lack of uniformity in the melting points of one of the firm's alloy filaments. Fifty filaments are selected and their melting points are determined. The following results are obtained:

320	326	325	318	322	320	329	317	316
331	320	320	317	329	316	308	321	319
322	335	318	313	327	314	329	323	327
323	324	314	308	305	328	330	322	310
324	314	312	318	313	320	324	311	317
325	328	319	310	324				

(a) Construct a frequency distribution and display the histogram.

*(b) Calculate the mean, median, quartiles, sample variance, and standard deviation. How many observations lie within one standard deviation from the mean? Within two standard deviations?

Hint: In this and most of the exercises that follow, we recommend that you use statistical software.

*1.3-3 The following 64 observations are a sample of daily weekday afternoon (3 to 7 P.M.) lead concentrations (in micrograms per cubic meter, or $\mu g/m^3$) recorded at an air monitoring station near the San Diego Freeway in Los Angeles during the fall of 1976:

6.7	5.4	5.2	6.0	8.7	6.0	6.4	8.3	5.3	5.9
7.6	5.0	6.9	6.8	4.9	6.3	5.0	6.0	7.2	8.0
8.1	7.2	10.9	9.2	8.6	6.2	6.1	6.5	7.8	6.2
8.5	6.4	8.1	2.1	6.1	6.5	7.9	15.1	9.5	10.6

8.4	8.3	5.9	6.0	6.4	3.9	9.9	7.6	6.8	8.6
8.5	11.2	7.0	7.1	6.0	9.0	10.1	8.0	6.8	7.3
9.7	9.3	3.2	6.4						

(a) Construct a frequency distribution and display the results in form of a histogram. Is this distribution symmetric?

*(b) Calculate the mean, median, quartiles, interquartile range, sample variance, and sample standard deviation.

(c) Construct a box-and-whisker plot.

1.3-4   Show that the sample variance can be calculated from

$$s^2 = \frac{1}{n-1}\left[\sum_{i=1}^{n} x_i^2 - \frac{[\sum_{i=1}^{n} x_i]^2}{n}\right].$$

Hint: Write

$$\sum_{i=1}^{n}(x_i - \bar{x})^2 = \sum_{i=1}^{n}(x_i^2 - 2\bar{x}x_i + \bar{x}^2),$$

and recall that

$$\sum_{i=1}^{n} 2\bar{x}x_i = 2\bar{x}\sum_{i=1}^{n} x_i, \quad \sum_{i=1}^{n} \bar{x}^2 = n\bar{x}^2.$$

1.3-5   Obtain the last 50 annual precipitation totals for the city you live in. Construct a histogram. Calculate the mean, the quartiles, and the standard deviation. Interpret your findings.

Can you think of other informative data displays? What about plotting the annual totals over time? What can you learn from such a plot?

*   **1.3-6**   Take 12 eggs from your refrigerator and measure their lengths. Describe your measurement procedure, and discuss the factors that may have caused the variability in these 12 observations. Construct a dot diagram of your measurements, and calculate their mean, sample variance, and sample standard deviation.

Of course, one source of the variability is actual size differences, despite the fact that all these eggs probably come from a carton of large (or medium, or extra-large) eggs. Another source of the variability could come from your measurement procedure; for example, you may not always line up the egg in exactly the same way when you measure its length. To learn more about this measurement variability, obtain another round of measurement with the same 12 eggs. You will then have two measurements for each egg. Note, however, that these two replicated measurements are not obtained by lining up each egg once and then taking two measurements. If that were done, we would obtain information only on the variability that is due to the measurement instrument.

Display the variability in the resulting 24 observations. You could make a histogram of these 24 measurements, but if you do, you are not separating the sources of the variability—that is, the variability due to actual size differences and the measurement variability. Think about other ways of displaying your data which reflect the fact that you have obtained two measurements on each egg, and display the variability that is due to the measurement procedure.

1.3-7 Discuss how you measure the breaking strength of ordinary household plastic wrap. Obtain 10 measurements, display their variability, discuss the factors that affect your measurements, and comment on how we could learn more about their contributions to the variability.

1.3-8 Snee has measured the thickness of the "ears" of paint cans. (The "ear" of a paint can is the tab that secures the lid of the can.) At periodic intervals, samples of five paint cans are taken from a hopper that collects the production from two machines, and the thickness of each ear is measured. The results (in inches $\times$ 1,000) of 30 such samples are as follows:

Sample	Measurements				
1	29	36	39	34	34
2	29	29	28	32	31
3	34	34	39	38	37
4	35	37	33	38	41
5	30	29	31	38	29
6	34	31	37	39	36
7	30	35	33	40	36
8	28	28	31	34	30
9	32	36	38	38	35
10	35	30	37	35	31
11	35	30	35	38	35
12	38	34	35	35	31
13	34	35	33	30	34
14	40	35	34	33	35
15	34	35	38	35	30
16	35	30	35	29	37
17	40	31	38	35	31
18	35	36	30	33	32
19	35	34	35	30	36
20	35	35	31	38	36
21	32	36	36	32	36
22	36	37	32	34	34
23	29	34	33	37	35
24	36	36	35	37	37
25	36	30	35	33	31
26	35	30	29	38	35
27	35	36	30	34	36
28	35	30	36	29	35
29	38	36	35	31	31
30	30	34	40	28	30

Construct a histogram. Since there are many observations, it makes sense to use about 15 intervals of equal length. Interpret the shape of the histogram. You will notice that this histogram has two "humps," or modes. We call such a histogram *bimodal,* compared with the unimodal histograms that we have seen, for example, in Figure 1.3-2. What could be the reason for this bimodal nature?

Experiment with a different number k of classes. For example, try $k = 10$ and $k = 6$. You will notice that the two humps will eventually collapse into one. This shows that the choice of k can change the appearance of the histogram. Be aware that searching for a shape that serves the experimenter's purpose in the best possible way and ignoring all other shapes could be a misuse of histograms. Can you think of any other useful displays of this information?

Hint: What could you learn from a plot of the 30 sample averages over time?

[R. D. Snee, "Graphical Analysis of Process Variation Studies," *Journal of Quality Technology,* 15:76–88 (April 1983).]

*1.3-9 The following data on the carbon monoxide emissions of $n = 794$ cars, grouped into nonoverlapping intervals of 24 grams per mile, were taken from an article by Snee and Pfeifer [R.D. Snee and C. G. Pfeifer, "Graphical Representation of Data," in *Encyclopedia of Statistical Sciences,* S. Kotz and N. L. Johnson, eds., Vol. 3 (New York: Wiley, 1983), pp. 488–511]:

Interval	Midpoint, M_i	Frequency, f_i
0–24	12	13
24–48	36	98
48–72	60	161
72–96	84	189
96–120	108	148
120–144	132	85
144–168	156	45
168–192	180	30
192–216	204	10
216–240	228	5
240–264	252	5
264–288	276	1
288–312	300	2
312–336	324	1
336–360	348	1

*(a) Calculate the relative frequencies, the cumulative frequencies, and the relative cumulative frequencies. Construct a histogram and plot the ogive. Determine the median and the interquartile range.

*(b) Calculate the sample mean, the sample variance, and the sample standard deviation.

Hint: Here, we have listed the grouped data; the individual observations are not available. In the calculation of the mean and variance from such grouped data, we usually assume that within each class the observations are distributed fairly evenly and that it is reasonable to represent each observation in a given class by the midpoint $M_i, i = 1, 2, \ldots, k$. Then we approximate the mean as

$$\bar{x} = \frac{\sum_{i=1}^{k} f_i M_i}{n}$$

and the variance as

$$s^2 = \frac{\sum_{i=1}^{k} f_i (M_i - \bar{x})^2}{n - 1}.$$

To calculate the $(100P)$th percentile, or the quantile of order P, we must locate the observation with rank $r = Pn + 0.5$. Obtain the cumulative frequencies, and determine the class that includes the percentile—the class whose cumulative frequency is the first to exceed r. Denote the lower and upper limits of this class by L and U, respectively, the frequency of this class by f, and the number of observations that are smaller than L by m; of course, $m < r$. Then the $(100P)$th percentile is given by

$$L + \frac{r - m}{f} (U - L).$$

As an illustration, let us calculate the 20th percentile. Since $n = 794$, we have to locate the observation with rank $r = (0.20)(794) + 0.5 = 159.3 \cong 159$. This percentage is in the third class, from 48–72, which has frequency $f = 161$. Because the cumulative frequency of the class with lower limit $L = 48$ is $m = 13 + 98 = 111$, we find that the 20th percentile is given by

$$48 + \frac{159 - 111}{161} (72 - 48) = 55.2.$$

Of course, we could also obtain this value from the ogive, at least approximately. It is the point where the ogive crosses the horizontal line that is drawn at a relative cumulative frequency of $P = 0.20$.

1.3-10 (a) An analysis of human mortality data has led to the following frequency distribution of the length of life:

Age Group	Number of Deaths (1,000s)	Age Group	Number of Deaths (1,000s)
0–5	39.3	55–60	76.4
5–10	12.0	60–65	99.9
10–15	9.5	65–70	123.3
15–20	10.8	70–75	138.6
20–25	12.3	75–80	134.2
25–30	14.6	80–85	103.6
30–35	18.1	85–90	56.6
35–40	23.2	90–95	18.6
40–45	30.8	95–100	3.0
45–50	41.7		
50–55	46.7		

Construct a histogram. Determine the mean, mode, and median. Use the hint in Exercise 1.3.9.

(b) The Human Mortality Database (http://www.mortality.org), developed by the Department of Demography at the University of California, Berkeley, U.S.A, and the Max Planck Institute for Demographic Research in Rostock, Germany, contains a wealth of demographic data for many countries and for many periods. Consider the number of deaths for your country of interest and for a certain specified year, and display the distribution of the length of life. Using the appropriate graphical displays and numeric summary statistics, compare the distributions of the length of life for England and Wales in 1950 and 2000.

1.3-11 Use the data on compressive strength of concrete blocks given in Table 1.3-1. Complete a stem-and-leaf display, using double leaves with stems 2•, 3*, 3t, 3f, 3s, 3•, 4*, and so on.

1.3-12 Consider the data in Exercises 1.3-2, 1.3-3, and 1.3-8.

(a) Construct ordered stem-and-leaf displays.

(b) Construct box-and-whisker plots.

1.3-13 The following data represent $n = 38$ hurricane losses (in millions of dollars). Because these hurricanes occurred over a period of 30 years, the figures have been adjusted for inflation. The observations are already ordered from the smallest to the largest.

2.73	3.08	4.47	6.77	7.12	10.56
14.47	15.35	16.98	18.38	19.03	25.30
29.11	30.15	33.73	40.60	41.41	47.91
49.40	52.60	59.92	63.12	77.81	102.94
103.22	123.68	140.14	192.01	198.45	227.34
329.51	361.20	421.68	513.59	545.78	750.39
863.88	1638.00				

(a) Calculate the sample mean and sample standard deviation.

(b) The losses range from about zero to $1,650 million. If we construct a histogram with 10 equal-sized intervals, we find that 27 of the 38 observations are in the first class, from 0 to 165, and that some cells are completely empty. In these kinds of situations, involving skewed data, statisticians often use transformations such as $\sqrt{x}$ or $\log x$ to make the distribution more symmetric. For each of these transformations, construct a stem-and-leaf display and a box-and-whisker plot for the transformed hurricane data.

*1.3-14 Consider the following yields of 30 consecutive batches (read across):

72.4	75.3	72.7	74.2	75.6
75.0	73.3	73.1	74.4	77.5
75.8	74.0	73.2	74.8	76.2
72.7	68.2	79.1	73.8	75.5
74.7	75.7	72.9	72.1	77.7
76.6	73.2	75.1	75.5	71.7

Construct a time-sequence plot, a stem-and-leaf display, and a box-and-whisker plot for these $n = 30$ observations. If you had a detailed account of the special circumstances that accompanied each batch (i.e., if you knew that certain batches were produced by the day shift while others were produced by the night shift, or if you knew that some batches used raw material from supplier A while others used material from supplier B, or if you knew that a machine malfunctioned during batch 17), how would you use this information to assign some of the variability to these causes?

1.3-15 A set of 10 data values has a mean and a median of $1,425 and ranges from a minimum value of $987 to a maximum value of $1,945. Later, we discover that the $1,945 value was mis-recorded and should have been $2,945. Find the corrected values for the mean and median.

1.3-16 The summary statistics of 54 measurements on height (in inches) are $\bar{x} = 69$ inches and $s = 3.75$ inches. We decide to express height in centimeters; that is, $y = (2.54)x$, where y is height in cm and x is height in inches. What are the mean and the standard deviation of height expressed in cm?

1.3-17 The purchasing department of a major company has 100 employees, 22 of whom have no children, 50 of whom have exactly one child, and 28 of whom have exactly two children. Calculate the mean number of children. Calculate the median number of children.

*1.3-18 Ledolter asked the 54 students in his evening MBA statistics course questions about their height, weight, gender, and number of children at home. The results are listed as follows:

Height (inches)	Weight (pounds)	Number of children	Gender
71	170	0	m
69	179	0	f
68	200	5	m
66	130	3	f
68	138	3	m
70	168	0	m
72	165	0	m
64	125	0	f
72	230	1	m
76	210	0	m
74	225	2	m
74	230	0	m
71	160	0	m
68	145	0	m
70	163	1	m
72	190	1	m
69	179	1	m
62	115	0	f
66	150	0	f

Height (inches)	Weight (pounds)	Number of children	Gender
75	190	1	m
72	190	2	m
72	150	3	m
73	165	0	m
71	175	0	m
76	170	0	m
72	200	1	m
73	185	0	m
65	135	1	f
66	150	0	f
69	180	0	m
64	180	1	f
76	200	0	m
63	135	0	f
70	175	0	m
71	183	0	m
66	130	0	f
61	110	0	f
75	255	2	m
70	145	0	f
70	190	2	m
72	225	1	m
69	150	0	m
66	140	0	f
68	185	2	m
67	190	2	m
71	155	0	m
64	148	0	f
73	185	2	m
74	190	2	m
69	165	0	m
74	278	2	m
74	210	2	m
68	198	0	m
73	195	3	m

Discuss the nature of the variables (i.e., whether they are categorical or measurement variables). For measurement variables, compute summary statistics (such as the mean, median, and standard deviation) and construct dot plots, histograms, and box-and-whisker displays. Calculate the summary statistics of the measurement variables (height, weight, number of children) for each group (male, female) separately. Also, calculate the frequency table for gender and the joint frequency table for gender and number of children. Display the information graphically with bar and pie charts, and interpret the results.

1.4 Comparisons of Samples: The Importance of Stratification

Stratification (i.e., showing the results for appropriate groupings separately) is very important, as it brings out relationships among variables that otherwise would go unnoticed. Unfortunately, the correct stratification factors for solving a problem are often not known at the outset. Hence, one should keep track of, and record, all factors that change while the data are collected. This information may turn out to be very helpful at a later stage of the analysis. Good graphical displays are important for the visualization of differences, and appropriate summary statistics are needed to quantify the differences.

1.4-1 Comparing Two Types of Wires

Let us begin our discussion with an illustration taken from a case study by J. H. Sheesley and reported in R. D. Snee, L. B. Hare, and J. R. Trout, *Experiments in Industry* (Milwaukee, WI: American Society for Quality Control, 1985). Sheesley was comparing the performance of two different types of lead wires that were needed in the production of ordinary household light bulbs. The company had noticed that, during assembly, a certain percentage of lead wires did not feed properly into machines that produced components of these light bulbs. Because this resulted in lost production, the company changed the way it produced the lead wires. It was then interested in evaluating whether the lead wires produced under the new process performed better than those produced under the old one. In Sheesley's report, there were several other important variables besides the two types of wires, such as different machines, shifts, plants, and so on. We ignore these variables and suppose that each type of wire, new and old, was "being treated fairly"; that is, there were no biases created by using one type of wire under better conditions than the other (e.g., with a better shift, or running one type on Wednesday and the other on Monday). An experiment was conducted, and for each production run, the average number of leads per hour that missed were recorded. The results of 24 production runs, 12 with the old and 12 with the new type of lead wire, are given in Table 1.4-1.

It is always instructive to compare the histograms (or the dot diagrams if the number of observations in each group is small, as is the case here). To facilitate the graphical comparison, one should draw these histograms (or dot diagrams) on

Table 1.4-1 Average Hourly Number of Misfeeding Lead Wires												
Old:	17.6	18.3	10.8	19.2	18.0	39.4	21.4	19.9	23.7	22.7	23.2	19.6
New:	12.4	28.1	11.5	7.8	16.7	16.8	25.6	23.7	26.9	11.2	21.5	18.9

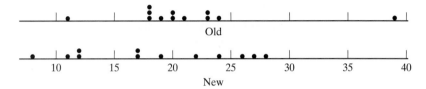

Figure 1.4-1 Dot diagrams for Sheesley's data in Table 1.4-1

the same scale. The two dot diagrams in Figure 1.4-1 suggest that the new process may have lowered the number of misfeeding leads somewhat. But the evidence from such few observations is certainly not strong, and this tentative conclusion is influenced by one large value for the old type of wire. In a later section, we study formal procedures that provide a quantitative assessment as to whether differences between two or more samples are "statistically significant." At any rate, it probably would be a good idea to collect more data.

1.4-2 Comparing Lead Concentrations from Two Different Years

Afternoon four-hour lead concentrations, in micrograms per cubic meter, were recorded next to the San Diego Freeway in Los Angeles during the fall of 1976 and the fall of 1977. Table 1.4-2 lists the weekday (Monday through Friday) concentrations. Weekend concentrations are not given here, as most of the lead originates from traffic, and traffic patterns for weekdays and weekends are quite different. Note that gasoline at that time was "leaded" and not lead free.

Table 1.4-2 Afternoon Four-Hour Lead Concentrations (in Micrograms per Cubic Meter) at the San Diego Freeway in Los Angeles during the Fall of 1976 and the Fall of 1977

Fall 1976										
6.7	7.6	8.1	8.5	8.4	8.5	9.7	5.4	5.0	7.2	6.4
8.3	11.2	9.3	5.2	6.9	10.9	8.1	5.9	7.0	3.2	6.0
6.8	9.2	2.1	6.0	7.1	6.4	8.7	4.9	8.6	6.1	6.4
6.0	6.0	6.3	6.2	6.5	3.9	9.0	6.4	5.0	6.1	7.9
9.9	10.1	8.3	6.0	6.5	15.1	7.6	8.0	5.3	7.2	7.8
9.5	6.8	6.8	5.9	8.0	6.2	10.6	8.6	7.3		
Fall 1977										
9.5	10.5	9.3	9.1	10.3	8.7	6.8	10.7	8.9	9.3	2.9
8.6	5.0	6.6	8.3	11.4	10.5	9.8	10.2	9.9	7.3	9.8
12.0	9.4	5.7	9.4	6.3	16.7	9.1	12.4	9.4	8.2	14.8
6.5	9.4	9.9	8.2	8.1	9.9	10.2	9.6	10.9	10.4	8.8
9.3	8.8	11.9	12.3	9.3	9.7	8.2	8.0	9.5	11.0	8.7
8.1	9.9	8.7	12.6	9.2	9.8	8.8	11.6	8.9		

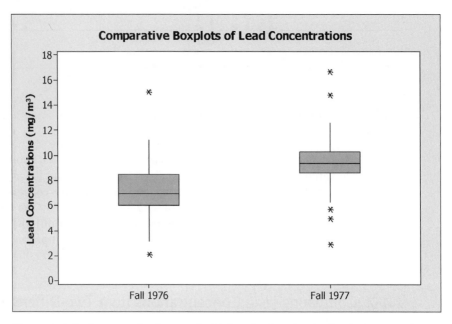

Figure 1.4-2 Comparative box-and-whisker displays for the lead concentrations (Minitab graph)

We are interested in a comparison of the 1976 and 1977 data. Is there a difference between the 1976 and the 1977 measurements? If so, how do they differ? To detect differences between the 1976 and 1977 data, we display the data on two dot diagrams, separately for the two years, but drawn on the same scale. Similarly, we can display the data through two box-and-whisker plots, one for each year. The box plots are shown in Figure 1.4-2. We also calculate, separately for each year, the relevant summary statistics, including the mean and the standard deviation; they are shown in Table 1.4-3.

The diagrams in Figure 1.4-2, as well as the summary statistics in Table 1.4-3, show that the 1977 observations tend to be higher than the ones for 1976. Initially, this was very surprising to us, as we expected the opposite to happen. Lead-free gasoline was introduced in the mid-1970s, and in 1977 more cars used lead-free gasoline than in 1976. Since virtually all lead roadside deposits come from automobile emissions, we expected the 1977 lead measurements to be lower than the ones for 1976. Why did we see a different result? The reason had to do with a traffic flow change that took place between 1976 and 1977. (See the 1979 paper by Ledolter and Tiao, listed in Exercise 1.4-1.) In 1976, the freeway consisted of four lanes in each direction.

Table 1.4-3 Summary Statistics for the Lead Concentration Data

Variable	Sample Size	Mean	Standard Deviation	Minimum	First Quartile	Median	Third Quartile	Maximum
Fall 1976	64	7.291	2.025	2.100	6.025	6.950	8.475	15.100
Fall 1977	64	9.422	2.081	2.900	8.625	9.400	10.275	16.700

In 1977, an additional, fifth, lane was added to the northbound direction. This change increased the traffic speed which, in turn, increased the lead emissions.

In addition to comparative histograms (dot diagrams) and box-and-whisker displays, there is another interesting graphical method, called a q–q plot, for comparing two samples or distributions. The reason for the name is that the $(100P)$th percentile is often called the *quantile of order P*, and in a q–q plot, quantiles of one sample (distribution) are plotted against corresponding quantiles of the other. By plotting pairs of corresponding quantiles (usually, it is sufficient to plot 10 to 15 pairs), we can get a good visual comparison of the two samples. To illustrate, consider the lead concentration data. Since both samples are of equal size, $n = 64$, it is convenient to use quantiles of order $(i - 0.5)/64, i = 1, 2, \ldots, 64$. That is, with equal sample sizes, we simply order the observations in each group and plot the order statistics of one group (here, 1976) on the vertical axis against the order statistics of the other (1977) on the horizontal axis.

A line with slope 1 is drawn through the origin to help with the comparison. If all points are on this 45-degree line, there is absolutely no difference between the two samples; in particular, their centers and their spreads are exactly the same, given that all their corresponding quantiles are equal. If all points are below the 45-degree line, however, the quantiles in the 1977 sample are larger than the corresponding ones in the 1976 sample. If, by contrast, all points are above the 45-degree line, the quantiles of the 1976 samples are larger than the ones in 1977. In Figure 1.4-3, all points are below the 45° line. Thus, there is strong evidence that the 1977 measurements are larger than the ones from 1976. This confirms the conclusion of the box-and-whisker plot.

From the q–q plot, we can also get information on the spreads of the two distributions. On the one hand, if the plotted points increase with a slope greater than 1, than the sample plotted on the horizontal axis is not as spread out as that plotted

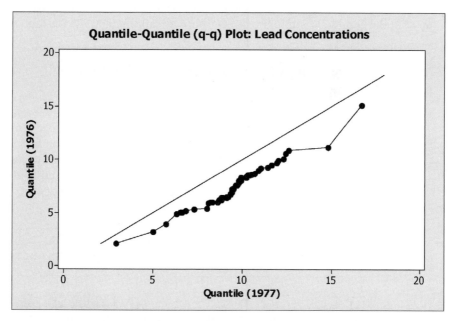

Figure 1.4-3 Quantile–quantile (q–q) plot for the lead concentrations

on the vertical axis. On the other hand, a slope of less than 1 means that the sample plotted on the horizontal axis is spread out more than that on the vertical axis. In Figure 1.4-3, we notice that the slope is roughly 1, confirming that the variability in the 1976 and 1977 data is about the same. This is consistent with the very similar standard deviations in Table 1.4-3.

The preceding discussion has illustrated a comparison of two samples. The examples that follow show how we can graphically assess the differences among three or more samples.

1.4-3 Number of Flaws for Three Different Products

A company divides its production of 4-foot by 8-foot wood-grained panels into three different quality groups. Ten samples of 100 panels each were taken from each group and the number of imperfections per 100 panels was counted. The results are listed in Table 1.4-4. The corresponding box-and-whisker displays are given in Figure 1.4-4.

It is quite obvious from this figure that there is a difference in the average number of imperfections among the three quality levels. Moreover, the variability (i.e., the spread of the distribution) becomes larger with poorer quality. It is interesting to note that the spreads, as measured by the respective interquartile ranges of 2, 5, and 7, are roughly proportional to the square root of the middle values, as measured by the medians 1, 4, and 10. From statistical theory, statisticians know that under such conditions a square-root transformation of these data should make the spreads about the same. In Figure 1.4-5, we show the box-and-whisker displays of $\sqrt{x}$, where x is the number of imperfections that are given in Table 1.4-4. The new respective interquartile ranges are 1.4, 1.4, and 1.1., which are now much more equal than the old ones. Reexpressing the data by taking a square-root transformation thus simplifies the comparison. On the square-root scale, the levels of the three groups differ, but the variations within the three groups are approximately constant.

1.4-4 Effects of Wind Direction on the Water Levels
of Lake Neusiedl

Lake Neusiedl is a large, shallow lake at the eastern border of Austria. The lake is long and narrow (extending 35 km in the north–south direction and 13 km in the east–west direction at its widest point). It covers about 285 square kilometers, and its deepest point is only 6 feet. The water level of the lake is affected by the prevailing winds. The water level at the city of Neusiedl, the northernmost point of the lake, is

Table 1.4-4 Number of Imperfections per 100 Panels

Quality Group	Observations									
I	0	2	1	1	0	1	0	2	3	0
II	4	1	7	1	4	0	2	5	6	9
III	6	7	18	4	14	10	10	15	11	9

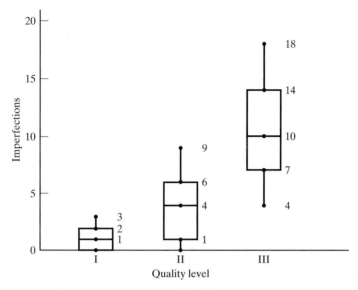

Figure 1.4-4 Comparative box-and-whisker displays for the number of flaws in Table 1.4-4
Note that the first and third quartiles are taken as the observation with ranks $(11)(0.25) = 2.75 \approx 3$ and $(11)(0.75) = 8.25 \approx 8$. Other software packages, such as Minitab, will interpolate linearly between the second- and third- (and eighth- and ninth-) largest observations. The basic conclusions will not change.

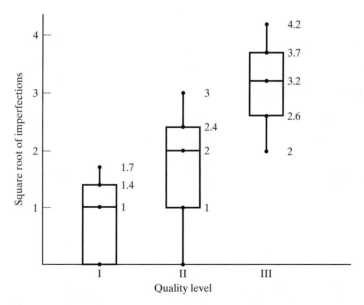

Figure 1.4-5 Comparative box-and-whisker displays for the transformed data in Table 1.4-4

Table 1.4-5 Summary Statistics for the Water-Level Differences at the City of Neusiedl

WD	N	Mean	StDev	Minimum	Q1	Q2	Q3	Maximum
4	36	−0.278	3.453	−13.0	−2.0	0.0	2.0	6.0
8 (East)	149	1.443	2.990	−10.0	0.0	2.0	3.0	9.0
12	803	3.501	3.260	−9.0	2.0	3.0	5.0	17.0
16 (South)	1091	3.856	4.162	−15.0	1.0	4.0	6.0	20.0
20	168	1.714	4.549	−9.0	−1.0	2.0	5.0	18.0
24 (West)	309	1.816	2.739	−8.0	0.5	2.0	3.0	8.0
28	2438	−1.292	4.268	−25.0	−3.0	−1.0	2.0	16.0
32 (North)	1039	−3.180	4.849	−31.0	−6.0	−2.0	0.0	10.0

thought to decrease with strong winds from the north as these winds push the water south. On the other hand, strong winds from the south push the water north, increasing the water level at Neusiedl.

Daily data from 1971 through 2004 are used in the analysis that follows. Water levels at Neusiedl and average water levels (determined from several stations throughout the lake) were obtained, and daily differences in water levels were calculated (in cm). Wind speed and wind direction at 2 p.m. were also available. We selected days with afternoon (2 p.m.) wind speeds between 12 and 38 km/h. Wind direction was coded as 4 (wind from northeast), 8 (from east), 12 (from southeast), 16 (from south), 20 (from southwest), 24 (from west), 28 (from northwest), and 32 (from north).

Summary statistics are shown in Table 1.4-5, and box plots (with the whiskers, but without the outliers) are shown in Figure 1.4-6. The data confirm that winds from

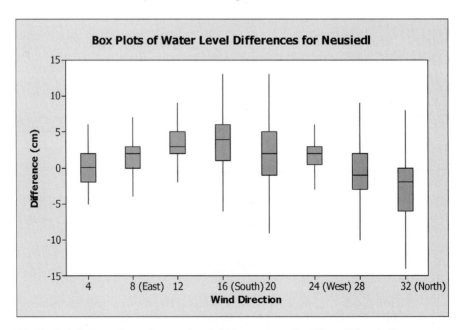

Figure 1.4-6 Box plots of water-level differences at the City of Neusiedl

the south and southeast increase the water level by 3–4 cm, while winds from the north and northwest decrease the water level by a somewhat smaller amount.

Exercises 1.4

*1.4-1   (Continuation of Exercise 1.3-3 and a check of the results in Section 1.4-2). During the fall of 1977, the weekday afternoon lead concentrations (in $\mu g/m^3$) at the measurement station near the San Diego Freeway in Los Angeles were as follows:

9.5	10.7	8.3	9.8	9.1	9.4	9.6	11.9	9.5	12.6
10.5	8.9	11.4	12.0	12.4	9.9	10.9	12.3	11.0	9.2
9.3	9.3	10.5	9.4	9.4	8.2	10.4	9.3	8.7	9.8
9.1	2.9	9.8	5.7	8.2	8.1	8.8	9.7	8.1	8.8
10.3	8.6	10.2	9.4	14.8	9.9	9.3	8.2	9.9	11.6
8.7	5.0	9.9	6.3	6.5	10.2	8.8	8.0	8.7	8.9
6.8	6.6	7.3	16.7						

*(a) Construct a histogram and a box-and-whisker display, and compare them with the ones you made for the 1976 data in Exercise 1.3-3. Make sure that you draw the 1976 and 1977 displays on the same scale. Interpret the displays.

(b) Using a q–q plot, compare the 1976 and 1977 samples. Here, the sample sizes for the two groups are the same, and you can plot the order statistics of one sample against the order statistics of the other. (You need not plot all pairs of corresponding order statistics.) How would you have proceeded if the sample sizes were not the same?

[J. Ledolter and G. C. Tiao, "Statistical Methods for Ambient Air Pollutants with Special Reference to the Los Angeles Catalyst Study (LACS) Data," *Environmental Science and Technology*, 13: 1233–1340 (1979).]

Remark: In the spring of 1977, a new traffic lane was added to the freeway. The new lane reduced traffic congestion, increased traffic speed, and thus increased lead concentrations.

1.4-2   The precipitation of platinum sulfide by three different methods led to the following results (already ordered by magnitude):

Method	Platinum Recovered from 10.12 Milligrams Pt						
1	10.28	10.29	10.31	10.33	10.33	10.33	10.34
2	10.22	10.23	10.23	10.24	10.27	10.28	10.29
3	10.22	10.25	10.30	10.33	10.36	10.38	10.38

Compare the quality of the three methods. Make the best graphical presentation and comparison you can. Use dot diagrams, box-and-whisker displays, q–q plots, and so on.

[J. W. Tukey, *Exploratory Data Analysis* (Reading, MA: Addison-Wesley, 1977).]

*1.4-3   Jaffe, Parker, and Wilson have investigated the concentrations of several hydrophobic organic substances (such as hexachlorobenzene, chlordane, heptachlor, aldrin, dialdrin, and endrin) in the Wolf River in Tennessee. Measurements were taken downstream of an abandoned dump site that had previously been used by the pesticide industry to dispose of its waste products.

It was expected that these hydrophobic substances might have a nonhomogeneous vertical distribution in the river because of differences in density between their compounds and water and because of the absorption of these compounds on sediments, which could lead to higher concentrations on the bottom. It is important to check this hypothesis because the standard procedure of sampling at six-tenths of the depth could miss the bulk of these pollutants.

Grab samples were taken with a La Motte–Vandorn water sampler of 1-liter capacity at various depths of the river. This sampler consists of a horizontal plexiglass tube of 7 centimeters diameter and a plunger on each side that shuts the sampler when the sampler is at the desired depth. Ten surface, 10 middepth, and 10 bottom samples were collected, all within a relatively short period. Until they were analyzed, the samples were stored in 1-quart mason jars at low temperatures.

In the analysis of the samples, a 250-milliliter water sample was taken from each mason jar and was extracted with 1 milliliter of either hexanes or petroleum ether. A sample of the extract was then injected into a gas chromatograph, and the output was compared against standards of known concentrations. The test procedure was repeated two more times, injecting different samples of the extract into the gas chromatograph. The average aldrin and hexachlorobenzene (HCB) concentration (in nanograms per liter) in these 30 samples was as follows:

Surface		Middepth		Bottom	
Aldrin	HCB	Aldrin	HCB	Aldrin	HCB
3.08	3.74	5.17	6.03	4.81	5.44
3.58	4.61	6.17	6.55	5.71	6.88
3.81	4.00	6.26	3.55	4.90	5.37
4.31	4.67	4.26	4.59	5.35	5.44
4.35	4.87	3.17	3.77	5.26	5.03
4.40	5.12	3.76	4.81	6.26	6.48
3.67	4.52	4.76	5.85	3.76	3.89
5.17	5.29	4.90	5.74	8.07	5.85
5.17	5.74	6.57	6.77	8.79	6.85
4.35	5.48	5.17	5.64	7.30	7.16

*(a) Consider aldrin. Compare the box-and-whisker displays for the surface, middepth, and bottom samples. Do you think that a transformation should be considered? If so, which transformation? Make appropriate plots to support your claim. Consider the logarithmic transformation and repeat your analysis. Do you think that the vertical distribution of this substance is homogeneous?

(b) Repeat the analysis with hexachlorobenzene.

[P. R. Jaffe, F. L. Parker, and D. J. Wilson, "Distribution of Toxic Substances in Rivers," *Journal of the Environmental Engineering Division*, 108: 639–649 (1982).]

I.4-4   Consider the data from experiments 4 and 5 in Project 1. Note that the target deviations are obtained from experiments that use slightly different versions of a feedback controller. Compare the histograms and box-and-whisker displays. Construct a q–q plot. Can you conclude that one controller leads to smaller target deviations than the other?

Hint: Use the Minitab calculator function "Percentile" to obtain the percentiles of order 0.1, 0.2, ..., 0.9 of each column (say, C1 for experiment 4 and C2 for experiment 5). That is, "Calc > Calculator > Percentile (C1,0.10)" for the 10th percentile, "Calc > Calculator > Percentile (C1,0.20)" for the 20th percentile, and so on. Store the results in two columns and execute the scatter plot function to create the *q–q* plot.

1.4-5   Consider the data in Project 3. Analyze the ratio (diesel use of test truck)/(diesel use of control truck) in part (d). Compare the ratios for baseline and test conditions. What conclusions do you reach?

1.4-6   A new emergency procedure was developed to reduce the time required to fix a certain manufacturing problem. Past data under the old system were available ($n = 25$). The staff was trained under the new procedure, and the response times for the next 15 occurrences of this manufacturing problem were recorded. The results are as follows:

Old procedure

    4.3   6.5   4.6   4.3   6.4   4.8   5.1   6.8   4.9   4.5   5.1   7.3   3.3
    5.0   4.6   7.0   5.1   3.8   5.2   4.1   5.7   4.6   5.9   3.1   6.2

New procedure

    6.2   4.0   3.3   4.5   2.3   3.0   3.2   6.0   3.7   4.5   5.3   4.0   5.4
    4.3   3.8

Compare the response times under the old and the new procedures. Are there differences in the mean responses? Discuss, using appropriate graphs and summary statistics. Would you switch to the new procedure?

1.4-7   Construct comparative dot diagrams and box-and-whisker displays for height and weight in Exercise 1.3-18, separately for the two gender groups.

1.5 Graphical Techniques, Correlation, and an Introduction to Least Squares

There are many statistical techniques that can help engineers and scientists make wise decisions, but one simple and always effective method is to display the relevant data in appropriate graphical form. We have illustrated the effectiveness of graphics in our previous discussion on time-sequence plots, dot diagrams, histograms, bar and pie charts, and box-and-whisker displays.

1.5-1 The *Challenger* Disaster

We now give a further illustration, using data from the *Challenger* explosion on January 28, 1986. The *Challenger* space shuttle was launched from Cape Kennedy in Florida on a very cold January morning. Meteorologists had predicted temperatures at launch to be around 30 degrees. The night before the launch, there was much debate among engineers and NASA officials as to whether a launch under

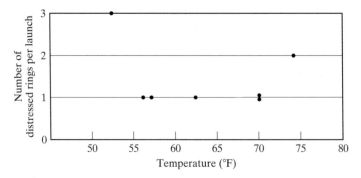

Figure 1.5-1 Scatter plot of the number of distressed rings per launch against temperature

such low-temperature conditions would be advisable. Several engineers advised against the launch because they thought that certain O-ring failures were related to temperature. Data on O-ring failures experienced in previous launches were available and were studied the night before the launch. There were seven known incidents of distressed O-rings. Figure 1.5-1 displays this information; it is a simple scatter plot of the number of distressed rings per launch against temperature at launch.

From this plot alone, there does not seem to be a strong relationship between the number of O-ring failures and temperature. On the basis of this information, along with many other technical and political considerations, it was decided to launch the *Challenger* space shuttle. As you all know, the launch resulted in disaster: the loss of seven lives and billions of dollars, and a serious setback to the space program.

One might argue that engineers had looked at the scatter plot of the number of failures against temperature, but could not see a relationship. However, this argument misses the fact that engineers did not display all the data that were relevant to the issue. They looked only at instances of failures and ignored the cases where there were no failures. In fact, there were 17 previous launches in which no failures occurred. A scatter plot of the number of distressed rings per launch against temperature using data from all previous shuttle launches is given in Figure 1.5-2.

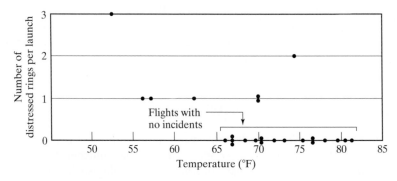

Figure 1.5-2 Scatter plot of the number of distressed rings per launch against temperature (all data)

It is difficult to look at these data and not see a relationship between failures and temperature. Moreover, one recognizes that an extrapolation is required and that an inference about the number of failures outside the observed range of temperature is needed. The temperature at *Challenger's* launch was only 31° F, while the lowest temperature recorded at a previous launch was 53° F. It is always very dangerous to extrapolate inferences to a region for which one does not have data. If NASA officials had looked at this plot, the launch would have been delayed.

The Rogers Commission, which investigated the *Challenger* disaster, saw the data in Figure 1.5-2 and reported that "a careful analysis of the flight history of the O-ring performance would have revealed the correlation of O-ring damage in low temperature." This example shows why it is so important to have statistically minded engineers involved in important decisions. Applied statisticians have told us time and time again, "In God we trust; others must have data."

This example raises two important points. First, it illustrates the importance of scatter plots in which we plot one variable against another. Earlier versions of such scatter plots are time-sequence plots, in which we plot the variable of interest against time. The other point that is made so well in the *Challenger* example is the importance of plotting *relevant data*. Yes, it is true that some data were used in making the decision. But not all the relevant data were utilized. To make good decisions, it takes knowledge of statistics, as well as subject knowledge, common sense, and an ability to question the relevance of information.

1.5-2 The Sample Correlation Coefficient as a Measure of Association in a Scatter Plot

Scatter plots are useful graphical tools, showing the nature and the strength of relationships between two variables. In this section, we give another example of a scatter plot and introduce a simple measure of association. Figure 1.5-3 shows a scatter plot

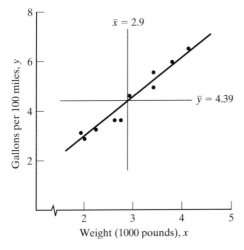

Figure 1.5-3 Scatter plot of fuel consumption (gallons per 100 miles traveled) against weight

Table 1.5-1 Weight x (1,000 Pounds) and Fuel Consumption y
(Gallons per 100 Miles)

Car	x	y	$(x - \bar{x})(y - \bar{y})$	$(x - \bar{x})^2$	$(y - \bar{y})^2$
AMC Concord	3.4	5.5	0.555	0.25	1.2321
Chevy Caprice	3.8	5.9	1.359	0.81	2.2801
Ford Country Squire Wagon	4.1	6.5	2.532	1.44	4.4521
Chevette	2.2	3.3	0.763	0.49	1.1881
Toyota Corona	2.6	3.6	0.237	0.09	0.6241
Ford Mustang Ghia	2.9	4.6	0.000	0.00	0.0441
Mazda GLC	2.0	2.9	1.341	0.81	2.2201
AMC Sprint	2.7	3.6	0.158	0.04	0.6241
VW Rabbit	1.9	3.1	1.290	1.00	1.6641
Buick Century	3.4	4.9	0.255	0.25	0.2601

Source: Data from H. V. Henderson and P. F. Velleman, "Building Multiple Regression Models Interactively," *Biometrics*, 37: 391–411 (1981).

of fuel consumption y (in gallons of fuel per hundred miles traveled) against weight x (in 1,000 pounds) of $n = 10$ automobiles. The data are presented in Table 1.5-1; the plot reveals several features:

1. It shows a *positive* association between the two variables: As expected, heavier cars require more fuel than lighter ones.

2. It also shows a relationship that is roughly *linear*: Over the range of weights studied, from 1,900 to 4,100 pounds, each added unit of weight increases fuel consumption by roughly the same amount. In Figure 1.5-3, we have added the "best-fitting" straight line; there is much more on that in Section 1.5-3 and in Chapter 8.

3. The plot also shows that the relationship between these two variables is fairly *strong*: The scatter around the best-fitting line is quite tight.

Scatter plots tell us graphically what these relationships are like. Sometimes, however, it is useful to attach to these plots a numeric value that assesses the association in terms of a single number. This is especially useful if one compares different data sets (e.g., scatter plots of fuel consumption against weight for domestic and imported cars, or scatter plots for different model years). The *sample correlation coefficient* is one such measure. It quantifies the degree of *linear* association among two variables. It is important to emphasize that it picks up only the linear association; it may miss more complicated associations whereby the variables are related in a non-linear fashion. There is always a danger when one characterizes a scatter plot with just a single summary statistic, as the statistic used may miss certain aspects of the data that are discernible to the eye; the sample correlation coefficient is no exception.

In Figure 1.5-3, we have also drawn in a vertical line at the average $\bar{x}$, and a horizontal line at the average $\bar{y}$. These lines divide the scatter into four quadrants. The deviations $x_i - \bar{x}$ and $y_i - \bar{y}$ are important components in developing this measure of association. Standardized deviations, $(x_i - \bar{x})/s_x$ and $(y_i - \bar{y})/s_y$, are in fact preferable, as we are looking for a measure of association that does not depend on the unit of measurements. Since $(x_i - \bar{x})$ and s_x are expressed in terms of the same unit, their ratio is dimensionless and is not affected by (linear) changes in the measurement unit. For example, it does not matter whether we measure weight in pounds, 1,000 pounds, kilograms, or tons, or fuel consumption in gallons, quarts, or liters.

The sample correlation coefficient depends on the products

$$[(x_i - \bar{x})/s_x][(y_i - \bar{y})/s_y].$$

In case of a positive association, these products are mostly positive, as larger (smaller) than average y-values usually occur together with larger (smaller) than average x-values. Negative products are not as common with positive association, because it is rarer to observe larger (smaller) than average y-values with smaller (larger) than average x-values. Of course, in case of a negative association, most of the products

$$[(x_i - \bar{x})/s_x][(y_i - \bar{y})/s_y]$$

are negative.

The sample correlation coefficient r is the "average" of these n products:

$$r = \frac{1}{n-1} \sum_{i=1}^{n} \left(\frac{x_i - \bar{x}}{s_x} \right) \left(\frac{y_i - \bar{y}}{s_y} \right)$$

$$= \frac{\sum (x_i - \bar{x})(y_i - \bar{y})}{(n-1)s_x s_y} = \frac{\sum (x_i - \bar{x})(y_i - \bar{y})}{\sqrt{\sum (x_i - \bar{x})^2} \sqrt{\sum (y_i - \bar{y})^2}}.$$

Note that, as with the definition of the sample variance and sample standard deviation, we have divided the sum of the n products by $(n-1)$. Adopting this particular definition, one can show that the sample correlation coefficient is always in the range $-1 \le r \le 1$.

The sample correlation coefficient measures the direction and the strength of a linear association. The *sign* of the coefficient tells us about the direction of the association; positive values imply a positive or direct association, and negative values imply a negative or indirect one. The *absolute value* of this coefficient tells us about the strength of the linear association. A coefficient of $+1$ or -1 implies that there is a perfect linear association; in such a situation, the scatter plot shows that all points lie exactly on a straight line. A coefficient of zero, by contrast, implies that there is no linear association.

As an illustration, we use the fuel consumption and weight of the 10 automobiles listed in Table 1.5-1 to calculate the sample correlation coefficient. We find that, rounded to three decimal digits, the sample means and sample standard deviations are given by $\bar{x} = 2.9$, $s_x = 0.759$, $\bar{y} = 4.39$, and $s_x = 1.273$. The last three columns in the table list the products $(x_i - \bar{x})(y_i - \bar{y})$ and the squares $(x_i - \bar{x})^2$ and $(y_i - \bar{y})^2$. Substituting sums of these columns into the expression for the sample correlation coefficient, we obtain

$$r = \frac{8.49}{\sqrt{(5.18)(14.589)}} = 0.977,$$

which implies a positive association that is very strong. This confirms the conclusions that we have drawn from the scatter plot in Figure 1.5-3.

Remark: There is yet another equivalent expression for the correlation coefficient, namely,

$$r = \frac{\sum x_i y_i - \left(\sum x_i\right)\left(\sum y_i\right)/n}{\sqrt{\left[\sum x_i^2 - \left(\sum x_i\right)^2/n\right]\left[\sum y_i^2 - \left(\sum y_i\right)^2/n\right]}}.$$

This expression is computationally useful, as it avoids some of the rounding errors that are made with the earlier formula. For the data in Table 1.5-3, we find that $\sum x_i = 29.0$, $\sum y_i = 43.9$, $\sum x_i^2 = 89.28$, $\sum y_i^2 = 207.31$, and $\sum x_i y_i = 135.8$. Accordingly,

$$r = \frac{135.80 - (29.0)(43.9)/10}{\sqrt{[89.28 - (29.0)^2/10][207.31 - (43.9)^2/10]}} = 0.977.$$

Fortunately, computer software is readily available to carry out these calculations. (See Section 1.7.)

Properties of the Correlation Coefficient The correlation coefficient has several useful properties:

1. The correlation coefficient is always between -1 and $+1$.

2. The *sign* of the correlation coefficient tells us about the direction of the association. A positive value indicates a positive (or direct) association; if one variable is larger than average, then the other variable also tends to be larger than average. A negative value indicates negative (or indirect) association; if one variable is larger than average, then the other tends to be smaller than average.

3. The *absolute value* of the correlation coefficient indicates the strength of the association. The correlation coefficient $r = +1$ implies that all points (x_i, y_i) lie exactly on a straight line with a positive slope. The correlation coefficient $r = -1$ implies that all points lie exactly on a straight line with negative slope. The further away the correlation coefficient is from -1 or $+1$, the weaker is the linear association. The correlation coefficient $r = 0$ means that there is no linear association between the two variables.

4. The correlation coefficient measures only the *linear association*. You can understand this statement by considering the situation in which equally spaced observations lie exactly on a circle of a certain fixed radius. Such a situation reflects a perfectly deterministic, but nonlinear, relationship. However, the numbers of observations in the four quadrants of the scatter plot of y against x are the same, and the (linear) correlation coefficient is zero. This is a limitation of the correlation coefficient, and we recommend that you always plot the data. A scatter plot includes more information than the correlation coefficient gives, as it is not always possible to characterize a scatter diagram by just a single number. A scatter plot can tell us whether the relationship between the two variables is more complicated than a linear association. Nonlinearity may be present.

5. Note that, for the correlation coefficient, it does not matter which variable is selected as the x or the y variable. The variables enter the correlation coefficient

through their product, and which variable is called x does not matter. Thus, the correlation between fuel consumption and weight is the same as the correlation between weight and fuel consumption.

6. Correlation does not immediately imply causation. The fact that two variables are correlated does not necessarily mean that a cause-and-effect relationship exists between them. Consequently, that y tends to be larger than average whenever x is larger than average cannot be interpreted as revealing that increases in x "cause" y to rise. Beware of the effects of "lurking" variables. Two variables may appear heavily correlated, but this correlation is not due to a direct link between them. There may be a third variable that affects both variables; we call this third variable the lurking variable. As an example, take one variable that increases with time and another that decreases with time. Then a scatter plot of the first variable against the second will show a strong negative association. Here, "time" is the lurking variable, and the apparent association between the two variables does not imply that there is a causal relationship between them. One can cite many other examples to demonstrate that correlation and causation should not be confused. One study showed that the consumption of alcohol is positively correlated with teachers' salaries: The higher the salaries, the greater is the consumption of alcohol. But it isn't that teachers rush out to cash in their latest pay rise at the nearest bar; rather it seems that underlying economic conditions affect both variables. During good times, people spend more on alcohol as well as on teachers.

7. In correlation studies, we should be aware of one other pitfall: The range of the variables that are being related may be very narrow, and with such limited-range data it may not be possible to obtain strong relationships. In fact, over a narrow range, it often looks like there is no relationship at all. Only if one could observe the variables over a wider range would relationships become evident. As an example, consider the relationship between high school and college grade point averages for students enrolled at your university. Obtain the relevant data from the registrar's office, and convince yourself that the correlation between these two variables is rather weak. Why is the correlation so low? One reason is that students accepted to your institution have fairly similar academic backgrounds. Your university probably does not admit many students with poor high school GPAs. The range of one of your variables is quite narrow, and as a result, the correlation coefficient is rather low.

1.5-3 Introduction to Least Squares

In Sections 1.5-1 and 1.5-2, we introduced scatter plots. Also, in Figure 1.5-3, we plotted the fuel consumption data in Table 1.5-1 along with the "best-fitting" straight line. We now discuss in what sense this line is best fitting among all possible straight lines that could be drawn through the points shown.

More generally, suppose that we wish to fit a curve $y = h(x; \alpha, \beta)$ that depends on two parameters α and β to n pairs of points, $(x_1, y_1), (x_2, y_2),...,(x_n, y_n)$. The height of the curve at x_i is $h(x_i; \alpha, \beta)$, and the height of the observed point is y_i. The distance between these heights is $|y_i - h(x_i; \alpha, \beta)|$, and the square of this distance measures, in some sense, how badly the curve misses the point. In the *method of least squares,* we select the parameters α and β so as to minimize the sum of the squares of these n distances — hence the name "least squares." That is, we find α and β to minimize

$$S(\alpha, \beta) = \sum_{i=1}^{n} [y_i - h(x_i; \alpha, \beta)]^2.$$

In the special case when $h(x; \alpha, \beta) = \alpha + \beta x$ is a linear function of x, this is a rather easy mathematical problem. Here,

$$S(\alpha, \beta) = \sum_{i=1}^{n} (y_i - \alpha - \beta x_i)^2.$$

Setting the two first partial derivatives equal to zero, we obtain

$$\frac{dS(\alpha, \beta)}{d\alpha} = \sum_{i=1}^{n} 2(y_i - \alpha - \beta x_i)(-1) = 0,$$

$$\frac{dS(\alpha, \beta)}{d\beta} = \sum_{i=1}^{n} 2(y_i - \alpha - \beta x_i)(-x_i) = 0.$$

These two linear equations in α and β are equivalent to

$$\sum_{i=1}^{n} y_i = n\alpha + \left(\sum_{i=1}^{n} x_i \right)\beta,$$

$$\sum_{i=1}^{n} x_i y_i = \left(\sum_{i=1}^{n} x_i \right)\alpha + \left(\sum_{i=1}^{n} x_i^2 \right)\beta.$$

If we multiply the first equation by $\sum x_i$ and the second by n and then subtract the two results (of course, this eliminates the α term), we obtain the solution, say, $\hat{\beta}$ for β, namely,

$$\hat{\beta} = \frac{n\sum x_i y_i - \left(\sum x_i\right)\left(\sum y_i\right)}{n\sum x_i^2 - \left(\sum x_i\right)^2} = \frac{\sum x_i y_i - \left(\sum x_i\right)\left(\sum y_i\right)/n}{\sum x_i^2 - \left(\sum x_i\right)^2/n} = \frac{\sum(x_i - \bar{x})(y_i - \bar{y})}{\sum(x_i - \bar{x})^2}.$$

This estimate of slope is closely related to the sample correlation coefficient $r = \sum(x_i - \bar{x})(y_i - \bar{y})/(n - 1)s_x s_y$ discussed in Section 1.5-2. Since $\sum(x_i - \bar{x})^2 = (n - 1)s_x^2$, it follows that

$$\hat{\beta} = \frac{(n - 1)rs_x s_y}{(n - 1)s_x^2} = r\frac{s_y}{s_x}.$$

Finally, if we substitute this result into the earlier equation

$$\sum y_i = n\alpha + \left(\sum x_i \right)\beta,$$

we obtain the estimate of the intercept:

$$\hat{\alpha} = \bar{y} - \hat{\beta}\bar{x}.$$

The best-fitting line in the sense of least squares is given by

$$\hat{y} = \left(\bar{y} - r\frac{s_y}{s_x}\bar{x}\right) + \left(r\frac{s_y}{s_x}\right)x$$

$$= \bar{y} + \left(r\frac{s_y}{s_x}\right)(x - \bar{x}),$$

where $\bar{x}$, $\bar{y}$, s_x, s_y, and r are the computed characteristics of the data.

For the data in Table 1.5-1, these minimizing values are

$$\hat{\beta} = 1.639, \quad \hat{\alpha} = -0.363$$

The slope 1.639 implies that for each additional 1,000 pounds of weight, driving 100 miles requires an additional 1.639 gallons of fuel. Since our data set includes only cars with weights between 1,900 and 4,100 pounds, it does not make sense to extrapolate the least-squares line to $x = 0$; thus, the intercept $\hat{\alpha} = -0.363$ by itself has no real meaning.

Remarks: Statisticians refer to the activity of fitting a curve of the form $h(x; \alpha, \beta) = \alpha + \beta x$ as *regression modeling,* and they talk about "regressing" the response variable y (here, fuel consumption) on the explanatory variable x (the weight of the car). Chapter 8 discusses regression modeling in great detail, including the fitting of equations that include more than one explanatory variable and the fitting of nonlinear relationships.

Computer software for fitting such regression models is widely available, and the user can ignore some of the details that have been explained here for the simplest (that is, linear) case. In your course, you will use packages such as Minitab, SPSS, SAS, Excel, or R, and it will be very easy to make scatter plots, calculate correlation coefficients, and obtain the fitted least-squares line. In Minitab, for example, you can use the commands under "Graph" (for scatter plots), "Stat > Descriptive" (for obtaining correlation coefficients), and "Stat > Regression" (for the least-squares fit). It is instructive to go through the (tedious) calculations once, at least for a small data set. But after you have done this, we expect you to use a program for the analysis.

Exercises 1.5

1.5-1   Consider the aldrin and hexachlorobenzene (HCB) concentrations of the 30 water samples listed in Exercise 1.4-3. Explore the relationship between these two variables. Make a scatter plot of aldrin against HCB. Do you think that there may be a problem with some of the observations? Which are the most likely candidates?

Calculate the sample correlation coefficient. Calculate r, with and without the suspicious observation(s).

*1.5-2   By the method of least squares, fit $y = \beta x$ to the following $n = 6$ points:

x	3	1	5	6	3	4
y	4	2	4	8	6	5

Note that here we are forcing the line through the origin, and the least squares estimate is different from the formula given in this section. Draw the fitted line on the scatter plot.

1.5-3   By the method of least squares, fit $y = \alpha + \beta x^2$ to the following $n = 8$ points:

x	−2	3	−1	0	−3	1	5	−3
y	7	15	3	1	11	6	30	16

*1.5-4   R.A. Fisher presents the following data, which give the logarithm to the base 10 of the volume occupied by algal cells on successive days, taken over a period for which the relative growth rate is approximately constant:

Day (x)	log Volume (log y)
1	3.592
2	3.823
3	4.174
4	4.534
5	4.956
6	5.163
7	5.495
8	5.602
9	6.087

Plot log y against x. Do you think that the logarithmic transformation is appropriate? If so, why? Calculate and interpret the sample correlation coefficient. By the method of least squares, fit log $y = \alpha + \beta x$ to the $n = 9$ data points.

[R. A. Fisher, *Statistical Methods for Research Workers* (Edinburgh: Oliver & Boyd, 1925).]

1.5-5   Tukey discusses the relationship between the vapor pressure of B-trimethyl-borazole (y, in millimeters of mercury) and temperature (x, in °C):

x	y	x	y
13.0	2.9	56.1	51.4
19.5	5.1	64.4	74.5
22.5	8.5	71.4	100.2
27.2	10.3	80.5	143.7
31.8	14.6	85.7	176.9
38.4	21.3	91.5	216.9
45.7	30.5		

Plot y against x. How can we straighten this plot? Try various transformations on y and/or x; for example, try square root and logarithmic transformations.

[J. W. Tukey, *Exploratory Data Analysis* (Reading, MA: Addison-Wesley, 1977).]

*1.5-6   In a statistical analysis of air pollution data, Ledolter and Tiao analyzed hourly carbon monoxide (CO) averages that were recorded on summer weekdays at a measurement station in Los Angeles. This particular station was established by the Environmental Protection Agency as part of a larger study to assess the effectiveness of the catalytic converter. The measurement station was located about 25 feet from the San Diego Freeway, which in this particular area is located at 145° north. Winds from 145° to 325° (which in the summer are the prevalent wind directions during the daylight hours) transport the CO emissions from the highway toward the measurement station. For each hour from 1 to 24, we have listed the average summer weekday CO concentration (in parts per million), the average weekday traffic density TD = traffic count/traffic speed, and the average perpendicular wind-speed component:

$$WS_p = \text{wind speed} \times \cos(\text{wind direction} - 235°).$$

Hour	CO	TD	WS_p	Hour	CO	TD	WS_p
1	2.4	50	−0.2	13	5.8	179	4.6
2	1.7	26	0.0	14	5.5	178	5.4
3	1.4	16	0.0	15	5.9	203	5.9
4	1.2	10	0.0	16	6.8	264	5.9
5	1.2	12	0.1	17	7.0	289	5.6
6	2.0	41	−0.1	18	7.4	308	4.9
7	3.4	157	−0.1	19	6.4	267	3.8
8	5.8	276	−0.2	20	5.0	190	2.5
9	6.8	282	0.2	21	3.8	125	1.4
10	6.6	242	1.0	22	3.5	120	0.6
11	6.6	200	2.3	23	3.3	116	0.4
12	6.3	186	3.8	24	3.1	87	0.1

(a) Make a map of the area and discuss why WS_p is a meaningful measure.

(b) Construct diurnal plots of CO, the traffic density TD, and the perpendicular wind-speed component. (Diurnal plots are plots of a variable against hour of the day.)

*(c) Interpret your graphs. For example, does it appear that CO is proportional to TD and that WS_p plays the role of transport as well as diffusion?

(d) Ignore traffic density for the moment, and assume that there is information on wind speed and direction. How can you change the diurnal CO plot such that it also shows wind speed and direction?

Hint: Represent wind direction by the vane direction on 360° with north on top, and represent wind speed by the length of the vane.

[J. Ledolter and G. C. Tiao, "Statistical Methods for Ambient Air Pollutants with Special Reference to the Los Angeles Catalyst Study (LACS) Data," *Environmental Science and Technology,* 13: 1233–1240 (1979).]

1.5-7 Consider the times it took Ledolter to ride his bicycle from his office in the Engineering School to the Princeton train station when he was a visitor during the academic year 1985–86. He usually left his office shortly after 5 p.m., as he had to catch a train at 5:20.

Date	Time (seconds)	Date		Time (seconds)
Sept. 3, 1985 (Tuesday)	290	Sept.	19	305
4	305		20	360
5	310		23	340
6	365		24	275
9	280		25	320
10	300		26	305
11	410		27	385
12	305		30	280
13	345	Oct.	1	315
16	260		2	320
17	295		3	290
18	290		4	330

Many factors affected his time. The most important were traffic conditions: He had to cross two major busy roads and at times had to wait, especially on Fridays, when the traffic was particularly heavy. Other factors were the weather conditions, his energy level, the condition of the bicycle, and whether he stopped along the way to chat with someone (e.g., this was the case on September 11).

Analyze the data. Display the variability. Check whether day of the week had an influence on the times. Check whether the times got faster (or slower) with experience.

Conduct a similar experiment and record the time it takes you to get to a certain class. Obtain measurements for the next 20 days. Keep a log of the special circumstances that affect your times, and analyze the 20 observations in any way you think is appropriate.

*1.5-8   The following $n = 15$ pairs of observations are the ACT Math and ACT Verbal test scores of 15 students:

(16, 19)	(18, 17)	(22, 18)	(20, 23)	(17, 20)
(25, 21)	(21, 24)	(23, 18)	(24, 18)	(31, 25)
(27, 29)	(28, 24)	(30, 24)	(27, 23)	(28, 24)

Construct a scatter plot of Math against Verbal test scores and calculate the correlation coefficient. Find the best-fitting straight line by the method of least squares.

*1.5-9   The following data set lists the annual 2005 salary (in $1,000) and the educational background of a sample of 25 employees at a large midwestern manufacturing company:

Educ	Salary	Educ	Salary
16	52.3	17	49.4
12	43.7	16	45.4
12	39.5	13	41.3
16	47.8	12	37.6

Educ	Salary	Educ	Salary
18	53.0	12	33.3
15	49.0	19	64.8
11	33.7	16	50.7
12	32.1	17	54.5
11	9.8	16	27.3
20	37.7	12	14.8
15	26.3	16	21.7
16	22.0	16	33.8
16	27.0		

Educational background is measured by the number of years of formal schooling (12 refers to a high school graduate, 16 to a college graduate, and 17 through 20 to an employee with a college degree plus the indicated number of years of graduate work).

Construct a scatter diagram of salary against educational achievement. Calculate the correlation coefficient. Determine the least-squares estimates of the coefficients in the linear model $h(x; \alpha, \beta) = \alpha + \beta x$. Interpret the results.

1.5-10 Explain, in your own terms and with your own examples, why a large correlation may not be a sign of causation. [What about annual numbers of storks and annual numbers of births for the post–World War II period?]

*1.5-11 Construct a scatter plot of weight against height for the data given in Exercise 1.3–18. Identify males and females on the scatter plot by different symbols (or color). Calculate the correlation coefficients for the combined data set, as well as for each group (male/female) separately.

1.6 The Importance of Experimentation

The information that one obtains by listening to ongoing processes gives a picture of what those processes are like. Such data provide the bases of *descriptive (enumerative)* studies. However, one must keep in mind that the analysis of such data represents a passive use of statistics, because the data occur without any active participation of the investigator. Although much can be learned by "listening" to a process, much more can be learned by active questioning, testing, and experimenting, much as a doctor does with a patient. It is through these *comparative (analytic)* studies that we learn how to improve our products and services.

Observing a patient's responses on cardiac tests with several different stress levels will provide more information than can be obtained from measurements taken during normal resting periods. The same is true for industrial processes: Much can be learned about a process by changing, according to a specified plan, the factors that are thought to have an influence. Measurements on how a process responds to such changes provide the investigator with valuable information on the effects of those factors. Such investigations involve the collection and analysis of data from carefully planned experiments. *Design of experiments,* a very important area of statistics, teaches engineers and physical scientists how to perform experiments that ensure the validity of experimental results and that lead, in relatively few experimental runs, to precise estimates of the effects of various factors on the response.

1.6–1 Design of Experiments

A poorly designed experiment may not shed light on a particular question that we would like to have answered. Thus, it is important to have an experimental plan *before* information is collected; trying to rescue the experiment at the data analysis stage *after* it has been run usually fails. Consulting a statistician after an experiment is finished often amounts simply to asking for a postmortem examination; the statistician cannot save the investigation at this point and can only suggest from which oversight it has failed. Consequently, it is important to design the experiment such that the validity of the experimental results is assured.

In good experimental designs, relatively few runs are needed to get precise estimates of the effects on the response of the factors studied. R.A. Fisher, an eminent statistician and scientist who developed this subject, said that a complete overhaul of an experimental design may increase the precision of the results ten- or twelve-fold for the same cost in time and labor.

Example 1.6-1 This example illustrates the importance of carefully planned experiments. Suppose that it is thought that one possible reason for the increased fraction of defectives in a manufacturing process is a change in the supplier of one of the raw materials needed in production—say, the ink in our ballpoint pen example of Section 1.2-2. To follow up on this hypothesis, the plant engineer produces a total of 40 lots of pens, 20 from the old raw material, *A,* and 20 from the new one, *B*. Since the plant can manufacture only 10 lots a day, it takes all day Thursday and Friday to produce the 20 lots from the old material and all day Saturday and Sunday to produce the 20 lots from the new material. Subsequent sample inspection shows that the old material includes 1.5 percent defectives, whereas the new one includes 5.8 percent. The engineer takes this difference as conclusive evidence that the new material is indeed worse than the old one. A worker on the assembly line, however, raises the point that on Thursday and Friday the line is staffed by experienced workers, whereas on Saturday and Sunday the company uses inexperienced part-time help. He documents his observation by pointing out that in the past the rejection rates for weekend production were usually higher than those on weekdays. Thus, he says, the difference in the percentage of defectives could just as well be due to the different experience levels of the workers. ∎

Of course, the worker is correct. In this experiment, we observed only experienced workers who use the old raw material and inexperienced workers who use the new one. We say that the two factors are *confounded*. From this experiment alone, it is not possible to separate the effects of these two factors. The observed difference in the percentage of defectives could be due to differences in raw materials, worker experience, or perhaps both.

The preceding example is an example of a poorly designed experiment. How could we have done better? If we believe that experience matters, each group of workers should have produced pens from both raw materials. Thus, the weekday group, as well as the weekend group, should have processed 10 lots from A and 10 lots from B. If we fear that the process varies somewhat from day to day, then each day we should produce 5 lots with A and 5 lots with B. Such a design could "block out" a possible day effect because comparisons could then be made *within* each day. The day represents a block, and the principle of allocating both raw materials (i.e., the different treatments) to each block is called *blocking* an experiment.

The original experiment has failed here because we have chosen a design in which our variable of interest is completely confounded with another variable that is believed to have an influence on the response. Usually, we can think of many variables besides the variable of interest that could influence the response. In our example, it was "day," but it could be batch, furnace run, location, plant, and so on. We refer to factors that can affect the response, but that are not of main interest, as the *blocking variables*. A good design makes sure that the levels of the variable of interest are not confounded with the levels of other variables.

If we know how these other factors change over the course of the experiment, we can always choose a design that avoids confounding. But what should we do if these factors and the way they change over the course of the experiment are unknown? How can we protect the validity of our results and guard against the effects of all these unspecified and unknown factors? The answer to this is *randomization,* which is another important principle in the design of experiments. In the context of our example, we would randomize the order of the experiments within each day. For example, the particular order of the 10 runs on Thursday (the 5 from A and 5 from B) would be randomized. We can get such a random arrangement by putting into a bowl five paper slips designated A and five slips designated B, mixing the slips, and drawing one after another. For example, the sequence $ABAABABBBAB$ would specify the order in which the experiments should be carried out. We start with a lot that uses material A, the next uses B, the next two use A, and so on. We would repeat such a drawing three more times; the resulting sequences would give us the order for the experiments on Friday, Saturday, and Sunday. This randomization spreads the risk of uncontrolled factors evenly over the levels of our factor of interest. A random arrangement would be much safer than a deterministic pattern in which, on each day, the five A's are run first and the five B's second; the deterministic arrangement would be particularly bad if there are trends in the quality of the production process.

In summary, the principles of *blocking* and *randomization* ensure the validity of the experimental results. If we know of certain factors that may affect the comparison, we should block the experiment with respect to those factors and make sure that the experiment is designed so that comparisons can be made within each block. To guard against all other unspecified factors, we should randomize the treatments within each block. The risk that this random order is confounded with some other important factor is quite small. Thus, we can say that randomization guarantees the validity of the inference in the face of unspecified disturbances. A simple rule of thumb is to *block what you can and randomize what you cannot block.*

Example 1.6-2   Let us consider another experiment. Suppose that you are an engineer for a major car manufacturer. Your job is to design a new disk for front disk brakes that is more durable than the one currently in use. After experimentation, you come up with a new component alloy that you think is more durable than the old one. You have already made several test runs in the laboratory and have found that the new material is indeed better than the old one. However, before management makes a decision to install these disks in the new cars, they want you to run an experiment that shows this advantage under general driving conditions.

For your study, you select a sample of 12 brand-new cars that are all of the same type. In 6 of the cars, you install the new experimental disks in the front brakes. You have just learned the importance of randomization. Even though you were told that the 12 cars are identical, you divide these 12 cars at random into two groups. You carry out the randomization by assigning a number to each car and drawing numbers at random. The first six numbers drawn identify the cars on which the new brake disks are installed. Next, you have to assign the 12 cars to 12 test drivers, who drive the cars for 5,000 miles. You know that there will be differences among drivers; there are always a few of us who drive faster and use the brakes more often than others. Thus, you allocate the drivers to the cars at random. You can do this by preparing a list of the 12 drivers, putting 12 slips of paper that identify the cars into a bowl, and selecting one slip after another. (These randomizations can be done with the computer.) The first car selected is driven by the first driver on your list, the second car by the next driver on your list, and so on. After you have made these assignments, the test drivers are instructed to use the cars for their normal day-to-day driving. After 5,000 miles, the cars are brought back to the factory and the loss of material on the disks is measured.

When the experiment is finished, you obtain 12 measurements, 6 on each material. Probably the first thing that you should do with this information is to plot the material loss on a simple line graph and construct a dot diagram. As in Figure 1.6-1, you may want to use two different symbols to distinguish between the new material, A (e.g., use ×) and the old material, B (say, ∘). You could also calculate the averages of these two groups of six, indicated by the arrows. Note that, on the basis of tests of 12 cars, the average loss of material from brake pads made of material A is slightly smaller than that for pads made of material B. However, does this really mean that material A is better than material B? Probably not, as the samples are small and there is large variability among the individual measurements. Referring to Figure 1.6-1, we see that the "×'s" and the "∘'s" are pretty well mixed up. We also notice that cars equipped with type A disks lead to the smallest, as well as the largest, material loss. Judging from this diagram—and thus from our experiment in general—we cannot say that this small difference in the averages is sufficient evidence that A is better

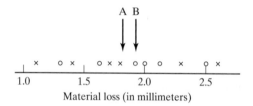

Figure 1.6-1 Loss of brake disk lining for 12 cars. Material A was used for the disks on 6 of the cars; their resulting losses are marked by ×. Material B was used on the remaining 6 cars, and their results are denoted by ∘. The arrows indicate the averages for A and B

than B. Of course, our engineer is very disappointed, since this evidence conflicts with his findings in the lab. What went wrong?

Although we have spread the possible driver effect evenly among the two treatments, the price that we paid is the rather large variability among our measurements. We could have done much better. Blocking with respect to driver and car would have eliminated a major source of variability. It is easy to block here, because we can install both types of disks in each car. We can put one type on the left front wheel and the other type on the right. Actually, we should make this assignment at random, assigning type A to the left wheels of six randomly selected cars and to the right wheels of the remaining six. Such random assignment would guard against right–left biases. At the conclusion of this experiment, we obtain two measurements on each of the 12 cars (drivers): one on A and one on B. The analysis of the difference eliminates the variability that is due to the cars and drivers. Because cars and drivers are major contributors to the variability, the blocking and the analysis of the differences together make the comparison between A and B much more precise. This is what R. A. Fisher meant when he said that a complete overhaul of the experimental design may often increase the precision of the results ten- or twelve-fold for the same cost in time and effort. Figure 1.6-2 shows this benefit very clearly. The two measurements on each car show that the material loss for A is less than that for B. This experiment provides strong evidence that material A is indeed better. However, if the experiment had not been blocked with respect to drivers, the conclusion would have been quite different: The summary diagram at the bottom of the figure shows little difference between A and B.

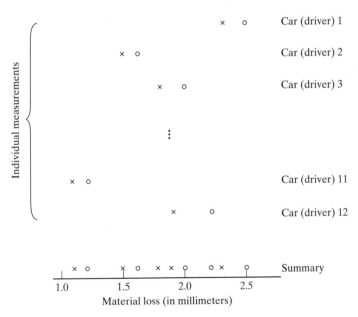

Figure 1.6-2 Loss of brake disk lining from an experiment with 12 cars. Both materials were mounted on each car ($\times$ corresponds to type A, $\circ$ to type B). To simplify the graphical comparison, we have shown only the results of five cars

Example 1.6-3   Suppose that we want to compare the fuel efficiencies of two specific trucks, A and B. A possible, but very bad, experimental design is to obtain fuel consumption measurements on n consecutive test runs of fixed length with truck A, followed by n consecutive runs with truck B. Since the runs are not carried out under the same weather, road, and traffic conditions, it is impossible to make a fair comparison. A design that runs the trucks in tandem, described in Project 3, is preferable, as the fuel consumption of both trucks is obtained under the same conditions. An analysis of the ratios (or the differences) in fuel consumption for the n runs will tell us which of the two trucks is the more efficient one.

A comparison of truck fuel efficiencies may not be the real goal of the investigation. In Project 3, it is the question whether a certain fuel additive improves fuel efficiency, and the question which of the two trucks is more fuel efficient is of minor interest. Nevertheless, we can use the "tandem-type" design to address this question, too. We can compare the ratio in fuel consumption that we obtained when both trucks use the same fuel with the ratio that is obtained when one of the trucks uses the additive. A difference in these ratios tells us whether the fuel additive has an effect.   ∎

1.6-2   Design of Experiments with Several Factors and the Determination of Optimum Conditions

There is usually a rather long list of factors (variables) that are thought to have an effect on the response variable. For example, the yield of a chemical process may depend on the reaction time, on the pressure and the temperature under which the reaction is completed, on the concentration of a certain additive, on the type of catalyst that is used in the reaction, and so on. The roughness of the finish of certain metal strips in a metallurgical process may depend on factors such as the solution temperature, the solution concentration, the roll size, and the roll tension. The impurity of a certain chemical product may depend on the type of polymer that is used, the polymer concentration, and the amount of a certain additive. How should we conduct experiments that allow us to estimate the effects of these factors, and how should we locate regions that lead to the largest (or smallest) possible response? In the case of yield as the response, we would look for a maximum; in the case of impurity or roughness, we would look for a minimum.

If there is only one factor, this problem is relatively straightforward. We would ask the operator in charge of the process, or a chemist, or an expert in metallurgy about the range over which the factor should be varied and would inquire into what level of the factor is currently being used. Then we would conduct several experiments at various levels of the factor, measure the responses, plot the responses on graph paper, and locate the direction in which we should move in order to increase (or decrease) the response.

But how should we proceed if there are $k = 5$ or $k = 10$ factors, particularly if there are constraints on the numbers of experimental runs? On the one hand, if the experiments are performed online (i.e., during the normal production schedule), the operator is going to be reluctant to make too many changes. Also, there is always the concern that with some of the changes recommended, the output may in fact be poor and substantial losses may result. On the other hand, experiments in the laboratory, away from the production process, are often time consuming and expensive. Of course, the cost of an experiment depends on the particular setting. Usually, however, the information from a well-designed experiment will outweigh the cost of experimentation.

Before you read further, think how you would design an experiment if you wanted to study the effects of $k = 5$ factors on the yield of a certain process. Assume that you know the feasible range of these five factors and the settings at which the process is currently run. In particular, think about how to specify the values of the five factors, called the design variables, and how you would analyze the results. By now, you are probably convinced that you need a plan or a design for choosing the levels of the design variables and a method for analyzing the results.

One way, but in many cases a very bad one, is to change the factors one at a time. To make our discussion simpler, assume that there are only $k = 2$ factors—say, temperature and pressure—that affect the yield of a certain chemical reaction. In the "*change one factor at time*" approach, the experimenter fixes one of the factors at a certain level—say, temperature at 220°C—and conducts runs at various levels of the second variable, pressure—say, at 80, 90, 100, 110, and 120 pounds per square inch (psi). She analyzes the yields at these levels, plots the responses on graph paper (yield against pressure), and locates the pressure that leads to the highest yield. Suppose that the maximum for a fixed temperature of 220°C is obtained when the pressure is 100 psi. She then fixes the pressure at this value and changes the temperature; suppose that she conducts experiments with temperatures of 180, 200, 220, 240, and 260°C. Again she analyzes the results, plots the yields against temperature, and finds the maximum; assume that it is somewhere between 220 and 240°C, say, 230°C. From these experiments, the engineer claims that the process runs optimally at a temperature of 230°C and a pressure of 100 psi.

Figure 1.6-3(a) shows that the engineer's conclusion is *wrong,* and it illustrates the problems with the "change one factor at a time" approach. In this graph, we have connected the settings of temperature (x_1) and pressure (x_2) that lead to the same response (y); these connecting curves are called *contours* and the plot is called a *contour plot.* We can obtain these contours by plotting the response y for given values of x_1 and x_2 in the three-dimensional (x_1, x_2, y) space and then slicing through this surface at various heights y; here, the y's are chosen as 70, 65, and 60. This contour plot shows that the point (temperature $= 230°C$, pressure $= 100$ psi) which we have obtained by changing the factors one at a time is *not* the maximum. The maximum is achieved at a temperature of 270°C and a pressure of about 85 psi.

The "change one factor at a time" approach has failed here because it assumes that the effects of changes in one factor are independent of the other factor; it assumes that the maximum with respect to one factor is independent of the settings of the other. This is not true in the illustration given in Figure 1.6-3; thus, the "change one factor at a time" approach fails.

To locate the optimal region, we have to change the factors *together.* For example, it is much better to start the experiment at the points (temperature $= 200°C$, pressure $= 100$ psi), (220°C, 100 psi), (200°C, 110 psi), and (220°C, 110 psi); see Figure 1.6-3(b). Observations at these four points would tell us to move in the northwest direction (i.e., increase temperature and lower pressure). Additional experiments could be performed on this path, and usually we would locate the maximum with very few runs.

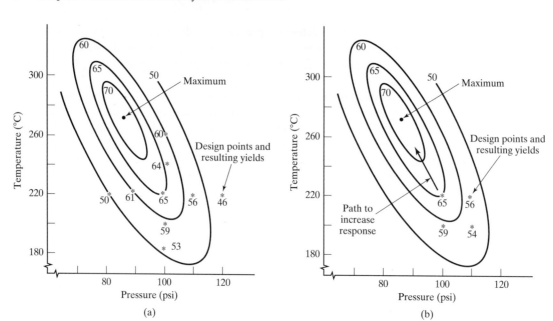

Figure 1.6-3 Contour plot of a response surface that involves two factors, temperature and pressure: (a) "Change one factor at a time" approach (the *incorrect* approach); (b) changing the factors together (the *correct* approach)

This simple example shows quite clearly that we should change the factors together, not one at a time. In this book, we discuss very simple, but extremely useful, arrangements that allow us to determine the effects of several factors on a response and to determine whether the effects of one factor depend on the levels of the other. [Note that the four runs given in Figure 1.6-3(b) show that the effect of a 20°C temperature increase, from 200°C to 220°C, is three times larger at a pressure of 100 psi than at 110 psi.] In these designs, we select a low and a high level for each factor and conduct experiments at every possible factor–level combination. The design in Figure 1.6-3(b) is one such example with $k = 2$ factors; there, we start with $2^2 = 4$ runs. If there are $k = 3$ factors, we have to conduct experiments at $2^3 = 8$ different factor–level combinations. The results of these experiments, properly analyzed, give us information on the effects of the various factors and tell us how to change the factors to increase the response.

Exercises 1.6

1.6-1   Purchase a bag of rubber bands of different lengths, widths, and colors. You are interested in the breaking strength of rubber bands. In particular, you would like to know whether length, width, or color affects the strength. Think about appropriate ways of measuring strength and discuss how you would carry out the experiment.

Carry out a small experiment and analyze your data in the form of graphs and simple summary statistics.

1.6-2 There are many other experiments with rubber bands of different sizes and colors that you could carry out.

   (a) For example, you could shoot rubber bands at a target some distance away and measure the accuracy of your shot (i.e., the deviation from the specified target). Study how the distance to the target and the diameter of the rubber band affect your shooting accuracy. What about the effects of color? Does it make a difference if several individuals shoot at the target?

   (b) Study the percent elongation of cut rubber bands that you obtain by adding a certain weight to one end of the band. How does this elongation depend on the length, width, and color of the rubber band? How does it depend on the weight?

   (c) Study the travel distance that you obtain when shooting rubber bands from a ruler. In particular, stretch the rubber bands to 8, 9, and 10 inches on your ruler, and let go. Do the width, the length, and the color of the rubber band make a difference? Have fun carrying out these experiments and discuss the results.

1.6-3 You would like to learn how the distance from the basket and the angle to the basket affect your shooting percentage in basketball. Design an experiment that allows you to study these two factors. Make sure that your results are not confounded by other factors, such as possible "fatigue" or "time of day" effects. Carry out the experiment and present your results.

1.6-4 Discuss the importance of the following experimental design principles: replication, randomization, blocking, and the sequential nature of experiments. Explain these concepts with simple illustrations (such as designs for the comparison of two types of tennis shoes, for the comparison of three methods of teaching statistics to engineers, and so on).

*1.6-5 In multifactor investigations, it is important that one study the response of a process at all possible factor–level combinations. Consider a process that depends on three factors, and suppose that you study each factor at two levels: low and high. How many different factor–level combinations are involved? Represent the experiments as the corner points of a cube. Select the responses at the eight levels to illustrate that a "change one factor at a time" approach to experimentation would miss the optimum of the process.

1.7 Available Statistical Computer Software and the Visualization of Data

1.7-1 Computer Software

You probably appreciate by now that the "hand calculation" of summary statistics, such as the mean and standard deviation in Section 1.3 and the correlation coefficient and the fitted least-squares line in Section 1.5, turns out to be a pretty cumbersome task, especially if the number of observations is large. Also, the construction of displays such as histograms, box plots, scatter plots, and contour diagrams becomes tedious if they have to be carried out from scratch. Fortunately, none of this is necessary, as many excellent statistical software packages are available for personal computers. Most programs work pretty much the same way, and it is not difficult to switch from one program to another. Most packages are spreadsheet programs. You enter data into columns of a spreadsheet, and you use simple commands for carrying out the desired operations. Commands for the various graphing procedures (such as time-sequence plots, scatter diagrams, and histograms) and for statistical analyses (such as describing the data through summary statistics and carrying out a regression least-squares fit) are executed through convenient pull-down windows.

Minitab, SPSS, SAS, and JMP are commonly used packages, and you may use one of them in your course. Many graphs and summaries presented in the first chapter were used with one or another of those packages, and in subsequent chapters we will utilize these packages even more.

Minitab is particularly easy to use. You enter the data into columns of a spreadsheet and use tabs on the command line to carry out the desired operations on the data. "Stat > Basic Statistics > Display Descriptive Statistics" will give you summary statistics such as the mean, median, and standard deviation. "Graph > Histogram" will result in a histogram of the observations that had been stored in specified columns. "Stat > Regression > Fitted Line Plot" (with the specified columns for the response and the explanatory variable) will result in the graph shown in Figure 1.5-3.

The software R is also very useful. It has two advantages over the aforementioned commercial packages: (1) Representing the joint, coordinated effort of hundreds of statisticians, R is both excellent technically and inexpensive. In fact, it is free and can be downloaded from http://cran.us.r-project.org. (2) The package is very general. It doesn't "box" you into the preprogrammed procedures that are available in the commercial packages and allows for a "richer" analysis. A disadvantage is that R is somewhat more difficult to learn, because the analysis in R is currently executed through a command language. Although the commands are fairly simple, an analysis with R involves a bit more than the dragging and clicking operations of the spreadsheet packages.

In general, these packages are computationally efficient and accurate. There is a certain area of statistics that focuses on computational aspects. The standard deviation is a good example for illustrating the ideas behind good statistical computing. The definition of the standard deviation in Section 1.3, namely,

$$s = \sqrt{\frac{\sum_{i=1}^{n}(x_i - \bar{x})^2}{n - 1}},$$

is informative, as it shows what s stands for. However, the definition is not the best equation for computation, especially if the data set is large. This is because rounding is commonly involved in the calculation of the average, and rounding gets carried into every one of the squared deviations from the average. Rounding errors can add up, especially if the data set is large. The alternative expression (see Exercise 1.3-4)

$$s = \sqrt{\frac{1}{n - 1}\left\{\sum_{i=1}^{n}x_i^2 - \frac{\left[\sum_{i=1}^{n}x_i\right]^2}{n}\right\}}$$

avoids the rounding and is better suited for calculation. Similar comments apply to the correlation coefficient discussed in Section 1.5.

1.7-2 The Visualization of Data

William S. Cleveland, in the preface to his book, *Visualizing Data,* writes,

Visualization is critical to data analysis. It provides a front line of attack, revealing intricate structure in data that cannot be absorbed in any other way. We discover unimagined effects, and we challenge imagined ones.

Graphs, charts, and diagrams are very important instruments for reasoning about quantitative information. Often, the most effective way to describe, explore, and summarize a set of numbers is through pictures of these numbers. Pictures force us to notice what we never expected to see. A picture may indeed be worth a thousand words.

Graphical methods are useful at all stages of statistical analysis, and data visualization tools need to be integrated into the analysis. Cleveland argues that relying exclusively on numerical statistical methods without appropriate data visualization is a crippling strategy that can lead to incorrect and missed conclusions. Graphical information, in combination with prior knowledge of the subject under investigation, becomes a powerful tool.

There are many variations of the graphical displays that we have presented in this chapter, and there are many more techniques that we haven't mentioned. We recommend that you study the book by John W. Tukey, *Exploratory Data Analysis,* the books by William S. Cleveland, *Visualizing Data* and *The Elements of Graphing Data,* and any of the marvelous books by Edward R. Tufte, *The Visual Display of Quantitative Information, Envisioning Information, Visual Explanations,* and *Beautiful Evidence.*

The focus of our discussion in this chapter is on univariate displays (dot diagrams, histograms, box plots, and stem-and-leaf plots, in all of which we display the distribution of a single variable) and on bivariate displays (scatter plots, which display relationships among two variables). Cleveland and Tufte also show how to display more than two variables. They discuss, in great detail and in much clarity and beauty, how to analyze multivariate data structures that involve three or more variables and include factors such as time and space. This type of analysis is very important, because data sets are nearly always multivariate. With three variables, one can stratify the scatter plot of two variables with respect to the values of the third variable—an approach that works especially well if the third variable is categorical. For example, imagine a scatter plot of tool wear against cutting speed. You may have data on two different tools: a standard tool and one that is reinforced with a certain alloy. It is quite informative to overlay the two scatter plots, using different symbols or colors for the two different tools.

Tufte pays great attention to the *design* of statistical graphics. He views excellence in graphics as the well-designed truthful presentation of interesting data. Excellence in graphics involves communicating complex ideas with clarity, precision, and efficiency. It gives the viewer the greatest number of ideas in the shortest time, with the least ink, and in the smallest place.

Graphs need to be effective. Cleveland and Tufte have much to say about the principles of good graph construction, and their books are full of useful practical advice. Let us list some of the important principles behind good graphical displays: Make the data stand out, and avoid superfluity. Show the data and avoid unneeded "chart junk." Use visually prominent graphical elements to show the data. Do not overdo the number of tick marks. Do not use too many data labels in the interior of the graph; that way, they don't interfere with the quantitative data shown. Add reference grids if you want to draw attention to certain values. Choose appropriate scales, as visual perception is affected by proportions and scale. For ready comparisons, use the same scale in comparing data from different groups or panels. Be aware of the effect of "zero," because the way the zero is located on a graph may change your perception of the data. Make sure that your graphs tend toward the horizontal and are greater in width than in height.

Graphs need to be truthful to the data. An important principle in graph construction is that the representation of the numbers, as physically displayed on the graph, should be proportional to the numerical quantities they represent. A common way to distort the truth is to deviate from that principle. A pie chart in which the areas depicted are not proportional to the frequencies being represented is one example of a misleading representation. Displaying the magnitude of numbers by squares whose length is proportional to the number is another example of a deceptive display. To see this, suppose that one number is twice as large as the other; then the larger square would be *four* times the size of the smaller one. Incorrect and nonuniform scales and unclear labeling are further sources for creating impressions that are not truthful to the data. Cutting off the bottom part of bars in a comparative bar chart creates a wrong impression.

Most graphs can be constructed with pencil and paper, as long as one understands the general principles of good graphing. In today's computer age, virtually all statistical software packages and spreadsheet programs include many different options for graphical displays. Computers have certainly changed the way we carry out the graphics; however, they have not affected the goals of the analysis. While modern spreadsheet software makes it quite easy to produce graphics, not all of the displays they create are good. Tufte is highly critical of graphs in Excel, which—as he writes—are loaded with "chart junk" such as unneeded cute ornamental hatching and three-dimensional perspectives.

Exercises 1.7

1.7-1   Familiarize yourself with the computer program that is being used in your course. Compare it with Excel, a commonly available Microsoft product. In Excel, look at the Excel commands under "Tools > Data Analysis"; if you don't see them listed, you need to load that macro first. The Excel pivot table feature (under "Data") is useful for analyzing categorical information. In Minitab, study the commands under "Graph" and "Stat" (especially Table, Basic Statistics, and Regression). In SPSS, look at "Graphs" and "Analyze" (especially Descriptive Statistics and Regression).

*1.7-2   Search the Web for one of Edward Tufte's books. Read sample pages and get a sense what he is trying to say. Summarize his main points.

Chapter 1 Additional Remarks

Understanding variability in processes is extremely important. The processes in question may be personal processes, such as making free throws in a pickup game of basketball or preparing the morning coffee before heading off to school; work processes, such as manufacturing certain widgets during your summer internship; service processes, such as answering user questions while staffing a consumer hotline; and so on. In all these cases, you try to reduce undesirable variation, move your process to a better level, and achieve this in a steady, continual fashion by listening and taking observations.

We are reminded of W. Edwards Deming, an esteemed statistician, who went to Japan after World War II and taught the Japanese the use of statistical methods and how to implement his philosophy of quality improvement. For his work, he was awarded the Emperor's Medal, and the Japanese established the *Deming Prize,* to be awarded each year to the company or individual contributing the most to quality improvement.

It was not until the 1980s that Americans paid much attention to Deming's ideas. At that time, Deming started to give short courses on his philosophy of quality, the highlight of which was his famous "red beads" experiment. He brought a box with over 1,000 beads, about 20 percent red (defective), and a paddle that held exactly 50 beads. His audience did not know the exact proportion of defectives. He selected about six "willing workers" from the audience, who then had to dip the paddle into the box and remove a set of beads. Deming gave detailed instructions on how to do this, emphasizing the "correct" way of scooping with just the right amount of "shaking" and the appropriate "angle" of the paddle. Two inspectors would count the number of defectives (i.e., the number of reds) and report the results to a secretary. The first worker, say, Mike, happened to have 9 defectives. "We can do better than that," Deming would declare and advise a little more shaking before drawing out the paddle. The next worker, Joe, had 13 defectives. Deming would "give him hell" because, in his words, the paddle came out at the wrong angle. This would continue until Mary ended up with 6 defectives. "That's more like it," Deming would praise, and, pointing to the next worker, he would say, "Tim, you saw how she did this." Tim might end up with 12 defectives and would get the question, "Didn't you learn anything from Mary?" Occasionally, someone would get 16 defectives and be fired, although Deming was known to encourage managers to retrain, not fire, those who needed help. The selection process would continue for well over an hour, with Deming playing the role of the tough boss encouraging the participants to "work harder and reduce the number of defects."

The "red beads" experiment was a good and also very entertaining illustration, and it communicated a very important point. Members of the audience, as well as the willing participants, knew that the only way to reduce the number of defectives, on average, was to improve the process—and this meant removing a number of red beads from the box. It is the managers and the engineers who can facilitate improvements, usually by redesigning and changing the underlying processes. Workers cannot make major improvements to the system by just trying harder; if the system is bad, such good efforts will not help. Workers may have good ideas for improvements, but it is the "owner" of the process (the one responsible for the process and in charge of changes) who can effect improvements. Deming also realized that few valuable suggestions will come to the surface if people are afraid to speak up and question the status quo. "Driving out fear" was another principle that Deming communicated to his audience.

There is a story that statisticians tell about Ralph Sampson, who was a great basketball player at the University of Virginia in the early 1980s and was drafted and signed to a lucrative contract by the National Basketball Association's Houston Rockets. Supposedly, Sampson was a major in communication studies, and that department reported that the average staring salary of its graduates was much higher than that of graduating engineers. This, of course, was due mainly to Sampson's high starting salary, which was included in calculating the average. Clearly, it would have been more appropriate for the department to report the *median* salary, which would have been considerably lower than that of engineers. So be careful, and know what type of typical value is being reported.

Of course, although the measure of the middle of the variation is an important characteristic, so is the measure of spread. For example, in Deming's red-bead experiment, how much variation around the typical value can we expect from numbers of defectives in repeated trials? If, in fact, 20 percent of the items in the box are defective, statisticians would say that they are quite confident that the number of reds in one trial of 50 beads will be somewhere between 2 and 18. We will learn how to compute the chance of this event in the next chapter.

We stress that it is important to design experiments properly. It is of interest to note that the famous (perhaps the *most* famous) statistician Ronald A. Fisher had much to do with the design of experiments. He was knighted for his contributions, as were other statisticians, such as Sir Maurice G. Kendall and Sir David R. Cox. This shows that society understands that statisticians do worthwhile things.

Everyone agrees that Sir Francis Galton was a great scientist. He contributed to many fields, and one of them was statistics. One of his rules was "Whenever you can, count." He followed this guiding principle in all his investigations as he truly believed in the importance of data. Some might say that he was the *first* statistician. While this is difficult to prove, he certainly made some of the early key discoveries in regression and correlation.

John Tukey died a few years ago, at the age of 85. Not only did he have many excellent ideas as a statistician, such as the stem-and-leaf display in exploratory data analysis, he also coined terms such as "software" (as opposed to hardware) and "bit" (for binary digit), and he rediscovered and made popular the fast Fourier transform. One of his many notable quotes should be remembered by all engineers: "Far better an approximate answer to the right question, which is often vague, than an exact answer to the wrong question, which can always be made precise." We think that he was taking a shot at mathematicians, although Tukey was excellent in that subject also.

Projects

Project 1

Machine tools in the metal-cutting industry have to machine parts within close tolerances, sometimes in the range of ±0.0005 inch. The tools involved in the cutting process wear with time. These changes cause drifts in the dimensional properties between the cutting tool and the work piece. The drift pattern is monitored and controlled.

In this project, we study several tool offset systems for a Cincinnati Milacron Cin-turn lathe. The study focuses on the bore size of various sizes of gear blanks for large tractors. Several different correction algorithms are developed and tested in the manufacturing environment. The algorithms estimate the drift in the measurement of the bore and recommend suitable tool offsets.

For boring operations, the drift may be in the positive direction, which means that the bore of the part has become oversized, or in the negative direction, implying that the bore of the part has become too small. In either case, it becomes necessary to offset the tool in the direction opposite that of the observed change and of such a magnitude that the dimensions of the parts are restored to acceptable limits.

Measurements of the finished bore size are taken by the operator, and on the basis of these measurements, the required tool offset for the next piece is calculated. The feedback control algorithm is implemented by a computer program. Note that adjustments are highly desirable in this context, as the observations come from a changing system. This is different from the "tampering" situation in a stable system.

The data sets that follow represent the outcomes of several experiments on a Cincinnati Milacron lathe. The first three experiments are conducted to obtain *process capability data*. This means that the tool is kept at its nominal position and is not changed over time. These data tell us about the capability of the process when it is not adjusted. The measurements (deviations from target, in inches $\times 10^5$) are as follows (read across):

Experiment 1 ($n = 22$)

15	0	10	5	5	10	−10	15	15	0
5	−5	−25	−15	−15	−15	−20	−15	−10	5
−20	−40								

Experiment 2 ($n = 30$)

70	75	70	65	65	50	55	40	35	10
10	10	20	−5	0	35	45	15	20	20
−10	−10	−5	−15	15	−15	−20	5	−15	−10

Experiment 3 ($n = 50$)

75	55	40	40	25	45	45	50	45	35
20	25	20	25	10	15	10	5	−5	−5
15	15	10	10	−10	−15	5	10	20	15
−5	5	5	5	0	15	5	15	10	0
0	15	20	5	20	0	5	−15	−5	−15

The results for experiments 4 and 5 that follow summarize the deviations that are obtained after implementing a feedback controller. The tool offset depends on the readings that are collected on previously machined parts. If the deviations from the target are within certain control limits (called the deadband), the readings are treated as good as nominal, and no offsets are made. If readings are outside these limits, the algorithm, in its simplest version, calculates the offset at a percentage of the (negative) deviation of the bore measurement. Two experiments with slightly different versions of this feedback controller are carried out, with the following results:

Experiment 4 ($n = 70$)

7.5	0	−2.5	−10	17.5	7.5	17.5	7.5	7.5	−2.5
−2.5	12.5	−10.0	2.5	−7.5	5.0	−7.5	2.5	−5.0	−7.5
−10	−5	7.5	7.5	7.5	7.5	7.5	5	17.5	−17.5

7.5	5	7.5	0	7.5	0	7.5	−2.5	7.5	7.5
7.5	5	7.5	7.5	−2.5	7.5	7.5	5	7.5	5
7.5	20	0	−2.5	0	0	0	2.5	5	0
20	0	2.5	−5	−7.5	−12.5	0	10	−2.5	−7.5

Experiment 5 ($n = 134$)

10	10	5	0	−15	5	15	−5	0	0
0	−10	−10	10	5	10	10	5	−5	10
5	0	0	0	−5	−10	−10	20	5	5
0	5	10	10	0	5	5	5	5	−5
0	−5	−5	5	0	−10	−15	10	10	10
5	10	0	0	10	5	5	20	−5	0
−5	−5	−15	10	0	10	−5	5	10	5
−5	5	5	10	0	0	−5	0	10	5
10	15	10	5	15	10	10	10	5	0
5	0	5	10	−15	20	10	−5	5	0
−10	−10	15	−10	5	0	−5	10	5	0
−5	5	5	−10	10	5	10	10	10	5
0	5	10	10	5	0	10	10	5	0
5	10	10	5						

Discuss the information that is contained in these data sets. Your analysis should contain time-sequence plots such as those in Figure 1.2-2. Discuss whether there are trend components in the data. If there are, describe their nature and discuss possible reasons for their occurrence. You may notice that the initial deviations in experiments 1 to 3 are always positive and quite large. Can you explain why?

Has feedback control made a difference? For each experiment, define and calculate a summary measure for "closeness to the target." Interpret your findings. In addition, use the graphical techniques discussed in Section 1.4 to compare the five distributions.

Project 2

The following table reports the monthly numbers of motor vehicle fatalities on Iowa roads from January 1950 through December 1979 (the data were provided by the Iowa Department of Public Safety):

1950:	31	40	32	39	41	56	59	60	59	57	60	53
1951:	45	39	46	39	42	56	48	57	60	68	46	63
1952:	29	23	39	32	46	45	53	71	58	47	50	38
1953:	29	29	37	38	48	45	76	70	54	54	48	73
1954:	51	35	40	46	52	40	62	60	71	49	59	47
1955:	48	35	35	48	56	40	56	60	63	59	56	53
1956:	63	50	41	45	53	57	79	71	55	63	62	61
1957:	35	54	41	59	68	62	57	72	62	56	54	70
1958:	55	48	38	44	39	34	46	63	40	72	61	58

1959:	57	40	52	60	58	40	49	54	65	87	48	69
1960:	43	26	35	49	38	59	53	54	58	85	56	63
1961:	40	37	46	54	57	55	57	72	54	69	43	52
1962:	30	26	38	38	37	73	58	78	50	69	49	72
1963:	30	35	47	46	53	58	86	56	77	58	70	79
1964:	49	60	61	49	63	63	85	89	75	83	75	82
1965:	53	44	43	54	63	72	53	92	83	76	86	77
1966:	45	57	57	68	70	66	91	91	98	92	89	80
1967:	53	37	65	44	63	73	98	80	75	89	81	60
1968:	56	39	70	64	85	77	77	100	81	69	77	75
1969:	40	42	54	63	79	66	61	82	81	88	74	51
1970:	64	61	57	60	84	63	84	92	90	107	70	80
1971:	48	49	53	65	57	70	72	87	86	112	63	66
1972:	44	44	62	43	71	72	85	97	82	91	91	92
1973:	50	48	60	67	79	83	67	104	63	77	67	48
1974:	41	46	41	49	46	56	74	73	71	60	67	68
1975:	38	28	56	48	67	76	83	57	56	52	61	52
1976:	49	50	46	45	75	88	93	91	72	71	50	55
1977:	49	39	47	48	52	50	73	57	55	65	43	62
1978:	29	29	33	45	72	77	61	73	60	55	75	41
1979:	36	42	37	45	53	66	54	61	60	69	56	53

Following are annual (1950–1979) and monthly (1978 and 1979) data on the number of vehicle miles of travel in Iowa (these data were provided by the Iowa Department of Transportation):

Annual traffic volume (millions of miles traveled)

1950:	9,434	1960:	11,255	1970:	16,053
1951:	9,350	1961:	11,471	1971:	16,581
1952:	9,721	1962:	11,687	1972:	17,127
1953:	9,970	1963:	12,036	1973:	17,690
1954:	9,811	1964:	12,384	1974:	17,250
1955:	10,244	1965:	12,733	1975:	17,853
1956:	10,382	1966:	13,439	1976:	18,440
1957:	10,227	1967:	14,145	1977:	19,028
1958:	10,609	1968:	15,047	1978:	19,466
1959:	10,971	1969:	15,542	1979:	18,959

Monthly traffic volume (millions of miles traveled)

1978:	1,319	1,295	1,520	1,598	1,774	1,778
	1,854	1,863	1,694	1,707	1,553	1,511
1979:	1,249	1,299	1,470	1,607	1,731	1,712
	1,757	1,792	1,648	1,653	1,510	1,531

Analyze the information. You should address the following issues:

(a) Discuss the time trend and the seasonal pattern in the numbers of traffic fatalities. Speculate on the factors that may be responsible for the trend and the seasonality. Discuss whether the change in the traffic speed limit from 70 mph to 55 mph in January 1974 has had an impact on fatalities.

   Hint: Construct time-sequence plots of the numbers of monthly fatalities, as well as the yearly totals. Compare the numbers of fatalities before and after the change in speed limit. To check for seasonality, you may want to calculate monthly averages of the numbers of fatalities (where the average for each month is taken over the 30 years) and plot them against the month.

(b) Has it become safer to drive? Support your answer with appropriate plots. Your analysis should take traffic volume into account. Would it make sense to consider the number of fatalities per million miles traveled?

(c) Update the information by obtaining recent data. Check whether the change in speed limit in May of 1987 (when the speed limit on Iowa rural interstates changed from 55 to 65 mph) has had an effect.

[Reference: Ledolter, J., and Chan, K.S., "Evaluating the Impact of the Increased 65 mph Speed Limit on Iowa Rural Interstates," *The American Statistician*, Vol. 50: 79–85 (1996).]

*Project 3

A producer of gasoline additives wants to learn whether a newly developed diesel additive increases fuel economy. The company hires an independent research institute to conduct a fuel consumption test of this new product.

   Two trucks are used in the experiment. The two trucks are identical International tractor/trailer combinations with the same type of engine, transmission, drive axle, tires, trailer, and identical 80,000-pound load of concrete blocks. The two trucks, with mileages of 250,000 and 275,000 miles, are leased from a brick manufacturer. Both trucks are first checked on a chassis dynamometer to ensure that the engines are performing properly.

   The test route represents a typical long-haul interstate highway operation. A low-density 20-mile-long traffic portion of I-80 between Iowa City and Amana is used in the experiment.

   Reference fuel consumption data are obtained from a control vehicle that is run in tandem with a test vehicle. The test procedure is divided into two segments: the baseline and the test segment. During the baseline, the test truck uses ordinary diesel; during the test segment, the additive is mixed at a ratio of 1 part additive and 500 parts fuel. Ordinary diesel without additive is used in the control truck for both the baseline and the test segments.

   The two trucks enter I-80 in Iowa City, with the second truck following the lead truck at a distance of approximately 1/2 mile. Fuel consumption is measured over a distance of exactly 40 miles—20 miles westbound and 20 miles eastbound on I-80—at a steady speed of 60 mph. After accelerating to 60 mph westbound, the driver starts the fuel meter upon reaching mile marker 262. At mile marker 242, the meter is stopped and the quantity of fuel used is recorded. The same procedure is used for the eastbound leg of the trip between markers 242 and 262. The same two drivers operate their assigned truck throughout the test program. The drivers take 16 test laps as practice to establish familiarity with the trucks and the route.

   After the baseline experiment is completed, the fuel tank of the test truck is drained and filled with fuel containing the diesel additive. The test truck is then driven 5,000 miles on the treated fuel. This is done to condition the engine to the new fuel supply and to maximize the additive's potential. After this conditioning period, 13 laps are run during the testing stage. The fuel consumption (in pounds) for laps during the baseline testing program (baseline laps 11–27) and for the test laps (test laps 1 to 13) are given in the following table:

	Baseline Laps			**Test Laps**	
Lap No.	**Control**	**Test**	**Lap No.**	**Control**	**Test**
11	50.1488	51.8912	1	50.0354	49.4885
12	49.2472	50.6840	2	50.3589	48.7868
13	50.7082	51.9655	3	49.8449	48.3581
14	49.6399	50.5968	4	50.8973	49.3284
15	50.4372	51.8736	5	51.2444	50.2403
16	50.2273	51.6166	6	55.9998	54.1003
17	50.0609	50.8950	7	55.0668	53.7085
18	50.2046	50.9899	8	52.0332	50.0265
19	50.4621	51.6013	9	51.3607	50.7168
20	50.1678	51.3997	10	51.0776	50.1817
21	49.6908	50.9830	11	50.8610	49.3742
22	49.6244	50.5118	12	50.5268	49.1016
23	47.5206	49.7935	13	49.1270	48.6476
24	49.7272	50.1476			
25	46.9486	48.8342			
26	47.9234	48.9613			
27	48.7489	49.1496			

(a) Consider the fuel consumption of the control truck during the baseline laps. Describe the variability by constructing a dot diagram of the 17 observations. Calculate the mean, median, standard deviation, and interquartile range.

(b) Repeat part (a) for the test truck during its baseline laps. To facilitate easy comparisons between the two trucks, display the test truck's dot diagram on the scale that you had selected for the control truck. What conclusions can you draw from the analyses in parts (a) and (b)?

(c) Discuss the factors (road conditions, weather, uncontrolled speed changes, and so on) that contribute to the variability in fuel consumption.

(d) Compare the fuel efficiencies of the two trucks during the baseline laps by considering, for each lap, the difference (or ratio) of the fuel consumptions of the test and the control truck. Analyze the 17 differences (or ratios) for the baseline laps. What conclusions can you draw from this analysis? Supplement your graphical analysis with appropriate summary statistics. Discuss why an analysis of differences or ratios is more effective than a comparison of the two dot diagrams (for the test and control truck) that you made in (a) and (b).

*(e) The main interest in this experiment is to compare the effectiveness of the diesel additive. Use the ratio (diesel use of test truck)/(diesel use of control truck) that we recommend in (d). Construct a dot diagram with the 13 ratios from the test

laps, and compare it against the dot diagram with the ratios from the 17 baseline laps. Again, to facilitate the comparison, you should make these plots on the same scale. Summarize your conclusions.

Project 4

W. E. Deming (*Out of the Crisis,* 1986, p. 327) asserts, "If anyone adjusts a stable process to try to compensate for a result that is undesirable, or for a result that is extra good, the output that follows will be worse than if he had left the process alone." Variation is a fact of life, and even stable processes exhibit some variation. Changing a stable process on the basis of a defective item or a complaint of a single customer will make things worse and will increase its variability. Improvements to a stable process (i.e., a reduction in its variability) can come only through fundamental changes in the system, and not through tampering with the process. The experiment described next (the "funnel experiment," taken from Deming) demonstrates the loss that is due to overadjustment.

(a) Conduct the following experiment to illustrate variability in a stable process: Designate a point on a table as the target. Assume that the coordinates of the target are $(0, 0)$. Take a funnel, suspend it about 3 feet off the table by constructing an appropriate holder, and place the spout of the funnel directly over the target. Take a light Styrofoam® ball, drop it through the funnel, and mark the spot on the table where it comes to rest. Repeat the experiment 50 times. Leave the funnel fixed, aimed perfectly at the target.

This experiment describes a stable system. The locations of impact vary, despite the fact that the funnel is located directly over the target. Comment briefly on the sources of variability. Analyze the results of your experiment. Note that each point is characterized by two coordinates. Define and use an appropriate distance measure in your analysis. Discuss.

(b) Repeat the same experiment, but now adjust the position of the funnel after each drop. If the ball at drop k ($k = 1, 2, \ldots$) comes to rest at point z_k, move the funnel a distance $-z_k$ from its previous position. More specifically, denote the position of the funnel at trial k by $(po_x(k), po_y(k))$ and the location of impact of the ball at trial k, relative to the target $(0, 0)$, by $(x(k), y(k))$. Let the position of the funnel at trial $k + 1$ be given by $(po_x(k + 1) = po_x(k) - x(k), po_y(k + 1) = po_y(k) - y(k))$. Analyze the resulting data and interpret the results. Does the experiment confirm Deming's claim?

(c) Consider a second adjustment procedure that moves the funnel at time $k + 1$ the distance $-z_k$ from its *origin* $(0, 0)$. That is, set $po_x(k + 1) = -x(k)$ and $po_y(k + 1) = -y(k)$. Generate 50 observations, analyze the data, and interpret your findings. Compare the results with the ones you obtained in part (b). [Note that the adjustment in (c) is different from the one in (b), where we move the funnel from its previous position.]

Project 5

Ask each of your fellow students to supply the following information (without, of course, revealing their identities): gender (male, female), height (inches), weight (pounds), age (years), current grade point average, current smoking status (no, yes), current drinking status (no, moderate, heavy), and television exposure (hours per day). Enter the data into a computer file and make the data accessible to all students. Data on categorical variables should be coded: for example, 0 (male) and 1 (female) for gender; 0 (no) and 1 (yes) for smoking; and 0 (no), 1 (moderate), and 2 (heavy) for drinking status.

Use simple graphical procedures and summary statistics, such as sample means, standard deviations, and correlation coefficients, to analyze the information you obtained. For example, discuss the relationship between weight and height (here, you should carry out the analysis for males and females combined, as well as for males and females separately). Explore differences in grade point averages and smoking and drinking patterns between females and males. Is there a relationship between grade point average and drinking? Is grade point average related to television exposure? Can you learn from this information whether extensive television exposure causes you to drink or whether drinking causes you to watch TV?

Project 6

Purchase a carton containing a dozen extra-large eggs and measure their lengths. Comment on measurement variability and process variability. Summarize your results numerically as well as graphically.

Note: Document your measurement procedure and describe your experimental protocol. To get at the measurement variability, you need to measure the same egg more than once. Obtain three measurements on each egg. Make sure that the person who measures an egg does not know its measurement from the previous round.

Project extension: How would you extend this experiment if you wanted to compare the accuracies of the measurements of two people? How would you carry out the experiment if you wanted to compare the accuracy of two measurement instruments—a simple ruler and an expensive caliper?

Project 7

Collect and analyze traffic data along a busy road of your choice. You may want to collect data on vehicle frequencies, car–truck mix, the number of passengers riding in the vehicle, and other factors that you may view useful.

    (a) Establish the appropriate operational definitions; for example, define what you mean by a car and by a truck. Develop a sampling strategy that allows you to learn about possible "time-of-the-day" and "day-of-the-week" effects.

    (b) Carry out your sampling strategy, collect the information, and analyze the data, using the appropriate statistical tools. Summarize your findings.

Project 8

Survey your fellow students about issues surrounding this engineering statistics course. Ask them how many hours they study for the course (hours per week), whether they like the course (yes, no), their average GPA in the required prerequisite calculus courses, their standing (sophomore, junior, senior), their gender (male, female), their area of specialization (mechanical, electrical, etc.), whether they believe that statistics should be a required course (yes, no), and so on. Feel free to add other questions that interest you.

Analyze the data numerically as well as graphically. Some of the information is categorical, and you will need tables to display the information. What do you learn? Are males more likely to enjoy the statistics course? Are students with better calculus grades more likely to enjoy the statistics course? Discuss.

Save your data, as we will need it in Chapter 4.

Note: When collecting survey data, you may want to allow for the anonymity of the responder. Discuss ways of achieving this.

Project 9

The data for this project come from Moberg, A., Sonechkin, D. M., Holmgren, K., Datsenko, N. M., and Karlén, W., "Highly Variable Northern Hemisphere Temperatures Reconstructed from Low- and High-Resolution Proxy Data," *Nature,* Vol. 433, No. 7026, 10 February 2005, pp. 613–617. Using a wavelet transform technique, the authors reconstruct Northern Hemisphere temperatures for the past 2,000 years (from A.D. 1 to 1979) by combining low-resolution proxies with tree-ring data. The data in the file Chapter1Project9NHTemp are given as temperature anomalies (in degrees C) from the Northern Hemisphere 1961–90 temperature average.

Study the time-series plot of the annual temperature anomalies. Discuss the authors' claim that high temperatures—similar to those observed in the 20th century before 1990—occurred around A.D. 1000 to 1100 and minimum temperatures that are about 0.7 degree C below the average of 1961–90 occurred around A.D. 1600. Discuss why this large natural multicentennial variability may be relevant to our understanding of global warming. Additional data on temperature and tree ring growth can be found on the websites http://www.ncdc.noaa.gov/paleo/treering.html and http://www.ncdc.noaa.gov/paleo/recons.html.

Project 10

The scatter diagrams in the first chapter of this book visualize the relationship between two variables. Imagine now that you have three variables. Discuss ways of displaying the relationships among three variables. Think about displays in three-dimensional space, as well as contour diagrams in two dimensions that fix the third variable at certain values. Investigate whether your computer software is able to construct such graphs. You will notice that good three-dimensional displays are still difficult to construct and interpret.

Consult the books by Edward Tufte (1983, 1990, 1997, 2006) for examples of exceptional displays of data in three or more dimensions. Visit his website if you don't have access to his books. For example, look at the map of the losses suffered by Napoleon's army in the Russian campaign of 1812, as originally drawn by C. J. Minard. Beginning at the Polish–Russian border, the thick band shows the size of the army at each position. The path of Napoleon's retreat from Moscow in the bitterly cold winter is depicted by the dark lower band, which is tied to temperature and time scales. Even though this example is not directly related to engineering, it gives a wonderful display of multidimensional data (in this case, the location in two-dimensions, the size of the army, the time, and the temperature).

Color can greatly help the display of data in higher dimensions. (Unfortunately, such color displays are not possible in this book.) Scan the engineering literature for examples that use color effectively. On our website, we show an illustration that is taken from the paper by Gaydos, Stanier, and Pandis, "Modeling of In Situ Ultrafine Atmospheric Particle Formation in the Eastern United States," in the *Journal of Geophysical Research,* Vol. 110 (2005), D07S12. There, the number of atmospheric aerosol particles (on the z-axis) is tracked as a function of time of day (x-axis), conditioned on particle size (y-axis). The top figure shows the prediction of a computational model from aerosol physics and chemistry, while the bottom graph plots the actual measurements at a surface site in Schenley Park, Pittsburgh, Pennsylvania, on July 27, 2001.

PROBABILITY MODELS AND DISCRETE DISTRIBUTIONS

2.1 Probability

In applied mathematics, we are usually concerned with either *deterministic* or *probabilistic* models, although in many instances these are intertwined. As an illustration, suppose that we are interested in the half-life of a certain substance, say, plutonium-241 (^{241} Pu). In many situations involving decay, we can assume that the rate of decay is proportional to the available amount x of the substance. That is, we have the simple differential equation

$$\frac{dx}{dt} = bx,$$

where b is the constant of proportionality. The solution of this equation is easily shown to be

$$x = ce^{bt},$$

where c is another constant. If we are given x at two different times, one of which is usually $t = 0$, the constants b and c can be found. To determine the half-life, we find that t for which the amount equals one-half of the original; it is $t = -(ln\,2)/b > 0$, because $b < 0$. We have thus found a solution to a problem described by what is commonly called a *deterministic model*—a model in which everything is known once the differential equation and the two boundary conditions are specified.

It is interesting to observe, however, that such a deterministic problem can easily involve a *probabilistic* or *stochastic* element. For instance, suppose that the observation x at time $t > 0$ is the true value (i.e., the signal) plus some random noise, which could be a measurement error or which could be due to slight changes in the experimental conditions. That is, if we repeated the experiment on several occasions, we could get slightly different values of x at a given time t. More generally, if we take n pairs of observations, say, $(t_1, x_1), (t_2, x_2), \ldots, (t_n, x_n)$, where some of the times $t_1, t_2, \ldots, t_n$ differ, we usually find that they do not lie exactly on one curve of the form $x = ce^{bt}$. Then the problem is to determine the constants b and c so that the curve $x = ce^{bt}$

seems to "fit" the n observed points $(t_1, x_1), (t_2, x_2), \ldots, (t_n, x_n)$ in some reasonable manner. To describe in more detail a situation that involves noise components, we need a *probabilistic model* involving random elements.

Remarks: Many engineers, scientists, and applied statisticians believe that an introductory course in statistics should not devote a lot of time to probability, as many other important areas need to be covered. We have tried to keep our discussion short and yet provide enough background so that we can develop the basic probability models. Also, there is often criticism of the use of coins, cards, chips, and dice in illustrating some of these concepts, because many believe that real engineering-type examples should be used exclusively. There is a great deal of truth in this, but we find that it requires much longer discussions to explain these real situations, and that defeats the attempt to minimize the amount of material on probability. Hence, in this and the next sections, we do use coins, cards, chips, and dice; but in most cases we find that students enjoy computing a few probabilities involving them. As a matter of fact, probability can be quite entertaining and educational; for example, many students like to be able to compute the probability of the caster winning in "craps" and the chances of winning Powerball.   ∎

Random experiments have outcomes that cannot be determined with certainty before the experiments are performed. To understand such experiments, we have to learn something about the basic concepts of probability. Let us begin with a very simple random experiment that we understand from everyday life: a flip of an unbiased coin. Clearly, the outcome is either heads (H) or tails (T), but it cannot be predicted with certainty. The collection of all possible outcomes, namely, $S = \{H, T\}$, is called the *sample space*. Suppose that we are interested in a subset A of our sample space; for example, let $A = \{H\}$ represent heads. Repeat this random experiment, say, n times, and count the number of times—say, f—that the outcome of the experiment was A. Here, f is called the *frequency* of the *event* A, and the ratio f/n is called the *relative frequency* of the event A in the n trials of the experiment.

We actually flipped a coin a large number of times and recorded in Table 2.1-1 the frequencies and relative frequencies for a number of different values of n. We note that the relative frequency, f/n, is unstable for small values of n, but tends to stabilize as n increases. We associate the number about which f/n stabilizes with the event A. Clearly, to arrive at a single number, we would need to repeat the experiment a very large number of times. Often, this is not feasible in practice, and we take

Table 2.1-1 The Results of Coin Flips		
n	f	f/n
10	3	0.300
20	11	0.550
50	23	0.460
100	52	0.520
200	96	0.480
500	241	0.482
1000	489	0.489

an approximate value of that number. For example, in the coin-tossing experiment, we might associate 1/2 with $A = \{H\}$ because we believe that the relative frequency of heads in a sequence of flips of an unbiased coin would stabilize around 1/2. We denote this association by $P(A) = 1/2$ and call $P(A)$ the *probability of the event A*.

In practice, $P(A) = 1/2$ is probably only an approximation of the probability of "heads" in a coin-tossing experiment, because of some bias in the coin or of some bias in the way the experiment is performed. Clearly, we know of experiments (with dice or cards) in which biases do exist (loaded dice or crooked dealers).

The preceding interpretation of probability is referred to as the *relative frequency approach,* and it obviously depends on the fact that an experiment can be repeated under essentially identical conditions. Many persons, however, extend probability to other situations, treating it as a rational measure of belief. For example, if B is the event that it will rain tomorrow, they might say that $P(B) = 2/5$ is their *personal* or *subjective probability* of the occurrence of that event. Hence, if they are not opposed to gambling, this could mean a willingness on their part to bet on the outcome B, so that the two possible payoffs are in the ratio $P(B)/[1 - P(B)] = (2/5)/(3/5) = 2/3$. Often, with this probability, they say that the *odds* of rain are 2 to 3. If they truly believe that $P(B) = 2/5$ is correct, they would be willing to accept either side of the bet: (1) Win 3 units if B occurs and lose 2 units if it does not occur, or (2) win 2 units if B does not occur and lose 3 units if it does.

With either interpretation of probability, the basic laws of probability are the same. Although we could develop these laws mathematically from certain axioms, we find that most of them appeal to the reader's intuition about probability anyway, whether we use the relative frequency or the subjective approach. Therefore, we simply list these laws using the well-known terminology from the algebra of sets. Recall the set operations intersection, "$\cap$" (and), and union "$\cup$" (or): The event $B_1 \cap B_2$ includes all elements that are in both B_1 and B_2. The event $B_1 \cup B_2$ includes all elements that are either in B_1 or in B_2. Most readers are familiar with an ordinary deck of 52 playing cards, so we give very simple illustrations involving the draw of a card at random from that deck. To check our intuition about probability at this point, consider the following probability assignments:

$P(B_1) = 13/52$, where B_1 represents the event that our random draw results in a spade.

$P(B_2) = 4/52$, where B_2 represents the event that we have drawn a king.

$P(B_1 \cap B_2) = 1/52$, as the intersection $B_1 \cap B_2$ represents a draw of the king of spades.

$P(B_1 \cup B_2) = 16/52$, as the union $B_1 \cup B_2$ represents the draw of one of the 13 spades or one of the other three nonspade kings.

2.1-1 The Laws of Probability

Because the frequency f of the event A in $n \geq 1$ trials of the experiment is such that $0 \leq f \leq n$, we have $0 \leq f/n \leq 1$. Hence, the probability $P(A)$ equals a number between zero and 1, possibly including those endpoints:

Law 2.1-1: $0 \leq P(A) \leq 1$.

Given that the frequency of the null set ϕ is zero and the frequency of the whole sample space S (the universal set) is n, the respective relative frequencies of $0/n = 0$ and $n/n = 1$ suggest the second law:

Law 2.1-2: $P(\phi) = 0$ and $P(S) = 1$.

If A_1 and A_2 are disjoint sets (i.e., their intersection has no elements, so that $A_1 \cap A_2 = \phi$), we say that events A_1 and A_2 are *mutually exclusive*. The probability that A_1 or A_2 occurs is then clearly the sum of the individual probabilities; that is,

$$P(A_1 \cup A_2) = P(A_1) + P(A_2),$$

where $A_1 \cup A_2$ is the union of the two disjoint sets. As an illustration, we have

$$P(\text{draw results in a spade or heart}) = \frac{13}{52} + \frac{13}{52} = \frac{26}{52}.$$

This law can be extended to several mutually disjoint sets, say, $A_1, A_2, \ldots, A_k$, where $A_i \cap A_j = \phi$, for $i \neq j$, to give the third law:

Law 2.1-3: If $A_1, A_2, \ldots, A_k$ are mutually exclusive events, then

$$P(A_1 \cup A_2 \cup \ldots \cup A_k) = P(A_1) + P(A_2) + \cdots + P(A_k).$$

The probability of not getting the event A is denoted by $P(A')$, where A' is the complement of A (everything that is in the universal set, but is not in A). It is obvious that

$$P(\text{draw does not result in a spade}) = \frac{39}{52} = 1 - \frac{13}{52}$$

$$= 1 - P(\text{draw results in a spade}),$$

which suggests the next law:

Law 2.1-4: $P(A') = 1 - P(A)$.

From Law 2.1-4, we note that the odds $P(B)/[1-P(B)]$ can be written as $P(B)/P(B')$. That is, the odds that event B occurs equals the ratio of the probability that B occurs to the probability that B does not occur. Of course, the odds against B equals the ratio $P(B')/P(B)$. For illustration, the odds against a spade in a random draw of a card from an ordinary deck of 52 cards are $(39/52)/(13/52)$—that is, 3 to 1.

Suppose now that we consider two events, A_1 and A_2, that are *not* mutually exclusive. Then the probability of the union $A_1 \cup A_2$ is less than $P(A_1) + P(A_2)$, because the probability of the intersection $A_1 \cap A_2$ is counted twice, once in $P(A_1)$ and once in $P(A_2)$. This consideration leads to the fifth law:

Law 2.1-5: $P(A_1 \cup A_2) = P(A_1) + P(A_2) - P(A_1 \cap A_2)$.

Actually, Law 2.1-5 can be extended to the probability of the union of several events: adding the probabilities of the "singles," subtracting the

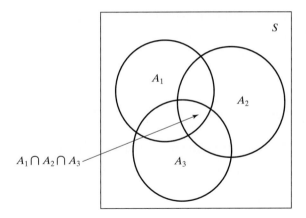

Figure 2.1-1 Venn diagram

probabilities of the "doubles," adding the probabilities of the "triples," and so on. In particular,

$$P(A_1 \cup A_2 \cup A_3) = P(A_1) + P(A_2) + P(A_3) - P(A_1 \cap A_2)$$
$$- P(A_1 \cap A_3) - P(A_2 \cap A_3) + P(A_1 \cap A_2 \cap A_3).$$

The reader can verify this equation by considering the Venn diagram in Figure 2.1-1, noting that the double intersections are counted twice, except for the common triple intersection, which is counted three times. Thus, subtracting the probabilities of the three doubles means that $A_1 \cap A_2 \cap A_3$ is not counted at all. Hence, we must add the term $P(A_1 \cap A_2 \cap A_3)$.

Example 2.1-1   Suppose that a large inventory of coated lenses includes 5 percent scratched lenses, 2 percent poorly coated lenses, and 1 percent lenses that are both scratched as well as poorly coated. If you pick one lens at random from this inventory, the probability of selecting a defective lens (one that is either scratched or poorly coated) is

$$P(A_1 \cup A_2) = P(A_1) + P(A_2) - P(A_1 \cap A_2) = \frac{5}{100} + \frac{2}{100} - \frac{1}{100} = \frac{6}{100},$$

or 6 percent.   ∎

We now consider a rule that allows us to assign probabilities to certain events under an "equally likely" assumption. Let $A_1, A_2, \ldots, A_k$ be *mutually exclusive* and *exhaustive* events. That is, not only do we have $A_i \cap A_j = \phi$, for $i \neq j$, but the union of the sets exhausts the sample space, or

$$A_1 \cup A_2 \cup \ldots \cup A_k = S.$$

Thus, from Laws 2.1–2 and 2.1–3, we have

$$P(A_1) + P(A_2) + \cdots + P(A_k) = P(S) = 1.$$

Suppose further that we can assume that each of the events $A_1, A_2, \ldots, A_k$ has the same probability; that is,

$$P(A_1) = P(A_2) = \ldots = P(A_k).$$

Then we have

$$kP(A_i) = 1 \text{ and } P(A_i) = \frac{1}{k}, i = 1, 2, \ldots, k.$$

If another event, B, is the union of r of these k mutually exclusive events, say,

$$B = A_1 \cup A_2 \cup \ldots \cup A_r, \quad r \le k,$$

then

$$P(B) = P(A_1) + P(A_2) + \cdots + P(A_r) = \frac{r}{k}.$$

Sometimes, for this particular partition of S, the integer k is said to be the *total number* of ways in which the random experiment can end, and the integer r is said to be the number of ways that are *favorable* to the event B. Thus, when we can assume these "equally likely" partitions, the probability of B is equal to the number of favorable ways divided by the total ways of ending an experiment. Note that the assumption of equal probability of $1/k$ must be checked before we can assign probabilities in this manner.

Example 2.1-2   It is really the rule about the assignment of probability that gave us the probabilities in the draw of a card. Let a card be drawn at random from an ordinary deck of 52 playing cards. Let us assume that each of the 52 outcomes in S has the same probability of 1/52. Let B_1 be the set of outcomes that are spades, so that $r_1 = 13$, and B_2 be the set of outcomes that are kings, so that $r_2 = 4$. Then

$$P(B_1) = \frac{13}{52} = \frac{1}{4} \quad \text{and} \quad P(B_2) = \frac{4}{52} = \frac{1}{13}.$$

Moreover, since $B_1 \cap B_2$ equals the king of spades with $r = 1$, we have

$$P(B_1 \cap B_2) = \frac{1}{52}$$

and

$$P(B_1 \cup B_2) = P(\text{draw resulting in either a spade or a king})$$

$$= \frac{13}{52} + \frac{4}{52} - \frac{1}{52} = \frac{16}{52} = \frac{4}{13}. \qquad \blacksquare$$

In Example 2.1-2, the probabilities were easy to compute because there was no difficulty in determining the appropriate values of r and k. However, instead of drawing only one card, suppose that five cards are taken, at random and without replacement, from the deck. We can think of each five-card hand as being an outcome in a sample space. It is reasonable to assume that each of these outcomes has the same probability. It is shown in many elementary mathematics books that the number of possible five-card hands is

$$k = \binom{52}{5} = \frac{52!}{5!47!} = 2{,}598{,}960,$$

where $n! = n(n-1) \ldots (3)(2)(1)$ and $0! = 1$. We refer to $n!$ as "n factorial."

In general, if n is a positive integer and y is a nonnegative integer with $y \leq n$, then the binomial coefficient

$$\binom{n}{y} = \frac{n!}{y!(n-y)!}$$

is equal to the number of combinations of n distinct items taken y at a time. In our illustration involving five cards, suppose that we want to determine the probability of the event B_1 that all five cards are spades. Because there are

$$r_1 = \binom{13}{5} = \frac{13!}{5!8!} = 1{,}287$$

different ways of selecting 5 spades out of 13, we find that

$$P(B_1) = \frac{r_1}{k} = \frac{\binom{13}{5}}{\binom{52}{5}} = 0.0005.$$

Next, let B_2 be that event that at least one of the five cards is a spade. Then the complement B_2' is the event that no card is a spade and all five cards are drawn from the remaining 39. Thus,

$$P(B_2') = \frac{\binom{39}{5}}{\binom{52}{5}} = 0.2215.$$

Of course, $P(B_2) = 1 - P(B_2') = 0.7785$.

Now suppose that B_3 is the event in which exactly three cards are kings and exactly two cards are queens. We can select the three kings in $\binom{4}{3}$ ways and the two queens in $\binom{4}{2}$ ways. By the multiplication rule in the study of counting, the number of ways favorable to B_3 is $r_3 = \binom{4}{3}\binom{4}{2} = (4)(6) = 24$. Multiplication is appropriate here, as we can combine each selection of three kings (there are four such selections) with each possible choice of two queens (there are six such choices). Thus,

$$P(B_3) = \frac{\binom{4}{3}\binom{4}{2}}{\binom{52}{5}} = 0.0000092.$$

Finally, in this illustration, let B_4 be the event in which there are exactly two kings, two queens, and one jack. Then

$$P(B_4) = \frac{\binom{4}{2}\binom{4}{2}\binom{4}{1}}{\binom{52}{5}} = 0.0000554,$$

because the numerator of this fraction is the number of ways that are favorable to B_4.

Example 2.1-3   A lot consisting of 100 fuses is inspected by the following procedure: Five of the fuses are selected at random and tested. If all five "blow" at the correct amperage, the lot is accepted. If, in fact, there are 20 defective fuses in the lot (and thus 80 good fuses), the probability of accepting the lot is

$$\frac{\binom{80}{5}}{\binom{100}{5}} = 0.32,$$

approximately. Note that this is a fairly high probability of acceptance for such a poor lot of fuses.   ∎

Example 2.1-4   In Powerball, the player selects 5 of the first 55 positive integers and a sixth number (the powerball) from the integers 1 through 42. Five numbers, as well as a powerball, are drawn without replacement. The player wins the "big" prize if all five numbers drawn and the powerball match the player's numbers (say, event B_1). Other prizes can be won also—for example, if the powerball does not match the player's selected powerball, but all five of the numbers drawn match the player's numbers (say, event B_2). If we suppose that the five numbers selected by the player (displayed on 55 balls in a big urn) are blue, the probability of drawing all blue balls and the correct powerball is

$$P(B_1) = \frac{\binom{5}{5}\binom{50}{0} \times \binom{1}{1}\binom{41}{0}}{\binom{55}{5} \times \binom{42}{1}} = \frac{(5)(4)(3)(2)(1)(1)}{(55)(54)(53)(52)(51)(42)}$$

$$= (6.844)(10^{-9}) = 1/(146{,}107{,}962),$$

because $\binom{5}{5} = \binom{50}{0} = \binom{1}{1} = \binom{41}{0} = 1$. Similarly,

$$P(B_2) = \frac{\binom{5}{5}\binom{50}{0} \times \binom{1}{0}\binom{41}{1}}{\binom{55}{5} \times \binom{42}{1}} = 41(6.844)(10^{-9})$$

$$= (2.806)(10^{-7}) = 1/(3{,}563{,}609).$$

The probabilities are quite low, and the odds of winning the big prize are not very good.   ∎

Exercises 2.1

*2.1-1 To test the quality of a shipment of crystal glasses, we selected 50 glasses at random. We found that one was scratched and chipped, three had only scratches, and two were only chipped. Consider the following events: A is the occurrence of a chipped glass and B is the occurrence of a scratched glass. Determine the relative frequencies of A, B, $A \cup B$, and $A \cap B$.

2.1-2 To 125 students that took last semester's engineering statistics course, the instructor gave 30 A's, 40 B's, 35 C's, 15 D's, and 5 F's. Calculate the relative frequencies of these five events. Calculate the relative frequency of getting a grade that is better than C.

*2.1-3 Consider rolling a single die (one of a pair of dice). Assume that each up side is equally likely. What is the probability that the experiment ends in rolling a "4"? What is the probability that it ends in an even number?

2.1-4 A person has purchased 5 of 1,000 tickets sold in a certain raffle. To determine the 10 prizewinners, 10 tickets are drawn at random and without replacement.

(a) Compute the probability that this person wins exactly one prize.

(b) Compute the probability that he wins at least one prize.

Hint: First compute the probability that he does not win a prize.

2.1-5 A bowl contains 20 chips, of which 9 are red, 8 are white, and 3 are blue. Six chips are taken at random and without replacement. Find the probability that

(a) each of the 6 chips is red;

(b) there are 3 red, 2 white, and 1 blue chip among the six chips selected;

(c) none of the 6 chips selected is blue; and

(d) there is at least 1 white and at least 1 blue chip among the 6 selected.

Hint: Consider the complement of at least 1 white and at least 1 blue chip.

*2.1-6 A lot of 100 fuses is accepted if at least 9 of 10 fuses taken at random "blow" at the correct amperage. If there are exactly 20 defectives among the 100 fuses, what is the probability of accepting the lot? If there are exactly 10 defectives among the 100 fuses, what is the probability of accepting the lot?

2.1-7 Compute the probability of being dealt, at random and without replacement, a 13-card bridge hand consisting of

(a) 4 spades, 4 hearts, 3 diamonds, and 2 clubs;

(b) 13 cards of the same suit.

2.1-8 If $S = A \cup B$, $P(A) = 0.7$, and $P(B) = 0.8$, find $P(A \cap B)$.

2.1-9 If $P(A) = 0.4$, $P(B) = 0.5$, and $P(A \cup B) = 0.7$, find

(a) $P(A \cap B)$;

(b) $P(A' \cup B')$.

*2.1-10 If $P(A) = 0.4$, $P(B) = 0.5$, and $P(A \cap B) = 0.3$, find

*(a) $P(A \cup B)$;

*(b) $P(A' \cap B)$;

*(c) $P(A' \cup B')$.

2.2 Conditional Probability and Independence

2.2-1 Conditional Probability

Let us start with a simple problem. Suppose that a bowl contains 10 chips of equal size: 5 red, 3 white, and 2 blue. We draw a chip at random and define the events

$$A = \text{the draw of a red or a blue chip}$$

and

$$B = \text{the draw of a red or a white chip}.$$

Of course, $A \cap B$ is the draw of a red chip. Assuming each chip to be equally likely, we would assign $P(A) = 7/10$, $P(B) = 8/10$, and $P(A \cap B) = 5/10$. Suppose, however, we are told that the draw resulted in either a red or a white chip, but not in a blue chip. This means that B has occurred. Intuitively, the probability of A, given the information that a blue chip has been ruled out, would be 5/8, as there are 5 red chips among the 8 chips that are red or white. That is, the conditional probability, $P(A|B)$, of event A given that event B has occurred is

$$P(A|B) = \frac{5}{8} = \frac{5/10}{8/10} = \frac{P(A \cap B)}{P(B)}.$$

This illustration suggests the next definition.

Definition 2.2-1 The *conditional probability of an event A given that event B has occurred* is defined by

$$P(A|B) = \frac{P(A \cap B)}{P(B)},$$

provided that $P(B) > 0$.

In this definition, we can think of B as *a reduced sample space*. It is called "reduced" because we consider only those outcomes that result in B. So $P(A|B)$ is the probability that event A occurs given that we are considering this new sample space B.

Example 2.2-1 Let us toss a nickel and a dime at random and record the result as an ordered pair: (result on nickel, result on dime). Thus, the sample space consists of the four pairs $(T, T), (T, H), (H, T)$, and (H, H), where T represents tails and H represents heads. Assuming that these four pairs are equally likely, we assign a probability of 1/4 to each. If the event A represents (H, H), and if the event B is the event of having at least one head, then

$$P(A|B) = \frac{P(A \cap B)}{P(B)} = \frac{P(A)}{P(B)} = \frac{1/4}{3/4} = \frac{1}{3}.$$

Here, A is a subset of B; thus, $A \cap B = A = \{(H, H)\}$, and B is the reduced sample space $\{(T, H), (H, T), (H, H)\}$. Further, if C is the event such that heads appears on the nickel, then

$$C \cap B = \{(H, T), (H, H)\}$$

and

$$P(C|B) = \frac{P(C \cap B)}{P(B)} = \frac{2/4}{3/4} = \frac{2}{3}.$$

But if D is the event such that tails appears on the nickel, then

$$D \cap B = \{(T,H)\}$$

and

$$P(D|B) = \frac{P(D \cap B)}{P(B)} = \frac{1/4}{3/4} = \frac{1}{3}. \qquad \blacksquare$$

Example 2.2-2 The following probabilities apply to the waiting time in front of a bank teller: $P(A_1$: no waiting time$) = 1/2$; $P(A_2$: minor waiting time$) = 1/3$; $P(A_3$: considerable waiting time$) = 1/6$. Assume that you are told that there is some waiting time. Then the conditional probability that there is considerable waiting time is

$$P(A_3|A_2 \cup A_3) = \frac{P(A_3)}{P(A_2 \cup A_3)} = \frac{1/6}{1/3 + 1/6} = \frac{1}{3}.$$

Here, we have used the fact that $A_3 \cap (A_2 \cup A_3) = A_3$. $\qquad \blacksquare$

Example 2.2-3 Let us draw 5 cards at random and without replacement from an ordinary deck of 52 cards. Let us obtain the conditional probability of an all-spade hand (event A) given that there are at least four spades in the hand (event B). Because $A \cap B = A$, it follows that

$$P(A|B) = \frac{P(A \cap B)}{P(B)} = \frac{P(A)}{P(B)}$$

$$= \frac{\binom{13}{5} / \binom{52}{5}}{\left[\binom{13}{4}\binom{39}{1} + \binom{13}{5}\binom{39}{0}\right] / \binom{52}{5}} = 0.0441. \qquad \blacksquare$$

The equation in Definition 2.2-1 of conditional probability is sometimes written as

$$P(A \cap B) = P(B)P(A|B)$$

or as

$$P(A \cap B) = P(A)P(B|A).$$

(The roles of A and B can be interchanged.) Written in either one of these forms, the equation is called the *multiplication rule* of probability. Many times, it is easier to assign the probabilities $P(B)$ and $P(A|B)$ and obtain $P(A \cap B)$ than it is to assign $P(A \cap B)$ and $P(B)$ to find $P(A|B)$.

Example 2.2-4   A bowl contains 6 white and 4 blue chips of the same size. Two chips are drawn at random in succession and without replacement. The probability that the first chip is white (event A) and the second chip is white (event B) is

$$P(A \cap B) = P(A)P(B|A) = \frac{6}{10} \cdot \frac{5}{9} = \frac{1}{3}.$$

Here, it is reasonable to assign $P(A) = 6/10$, because there are 6 white chips among the 10, and $P(B|A) = 5/9$, because after it is given that the first draw is white, there are only 5 white and 4 blue chips.   ∎

The multiplication rule can be extended to three or more events. We have

$$P(A_1 \cap A_2 \cap A_3) = P(A_1 \cap A_2)P(A_3|A_1 \cap A_2)$$
$$= P(A_1)P(A_2|A_1)P(A_3|A_1 \cap A_2),$$
$$P(A_1 \cap A_2 \cap A_3 \cap A_4) = P(A_1 \cap A_2 \cap A_3)P(A_4|A_1 \cap A_2 \cap A_3)$$
$$= P(A_1)P(A_2|A_1)P(A_3|A_1 \cap A_2)P(A_4|A_1 \cap A_2 \cap A_3),$$

and so on.

Example 2.2-5   Three cards are dealt at random in succession and without replacement. The probability of three spades (i.e., spade first, A_1; spade second, A_2; and spade third, A_3) is

$$\left(\frac{13}{52}\right)\left(\frac{12}{51}\right)\left(\frac{11}{50}\right) = 0.0129.$$

The first factor occurs because there are 13 spades among 52 equally likely cards. If a spade occurs first, there are 12 spades left in the 51 remaining cards; this gives us the factor 12/51. Finally, if spades occur on the first two draws, there are 11 spades left in the remaining 50 cards—hence the factor 11/50. Note that this probability can also be calculated from

$$\binom{13}{3} \bigg/ \binom{52}{3}.$$ ∎

2.2-2 Independence

It may happen that $P(A|B) = P(A)$ and $P(B|A) = P(B)$. That is, the fact that B has occurred does not change the probability of A, and the fact that A has occurred does not change the probability of B. In this case, the multiplication rule becomes

$$P(A \cap B) = P(A)P(B),$$

and we say that A and B are *independent* events.

> **Definition 2.2-2** Events A and B are independent if and only if
> $$P(A \cap B) = P(A)P(B).$$
> Otherwise, A and B are *dependent* events.

Example 2.2-6 When we toss the nickel and the dime at random, it is reasonable to assume that the two resulting events are independent. Thus,

$$P(H \text{ on nickel and } H \text{ on dime}) = \left(\frac{1}{2}\right)\left(\frac{1}{2}\right) = \frac{1}{4}.$$

In an obvious notation, we also have

$$P(T, T) = P(T, H) = P(H, T) = \left(\frac{1}{2}\right)\left(\frac{1}{2}\right) = \frac{1}{4},$$

which was our assumption in Example 2.2-1. ∎

Example 2.2-7 The probability of the event A_1 that a certain part of a spaceship works during a flight is determined to be 0.99. Since the engineers are reluctant to risk the probability 0.01 of failure, they insert another part in parallel. This means that failure occurs if and only if both parts fail. Let the event A_2 stand for the success of the second part. Assume that it has been determined that $P(A_2) = 0.99$ and that A_1 and A_2 are independent events; thus,

$$P(A_1 \cap A_2) = P(A_1)P(A_2) = 0.9801.$$

Then the probability of failure (which means that both parts fail) is

$$P[(A_1 \cup A_2)'] = 1 - P(A_1 \cup A_2)$$
$$= 1 - [P(A_1) + P(A_2) - P(A_1 \cap A_2)]$$
$$= 1 - (0.99 + 0.99 - 0.9801) = 0.0001.$$

Alternatively, we could calculate the probability that both parts fail as

$$P(A_1' \cap A_2') = P(A_1')P(A_2') = (0.01)(0.01) = 0.0001.$$ ∎

Remark In our alternative calculation in Example 2.2-7, we used the fact (but had not proved it) that if A_1 and A_2 are independent, so are A_1' and A_2'. Furthermore, it is also true that A_1 and A_2' are independent, as are A_1' and A_2. We prove the latter, and the other two may be shown in a similar manner. Accordingly, we show that, for independent events A_1 and A_2,

$$P(A_1' \cap A_2) = P(A_1')P(A_2).$$

$A_1 \cup A_1' = S$, so

$$P(A_2) = P[(A_1 \cup A_1') \cap A_2] = P[(A_1 \cap A_2) \cup (A_1' \cap A_2)]$$
$$= P(A_1 \cap A_2) + P(A_1' \cap A_2),$$

because $A_1 \cap A_2$ and $A_1' \cap A_2$ are mutually exclusive. Thus, by the independence of A_1 and A_2,

$$P(A_2) = P(A_1)P(A_2) + P(A_1' \cap A_2)$$

and

$$P(A_1' \cap A_2) = P(A_2) - P(A_1)P(A_2) = [1 - P(A_1)]P(A_2) = P(A_1')P(A_2). \quad \blacksquare$$

The concept of independence can be extended to more than two events. Events A_1, $A_2, \ldots, A_k$ are *mutually independent* if and only if

$$P(A_{r_1} \cap A_{r_2} \cap \ldots \cap A_{r_i}) = P(A_{r_1})P(A_{r_2}) \ldots P(A_{r_i}),$$

where $r_1, r_2, \ldots, r_i$ are any i distinct integers from $\{1, 2, \ldots, k\}$ with $i = 2, 3, \ldots, k$. In particular, note that with $i = 2$, the events must be *pairwise independent;* that is,

$$P(A_{r_1} \cap A_{r_2}) = P(A_{r_1})P(A_{r_2}), \quad r_1 \neq r_2.$$

More important, with $i = k$, we must have

$$P(A_1 \cap A_2 \cap \ldots \cap A_k) = P(A_1)P(A_2) \ldots P(A_k).$$

Example 2.2-8   The probability that a certain type of component works successfully is 0.99. However, this part is so critical that three such components are placed in parallel; with this system, at least one of the components must work to ensure the success of the operation. If we can assume that the outcomes associated with the three components are independent, the probability that all fail is

$$(0.01)(0.01)(0.01) = 0.000001.$$

Thus, the probability of the complement of the event that all fail (i.e., that at least one works successfully) is

$$1 - 0.000001 = 0.999999.$$

That is, by placing three such components in parallel, we increase the probability of success to 0.999999.   ∎

2.2-3 Bayes' Theorem

We start our discussion of Bayes' theorem with a simple example. Suppose that we produce components for a very complicated piece of machinery, and suppose further that 5 percent of the components are defective. Suppose also that, prior to a lengthy and expensive assembly process, it is impossible to determine for sure whether a component is defective. However, suppose that we have developed a quick, but not totally reliable, method for screening out defective items. We know that this test rejects good parts as defective in 1 percent of the cases and accepts defective parts as good ones in 10 percent of the cases.

We apply the new method to a newly produced item, and the test indicates that the item is good. What is the probability that this item is, in fact, defective?

Let us introduce some notation. Let D denote the event that the item is defective. Then $P(D) = 0.05$ and $P(D') = 0.95$. Furthermore, let TD be the event that the test indicates a defect; then, according to our assumptions, $P(\mathrm{TD}|D') = 0.01$ and $P(\mathrm{TD}'|D) = 0.10$, where TD' is the event that the test indicates that the item is good. This implies that the probabilities of the complements are $P(\mathrm{TD}'|D') = 1 - P(\mathrm{TD}|D') = 0.99$ and $P(\mathrm{TD}|D) = 1 - P(\mathrm{TD}'|D) = 0.90$. The objective is to find

$$P(D|\mathrm{TD}') = \frac{P(D \cap \mathrm{TD}')}{P(\mathrm{TD}')} = \frac{P(\mathrm{TD}'|D)P(D)}{P(\mathrm{TD}')},$$

using Definition 2.2-1. Good, as well as bad items, can pass our test; thus, $\mathrm{TD}' = (\mathrm{TD}' \cap D) \cup (\mathrm{TD}' \cap D')$. Because the events denoted in this union are disjoint (an item cannot be good and defective at the same time), we can use Law 2.1–3 and the multiplication rule to write

$$P(\mathrm{TD}') = P(\mathrm{TD}' \cap D) + P(\mathrm{TD}' \cap D')$$
$$= P(\mathrm{TD}'|D)P(D) + P(\mathrm{TD}'|D')P(D').$$

Therefore, our desired probability is

$$P(D|\mathrm{TD}') = \frac{P(\mathrm{TD}'|D)P(D)}{P(\mathrm{TD}'|D)P(D) + P(\mathrm{TD}'|D')P(D')}$$

$$= \frac{(0.10)(0.05)}{(0.10)(0.05) + (0.99)(0.95)} = 0.0053.$$

If we ship without any screening, we ship 5 percent defectives. The probability $P(D) = 0.05$ that a randomly selected item is defective is called the *prior probability* of being defective. If we instead apply our imperfect screening test and ship only items that pass the test, we reduce the probability of shipping defective items to 0.0053 (or 0.53 percent). The probability $P(D|\mathrm{TD}') = 0.0053$ is called the *posterior probability* of being defective, as it is calculated after obtaining the additional information that the item has passed the test.

In general, we can state

Bayes' Theorem   Suppose that a random experiment can result in k mutually exclusive and exhaustive outcomes $A_1, A_2, \ldots A_k$, with *prior probabilities* $P(A_1), P(A_2), \ldots, P(A_k)$. Suppose also that there is another event B for which the conditional probabilities $P(B|A_1), P(B|A_2), \ldots, P(B|A_k)$ can be given. Then the *posterior probability* of A_i given B is

$$P(A_i|B) = \frac{P(A_i \cap B)}{P(B)} = \frac{P(B|A_i)P(A_i)}{\sum_{j=1}^{k} P(B|A_j)P(A_j)}.$$

The proof is straightforward, because, first,

$$P(A_i \cap B) = P(A_i)P(B|A_i).$$

Furthermore, because $B = (A_1 \cap B) \cup (A_2 \cap B) \cup \ldots \cup (A_k \cap B)$ partitions B into k mutually exclusive events, we have

$$P(B) = P(A_1)P(B|A_1) + P(A_2)P(B|A_2) + \cdots + P(A_k)P(B|A_k).$$

Example 2.2-9   Items in your inventory are produced at three different plants, 50 percent from plant A_1, 30 percent from plant A_2, and 20 percent from plant A_3. You are aware that your plants produce at different levels of quality: A_1 produces 5 percent defectives, A_2 produces 7 percent defectives, and A_3 yields 8 percent defectives. You select an item from your inventory and it turns out to be defective. Then the probability that this item is from plant A_1 is

$$P(A_1|D) = \frac{(0.05)(0.50)}{(0.05)(0.50) + (0.07)(0.30) + (0.08)(0.20)} = 0.403.$$

Note that the prior probability $P(A_1) = 0.50$ is larger than the posterior probability $P(A_1|D) = 0.403$ because A_1 tends to produce fewer defectives than do A_2 and A_3. Similarly, we can calculate the probability that the defective part comes from plant 2 as $P(A_2|D) = 0.339$ and the probability that it comes from plant 3 as $P(A_3|D) = 0.258$. These two posterior probabilities are larger than their corresponding prior probabilities.   ∎

Exercises 2.2

2.2-1   A bowl contains 10 chips: 6 red and 4 blue. Three chips are drawn at random and without replacement. Compute the conditional probability that

(a) 2 are red and 1 is blue, given that at least 1 red chip is among the three selected;

(b) all 3 are red, given that at least 2 red chips are in the sample of 3 chips.

2.2-2   A hand of 13 cards is dealt at random and without replacement from an ordinary deck of 52 playing cards. Find the conditional probability that there are at least three aces in the hand, given that there are at least two aces.

*2.2-3   Five cards are to be drawn successively at random and without replacement from an ordinary deck of cards. Compute the probability that the sequence "spade, club, club, heart, spade" is observed in that particular order.

2.2-4   Consider flipping at random two coins, one a nickel and the other a dime, and recording either H or T for each. What is the sample space of this random experiment? Assume that the flips are independent. What is the probability of getting two heads? What is the probability of getting exactly one head? What is the probability of getting at least one head?

*2.2-5   Consider flipping one unbiased coin at random seven times, recording the sequence of H's and T's. How many different outcomes are in the sample space? That is, how many different sequences are possible? If the flips are independent, what is the probability of getting seven heads? What is the probability of getting exactly six heads?

Hint: How many outcomes in the sample space have exactly six heads, and what is the probability of each?

2.2-6   You are rolling two unbiased dice, one red and one blue, and recording the ordered pair (spots up on red die, spots up on blue die). Determine the possible outcomes of this experiment. If the results on the dice are independent, determine the probabilities of the following events:

(a) a 1 on the red die;

(b) a 1 on each die;

(c) a 1 on at least one of the two dice;

(d) the sum of the two values equals 7;

(e) the sum of the two values is an even number;

(f) the number on the blue die is at least as large as the number on the red die;

(g) getting equal numbers on the two dice.

*2.2-7 At each of five defense stations, the probability of downing an attacking airplane is 0.10. If a plane has to pass all five stations before arriving at its target, what is the probability that it gets shot down before it reaches the target?

Hint: Assume independence, and compute the probability that the plane will successfully pass all defense stations.

*2.2-8 Suppose that the numbers 0 through 9 constitute the possible outcomes of a random experiment, and assume that each number is equally likely. Define the following events: A_1, the number is even; A_2, the number is between 4 and 7, inclusive.

*(a) Are A_1 and A_2 mutually exclusive events?

*(b) Calculate $P(A_1), P(A_2), P(A_1 \cap A_2)$, and $P(A_1 \cup A_2)$.

*(c) Are A_1 and A_2 independent events?

*2.2-9 A device for checking welds in pipes is designed to signal if the weld is defective. The assignment of probabilities for the status of the weld and the response of the device is as follows:

	Device response	
	Signal	**No signal**
Defective	0.05	0.01
Not defective	0.02	0.92

The probabilities given mean, for example, that $P(\text{defective} \cap \text{signal}) = 0.05$. Choose a pipe at random.

*(a) What is the probability that the pipe is defective?

*(b) What is the probability that it is not defective?

*(c) If the device signals a defect, what is the conditional probability that the pipe is actually not defective?

*(d) If the device does not signal a defect, what is the conditional probability that the device is, in fact, defective?

2.2-10 A company's management made a certain proposal to its sales representatives in different sales regions. Questionnaires were sent to each representative, and the results obtained were as follows:

	Region		
	East	**Midwest**	**West**
Opposed	50	80	50
Not opposed	100	120	200
Total	150	200	250

(a) What is the probability that a questionnaire selected at random is that of a Western sales representative in favor of the proposal?

(b) What is the probability that it is that of a Midwestern sales representative?

(c) If a questionnaire is selected at random from the group that responded unfavorably to the proposal, what is the conditional probability that the respondent comes from the East?

(d) Are regional district and opinion on the proposal independent? If yes, prove your assertion. If no, with the same marginal totals, specify what the numbers in the cells of the table would have been had the two factors been independent.

Hint: If the regions are denoted by A_1, A_2, A_3, and opinions are called B_1, B_2, respectively, is it true that $P(A_i \cap B_j) = P(A_i)P(B_j)$, for all $i = 1, 2, 3$ and $j = 1, 2$? If so, regions and opinions would be independent.

2.2-11 A survey organization asked respondents what their views were on the probable future direction of the economy and how they voted for president in the last election. The following table shows the fractions of respondents in nine classifications:

	View on Economy		
	Optimistic	**Pessimistic**	**Neutral**
Voted for the president	0.20	0.08	0.12
Voted against the president	0.08	0.15	0.12
Did not vote	0.07	0.08	0.10

(a) What is the probability that a randomly chosen respondent voted for the president?

(b) What is the probability that a randomly chosen respondent is pessimistic about the economy?

(c) What is the conditional probability that a respondent who voted for the president will be pessimistic about the economy?

(d) What is the conditional probability that a respondent who is pessimistic about the economy voted for the president?

(e) Are views on the economy independent of how respondents voted?

Hint: See Exercise 2.2-10.

2.2-12 Two inspectors review the quality of a manufactured component and classify the items as "good," "repairable," and "scrap." Each inspector looks at every item, but they do not always classify the items in the same way. Describe the sample space, and assign probabilities to the nine possible outcomes. Make certain that they are nonnegative and add to 1; the probability of the sample space has to equal 1. For your chosen assignment, calculate the marginal probabilities for each inspector. Do the inspectors act independently?

2.2-13 A popular gambling game called craps is played as follows: A player (called the "caster") rolls two dice, and the sum of the two numbers that appear is observed. If that sum is 7 or 11, the caster wins the game immediately. If the sum equals 2, 3, or 12, the player loses immediately and that collection of numbers is usually referred to as "craps." If, however, the sum on the first roll is 4, 5, 6, 8, 9, or 10, the game continues and that value is called the caster's "point." The caster can then win by rolling again and again until that sum (the caster's point) occurs a second time before the sum of 7 appears once. If the sum of 7 is obtained before the caster's point is observed a second time, the caster loses.

(a) Refer to Exercise 2.2-6 and establish the following probabilities for the sum resulting from a single roll of a pair of unbiased dice (clearly, the fact that they were colored in that exercise does not influence these probabilities):

Sum	2	3	4	5	6	7	8	9	10	11	12
Probability	1/36	2/36	3/36	4/36	5/36	6/36	5/36	4/36	3/36	2/36	1/36

(b) The probability of winning with the sum of 7 or 11 on the first roll is 8/36. Why?

(c) The probability of winning with the sum of 4 on the first roll is computed as

$$P(4 \text{ on 1st})P(4 \text{ the 2nd time before } 7|4 \text{ on 1st}).$$

The first factor is, of course, 3/36. The second is the conditional probability of 4 given that the sequence of rolls ends in either 4 or 7, which have the respective probabilities of 3/36 and 6/36. Thus, the answer is

$$\left(\frac{3}{36}\right)\left(\frac{3/36}{3/36 + 6/36}\right) = \frac{1}{36}.$$

Using similar arguments, show that the probabilities of winning with a 5, 6, 8, 9, and 10 on the first roll are 2/45, 25/396, 25/396, 2/45, and 1/36, respectively.

(d) Thus, the probability that the caster wins at craps is

$$\frac{8}{36} + 2\left(\frac{1}{36} + \frac{2}{45} + \frac{25}{396}\right) \approx 0.493.$$

*2.2-14 Bowl A_1 contains three red and two white chips, and bowl A_2 contains two red and five white chips. A fair die is cast. If the outcome is "five or six," a chip is taken from A_1; otherwise a chip is taken from A_2.

*(a) Compute the probability that a white chip (say, B) is taken.

Hint: Note that $B = (B \cap A_1) \cup (B \cap A_2)$ and $(B \cap A_1) \cap (B \cap A_2) = \phi$.

*(b) Given that a white chip is observed, compute the conditional (posterior) probability that it came from A_1.

(c) Compare $P(A_1 \mid B)$ determined in part (b) with the prior probability $P(A_1)$. Does the result agree with your intuition?

2.2-15 Suppose that we wish to determine whether a rare, but very costly, flaw is present (event F). Let us assume that $P(F) = 0.01$; thus, $P(F') = 0.99$. A fairly simple procedure is proposed to test for this flaw. However, the test is preliminary, as the probabilities of reaching the wrong conclusion are large. About 5 percent of the time, the test indicates a flaw (TF) when no flaw is present; and about 3 percent of the time, it indicates the absence of a flaw when a flaw is present. That is, $P(TF \mid F') = 0.05$, but $P(TF' \mid F) = 0.03$.

(a) Show that the probability that the test indicates a flaw is $P(TF) = 0.0592$.

(b) Show that the posterior probability that there is no flaw, given that the test has indicated a flaw, is $P(F'|TF) = 0.8361$.

Remark: Armed with these probabilities, a statistician would argue very strongly for better test procedures. The error rates of the present test, in particular, $P(TF|F') = 0.05$, are much too large relative to the proportion of flaws in the population [$P(F) = 0.01$]. More reliable test procedures must be found.

*2.2-16 A Pap smear is a screening procedure to detect cervical cancer. For women with this cancer, 16 percent of the tests are false negatives. For women without this cancer, about 19 percent are false positives. In the United States, there are about 8 women in 100,000 who have this cancer.

*(a) A woman tests positive on this screening test. Calculate her probability of having cancer.

*(b) Questions have been raised about the value of the Pap smear procedure. Comment on this fact. How many cases among every one million positive Pap smears represent true cases of cervical cancer? Why is this number so low?

2.2-17 An engineer gathered data on failures of a circuit that has been considered for use in a satellite. After several experiments, the engineer decided that failures could be divided into the following mutually exclusive and exhaustive categories: A = integrated circuit failure; B = PC board failure; C = power supply failure; and D = other failure.

The engineer also isolated a common problem M that is thought to contribute to all four types of failure. He established the following connections: $P(A) = 0.40; P(M|A) = 0.05;\ \ P(B) = 0.25;\ \ P(M|B) = 0.05;\ \ P(C) = 0.30;$ $P(M|C) = 0.04; P(D) = 0.05; P(M|D) = 0.01$
Find $P(M)$, $P(M')$, $\mathrm{P}(C|M)$, $\mathrm{P}(C|M')$. Interpret these events within their engineering context.

2.2-18 Thirty percent of the students in a calculus course and 20 percent of students in a statistics course receive A's. Furthermore, 60 percent of the students with an A in calculus receive an A in the statistics course. John received an A in the statistics course. Calculate the probability that he also received an A in the calculus course.

* 2.2-19 A certain disease is quite rare; only 10 percent of the population are aflicted by it. Your company has developed a test that can be used to recognize whether that disease is present. But your test is not perfect: The probability of failing to recognize the presence of the disease is 20 percent, and the probability of falsely concluding that the disease is present when in fact it is not is 10 percent.

You are testing a person and find that the test is positive. Determine the probability that the disease is actually present.

2.2-20 Suppose that 65% of all those who enroll at the University of Iowa finish in six years. What is the probability that five freshman selected at random will all finish in six years?

2.3 Random Variables and Expectations

2.3-1 Random Variables and Their Distributions

Engineers and scientists are interested in quantities that cannot be predicted with certainty; for example, consider (1) the number of defective fuses in a shipping lot of fuses, (2) the thickness of a lens, (3) the compression strength of a concrete block, (4) the tear strength of paper, and (5) the number of flaws in 1,000 feet of wire. Measurements of these quantities can be thought to depend on the outcomes of experiments that cannot be predicted with certainty. For example, suppose that we are interested in the compressive strength of certain concrete blocks. Let us select a concrete block from a lot of concrete blocks and measure its compressive strength. This gives us a certain number. However, we could repeat the entire experiment, select and measure a different block, and obtain a different value. Not all blocks are identical; process variability is usually present, and not all items that leave the production line are the same. Measurement equipment and procedures may also influence the outcome; often, repeated measurements on the same item lead to different results.

Measurements can be thought of as functions of the outcomes of random experiments. Such functions are called *random variables,* because their values cannot be predicted with certainty before the experiment is performed. That is, random variables are functions of the outcomes of random experiments. Engineers want to know something about the characteristics of random variables to be able to deal with the processes that produce them. It is the purpose of this section to introduce some useful terminology. We begin with an example.

Example 2.3-1 Let an unbiased coin be tossed at random three independent times. Let the random variable X be the number of heads observed in the three tosses. Here, the random variable is a function of the outcomes of the three trials, and it can take on one of the values $0, 1, 2,$ or 3. We say that the space of all possible values of X is $R = \{0, 1, 2, 3\}$. A random variable that can take on distinct values is called a *discrete* random variable, so here X is a discrete random variable. In Chapter 3, we consider *continuous* random variables, which can take on any value in a continuum. It is easy to show that in this particular example the probabilities associated with the events $X = x, x = 0, 1, 2, 3$, are given by

$$P(X = x) = \binom{3}{x}\left(\frac{1}{2}\right)^3, \quad x = 0, 1, 2, 3.$$

Recall that the number of combinations of n distinct objects taken r at a time is

$$\binom{n}{r} = \frac{n!}{r!(n-r)!}, \quad r = 0, 1, 2, \ldots, n,$$

where $0! = 1$ and $k! = (k)(k-1)(k-2)\ldots(3)(2)(1)$. In our case, we have

$$\binom{3}{0} = \frac{3!}{0!3!} = 1, \quad \binom{3}{1} = \frac{3!}{1!2!} = 3, \quad \binom{3}{2} = 3, \quad \binom{3}{3} = 1.$$

To illustrate the calculation of these probabilities, consider $P(X = 2)$. We can get exactly two heads in three trials in three different ways: $(T, H, H), (H, T, H)$, and (H, H, T). Thus,

$$P(X = 2) = P(T, H, H) + P(H, T, H) + P(H, H, T).$$

Due to the independence of the tosses, we have

$$P(X = 2) = \left(\frac{1}{2}\right)^3 + \left(\frac{1}{2}\right)^3 + \left(\frac{1}{2}\right)^3 = 3\left(\frac{1}{2}\right)^3 = \frac{3}{8}.$$

In similar fashion, we find that

$$P(X = 0) = \frac{1}{8}, \quad P(X = 1) = \frac{3}{8}, \quad P(X = 3) = \frac{1}{8}. \qquad \blacksquare$$

Often, the probability $P(X = x)$ is denoted by $f(x)$ and is called the *probability function,* the *frequency function,* the *probability mass function,* or the *probability density function* of a random variable of the *discrete* type. We use the last of these expressions and frequently abbreviate *probability density function* by p.d.f., or simply *density.*

Because $f(x) = P(X = x)$ is a probability density function defined on the discrete space R, it has the following properties:

1. $f(x) \geq 0, x \in R$ (probability is nonnegative).
2. $\sum_{x \in R} f(x) = 1$ (the sum of the probabilities over all possible values x must equal 1).
3. $P(X \in A) = \sum_{x \in A} f(x)$, where A is a subset of R.

Example 2.3-2   Toss an unbiased coin a number of independent times until a head appears. Let X be the number of trials needed to obtain that first head. Here, X is a random variable of the discrete type, and the space of X is $R = \{1, 2, 3, 4, \ldots\}$. To achieve $X = x$, there must be tails on the first $x - 1$ trials and then a head on the xth trial. The probability of this sequence of events is

$$P(\overbrace{T, T, \ldots T}^{x-1}, H) = \overbrace{\left(\frac{1}{2}\right)\left(\frac{1}{2}\right) \cdots \left(\frac{1}{2}\right)}^{x-1}\left(\frac{1}{2}\right) = \left(\frac{1}{2}\right)^x.$$

That is,

$$f(x) = \left(\frac{1}{2}\right)^x, \quad x \in R = \{1, 2, 3, 4, \ldots\}.$$

Of course, $f(x) = (1/2)^x > 0$, provided that $x = 1, 2, 3, \ldots$. Furthermore,

$$\sum_{x=1}^{\infty} f(x) = \sum_{x=1}^{\infty} \left(\frac{1}{2}\right)^x = \frac{1/2}{1 - 1/2} = 1,$$

because this is the sum of an infinite geometric series whose first term is $a = 1/2$ and whose common ratio is $r = 1/2$. Also, the probability that we need, for example, between three and five tosses of the coin to get that first head is

$$P(X = 3, 4, 5) = \sum_{x=3}^{5} \left(\frac{1}{2}\right)^x = \left(\frac{1}{2}\right)^3 + \left(\frac{1}{2}\right)^4 + \left(\frac{1}{2}\right)^5 = \frac{7}{32}. \qquad \blacksquare$$

We let the function $F(x)$ represent the cumulative sum of all the probabilities less than or equal to x. That is,

$$F(x) = P(X \leq x) = \sum_{t \in A} f(t),$$

where $A = \{t; t \in R \text{ and } t \leq x\}$. Because $F(x)$ cumulates all the probabilities less than or equal to x, it is called the *cumulative distribution function (c.d.f.)* of X or, more simply, the *distribution function* of X. For $x = 1, 2, 3, \ldots$, the c.d.f. for the random variable in Example 2.3-2 is

$$F(x) = \sum_{t=1}^{x} \left(\frac{1}{2}\right)^t = \frac{(1/2) - (1/2)(1/2)^x}{1 - 1/2} = 1 - \left(\frac{1}{2}\right)^x.$$

To find $F(x)$ here, we used the result for the sum of x terms of a geometric series with $a = r = 1/2$. Note that the c.d.f. is defined for any x value, not just the ones in the space R. For x values that are different from the integers $1, 2, 3, \ldots$, the c.d.f. is $F(x) = 1 - (1/2)^{[x]}$, where $[x]$ (called "the greatest integer in x") is the largest nonnegative integer that is smaller than (or equal to) x. More specifically, $F(x) = 0$ for $x < 1$, $F(x) = 1 - (1/2)$ for $1 \le x < 2$, $F(x) = 1 - (1/2)^2$ for $2 \le x < 3$, and so on.

Using the $F(x)$ of our illustration, we note that

$$P(X = 3, 4, 5) = P(X \le 5) - P(X \le 2) = F(5) - F(2)$$

$$= \left[1 - \left(\frac{1}{2}\right)^5 \right] - \left[1 - \left(\frac{1}{2}\right)^2 \right] = \frac{1}{4} - \frac{1}{32} = \frac{7}{32}.$$

More generally, if X is a random variable of the discrete type with an outcome space R consisting of integers, then

$$P(a \le X \le b) = F(b) - F(a - 1),$$

where $a \in R$ and $b \in R$. Most tables of well-known discrete probability distributions are given in terms of their c.d.f. $F(x)$, and they are accessible through computer software and most scientific calculators.

2.3-2 Expectations of Random Variables

Suppose that we refer once again to Example 2.3-1, where X is the number of heads in three independent tosses of an unbiased coin and

$$f(x) = \binom{3}{x} \left(\frac{1}{2}\right)^3, \quad x = 0, 1, 2, 3.$$

Let us play a game and award X^2 dollars to the person flipping the coin. Before we consider a charge for playing such a game, we observe that the player receives $0, 1, 4$, or 9 dollars for the respective x values $0, 1, 2, 3$. The probabilities $1/8, 3/8, 3/8$, and $1/8$ are associated with the respective amounts $0, 1, 4$, and 9. That is, if the game is played a large number of times, about one-eighth of the trials lead to a payment of zero dollars, about three-eighths to a payment of one dollar, about three-eighths to a payment of four dollars, and about one-eighth to a payment of nine dollars. Thus, the average payment is

$$(0)\left(\frac{1}{8}\right) + (1)\left(\frac{3}{8}\right) + (4)\left(\frac{3}{8}\right) + (9)\left(\frac{1}{8}\right) = 3.$$

This is a weighted average of the possible outcomes of the random variable X^2, namely, $0, 1, 4, 9$, with respective weights $1/8, 3/8, 3/8$, and $1/8$. Note that this average of 3 does not equal any of the possible outcomes. If we let $u(X) = X^2$, we denote the weighted average by

$$E[u(X)] = \sum_{x \in R} u(x) f(x)$$

and call it the *mathematical expectation*, or the *expected value*, of $u(X)$. Incidentally, because in this illustration $E[X^2] = 3$, it is likely that the person would be charged something like $3.25 to play the game, to provide the organizer of the game a profit of 25 cents per play, on the average.

Remark: It should be pointed out that $u(X)$ itself is a random variable, say, Y, taking on values in some space R_1. Suppose we find that the p.d.f. of Y is $g(y)$. Then $E(Y)$ is given by

$$\sum_{y \in R_1} y g(y).$$

But does this equal

$$\sum_{x \in R} u(x) f(x)?$$

It seems that it should, as these two different summations are seemingly computing the average of $Y = u(X)$. In fact, the two summations are equal, and this equality is illustrated in Exercise 2.3-8. ■

It is easy to show (Exercise 2.3-7) that the mathematical expectation E is a *linear* or *distributive operation* with the following properties:

1. $E(c) = c$, where c is a constant.

2. $E\left[\sum_{i=1}^{k} c_i u_i(X)\right] = \sum_{i=1}^{k} c_i E[u_i(X)],$

   where $c_1, c_2, \ldots, c_k$ are constants. In particular,

$$E[c_1 u_1(X) + c_2 u_2(X)] = c_1 E[u_1(X)] + c_2 E[u_2(X)].$$

Example 2.3-3   Using the setup of Example 2.3-1, we let $u_1(X) = X$ and $u_2(X) = X^2$. We already know that $E[u_2(X)] = 3$. Thus, if we had a payment equal to $3X + 2X^2$, then

$$E[3u_1(X) + 2u_2(X)] = (3)\left(\frac{3}{2}\right) + (2)(3) = \frac{21}{2},$$

because

$$E[u_1(X)] = \sum_{x=0}^{3} x f(x) = (0)\left(\frac{1}{8}\right) + (1)\left(\frac{3}{8}\right) + (2)\left(\frac{3}{8}\right) + (3)\left(\frac{1}{8}\right) = \frac{3}{2}. \quad ■$$

In Chapter 1, we defined the (sample) mean of a set of numbers by their average, with each number given the same weight. Here, we define the mean of a random variable X as the expected value of $u(X) = X$. It is a weighted average of all possible outcomes; each outcome $x \in R$ is weighted by its probability $f(x)$. We denote this expectation by the Greek letter mu:

$$\mu = E(X) = \sum_{x \in R} x f(x).$$

We call $\mu = E(X)$ the *mean* of X or the mean of the distribution of X. In Example 2.3-3, we note that the mean of X of Example 2.3-1 is $\mu = E(X) = 3/2$.

Example 2.3-4   Using the number of trials needed to obtain the first head as the random variable X with p.d.f. $f(x) = (1/2)^x, x = 1, 2, 3, \ldots$, we have

$$\mu = E(X) = \sum_{x=1}^{\infty} x\left(\frac{1}{2}\right)^x = (1)\left(\frac{1}{2}\right) + (2)\left(\frac{1}{2}\right)^2 + (3)\left(\frac{1}{2}\right)^3 + \ldots = 2.$$

To see that this infinite series sums to 2, note that

$$\left(\frac{1}{2}\right)\mu = (1)\left(\frac{1}{2}\right)^2 + (2)\left(\frac{1}{2}\right)^3 + (3)\left(\frac{1}{2}\right)^4 + \ldots.$$

The difference of the expressions for μ and $(1/2)\mu$ equals

$$\frac{\mu}{2} = (1)\left(\frac{1}{2}\right) + (1)\left(\frac{1}{2}\right)^2 + (1)\left(\frac{1}{2}\right)^3 + \ldots = \frac{1/2}{1 - 1/2} = 1.$$

Thus, $\mu = 2$. Incidentally, it is interesting to observe that $\mu = 2$ is the weighted average of the infinite set of numbers $1, 2, 3, 4, \ldots$. Clearly, the weight associated with a large value of x, such as 1,000, must be small, and indeed it is, as $f(1000) = (1/2)^{1000}$. ∎

A (sample) variance of a set of numbers with equal weights was also defined in Chapter 1. Here we define the *variance* of X (or the variance of the distribution of X) as the weighted average of the outcomes $(x - \mu)^2$; that is,

$$\sigma^2 = E[(X - \mu)^2] = \sum_{x\in R}(x - \mu)^2 f(x),$$

where σ is the Greek lowercase letter sigma. Sometimes we use the expression var $(X) = \sigma^2$. Since $f(x) \geq 0$, the variance is nonnegative. The positive square root of the variance is called the *standard deviation* of X:

$$\sigma = \sqrt{\operatorname{Var}(X)} = \sqrt{E[(X - \mu)^2]}.$$

The standard deviation σ is expressed in the same units as the random variable X. Because, on the average, $(X - \mu)^2$ equals σ^2, we can think of the standard deviation σ as the *approximate* average distance from the outcomes of the random variable X to the mean μ.

There is another way of computing the variance. We have

$$\sigma^2 = E[(X - \mu)^2] = E[X^2 - 2\mu X + \mu^2].$$

However, the expectation is a linear operation and $E(\mu^2) = \mu^2$, so we obtain

$$\sigma^2 = E(X^2) - 2\mu E(X) + \mu^2 = E(X^2) - 2\mu^2 + \mu^2 = E(X^2) - \mu^2.$$

Thus, $\sigma^2 = E(X^2) - \mu^2$, and the standard deviation equals

$$\sigma = \sqrt{E(X^2) - \mu^2}.$$

Example 2.3-5   Let X have a p.d.f given by

x	1	2	3	4
$f(x)$	0.4	0.2	0.3	0.1

Hence,

$$\mu = E(X) = (1)(0.4) + (2)(0.2) + (3)(0.3) + (4)(0.1) = 2.1,$$
$$E(X^2) = 1^2(0.4) + 2^2(0.2) + 3^2(0.3) + 4^2(0.1) = 5.5,$$
$$\sigma^2 = E(X^2) - \mu^2 = 5.5 - 4.41 = 1.09,$$

and

$$\sigma = \sqrt{1.09} = 1.044. \qquad \blacksquare$$

In the next two sections, we determine the mean μ and the standard deviation σ for each of two important discrete distributions. Those who have studied mechanics recognize the mean μ as the centroid of a system of weights with density $f(x)$, while σ is the radius of gyration. It is also helpful here to note for later consideration that if $Y = cX$, where c is a constant, then

$$\mu_Y = E(Y) = E(cX) = c\mu_X,$$
$$\sigma_Y^2 = E[(Y - \mu_Y)^2] = E[(cX - c\mu_X)^2] = c^2 E[(X - \mu_X)^2] = c^2 \sigma_X^2,$$

and

$$\sigma_Y = |c|\sigma_X.$$

For example, if Y is expressed in inches and X is the same measurement in feet, so that $Y = 12X$, then, $\mu_Y = 12\mu_X, \sigma_Y^2 = 144\sigma_X^2$, and $\sigma_Y = 12\sigma_X$.

Exercises 2.3

* 2.3-1 Let the p.d.f of X be defined by $f(x) = x/10, x = 1, 2, 3, 4$. Determine
*(a) $P(X \le 2)$,
*(b) $P(2 \le X \le 4)$,
*(c) $\mu = E(X)$, and
*(d) $\sigma^2 = E[(X - \mu)^2]$.

* 2.3-2 For each of the following, determine the constant c so that $f(x)$ enjoys the properties of a p.d.f.:
*(a) $f(x) = cx, x = 1, 2, \ldots, 25$.
*(b) $f(x) = c(x + 1)^3, x = 0, 1, 2$.
*(c) $f(x) = c(1/3)^x, x = 1, 2, 3, \ldots$.

* 2.3-3 Let X have the p.d.f. $f(x) = (1/6)(5/6)^{x-1}, x = 1, 2, 3, \ldots$.
(a) Show that $f(x)$ is a valid p.d.f.
*(b) Determine $F(x) = P(X \le x), x = 1, 2, 3, \ldots$.
*(c) Compute $P(4 \le X \le 7)$.
*(d) Evaluate $\mu = E(X)$.

2.3-4 Suppose that the probability density function $f(x)$ of the length X of an international telephone call, rounded up to the next minute, is given by

x	1	2	3	4
$f(x)$	0.2	0.5	0.2	0.1

(a) Calculate $P(X \le 2), P(X < 2),$ and $P(X \ge 1)$.

(b) Calculate and plot the cumulative distribution function $F(x)$ against x. Note that this function is a step function: At each $x = 1, 2, 3, 4$, the function will jump to a new level. The height of the step at x is $f(x)$.

(c) Calculate the mean $\mu = E(X)$.

(d) Calculate $E(X^2)$, and use the result to determine the variance $\sigma^2 = E(X^2) - [E(X)]^2$.

(e) Use the formula $\sigma^2 = E[(X - \mu^2)]$ to calculate the variance, and show that it is the same as that found in part (d).

*2.3-5 The yearly number X of large contracts that are awarded to a company is a random variable with p.d.f.

x	0	1	2	3
$f(x)$	0.2	0.4	0.3	0.1

Calculate the mean, variance, and standard deviation of this distribution.

*2.3-6 Let X have the p.d.f. $f(x) = x^2/30, x = 1, 2, 3, 4$. Suppose that a game has payoff $u(X) = (4 - X)^3$. Compute the expected value of the payoff, $E[(4-X)^3]$.

2.3-7 Using the definition of the mathematical expectation, show that

$$E\left[\sum_{i=1}^k c_i u_i(X)\right] = \sum_{i=1}^k c_i E[u_i(X)],$$

where $c_1, c_2, \ldots, c_k$ are constants.

Hint: Write

$$E\left[\sum_{i=1}^k c_i u_i(X)\right] = \sum_{x \in R}\left[\sum_{i=1}^k c_i u_i(x)\right] f(x) = \sum_{i=1}^k \sum_{x \in R} c_i u_i(x) f(x)$$

after interchanging the order of summation.

2.3-8 With the random variable X of Example 2.3-1, argue that the random variable $Y = X^2$ has the p.d.f. $g(y)$ given by

y	0	1	4	9
$g(y)$	1/8	3/8	3/8	1/8

(a) Evaluate $E(Y) = \sum_{y \in R_1} y g(y)$, where $R_1 = \{0, 1, 4, 9\}$.

(b) Evaluate $E(X^2) = \sum_{x \in R} x^2 f(x)$, where $f(x)$ and R are as described in Example 2.3-1.

(c) Note from parts (a) and (b) that

$$\sum_{y \in R_1} y g(y) = \sum_{x \in R} x^2 f(x).$$

More generally, if $Y = u(X)$, then

$$E(Y) = \sum_{x \in R} u(x) f(x) = \sum_{y \in R_1} y g(y),$$

where $g(y)$ is the p.d.f. of $Y = u(X)$ with space R_1.

2.3-9   Let X have the p.d.f. $f(x) = x^3/36, x = 1, 2, 3.$
(a) Show that $Y = X^3$ has p.d.f. $g(y) = y/36, y = 1, 8, 27.$
(b) Evaluate the two summations

$$\sum_{x=1}^{3} x^3 f(x) \text{ and } \sum_{y \in R_1} y g(y),$$

where $R_1 = \{1, 8, 27\}$.

2.4 The Binomial and Related Distributions

2.4-1 Bernoulli Trials

Consider a random experiment, the outcome of which can be classified into one of two mutually exclusive and exhaustive ways. We frequently call these two ways *success* and *failure,* but these words could stand for heads and tails, female and male, death and life, defective and nondefective, and so on. If this random experiment is repeated a number of times, so that

1. the outcomes of the trials are mutually independent and
2. the probability p of success is the same in each trial,

then this sequence of trials is called a sequence of *Bernoulli trials.*
     Let X be a random variable associated with one Bernoulli trial in such a way that $X = 1$ represents success and $X = 0$ represents failure. Because

$$P(X = 1) = p \text{ and } P(X = 0) = 1 - p = q,$$

we can write the p.d.f. of X as

$$f(x) = P(X = x) = p^x(1 - p)^{1-x} = p^x q^{1-x}, x = 0, 1.$$

The mean and the variance of X are, respectively,

$$\mu = (0)(1 - p) + (1)(p) = p$$

and

$$\sigma^2 = E(X^2) - p^2 = [(0)^2(1 - p) + (1)^2 p] - p^2 = p(1 - p) = pq.$$

Let X_i be the random variable associated with the ith Bernoulli trial, $i = 1, 2, \ldots, n$. Thus, each X_i has the same p.d.f., mean, and variance, namely,

$$f(x) = p^x q^{1-x}, x = 0, 1; \mu = p; \sigma^2 = pq.$$

The fact that the outcomes of the n trials are mutually independent means that all events like

$$(X_1 = 1), (X_2 = 0), (X_3 = 0), \ldots, (X_{n-1} = 1), (X_n = 0)$$

are mutually independent (see Section 2.2-2); thus, we call $X_1, X_2, \ldots, X_n$ mutually independent random variables. In general, a collection of n mutually independent random variables, each having the same distribution, is called a *random sample* from that distribution.

The probability of getting a particular sequence of successes and failures is easily calculated with mutually independent random variables. If $n = 5$, the probability of the sequence (success, success, failure, success, failure) is

$$P(X_1 = 1, X_2 = 1, X_3 = 0, X_4 = 1, X_5 = 0) = ppqpq = p^3 q^2.$$

More generally, if x_i equals zero or 1, for $i = 1, 2, \ldots, n$, then, from the independence, we have

$$
\begin{aligned}
P(X_1 = x_1, X_2 &= x_2, \ldots, X_n = x_n) \\
&= f(x_1)f(x_2)\ldots f(x_n) \\
&= p^{x_1}(1 - p)^{1-x_1} p^{x_2}(1 - p)^{1-x_2} \ldots p^{x_n}(1 - p)^{1-x_n} \\
&= p^{\sum x_i}(1 - p)^{n - \sum x_i} = p^y(1 - p)^{n-y} = p^y q^{n-y},
\end{aligned}
$$

where $y = \sum_{i=1}^{n} x_i$ is the number of 1's (successes) among $x_1, x_2, \ldots, x_n$. This agrees with our result when $n = 5$ and the number of successes is $y = 3$.

In practice, the common probability p associated with the Bernoulli trials is usually unknown. For example, we often do not know the exact probability p of producing a defective part in a certain manufacturing process. How, then, can we estimate p once the random sample is observed to be $x_1, x_2, \ldots, x_n$, where each x_i is equal to zero or 1? The intuitive answer to this question is to take the relative frequency of defectives, $\hat{p} = \sum_{i=1}^{n} x_i / n$. The numerator, $\sum_{i=1}^{n} x_i$, is the number of defectives; the "hat" above p indicates that we have an estimate of p.

2.4-2 The Binomial Distribution

Because $Y = \sum_{i=1}^{n} X_i$, the number of successes in n Bernoulli trials, is an important statistic in estimating p, we should learn more about its distribution. Clearly, the space of Y is $R = \{0, 1, 2, \ldots, n\}$. If exactly y successes occur, where $y = 0, 1, 2, \ldots, n$, then there must be $n - y$ failures. The trials are independent, so the probability of one such sequence is $p^y (1 - p)^{n-y}$. There are, however,

$$\binom{n}{y} = \frac{n!}{y!(n - y)!}$$

ways in which exactly y ones can be assigned to y of the variables $X_1, X_2, \ldots, X_n$. Thus, $g(y) = P(Y = y)$ is the sum of the probabilities of these $\binom{n}{y}$ mutually exclusive events, each with probability $p^y (1 - p)^{n-y}$. That is,

$$g(y) = \binom{n}{y} p^y (1 - p)^{n-y}, y = 0, 1, 2, \ldots, n.$$

This is the p.d.f of the *binomial distribution* with *parameters n and p*, often abbreviated $b(n, p)$. The distribution is called binomial because these $n + 1$ probabilities given by $g(y)$, $y = 0, 1, 2, \ldots, n$, are equal to the respective $n + 1$ terms in the expansion of a certain binomial raised to the n th power, namely,

$$1 = [(1 - p) + p]^n = \sum_{y=0}^{n} \binom{n}{y} p^y (1 - p)^{n-y}.$$

Incidentally, this equation shows that the binomial probabilities $g(y)$, $y = 0, 1, 2, \ldots, n$, sum to 1.

We find the mean of Y by considering

$$\mu = E(Y) = \sum_{y=0}^{n} y \frac{n!}{y!(n-y)!} p^y (1-p)^{n-y}$$

$$= \sum_{y=1}^{n} \frac{n!}{(y-1)!(n-y)!} p^y (1-p)^{n-y}$$

because $y/y! = 1/(y - 1)!$ when $y = 1, 2, \ldots, n$. We rewrite the equation for μ as follows:

$$\mu = (np) \sum_{y=1}^{n} \frac{(n-1)!}{(y-1)!(n-y)!} p^{y-1}(1-p)^{n-y}$$

$$= (np)[(1-p)^{n-1} + (n-1)p(1-p)^{n-2} + \cdots + p^{n-1}]$$

$$= (np)[(1-p) + p]^{n-1} = np.$$

That is, $\mu = E(Y) = np$ is the mean of the number of successes in n Bernoulli trials, each with probability of success p. As an illustration, if $n = 100$ and $p = 1/4$, the mean is $np = (100)(1/4) = 25$. We expect 25 successes, a result that agrees with our intuition.

The variance of Y is found by noting that

$$E(Y^2) = E[Y(Y - 1)] + E(Y).$$

It can be shown (Exercise 2.4-5) that $E[Y(Y - 1)] = n(n - 1)p^2$; thus,

$$\sigma^2 = \text{var}(Y) = E(Y^2) - (np)^2 = n(n-1)p^2 + np - (np)^2 = np(1 - p).$$

Hence, we have demonstrated that if Y is $b(n, p)$, then

$$E(Y) = np \quad \text{and} \quad \text{var}(Y) = E[(Y - np)^2] = np(1 - p).$$

With these expectations, it is easy to find the mean and the variance of the estimator $\hat{p} = Y/n$, of p. The mean of $\hat{p}$ is

$$E(\hat{p}) = E\left(\frac{Y}{n}\right) = \left(\frac{1}{n}\right)E(Y) = \frac{1}{n}np = p,$$

and $\text{var}(\hat{p})$ equals

$$E[(\hat{p} - p)^2] = E\left[\left(\frac{Y}{n} - p\right)^2\right] = \left(\frac{1}{n^2}\right)E[(Y - np)^2]$$

$$= \frac{1}{n^2}[np(1-p)] = \frac{p(1-p)}{n}.$$

Remark: Here we call $\hat{p} = Y/n$ an *estimator* of the parameter p. Say that we actually performed the experiment n independent times and observed y successes; that is, y is a known number. As an illustration, suppose that $n = 50$ and $y = 17$; then we call $y/n = 17/50 = 0.34$ an *estimate* of p. That is, the random variable Y/n is an estimator of p *before* the experiment is performed; but *after* the experiment results in a given number of successes, say, y, the known value, y/n is called an estimate of p. In general, an estimator is a random variable and an estimate is an observation of that random variable. ∎

Statistical software (and also most scientific calculators) will compute the binomial probability density function (p.d.f.) $f(y) = P(Y = y)$ and the cumulative probability distribution function (c.d.f.) $F(y) = P(Y \leq y)$, for specified values n and p. For example, in Minitab, you would use "Calc > Probability Distributions > Binomial." Specifying "Probability" would give you the p.d.f., while "Cumulative Probability" would give you the c.d.f.. Tables of cumulative binomial probabilities, for selected values of n and p, are also given in Table C.2 of Appendix C. The binomial probability density functions for $n = 10$ and $p = 0.3, 0.5,$ and 0.9, are plotted in Figure 2.4-1.

Example 2.4-1   Let Y be $b(n = 14, p = 0.15)$. Then

$$P(Y = 3, 4, 5) = P(Y = 3) + P(Y = 4) + P(y = 5) = \sum_{y=3}^{5} \binom{14}{y}(0.15)^y(0.85)^{14-y}$$

$$= 0.2056 + 0.0998 + 0.0352 = 0.3406,$$

or

$$P(Y = 3, 4, 5) = P(Y \leq 5) - P(Y \leq 2) = F(5) - F(2)$$

$$= 0.9885 - 0.6479 = 0.3406.$$

The preceding probabilities can be looked up in Table C.2, or they can be obtained from statistical computer software such as Minitab. ∎

Example 2.4-2   Let Y be $b(n = 8, p = 0.75)$, so that $E(Y) = (8)(0.75) = 6$ and $var(Y) = 8(0.75)(0.25) = 1.5$. Then

$$P(Y = 6, 7) = P(Y = 6) + P(Y = 7) = 0.3114 + 0.2670 = 0.5784$$

or

$$P(Y = 6, 7) = P(Y \leq 7) - P(Y \leq 5) = 0.6785 - 0.1001 = 0.5784. \quad ∎$$

Example 2.4-3   The probability that a certain type of transformer fails in the first 10 years of operation is $p = 0.05$. If we observe $n = 20$ such transformers, and if we can assume independence, we know that the number Y of transformers failing in the first 10 years is $b(n = 20, p = 0.05)$. Therefore, $E(Y) = (20)(0.05) = 1$ is the expected number that fail in the first 10 years. The probability that between 1 and 4 transformers, inclusive, fail is

$$P(Y = 1, 2, 3, 4) = P(Y \leq 4) - P(Y \leq 0) = 0.9974 - 0.3585 = 0.6389. \quad ∎$$

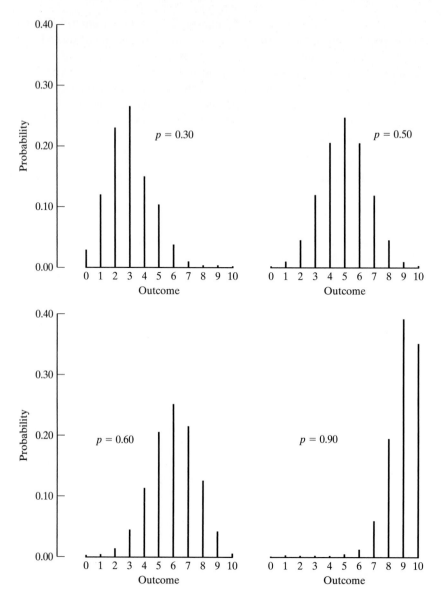

Figure 2.4-1 Probability densities of various $b(10, p)$ distributions

Remarks: Independence is important here. Intuitively, independence means that the status of one transformer does not depend on the status of another. Independence, for example, would probably not be present if the transformers are taken sequentially from the production line and if we believe that there is some carryover in the quality of consecutive items.

The application of the binomial distribution also requires that the probability of failure be the same for all transformers. Suppose that you take the 20 transformers from two groups, say, a certain fraction from lots that come from machine A and the rest from lots that were produced by machine B. If there is reason to believe that these two machines produce transformers of different quality, it would be inappropriate to combine the transformers into one group. This would certainly violate the assumption that the probabilities of failure are the same. In most cases, however, violation of the independence assumption causes a greater change in the distribution of Y than do slight changes of p from trial to trial.

2.4-3 The Negative Binomial Distribution

Let us consider the problem of observing a sequence of Bernoulli trials until exactly r successes occur, where r is a fixed positive integer. Let the random variable Z denote the number of trials needed to observe the rth success. Of course, the space of Z is $R = \{r, r + 1, r + 2, \ldots\}$, the collection of an infinite, but countable, number of values. If z is an element of R, then the event $Z = z$ occurs when there are exactly $r - 1$ successes in the first $z - 1$ trials, followed by a success on the zth trial. Because the latter two events are independent, we obtain $P(Z = z)$ by multiplying their probabilities. That is, the p.d.f of Z is

$$h(z) = P(Z = z) = \left[\binom{z - 1}{r - 1}p^{r-1}(1 - p)^{z-r}\right](p)$$

$$= \binom{z - 1}{r - 1}p^{r}(1 - p)^{z-r} \quad \text{for } z = r, r + 1, r + 2, \ldots.$$

We say that Z has a *negative binomial distribution* with parameters r and p.

Example 2.4-4 A process is continued until the first defective is produced. Assume that the trials are independent (which may not always be a good assumption in practice). If the process produces $p = 0.05$ defectives, the probability that the first defective appears on the fifth trial is

$$P(Z = 5) = \binom{4}{0}(0.05)(0.95)^4 = 0.0407.$$

The probability that the first defective occurs on or before the fifth trial is

$$P(Z \leq 5) = \sum_{z=1}^{5}\binom{z - 1}{0}(0.05)(0.95)^{z-1} = 0.2262.$$

The negative binomial distribution with parameters $r = 1$ and p is also known as the *geometric distribution*. It is the distribution of the number of trials that are needed to achieve the first success. Example 2.3-2 in Section 2.3 discussed the special case with $p = 0.5$. ∎

2.4-4 The Hypergeometric Distribution

Consider a collection of $N = N_1 + N_2$ objects, N_1 of them belonging to one of two distinct classes (say, defective items) and N_2 of them to the other class (say, good items). Take n of these objects at random and without replacement from the N objects.

We wish to find the probability that exactly w of these n sampled objects belong to the first class (i.e., are defective). Here, w is a nonnegative integer satisfying the inequalities $w \leq n$, $w \leq N_1$, and $n - w \leq N_2$. Of course, if $n \leq N_1$ and $n \leq N_2$, then the space of W, the number of items in the first class, is $R = \{0, 1, 2, \cdots, n\}$. The p.d.f. of W is

$$k(w) = P(W = w) = \frac{\binom{N_1}{w}\binom{N_2}{n - w}}{\binom{N_1 + N_2}{n}}, \quad w = 0, 1, \ldots, n,$$

and it is said to be the p.d.f. of a *hypergeometric distribution*.

The hypergeometric distribution and the binomial distribution are closely related. Their difference lies in the sampling scheme that is adopted. The binomial distribution assumes independent trials and constant probabilities of success. It assumes a sampling scheme with replacement, so that trials are independent and the probability of success is constant across trials. In the hypergeometric distribution, by contrast, the trials are no longer independent. Because sampling is without replacement, the probability of success at the ith trial depends on the outcomes of the previous ones.

However, if N_1 and N_2 are large compared with n, the hypergeometric distribution can be approximated by the binomial distribution. This is intuitive, as in this case the probability of a success does not change much from one trial to the next and is close to $p = N_1/(N_1 + N_2)$.

Example 2.4-5   A lot (that is, collection) of 1,000 items consists of $N_1 = 100$ defective and $N_2 = 900$ good items. A sample of $n = 20$ items is taken at random and without replacement from this lot. The probability that the sample contains two or fewer defective items is

$$P(W \leq 2) = \sum_{w=0}^{2} k(w) = \sum_{w=0}^{2} \frac{\binom{100}{w}\binom{900}{20 - w}}{\binom{1000}{20}} = 0.6772.$$

This result can be closely approximated with a binomial distribution $b(n = 20, p = 100/1000 = 0.1)$; that is,

$$P(W \leq 2) \approx \sum_{w=0}^{2} \binom{20}{w}(0.1)^w(0.9)^{20-w} = 0.6769.$$

You can get these probabilities with the Minitab instructions "Calc > Probability Distributions > Hypergeometric" and "Calc > Probability Distributions > Binomial."   ∎

Exercises 2.4

*2.4-1   The probability of producing a defective item is 0.02.

   *(a) What is the probability that zero out of 10 items that are selected independently are defective?

   *(b) How many defective items do you expect in a sample of 100 items?

2.4-2 The university football team has 11 games on its schedule. Assume that the probability of winning each game is 0.40 and that there are no ties. Assuming independence (which may not be a good assumption here because there are streaks of wins and losses), calculate the probability that this year's team will have a winning season (i.e., that the team will win at least six games).

*2.4-3 The probability of producing a high-quality color print is 0.10. How many prints do we have to produce such that the probability of producing at least one quality print is larger than 0.90?

*2.4-4 A roofing contractor offers an inexpensive method of fixing leaky roofs. However, the method is not foolproof. The contractor estimates that after the job is done, 10 percent of the roofs will still leak.

*(a) The resident manager of an apartment complex hires this contractor to fix the roofs on six buildings. What is the probability that after the work has been done, at least two (that is, two or more) of these roofs will still leak?

*(b) The contractor offers a guarantee that pays $100 if a roof still leaks after he has worked on it. If he works on 81 roofs, find the mean and standard deviation of the amount of money that he will have to pay as a result of this guarantee.

2.4-5 Show that if Y follows a $b(n, p)$ distribution, then $E[Y(Y - 1)] = n(n - 1) p^2$.

Hint: Write

$$E[Y(Y - 1)] = \sum_{y=0}^{n} y(y - 1)\frac{n!}{y!(n - y)!}p^y(1 - p)^{n-y}$$

$$= \sum_{y=2}^{n} y(y - 1)\frac{n!}{y!(n - y)!}p^y(1 - p)^{n-y},$$

because the $y = 0$ and $y = 1$ terms equal zero. Use $y(y - 1)/y! = 1/(y - 2)!$ when $y \geq 2$. Factor out $n(n-1)p^2$ and note that the remaining summation equals $[(1 - p) + p]^{n-2} = 1$.

2.4-6 If a student answers questions on a true–false test randomly (i.e., assume that $p = 0.5$) and independently, determine the probability that

(a) the first correct answer is in response to question 4,

(b) at most four questions (that is, four or fewer) must be answered to get the first correct answer.

2.4-7 The probability that a machine produces a defective item is $p = 0.05$. Each item is checked as it is produced. Assuming independence among items, compute the probability that the third defective item is found

(a) on the 20th trial,

(b) on or before the 20th trial.

2.4-8 Of 10 fish in a pond, 4 are tagged. If $n = 3$ fish are caught at random and not replaced, what is the probability that

(a) 2 of the 3 fish are tagged,

(b) 2 or fewer are tagged?

2.4-9 Suppose that there are 100 defectives in a lot of 2,000 items. If a sample of size $n = 10$ is taken at random and without replacement,

(a) Determine the probability that there are two or fewer defectives in the sample.

(b) Use the binomial distribution to approximate the probability in part (a).

*2.4-10 Assume that the probability of a male birth is 0.5 and that the binomial distribution is applicable. Determine the probability that a couple with

*(a) three children has at least two days,

(b) five children has at least two boys.

(c) What are the corresponding probabilities if the probability of a male birth is 0.53?

*2.4-11 Electric switches are shipped in packages of eight items. The probability that an item is defective is 0.1. What is the probability that a package contains

*(a) at least one defective switch?

*(b) fewer than two defective switches? Apply the binomial distribution.

(c) Calculate the expected number of defectives in a package.

2.4-12 A customer is supplied with four randomly selected packages. (See Exercise 2.4-11.) What is the probability that

(a) each package contains at least one defective?

(b) at least one package contains fewer than two defectives?

2.4-13 Assume that the probability that a switch is defective (see Exercise 2.4-11) has been reduced to 0.05. Now what are the probabilities in parts (a) and (b) of Exercise 2.4-12? What are the probabilities of having (c) at least 3 and (d) fewer than 5 defectives in a single package of 40 items?

*2.4-14 Your stockbroker has a 65 percent probability of success in picking stocks that appreciate. You are investing in 20 securities that he has suggested. Calculate the probability that 9, 10, or 11 stocks will appreciate; that fewer than 14 will appreciate. Calculate the mean and the standard deviation of the number of stocks that will appreciate. State the assumptions that allowed you to make these calculations.

*2.4-15 Your sampling inspection plan works as follows: From a very large lot, you select a random sample of 10 items. If all 10 items pass inspection, you ship the entire lot. Otherwise you reject it. Assuming that your defect rate is 3 percent, calculate the probability that you will reject the lot.

2.4-16 Use your computer software to obtain the p.d.f. and the c.d.f. of the

(a) binomial distribution with parameters $n = 10$ and $p = 0.6$;

(b) binomial distribution with parameters $n = 20$ and $p = 0.6$;

(c) negative binomial distribution with $r = 1$ and $p = 0.6$;

(d) negative binomial distribution with $r = 2$ and $p = 0.6$.

Sketch the p.d.f. and the c.d.f. of each distribution.

2.4-17 Consider the geometric distribution with p.d.f. $h(z) = (1 - p)^{z-1}p$ for $z = 1, 2, \ldots$. Graph the density for different values of p ($p = 0.2, 0.5, 0.8$) and show that the mean of this distribution is given by $\mu = 1/p$.

2.5 Poisson Distribution and Poisson Process

2.5-1 The Poisson Distribution

A binomial probability can be approximated by another important probability when the number of trials, n, is large and the probability of success, p, is small. To find this approximating probability distribution, we investigate the limit of a binomial probability when $n \to \infty$ and $p \to 0$ such that $np = \lambda$, where the Greek lowercase letter lambda represents a positive constant. Replacing p by λ/n and keeping y fixed, we have

$$
\begin{aligned}
\lim_{n \to \infty} P(Y = y) &= \lim_{n \to \infty} \left[\frac{n!}{y!(n - y)!} \left(\frac{\lambda}{n}\right)^y \left(1 - \frac{\lambda}{n}\right)^{n-y} \right] \\
&= \lim_{n \to \infty} \left[\frac{n(n - 1) \dots (n - y + 1)}{n^y} \left(\frac{\lambda^y}{y!}\right) \frac{(1 - \lambda/n)^n}{(1 - \lambda/n)^y} \right] \\
&= \lim_{n \to \infty} \left[(1)\left(1 - \frac{1}{n}\right) \dots \left(1 - \frac{y - 1}{n}\right) \left(\frac{\lambda^y}{y!}\right) \frac{(1 - \lambda/n)^n}{(1 - \lambda/n)^y} \right] \\
&= \frac{\lambda^y e^{-\lambda}}{y!},
\end{aligned}
$$

because we know from calculus that

$$
\lim_{n \to \infty} \left(1 - \frac{\lambda}{n}\right)^n = e^{-\lambda},
$$

and because the limits of the rest of the factors, other than $\lambda^y/y!$ (which does not depend on n), equal 1. The space of y, which is $\{0, 1, 2, \dots, n\}$ in the binomial distribution, now includes all nonnegative integers, $\{0, 1, 2, 3, \dots\}$, in this limiting process.

We show that

$$
h(y) = \frac{\lambda^y e^{-\lambda}}{y!}, \quad y = 0, 1, 2, \dots,
$$

satisfies the properties of a p.d.f. Since $\lambda > 0$, it follows that $h(y) \geq 0$. In addition,

$$
\sum_{y=0}^{\infty} h(y) = \sum_{y=0}^{\infty} \frac{\lambda^y e^{-\lambda}}{y!} = e^{-\lambda} \left[1 + \frac{\lambda}{1!} + \frac{\lambda^2}{2!} + \frac{\lambda^3}{3!} + \cdots \right].
$$

However, the infinite sum in this last expression is the MacLaurin's expansion (special case of Taylor's expansion) of e^{λ}. Thus,

$$
\sum_{y=0}^{\infty} h(y) = e^{-\lambda} e^{\lambda} = 1.
$$

So $h(y)$ satisfies the properties of a p.d.f. of a random variable of the discrete type. $h(y)$ is called a *Poisson* p.d.f. with parameter λ.

Since np is the mean of the binomial distribution, we suspect that the parameter λ is the mean of the Poisson distribution. We investigate this conjecture by evaluating

$$\mu = E(Y) = \sum_{y=0}^{\infty} yh(y) = (0)e^{-\lambda} + (1)\frac{\lambda e^{-\lambda}}{1!} + (2)\frac{\lambda^2 e^{-\lambda}}{2!} + (3)\frac{\lambda^3 e^{-\lambda}}{3!} + \cdots$$

$$= e^{-\lambda}\left(\lambda + \frac{\lambda^2}{1!} + \frac{\lambda^3}{2!} + \cdots\right) = \lambda e^{-\lambda}\left(1 + \frac{\lambda}{1!} + \frac{\lambda^2}{2!} + \cdots\right) = \lambda e^{-\lambda}e^{\lambda} = \lambda.$$

Thus, the mean of the Poisson distribution is indeed λ, as we suspected.

The variance of the binomial distribution is $np(1 - p) = \lambda(1 - \lambda/n)$, which equals λ in the limit as $n \to \infty$. Could it be that λ is also the variance of the Poisson distribution? That this is indeed the case may be seen by noting that

$$E(Y^2) = E[Y(Y - 1)] + E(Y).$$

But $E[Y(Y - 1)] = \lambda^2$, as demonstrated in Exercise 2.5-7. Thus, we have

$$\sigma^2 = E(Y^2) - \lambda^2 = \lambda^2 + \lambda - \lambda^2 = \lambda.$$

We observe, then, that the Poisson distribution is a distribution in which the mean and the variance are equal to a common value λ. This is a very important characteristic of the Poisson distribution.

The p.d.f. $f(y) = P(Y = y)$ and the c.d.f. $F(y) = P(Y \le y)$ can be obtained from your computer software or your scientific calculator. All you need to do is provide the value for λ. In Minitab, you would use "Calc > Probability Distributions (Poisson)." Specifying "Probability" gives you the p.d.f., while "Cumulative Probability" provides the c.d.f. Tables of cumulative Poisson probabilities for selected values of λ are also given in Table C.3 of Appendix C. The densities of four Poisson distributions, with parameters $\lambda = 1, 2, 5$, and 10, are shown in Figure 2.5-1.

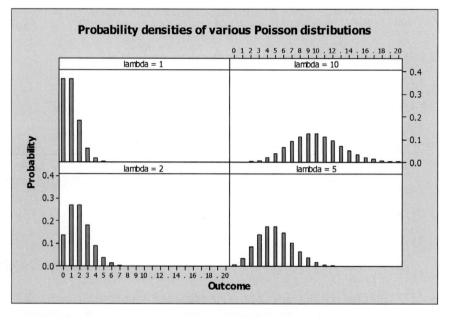

Figure 2.5-1 Probability densities of various Poisson distributions

Example 2.5-1   Let Y be the number of the $n = 20$ transformers of Example 2.4-3 that fail in the first 10 years. That is, Y is $b(n = 20, p = 0.05)$. Here, $n = 20$ is not too large, but it is still interesting to see how various values of the binomial c.d.f. with $n = 20$ and $p = 0.05$, compare with the Poisson c.d.f. with $\lambda = (20)(0.05) = 1$. (See Table 2.5-1.) Even with a small sample value of n, such as 20, the approximation is quite good; it is extremely good with much larger values of n.

Table 2.5-1 Two Cumulative Distribution Functions $F(y) = P(Y \leq y)$		
y	Binomial c.d.f. $n = 20, p = 0.05$	Poisson c.d.f. $\lambda = (20)(0.05) = 1$
0	0.3585	0.3679
1	0.7358	0.7358
2	0.9245	0.9197
3	0.9841	0.9810
4	0.9974	0.9963

∎

Example 2.5-2   Let Y be $b(n = 50, p = 0.96)$. Obviously, p is not small, but the Poisson approximation can still be used if the roles of *success* and *failure* are interchanged. That is, the number of failures $W = 50 - Y$ is $b(n = 50, p = 0.04)$. Thus,

$$P(Y = 46, 47, 48, 49, 50) = P(W = 0, 1, 2, 3, 4) \approx F(4) = 0.947,$$

where $F(4)$ comes from the c.d.f. of the Poisson distribution with $\lambda = (50)(0.04) = 2$.

∎

2.5-2 The Poisson Process

We have discovered that the Poisson distribution is a good approximation to the binomial distribution when n is large and p is small (or close to 1). However, the Poisson distribution is also an appropriate model for many other random phenomena. To illustrate, the number of flaws on a casting or the number of breakdowns of a certain machine over a given period may very well be random variables that have (at least approximate) Poisson distributions. To see this intuitively, consider the casting and divide its surface into n equal parts. Let n be large, so that each part is small and not likely to have more than one flaw. Then λ is the average number of flaws per casting and since we have divided a casting into n equal parts, it is reasonable to assign to each part a probability λ/n that there is one flaw. Moreover, since each part is so small, suppose that the probability of two or more flaws in it is zero. Since the probability of a flaw in a given part is λ/n, the probability of the part not having a flaw is $1 - (\lambda/n)$. If we can assume that the outcomes (either flaw or nonflaw in each part) are mutually independent with common probability λ/n of a flaw, we have, in essence, n Bernoulli trials. The probability of y defects in these n trials is then

$$\frac{n!}{y!(n-y)!}\left(\frac{\lambda}{n}\right)^{y}\left(1 - \frac{\lambda}{n}\right)^{n-y}.$$

However, we have proved that this expression approaches $h(y) = \lambda^y e^{-\lambda}/y!$ as $n \to \infty$.

Let us describe what we have just derived more formally by considering the assumptions of a Poisson process. The process generates a number of "changes" (occurrences, accidents, flaws, claims, etc.) in a fixed "interval" (time, space, length, etc.). Suppose that the number of changes, Y, in a fixed interval satisfies the following assumptions:

1. The probability of exactly one change in a very short interval of length h is approximately proportional to h; that is, it equals λh, where λ is the constant of proportionality.

2. The probability of two or more changes in this short interval of length h is approximately equal to zero.

3. The numbers of changes in nonoverlapping intervals are mutually independent.

If we then count the number of changes, Y, in a fixed interval of length w, we can show that Y has a Poisson distribution with mean λw. In particular, if $w = 1$, then the mean is λ. Note that in our example with $w = 1$ casting, we assumed independence from one part of the casting to another, that the probability of exactly *one* flaw in one such part is proportional to λ, and that the probability of more than one flaw is zero. Before any model is used in practice, the assumptions must be checked to see if they are appropriate. Certainly, before using the Poisson model, one should assess whether the assumption of independence from part to part appears reasonable. That is, is it a reasonable assumption that the quality of the casting is more or less uniform, or are there regions where the flaws occur in groups?

Example 2.5-3 Let the number Y of traffic deaths per week in a certain state have an approximate Poisson distribution with mean $\lambda = 7$. It is reasonable to assume a Poisson distribution in this case, as one can use either one of the following two justifications: First, we can note that there are many drivers (n is large) and the probability of someone being involved in a fatal traffic accident is quite small (p is small). Alternatively, we can use the results of the Poisson process to argue that the distribution of weekly traffic deaths follows a Poisson distribution. The week can be divided into short periods, say, hours. The probability of observing exactly one accident during such a short period should be proportional to the length of the period, and the probability of observing more than one should be negligible. Assuming, furthermore, that the accidents occur more or less independently over time would also lead us to a Poisson distribution.

For illustration, the probability of more than nine accidents is

$$P(Y \geq 10) = 1 - P(Y \leq 9) = 1 - 0.830 = 0.170,$$

where the cumulative probability $F(9) = 0.830$ is obtained from the Poisson distribution with $\lambda = 7$. You can use Minitab or any other statistical software package for this calculation. The probability of exactly five weekly traffic deaths is

$$P(Y = 5) = P(Y \leq 5) - P(Y \leq 4) = F(5) - F(4)$$

$$= 0.301 - 0.173 = 0.128.$$

Example 2.5-4

For 10,000 feet of a particular type of power-line wire, the number Y of failures has a Poisson distribution with $\lambda = 0.2$. The probability that there is more than one failure is given by

$$P(Y > 1) = 1 - P(Y \leq 1) = 1 - 0.982 = 0.018.$$ ∎

Like the parameter p of the binomial distribution, the parameter λ of a Poisson distribution is usually unknown. Since λ is the mean of this distribution, it is reasonable to estimate it by observing several values of Y, say, $Y_1, Y_2, \ldots, Y_n$, and determine the mean of these observations, say, $\overline{Y}$. That is, $\hat{\lambda} = \overline{Y}$ is an estimator of the unknown quantity λ. Here, as with the estimator of the parameter p of a binomial, we use the "hat" above λ to denote an estimator.

As an example, assume that your company manufactures wood panels and that there are good reasons to believe that the number of flaws on a panel follows a Poisson distribution. But you do not know λ. To estimate it, you are planning to take a sample of, say, $n = 20$ panels. Before you actually look at the sample items, you note that the number of flaws, $Y_i, i = 1, 2, \ldots, 20$, is a random variable; it could be zero with probability $e^{-\lambda}$, 1 with probability $\lambda e^{-\lambda}$, and so on. Thus, the estimator $\hat{\lambda} = \overline{Y} = \sum Y_i / 20$ is a random variable and has a distribution with a certain mean and a certain variance. (We will study such distributions in a later section.) But once you have inspected the 20 panels and determined their numbers of flaws, say,

$$2 \quad 4 \quad 1 \quad 0 \quad 0 \quad 3 \quad 2 \quad 1 \quad 5 \quad 3 \quad 0 \quad 1 \quad 1 \quad 0 \quad 4 \quad 3 \quad 2 \quad 2 \quad 1 \quad 0,$$

you can calculate the average $\overline{y} = 35/20 = 1.75$ and take this number as an estimate of the unknown parameter λ in the Poisson distribution. For this to be a good estimate, we have to take a representative sample. Ideally, we would like a *random sample* from the population in which we are interested. If our interest is in today's production, we should give each item produced an equal chance of being in our sample; we should not simply go to machine 1, or worker B, or pile H and take all items from these subgroups. It may turn out that the average number of flaws in the random sample, and thus the parameter λ in the Poisson distribution, is large. Then the next logical step would be to ask why. For example, is it due to machine 1, or because of worker B, or because of variability in the raw material? To learn more about that, one could take random samples from each machine and compare their numbers of flaws. Doing so brings us back to the techniques of Section 1.4, where we compared different samples.

Exercises 2.5

*2.5-1 If the probability that a single item is defective is 0.04, what are the probabilities that a sample of 100 will contain

*(a) exactly zero defectives,

*(b) exactly 4 defectives, and

*(c) more than 5 defectives.

Use the Poisson approximation.

2.5-2 On average, 2.5 telephone calls per minute are received at a corporation's switchboard. Making appropriate assumptions about the distribution (provide justification), find the probability that at any given minute there will be more than two calls.

*2.5-3   A telephone switchboard receives, on average, 300 calls an hour, but can make only 10 connections during a given minute. Determine the probability that the switchboard cannot handle all incoming calls in a given minute.

   Hint: Assume that the number of incoming calls per minute follows a Poisson distribution with parameter $\lambda = 300/60 = 5$.

*2.5-4   The daily number of plant shutdowns follows a Poisson distribution with mean 2.0. What is the probability that there are

   *(a) more than three shutdowns in a day, and

   *(b) at least one in a day?

   *(c) Assume that the company loses \$1,000 with each shutdown. Calculate the expected daily loss.

*2.5-5   Suppose that in one year the number of industrial accidents X follows a Poisson distribution with mean 3.0. If each accident leads to an insurance claim of \$5,000, how much money would an insurance company need to keep in reserve to be 95 percent certain that the claims are covered?

2.5-6   Bortkiewicz (1898, *Das Gesetz der kleinen Zahlen*) collected data on the number of horsemen, Y, that were killed by kicks from horses in each of 10 Prussian cavalry regiments. Data for 20 years (thus, 200 observations in total) are as follows:

Number of Fatalities	Observed Frequency
0	109
1	65
2	22
3	3
4	1
5	0

   Assume that the data come from a Poisson distribution. Note that this makes sense because fatalities due to kicks from horses are very rare events. Calculate an estimate of the parameter λ. Using this estimate, calculate the probabilities and expected frequencies of $Y = 0, 1, 2, 3, 4$, and ≥ 5, and compare them with the observed frequencies. Comment on the "fit."

2.5-7   Show that, for the Poisson distribution,

$$E[Y(Y - 1)] = \lambda^2.$$

   Hint: Note that

$$E[Y(Y - 1)] = \sum_{y=0}^{\infty} y(y - 1)\frac{\lambda^y e^{-\lambda}}{y!} = \sum_{y=2}^{\infty} \frac{\lambda^y e^{-\lambda}}{(y - 2)!}$$

   and factor λ^2 out of the last summation.

2.5-8   A furniture factory found that the number of reclamations concerning wood delivered by a certain supplier was six per year, on average. Making the appropriate distributional assumption, calculate the probability of having no reclamation in

   (a) all of next year, and

   (b) the next quarter.

2.5-9   Use your computer software to obtain the p.d.f. and the c.d.f. of

(a) the Poisson distribution with parameter (mean) $\lambda = 3$, and

(b) the Poisson distribution with parameter (mean) $\lambda = 6$.

Sketch the p.d.f. and the c.d.f. of each distribution. What are the means and standard deviations of these distributions?

2.6 Multivariate Distributions

Engineers often take more than one measurement on each item selected. For example, they may measure the percentage X of cotton in a fabric specimen, as well as its breaking strength Y. Or they may take measurements on the length X, width Y, and height Z of a certain item. Or, for each production shift, they may count the number of defective items, X, as well as the number of problems on the assembly line, Y.

2.6-1 Joint, Marginal, and Conditional Distributions

Consider the probabilities that are associated with two *discrete random variables* X and Y. We call the probability function

$$f(x, y) = P(X = x, Y = y), (x,y) \in R,$$

where R is the space of (X, Y), the *joint probability density function (p.d.f.)* of X and Y, or simply the *joint density* of X and Y.

Example 2.6-1   Twenty chips of the same size are placed in a bowl. Each chip has a pair (x, y) of numbers written on it. Assume that there are 4 chips with $(1, 1)$ written on them, 3 with $(2, 1)$, 1 with $(3, 1)$, 2 with $(1, 2)$, 4 with $(2, 2)$, and 6 with $(3, 2)$. In other words, the pairs of numbers are distributed among the 20 chips in accordance with Table 2.6-1.

If a chip is drawn at random, the probability of obtaining a chip with, for example, $(2, 1)$, is 3/20 under the assumption that each chip has the same probability of being picked. Also, the probability of drawing a chip with $(1, 2)$ is 2/20. Let us denote the coordinates of the chip that is to be selected at random by (X, Y), so that (X, Y) is a pair of discrete random variables. With this convention, the probability of obtaining a chip with coordinates $(2, 1)$ can be written

$$P(X = 2, Y = 1) = \frac{3}{20},$$

Table 2.6-1   Classification of 20 Chips

	x			
y	1	2	3	Total
1	4	3	1	8
2	2	4	6	12
Total	6	7	7	20

where $(X = 2, Y = 1)$ means $(X = 2)$ *and* $(Y = 1)$. Other examples are

$$P(X = 3, Y = 2) = \frac{6}{20} \text{ and } P(X = 1, Y = 1) = \frac{4}{20}.$$

Moreover, we can ask for probabilities about X and Y alone. For example,

$$P(X = 2) = P(X = 2, Y = 1) + P(X = 2, Y = 2) = \frac{3}{20} + \frac{4}{20} = \frac{7}{20},$$

because the event $X = 2$ is the union of the mutually exclusive events $(X = 2, Y = 1)$ and $(X = 2, Y = 2)$.
Probabilities such as

$$P(X = 1) = \frac{4}{20} + \frac{2}{20} = \frac{6}{20}$$

and

$$P(Y = 2) = \frac{2}{20} + \frac{4}{20} + \frac{6}{20} = \frac{12}{20}$$

are called *marginal probabilities* because they are usually recorded in the margins of the joint probability table. Table 2.6-2 lists the joint probabilities $P(X = x, Y = y)$, as well as the marginal probabilities $f_1(x) = P(X = x)$ and $f_2(y) = P(Y = y)$, for Example 2.6-1.

Table 2.6-2 $f(x, y) = P(X = x, Y = y)$

y	x			$P(Y = y)$
	1	2	3	
1	4/20	3/20	1/20	8/20
2	2/20	4/20	6/20	12/20
$P(X = x)$	6/20	7/20	7/20	

From the table, we note that the joint probability density function $f(x, y)$ defined on a discrete two-dimensional space R has the following properties:

1. $0 \leq f(x, y) = P(X = x, Y = y) \leq 1.$
2. $\sum\sum_{(x, y) \in R} f(x, y) = 1.$
3. $P[(X, Y) \in A] = \sum\sum_{(x, y) \in A} f(x, y),$ where A is a subset of R.

As an illustration of the third property, note in Example 2.6-1 that we have

$$P(X + Y = 3) = P(X = 1, Y = 2) + P(X = 2, Y = 1) = f(1, 2) + f(2, 1) = \frac{2}{20} + \frac{3}{20} = \frac{5}{20},$$

because $X + Y = 3$ happens when and only when $(X = 1, Y = 2)$ or $(X = 2, Y = 1)$.

Our example illustrated the computation of the marginal probabilities

$$f_1(x) = P(X = x) = \sum_y f(x, y) = \sum_y P(X = x, Y = y),$$

$$f_2(y) = P(Y = y) = \sum_x f(x, y) = \sum_x P(X = x, Y = y).$$

We call $f_1(x)$ and $f_2(y)$ the *marginal probability density functions* of X and Y, respectively.

   If we are now told that we will consider only chips with $X = 1$, what would be the probability of $Y = 2$? Certainly not $P(Y = 2) = 12/20$, as we now have the additional information that we are considering only the six chips in the $x = 1$ column of Table 2.6-1. The correct answer is 2/6, as there are two $Y = 2$ chips among those six chips. This conditional probability of $Y = 2$ given that $X = 1$ is written as

$$P(Y = 2 | X = 1) = \frac{P(X = 1, Y = 2)}{P(X = 1)} = \frac{2/20}{6/20} = \frac{2}{6}.$$

Conditional probabilities such as

$$P(Y = y | X = x) = \frac{P(X = x, Y = y)}{P(X = x)} = \frac{f(x, y)}{f_1(x)},$$

provided that $f_1(x) = P(X = x) > 0$, give us the *conditional probability density function* of Y given that $X = x$. We write this function as

$$g(y|x) = \frac{f(x, y)}{f_1(x)}.$$

Summing its values over y for fixed x gives 1, as it is a (conditional) discrete p.d.f. Similarly, the conditional p.d.f. of X given $Y = y$ is

$$h(x|y) = \frac{f(x, y)}{f_2(y)}.$$

For the example in Table 2.6-2, we find

$$g(y|x = 1) = \begin{cases} \dfrac{4/20}{6/20} = \dfrac{2}{3}, & y = 1 \\[2mm] \dfrac{2/20}{6/20} = \dfrac{1}{3}, & y = 2 \end{cases}$$

$$g(y|x = 2) = \begin{cases} \dfrac{3/20}{7/20} = \dfrac{3}{7}, & y = 1 \\[2mm] \dfrac{4/20}{7/20} = \dfrac{4}{7}, & y = 2 \end{cases}$$

$$g(y|x = 3) = \begin{cases} \dfrac{1/20}{7/20} = \dfrac{1}{7}, & y = 1 \\[2mm] \dfrac{6/20}{7/20} = \dfrac{6}{7}, & y = 2 \end{cases}$$

and

$$h(x|y = 1) = \begin{cases} \dfrac{4/20}{8/20} = \dfrac{4}{8}, & x = 1 \\[2ex] \dfrac{3/20}{8/20} = \dfrac{3}{8}, & x = 2 \\[2ex] \dfrac{1/20}{8/20} = \dfrac{1}{8}, & x = 3 \end{cases}$$

$$h(x|y = 2) = \begin{cases} \dfrac{2/20}{12/20} = \dfrac{1}{6}, & x = 1 \\[2ex] \dfrac{4/20}{12/20} = \dfrac{2}{6}, & x = 2 \\[2ex] \dfrac{6/20}{12/20} = \dfrac{3}{6}, & x = 3 \end{cases}$$

Both $g(y|x)$ and $h(x|y)$ satisfy the properties of being a p.d.f., so we can compute *conditional probabilities* such as

$$P(a < Y < b | X = x) = \sum_{\{y:a<y<b\}} g(y|x)$$

and *conditional expectations* like

$$E[u(Y)|X = x] = \sum_y u(y)g(y|x).$$

In particular, the conditional mean and conditional variance of Y, given that $X = x$, are, respectively,

$$\mu_{Y|x} = E(Y|x) = \sum_y y g(y|x)$$

and

$$\sigma^2_{Y|x} = E[(Y - \mu_{Y|x})^2 | x] = \sum_y (y - \mu_{Y|x})^2 g(y|x).$$

As an illustration, in the previous example, when given $x = 1$, we have

$$\mu_{Y|x=1} = E(Y|x = 1) = (1)(2/3) + (2)(1/3) = 4/3,$$
$$\sigma^2_{Y|x=1} = (1 - (4/3))^2(2/3) + (2 - (4/3))^2(1/3) = 2/9.$$

Example 2.6-2   Let X and Y have the joint p.d.f.

$$f(x, y) = \frac{x + 2y}{18}, \quad x = 1, 2, y = 1, 2.$$

Then the marginal densities of X and Y are, respectively,

$$f_1(x) = \sum_{y=1}^{2} \frac{x + 2y}{18} = \frac{2x + 6}{18}, \quad x = 1, 2$$

and

$$f_2(y) = \sum_{x=1}^{2} \frac{x + 2y}{18} = \frac{3 + 4y}{18}, \quad y = 1, 2.$$

The conditional densities are, respectively,

$$g(y|x) = \frac{x + 2y}{2x + 6}, \quad y = 1, 2$$

and

$$h(x|y) = \frac{x + 2y}{3 + 4y}, \quad x = 1, 2.$$ ∎

Example 2.6-3 Let

$$f(x, y) = \frac{xy^2}{30}, \quad x = 1, 2, 3, \ y = 1, 2$$

be the joint p.d.f. of X and Y. Hence,

$$f_1(x) = \sum_{y=1}^{2} \frac{xy^2}{30} = \frac{x}{6}, \quad x = 1, 2, 3$$

and

$$f_2(y) = \sum_{x=1}^{3} \frac{xy^2}{30} = \frac{y^2}{5}, \quad y = 1, 2$$

are the marginal densities of X and Y, respectively. ∎

2.6-2 Independence and Dependence of Random Variables

If

$$f(x, y) = P(X = x, Y = y) = P(X = x)P(Y = y) = f_1(x)f_2(y)$$

for all x and y, then X and Y are said to be *independent random variables*. That is, the joint p.d.f. of independent random variables is the product of the marginal densities. If random variables are not independent, they are said to be *dependent*.

The random variables in Example 2.6-1 are dependent. You can check that products of the marginal probabilities do not result in joint probabilities. Also, the conditional distributions depend on the conditioning argument; the densities $h(x|y = 1)$ and $h(x|y = 2)$ are not the same. The random variables in Example 2.6-3, on the other hand, are independent: All joint probabilities are products of the respective marginal probabilities.

Example Let X and Y have the joint p.d.f. $f(x, y)$ given by the display in Table 2.6-3.
2.6-4

Table 2.6-3 Joint p.d.f $f(x, y)$

			x		
y	1	2	3	4	$f_2(y)$
1	1/30	2/30	4/30	3/30	1/3
2	2/30	4/30	5/30	9/30	2/3
$f_1(x)$	1/10	2/10	3/10	4/10	

Here, the events $X = 2$ and $Y = 1$ are independent, because

$$\frac{2}{30} = f(2,1) = P(X = 2, Y = 1) = P(X = 2)P(Y = 1) = f_1(2)f_2(1) = \left(\frac{2}{10}\right)\left(\frac{1}{3}\right).$$

However, X and Y are not independent random variables, because $f(x, y)$ does not equal $f_1(x)f_2(y)$ for *all* possible x and y values; for example,

$$\frac{5}{30} = f(x = 3, y = 2) \neq f_1(x = 3)f_2(y = 2) = \left(\frac{3}{10}\right)\left(\frac{2}{3}\right).$$

That is, X and Y are *dependent* random variables.
In this example, it is easy to make X and Y independent random variables by slightly modifying the probabilities in the $x = 3$ and $x = 4$ columns, as in Table 2.6-4.

Table 2.6-4 Joint p.d.f. of Independent Random Variables

			x		
y	1	2	3	4	$f_2(y)$
1	1/30	2/30	3/30	4/30	1/3
2	2/30	4/30	6/30	8/30	2/3
$f_1(x)$	1/10	2/10	3/10	4/10	

Now the rows of $f(x, y)$ are proportional to each other and so are the columns of $f(x, y)$. Thus,

$$f(x, y) = f_1(x)f_2(y), x = 1, 2, 3, 4, \text{ and } y = 1, 2,$$

which shows that X and Y are independent random variables.   ■

2.6-3 Expectations of Functions of Several Random Variables

The concept of mathematical expectation can be extended to functions of two or more random variables. Let $u(X, Y)$ be a function of two discrete random variables X and Y having p.d.f. $f(x, y)$ and space R. Then

$$E[u(X, Y)] = \sum_{(x,y) \in R} \sum u(x, y) f(x, y)$$

is called the *expectation*, or the *expected value*, of $u(X, Y)$.

In particular, the respective expectations (*means*) of X and Y are

$$\mu_X = E(X) = \sum_{(x, y) \in R} \sum x f(x, y) = \sum_x x f_1(x)$$

and

$$\mu_Y = E(Y) = \sum_{(x, y) \in R} \sum y f(x, y) = \sum_y y f_2(y).$$

That shows that the means of X and Y can be computed with either the joint p.d.f. or the appropriate marginal p.d.f. The same is true for the *variances*, which are given by

$$\text{var}(X) = E[(X - \mu_X)^2] = \sum_{(x, y) \in R} \sum (x - \mu_X)^2 f(x, y) = \sum_x (x - \mu_X)^2 f_1(x)$$

and

$$\text{var}(Y) = E[(Y - \mu_Y)^2] = \sum_{(x, y) \in R} \sum (y - \mu_Y)^2 f(x, y) = \sum_y (y - \mu_Y)^2 f_2(y).$$

A new expression, the *covariance* of X and Y, requires knowledge of the joint p.d.f.; the covariance is given by

$$\sigma_{XY} = \text{cov}(X, Y) = E[(X - \mu_X)(Y - \mu_Y)] = \sum_{(x, y) \in R} \sum (x - \mu_X)(y - \mu_Y) f(x, y).$$

Once these expressions are found, we can compute the *correlation coefficient* of X and Y, denoted by the Greek lowercase letter rho and given by

$$\rho = \frac{\text{cov}(X, Y)}{\sqrt{\text{var}(X)\text{var}(Y)}} = \frac{\sigma_{XY}}{\sigma_X \sigma_Y},$$

where σ_X and σ_Y are the respective standard deviations.

Of course, we can continue to calculate the variances from

$$\sigma_X^2 = E(X^2) - \mu_X^2 \text{ and } \sigma_Y^2 = E(Y^2) - \mu_Y^2.$$

There is a similar method for computing the covariance. Because the expectation is a linear operation, we have

$$\begin{aligned}
\text{cov}(X, Y) &= E[(X - \mu_X)(Y - \mu_Y)] = E(XY - \mu_X Y - \mu_Y X + \mu_X \mu_Y) \\
&= E(XY) - \mu_X E(Y) - \mu_Y E(X) + \mu_X \mu_Y \\
&= E(XY) - \mu_X \mu_Y - \mu_Y \mu_X + \mu_X \mu_Y \\
&= E(XY) - \mu_X \mu_Y.
\end{aligned}$$

∎

Example 2.6-5 Let X and Y have the joint p.d.f. $f(x, y)$ given in Table 2.6-5.

Table 2.6-5	Joint p.d.f. $f(x, y)$		
	x		
y	1	2	$f_2(y)$
1	0.4	0.1	0.5
2	0.1	0.4	0.5
$f_1(x)$	0.5	0.5	

Here,

$$\mu_X = \mu_Y = (1)(0.5) + (2)(0.5) = 1.5$$

and

$$\sigma_X^2 = \sigma_Y^2 = (1)^2(0.5) + (2)^2(0.5) - (1.5)^2 = 0.25.$$

The covariance is equal to

$$\text{cov}(X,Y) = E(XY) - \mu_X\mu_Y$$
$$= (1)(1)(0.4) + (1)(2)(0.1) + (2)(1)(0.1) + (2)(2)(0.4) - (1.5)(1.5)$$
$$= 0.15.$$

This means that the correlation coefficient is

$$\rho = \frac{0.15}{\sqrt{(0.25)(0.25)}} = \frac{0.15}{0.25} = 0.6.$$

In Example 2.6-5, there is a larger probability, 0.4, of a smaller X occurring with a smaller Y and of a larger X occurring with a larger Y than is the probability, 0.1, of having a small value of one random variable occurring with a large value of the other. Accordingly, we find a positive correlation coefficient of 0.6. It is interesting to note that if we increased the probability of obtaining a large X with a large Y and a small X with a small Y, then ρ would increase. For example, suppose that

$$f(1, 1) = f(2, 2) = 0.5,$$

so that $f(1, 2) = f(2, 1) = 0$; then it is easy to show [Exercise 2.6-2(a)] that $\rho = 1$. However, if

$$f(1, 1) = f(2, 2) = f(1, 2) = f(2, 1) = 0.25,$$

so that X and Y are independent random variables, then $\rho = 0$ [Exercise 2.6-2(b)]. Continuing in this manner, if we shift more probability to the "small–large" situation, say,

$$f(1, 2) = f(2, 1) = 0.4$$

with
$$f(1, 1) = f(2, 2) = 0.1,$$

then the correlation coefficient becomes negative, namely, $\rho = -0.6$. [See Exercise 2.6-2(c).] These illustrations should give the reader some intuition about the meaning of the correlation coefficient. Also, note that the correlation coefficient in Section 1.5 is the sample estimate of ρ.

Remark: The covariance depends critically on the units of measurement. For example, if we multiply each possible value of X by a constant c_1 and each value of Y by a constant c_2, the covariance has to be multiplied by the factor c_1c_2. Obviously, this makes it difficult to assess the strength of a relationship from $cov(X,Y)$, since a change from meters to yards, kilograms to pounds, or centimeters to millimeters affects its magnitude. However, a change in the units of measurements does not change the correlation coefficient ρ. As will be shown in Exercise 2.6-7, ρ is always between -1 and $+1$.   ∎

As we will see in the discussion that follows, the fact that $\rho = 0$ for the independent situation in our illustration is not just a coincidence. In general, if X and Y are independent random variables [i.e., $f(x, y) = f_1(x)f_2(y)$], and if we are computing the expected value of a product of a function of X alone and a function of Y alone—say, $u(X)v(Y)$—then we obtain

$$E[u(X)v(Y)] = \sum_x \sum_y u(x)v(x)f_1(x)f_2(y)$$

$$= \left[\sum_x u(x)f_1(x)\right]\left[\sum_y v(y)f_2(y)\right] = E[u(X)]E[v(Y)].$$

Thus, when the product is $(X - \mu_X)(Y - \mu_Y)$, we find, for independent random variables X and Y, that

$$cov(X, Y) = E[(X - \mu_X)(Y - \mu_Y)] = [E(X - \mu_X)][E(Y - \mu_Y)]$$
$$= (\mu_X - \mu_X)(\mu_Y - \mu_Y) = 0.$$

Thus, $\rho = cov(X, Y)/\sigma_X\sigma_Y = 0$ if the random variables X and Y are independent. The converse is not necessarily true, however; that is, $\rho = 0$ does not automatically imply independence. This can be seen from the next example.

Example 2.6-6 Let X and Y have the joint p.d.f. $f(x, y)$ given in Table 2.6-6.

Table 2.6-6 Joint p.d.f. $f(x, y)$

y	1	2	3	$f_2(y)$
1	0.1	0.3	0.1	0.5
2	0.2	0.1	0.2	0.5
$f_1(x)$	0.3	0.4	0.3	

It is easy to see that $\mu_X = 2$ and $\mu_Y = 1.5$. Thus,

$$\text{cov}(X, Y) = (1)(1)(0.1) + (1)(2)(0.2) + (2)(1)(0.3) + (2)(2)(0.1)$$
$$+ (3)(1)(0.1) + (3)(2)(0.2) - (2)(1.5) = 0,$$

and of course, $\rho = 0$. But clearly, X and Y are dependent; for example, $P(X = 2, Y = 1) \neq P(X = 2)P(Y = 1)$ because $0.3 = f(2, 1) \neq f_1(2)f_2(1) = (0.4)(0.5)$. That is, X and Y must be dependent because the columns in Table 2.6-6 are not proportional, yet $\rho = 0$. ∎

Example 2.6-7 (Continuation of Example 2.6-2)
From the marginal densities of X and Y, it is easy to show that

$$\mu_X = \frac{14}{9}, \sigma_X^2 = \frac{20}{81}; \mu_Y = \frac{29}{18}, \sigma_X^2 = \frac{77}{324}.$$

Because

$$E(XY) = \sum_{x=1}^{2}\sum_{y=1}^{2} xy \frac{x + 2y}{18} = (1)(1)(3/18) + (1)(2)(5/18) + (2)(1)(4/18)$$
$$+ (2)(2)(6/18) = 45/18,$$

the covariance of X and Y is

$$\sigma_{XY} = \text{cov}(X, Y) = \frac{45}{18} - \left(\frac{14}{9}\right)\left(\frac{29}{18}\right) = -\frac{1}{162},$$

and the correlation coefficient is

$$\rho = \frac{\text{cov}(X,Y)}{\sigma_X \sigma_Y} = \frac{-1/162}{\sqrt{(20/81)(77/324)}} = -0.025.$$ ∎

2.6-4 Means and Variances of Linear Combinations of Random Variables

Consider a situation in which we have two random variables X and Y with joint p.d.f. $f(x, y)$. Suppose that we have calculated marginal means μ_X and μ_Y, marginal variances σ_X^2 and σ_Y^2, and the covariance σ_{XY}. Consider the linear combination $W = a_0 + a_1X + a_2Y$. What is its mean and what is its variance?
Using the definition of expectation from the previous section, we find that

$$\mu_W = \sum_{(x, y)\in R}\sum (a_0 + a_1x + a_2y)f(x, y)$$

$$= a_0 \sum_{(x, y)\in R}\sum f(x, y) + a_1 \sum_{(x, y)\in R}\sum xf(x, y) + a_2 \sum_{(x, y)\in R}\sum yf(x, y)$$

$$= a_0 + a_1\mu_X + a_2\mu_Y$$

and

$$\sigma_W^2 = E[(W - \mu_W)^2] = E\{[a_0 + a_1X + a_2Y - (a_0 + a_1\mu_X + a_2\mu_Y)]^2\}$$
$$= E\{[a_1(X - \mu_X) + a_2(Y - \mu_Y)]^2\}$$
$$= a_1^2E[(X - \mu_X)^2] + a_2^2E[(Y - \mu_Y)^2] + 2a_1a_2E[(X - \mu_X)(Y - \mu_Y)]$$
$$= a_1^2\sigma_X^2 + a_2^2\sigma_Y^2 + 2a_1a_2\sigma_{XY}.$$

Note that the constant a_0 does not affect the variance of the linear combination. If X and Y are independent, the covariance is zero and the variance simplifies to

$$\sigma_W^2 = a_1^2\sigma_X^2 + a_2^2\sigma_Y^2.$$

These results can be generalized to n random variables. Of particular interest is the case when the n random variables $X_1, X_2, \ldots, X_n$ are mutually independent with means $\mu_1, \mu_2, \ldots, \mu_n$, variances $\sigma_1^2, \sigma_2^2, \ldots, \sigma_n^2$, and zero covariances. Then the linear combination

$$W = a_0 + a_1X_1 + a_2X_2 + \cdots + a_nX_n$$

is a random variable with mean

$$\mu_W = a_0 + a_1\mu_1 + a_2\mu_2 + \cdots + a_n\mu_n$$

and variance

$$\sigma_W^2 = a_1^2\sigma_1^2 + a_2^2\sigma_2^2 + \cdots + a_n^2\sigma_n^2.$$

Example 2.6-8 Consider the random variables X and Y with joint p.d.f. $f(x, y)$ given in Table 2.6-5. We know from Example 2.6-5 that $\mu_X = \mu_Y = 1.5$, $\sigma_X^2 = \sigma_Y^2 = 0.25$, and covariance $\sigma_{XY} = 0.15$. Consider the random variable $W = 10 - 3X + 4Y$. Then $\mu_W = 10 - (3)(1.5) + (4)(1.5) = 11.5$ and $\sigma_W^2 = (-3)^2(0.25) + (4)^2(0.25) + (2)(-3)(4)(0.15) = 2.65.$ ∎

Exercises 2.6

*2.6-1 Let the joint p.d.f. of X and Y, $f(x, y)$, be given by

		x	
y	1	2	3
1	0.3	0.2	0.1
2	0.1	0.1	0.2

*(a) Determine the marginal densities.

*(b) Compute the means and the variances of X and Y.

*(c) Calculate $\sigma_{XY} = \text{cov}(X, Y)$ and the correlation coefficient ρ. Are X and Y independent?

*(d) Let $Z = X + Y$. Determine $k(z) = P(Z = z), z = 2, 3, 4, 5$. Determine the mean and the variance of Z.

Hint: For example, $P(Z = 4) = P(X = 2, Y = 2) + P(X = 3, Y = 1)$.

(e) Determine the mean and the variance of Z by applying the rules in Section 2.6-4 about the mean and the variance of a linear combination, and show that your results agree with those in (d).

2.6-2 If X and Y have the following three joint densities, compute ρ in each case:

(a)

		x	
y	1	2
1	0.5	0
2	0	0.5

(b)

		x	
y	1	2
1	0.25	0.25
2	0.25	0.25

(c)

		x	
y	1	2
1	0.1	0.4
2	0.4	0.1

* 2.6-3 Let X and Y have the joint p.d.f. given by

		x	
y	1	2	3
1	0.05	0.15	0.20
2	0.10	0.10	0.10
3	0.15	0.15	0.00

*(a) Determine the marginal densities. Are X and Y independent?

*(b) Determine the means, the variances, and the correlation coefficient.

*(c) Find the conditional mean $E(Y \mid x)$ and the conditional variance $\text{var}(Y \mid x)$, for $x = 1, 2, 3$.

2.6-4 Change Exercise 2.6-3 slightly by having the probabilities in the $x = 3$ column read 0.15, 0.10, and 0.05, instead of 0.20, 0.10, and 0.00.

2.6-5 Let X and Y have the joint p.d.f. given by

		x	
y	1	2	3
1	0.5	0	0
2	0	0.3	0
3	0	0	0.2

Show that $\rho = 1$ and $E(Y \mid x) = x$; note that all points of positive probability lie on this line.

2.6-6 The joint p.d.f. of grades in the calculus course (X) and grades in the engineering statistics course (Y) is as follows (here, X and Y range from 0, which is F, to 4, which is A):

			x		
y	0	1	2	3	4
0	0.02	0.02	0.03	0.01	0.00
1	0.02	0.03	0.06	0.03	0.03
2	0.05	0.06	0.10	0.05	0.04
3	0.05	0.08	0.08	0.05	0.05
4	0.01	0.04	0.04	0.03	0.02

(a) Calculate the marginal densities and check whether X and Y are independent. Find the conditional density $P(Y = y|X = 4)$ and determine the conditional mean.

(b) Calculate the covariance and the correlation coefficient.

2.6-7 Show that if the correlation coefficient ρ of X and Y exists, then $-1 \le \rho \le 1$.

Hint: The discriminant of the nonnegative quadratic function in v,

$$h(v) = E\{[(X - \mu_X) + v(Y - \mu_Y)]^2\},$$

must be less than or equal to zero.

2.6-8 In many instances, the mean of the conditional distribution is a linear function, say, $E(Y|x) = a + bx$. Show that in this case it must be of the form

$$E(Y|x) = \mu_Y + \rho \frac{\sigma_Y}{\sigma_X}(x - \mu_X).$$

Hint: Show that the following equations are true:

1. $E(Y) = a + bE(X)$, or $\mu_Y = a + b\mu_X$
2. $E(XY) = aE(X) + bE(X^2)$, or $\mu_X\mu_Y + \rho\sigma_X\sigma_Y = a\mu_X + b(\mu_X^2 + \sigma_X^2)$

Solve the two equations for a and b.

Remark: This result provides additional insight into the correlation coefficient ρ. The slope of the conditional mean line is positive or negative, depending on the sign of ρ. Also, if $\rho = 1$ or $\rho = -1$, then all discrete points having positive probability must lie on the conditional mean line. [See Exercise 2.6-5.]

*2.6-9 Let X_1, X_2 be a random sample of size $n = 2$ from a discrete uniform distribution with p.d.f. $f(x) = 1/3, x = 1, 2, 3$. Let $Y = X_1 + X_2$.

*(a) Compute the mean μ_X and variance σ_X^2 of the underlying distribution, and use them to determine μ_Y and σ_Y^2 by applying the rules in Section 2.6-4 about means and variances of linear combinations.

*(b) Find the p.d.f. $g(y) = P(Y = y), y = 2, 3, 4, 5, 6$, and use it to compute μ_Y and σ_Y^2.

*(c) Find the p.d.f. for $W = X_1X_2$, and calculate the mean and the variance of W.

2.6-10   Consider the random variables X and Y with distribution given in Exercise 2.6-3. Obtain the mean and variance of the following linear combinations:

(a) $Z = 2X + 3Y$.

(b) $W = 2X - 3Y$.

2.6-11   Consider three mutually independent random variables X, Y, and Z with respective means $\mu_X = 10$, $\mu_Y = 5$, $\mu_Z = 2$, and standard deviations $\sigma_X = 3$, $\sigma_Y = 1$, $\sigma_Z = 2$. Obtain the mean and variance of the following linear combinations:

(a) $V = 5 + 2X + 3Y - Z$.

(b) $W = 10 + X - 3Y + 2Z$.

2.7 The Estimation of Parameters from Random Samples

Distributions depend on key parameters. For the Bernoulli distribution in Section 2.4, the parameter is the probability of success in a single trial, p. For the Poisson distribution in Section 2.5, the parameter is the mean number of successes, λ. Parameters are usually not known and must be estimated from the realizations $x_1, x_2, \ldots, x_n$ of a random sample.

Estimates judged reasonable and intuitive were suggested in previous sections. The sample proportion $\hat{p} = \sum x_i/n$ was the recommended estimate for p in a Bernoulli trial (Section 2.4.2), and the average number of successes in n trials, $\hat{\lambda} = \sum x_i/n$, was the recommended estimate for the parameter λ in the Poisson distribution (Section 2.5.2).

In this section, we discuss a formal way of estimating a parameter of a distribution. We introduce *maximum likelihood estimates* of the unknown parameters (in general, there can be more than one parameter), and we look at properties of the resulting estimates. As one would expect, the uncertainty in estimates decreases with increasing sample size.

2.7-1 Maximum Likelihood Estimation

Let us denote the probability density function with $f(x;\theta)$, where θ is the parameter that needs to be estimated. In estimation, we take a random sample $X_1, X_2, \ldots, X_n$ from the distribution. That is, we repeat the experiment n independent times, observe the sample $X_1, X_2, \ldots, X_n$ in which X_i has the same distribution $f(x;\theta)$, and estimate the value of θ using the realizations $x_1, x_2, \ldots, x_n$. The function of $X_1, X_2, \ldots, X_n$ used to estimate the parameter θ—say, the statistic $u(X_1, X_2, \ldots, X_n)$—is called the *estimator* of θ. We want it to be such that the computed *estimate* $u(x_1, x_2, \ldots, x_n)$ is close to θ. Since the estimate is a single number, we refer to it as a point estimate. The corresponding estimator is called the point estimator.

Because of the independence of the random variables $X_1, X_2, \ldots, X_n$, we can write the joint probability density function as

$$P(X_1 = x_1, X_2 = x_2, \ldots, X_n = x_n) = P(X_1 = x_1)P(X_2 = x_2)\ldots P(X_n = x_n)$$

$$= f(x_1; \theta)f(x_2; \theta)\ldots f(x_n; \theta).$$

A reasonable way to proceed toward finding a good estimate of θ is to regard the joint probability density as a function of θ (note that the data $x_1, x_2, \ldots, x_n$ have been observed and are known) and find the value of θ that maximizes this function. In other words, we find the value of θ that is most likely to have produced the sample values $x_1, x_2, \ldots, x_n$. The joint probability density function, when regarded as a function of the parameter, is called the *likelihood function*, and the value of θ that maximizes this function is called the *maximum likelihood estimate* $\hat{\theta}$. We use the "hat" to distinguish estimates of parameters from parameters.

2.7-2 Examples

Let us consider three examples.

Example 2.7-1

Estimation of p from a sequence of n Bernoulli trials

The probability density function (p.d.f.) of X, the outcome of a Bernoulli trial with probability of success p, is given by

$$f(x; p) = p^x (1 - p)^{1-x} \quad x = 0,1; \quad 0 < p < 1.$$

Given a random sample $X_1, X_2, \ldots, X_n$, the problem is to find $u(X_1, X_2, \ldots, X_n)$ such that $u(x_1, x_2, \ldots, x_n)$ is a good point estimate of p. The joint p.d.f. is given by

$$P(X_1 = x_1, X_2 = x_2, \ldots, X_n = x_n) = \prod_{i=1}^{n} p^{x_i}(1 - p)^{1-x_i} = p^{\sum x_i}(1 - p)^{n - \sum x_i}.$$

Treating this equation as a function of p (for given sample realizations), we obtain the likelihood function

$$L(p) = p^{\sum x_i}(1 - p)^{n - \sum x_i}.$$

This function needs to be maximized. In finding the maximum likelihood estimator, it is often easier to find the value of the parameter that maximizes the logarithm of the likelihood function. Because the logarithm is an increasing function, the solutions will be the same. Hence, we write

$$\ln L(p) = \left(\sum x_i\right)\ln(p) + [n - \sum x_i]\ln(1 - p).$$

Setting the first derivative to zero leads to

$$\frac{d \ln L(p)}{dp} = \frac{\sum x_i}{p} - \frac{n - \sum x_i}{1 - p} = 0,$$

$$\sum x_i - np = 0,$$

and

$$p = \frac{\sum x_i}{n} = \bar{x}.$$

Convince yourself that the second-order conditions for a maximum are satisfied, confirming that this value of p, namely, $\hat{p} = \bar{x}$, maximizes $\ln L(p)$. The statistic

$$\hat{p} = \frac{\sum X_i}{n} = \bar{X}\text{—the proportion of successes in a random sample of } n \text{ trials—is the}$$

maximum likelihood estimator of p. Once sample realizations have been obtained, the observed statistic is called the maximum likelihood estimate. Note that this was the estimate that we had suggested in Section 2.4-2. ∎

Example 2.7-2

Estimation of λ in the Poisson distribution

Let X have a Poisson distribution with p.d.f.

$$f(x; \lambda) = \frac{\lambda^x}{x!}e^{-\lambda}, \quad x = 0, 1, 2, \ldots; \quad \lambda > 0.$$

The likelihood function—that is, the joint probability density function (p.d.f.) $P(X_1 = x_1, X_2 = x_2, \ldots, X_n = x_n) = f(x_1; \lambda) f(x_2; \lambda) \ldots f(x_n; \lambda)$ treated as a function of the parameter λ—is given by

$$L(\lambda) = \left(\frac{\lambda^{x_1}}{x_1!}e^{-\lambda}\right)\left(\frac{\lambda^{x_2}}{x_2!}e^{-\lambda}\right)\ldots\left(\frac{\lambda^{x_n}}{x_n!}e^{-\lambda}\right) = \frac{1}{x_1! x_2! \ldots x_n!}\lambda^{\sum x_i}e^{-n\lambda},$$

and its logarithm is

$$\ln L(\lambda) = \left(\sum x_i\right)\ln(\lambda) - n\lambda - \ln(x_1! x_2! \ldots x_n!).$$

Now we set the derivative with respect to λ equal to zero:

$$\frac{d[\ln L(\lambda)]}{d\lambda} = \frac{\sum x_i}{\lambda} - n = 0.$$

This leads to

$$\lambda = \frac{\sum x_i}{n} = \bar{x}.$$

The second derivative, evaluated at $\lambda = \bar{x}$, is

$$-\frac{\sum x_i}{\lambda^2} = -\frac{n\bar{x}}{\bar{x}^2} = -\frac{n}{\bar{x}} < 0,$$

confirming that this value of λ, namely, $\hat{\lambda} = \bar{x}$, maximizes $\ln L(\lambda)$. The maximum likelihood estimator of λ is $\hat{\lambda} = \bar{X}$. Note that this was the estimate we had suggested in Section 2.5-3. ∎

Example 2.7-3

Estimation of p in the geometric distribution

Let $X_1, X_2, \ldots, X_n$ be a random sample from the geometric distribution with p.d.f. $f(x; p) = p(1 - p)^{x-1}$, for $x = 1, 2, 3, \ldots$. This is a special case of the negative binomial distribution in Section 2.4-3 with $r = 1$. The random variable X expresses the number of Bernoulli trials that are needed to reach the first success. The likelihood function is given by

$$L(p) = \prod_{i=1}^{n} p(1 - p)^{x_i - 1} = p^n(1 - p)^{(\sum x_i) - n} \quad \text{for } 0 < p < 1.$$

Setting the first derivative of

$$\ln L(p) = n \ln p + \left(\sum x_i - n\right)\ln(1 - p)$$

equal to zero leads to

$$\frac{d \ln L(p)}{dp} = \frac{n}{p} - \frac{\sum x_i - n}{1 - p} = 0.$$

Solving for p, we obtain

$$p = \frac{n}{\sum x_i} = \frac{1}{\bar{x}}$$

and this solution provides a maximum. Hence, the maximum likelihood estimator of p is

$$\hat{p} = \frac{n}{\sum X_i} = \frac{1}{\bar{X}}.$$

This estimator agrees with our intuition. There are n successes among $\sum x_i$ trials, for a proportion of $n / \sum x_i = 1/\bar{x}$ successes. ∎

2.7-3 Properties of Estimators

Estimates of unknown parameters are functions of the realizations of a random sample. Once the data are in, the estimates are certain numbers. But prior to sampling, the estimators are functions of random variables and they themselves have a distribution. We call this distribution the *sampling distribution* of the estimators. What are the properties of sampling distributions? Is it possible to attach error bounds to the estimates?

Let us consider the estimate of the probability of success from n independent Bernoulli trials. Example 2.7-1 of the previous section has shown that the maximum likelihood estimator of p is given by

$$\hat{p} = u(X_1, X_2, \ldots, X_n) = \frac{1}{n}(X_1 + X_2 + \cdots + X_n).$$

The estimator $\hat{p}$ is a linear combination of n independent random variables X_i that follow the same Bernoulli distribution with mean $E(X_i) = p$ and variance var $(X_i) = p(1 - p)$. Applying the results on means and variances of linear combinations of independent random variables presented in Section 2.6-4, we find that

$$E(\hat{p}) = \frac{1}{n}p + \frac{1}{n}p + \cdots + \frac{1}{n}p = p$$

and

$$\text{var}(\hat{p}) = \frac{1}{n^2}p(1 - p) + \frac{1}{n^2}p(1 - p) + \cdots + \frac{1}{n^2}p(1 - p) = \frac{p(1 - p)}{n}.$$

The first result, $E(\hat{p}) = p$, implies that the estimate $\hat{p}$ varies around the true value p. For some sample realizations, the estimate will be larger than p; for others, it will be smaller. But in the long run, the estimate is right on target. We call the property $E(\hat{p}) = p$ *unbiasedness,* and we refer to the estimator $\hat{p}$ as an *unbiased estimator* of p. Most of the time, we want our estimators to be unbiased.

The second result, $\mathrm{var}(\hat{p}) = \dfrac{p(1 - p)}{n}$, implies that the variability of the estimator around the true value p (in other words, the variability of our estimation error) shrinks with the sample size n. The standard deviation of the estimator,

$$\sigma_{\hat{p}} = \sqrt{\mathrm{var}(\hat{p})} = \sqrt{\frac{p(1 - p)}{n}},$$

is also called the *standard error* of the estimator $\hat{p}$. It represents the standard deviation of the estimation error that one makes when using $\hat{p}$ as an estimate of the unknown p. Our second result shows that the standard error shrinks with the square root of the sample size. Large samples will give you an estimate that is close to the true parameter, while, with a small sample, you can be far from the true value. Since p is not known, we can calculate only an estimate of the standard error:

$$\hat{\sigma}_{\hat{p}} = \sqrt{\frac{\hat{p}(1 - \hat{p})}{n}}.$$

We will see in a subsequent chapter that the interval

$$\hat{p} \pm 2\hat{\sigma}_{\hat{p}}, \quad \text{or} \quad \hat{p} \pm 2\sqrt{\frac{\hat{p}(1 - \hat{p})}{n}},$$

covers the unknown true parameter p with a probability of about 95 percent. We will refer to this interval as an approximate 95 percent *confidence interval* for p. We don't know for sure whether the calculated interval covers the unknown p; however, in repeated random samples of size n, these intervals have an approximate 95 percent chance of covering p.

Similar comments can be made about the estimator $\hat{\lambda} = \overline{X}$ of the parameter λ of a Poisson distribution. Of course, $\hat{\lambda} = \overline{X}$ is a linear function of $X_1, X_2, \ldots, X_n$, namely,

$$\hat{\lambda} = \frac{1}{n}(X_1 + X_2 + \cdots + X_n),$$

with mean

$$E(\hat{\lambda}) = \lambda$$

and variance

$$\mathrm{var}(\hat{\lambda}) = \left(\frac{1}{n}\right)^2 \lambda + \left(\frac{1}{n}\right)^2 \lambda + \cdots + \left(\frac{1}{n}\right)^2 \lambda = \frac{\lambda}{n}.$$

Since λ is unknown, the standard deviation of $\hat{\lambda}$ (i.e., its standard error) is estimated by $\sqrt{\hat{\lambda}/n}$. Again, the interval

$$\hat{\lambda} \pm 2\sqrt{\hat{\lambda}/n}$$

serves as an approximate 95 percent confidence interval for λ.

Here we have introduced the important concepts of estimation, sampling distributions, unbiasedness, standard errors, and confidence intervals for the special case of estimating the unknown p from a sequence of Bernoulli trials and the unknown λ

in the Poisson distribution. More generally, if a random sample $X_1, X_2, \ldots, X_n$ arises from a distribution with unknown mean μ and standard deviation σ, then

$$\overline{X} = \frac{1}{n}(X_1 + X_2 + \cdots + X_n)$$

has mean

$$E(\overline{X}) = \frac{1}{n}(\mu + \mu + \cdots + \mu) = \mu$$

and variance

$$\sigma_{\overline{X}}^2 = \text{var}(\overline{X}) = \left(\frac{1}{n}\right)^2 \sigma^2 + \left(\frac{1}{n}\right)^2 \sigma^2 + \cdots + \left(\frac{1}{n}\right)^2 \sigma^2 = \frac{\sigma^2}{n}.$$

It is an important fact that $\overline{X}$ has an approximate normal distribution. The normal distribution will be introduced in the next chapter, and in Chapter 4 we will learn that

$$\overline{x} \pm 2\sigma/\sqrt{n}$$

serves as an approximate 95% confidence interval for the unknown mean μ. If σ is also unknown, it can be estimated by the sample standard deviation s; thus,

$$\overline{x} \pm 2s/\sqrt{n}$$

provides an approximate 95% confidence interval for μ.

Exercises 2.7

2.7-1 We wish to estimate the size N of a population of animals (such as ducks, ants, or fish). We use the following procedure: We sample and tag n_1 animals and release them. We wait for some time to allow for fairly good "mixing" of the animals with and without tags. Then we sample n_2 animals and determine the number m of animals that had been tagged.

(a) Convince yourself that $\hat{N} = n_1 n_2/m$ is the appropriate estimator of the unknown population size N.

Note: Discuss why it is reasonable to assume that $n_1/N \approx m/n_2$, and obtain your estimator of N from this equality. Also, one can show that $\hat{N} \approx n_1 n_2/m$ is the maximum likelihood estimator. We don't require a proof, as that would go beyond the scope of this course. But if you are interested in learning what such a proof looks like, consult Tanis, E. A., and Hogg, R. V., *A Brief Course in Mathematical Statistics* (Upper Saddle River, NJ: Prentice Hall, 2006), pp. 73, 74.

(b) A slightly better (but quite similar) estimate is given by $N^* = \dfrac{(n_1 + 1)(n_2 + 1)}{m + 1} - 1$, and an approximate 95 percent confidence interval for the population size N can be obtained from $N^* \pm 2\sqrt{\text{var}(N^*)}$, where

$$\text{var}(N^*) = \frac{(n_1 + 1)(n_2 + 1)(n_1 - m)(n_2 - m)}{(m + 1)^2(m + 2)} - 1.$$

Consider the recapture data for the six ant colonies given below.

Colony	n_1	n_2	m
1	500	149	7
2	500	159	11
3	500	189	17
4	1,000	243	21
5	500	437	68
6	500	321	89

Estimate the sizes of the populations, and calculate approximate 95 percent confidence intervals for those sizes.

(c) Prior to the shooting season, ducks are trapped and banded before being released. Shooters return 12 percent of all bands. The total duck kill is 5 million ducks. Estimate the total duck population.

Note: Here, $n_2 = 5$ million and $m/n_1 = 0.12$.

2.7-2 Assume a discrete distribution with space $R = \{1, 2, 3\}$ and probabilities $P(X = 1) = P(X = 3) = p$ and $P(X = 2) = 1 - 2p$, with $0 < p < 0.5$.

(a) Graph the distribution, and determine its mean and variance.

(b) From a random sample of $n = 100$, you observe n_1 realizations of $X = 1$, n_2 realizations of $X = 2$, and $n_3 = n - n_1 - n_2$ realizations of $X = 3$. Determine the maximum likelihood estimator of p.

Chapter 2 Additional Remarks

Most statisticians believe that the serious study of probability began in 1654, when Chevalier de Méré, a French nobleman who liked to gamble, challenged the mathematician Blaise Pascal to consider the following problem: Compare the chances of (a) getting at least one six in 4 rolls of a single die and (b) getting at least one double six in 24 rolls of a pair of dice. The gambler, de Méré, thought that the chances should be the same, because four is to six, the number of ways in which a die can result, is the same ratio as 24 is to 36, the number of ways in which a pair of dice can result. However, he was winning betting on (a) and losing betting on (b). Pascal, not wanting to work on this problem alone, asked another great mathematician, Pierre de Fermat, to partner with him on solving the problem. Although these two great mathematicians solved this problem first, the solution can be explained to students after they have had very little probability. Because 5/6 is the probability of not getting a six on a single roll, $(5/6)^4$ is the probability of not getting a six on 4 rolls; thus, the probability of (a) is $1 - (5/6)^4 = 0.518$ approximately. The corresponding probabilities with a pair of dice are 35/36 and $(35/36)^{24}$; hence, the probability of (b) is $1 - (35/36)^{24} = 0.491$ approximately. It seems amazing to us that de Méré could observe enough trials of those two events to detect the slight differences in these two probabilities.

The two mathematicians considered another related problem that had been around for a long time: "A and B are playing a fair game of balla. The players agree to continue until one has won six games. However, they are interrupted when A had won five and B three. How should the stakes be divided?" Some may answer three to one, with A getting three-fourths of the stakes. However, note that B can be the victor only by winning the next three games, and that has probability $(1/2)^3 = 1/8$. Thus, A's probability of winning is $1 - (1/2)^3 = 7/8$, and hence the stakes should be divided *seven* to one.

We believe that Pascal continued to work on the generalization of this latter problem; to illustrate, what should be done if A has won four games and B two? Then, to come out ahead, B must win the next four or must win three of the next four and then also win on the fifth trial. The sum of these two probabilities is $(1/2)^4 + \binom{4}{3}(1/2)^3(1/2)(1/2) = 3/16$. In any case, Pascal continued long enough to discover *Pascal's triangle,* involving binomial coefficients, namely,

$$
\begin{array}{ccccccccccc}
 & & & & & 1 & & & & & \\
 & & & & 1 & & 1 & & & & \\
 & & & 1 & & 2 & & 1 & & & \\
 & & 1 & & 3 & & 3 & & 1 & & \\
 & 1 & & 4 & & 6 & & 4 & & 1 & \\
1 & & 5 & & 10 & & 10 & & 5 & & 1 \\
 & & & & & \vdots & & & & &
\end{array}
$$

The relationships can be summarized with the equation

$$
\binom{n-1}{r} + \binom{n-1}{r-1} = \binom{n}{r},
$$

which can be proven quite easily by expressing the binomial coefficients in terms of the corresponding factorials.

The next major contributions to probability were made by the Bernoullis, a remarkable Swiss family of eight mathematicians from the late 1600s to the early 1700s. We will mention only two of them. Jacob, who was the oldest, discovered that Y/n, where Y is the number of successes in n independent trials, approaches, in some probabilistic sense, the probability p of success of each trial. This is called the *law of large numbers.* We used this relative frequency idea as the motivation of the axioms of probability.

The second was Daniel, who was a nephew of Jacob. He understood that expected value was the sum of the products of payoffs and corresponding probabilities. He considered the following problem, which is sometimes called the St. Petersburg paradox (St. Petersburg is a city in Russia, located at the east end of the Gulf of Finland on the Baltic Sea): "Player A continues to toss a coin until a head first appears, say on the xth trial. Then player A gives player B 2^{x-1} dollars (originally "ducats," but we use dollars). How much should B pay A to enter the game?" Clearly, from the geometric distribution in Section 2.4-3,

$$
\sum_{x=1}^{\infty}(2^{x-1})(1/2)(1/2)^{x-1} = \sum_{x=1}^{\infty}(1/2) = \infty;
$$

that is, the expected return for player B is unbounded. A rational gambler would enter a game if the price of entry was less than the expected value. But would B give A something like \$1,000 to play? If so, then player B wins only if $x \geq 11$, although the amount of his winnings (if he does win) would be substantial. Now, would A be willing to accept the $1,000 payment for B to play, knowing that there is a small probability that player B could ruin him?

This problem made Daniel think about the *utility* of money. For most of us, $2 is worth twice as much as $1. However, is $2 million worth twice as much as $1 million? Possibly to Bill Gates it is, but to us (remember, Hogg is already retired), it isn't. Suppose Hogg has just $1 million to use in retirement and Gates offers him the following deal: Gates will put up $2 million against that $1 million, flip a fair coin, and have the winner take all (i.e., the $3 million). Hogg will have either 0 or $3 million, each with probability 1/2, and the expected value, $(0)(1/2) + (3)(1/2) = 1.5$, is much more than his $1 million. This is a great bet! However, Hogg would never take it, because the additional utility of the extra $2 million is not worth the utility of the $1 million he already has. Keeping the same ratio, would Hogg bet $100 against $200? Yes. Would he bet $1,000 against $2,000? Yes, although he would have to run it by his wife first. Would he bet $10,000 against $20,000? Probably not. So Hogg's utility function is starting to bend somewhere between $1,000 and $10,000.

The concept of utility is useful in understanding insurance. When you pay an insurance company a premium, it is always more than what they expect to pay you, on average. On average, it is a bad bet for you, and you should not insure something if the insured amount reflects your true utility. As a very rough rule, we would say never insure anything less than two month's salary. This is not a theorem, but only a rough guide, and each of us must evaluate his or her own situation.

R.A. Fisher not only was the architect of key concepts in the design of experiments, but also contributed a great deal to the estimation of parameters—in particular, maximum likelihood estimation. Another group of statisticians, called Bayesians, has a different approach to estimation. A Bayesian treats a parameter θ as a random variable. Say that there is another random variable X that can be observed and that has p.d.f. $f(x|\theta)$; that is, $f(x|\theta)$ is the model for the conditional p.d.f. of X, given θ. Suppose θ has the prior probabilities given by p.d.f. $g(\theta)$. Then the joint p.d.f. of X and θ is given by $g(\theta)f(x|\theta)$, and the marginal p.d.f. of X is given by the sum $h(x) = \sum_\theta g(\theta)f(x|\theta)$. So the conditional p.d.f. of θ, given that $X = x$, is

$$k(\theta|x) = \frac{g(\theta)f(x|\theta)}{h(x)} = \frac{g(\theta)f(x|\theta)}{\sum_\theta g(\theta)f(x|\theta)},$$

which, after a little bit of thought, you should recognize as Bayes' theorem. So the posterior probabilities about the parameter θ, $k(\theta|x)$, are different from the prior probabilities, $g(\theta)$, after X is observed to be x. Repeating the experiment will result in many x values, say, $x_1, x_2, \ldots, x_n$, and the Bayesians use the posterior distribution $k(\theta | x_1, x_2, \ldots, x_n)$ to make their inferences about the parameter θ. It is interesting to note that the Reverend Thomas Bayes, the mathematician and Presbyterian minister who started this line of thinking, never published a mathematical article during his lifetime; his famous paper was published about two years after his death.

Projects

*Project 1: Decision trees

[This project is adapted from Evans, James R., *Statistics, Data Analysis, and Decision Modeling*, 3d ed. (Upper Saddle River, NJ: Prentice Hall, 2007).]

You expect to reduce travel costs by developing software for conducting meetings electronically. The following three scenarios have been suggested:

- In-house development of the software: Cost 112.0 K + yearly running costs. The chance of completing the system is 0.70. If the system is successful, its availability is 84 percent.

- Step-up in-house development: Cost 175.6 K + yearly running costs. The chance of success is 0.90. If the system is successful, its availability is 95 percent.

- Buy-down development: 77.4 K + yearly running costs. The chance of success is 0.50. If the system is successful, its availability is 60 percent.

The yearly operating costs are the same for all three systems: $81,600 in year 1, $85,680 in year 2, and $89,964 in year 3.

Assume two different, equally likely, scenarios for demand:

- Low usage: 50 meetings (of five participants with savings of $500 each), for a total of 125 K.

- High usage: 100 meetings (of five participants with savings of $500 each), for a total of 250 K.

The project is to be evaluated over a three-year basis, using a discount rate of 12 percent per year. Which decision maximizes the expected benefit?

Notes: You need to determine the cost savings for all possible outcomes. For each strategy, there are three possibilities: (success and high demand), (success and low demand), and (failure). In the latter case, the demand does not matter, as the program is not operational. You know the probabilities of these states. The benefit from successful in-house development and high usage is calculated as follows: Each year, the expected reduction in travel costs is $(0.84)(\$250,000) = \$210,000$. Savings in year 1 are $\$210,000 - \$81,600 = \$128,400$; in year 2, $\$210,000 - \$85,680 = \$124,320$; and in year 3, $\$210,000 - \$89,964 = \$120,036$. The discounted benefit with a 12 percent discount rate is $\$128,400/(1.12) + \$124,320/(1.12)^2 + \$120,036/(1.12)^3 - \$112,000$ (the development cost) $= \$187,190$. For low usage, the expected reduction in travel costs is $(0.84)(\$125,000) = \$105,000$. Savings in year 1 are $\$105,000 - \$81,600 = \$23,400$; in year 2, $\$105,000 - \$85,680 = \$19,320$; and in year 3, $\$105,000 - \$89,964 = \$15,036$. The discounted benefit with a 12 percent discount rate is $\$23,400/(1.12) + \$19,320/(1.12)^2 + \$15,036/(1.12)^3 - \$112,000$ (the development cost) $= -\$65,000$. If the in-house program fails (i.e., the program is never operational and hence no operating costs arise), we lose the development costs; that is, the benefit is $-\$112,000$.

Calculate the benefits for the step-up in-house development. Show that the cost savings are $189,640 (high usage), −$95,580 (low usage), and −$175,600 (failure). For the buy-down development, the benefits are $77,680 (high usage), −$102,460 (low usage), and −$77,400 (failure).

You can calculate the expected benefit for each of the three possible strategies and make your decision that way. Alternatively—and this is recommended for more complicated situations—you can use "decision tree" analysis. Here, we have one *decision node* (with three plans) and two *event nodes* (with probabilities governing the success of the system and the uncertain demand). Situations with more than one decision node and several event nodes can be handled quite easily with "decision tree" software. You will find convenient EXCEL decision tree add-ins by searching the Web; many of these packages can be tried for free.

Project 2

The number of work stoppages per shift has a Poisson distribution, with a parameter that depends on the workforce that is available. For shifts staffed by the fully trained regular workforce, the parameter is λ_1; for shifts staffed by part-time workers, the parameter is $\lambda_2 > \lambda_1 > 0$. Assume that the fraction of shifts that are staffed by the fully trained workforce is p ($0 < p < 1$).

(a) Obtain the distribution of X, the number of work stoppages per shift. Determine its mean and variance. Graph the distribution for various combinations, such as $(\lambda_1 = 2; \lambda_2 = 3; p = 0.8)$, $(\lambda_1 = 2; \lambda_2 = 7; p = 0.8)$, and $(\lambda_1 = 2; \lambda_2 = 7; p = 0.5)$.

Note: Because we are mixing two distributions, we say that X has a *mixture distribution*.

(b) Assume that you observe the number of work stoppages, $x_1, x_2, \ldots, x_n$, for n shifts. How would you estimate the parameters $\lambda_2 > \lambda_1 > 0$ and $0 < p < 1$?

No actual derivations are necessary, but try to sketch a possible approach. Write down the log-likelihood function that needs to be maximized. You will notice that it is impossible to determine explicit solutions; iterative optimization methods need to be used. In general, will it be easy to estimate the three parameters?

*Project 3

Consider a digital communication receiver with response $x = s + n$, where s is the signal, taking on values -1 and $+1$ with equal probability. The noise n is independent of the signal and follows a discrete distribution with values $-2, -1, 0, 1, 2$ and probabilities 0.1, $0.3, 0.3, 0.2, 0.1$. Develop a detector for the signal that minimizes the probability of making an error.

3

CONTINUOUS PROBABILITY MODELS

3.1 Continuous Random Variables

In Chapter 2, we considered discrete random variables; for example, the binomial random variable can take only one of the values $0, 1, 2, \ldots, n$, and the Poisson random variable must equal a nonnegative integer. We recognize, however, that many random variables, such as the weight of an item, the tear strength of a piece of paper, and the length of life of a motor, can assume any value in certain intervals. These random variables are said to be *continuous random variables,* because their realizations fall into a given continuum, or interval.

3.1-1 Empirical Distributions

Let us suppose that we have a random sample from a distribution of either the discrete or continuous type. Before the sample is drawn, the future observations, $X_1, X_2, \ldots, X_n$, are treated as mutually independent random variables, each coming from the same underlying distribution. Once the sample is observed, the resulting observations are denoted by the lowercase letters $x_1, x_2, \ldots, x_n$, respectively. If we now assign a weight of $1/n$ to each of these n observations, we have, in effect, created a distribution of the discrete type, because the weights are nonnegative and sum to 1. This kind of distribution is called an *empirical distribution.*

We can find the mean and the variance of an empirical distribution by the same procedure that we use with any other discrete distribution. Since each value x_i has the "probability" (weight) $1/n$, we find that the mean and the variance are, respectively,

$$\sum_{i=1}^{n} x_i \left(\frac{1}{n} \right) = \frac{1}{n} \sum_{i=1}^{n} x_i = \bar{x}$$

and

$$\sum_{i=1}^{n} (x_i - \bar{x})^2 \left(\frac{1}{n} \right) = \frac{1}{n} \sum_{i=1}^{n} (x_i - \bar{x})^2 = \frac{1}{n} \sum_{i=1}^{n} x_i^2 - \bar{x}^2 = v.$$

Note that the mean of the empirical distribution and the sample mean are the same, namely, $\bar{x}$. Moreover, the variance, v, of the empirical distribution and the sample variance, s^2, are related by the expression $v = (n - 1)s^2/n$; thus, they are almost equal for large or moderately large n.

An empirical distribution can help us model the underlying distribution from which the sample is taken. To illustrate this fact, let us refer to the compressive strengths of the $n = 90$ concrete blocks discussed in Section 1.3. There, we described the variability of these measurements with a histogram. In the histogram in Figure 1.3-1(b), the height above each class is either the frequency of the class, say, f_i, or the relative frequency f_i/n. Now let us go one step further and choose the height of the bar such that the *area* associated with that class is equal to the relative frequency f_i/n. Since the length of each class interval in that example is 4 units, the heights become $f_i/4n$. Let us call the resulting histogram a *normalized relative frequency histogram* and denote it by $h(x)$. This histogram is called "normalized" because the area underneath it is 1. It is clear that we can obtain the relative frequency of a number of classes simply by integrating $h(x)$ over those classes. For example, the relative frequency of the three classes from 48.0 to 60.0 is equal to

$$\int_{48}^{60} h(x)dx = 4 \cdot \frac{17}{(90)(4)} + 4 \cdot \frac{9}{(90)(4)} + 4 \cdot \frac{3}{(90)(4)} = \frac{29}{90}.$$

One might think that the integration is too complicated; wouldn't it be much easier to sum the frequencies 17, 9, and 3 and divide by $n = 90$, obtaining 29/90? The more difficult integration, however, leads to a generalization that proves very valuable when we have histograms with unequal class widths.

Say that the *class boundaries* for the k classes are

$$[c_0, c_1), [c_1, c_2), \ldots, [c_{k-1}, c_k).$$

We use this notation because we remember that the class intervals are closed on the left; that is, a value on a boundary point is assigned to the class that has that value as its lower boundary. The (normalized) *relative frequency histogram* is defined by

$$h(x) = \frac{f_i/n}{c_i - c_{i-1}} \quad \text{for } c_{i-1} \le x < c_i, \quad i = 1, 2, \ldots, k,$$

where n is the number of observations and f_i is the frequency of the class $[c_{i-1}, c_i)$. In this definition, it is *not* necessary for the classes to be of equal length, and in some cases we can describe the distribution of the data better by making the class lengths unequal. Note that the relative frequency of an interval $[a, b)$, where $c_0 \le a < b \le c_k$, can always be approximated by the integral

$$\int_a^b h(x)dx.$$

If $a < b$ are boundary points associated with the classes, this integral gives the exact relative frequency of the interval $[a, b)$. If either a or b is not a boundary point, the integral provides an approximation to the relative frequency of the interval $[a, b)$. Since relative frequency is an approximation to probability, this integral can be thought of as an approximation to the probability $P(a \le X < b)$, where X is the random variable under consideration.

Example 3.1-1   W. Nelson, *Applied Life Data Analysis* (New York: Wiley, 1982), lists $n = 19$ times to breakdown of an insulating fluid between electrodes recorded at the voltage of 34 kilovolts. These times (in minutes), already ordered from smallest to largest, are

0.19	0.78	0.96	1.31	2.78	3.16	4.15	4.67	4.85	6.50
7.35	8.01	8.29	12.06	31.75	32.52	33.91	36.71	72.89	

Since $n = 19$, $(5 - 0.5)/19 = 0.24$, $(10 - 0.5)/19 = 0.50$, and $(15 - 0.5)/19 = 0.76$, we find that the order statistics $x_{(5)} = 2.78$, $x_{(10)} = 6.50$, and $x_{(15)} = 31.75$, equal the 24th, 50th, and 76th percentiles, respectively. An approximate box-and-whisker diagram is as given in Figure 3.1-1. The plot clearly shows that these data are highly skewed to the right. Hence, a histogram with equal class intervals is inappropriate in this situation, as many classes would be empty. Clearly, there is no unique way to construct a relative frequency histogram, and an analyst must use his or her judgment in the selection of the number k and the lengths of the class intervals. For example, with these data, we question whether we want to show a second, smaller mode that seemingly appears around 33. We think not, because there is no apparent reason in this experiment for that second mode to appear; thus, we would rather smooth it out. As a matter of fact, we believe that these data can be described well with only $k = 2$ classes: $[0, 10)$ and $[10, 75)$. However, other persons might disagree with our selection of

$$
h(x) = \begin{cases} \dfrac{13/19}{10} = 0.0684, & 0 \le x < 10, \\[2mm] \dfrac{6/19}{65} = 0.0049, & 10 \le x < 75. \end{cases}
$$

We feel, however, that $h(x)$, as depicted in Figure 3.1-2, is a relatively accurate description of the situation. Better than other relative frequency histograms, it suggests an underlying model described by some $f(x)$. We could be wrong, however, and in Exercise 3.1-1 we ask readers to find other relative frequency histograms that, in their opinion, achieve a better balance between smoothness and providing an accurate description of these data. In any case, one should always make certain that the total area under $h(x)$ is equal to 1; in this case, it is

$$
\int_0^{75} h(x)\,dx = (10)\left(\frac{13/19}{10}\right) + (65)\left(\frac{6/19}{65}\right) = \frac{13}{19} + \frac{6}{19} = 1.
$$

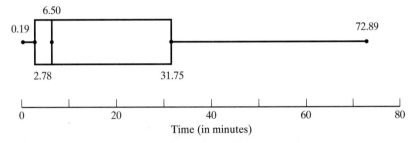

Figure 3.1-1 Box-and-whisker plot of breakdown times

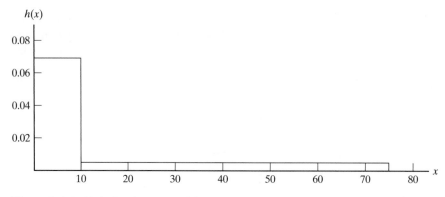

Figure 3.1-2 Relative frequency histogram of breakdown times   ∎

3.1-2 Distributions of Continuous Random Variables

In order to describe the variability of a continuous random variable X, we must find a *probability density function (p.d.f.)* $f(x)$ that gives us the probabilities $P(a \leq X < b)$ through the integral

$$P(a \leq X < b) = \int_a^b f(x)dx.$$

That is, the probability $P(a \leq X < b)$ is the area between the graph of $f(x)$, the x-axis, and the vertical lines $x = a$ and $x = b$, as depicted in Figure 3.1-3. In some sense, $f(x)$ is the limit of the normalized relative frequency histogram $h(x)$ as n increases and the lengths of the class intervals go to zero.

Since areas, such as $P(a \leq X < b)$ for all $a < b$, represent probabilities, we require that the total area between the graph of $f(x)$ and the x axis be equal to 1. That is,

$$\int_R f(x)dx = 1,$$

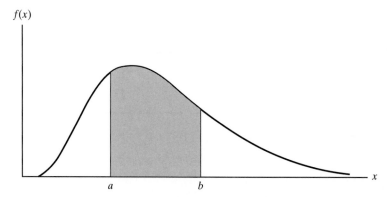

Figure 3.1-3 Plot of a probability density $f(x)$. The shaded area represents $P(a \leq X < b)$

where the integral is taken over R, the outcome space of the random variable X. Moreover, since probabilities are nonnegative, we also need

$$f(x) \geq 0, \quad x \in R.$$

Then, if A is a subset of R, the probability of the event $\{X \in A\}$ is given by

$$P(X \in A) = \int_A f(x)dx.$$

It is interesting to observe that if A consists of a single point, say, $x = b$, then

$$P(X = b) = \int_b^b f(x)dx = 0,$$

because the integral over a single point is equal to zero. This agrees with our intuition, because if the space R is really an interval with an uncountable infinity of points, the probability of one particular point is certainly zero. Of course, in practice, we usually record an observation only to so many decimal points. (For instance, three, and thus the number $\sqrt{2}$ is recorded as 1.414.) This "rounding" really forces us back into a discrete situation in which we have a finite, but very large, number of possible outcomes. Nevertheless, it is often more convenient to approximate their probabilities by an appropriate continuous probability model. So, in a practical situation, we look for a p.d.f. $f(x)$ from which we can compute the probabilities, at least approximately. Let us consider a simple example.

Example 3.1-2 Let us consider a balanced spinner that, after a spin, will point to a number X between zero and 1, as depicted in Figure 3.1-4. A reasonable model for this random variable X is $f(x) = 1, x \in R = \{x; 0 \leq x < 1\}$, which we usually write more simply as

$$f(x) = 1, \quad 0 \leq x < 1.$$

Such a p.d.f., which is constant on its space R, is called a *rectangular* or *uniform* distribution. With this model, we obtain

$$P\left(\frac{1}{4} \leq X < \frac{1}{2}\right) = \int_{1/4}^{1/2} 1 \cdot dx = 0.25$$

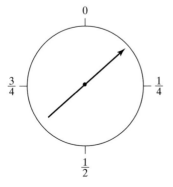

Figure 3.1-4 Balanced spinner

and

$$P(0.1 \le X < 0.7) = \int_{0.1}^{0.7} 1 \cdot dx = 0.6.$$

Since the spinner is balanced, probabilities such as these two certainly are in agreement with our intuition.   ∎

There are certain conventions that are commonly used in the context of continuous random variables. Since, in the continuous case, $P(X = x) = 0$ for all $x \in R$, we find that

$$P(a < X < b) = P(a \le X < b) = P(a < X \le b) = P(a \le X \le b).$$

That is, we can either include or exclude the equality sign in these expressions without changing the probability. In addition, so that we do not need to refer to R in each illustration, we tacitly extend the definition of $f(x)$ to be equal to zero when $x \notin R$. That is, in Example 3.1-2, we can think of

$$f(x) = 1, \quad x \in R = \{x; 0 \le x < 1\}$$

as being

$$f(x) = \begin{cases} 1, & 0 \le x < 1, \\ 0, & \text{elsewhere.} \end{cases}$$

It is even simpler to write it as

$$f(x) = 1, \quad 0 \le x < 1;$$

this means that $f(x) = 1$ when $0 \le x < 1$ and $f(x) = 0$ elsewhere. With this convention, we could write

$$P(X < 0.6) = \int_{-\infty}^{0.6} f(x)dx = \int_{0}^{0.6} f(x)\,dx,$$

because, in this illustration, $f(x) = 0$ when $-\infty < x < 0$.

Example 3.1-3   Suppose that the length X of the life (in years) of the field winding in a generator has a distribution that can be described by the p.d.f.

$$f(x) = \frac{(1.8)x^{0.8}}{8^{1.8}} \exp\left[-\left(\frac{x}{8}\right)^{1.8}\right], \quad 0 \le x < \infty,$$

where $\exp(b)$ means e^b. The fraction of such windings that fail before the one-year warranty expires is given by

$$P(X < 1) = \int_{0}^{1} f(x)dx = \int_{0}^{1} \frac{(1.8)x^{0.8}}{8^{1.8}} \exp\left[-\left(\frac{x}{8}\right)^{1.8}\right]dx$$

$$= -\exp\left[-\left(\frac{x}{8}\right)^{1.8}\right]\Bigg|_{0}^{1} = 1 - \exp\left[-\left(\frac{1}{8}\right)^{1.8}\right] = 0.0234.$$

The p.d.f. $f(x)$ and the probability of interest are shown in Figure 3.1-5.

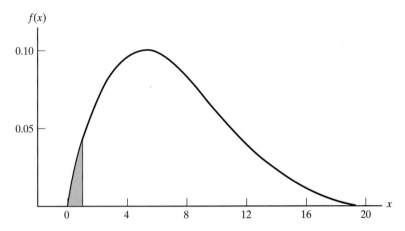

Figure 3.1-5 Plot of the p.d.f. of Example 3.1-3. The shaded area represents $P(X < 1)$

■

As in the discrete case, it is frequently beneficial to use the *cumulative distribution function (c.d.f.)*:

$$F(x) = P(X \le x) = P(X < x) = \int_{-\infty}^{x} f(w)\, dw.$$

$F(x)$ is the area between the graph of the p.d.f. and the x-axis, up to the vertical line through the general point x. Of course, from the fundamental theorem of calculus, we know that

$$F'(x) = f(x),$$

for x values where the derivative $F'(x)$ exists. Note that since $F(x)$ cumulates the probability up to and including x, it is a nondecreasing function such that

$$\lim_{x \to -\infty} F(x) = 0 \quad \text{and} \quad \lim_{x \to \infty} F(x) = 1.$$

The point, x_p, $0 < p < 1$, for which

$$F(x_p) = P(X \le x_p) = \int_{-\infty}^{x_p} f(x)\, dx = p$$

is called the *(100p)th percentile* of the distribution. It is also called the *quantile of order p*. The 50th percentile is the median; the 25th and 75th percentiles are, respectively, the first and third quartiles of the distribution. These three percentiles are also called *quantiles* of orders 0.5, 0.25, and 0.75, respectively.

Example 3.1-4 Let us hit the spinner of Example 3.1-2 $n = 3$ independent times, obtaining three independent observations X_1, X_2, X_3. Let Y equal the maximum value of these three observations. Since $Y \le y, 0 \le y < 1$, if and only if each observation is less than or equal to y—that is $X_1 \le y, X_2 \le y$, and $X_3 \le y$—the distribution function of Y is

$$G(y) = P(Y \le y) = P(X_1 \le y, X_2 \le y, X_3 \le y)$$
$$= P(X_1 \le y)P(X_2 \le y)P(X_3 \le Y).$$

The last step follows from the independence of the events

$$(X_1 \le y), (X_2 \le y), \text{ and } (X_3 \le y).$$

Furthermore,

$$P(X_i \le y) = \int_0^y 1 \cdot dx = y, \quad i = 1, 2, 3,$$

implies that

$$G(y) = y^3, \quad 0 \le y < 1.$$

Accordingly, the p.d.f. of Y is

$$g(y) = G'(y) = 3y^2, \quad 0 \le y < 1.$$

For example, the probability that the maximum spin exceeds 0.8 is

$$P(Y > 0.8) = \int_{0.8}^1 3y^2 dy = 1 - (0.8)^3 = 0.488.$$

It is also easy to calculate the percentiles of the distribution of $Y = \max(X_1, X_2, X_3)$. The percentiles y_p satisfy $G(y_p) = (y_p)^3 = p$ or $y_p = p^{1/3}$. For example, the median is 0.794, and the first and third quartiles are 0.630 and 0.909, respectively. ∎

For random variables of the continuous type, the definition and the rules associated with *mathematical expectation* are the same as in the discrete case, except that integrals replace summations. That is,

$$E[u(x)] = \int_R u(x)f(x)dx = \int_{-\infty}^\infty u(x)f(x)dx$$

is the expected value of $u(X)$. The *mean* and *variance* of X are, respectively,

$$\mu = E(X) = \int_{-\infty}^\infty xf(x)dx$$

and

$$\sigma^2 = E[(X - \mu)^2] = \int_{-\infty}^\infty (x - \mu)^2 f(x)dx$$

$$= \int_{-\infty}^\infty x^2 f(x)dx - \mu^2 = E(X^2) - \mu^2.$$

Example 3.1-5   Let X have the p.d.f. of Example 3.1-2, namely,

$$f(x) = 1, \quad 0 \le x < 1.$$

Then

$$\mu = E(X) = \int_0^1 x \cdot 1 \cdot dx = \frac{x^2}{2}\Big|_0^1 = \frac{1}{2}$$

and

$$\sigma^2 = E(X^2) - \left(\frac{1}{2}\right)^2 = \int_0^1 x^2 \cdot 1 \cdot dx - \frac{1}{4} = \frac{1}{3} - \frac{1}{4} = \frac{1}{12}.$$ ∎

Example 3.1-6 For $Y = \max(X_1, X_2, X_3)$ of Example 3.1-4,

$$\mu = E(Y) = \int_0^1 y(3y^2)dy = \frac{3}{4}$$

and

$$\sigma^2 = \mathrm{var}(Y) = \int_0^1 y^2(3y^2)dy - \left(\frac{3}{4}\right)^2 = \frac{3}{5} - \frac{9}{16} = \frac{3}{80}.$$

Thus, the standard deviation of Y is

$$\sigma = \sqrt{\mathrm{var}(Y)} = \sqrt{\frac{3}{80}} = 0.19.$$ ∎

Exercises 3.1

3.1-1 For the data in Example 3.1-1, construct other normalized relative frequency histograms by trying other values of k and different boundary values.

Hint: As a start, take $k = 4$ with classes of $[0, 10)$, $[10, 30)$, $[30, 40)$, and $[40, 80)$. Compare your plots with that in Figure 3.1-2.

3.1-2 Nelson has recorded the operating hours until failure on $n = 22$ transformers, where the data are already ordered from smallest to largest as follows:

10	314	730	740	990	1046	1570	1870
2020	2040	2096	2110	2177	2306	2690	3200
3360	3444	3508	3770	4042	4186		

(a) Draw a box-and-whisker diagram.

(b) Construct a normalized relative frequency histogram.

[W. Nelson, *Applied Life Data Analysis* (New York: Wiley, 1982), p. 137.]

3.1-3 Hogg and Tanis have reported on 40 losses due to wind-related catastrophies in 1977. The data include only those losses of \$2,000,000 or more, and are recorded as follows to the nearest \$1,000,000 (in units of \$1,000,000):

2	2	2	2	2	2	2	2	2	2
2	2	3	3	3	3	4	4	4	5
5	5	5	6	6	6	6	8	8	9
15	17	22	23	24	24	25	27	32	43

(a) Draw a box-and-whisker diagram.

(b) Construct a normalized relative frequency histogram.

[R. V. Hogg and E. A. Tanis, *Probability and Statistical Inference*, 7th ed. (Upper Saddle River, NJ: Prentice Hall, 2006), p. 129].

Hint: The authors used $k = 4$ with classes [1.5, 2.5), [2.5, 6.5), [6.5, 29.5), and [29.5, 49.5). However, try others, such as $k = 5$ and [1.5, 2.5), [2.5, 9.5), [9.5, 20.5), [20.5, 28.5), and [28.5, 45.5). Compare the results.

*3.1-4 Let X have the p.d.f. $f(x) = 3(1 - x)^2, 0 \le x < 1$. Graph the p.d.f. and compute

*(a) $P(0.1 < X < 0.5)$.

*(b) $P(X > 0.4)$.

*(c) $P(0.3 < X < 2)$.

Hint: In parts (b) and (c), recognize that the outcome space of X is $0 \le x < 1$.

*3.1-5 Find the mean and the variance, and determine the 90th percentile, of each of the distributions given by the following densities:

*(a) $f(x) = 2x, 0 \le x < 1$.

*(b) $f(x) = 6x(1 - x), 0 \le x < 1$.

*3.1-6 Find the 50th percentile (*median*), the 25th percentile (*first quartile*), the 75th percentile (*third quartile*), and the 90th percentile (also called the *ninth decile*) for the following densities:

*(a) $f(x) = 4x^3, 0 \le x < 1$.

*(b) $f(x) = e^{-x}, 0 \le x < \infty$.

*3.1-7 Consider the uniform (rectangular) distribution on the space $[a, b)$, where $a < b$, with p.d.f.

$$f(x) = \frac{1}{b - a}, \quad a \le x < b.$$

*(a) Obtain the cumulative distribution function $F(x)$. Plot $f(x)$ and $F(x)$. Determine the median and the first and third quartiles.

*(b) Calculate the mean and the variance.

3.1-8 Suppose that the p.d.f. of the life (in weeks) of a certain part is $f(x) = 3x^2/(400)^3, 0 \le x < 400$.

(a) Graph the p.d.f. and show that the probability of a part failing in the first 200 weeks is $1/8$.

(b) To decrease this probability, four independent parts are placed in parallel so that all must fail if the system is to fail. Show that the p.d.f. of the life Y of this parallel system is $g(y) = 12y^{11}/(400)^{12}, 0 \le y < 400$.

Hint: $P(Y \le y) = P(X_1 \le y)P(X_2 \le y)P(X_3 \le y)P(X_4 \le y)$, where X_1, X_2, X_3, X_4 are the lives of the respective parts.

(c) Determine $P(Y \le 200)$ and compare it with the answer in part (a).

3.1-9 An insurance agent receives a bonus if the loss ratio $L =$ (total losses/total premiums) on his business is less than 0.5. Represent the total losses by a random variable X, and let the bonus equal $(0.5 - L)(T/30)$ if $L < 0.5$ and zero otherwise. If X (in \$100,000) has the p.d.f.

$$f(x) = \frac{3}{x^4}, \quad x \geq 1,$$

and if T (also in \$100,000) equals 3, determine the expected value of the bonus.

3.1-10 The life X in years of a voltage regulator of a car has the p.d.f.

$$f(x) = \frac{3x^2}{7^3} \exp[-(x/7)^3], \quad 0 \leq x < \infty.$$

(a) Graph the p.d.f.

(b) What is the probability that this regulator will last at least 7 years?

(c) Given that it has lasted at least 7 years, what is the conditional probability that it will last at least another 3.5 years?

3.2 The Normal Distribution

The normal distribution is the most important distribution in the study of statistics, not only because many data sets are about normally distributed, but also because many estimators, such as $\hat{p}$ from the binomial model and $\hat{\lambda}$ from the Poisson model, have approximate normal distributions. We comment much more on the importance of the normal distribution for estimation in Section 4.1.

If X has a *normal distribution* with mean μ and variance σ^2, its p.d.f. is

$$f(x) = \frac{1}{\sqrt{2\pi}\sigma} \exp\left[-\frac{(x - \mu)^2}{2\sigma^2}\right], \quad -\infty < x < \infty.$$

We abbreviate this property by saying that X is $N(\mu, \sigma^2)$. The graph of $f(x)$ is the well-known bell-shaped curve displayed in Figure 3.2-1. The curve is symmetric about $x = \mu$ and reaches its highest value at that point. The more mathematical reader may want to verify that

$$\int_{-\infty}^{\infty} f(x)\,dx = 1, \quad E(X) = \int_{-\infty}^{\infty} xf(x)\,dx = \mu,$$

$$\text{var}(X) = \int_{-\infty}^{\infty} (x - \mu)^2 f(x)\,dx = \sigma^2,$$

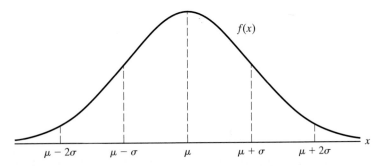

Figure 3.2-1 Probability density function of the $N(\mu, \sigma^2)$ distribution

and that the points of inflection of $f(x)$ are at $\mu - \sigma$ and $\mu + \sigma$. (See Exercise 3.2-9, which gives helpful hints.)

Let us accept the fact that $f(x)$ is a p.d.f. with mean μ and variance σ^2. Then, if the random variable Z has a normal distribution with mean zero and variance 1 [i.e., Z is $N(0, 1)$], its p.d.f. is

$$f(z) = \frac{1}{\sqrt{2\pi}} e^{-z^2/2}, \quad -\infty < z < \infty,$$

and Z is said to have a *standard normal distribution*. For such a Z, let us try to compute some probabilities, such as

$$P(-1 < Z < 1.5) = \int_{-1}^{1.5} \frac{1}{\sqrt{2\pi}} e^{-z^2/2} dz.$$

We notice quickly that we cannot find an antiderivative of $f(z)$; that is, we cannot find a simple function $F(z)$ such that $F'(z) = f(z)$. Thus, we are not able to use the fundamental theorem of calculus to evaluate this integral, and we are forced to use numerical methods. Fortunately, statisticians have faced these integrals for years and have produced a table of the (cumulative) distribution function (c.d.f.) of the $N(0, 1)$ distribution, or

$$\Phi(z) = P(Z \leq z) = \int_{-\infty}^{z} \frac{1}{\sqrt{2\pi}} e^{-w^2/2} dw.$$

We give one such tabulation in Table C.4 in Appendix C, for both positive and negative values of z. For example, the probability $\Phi(1.5) = 0.9332$ in this table corresponds to the shaded area in Figure 3.2-2(a). Similarly, $\Phi(-1.0) = 0.1587$. You can also use statistical software such as Minitab (and even scientific calculators) to obtain these values. In Minitab, for example, simply use "Calc > Probability Distributions > Normal," and check the "Cumulative Probability" tab.

A plot of $\Phi(z)$, the c.d.f. of the $N(0, 1)$ distribution, is given in Figure 3.2-2(b). This plot has a characteristic "S-shaped" pattern. Note that the shaded area in Figure 3.2-2(a) is the ordinate of the c.d.f. at z in Figure 3.2-2(b). Although the c.d.f. and p.d.f. can supply the same information, we find the p.d.f. more descriptive of the distribution.

With this notation, we can write our required probability as

$$P(-1 < Z < 1.5) = \Phi(1.5) - \Phi(-1);$$

that is, we determine the area under the p.d.f. up to 1.5 and subtract the area under the curve up to -1. We find, from Table C.4, that

$$P(-1 < Z < 1.5) = \Phi(1.5) - \Phi(-1)$$

$$= 0.9332 - 0.1587 = 0.7745.$$

Note that, due to the symmetry of $f(z)$ about $z = 0$, it follows that $\Phi(-1) = 1 - \Phi(1)$, or, more generally,

$$\Phi(-z) = 1 - \Phi(z).$$

The entries in Table C.4 can be used to find the percentiles of the $N(0, 1)$ distribution. The $(100p)$th percentile (or the quantile of order p) is the value z_p for which

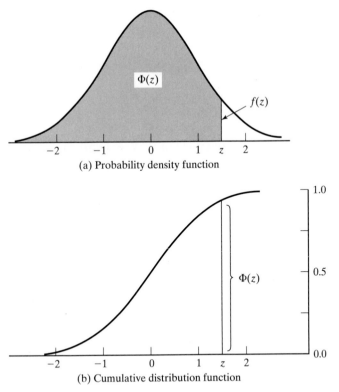

(a) Probability density function

(b) Cumulative distribution function

Figure 3.2-2 Plot of the p.d.f. $f(z)$ and the c.d.f. $\Phi(z) = P(Z \leq z)$ of the $N(0, 1)$ distribution

$\Phi(z_p) = P(Z \leq z_p) = p$. For example, we find that the 95th percentile is $z_{0.95} = 1.645$. The symmetry of the p.d.f. around zero implies that $z_{1-p} = -z_p$; thus, the 5th percentile is $z_{0.05} = -1.645$. Similarly, you can check that the 1st, 25th, 50th, 75th, and 99th percentiles are given by $-2.326, -0.675, 0.0, 0.675$, and 2.326, respectively. You obtain the percentiles with the Minitab command "Calc > Probability Distributions > Normal," clicking the "Inverse Cumulative Probability" tab.

So far, we have learned how to find probabilities associated with the standard $N(0, 1)$ distribution. Suppose, however, that X is $N(\mu = 75, \sigma^2 = 100)$ and we want a probability such as $P(70 < X < 90)$. In general, say that X is $N(\mu, \sigma^2)$, and we wish to determine

$$P(a < X < b) = \int_a^b \frac{1}{\sqrt{2\pi}\sigma} \exp\left[-\frac{(x - \mu)^2}{2\sigma^2}\right] dx.$$

If we change the variable of integration by letting $w = (x - \mu)/\sigma$, so that $x = \mu + \sigma w$ and $dx/dw = \sigma$, we find that

$$P(a < X < b) = \int_{(a-\mu)/\sigma}^{(b-\mu)/\sigma} \frac{1}{\sqrt{2\pi}} e^{-w^2/2} dw.$$

Since the integrand in this integral is the p.d.f. of the $N(0, 1)$ distribution, we can use the entries in Table C.4 to evaluate these probabilities. Hence, we have

$$P(a < X < b) = \Phi\left(\frac{b - \mu}{\sigma}\right) - \Phi\left(\frac{a - \mu}{\sigma}\right).$$

That is, to compute this probability, we use the c.d.f. of the $N(0, 1)$ distribution after standardizing the bounds of the interval, namely, a and b. The standardization is carried out by subtracting the mean μ and dividing by the standard deviation σ.

Note that the values of X that satisfy $(a < X < b)$ are the same as the ones that satisfy

$$\left(\frac{a - \mu}{\sigma} < \frac{X - \mu}{\sigma} < \frac{b - \mu}{\sigma}\right)$$

and

$$P(a < X < b) = P\left(\frac{a - \mu}{\sigma} < \frac{X - \mu}{\sigma} < \frac{b - \mu}{\sigma}\right) = \Phi\left(\frac{b - \mu}{\sigma}\right) - \Phi\left(\frac{a - \mu}{\sigma}\right).$$

Thus, we have actually proven that if X is $N(\mu, \sigma^2)$, then the random variable

$$Z = \frac{X - \mu}{\sigma} \quad \text{is} \quad N(0, 1).$$

The percentiles of the $N(\mu, \sigma^2)$ distribution are linearly related to the percentiles of the $N(0, 1)$ distribution. Using the transformation from the $N(\mu, \sigma^2)$ to the $N(0, 1)$ distribution, we find that

$$p = P(X \le x_p) = P\left(\frac{X - \mu}{\sigma} \le \frac{x_p - \mu}{\sigma}\right).$$

Therefore,

$$\frac{x_p - \mu}{\sigma} = z_p \quad \text{and} \quad x_p = \mu + \sigma z_p,$$

where z_p is the percentile of the $N(0, 1)$ distribution.

Example 3.2-1 If X is $N(\mu = 75, \sigma^2 = 100)$, then

$$P(70 < X < 90) = \Phi\left(\frac{90 - 75}{10}\right) - \Phi\left(\frac{70 - 75}{10}\right)$$

$$= \Phi(1.5) - \Phi(-0.5)$$

$$= 0.9332 - 0.3085 = 0.6247,$$

$$P(80 < X < 95) = \Phi\left(\frac{95 - 75}{10}\right) - \Phi\left(\frac{80 - 75}{10}\right)$$

$$= \Phi(2.0) - \Phi(0.5)$$

$$= 0.9772 - 0.6915 = 0.2857,$$

and

$$P(55 < X < 65) = \Phi\left(\frac{65 - 75}{10}\right) - \Phi\left(\frac{55 - 75}{10}\right)$$

$$= \Phi(-1) - \Phi(-2) = 0.1587 - 0.0228 = 0.1359.$$

The 95th percentile of this distribution is given by $75 + (10)(1.645) = 91.45$. Similarly, the three quartiles are given by 69.02, 75.0, and 80.98, respectively. ■

Remark: We calculate the probability by standardizing the limits and using the c.d.f. of the standard normal distribution. The Minitab command "Calc > Probability Distributions > Normal" avoids the standardization by having us enter the mean and standard deviation directly.

Example 3.2-2 Say that the time X required to assemble an item is a normally distributed random variable with mean $\mu = 15.8$ minutes and standard deviation $\sigma = 2.4$ minutes. Then the probability that the next item will take more than 17 minutes to assemble is

$$P(X > 17) = P\left(\frac{X - 15.8}{2.4} > \frac{17 - 15.8}{2.4}\right) = P(Z > 0.5)$$

$$= 1 - \Phi(0.5) = 1 - 0.6915 = 0.3085.$$ ■

To gain a greater appreciation of the normal distribution and the relationship of its probabilities to the parameters μ and σ, consider the following discussion: If X is $N(\mu, \sigma^2)$, then, with $k > 0$, we have

$$P(\mu - k\sigma < X < \mu + k\sigma) = \Phi\left(\frac{\mu + k\sigma - \mu}{\sigma}\right) - \Phi\left(\frac{\mu - k\sigma - \mu}{\sigma}\right)$$

$$= \Phi(k) - \Phi(-k) = \Phi(k) - [1 - \Phi(k)]$$

$$= 2\Phi(k) - 1.$$

For selected values of k, we obtain the following probabilities:

k	1	1.282	1.645	1.96	2	2.576	3
$P(\mu - k\sigma < X < \mu + k\sigma)$	0.6826	0.80	0.90	0.95	0.9544	0.99	0.9974

These results imply that about 68 percent of normally distributed observations are between $\mu - \sigma$ and $\mu + \sigma$, about 95 percent are in the interval given by $\mu \pm 2\sigma$, and almost all (99.74%) are within three standard deviations of the mean μ. We call the bounds of the interval, from $\mu - k\sigma$ to $\mu + k\sigma$, the $100[2\Phi(k) - 1]$ percent *tolerance limits* of a normal distribution.

What do these results imply in a real situation? Say that we find that the inside diameter of a washer that our company produces has a mean of 0.503 and a standard deviation of 0.005. If we can assume that these diameters are normally distributed, then 99 percent of them are between $0.503 \pm 2.576(0.005)$—that is, roughly between 0.490 and 0.516. If the specifications for these diameters were 0.500 ± 0.005 (i.e., the diameter of a useful washer should be between 0.495 and 0.505), this finding causes some concern, because we would probably want at least 99 percent of our washers to be within specifications. Questions should be asked: Are the specifications realistic and necessary? If so, what changes should be made to the process to get a larger number of washers within the specifications? In this particular example, it would be desirable to shift the mean to 0.500 and reduce the standard deviation as much as possible. The shifting of the mean may not be too difficult to achieve, but the reduction of the variance could require very expensive modifications of our production process.

Exercises 3.2

*3.2-1   Let Z be $N(0, 1)$. Graph its p.d.f. and determine the following probabilities:

*(a) $P(-0.7 < Z < 1.3)$.

*(b) $P(0.2 < Z < 1.1)$.

*(c) $P(-1.9 < Z < -0.6)$.

3.2-2   Let X be $N(\mu = 25, \sigma^2 = 16)$. Graph its p.d.f. and determine the following probabilities:

(a) $P(22 < X < 26)$.

(b) $P(19 < X < 24)$.

(c) $P(X > 23)$.

(d) $P(|X - 25| < 5)$.

*3.2-3   If X is $N(\mu = 5, \sigma^2 = 4)$, find each c so that

*(a) $P(X < c) = 0.8749$.

*(b) $P(c < X) = 0.6406$.

*(c) $P(X < c) = 0.95$.

*(d) $P(|X - 5| < c) = 0.95$.

3.2-4   The weight of the cereal in a box is a random variable with mean 12.15 ounces and standard deviation 0.2 ounce. What percentage of the boxes have contents that weigh under 12 ounces? If the manufacturer claims that there are 12 ounces in a box, does the percentage you just found cause concern? If so, what two things could be done to correct the situation? Which would probably be cheaper immediately, but might cost more in the long run?

3.2-5   Suppose that the specifications on the diameter of a rotor shaft are 0.25 ± 0.002. If the diameters are distributed normally with mean $\mu = 0.251$ and standard deviation $\sigma = 0.001$, what percentage of them are within specifications?

*3.2-6   Due to many factors, such as slight variations in the metal, different operators, and different machines, the thickness of manufactured metal plates varies and can be considered a normal random variable with mean $\mu = 20$ mm and standard deviation $\sigma = 0.04$ mm. How much scrap can be expected if the thickness of the metal plates

(a) Has to be at least 19.95 mm?

(b) Can be at most 20.10 mm?

(c) Can differ at most by 0.05 mm from the target 20 mm?

(d) How would one have to set the tolerance limits $20 - c$ and $20 + c$ such that one produces at most 5 percent scrap?

(e) Assume that the mean has shifted to $\mu = 20.10$. Calculate the percentage of metal plates that exceed the tolerance limits in part (d).

*3.2-7   A machine fills 100-pound bags of dry concrete mix. The actual weight of the mix that is put in each bag is a normal random variable with standard deviation $\sigma = 0.5$ pound. The mean of the distribution can be set by the operator. At what mean weight should the machine be set such that only 5 percent of the bags are underweight?

3.2-8   The average weight of a certain brand of refrigerator is 31 pounds. Due to variability in raw materials and production conditions, weight is a random variable; assume that the distribution is normal with a standard deviation of $\sigma = 0.5$ pound.

(a) What is the probability that a randomly selected refrigerator is heavier than 32.0 pounds?

(b) What is the probability that the weight of a randomly selected refrigerator is between 30.0 and 30.5 pounds?

3.2-9   **[Difficult]** Let $f(x)$ be a p.d.f. associated with $N(\mu, \sigma^2)$.

(a) Show that

$$I = \int_{-\infty}^{\infty} f(x)\,dx = 1.$$

Hint: Change the variables by letting $y = (x - \mu)/\sigma$. Then

$$I = \int_{-\infty}^{\infty} \frac{1}{\sqrt{2\pi}} e^{-y^2/2}\,dy.$$

Write

$$I^2 = \int_{-\infty}^{\infty} \int_{-\infty}^{\infty} \frac{1}{2\pi} e^{-(y^2+z^2)/2}\,dy\,dz,$$

and use polar coordinates $y = r\cos\theta$ and $z = r\sin\theta$ to evaluate I^2.

(b) Show that

$$\int_{-\infty}^{\infty} xf(x)\,dx = \mu.$$

Hint: Prove that

$$\int_{-\infty}^{\infty} \left(\frac{x - \mu}{\sigma}\right) f(x)\,dx = 0$$

by finding an antiderivative of the integrand; then use the result of part (a).

(c) Show that

$$\int_{-\infty}^{\infty} (x - \mu^2) f(x)\,dx = \sigma^2,$$

using integration by parts.

(d) Prove that the points of inflection of $f(x)$ are $\mu - \sigma$ and $\mu + \sigma$.

Hint: Set $f''(x) = 0$.

3.2-10 The "fill problem" is important in many industries, including those making toothpaste, cereal, beer, and so on. If such an industry claims that it is selling 12 ounces of its product in a certain container, it must have a mean fill weight greater than 12 ounces, or else the Food and Drug Administration (FDA) will crack down. The FDA will allow only a very small percentage of the containers to contain less than 12 ounces.

(a) If the contents of a container have a $N(\mu = 12.1, \sigma^2)$ distribution, find σ so that $P(X < 12) = 0.01$.

(b) If $\sigma = 0.05$, find μ so that $P(X < 12) = 0.01$.

*3.2-11 Y has a normal distribution with mean 1 and standard deviation 2. Determine $P(Y^2 < 9)$.

3.2-12 The times necessary to complete a certain service for a class of bank customers are described by a normal distribution with mean 15 minutes and standard deviation 2.1 minutes. Service times are considered excessive if they exceed 20 minutes. Over the long run, what percentage of customers will experience excessive service times?

3.2-13 Measurements associated with the cap torque are normally distributed with $\mu = 35$ psi (pounds per square inch) and $\sigma = 5$ psi. The required specification limits for torque are a target of 30 and lower and upper limits of 20 and 40, respectively. What percentage of caps are between 20 and 40 psi?

3.2-14 Car insurance claims are normally distributed with a mean of $2,100 and a standard deviation of $100. A new claim comes in at $2,250. The claimant is told by the insurance company that he belongs to a high-risk group. Calculate the proportion of claims that exceed $2,250.

3.2-15 On average, a vacuum wears out after 2,000 hours, with a standard deviation of 400 hours. Assume that the distribution of its life is normal. The manufacturer wants to set the warranty such that only 5 percent of the vaccums are returned for warranty work. Determine the length of the warranty.

*3.2-16 The dimensions of steel flats vary according to a normal distribution with a mean of 5.1 cm and a standard deviation of 0.10 cm.

*(a) Obtain the 95th and 99th percentile of the distribution.

*(b) The lower and upper specification limits for this product are 4.8 and 5.2 cm, respectively. Calculate the probability that the dimension of a product is outside the specification limits.

*(c) It is thought that adjusting the process so that the width distribution is centered at 5.00 cm (with the same standard deviation of 0.10 cm) will reduce the probability of being outside the specification limits. Explain.

3.3 Other Useful Distributions

Let X be a continuous random variable that represents the length of life of a component (such as a light bulb, a set of windings, another kind of part, or even a human being). Say the p.d.f. and c.d.f. of X are $f(x)$ and $F(x)$, respectively, with outcome space $0 \leq x < \infty$. We are often interested in the probability that a component fails in the interval $(x, x + \Delta x)$, given that it has lasted at least x units. Since the event $(x < X < x + \Delta x)$ is a subset of $(X > x)$, this conditional probability equals

$$P(x < X < x + \Delta x \mid X > x) = \frac{P(x < X < x + \Delta x)}{P(X > x)} \approx \frac{f(x)\Delta x}{1 - F(x)}.$$

The right side of this equation is an approximation, because, for Δx small, the probability of failing in the interval $(x, x + \Delta x)$ is approximately equal to $f(x)\Delta x$. If Δx denotes one unit, the conditional probability is approximately

$$\lambda(x) = \frac{f(x)}{1 - F(x)},$$

and $\lambda(x)$ is called the *failure rate function*. That is, the failure rate at x is the height of the p.d.f. at x, divided by the probability of exceeding x. The failure rate is the approximate conditional probability of failing in the next unit, given that the component has already lasted x units of time. Frequently, since parts (and people) "wear out," the failure rate tends to be an increasing function of x, but there are cases in which $\lambda(x)$ is a constant or decreasing function of x ("old is as good or better than new").

Of course, with $\lambda(t) = f(t)/[1 - F(t)]$ and $F'(t) = f(t)$, we have

$$\int_0^x \lambda(t)\,dt = \int_0^x \frac{f(t)}{1 - F(t)}\,dt$$

$$= -\ln[1 - F(t)]\Big|_0^x$$

$$= -\ln[1 - F(x)] + \ln[1 - F(0)].$$

However, X is the length of life, so $P(X \le 0) = F(0) = 0$, and it follows that

$$\int_0^x \lambda(t)\,dt = -\ln[1 - F(x)]$$

and

$$F(x) = 1 - \exp\left[-\int_0^x \lambda(t)\,dt\right], \quad 0 \le x < \infty.$$

Accordingly, from the fundamental theorem of calculus,

$$f(x) = F'(x) = \lambda(x)\exp\left[-\int_0^x \lambda(t)\,dt\right], \quad 0 \le x < \infty.$$

By considering various failure rate functions, we can generate a number of continuous-type distribution functions and their corresponding density functions.

3.3-1 Weibull Distribution

Let the failure rate be proportional to a power of x, say, $\lambda(x) = \alpha x^{\alpha-1}/\beta^{\alpha}$, $0 \le x < \infty$, where the parameters α and β are positive. The failure rate function increases linearly when $\alpha = 2$. The increase in the failure rate function is faster (slower) than linear when $\alpha > 2$ ($\alpha < 2$). Then

$$f(x) = \frac{\alpha x^{\alpha-1}}{\beta^{\alpha}} \exp\left[-\left(\frac{x}{\beta}\right)^{\alpha}\right], \quad 0 \le x < \infty.$$

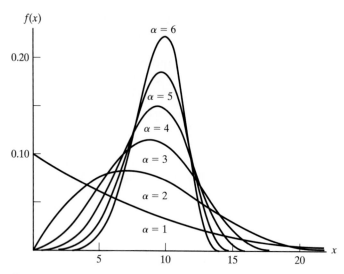

Figure 3.3-1 Graphs of several Weibull densities with $\beta = 10$

This was the p.d.f. used in Example 3.1-3 with $\beta = 8$ and $\alpha = 1.8$. The p.d.f. $f(x)$ is said to be that of a *Weibull distribution* with parameters α and β. In particular, if $\alpha = 1$, we obtain the p.d.f. of an *exponential distribution*:

$$f(x) = \frac{1}{\beta} e^{-x/\beta}, \quad 0 \le x < \infty.$$

The Weibull family of distributions is useful in many engineering applications. Several examples of Weibull densities with $\beta = 10$ are given in Figure 3.3-1.

The calculation of the Weibull c.d.f., $F(x) = P(X \le x) = \int_0^x f(w)\,dw$, involves integration. Fortunately, computer programs such as Minitab include convenient look-up functions that provide these probabilities without difficulties. All one needs to do is enter "Calc > Probability Distributions > Weibull" and ask for "cumulative probability." Minitab calls α and β the shape and scale parameters, respectively. Check, for example, that $P(X \le 10) = 0.6321$ for the Weibull distribution with $\alpha = 2$ and $\beta = 10$. With the "inverse probability" feature of this Minitab command, we find the 90th percentile $x_{0.90} = 15.17$. With the "probability density" feature, we find the density $f(x = 10) = 0.0736$. Relate these numbers to the curve in Figure 3.3-1.

3.3-2 Gompertz Distribution

Human mortality (a kind of failure rate, to be sure) increases about exponentially once a person reaches his or her middle twenties. As a matter of fact, the increase equals roughly 10 percent each year; hence,

$$\lambda(x) \approx c(1.1)^x = ce^{bx},$$

where $b = \ln(1.1)$. The corresponding p.d.f., with parameters $b > 0$ and $c > 0$, is

$$f(x) = ce^{bx}\exp\left(-\frac{c}{b}e^{bx} + \frac{c}{b}\right), \quad 0 \le x < \infty.$$

The function $f(x)$ is known as the p.d.f. of the *Gompertz distribution*. It is widely used in actuarial science because it provides an excellent description of the distribution of the length of human life. In Figure 3.3-2, we display the failure rates and density functions of an exponential, a Weibull, and a Gompertz distribution.

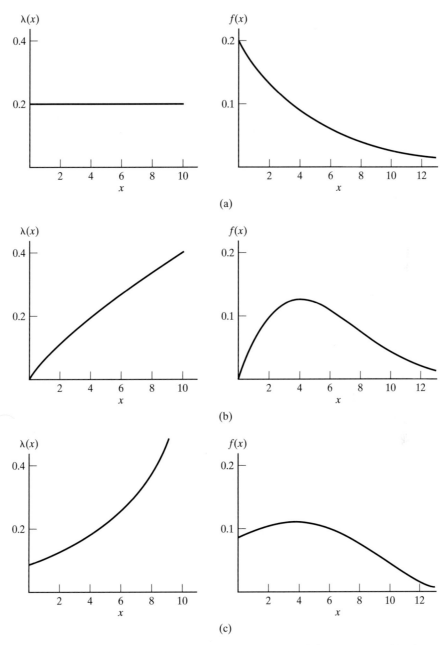

Figure 3.3-2 Plot of the failure rate $\lambda(x)$ and the p.d.f. $f(x)$ of the following three distributions: (a) exponential with $\beta = 5$; (b) Weibull with $\beta = 6.4$ and $\alpha = 1.8$; (c) Gompertz with $b = \ln(1.2)$ and $c = 0.087$

Engineers study the same p.d.f. $f(x)$ as that of Gompertz, but they usually replace the coefficient b by $1/\delta$ and c/b by $e^{-\lambda/\delta}$. With this change, which only amounts to a relabeling or reparameterization of the coefficients, the p.d.f. becomes

$$f(x) = \frac{1}{\delta}\exp\left(\frac{x-\lambda}{\delta}\right)\exp\left[-\exp\left(\frac{x-\lambda}{\delta}\right) + \exp\left(\frac{-\lambda}{\delta}\right)\right],$$

for $0 \leq x < \infty$. The new parameters are $\delta > 0$ and $-\infty < \lambda < \infty$.

3.3-3 Extreme Value Distribution

The p.d.f. of this last (modified Gompertz) distribution is zero for $x < 0$. If we want to extend the space of a new random variable Y to $-\infty < y < \infty$, but use the same functional form as $f(x)$, we have to divide the p.d.f. by $\int_{-\infty}^{\infty} f(x)\, dx$, because the area under the new density over the space $-\infty < y < \infty$ has to be 1. Since

$$\int_{-\infty}^{\infty} f(x)dx = \int_{-\infty}^{0} f(x)dx + \int_{0}^{\infty} f(x)dx$$

$$= \left\{-1 + \exp\left[\exp\left(-\frac{\lambda}{\delta}\right)\right]\right\} + 1$$

$$= \exp\left[\exp\left(-\frac{\lambda}{\delta}\right)\right],$$

the new random variable Y, which now has the space $-\infty < y < \infty$, is given by

$$g(y) = \frac{1}{\delta}\exp\left(\frac{y-\lambda}{\delta}\right)\exp\left[-\exp\left(\frac{y-\lambda}{\delta}\right)\right], \quad -\infty < y < \infty.$$

The function $g(y)$ is known as the p.d.f. of the *extreme value distribution*. As the name already indicates, this function provides a useful description of the variability of extreme phenomena, such as floods, temperature extremes, monthly maxima of air pollutant concentrations, and so on. The p.d.f. depends on two parameters; λ is usually called the location parameter and δ is known as the scale parameter. (See Exercise 3.3-9 for further discussion.)

3.3-4 Gamma Distribution

Another distribution of the continuous type that arises in engineering applications is the *gamma distribution* with p.d.f.

$$f(y) = \frac{1}{\Gamma(\alpha)\beta^{\alpha}}y^{\alpha-1}e^{-y/\beta}, \quad 0 \leq y < \infty,$$

and parameters $\alpha > 0$ (also called the shape parameter) and $\beta > 0$ (referred to as the scale parameter). The gamma function,

$$\Gamma(\alpha) = \int_0^\infty w^{\alpha-1} e^{-w}\, dw,$$

is a generalized factorial, as it can be shown that $\Gamma(\alpha) = (\alpha - 1)!$ if α is a positive integer. Moreover, in Exercise 3.3-11, it is shown that $\Gamma(\alpha) = (\alpha - 1)\Gamma(\alpha - 1)$, provided that $\alpha > 1$. This implies that, for $\alpha > 1$, we can express the gamma function as $\Gamma(\alpha) = (\alpha - 1)(\alpha - 2) \cdots (\alpha - r)\Gamma(\alpha - r)$, where $\alpha - r$ is a number between 0 and 1. The gamma function for arguments between 0 and 1 can be looked up in standard mathematical tables; for example, $\Gamma(1) = 1$ and $\Gamma(0.5) = \sqrt{\pi}$.

The gamma distribution arises as the distribution of waiting times. In Section 2.5, we have learned that, for a Poisson process, the number of occurrences in an interval of length w follows a Poisson distribution with parameter λw, where λ is the mean occurrence in an interval of length 1. Similarly, one can show that the time it takes to obtain exactly k occurrences has a gamma distribution with parameters $\alpha = k$ and $\beta = 1/\lambda$.

As a special case of the gamma distribution with $k = \alpha = 1$, we obtain the exponential distribution with parameter $\beta = 1/\lambda$. This is the distribution of the waiting time to the first occurrence. To see the effect of the parameters on the shape of the gamma p.d.f., several combinations of α and β have been used for the graphs displayed in Figure 3.3-3.

3.3-5 Chi-Square Distribution

Another special case of the gamma distribution that statisticians frequently use is that in which $\alpha = r/2$, where r is a positive integer and $\beta = 2$. A random variable Y that has the p.d.f.

$$f(y) = \frac{1}{\Gamma(r/2)2^{r/2}} y^{(r/2)-1} e^{-y/2}, \quad 0 \le y < \infty,$$

is said to have a *chi-square distribution* with parameter r. For no obvious reason at this time, we call the parameter r the *number of degrees of freedom* of this chi-square distribution, which is written, for brevity, $\chi^2(r)$.

3.3-6 Lognormal Distribution

Let the random variable X be $N(\mu, \sigma^2)$, and consider the random variable $Y = \exp(X)$, or equivalently, $X = \ln(Y)$. The distribution function of Y is

$$G(y) = P(Y \le y) = P(X \le \ln y), \quad 0 \le y < \infty.$$

That is,

$$G(y) = \int_{-\infty}^{\ln y} \frac{1}{\sqrt{2\pi}\sigma} \exp\left[-\frac{(x - \mu)^2}{2\sigma^2}\right] dx.$$

The p.d.f. of Y is

$$g(y) = G'(y) = \frac{1}{\sqrt{2\pi}\sigma y} \exp\left[-\frac{(\ln y - \mu)^2}{2\sigma^2}\right], \quad 0 \le y < \infty.$$

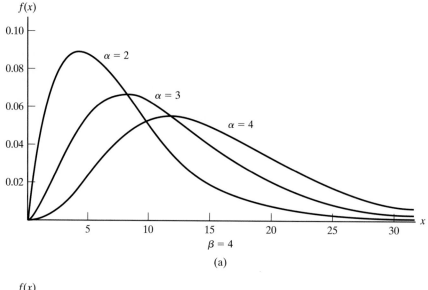

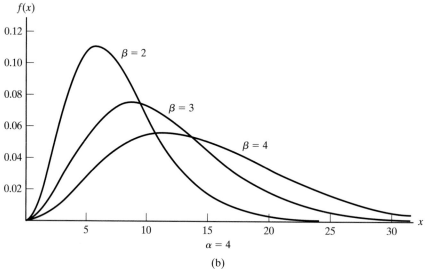

Figure 3.3-3 Graphs of several gamma densities

Because this distribution results from the transformation $X = \ln(Y)$, the random variable Y is said to have a *lognormal distribution* with parameters μ and σ. Figure 3.3-4 shows the graphs of several lognormal densities. Incidentally, we now see why a log transformation of a positive random variable that is skewed to the right frequently creates a distribution that resembles a normal distribution.

In Table 3.3-1, we summarize the most important continuous distributions by recording their densities and the corresponding means and variances. In Exercises 3.2-9, 3.3-12, and 3.3-13, hints are given on how to compute the means and variances of the normal, gamma, and Weibull distributions, respectively.

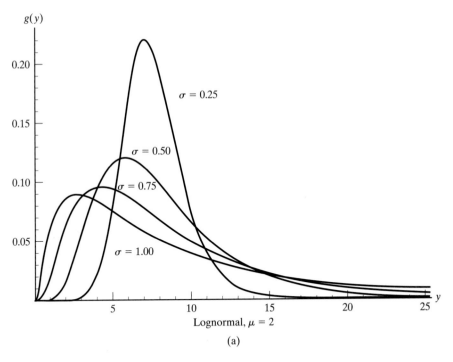

Lognormal, $\mu = 2$

(a)

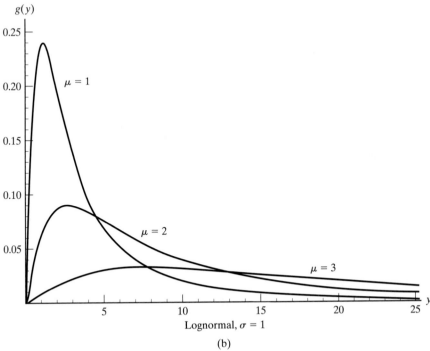

Lognormal, $\sigma = 1$

(b)

Figure 3.3-4 Graphs of several lognormal densities

Table 3.3-1 Important Continuous Distributions

Distribution	p.d.f.	Mean	Variance
Normal	$\dfrac{1}{\sqrt{2\pi}\,\sigma}\exp\left[-\dfrac{(x-\mu)^2}{2\sigma^2}\right],\ -\infty < x < \infty$	μ	σ^2
Gamma	$\dfrac{1}{\Gamma(\alpha)\beta^\alpha}x^{\alpha-1}e^{-x/\beta},0\le x<\infty$	$\alpha\beta$	$\alpha\beta^2$
Exponential ($\alpha=1$)	$\dfrac{1}{\beta}e^{-x/\beta},0\le x<\infty$	β	β^2
Chi-square ($\alpha=r/2$; $\beta=2$)	$\dfrac{1}{2^{r/2}\Gamma(r/2)}x^{(r/2)-1}e^{-x/2},0\le x<\infty$	r	$2r$
Weibull	$\dfrac{\alpha x^{\alpha-1}}{\beta^\alpha}\exp\left[-\left(\dfrac{x}{\beta}\right)^\alpha\right],0\le x<\infty$	$\beta\Gamma\left(\dfrac{1}{\alpha}+1\right)$	$\beta^2\left\{\Gamma\left(\dfrac{2}{\alpha}+1\right)-\left[\Gamma\left(\dfrac{1}{\alpha}+1\right)\right]^2\right\}$
Lognormal	$\dfrac{1}{\sqrt{2\pi}\,\sigma y}\exp\left[-\dfrac{(\ln y-\mu)^2}{2\sigma^2}\right],0\le y<\infty$	$e^{\mu+\sigma^2/2}$	$e^{2\mu+\sigma^2}(e^{\sigma^2}-1)$

Note on computation: Computer software makes it easy to work with these distributions. Minitab, with its "Calc > Probability Distributions" command, provides the p.d.f. (probability density), c.d.f. (cumulative probability), and percentiles (inverse cumulative probability) for a wide selection of distributions. Note that the gamma distribution in Minitab refers to α and β as the shape and scale parameters, respectively. The (largest) extreme value distribution refers to λ and δ as the location and scale parameters, respectively, and the lognormal uses the same terminology for its parameters μ and σ.

Check, for example, that the following are true:

- For the gamma distribution with $\alpha=2$ and $\beta=4$, $P(Y\le 5)=0.3554$ and the 90th percentile $y_{0.90}=15.56$;

- For the extreme value distribution with $\lambda=2$ and $\delta=5$, $P(Y\le 7)=0.6922$ and the 90th percentile $y_{0.90}=13.25$;

- For the lognormal distribution with location $\mu=2$ and scale $\sigma=1$, $P(Y\le 5)=0.3481$ and the 90th percentile $y_{0.90}=26.62$.

Furthermore, you can determine the ordinates in Figures 3.3-3 and 3.3-4 by using the "probability density" feature of the "Calc > Probability Distributions" command. For example, for the lognormal distribution with $\mu=2$ and $\sigma=1$, $g(y=5)=0.0739$.

Other statistical software packages include the same functions, but with slightly different terminology.

Exercises 3.3

3.3-1 Let X have the Weibull p.d.f. with $\alpha = 4$ and $\beta = 60$. Draw the p.d.f. and compute probabilities such as $P(60 < X < 80)$ and $P(20 < X < 40)$. Discuss whether this model could serve as a p.d.f. for the length of a human life.

★3.3-2 Let X have an exponential p.d.f. with $\beta = 500$; that is, $f(x) = (1/500)\exp(-x/500), 0 \leq x < \infty$.

*(a) Draw the p.d.f. and compute $P(X > 500)$.

(b) Compute the conditional probability

$$P(X > 1{,}000 | X > 500) = P(X > 1{,}000)/P(X > 500).$$

(c) Note that the answers in parts (a) and (b) are the same. Show, in general, that

$$P(X > x + c | X > x) = P(X > c).$$

Since the exponential distribution has a constant failure rate, we might have expected this result. That is, for this model, "old is as good as new."

3.3-3 Let X_1, X_2, and X_3 be $n = 3$ independent observations from a Weibull distribution with parameters α and β. Define Y as the minimum value of these three observations. Note that if $0 \leq y < \infty$, then $Y > y$ is equivalent to $X_1 > y$, $X_2 > y$, and $X_3 > y$, so that

$$G(y) = 1 - P(Y > y) = 1 - P(X_1 > y, X_2 > y, X_3 > y).$$

Show that $g(y) = G'(y)$ is a Weibull p.d.f. with parameters α and $\beta/3^{1/\alpha}$.

★3.3-4 Telephone calls enter a college switchboard according to a Poisson process on the average of three calls every 4 minutes (i.e., at a rate of $\lambda = 0.75$ per minute). Let W denote the waiting time in minutes until the second call.

*(a) What is the p.d.f. of W?

*(b) Compute $P(W > 1.5 \text{ minutes})$.

★3.3-5 Let X equal the number of alpha particle emissions of carbon-14 per second that are counted by a Geiger counter. Assume that X has a Poisson distribution with mean 8. Let W be the time in seconds before the second count is made. Determine $P(W \leq 0.5)$ and $P(W > 0.75)$.

3.3-6 Let Y have a lognormal distribution with parameters $\mu = 5$ and $\sigma = 1$. Obtain the mean, variance, and standard deviation of Y. Sketch its p.d.f. Compute $P(Y < 91)$.

Hint: $P(Y < 91) = P(\ln Y < \ln 91)$.

★3.3-7 Nelson reports that the life, in thousands of miles, of a certain type of electronic control for locomotives has an approximate lognormal distribution with parameters $\mu = 2.236$ and standard deviation $\sigma = 0.320$, where he has taken common (base-10) logarithms. Show that the fraction of these controls that would fail on an 80,000-mile warranty is 0.15. Nelson notes that this fraction was so high that the control had to be redesigned.

Hint: $P(X < 80) = P(\log_{10} X < \log_{10} 80)$; use the fact that $\log_{10} X$ has a normal distribution with mean 2.236 and standard deviation $\sigma = 0.320$.

[W. Nelson, *Applied Life Data Analysis* (New York: Wiley, 1982)]

3.3-8 Start with a uniform random variable U on the interval $[0, 1]$. Let $X = -b \ln(1 - U)$. Show that X has an exponential distribution.

Hint: Find the cumulative distribution function of X, and use the fact that $X \le x$ if and only if $U \le 1 - \exp(-x/b)$.

3.3-9 Consider the p.d.f. of the extreme value distribution

$$g(y) = \frac{1}{\delta} \exp\left(\frac{y - \delta}{\delta}\right) \exp\left[-\exp\left(\frac{y - \lambda}{\delta}\right)\right], \quad -\infty < y < \infty.$$

(a) Show that $g(y)$ is a valid p.d.f. and that it reaches its maximum at λ; that is, λ is the *mode*. This explains why we refer to λ as the location parameter.

(b) Show that the percentiles are given by $y_p = \lambda + \delta \ln[-\ln(1 - p)]$. This explains why we refer to δ as the scale parameter. An increase in δ will stretch the distribution and thus change the scale.

3.3-10 Let three parts be placed in *series*, so that the failure of any one of them results in the failure of the system. Assume that the lengths X_1, X_2, and X_3 of lives of the parts have independent distributions with failure rates $\lambda_1(x), \lambda_2(x)$, and $\lambda_3(x)$, respectively.

(a) Show that $Y = \min(X_1, X_2, X_3)$, the length of life of the system, has distribution function

$$G(y) = 1 - \exp\left\{-\int_0^y [\lambda_1(x) + \lambda_2(x) + \lambda_3(x)]\,dx\right\},$$

so that the failure rate of Y is $\lambda_1(y) + \lambda_2(y) + \lambda_3(y)$.

Hint:

$$P(Y \le y) = 1 - P(Y > y)$$
$$= 1 - P(X_1 > y)P(X_2 > y)P(X_3 > y).$$

(b) If $\lambda_i(x) = x^2/9, i = 1, 2, 3$, find the p.d.f. of Y. What type of distribution is this?

3.3-11 Show that $\Gamma(\alpha) = (\alpha - 1)\Gamma(\alpha - 1)$, provided that $\alpha > 1$.

Hint: Integrate

$$\Gamma(\alpha) = \int_0^\infty w^{\alpha-1} e^{-w}\,dw$$

by parts one time.

3.3-12 Let Y have a gamma distribution with parameters α and β. Show that the mean and the variance are $\alpha\beta$ and $\alpha\beta^2$, respectively.

Hint: In the integrals representing $E(Y)$ and $E(Y^2)$, change variables by writing $w = y/\beta$ and use the result of Exercise 3.3-11 to obtain $E(Y) = \beta\Gamma(\alpha + 1)/\Gamma(\alpha) = \alpha\beta$ and $E(Y^2) = \beta^2\Gamma(\alpha + 2)/\Gamma(\alpha) = (\alpha + 1)\alpha\beta^2$.

3.3-13 Let Y have a Weibull distribution with parameters α and β. Show that the mean and the variance are $\beta\Gamma((1/\alpha) + 1)$ and $\beta^2\{\Gamma((2/\alpha) + 1) - [\Gamma((1/\alpha) + 1)]^2\}$, respectively.

Hint: In the integrals representing $E(Y)$ and $E(Y^2)$, let $w = (y/\beta)^\alpha$.

3.3-14 Experiment with the available statistical software packages, and practice looking up cumulative probabilities and percentiles. In Minitab, you use the command "Calc > Probability Distributions" and you specify the desired distribution and its parameters. Asking for the "cumulative probability" gives you the probability $F(x) = P(X \le x)$ for a specified value x. Asking for the "inverse cumulative probability" gives you the percentile $x_p = F^{-1}(p)$ for a specified proportion p. Minitab covers most discrete and continuous distributions that we discuss in Chapters 2 and 3.

    (a) For the gamma distribution with shape $\alpha = 3$ and scale $\beta = 4$, determine the percentiles of order $0.1, 0.5$ (median), and 0.95, and find the probabilities $P(3 < X < 10)$ and $P(X > 10)$.

    (b) For the lognormal distribution with location $\mu = 2$ and scale $\sigma = 1$, determine the first, second (median), and third quartiles, and find the probabilities $P(1 < X < 10), P(X > 10)$, and $P(X < 2)$.

    (c) For the Weibull distribution with shape $\alpha = 2$ and scale $\beta = 10$, determine the first, second (median), and third quartiles, and find the probabilities $P(3 < X < 15), P(X > 10)$, and $P(X < 2)$.

*3.3-15 We usually have about five power outages in a four-week period. Making appropriate distributional assumptions:

    *(a) What is the chance that we can get through a week without an outage?

    *(b) With probability 75 percent, how long can we go without an outage?

3.3-16 **(Maxwell–Boltzmann distribution)** Consider three independent normal random variables X_1, X_2, X_3, with zero means and common standard deviation a. Then the quantity $W = \sqrt{X_1^2 + X_2^2 + X_3^2}$ has a *Maxwell–Boltzmann distribution* with parameter a. The density of this continuous distribution is given by

$$f(w) = (1/a)(2/\pi)^{1/2}(w/a)^2\exp[-(1/2)(w/a)^2], \quad \text{for } 0 \le w < \infty.$$

Boltzmann showed that this distribution describes the speed of molecules in a gas with molecular mass m at temperature T. The parameter $a = \sqrt{kT/m}$ also includes the Boltzmann constant k (which relates temperature to energy).

    (a) Consider the Maxwell–Boltzmann distribution with parameter $a = 1$. Determine the mean and the standard deviation of that distribution.
       Note: Later, we find that the sum of squares of independent standard normal random variables follows a chi-square distribution.

    (b) Draw the densities of the three Boltzmann distributions with $a = 0.5, a = 1$, and $a = 2$.

3.4 Simulation: Generating Random Variables

3.4-1 Motivation

Engineers design structures such as dams, bridges, and skyscrapers, and these structures need to withstand extreme floods and strong earthquakes. Network engineers design the flow of work processes in a way such that the waiting times of arriving customers are kept short. Airlines and rental car companies face uncertain demand for their products and services, and they study ways of overbooking reservations that are financially beneficial to them.

The magnitudes of earthquakes, the hourly volume of water rushing down a river, the number of customers arriving within a certain time interval, the waiting time until service is rendered, the time required to complete a task (service, assembly operation, etc.), and the demand for a certain flight from city A to city B on day C are random variables with certain distributions. The number of customers arriving may be described by a Poisson distribution, the waiting times until service may be exponential, the hourly volume of water may be lognormal, and the demand for airplane seats may follow a normal distribution.

For certain simple and well-structured problems, it is possible to obtain explicit analytic solutions. The "News vendor problem" is a good example. In this problem, a vendor is facing an uncertain demand for a perishable product that can only be sold at a loss once a certain target date has passed. In Project 1 of this chapter, we ask you to derive the volume of orders that maximizes the expected profit. You will learn that the solution can be expressed in terms of the cumulative probability distribution of the uncertain demand, and the applicable "overage" and "underage" costs which represent the losses due to orders above and below the demand. Similarly, queuing systems with simple distributions for the number of arriving customers and the time of service and with simple priority rules can be solved theoretically, and it is possible to derive explicit expressions for key characteristics, such as the mean waiting time. However, for many realistic situations that go beyond textbook formulations (e.g., several service providers, general distributions, complicated protocols of who gets served first), analytical solutions are no longer possible.

In these situations, simulation becomes a practical and useful approach for eliciting how a system responds to various random inputs. In simulation, we draw realizations from specified distributions that govern the random variables (such as the number of arriving customers, the service times, etc.), pass these realizations through the system, and assess the state of the system. We repeat the process many times and evaluate the variable of interest (such as the mean profit, the average number of people waiting in the queue, etc.).

Simulation is also useful in a number of other areas. Many decision models contain variables that are affected by uncertainty. For example, in a certain financial model, we may be interested in the distribution of the cumulative discounted cash flow over several years during which sales, sales growth, operating expenses, and inflation are uncertain, with their uncertainty described by probability distributions. Or consider a project management model dealing with issues of identifying expected completion times when the completion times of individual components are uncertain. Although analytical methods such as the Program Evaluation and Review Technique (PERT) are available, some of their assumptions may not be valid, and it may be preferable to simulate the system. An important benefit of simulation is its ability to conduct sensitivity analyses to understand the impact of changed assumptions on the output.

For simulation, we need to generate outcomes from different distributions — discrete as well as continuous ones. The question becomes, "How do we generate random variables from specified distributions?"

3.4-2 Generating Discrete Random Variables

Let us start our discussion by generating uniform random integers between 0 and 9. This is quite easy, as we can use an urn with 10 slips of paper, numbered 0 to 9, and draw from it consecutive slips at random with replacement. Of course, we should shuffle the slips each time before we draw a slip, to make sure that consecutive draws are independent and that there is no carryover from one draw to the next. We may get the following sequence: 1, 9, 0, 2, 2, 8, 7, 5, 6, and so on. Think about doing this a million times, and we would get a million random digits. Of course, we wouldn't do this in practice. Instead, we would use a computer to generate "pseudo-random" numbers that have the same properties as the random numbers one gets by drawing slips from an urn.

If we want random numbers with three digits (i.e., random numbers between 000 and 999), we select nonoverlapping triples from our string of single digits. In our example, this gives us 190, 228, 756, and so on. (Note that it is important that these numbers be nonoverlapping, because overlapping numbers introduce a carryover from one number to the next, making the subsequent number somewhat predictable.)

Once we have generated the (discrete) uniform random numbers, we can generate realizations from any discrete distribution. Take, for example, the binomial distribution with $n = 3$ and $p = 0.8$. We know from Section 2.4 that the p.d.f. and the c.d.f. of X are

$$P(X = 0) = 0.008, \quad P(X \leq 0) = 0.008;$$
$$P(X = 1) = 0.096, \quad P(X \leq 1) = 0.104;$$
$$P(X = 2) = 0.384, \quad P(X \leq 2) = 0.488;$$
$$P(X = 3) = 0.512, \quad P(X \leq 3) = 1.000.$$

We use uniform random digits between 000 and 999, as the probabilities are expressed with three decimal positions. Using the three-digit number D, we define

$$X = 0 \quad \text{if} \quad 000 \leq D \leq 007,$$
$$X = 1 \quad \text{if} \quad 008 \leq D \leq 103,$$
$$X = 2 \quad \text{if} \quad 104 \leq D \leq 487,$$
$$X = 3 \quad \text{if} \quad 488 \leq D \leq 999.$$

With the three-digit numbers we generated, we get $X = 2$ (for $D = 190$), $X = 2$ (for $D = 228$), $X = 3$ (for $D = 756$), and so on. Repeating this 1,000 times, we will have generated 1,000 independent realizations from the binomial distribution with $n = 3$ and $p = 0.8$.

Most computer packages include routines that allow us to generate realizations of specified discrete distributions directly, without having to go through uniform random numbers first. For example, the Minitab command "Calc > Random Data > Binomial" generates realizations from the binomial distribution. To do this, you enter the number of realizations desired, as well as the parameters n and p. Similarly, you can use the Minitab command "Calc > Random Data > Poisson" to generate 10,000 realizations from the Poisson distribution with parameter, say, $\lambda = 1$. Draw the frequency distribution of the resulting observations and check that it looks like the p.d.f. of the Poisson distribution with $\lambda = 1$ by comparing the relative frequencies with the Poisson probabilities. Note that every once in a while you will get a large value; $P(X = 10) = \frac{1}{10!}e^{-1}1^{10} = (1.0138)10^{-7}$, for example, allows one such realization in about every 10 million trials.

3.4-3 Generating Continuous Random Variables

In the previous section, we generated realizations from a discrete uniform distribution by using the conceptual model of drawing slips from an urn. From these uniform random digits between 0 and 9, we can form very long tuples of nonoverlapping digits (say, 10 or 20 digits long) and divide these digits by 10^{10} (or 10^{20} if we use strings of 20). We will then have generated (almost) continuous random variables between 0 and 1. Continuous uniform numbers are also available from statistical software packages. In Minitab, you use "Calc > Random Numbers > Uniform"; all you have to do is specify the lower and upper endpoints of the uniform distribution (here, 0 and 1). In Exercise 3.4-2, we ask you to generate 10,000 uniform random numbers between 0 and 1 and draw the resulting histogram. This exercise should convince you that your software is doing the right thing.

The following result shows that these continuous $U(0, 1)$ random numbers can be used to generate random numbers from *any other* continuous distribution. The result also tells us how to do this operationally.

Result: Let $F(x)$ be the cumulative distribution of a continuous random variable X. We assume that $F(x)$ is strictly increasing on the interval $a < x < b$, with $F(a) = 0$ and $F(b) = 1$, where a and b can be $-\infty$ and $+\infty$, respectively. Let U have a continuous uniform distribution on the interval from 0 to 1, with p.d.f. $g(u) = 1$, for $0 < u < 1$. Then $X = F^{-1}(U)$ is a continuous-type random variable with cumulative probability distribution (c.d.f.) $F(x)$.

Proof: Consider

$$P(X \leq x) = P(F^{-1}(U) \leq x), \quad \text{for} \quad a < x < b.$$

$F(x)$ is strictly increasing. Hence, the event $[F^{-1}(U) \leq x]$ is equivalent to the event $[U \leq F(x)]$. Therefore,

$$P(X \leq x) = P(U \leq F(x)), \quad \text{for} \quad a < x < b.$$

But U is uniform on the interval from 0 to 1, and

$$P(U \leq u) = u \quad \text{for} \quad 0 < u < 1.$$

Thus,

$$P(X \leq x) = P(U \leq F(x)) = F(x), \quad \text{for} \quad 0 < F(x) < 1.$$

That is, $F(x)$ is the c.d.f. of the random variable X.

Notes:

1. The assumption of a strictly increasing $F(x)$ rules out distributions that give zero probability to sections in the interior of the interval.

2. Equivalence between the event $[F^{-1}(U) \leq x]$ and the event $[U \leq F(x)]$ means that the same set of real numbers U satisfy $[F^{-1}(U) \leq x]$ and $[U \leq F(x)]$.

This result gives us a simple recipe for generating random variables from a distribution with c.d.f. $F(x)$: Generate a realization u from the $U(0, 1)$ distribution, and compute $x = F^{-1}(u)$. Of course, this requires that we be able to evaluate $F^{-1}(u)$, or at least have a way to look it up from a table.

Example 3.4-1   **Generating realizations from an exponential distribution**

Let us obtain a random realization from the exponential distribution with mean $\beta = 10$. (See Section 3.3.) The p.d.f. and the c.d.f. of this distribution are respectively given by

$$f(x) = (1/\beta)\exp(-x/\beta) \text{ and } F(x) = 1 - \exp(-x/\beta), \quad \text{for } 0 \leq x < \infty.$$

Solving $u = F(x)$ for x yields the inverse $x = F^{-1}(u) = -\beta\ln(1 - u) = -10\ln(1 - u)$. We generate the $U(0, 1)$ realization u (say, $u = 0.14$) and obtain $x = -10\ln(1 - 0.14) = 1.508$ as our first realization, and so on. Of course, you can use the Minitab function "Calc > Random Numbers > Exponential" directly, without having to go through the preceding steps; all you have to do is specify the number of realizations that you want and the mean β of the exponential distribution. ■

Example 3.4-2   **Generating realizations from a normal distribution**

The c.d.f. of the standard normal distribution, given in Section 3.2, is

$$F(x) = \frac{1}{\sqrt{2\pi}} \int_{-\infty}^{x} \exp(-w^2/2)dw, \quad \text{for } -\infty < x < \infty.$$

Earlier, we noted that this integral can be evaluated only numerically. Hence, there is no algebraic expression for the inverse cumulative probability distribution function $x = F^{-1}(u)$. However, a look-up function exists in most computer packages and even calculators. For example, we can evaluate $F^{-1}(u)$, for a given value of u, by using the Minitab command "Calc > Probability Distribution > Normal" and checking the inverse cumulative probability option. For instance, from $u = 0.14$, we obtain $x = -1.08032$. If we want a realization from a general normal distribution with specified mean μ and standard deviation σ, the realization is $\mu + (-1.08032)\sigma$.

Of course, you can use Minitab's random-number generator "Calc > Random Numbers > Normal" directly; all you have to do is specify the number of realizations you want and the mean and standard deviation. ■

Exercises 3.4

3.4-1   Generate four-digit uniform random numbers by considering nonoverlapping strings of four random digits between 0 and 9. Use the four-digit uniform random numbers to generate random variables from a discrete distribution. For example, use them to generate 1,000 realizations from the binomial distribution with parameters $n = 4$ and $p = 0.20$, with probabilities $P(X = 0) = 0.4096$, $P(X = 1) = 0.4096, P(X = 2) = 0.1536, P(X = 3) = 0.0256,$ and $P(X = 4) = 0.0016$.

It is instructive to generate the binomial random variables from the initial uniform random numbers, even though most statistical software packages allow you to generate the observations directly. In Minitab, for example, you can use "Calc > Random Data > Binomial."

*3.4-2 (a) Consider the continuous uniform distribution on the interval from 0 to 1, with p.d.f. $g(u) = 1$, for $0 < u < 1$. Determine (theoretically) its mean, variance, standard deviation, 95th percentile, and interquartile range. (See Section 3.1.)

(b) Use any available statistical software program to generate 1,000 observations from a continuous uniform distribution between 0 and 1. For example, in Minitab, you use the command "Calc > Random Data > Uniform." In Excel, you use "Tools > Data Analysis > Random Number Generation (uniform distribution)." Draw a histogram of the resulting 1,000 observations.

Note: Make sure that the bins of your histogram have been selected appropriately. A frequent problem in software-generated histograms is that the first and last bins extend beyond the limits (here, 0 and 1). For example, a histogram with bin width 0.10 and starting midpoint 0 would lead to bins extending from -0.05 to 0.05 and from 0.95 to 1.05. You should not be surprised that the observed frequencies in these two bins are about half the frequencies in all others, as the uniform distribution between 0 and 1 leaves no chance that an observation is smaller than 0 or larger than 1.

(c) Calculate the sample statistics of the observations you generated, and compare them with the population values derived in (a).

(d) Search the literature and describe how such uniform random numbers are actually generated. [No, it is *not* true that graduate students draw these digits with replacement from 10 digits, numbered 0 to 9.]

(e) Among your simulated data, there should be no carryover from one random number to the next. Investigate whether this is true for the numbers you have generated.

Note: You could construct a scatter plot of the result of draw i against the result of draw $i - 1$ and calculate the correlation coefficient. Discuss whether this simple check is sufficient to show that there is no carryover from one number to the next.

3.4-3 Use the 1,000 uniform random numbers you generated in Exercise 3.4-2, and consider the exponential distribution in Section 3.3 with parameter $\beta = 1$, probability density $f(x) = \exp(-x)$, and cumulative distribution function $F(x) = 1 - \exp(-x)$.

Generate 1,000 observations from this distribution, draw the histogram, obtain the sample mean and variance, and relate the sample statistics to the population mean and variance.

Note: Even without special software, it is easy to generate random numbers from this distribution. Use the uniform number generated between 0 and 1 (say, u), and solve the equation $F(x) = 1 - \exp(-x) = u$ for x. The result, $x = -\log(1 - u)$, is a realization from the exponential distribution with parameter $\beta = 1$.

3.4-4 (a) Use the uniform random variables $u_1, u_2, \ldots, u_{1,000}$ from Exercise 3.4-2, and generate 1,000 observations from a standard normal distribution.

Note: You need to solve the equation $F(x) = \dfrac{1}{\sqrt{2\pi}} \displaystyle\int_{-\infty}^{x} \exp(-w^2/2)\,dw = u$

for $x = F^{-1}(u)$. In Section 3.2, you learned that the integral on the right-hand side of this equation can be evaluated only numerically. Fortunately, the inverse function $F^{-1}(u)$ has been tabulated in most statistical packages (as well as on scientific calculators), and for any given value u, it is easy to look up $x = F^{-1}(u)$. In Minitab, for example, one stores the u's in one column and obtains the column of x's by applying the function "Calc > Probability Distributions > Normal" to the column of u's and checking the Inverse Cumulative Probability option.

(b) Computer packages generate normal random variables directly, so there is no need to generate uniform random variables first and then transform the uniform numbers by applying the inverse cumulative distribution function of the normal (target) distribution to them. For example, in Minitab, you use the command "Calc > Random Data > Normal". Generate the random numbers this way.

Note: Many computer packages use the Box–Muller method for generating normal random numbers [Box, G. E. P., and Muller, M. E.: "A note on the generation of random normal deviates," *Annals Math. Stat,* Vol. 29 (1958), pp. 610–611].

(c) Draw a histogram of the normal data generated and obtain the sample statistics (mean, variance).

3.4-5   Experiment with your statistical software. Generate sets of random variables (from distributions such as the normal, gamma, and Weibull). In Minitab, you can use the command "Calc > Random Data" and specify the distribution of interest.

Obtain histograms of the data generated, and check whether the histograms resemble the probability density functions of the selected distributions you studied in Sections 3.2 and 3.3.

3.5 Distributions of Two or More Continuous Random Variables

3.5-1 Joint, Marginal, and Conditional Distributions, and Mathematical Expectations

In Section 2.6, we discussed the joint probability density function of two discrete random variables X and Y. We can do the same for two continuous random variables and define the joint p.d.f. as

$$f(x, y) \geq 0, \quad \text{for } -\infty < x, y < \infty.$$

In order to make this a genuine density, we require that the integral of this function, taken over the two-dimensional space R^2, equals 1. That is, we need

$$\int_{-\infty}^{\infty} \int_{-\infty}^{\infty} f(x, y)dxdy = 1.$$

The probability that $(X, Y) \in A$, where A is a certain specified region in the two-dimensional space R^2, is calculated from

$$P[(X, Y) \in A] = \iint_{(x, y) \in A} f(x, y)dxdy.$$

The only difference from the discrete situation is that we use integrals instead of sums.

Marginal probability density functions (p.d.f.'s) of X and Y are obtained from

$$f_1(x) = \int_{-\infty}^{\infty} f(x, y)dy \quad \text{and} \quad f_2(y) = \int_{-\infty}^{\infty} f(x, y)dx,$$

and conditional probability density functions are given by

$$g(y|x) = \frac{f(x, y)}{f_1(x)} \quad \text{and} \quad h(x|y) = \frac{f(x, y)}{f_2(y)}.$$

The expectation of $u(X, Y)$, a function of the random variables X and Y, is obtained from

$$E[u(X, Y)] = \int_{-\infty}^{\infty} \int_{-\infty}^{\infty} u(x, y)f(x, y)dxdy.$$

Again, the only difference from the discrete case is that integrals replace sums. The covariance between X and Y, namely,

$$\text{cov}(X,Y) = \sigma_{XY} = E[(X - \mu_X)(Y - \mu_Y)] = \int_{-\infty}^{\infty} \int_{-\infty}^{\infty} (x - \mu_X)(y - \mu_Y)f(x, y)dxdy,$$

is a special case.

The results for the mean and variance of a linear combination of discrete random variables also carry over to the continuous case. That is, for $W = a_0 + a_1 X + a_2 Y$, we find that

$$\mu_W = E(W) = a_0 + a_1\mu_X + a_2\mu_Y$$

and

$$\sigma_W^2 = \text{var}(W) = a_1^2\sigma_X^2 + a_2^2\sigma_Y^2 + 2a_1a_2\sigma_{XY}.$$

It is easy to generalize this result to linear functions of n random variables, $W = a_0 + \sum_{i=1}^{n}a_iX_i$. One can show that

$$\mu_W = a_0 + \sum_{i=1}^{n}a_i\mu_i \quad \text{and} \quad \sigma_W^2 = \sum_{i=1}^{n}a_i^2\sigma_i^2 + 2\sum\sum_{i<j}a_ia_j\sigma_{ij},$$

where $\mu_i = E(X_i)$, $\sigma_i^2 = \text{var}(X_i)$, $i = 1, 2, \ldots, n$, and $\text{cov}(X_i, X_j) = \sigma_{ij} = \rho_{ij}\sigma_i\sigma_j$. The mean of W is not affected by the covariance (correlation) among the random variables; correlation enters only into the calculation of the variance. Of course, for mutually independent random variables,

$$\sigma_W^2 = \sum_{i=1}^{n}a_i^2\sigma_i^2.$$

3.5-2 Propagation of Errors

Many engineering relationships involve nonlinear functions of several components that are subject to variability. Usually, engineers have a pretty good idea about variabilites of individual components, and they want to know how they "propagate" into the variabilities of the combined result. Following are a few motivating examples.

The density of a rock is determined by putting the rock into a graduated cylinder partially filled with water and measuring the volume of water that is being displaced. The density of the rock is given by $D = m/(V_1 - V_0)$, where m is the mass of the rock, V_0 is the initial volume of water, and V_1 is the volume of the water plus the rock. Assume that the mass of the rock is known. (Suppose it is 700 grams.) Your measurements of the water volumes are subject to error; assume, for example, that the volumes V_0 and V_1 are independent random variables with means 400 and 600 ml and standard deviations of 0.2 ml each. What can you say about the rock's density?

Two resistors with resistances R_1 and R_2 are connected in parallel. The combined resistance is given by $R = (R_1 R_2)/(R_1 + R_2)$. The first resistance (R_1) is measured as $50 \pm 10\,\Omega$; the second (R_2) is measured as $25 \pm 5\,\Omega$, where the numbers after the $\pm$ sign are the standard deviations of the measurement errors (expressed in ohms, Ω). Furthermore, you have good reason to believe that the two measurement errors are independent. How can you determine the mean and the standard deviation of R?

Consider two gases with molar masses M_1 and M_2. Graham's law states that the ratio of their rates of effusion is given by

$$R = \frac{\text{Rate}_1}{\text{Rate}_2} = \sqrt{\frac{M_1}{M_2}};$$

if the molecular weight of one gas is four times that of another, then the first gas would diffuse through a porous plug or escape through a small pinhole in a vessel at half the rate of the other. Suppose you find that the effusion rate of an unknown gas is 1.70 ± 0.10 greater than the effusion rate of carbon dioxide (which, as you know from your chemistry course, has molar mass $44\,\text{g/mol}$). How can you determine the mean molar mass of the unknown gas and how can you find its variability?

An Atwood machine consists of two bodies X and Y of unequal mass that are attached to a thin cord that passes over a lightweight pulley. When the masses are released, the larger mass X accelerates down with acceleration

$$A = g\frac{X - Y}{X + Y},$$

where g is the constant of acceleration due to gravity. Suppose that the two bodies have masses X and Y measured as 100 ± 2 grams and 50 ± 2 grams, respectively. How can you find the relative uncertainty of the acceleration (i.e., the standard deviation of A divided by the mean of A)?

The general approach Let us define the variable of interest as $W = u(X, Y)$, a general nonlinear function of X and Y. We do not know the joint p.d.f. of (X, Y), but we know the means (μ_X, μ_Y), standard deviations (σ_X, σ_Y), and covariance σ_{XY}. We expand $W = u(X, Y)$ in a first-order Taylor series expansion around the means μ_X and μ_Y. That is,

$$W = u(X, Y) \approx a_0 + a_1(X - \mu_X) + a_2(Y - \mu_Y),$$

where

$$a_0 = u(\mu_X, \mu_Y); \quad a_1 = \left.\frac{du(x, y)}{dx}\right|_{x=\mu_X, y=\mu_Y}; \quad a_2 = \left.\frac{du(x, y)}{dy}\right|_{x=\mu_X, y=\mu_Y}.$$

The coefficients a_1 and a_2 are the first derivatives evaluated at the means μ_X and μ_Y. Applying our results about the mean and variance of a linear combination of random variables, we find that

$$\mu_W = E(W) \approx a_0,$$

$$\sigma_W^2 = \text{var}(W) \approx a_1^2 \sigma_X^2 + a_2^2 \sigma_Y^2 + 2a_1 a_2 \sigma_{XY}.$$

The two expressions provide fairly good approximations for the mean and the variance, respectively, of W. Note that the approximation of the mean could be improved by considering a second-order Taylor series expansion of $u(X, Y)$. However, the benefit is usually minor.

Let us consider again several of the examples we mentioned in the introductory paragraphs of this section.

Example 3.5-1

Consider the two resistors connected in parallel with resistances R_1 and R_2, and combined resistance $R = (R_1 R_2)/(R_1 + R_2)$. We are told that $\mu_{R1} = 50$, $\mu_{R2} = 25$, $\sigma_{R1} = 10$, and $\sigma_{R2} = 5$, all in units of ohms, and that the resistances R_1 and R_2 are independent. For this example, we find that

$$a_0 = \frac{\mu_{R1}\mu_{R2}}{\mu_{R1} + \mu_{R2}} = \frac{50}{3}; a_1 = \left[\frac{\mu_{R2}}{\mu_{R1} + \mu_{R2}}\right]^2 = \frac{1}{9}; a_2 = \left[\frac{\mu_{R1}}{\mu_{R1} + \mu_{R2}}\right]^2 = \frac{4}{9}.$$

Since $\text{cov}(R_1, R_2) = 0$, it follows that

$$\sigma_W^2 = \text{var}(W) \approx \frac{1}{81}100 + \frac{16}{81}25 = \frac{500}{81} = 6.1728$$

and

$$\mu_W = E(W) \approx \frac{50}{3} = 16.6667.$$

In sum, we learn that the mean of the combined resistance is about 16.67, and its standard deviation is $\sigma_W \approx \sqrt{6.1728} = 2.48$. ∎

Example 3.5-2

In the context of the Atwood machine, the acceleration of the larger mass relative to the constant acceleration due to gravity, g, is given by $W = A/g = (X - Y)/(X + Y)$. We are told that $\mu_X = 100, \mu_Y = 50, \sigma_X = \sigma_Y = 2$, and $\sigma_{XY} = 0$.

For this nonlinear function, we find that

$$a_0 = \frac{\mu_X - \mu_Y}{\mu_X + \mu_Y} = \frac{1}{3}; a_1 = \frac{2\mu_Y}{[\mu_X + \mu_Y]^2} = \frac{2}{450}; a_2 = \frac{-2\mu_X}{[\mu_X + \mu_Y]^2} = \frac{-4}{450}.$$

We then obtain the variance

$$\sigma_W^2 = \text{var}(W) \approx \left[\frac{2}{450}\right]^2 4 + \left[\frac{-4}{450}\right]^2 4 = \frac{4}{10,125} = 0.0004,$$

standard deviation

$$\sigma_W \approx \sqrt{0.0004} = 0.02,$$

and mean

$$E(W) = \mu_W \approx \frac{1}{3}.$$ ∎

Example 3.5-3   The effusion rate of an unknown gas is 1.70 ± 0.10 greater than the effusion rate of carbon dioxide, which has molar mass 44 g/mol. Solving the equation implied by Graham's law, we find that the molar mass M_1 of the unknown gas is given by $M_1 = M_2R^2$, where $M_2 = 44$ g/mol is the molar mass of carbon dioxide and R is the ratio of effusion rates with mean $\mu_R = 1.70$ and standard deviation $\sigma_R = 0.10$. Here, the nonlinear function includes a single random variable. The first and half of the second derivatives of $M_2R^2 = 44R^2$, evaluated at $\mu_R = 1.70$, are, respectively,

$$a_1 = \frac{d(44R^2)}{dR} = (44)(2)\mu_R = 149.6 \quad \text{and} \quad a_2 = \frac{1}{2}\frac{d^2(44R^2)}{dR^2} = \frac{(44)(2)}{2} = 44.$$

The constant $a_0 = 44(\mu_R)^2 = (44)(1.7)^2 = 127.16$. Using our first-order Taylor series expansion, we find that

$$\mu_{M1} \approx a_0 = 127.16,$$
$$\sigma_{M1}^2 = \text{var}(M_1) \approx (149.6)^2(0.1)^2 = 223.8 \quad \text{and} \quad \sigma_{M1} = \sqrt{223.8} = 14.96.$$

Using the second-order Taylor series expansion $M_1 = 44R^2 \approx a_0 + a_1(R - \mu_R) + a_2(R - \mu_R)^2$, we find a more accurate value for the mean:

$$\mu_{M1} \approx 127.16 + (44)(0.10)^2 = 127.60.$$

In sum, the molecular weight of the first gas has mean 127.6 g/mol, with a standard deviation of about 15 g/mol.

Note: In this special case, we can calculate the mean exactly, without using the Taylor series expansion. We learned earlier that a variance can be expressed as $\sigma_R^2 = E(R^2) - (\mu_R)^2$. Hence, $E(R^2) = \sigma_R^2 + (\mu_R)^2$ and $E(M_1) = 44E(R^2) = 44[\sigma_R^2 + (\mu_R)^2]$ $= 44[(0.10)^2 + (1.70)^2] = 127.6$, the latter of which is identical to the expression that we obtained from our second-order Taylor series expansion. Observe that derivatives of order 3 or higher are zero for this particular nonlinear (quadratic) function. ∎

Remark: The general approach presented in Section 3.5-2 approximates the mean and the variance of a very general function of two random variables without making assumptions about the distributions of the random variables. Of course, if we knew more about the distributions of the random variables (i.e., their densities, and not just their means, variances, and covariances), we could simulate the distribution of the resulting random variable by using the approach in Section 3.4. The histogram of the simulations would tell us about the distribution, and the summary statistics of the simulations would tell us about its mean and variance.

Exercises 3.5

*3.5-1   Consider the bivariate density $f(x, y) = 4xy$ for $0 \le x < 1, 0 \le y < 1$.

*(a) Check to make sure that this is a genuine density.

*(b) Obtain the marginal densities, and check whether the random variables are independent.

*3.5-2 Consider the bivariate density $f(x, y) = c(x + y)$ for $0 \leq x < 1, 0 \leq y < 1$.

*(a) Obtain the appropriate normalization constant c.

*(b) Obtain the marginal densities for X and Y, and calculate their means and variances.

(c) Obtain the covariance between X and Y, and check whether the random variables are independent.

(d) Obtain the conditional distribution of Y, given that $X = x$.

*3.5-3 The density of a rock is determined by putting the rock into a graduated cylinder partially filled with water and measuring the volume of water displaced. (A graduated cylinder is a cylinder with graduation marks, as commonly found in labs.) The density of the rock is given by $D = m/(V_1 - V_0)$, where m is the mass of the rock, V_0 is the initial volume of water, and V_1 is the volume of the water plus the rock. Assume that the mass of the rock is 700 grams, V_0 is 400 ± 0.2 ml, and V_1 is 600 ± 0.2 ml. The first numbers in these expressions represent the means; the quantities after the $\pm$ signs are the standard deviations. Assume that the two measurements are independent.

*(a) Obtain the mean and standard deviation of the density of the rock.

*(b) Assume that the mass of the rock itself is subject to measurement variability, and suppose that the mass is measured as 700 ± 5 grams. Assume that the three measurements are independent. Obtain the mean and standard deviation of the density of the rock.

Note: For a function $W = u(X, Y, Z)$ of three random variables, the first-order Taylor series expansion leads to the variance

$$\sigma_W^2 = \text{var}(W) \approx a_1^2 \sigma_X^2 + a_2^2 \sigma_Y^2 + a_3^2 \sigma_Z^2 + 2a_1 a_2 \sigma_{XY} + 2a_1 a_3 \sigma_{XZ} + 2a_2 a_3 \sigma_{YZ},$$

with the appropriate simplification for zero covariances. The coefficients a_1, a_2, a_3 are the first derivatives $u(X, Y, Z)$, evaluated at the means. The first-order approximation of the mean is given by $E(W) = \mu_W \approx a_0 = u(\mu_X, \mu_Y, \mu_Z)$.

3.5-4 Two resistors with resistances R_1 and R_2 are connected in parallel. The combined resistance is given by $R = (R_1 R_2)/(R_1 + R_2)$. The first resistance (R_1) is measured as $50 \pm 10 \, \Omega$; the second (R_2) is measured as $25 \pm 5 \, \Omega$. Assume that the two measurements are independent.

(a) Obtain the mean and standard deviation of the combined resistance. This is what we did in Example 3.5-1.

(b) Suppose that it is possible to replace the first (and rather variable) resistor with a more accurate (but more expensive) resistor with standard deviation $2 \, \Omega$. Discuss whether this switch would be worthwhile.

(c) Assume that the two measurements in (a) are correlated with correlation $+0.3$. Determine the standard deviation of the combined resistance.

3.5-5 (a) Your goal is to determine the area of a rectangle, $A = XY$. Measurements of the sides, X and Y, are accurate to within 1 cm (i.e., $\sigma = 1$ cm). Assume that your measurements are independent. Evaluate the *relative* uncertainty in the estimate of the area (i.e., the standard deviation of the area, divided by the mean of the area).

(b) When calculating the mean of the area of a square, $A = X^2$, use the fact that $E(X^2) = \sigma_X^2 + (\mu_X)^2$ directly. Discuss how this fact relates to finding the mean from a second-order Taylor series expansion of $A = X^2$. Also, note that you get the same result from a first-order expansion of $A = X^2$ in terms of X^2, evaluated at $E(X^2) = \sigma_X^2 + (\mu_X)^2$.

(c) Foresters estimate the volume of a tree from the formula $V = (\pi/4)D^2H$, where H is the length of the tree and D is the diameter of the tree, measured at breast height, namely, 4.5 feet. Assume that the length of the tree is measured as 30 ± 0.3 meters and the diameter as 50 ± 2 cm. Assume that the measurement errors are uncorrelated. Find the mean and standard deviation of V, and discuss the relative uncertainty in the estimate of the volume (i.e., the standard deviation of V, divided by the mean of V).

Note: Make certain that D and H are expressed in the same units. Also, when determining the mean of V, you may want to use a first-order Taylor expansion of $V = (\pi/4)D^2H$ in terms of the variables D^2 and H, evaluated at $E(D^2) = \sigma_D^2 + (\mu_D)^2$ and μ_H.

Here, we have given the simplest formula for determining the volume of a tree. Many other more accurate, but also more complicated, approximations have been proposed. Search the Web if you are interested in this subject.

3.5-6 The Darcy–Weisbach equation is an important and widely used equation in hydraulics that enables the calculation of the head loss due to friction within a pipe of given length and diameter:

$$H_f = f\frac{L}{D}\frac{V^2}{2g}.$$

The head loss due to friction (H_f) is a function of the Darcy friction factor (f), the ratio of the length to the diameter of the pipe (L/D), the velocity of the flow (V), and the standard constant for the acceleration due to gravity (g). (Refer to your physics course or to Web references for a description.)

Assume that, for a certain pipe, the flow rate V is measured as $60 \pm 1\,\mathrm{m^3/sec}$, the length L as $500 \pm 1\,\mathrm{cm}$, and the diameter D as $10 \pm 0.03\,\mathrm{cm}$. Assume that the random variables are independent. Determine the mean and the standard deviation of the head loss due to friction. Determine the relative accuracy of your estimate of the head loss due to friction.

Note: When determining the mean of H_f, you may want to use a first-order Taylor expansion of H_f in terms of the variables V^2, L, and D, evaluated at $E(V^2) = \sigma_V^2 + (\mu_V)^2$, μ_L, and μ_D, respectively.

*3.5-7 The total time required to complete a building project depends on the completion times of several individual components. Assume that there are three components, with respective means of 2, 5, and 4 days, and standard deviations of 1, 2, and 1.5 days.

*(a) Assuming independence, determine the mean and the standard deviation of the total time to completion.

*(b) Assume that completion times are positively correlated, with constant correlation $+0.8$. Determine the mean and the standard deviation of the total time to completion.

3.6 Fitting and Checking Models

In this chapter, we have discussed several continuous distributions characterized by their probability density functions. Each distribution depends on certain key characteristics, which we have called parameters. The exponential distribution, for example, depends on a single parameter: the mean $\beta > 0$. The gamma distribution depends on two parameters: the shape parameter $\alpha > 0$ and the scale parameter $\beta > 0$. The normal distribution depends on the mean $-\infty < \mu < \infty$ and the variance $\sigma^2 > 0$.

Two questions arise, and they are addressed in this section: Given sample realizations $x_1, x_2, \ldots, x_n$,

1. How can we estimate the parameters of the distribution?
2. How can we check whether the observations are consistent with a certain distribution?

3.6-1 Estimation of Parameters

Perhaps the simplest way of estimating parameters is to equate the moments of the empirical distribution in Section 3.1-1 to the theoretical moments of the distribution that is being considered for the data. We call this approach to estimation the *method of moments*. The mean and the variance of the empirical distribution are given by $\bar{x}$ and $v = \dfrac{1}{n}\sum_{i=1}^{n}(x_i - \bar{x})^2$, respectively.

For the exponential distribution with mean $\mu = \beta$, the estimate based on the method of moments is $\hat{\beta} = \bar{x}$. For the gamma distribution with mean $\mu = \alpha\beta$ and variance $\sigma^2 = \alpha\beta^2$ (see Table 3.3-1), we obtain the solutions $\alpha = \mu^2/\sigma^2$ and $\beta = \sigma^2/\mu$. Hence, the method-of-moments estimates are given by $\hat{\alpha} = \bar{x}^2/v$ and $\hat{\beta} = v/\bar{x}$. For the normal distribution, the method-of-moments estimates are $\hat{\mu} = \bar{x}$ and $\hat{\sigma}^2 = v$.

Maximum likelihood estimation represents another useful, general approach. We used this approach in Section 2.6 to estimate parameters of discrete distributions. A random sample implies that the realizations of the random variables $X_1, X_2, \ldots, X_n$ are independent and from the same distribution, with p.d.f. $f(x;\theta)$, where θ denotes the parameter of the distribution. Hence, the joint p.d.f. is given by the product of the marginal distributions:

$$g(x_1, x_2, \ldots, x_n; \theta) = f(x_1; \theta)f(x_2; \theta)\cdots f(x_n; \theta).$$

Treating this equation as a function of the parameter θ leads to the likelihood function $L(\theta)$. The maximum likelihood estimate maximizes the likelihood function $L(\theta)$ or, equivalently, its logarithm $\ln L(\theta)$. The treatment can be generalized to two or more parameters.

Example 3.6-1 Exponential distribution with mean $\beta > 0$:

From its p.d.f. $f(x; \beta) = \dfrac{1}{\beta}e^{-x/\beta}$ and the resulting joint p.d.f., we obtain the likelihood function

$$L(\beta) = (1/\beta)^n \exp\left(-\sum_{i=1}^{n} x_i/\beta\right)$$

and its logarithm

$$\ln L(\beta) = -n \ln \beta - \sum_{i=1}^{n} x_i/\beta.$$

Setting the first derivative equal to zero, that is,

$$\frac{d \ln L(\beta)}{d\beta} = -\frac{n}{\beta} + \frac{\sum_{i=1}^{n} x_i}{\beta^2} = 0,$$

leads to the maximum likelihood estimate

$$\hat{\beta} = \frac{\sum_{i=1}^{n} x_i}{n} = \bar{x}.$$

This estimate maximizes the log-likelihood function, because the second derivative of $\ln L(\beta)$, evaluated at $\hat{\beta} = \bar{x}$, is negative. The maximum likelihood estimate is identical to the method-of-moments estimate discussed earlier.   ■

Example 3.6-2   Normal distribution with mean μ and variance $\sigma^2 > 0$:

Using the p.d.f. $f(x;\mu, \sigma^2) = \dfrac{1}{\sqrt{2\pi}\sigma}\exp\left[-\dfrac{1}{2\sigma^2}(x - \mu)^2\right]$, we obtain the likelihood function

$$L(\mu, \sigma^2) = (2\pi)^{-n/2}(\sigma^2)^{-n/2}\exp\left[-\frac{1}{2\sigma^2}\sum_{i=1}^{n}(x_i - \mu)^2\right]$$

and its logarithm

$$\ln L(\mu, \sigma^2) = (-n/2)\ln(2\pi) - (n/2)\ln(\sigma^2) - \frac{1}{2\sigma^2}\sum_{i=1}^{n}(x_i - \mu)^2.$$

The mean μ enters only into the last term. The log-likelihood function is maximized when $\sum_{i=1}^{n}(x_i - \mu)^2$ is minimized, implying the maximum likelihood estimate $\hat{\mu} = \bar{x}$. Setting the derivative of $\ln L(\bar{x}, \sigma^2)$ with respect to σ^2 equal to zero leads to

$$\frac{d \ln L(\bar{x}, \sigma^2)}{d\sigma^2} = -\frac{n}{2\sigma^2} + \frac{1}{2\sigma^4}\sum_{i=1}^{n}(x_i - \bar{x})^2 = 0,$$

and this implies that

$$n\sigma^2 = \sum_{i=1}^{n}(x_i - \bar{x})^2 \quad \text{and} \quad \sigma^2 = \frac{1}{n}\sum_{i=1}^{n}(x_i - \bar{x})^2 = v.$$

The maximum likelihood estimates $\hat{\mu} = \bar{x}$ and $\hat{\sigma}^2 = \dfrac{1}{n}\sum_{i=1}^{n}(x_i - \bar{x})^2 = v$ are again identical to the method-of-moments estimates discussed earlier. Note that, for large n, there is little difference between the sample variance $s^2 = \dfrac{1}{n-1}\sum_{i=1}^{n}(x_i - \bar{x})^2 = \dfrac{n}{n-1}v$ and the variance of the empirical distribution, v. Furthermore, as we will explain in a later section, there is an advantage to using s^2, so we will use s^2 instead of v. ∎

Remark: For given data, $\hat{\mu} = \bar{x}$ is the estimate of the unknown mean in each of the two distributions. Prior to sampling, the estimator $\hat{\mu} = \overline{X}$ is a random variable with a sampling distribution that is approximately normal (exactly normal in the case of a normal distribution). We can use the result obtained for the variance of a linear combination of independent random variables in Section 3.5-1 and show that its variance is given by $\mathrm{var}(\overline{X}) = \sigma^2/n$. Hence, $\bar{x} \pm 2\sigma/\sqrt{n}$ could be thought of as an approximate 95 percent confidence interval for the unknown means β and μ, respectively. However, because σ is unknown in each case, that interval could be approximated by $\bar{x} \pm 2s/\sqrt{n}$. This type of inference is studied in much more detail in the next chapter.

3.6-2 Checking for Normality

Suppose that we are given a sample of n observations $x_1, x_2, \ldots, x_n$. How can we check whether the normal distribution provides a good representation (model) of our data?

Probably the simplest way to check whether observations come from a normal distribution is to plot the histogram of the observations. Because the relative frequency histogram is an estimate of the probability density, we have to check whether it resembles the bell-shaped curve of the normal distribution in Figure 3.2-1. The histogram of the 90 measurements in Figure 1.3-1(b) looks more or less bell shaped. We could say that the distribution is roughly normal. However, it is quite difficult to assess normality from a histogram if there are only a few observations. For example, it is difficult to say whether the 10 observations in the dot diagram in Figure 1.3-1(a) come from a normal distribution.

In addition to the histogram, which, when scaled correctly, is an estimate of the p.d.f., we can calculate an estimate of the c.d.f. and assess whether it looks like the c.d.f. of a normal distribution. The empirical distribution was defined in Section 3.1; associated with it is the empirical c.d.f., which is given by

$$F_n(x) = \frac{\#(\{x_i : x_i \le x\})}{n},$$

where $\#(\{x_i : x_i \le x\})$ is the number of observations that are less than or equal to x. Note that $F_n(x)$ is the relative frequency of measurements that are smaller than or equal to x. As an illustration, we have used the first 10 observations in Table 1.3-1 to construct the empirical c.d.f. in Figure 3.6-1. The 10 observations, ordered from smallest to largest, are

31.3   32.3   42.2   42.3   44.5   47.5   49.2   50.0   53.9   60.9

The ordered observations are written as $x_{(i)}$, $i = 1, 2, \ldots, n$, and are called the *order statistics* of the sample. (See Section 1.3.) For example, $x_{(1)}$ is the smallest

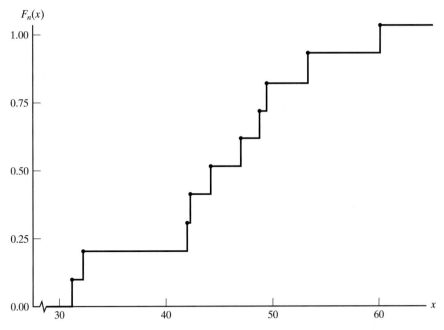

Figure 3.6-1 Empirical c.d.f. (step function) $F_n(x)$

observation and is the largest; the number in parentheses denotes the rank of the observation.

The empirical c.d.f. is defined for any real-valued x and is a nondecreasing step function of x, with jumps of magnitude $1/n$ at each of the n observations. Now, to determine whether the observations are normally distributed, we can check whether the empirical c.d.f. resembles the "S-shaped" curve in Figure 3.2-2(b) that we expect from a normal distribution.

It is usually difficult for us to recognize nonlinear patterns. The human eye is much better at recognizing *linear* tendencies. Thus, it would be more convenient if we could stretch the y-axis and create a special graph paper such that, under normality, the points $(x_{(i)}, P_i = (i - 0.5)/n)$ would lie on a straight line. Recall that $x_{(i)}$ is the sample quantile of order $100(i - 0.5)/n$. Such special graph paper is, in fact, available and is known as *normal probability paper*. It looks like ordinary graph paper, except that the distances on the y-axis, where we plot the values $P_i = (i - 0.5)/n, i = 1, 2, \ldots, n$, become larger as one moves away from the center-line at $P = 0.5$. For example, the distance between $P = 0.3$ and $P = 0.4$ (which, due to the symmetry of the normal distribution, is the same as the distance from 0.6 to 0.7) is larger than the distance between $P = 0.4$ and $P = 0.5$. The graph is known as a *normal probability plot*. Fortunately, computer programs provide such a graph, and the special graph paper is not needed.

There is another, and equivalent, way to think about this transformation of the y-axis. The ith-largest observation in our sample, $x_{(i)}$, is the sample quantile of order $P_i = (i - 0.5)/n$. One can also calculate the quantile of order P_i of the $N(\mu, \sigma^2)$ distribution. This distributional quantile is given by $q_i = \mu + \sigma z_i$, where z_i is the corresponding quantile of the $N(0, 1)$ distribution. The quantiles $z_1, z_2, \ldots, z_n$, which satisfy $P_i = P(Z \leq z_i) = \Phi(z_i)$, for $i = 1, \ldots, n$, are called the *standardized normal scores* (quantiles) associated with the n ordered observations $x_{(1)}, x_{(2)}, \ldots, x_{(n)}$. Now, if the data do in fact come from the $N(\mu, \sigma^2)$ distribution, then $q_i \cong x_{(i)}$ and the points on the scatter plot of q_i against $x_{(i)}$ should about fall on a 45-degree line through the origin. Such a plot is called a *quantile–quantile* (or *q–q*) plot, with the quantile of one (theoretical) distribution plotted against the corresponding quantile of another (empirical) distribution. That is, we plot two quantiles against each other, an observed quantile and one that is implied by a certain theoretical distribution.

The normal quantiles q_i depend on the parameters μ and σ^2. In practice, we can replace those parameters by their estimates $\bar{x}$ and s^2. Alternatively, we can simply plot the standardized normal scores z_i directly against the observed quantiles $x_{(i)}$. Because $q_i = \bar{x} + s z_i$ and because, under normality, $x_{(i)} \cong q_i$, we find that the points on the scatter plot of z_i against $x_{(i)}$ should lie approximately on a straight line with slope $1/s$ and that goes through the point $(\bar{x}, 0)$. Deviations from a linear pattern provide evidence that the underlying distribution is not normal. It is also easy to estimate from this plot, at least approximately, the parameters μ and σ of the normal distribution. The reciprocal of the slope gives us an estimate of the standard deviation; the intersection of a horizontal line at $z_i = 0$ with the approximate straight line gives us the estimate of μ.

We illustrate this plot with the first 10 observations from Table 1.3-1. We determine the ranks of the observations (if there were ties, we would assign the average rank) and calculate $P_i = (i - 0.5)/n$. The standardized normal scores satisfy $P_i = P(Z \leq z_i) = \Phi(z_i)$, for $i = 1, 2, \ldots, n$, and are determined from Table C.4. For example, z_1 satisfies $P_1 = (1 - 0.5)/10 = 0.05 = P(Z \leq z_1)$ and is given by $z_1 = -1.645$; the second normal score, z_2, satisfies $P_2 = (2 - 0.5)/10 = 0.15 = P(Z \leq z_2)$ and is given by $z_2 = -1.04$; and so on. These normal scores are listed in the last column of Table 3.6-1. A plot of the normal scores against the observations is given in Figure 3.6-2. This plot exhibits a fairly good linear relationship, so we have reason to believe that these data come from a normal distribution. We can also get a quick estimate of the standard deviation. Note that the slope in Figure 3.6-2 is about $(2)(1.6)/30 = 0.11$, and its reciprocal is 9.4. The latter is very close to the sample standard deviation $s = 9.1$.

Table 3.6-1 First 10 Observations from Table 1.3-1 and Their Normal Scores

Observation	Rank i	$P_i = \dfrac{i - 0.5}{n}$	Normal Score z_i
49.2	7	0.65	0.39
53.9	9	0.85	1.04
50.0	8	0.75	0.67
44.5	5	0.45	−0.13
42.2	3	0.25	−0.67
42.3	4	0.35	−0.39
32.3	2	0.15	−1.04
31.3	1	0.05	−1.64
60.9	10	0.95	1.64
47.5	6	0.55	0.13

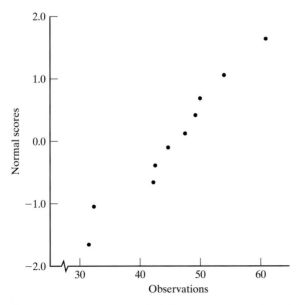

Figure 3.6-2 Normal probability plot. Scatter plot of the normal scores against the observations

Remark: Here, we have taken the ith-order statistic $x_{(i)}$ as the quantile of order $P_i = (i - 0.5)/n$. In the statistics literature, there are some slight differences in the definition of sample quantiles; this fact was mentioned in Section 1.3. Some authors define $x_{(i)}$ as the quantile of order $i/(n + 1)$, and some (the default specification in the computer software Minitab) define it as the quantile of order $(i - 0.3)/(n + 0.4)$. Thus, the standardized normal scores (quantiles) could vary somewhat with the particular definition one has adopted. Interestingly, each of the three definitions can be justified on different theoretical grounds. For our purposes, however, it suffices to say that, for reasonably large samples, these three definitions give similar results.

A normal probability plot for all 90 observations is given in Figure 3.6-3. We have used the Minitab command "Graph > Probability Plot (Normal)." All that is required for the user is to enter the data into a column and execute this command. We find that the data are in excellent agreement with a normal distribution.

Remark: Minitab plots the normal scores on the y-axis against the observations on the x-axis, exactly as we have illustrated in Figure 3.6-2. Normal scores are shown on the y-axis if one has requested "Score" in the "Scale > Y−Scale Type" window. Cumulative probabilities of the scores, $P_i = \Phi(z_i)$, are listed as the tick marks on the y-axis if "Probability" is specified in that window. In this case, the score $z_i = -1.96$ gets labeled as 0.025, the score 0 as 0.50, the score 1.645 as 0.95, and so on. Minitab adds to these graphs the line that is expected under normality. Deviations from the linear pattern are an indication of nonnormality.

Note that, through its "Tools > Options > Individual Graphs > Probability Plot" command, Minitab allows the user to change the default definition of $x_{(i)}$ from a

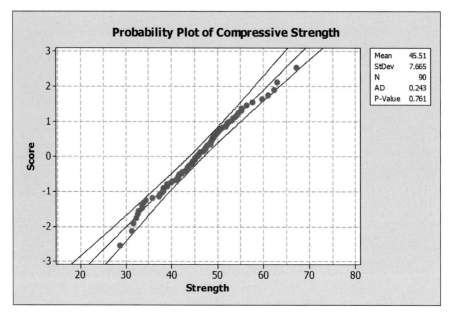

Figure 3.6-3 Minitab normal probability plot for the strength of concrete blocks (90 observations)

quantile of order $(i - 0.3)/(n + 0.4)$ (known as the median rank method) to a quantile of order $i/(n + 1)$ (the mean rank method) or $(i - 0.5)/n$ (the modified Kaplan–Meier method), the method we use here.

3.6-3 Checking Other Models through Quantile–Quantile Plots

So far, we have checked whether the data follow a normal distribution. But quantiles can be obtained for any other distribution, and q–q plots can check the appropriateness of other models.

To illustrate, suppose that by studying the histogram of a skewed set of data, we believe that a gamma distribution will provide a fairly good fit. Then, equating the first two theoretical moments (i.e., $\mu = \alpha\beta$ and $\sigma^2 = \alpha\beta^2$; see Table 3.3-1) to the corresponding sample moments, we obtain the method-of-moments estimates

$$\hat{\alpha} = \frac{\bar{x}^2}{s^2} \text{ and } \hat{\beta} = \frac{s^2}{\bar{x}}.$$

Example 3.6-3 Referring to Exercise 3.1-3, in which 40 losses due to wind-related catastrophies are given, we find that $\bar{x} = 9.225$ and $s_x = 10.237$. Because only losses above 1.5 (in millions) are considered, we subtract 1.5 from each observation, say, $y = x - 1.5$, to obtain $\bar{y} = 7.725$ and $s_y = 10.237$; note that the standard deviation is not affected if one subtracts a constant from each observation. The modified observations are as follows:

0.5	0.5	0.5	0.5	0.5	0.5	0.5	0.5	0.5	0.5
0.5	0.5	1.5	1.5	1.5	1.5	2.5	2.5	2.5	3.5
3.5	3.5	3.5	4.5	4.5	4.5	4.5	6.5	6.5	7.5
13.5	15.5	20.5	21.5	22.5	22.5	23.5	25.5	30.5	41.5

If we believe that a gamma distribution provides a reasonable model for the distribution of y values, we estimate the parameters by

$$\hat{\alpha} = \left(\frac{7.725}{10.237}\right)^2 = 0.569 \quad \text{and} \quad \hat{\beta} = \frac{(10.237)^2}{7.725} = 13.566.$$

It is interesting to compare the fitted gamma p.d.f.,

$$g(y) = \frac{y^{0.569-1} e^{-y/13.566}}{\Gamma(0.569)(13.566)^{0.569}}, \quad 0 \le y < \infty,$$

with the relative frequency histogram using the $k = 4$ classes $[0, 1), [1, 5), [5, 28)$, and $[28, 48)$; see Figure 3.6-4.

We can also use a quantile–quantile plot, similar to the normal probability plot discussed earlier, to check whether a set of observations comes from a specified theoretical distribution function $F(x) = P(X \le x)$. Let $x_{(1)} \le x_{(2)} \le \ldots \le x_{(n)}$ be the

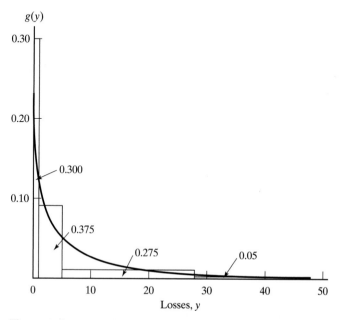

Figure 3.6-4 Relative frequency histogram of losses due to wind-related catastrophies and the p.d.f. of the corresponding gamma distribution

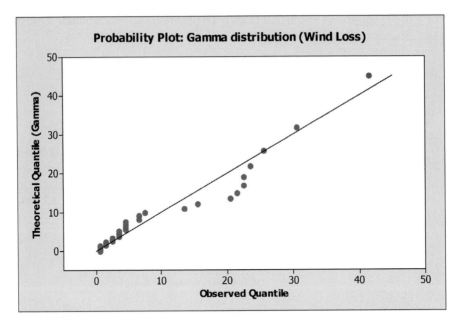

Figure 3.6-5 Probability plot: Gamma distribution (wind loss)

order statistics. For values of i ($1 \le i \le n$), determine $q_i = F^{-1}\left(\dfrac{i - 0.5}{n}\right)$, the quantile
of this distribution. Of course, here we assume that the c.d.f., $F(x)$, is of a form that
allows easy calculation of the theoretical quantiles. We then plot the implied theoretical quantiles q_i against the observed quantiles (order statistics) $x_{(i)}$, giving us a q–q
graph. If these points lie fairly close to a line with slope 1 and passing through the
origin, we say that $F(x)$ provides a reasonable model for the underlying distribution.

The quantile–quantile graph, plotted against a gamma distribution, is shown in
Figure 3.6-5. We have added to the graph a line with slope 1 and passing through the
origin. For the calculation of the quantiles, we have used the Minitab command
"Calc > Probability Distribution > Gamma," checking the inverse cumulative
probability tab and entering the method-of-moments estimates for α and β. The
result shows that the gamma distribution provides a fairly good fit to the data.   ∎

Example 3.6-4   Suppose someone thought that the data in Example 3.6-3 could be fit with an exponential distribution (once 1.5 had been subtracted from each of the $n = 40$ observations). From Example 3.6-3, we know that the mean of the 40 values is $\bar{y} = 7.725$.
Equating this mean to the mean, β, of the exponential distribution (see Table 3.3-1),
we find that the fitted exponential density is

$$g(y) = \frac{1}{7.725} e^{-y/7.725}, \quad 0 \le y < \infty.$$

The corresponding distribution function is

$$G(y) = \int_0^y g(w)\, dw = 1 - e^{-y/7.725}, \quad 0 \le y < \infty,$$

Table 3.6-2 Observed and Theoretical Quantiles

	Quantile	
p	Observed	Theoretical
0.0625	0.5	0.50
0.1375	0.5	1.14
0.2125	0.5	1.85
0.2875	0.5	2.62
0.3625	1.5	3.48
0.4375	2.5	4.44
0.5125	3.5	5.55
0.5875	4.5	6.84
0.6625	4.5	8.39
0.7375	7.5	10.33
0.8125	20.5	12.93
0.8875	22.5	16.88
0.9625	30.5	25.36

and the quantile of order p, q_p, found by solving

$$p = 1 - e^{-q_p/7.725},$$

is

$$q_p = -7.725 \ln(1 - p).$$

Let us consider every third order statistic. For example, the third order statistic, 0.5, is the $100(3 - 0.5)/40 = 6.25$th percentile; the sixth order statistic, 0.5, is the $100(6 - 0.5)/40 = 13.75$th percentile; and so on until the 39th order statistic, 30.5, which is the $100(39 - 0.5)/40 = 96.25$th percentile of the data. These observed quantiles, together with the implied theoretical quantiles $q_p = -7.725 \ln(1 - p)$, are listed in Table 3.6-2.

Figure 3.6-6 shows the scatter plot of the theoretical quantiles implied by the exponential distribution against the empirical quantiles. The points on this q–q plot do not follow a line with slope 1 and passing through the origin. Although it is not an impossible fit, a better one would have been achieved if the smaller quantiles of the theoretical distribution were yet smaller and the larger quantiles yet larger. That is exactly what was achieved in Example 3.6-3 by fitting a gamma distribution.   ■

Remarks:

1. Several observations in the example have the same magnitude and are tied in rank. For example, 12 observations have the value 0.5. We have assigned ranks 1 through 12 to these observations arbitrarily. Alternatively, we could have assigned all of them the average rank (6.5). This would replace the 12 different implied quantiles at the empirical quantile 0.5 with a single quantile–quantile pair.

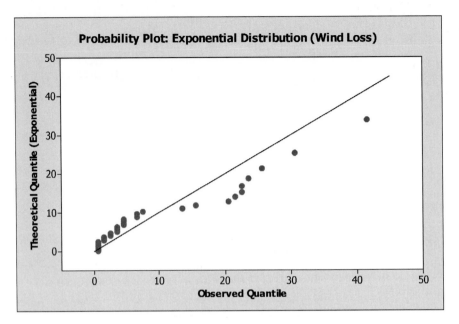

Figure 3.6-6 Probability plot: Exponential distribution (wind loss)

2. The implied quantiles can be obtained from statistical software. In Minitab, we use "Calc > Probability Distributions > Exponential," checking the tab for inverse cumulative probability and entering the estimate for the mean.

3. Instead of creating the q–q plot from first principles (as we did in Figure 3.6-6), we could have used the probability plotting routines of the available statistics software package. The Minitab routine "Graph > Probability Plots > Exponential" will do this. Note however, that both the y-axis and the x-axis are then logarithmic and that the plot will not look exactly like the one we created. Apart from these plotting issues, it is the assessment of linearity that is of central importance in interpreting these graphs. Deviations from linearity are indications that the adopted model does not fit.

Exercises 3.6

*3.6-1 Consider the following 25 observations:

10.1	37.5	8.4	99.0	12.8
66.0	31.0	85.3	63.6	73.8
98.5	11.8	83.5	88.7	99.6
65.5	80.1	74.4	69.9	9.9
91.5	80.3	44.1	12.6	63.6

(a) Construct a relative frequency histogram and sketch the empirical c.d.f.

(b) Determine the normal scores (implied theoretical quantiles) of the observations, and plot the normal scores against the observations. From this plot, and from the graphs in part (a), comment on whether these observations are from a normal distribution. If you find that they are not, what other models (distributions) are suggested by the data?

*3.6-2 In this section, we suggested a quantile plot to check for normality. Similar quantile plots can be made to check whether the observations come from certain other distributions. Consider the exponential distribution with the density $g(y) = (1/\beta)\exp(-y/\beta), 0 \le y < \infty$.

*(a) Show that $y_p = -\beta \ln(1 - p)$ is the quantile of order p.

(b) A plot of $-\ln(1 - P_i)$ against the order statistics $y_{(i)}$, where i is the rank of the observation and $P_i = (i - 0.5)/n$, is suggested as a check on whether the data are from an exponential distribution. Explain why this plot is appropriate, and discuss how one could estimate the parameter β from such a plot.

(c) It is hypothesized that the times to failure of certain welds that are exposed to continuous vibrations of a specified frequency and amplitude follow an exponential distribution. The times to failure (in units of 1,000 cycles) of 10 welded pieces are as follows:

580	300	990	240	1,900
480	1,040	1,500	50	280

Check whether this hypothesis is valid.

*3.6-3 Consider the $n = 90$ data in Table 1.3-1. Check whether these observations come from a normal distribution. Estimate μ and σ from the normal probability plot.

3.6-4 Check whether the melting points in Exercise 1.3-2 could come from a normal distribution. Obtain rough estimates of μ and σ from the normal probability plot.

3.6-5 Consider the carbon monoxide emissions in Exercise 1.3-9. Construct and interpret a normal probability plot. Because the data are already grouped, you may want to use the upper class limits as the observed order statistics in your quantile–quantile plot.

3.6-6 Assume that Y follows a lognormal distribution with parameters μ and σ. (See Section 3.3-6.)

(a) Show that the quantile of order p is given by $y_p = \exp(\mu + \sigma z_p)$, where z_p is the quantile of the $N(0, 1)$ distribution.

(b) Consider a scatter plot of the logarithms of the order statistics, $\ln y_{(i)}$, against z_i, where z_i is the normal score satisfying $P_i = (i - 0.5)/n = \Phi(z_i)$. Why is this a useful plot for checking whether the observations are from a lognormal distribution?

(c) Apply the plot from part (b) to the carbon monoxide emissions that were studied in Exercise 3.6-5. Which of the two models, normal or lognormal, is more plausible? Use your computer software to generate the graphs.

3.6-7 In Exercise 3.4-2 of Section 3.4, you generated observations from a continuous uniform distribution on the interval from 0 to 1. Using $n = 100$ observations, construct a quantile–quantile (probability) plot and check the appropriateness of the uniform distribution.

Note: Order the observations from smallest to the largest. The observation with rank i is the empirical quantile of order $(i - 0.5)/n$, where $n = 100$ is the sample size. Obtain the quantile that is implied by this target distribution; for the uniform distribution, this quantile is $(i - 0.5)/n$. Plot the empirical quantiles against the implied theoretical quantiles, and check whether the points scatter around a line with slope 1 and passing through the origin.

Given that you generated the data from the uniform distribution, the points should scatter around the aforesaid line. Deviations from the expected pattern are due to randomness. Repeat the probability plot for several samples of size 100, to gain an appreciation of the randomness and the differences that one can expect.

You can also use your computer software and obtain the probability plot directly. This will simplify your task.

3.6-8 In Exercise 3.4-3 of Section 3.4, you generated observations from an exponential distribution with parameter (mean) $\beta = 1$. Carry out a quantile–quantile (probability) plot on $n = 100$ observations, and check whether the data can be described by an exponential distribution with $\beta = 1$. Furthermore, obtain a quantile–quantile plot to check for a uniform distribution. Discuss your findings. Construct your graphs in two different ways:

(a) Construct them from scratch, using the inverse cumulative probability of the exponential and uniform distributions. The ith-largest (generated) observation is the empirical quantile of order $(i - 0.5)/n$. The corresponding theoretical quantile is $-\ln(1 - (i - 0.5)/n)$ for the exponential distribution and $(i - 0.5)/n$ for the uniform distribution.

(b) Confirm your results with probability plots that you obtain from your statistical computer software (e.g., Minitab's "Plot > Probability Plot" function).

3.6-9 Use your computer software to generate 150 observations from a lognormal distribution with parameters $\mu = 1$ and $\sigma = 1$.

(a) Construct a quantile–quantile (probability) plot for $n = 150$ observations, and check whether the observations are from a normal distribution.

Note: The implied normal quantile of order $(i - 0.5)/n$ from the standard normal distribution is $F^{-1}((i - 0.5)/n)$ and can be obtained from the inverse cumulative distribution function of the normal distribution. Fortunately, statistical computer software can construct probability plots directly and thus avoid the explicit evaluation of the implied theoretical quantiles. The Minitab function under "Graph > Probability Plot" will construct probability plots for a variety of target distributions, including the normal distribution. Try this command with a normal target distribution, and check whether the scatter plot exhibits linearity.

(b) Check whether the generated observations are from an exponential distribution.

Note: The c.d.f. of the exponential distribution with parameter β is $F(x) = 1 - \exp(-x/\beta)$, and its quantile of order p is $y_p = -\beta \log(1 - p)$. A plot of $-\ln(1 - P_i)$ against the order statistics $y_{(i)}$, where i is the rank of the observation and $P_i = (i - 0.5)/n$, should be linear with slope $1/\beta$. Or, in an even simpler procedure, avoid the evaluation of the implied theoretical quantiles by using "Graph > Probability Plot" and specifying the exponential as the target distribution. Minitab will estimate the parameter β from the data, obtain the implied quantiles of the (estimated) exponential distribution through its inverse cumulative probability function, and construct the plot. All a user has to do is assess the linearity of the scatter plot.

(c) Check whether the observations are from a Weibull distribution.

Hint: Use the Minitab command "Graph > Probability Plot" and specify the Weibull as the target distribution. The Weibull distribution, discussed in Section 3.3, includes two parameters (shape and scale). Minitab will estimate these parameters from the data, obtain the implied quantiles of the (estimated) Weibull distribution through its inverse cumulative probability function, and construct the plot. This approach is easier than constructing the graph from first principles.

3.6-10 Experiment with your statistical software. Generate sets of random variables (from distributions such as the normal, gamma, Weibull, etc.). In Minitab, you can use "Calc > Random Data."

Obtain histograms of the data you generated, and construct quantile–quantile plots against various target distributions. In Minitab, you use "Graph > Probability Plot" and specify the distribution. Note that, for some distributions (e.g., exponential and gamma distributions), Minitab will plot points on a log–log scale. Minitab adds the expected line to these graphs, helping the user assess deviations from linearity.

Comment on your findings. How easy is it to decide whether a given data set has been generated from a certain distribution? Experiment with several sample sizes (50, 100, ..., 500).

*3.6-11 (a) The $n = 105$ operating times (in hours) for a single surgeon's strabismus cases (involving a recession or resection procedure of one horizontal muscle) are as follows (the data are from F. Dexter and J. Ledolter, "Bayesian Prediction Bounds and Comparisons of Operating Room Times Even for Procedures with Few or No Historic Data," *Anesthesiology,* 103(2005), 1259–1267.

0.70	0.90	0.95	0.95	1.02	1.07	1.07	1.08	1.12	1.15	1.15
1.17	1.17	1.18	1.18	1.22	1.22	1.23	1.23	1.25	1.25	1.25
1.27	1.28	1.28	1.30	1.30	1.32	1.32	1.32	1.32	1.33	1.35
1.37	1.37	1.37	1.38	1.38	1.38	1.40	1.40	1.40	1.40	1.42
1.45	1.45	1.47	1.48	1.48	1.48	1.48	1.48	1.50	1.52	1.52
1.52	1.53	1.53	1.53	1.55	1.55	1.57	1.57	1.58	1.60	1.60
1.60	1.60	1.62	1.62	1.63	1.63	1.63	1.67	1.67	1.68	1.70
1.70	1.72	1.72	1.72	1.73	1.75	1.78	1.80	1.82	1.83	1.87
1.88	1.90	1.90	1.90	1.92	1.92	1.93	1.98	1.98	2.03	2.03
2.03	2.07	2.10	2.12	2.20	2.30					

Construct a histogram of the observations, and display quantile–quantile (probability) plots to check for a normal and a lognormal distribution. Interpret the results. Which of the two distributions provides a better fit?

*(b) The operating turnover time represents the time it takes to prepare an operating room for the next patient. Consider the following average turnover times (in minutes) of 31 U.S. hospitals (the data are from F. Dexter, R. Epstein, E. Marcon, and J. Ledolter, "Estimating the Incidence of Prolonged Turnover Times and Delays by Time of Day, " *Anesthesiology,* 102 (2005), 1242–1248):

22.6	24.0	24.1	25.0	25.0	25.6	27.9	29.0	29.0	30.0	31.0
31.0	31.0	31.0	33.0	33.1	33.6	34.0	35.0	35.4	36.0	36.0
37.0	37.0	38.0	42.3	47.0	49.0	49.0	55.0	58.0		

Construct a dot plot of the information, and display quantile–quantile (probability) plots to check for a normal and a lognormal distribution. Interpret the results. Which of the two distributions provides a better fit?

3.6-12 Write down the log-likelihood function for estimating the parameters α and β of the gamma distribution. Discuss how this would help you estimate the unknown parameters. Note that we are not asking you to actually determine the maximum likelihood estimates of *both* parameters, because it is not possible to write down explicit expressions; iterative numerical optimization approaches must be used. However, if you were given the estimate of α, how would you obtain the maximum likelihood estimate of β?

3.7 Introduction to Reliability

The *reliability* of a product or a system is related to the probability that it will function properly over a certain period. In Section 3.3, certain basic concepts associated with reliability were introduced. In particular, the *failure rate function* of a system was defined by

$$\lambda(x) = \frac{f(x)}{1 - F(x)},$$

where $f(x)$ is the p.d.f. and $F(x) = P(X \le x)$ is the c.d.f. of the random variable X, the time to failure. We learned that $\lambda(x)\Delta x$ is approximately the conditional probability that an item fails in the interval $(x, x + \Delta x)$, given that its lifetime (i.e., the time to failure) X exceeds x.

Frequently, $R(x) = 1 - F(x) = P(X > x)$ is called the *reliability function*, because it expresses the probability that an item will function longer than x. The failure rate function can then be written as $\lambda(x) = f(x)/R(x)$.

We also learned in Section 3.3 how to obtain the p.d.f. $f(x)$ from a given failure rate function $\lambda(x)$:

$$f(x) = \lambda(x)\exp\left[-\int_0^x \lambda(t)dt\right], \quad 0 \le x < \infty.$$

For the failure rate function $\lambda(x) = \alpha x^{\alpha-1}/\beta^\alpha$, with $\alpha > 0$ and $\beta > 0$, we obtained the *Weibull* p.d.f.:

$$f(x) = \frac{\alpha x^{\alpha-1}}{\beta^\alpha} \exp\left[-\left(\frac{x}{\beta}\right)^\alpha\right], \quad 0 \le x < \infty.$$

This is a commonly used distribution model in reliability. Of course, if $\alpha = 1$, then $f(x)$ is the exponential p.d.f. with mean $\beta = 1/\lambda$, namely,

$$f(x) = \frac{1}{\beta}e^{-x/\beta} = \lambda e^{-\lambda x}, \quad 0 \le x < \infty.$$

The failure rate function for this distribution is constant; that is, $\lambda(x) = \lambda$. The parameter $\lambda = 1/\beta$ is known as the failure rate of the exponential distribution.

Let us now consider a system of components that are arranged in *series*, such as that depicted in Figure 3.7-1. For this system to function, all components,

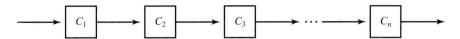

Figure 3.7-1 A series system

$C_1, C_2, \ldots, C_n$, must function properly. Thus, if the random variable X_i represents the lifetime of component $C_i, i = 1, 2, \ldots, n$, then the lifetime, say, Y, of the system is the minimum of $X_1, X_2, \ldots, X_n$. That is,

$$Y = \min(X_1, X_2, \ldots, X_n).$$

The lifetimes $X_1, X_2, \ldots, X_n$ of the components are often mutually independent, in which case the reliability function of the series system is given by

$$R(y) = P(Y > y) = P(X_1 > y)P(X_2 > y)\ldots P(X_n > y)$$
$$= R_1(y)R_2(y)\ldots R_n(y),$$

where $R_i(y) = P(X_i > y), i = 1, 2, \ldots, n$. That is, the reliability function of the series system is equal to the product of the individual reliability functions. Of course, we have to assume that the lifetimes of the individual components are independent.

If each individual lifetime X_i has an exponential distribution with failure rate $\lambda_i = 1/\beta_i, i = 1, 2, \ldots, n$, then

$$R_i(y) = \int_y^\infty \lambda_i e^{-\lambda_i x}\, dx = e^{-\lambda_i y}$$

and

$$R(y) = e^{-\lambda_1 y} e^{-\lambda_2 y} \cdots e^{-\lambda_n y} = e^{-(\Sigma \lambda_i)y}.$$

The p.d.f. of Y is $g(y) = G'(y) = -R'(y) = (\Sigma \lambda_i)e^{-(\Sigma \lambda_i)y}$. Hence, the failure rate of the system, $g(y)/R(y) = \Sigma \lambda_i = \lambda_Y$, is the sum of the individual failure rates, and the lifetime Y of the system has an exponential distribution with mean $1/\lambda_Y$.

Example 3.7-1 Suppose that an electronic circuit has four transistors arranged in series. Assume that the lifetimes of these transistors are independent and that they have exponential distributions with respective failure rates $\lambda_1 = 0.00006$, $\lambda_2 = 0.00003$, $\lambda_3 = 0.00012$, and $\lambda_4 = 0.000018$, expressed in terms of failures per minute. A failure rate of $\lambda_1 = 0.00006$, for example, implies that the average time to failure is $1/0.00006 = 16{,}667$ minutes, or about 278 hours. For our system with these four transistors arranged in series, we can expect $\lambda_Y = \sum_{i=1}^{4} \lambda_i = 0.000228$ failure per minute. Moreover, the lifetime Y of the system has an exponential distribution with mean $1/\lambda_Y = 1/0.000228 = 4{,}386$ minutes, or about 73 hours. ∎

If the proper functioning of a system is extremely important (e.g., failure could mean loss of life or great economic loss), a system is often constructed by placing components in *parallel,* as depicted in Figure 3.7-2. Here, all components are active simultaneously. For this system to function properly, *at least* one of the components

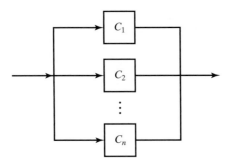

Figure 3.7-2 A parallel system

must function. Thus, the lifetime, say, Z, of such a system is the maximum of the lifetimes, $X_1, X_2, \ldots, X_n$, of the individual components. That is,

$$Z = \max(X_1, X_2, \ldots, X_n).$$

If $X_1, X_2, \ldots, X_n$ are mutually independent, then the reliability function of the parallel system is

$$\begin{aligned} R(z) = P(Z > z) &= 1 - P(Z \le z) \\ &= 1 - P(X_1 < z)P(X_2 < z)\ldots P(X_n < z) \\ &= 1 - [1 - R_1(z)][1 - R_2(z)]\ldots[1 - R_n(z)], \end{aligned}$$

where $R_i(z) = P(X_i > z) = 1 - P(X_i \le z)$ is the reliability function of the ith component, $i = 1, 2, \ldots, n$.

Example 3.7-2   Suppose that the four transistors of Example 3.7-1 are arranged in parallel. Then the reliability function of the system is

$$R(z) = 1 - [1 - e^{-(0.00006)z}][1 - e^{-(0.00003)z}][1 - e^{-(0.00012)z}][1 - e^{-(0.000018)z}].$$

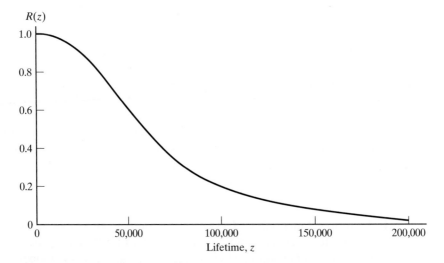

Figure 3.7-3 Plot of $R(z) = P(Z > z)$ in Example 3.7-2

This expression cannot be simplified much more, and we would need to investigate $R(z)$ further by substituting various values of z. The graph of $R(z) = P(Z > z)$, plotted in Figure 3.7-3, shows that, for this particular system, there is a 20 percent chance that the lifetime exceeds 100,000 minutes and about a 5 percent probability that it exceeds 170,000 minutes. The median lifetime is somewhere between 50,000 and 60,000 minutes. ∎

In the formulation of a parallel system, all components are active simultaneously; thus, the lifetime of this system equals the maximum lifetime of the individual components. Clearly, it would be better if we could devise a system in which only one component were active at a time and another component started only when the current component failed. In this *standby* system, the procedure continues until all n components have failed, one at a time. Clearly, the lifetime, say, W, of the system is equal to the sum of the individual lifetimes. That is, $W = X_1 + X_2 + \cdots + X_n$. If the individual X_i, for $i = 1, 2, \ldots, n$, are mutually independent with means μ_i and variances σ_i^2, then

$$\mu_W = E(W) = \sum_{i=1}^{n} \mu_i$$

and

$$\sigma_W^2 = E[(W - \mu_W)^2] = \sum_{i=1}^{n} \sigma_i^2.$$

However, to determine the reliability function of this system, we must find the distribution of W and then evaluate the probability

$$R(w) = P(W > w).$$

For example, suppose that X_i, for $i = 1, 2, \ldots, n$, has a gamma distribution with parameters α_i and common β; that is,

$$f(x) = \frac{1}{\Gamma(\alpha_i)\beta^{\alpha_i}} x^{\alpha_i - 1} e^{-x/\beta}, \quad 0 \le 0 < \infty.$$

(See Section 3.3.) Then the sum W also has a gamma distribution, with parameters $\sum_{i=1}^{n} \alpha_i$ and β. In particular, if $\alpha_1 = \alpha_2 = \ldots = \alpha_n = 1$, so that each X_i has the same exponential distribution with $\beta = 1/\lambda$, then W has a gamma distribution with parameters n and $\beta = 1/\lambda$. Thus, in the latter case, we have

$$R(w) = \int_{w}^{\infty} \frac{1}{(n-1)!\beta^n} y^{n-1} e^{-y/\beta} dy,$$

and it follows that $\mu_W = n\beta = n/\lambda$ and $\sigma_W^2 = n\beta^2 = n/\lambda^2$.

Exercises 3.7

3.7-1   Let X have the p.d.f. $f(x) = 1/x^2$, $1 \le x < \infty$. Determine and sketch the reliability function $R(x) = P(X > x)$ and the failure rate function $\lambda(x) = f(x)/R(x)$.

*3.7-2   Let X have the p.d.f. $f(x) = (x/\beta^2)e^{-x/\beta}$, $0 \le x < \infty$. Determine and sketch the reliability function $R(x)$ and the failure rate function $\lambda(x)$.

*3.7-3   Suppose that the respective lifetimes of three components arranged in series are mutually independent with exponential distributions having failure rates $\lambda_1 = 0.003$, $\lambda_2 = 0.010$, and $\lambda_3 = 0.008$.

    *(a) Determine the reliability function $R(y)$ and the failure rate function of the series system.

    *(b) What is μ_Y, the mean time to failure of this system?

    *(c) Evaluate $R(\mu_Y)$.

3.7-4   Repeat Exercise 3.7-3(a), but assume that the components are now placed in parallel.

3.7-5   Three identical components are connected in parallel. Their lifetimes are mutually independent, and each lifetime has an exponential distribution with $\lambda = 0.02$.

    (a) If at least 1 component must operate for the system to function properly, evaluate the reliability function at time 15.

    (b) Suppose the system is designed so that at least 2 components must operate for the system to function properly. Evaluate the reliability function of this system at time 15. Compare this answer with that of part (a).

    (c) If the 3 components can be placed in a standby system, evaluate the reliability function of this system at time 15. Does this make sense in light of the answers to parts (a) and (b)?

*3.7-6   Consider a parallel system with $k > 1$ components. Assume that the lifetime of each of these components follows an exponential distribution with failure rate λ, and suppose that the k lifetimes are mutually independent. Obtain an expression for the $(100p)$th percentile of the lifetime distribution for the parallel system.

*3.7-7   An army unit consisting of six soldiers is outfitted with inferior weapons. The mean time to failure of a weapon is 5 minutes, and it is assumed that the distribution of the failure time is exponential. The first soldier fires his weapon until the weapon fails; then the next soldier fires and continues firing until his weapon fails, and so on. What is the probability of providing firepower for 25 minutes? 35 minutes?

    Hint: Use the fact that the sum of independent exponential random variables has a gamma distribution. Use your computer software to calculate the probability.

Chapter 3 Additional Remarks

In this chapter, we studied several continuous distributions and introduced the very important *normal distribution*. Why does this distribution come up so often? It has to do with the fact that a measurement is usually affected by many random disturbances. For example, measurements on the lengths of eggs depend on the particular egg (not all eggs are the same), the person who measures the egg, the way the egg is being measured (how it is placed in a holder, which caliper is used, how the measurement tool is applied), and so on. The measurement "length of an egg" is subject to many random influences that contribute to the final result, without any of them dominating. It turns out that the measurement has an approximate normal distribution. The *central limit theorem,* discussed in the next chapter, will explain this phenomenon theoretically.

Course grades are often normally distributed, usually by design, as instructors "grade on a (normal) curve." We believe that the normal distribution is used way too often for this purpose. *All* students should be able to earn "A" grades, and grades should not be restricted to a "normal curve."

The normal distribution is symmetric. The other important continuous distributions in this chapter are skewed. Engineers find the Weibull distribution highly useful, as it implies an increasing failure rate $\lambda = \alpha x^{\alpha-1}/\beta^{\alpha}$ (for $\alpha > 1$), a force of mortality that is being followed by many manufactured parts. Of course, for $\alpha = 1$, the failure rate is constant. In this case, an old part would be as good as a new one. Or, applied to the human body, an old man would have the same chance of living another 50 years as a young man; we could call this the "mathematical fountain of eternal youth." Unfortunately, the human body has an exponential failure rate that increases about 10 percent each year, and this implies that our failure rate doubles about every 7.3 years. Thank goodness, it starts low.

Why do we study distributions? One reason is that we want to know answers to questions such as "What is the typical value," "what percent of the observations are smaller than *x*, or larger than *y*, or in between *x* and *y*," "what is the 95th percentile," and so on. Knowing answers to these and similar questions helps us make decisions.

Engineers must understand how the variability in certain input variables of an engineering system propagates to output signals. In certain special cases, like the news vendor problem in Project 1, it is possible to solve such problems analytically. But in many realistic situations in which the engineering system is described by a complicated set of equations containing many engineering parameters, this is no longer feasible, and we must resort to *simulation* methods. That is, we simulate the input variables to the engineering system by drawing realizations from certain distributions, pass the simulated values through the model, and record the values of the output variables of interest. Doing this over and over again results in a distribution for those variables. The parameters in such models represent the characteristics of the system, and they can be altered by design changes. By changing the values of the parameters, we can assess the sensitivity of the output to those changes. For example, consider the design of a dam that needs to withstand floods. Using simulated data on the volume of water in the river, we can assess how various dam designs affect the probabilities of catastrophic failure.

"Monte Carlo" is often attached to the term "simulation," and one speaks of *Monte Carlo simulations.* The term was first used during the development of the atomic bomb as a code name for computer simulations of nuclear fission. Researchers coined this term because of the similarity of the simulations to random sampling in games of chance, such as roulette in the casinos of Monte Carlo.

Projects

*Project 1: Newsvendor problem in operations research

Suppose you are selling a perishable product (say, a newspaper or a service in which time is of essence) and you have to order a certain number Q (the "stock") to cover the uncertain demand for your product. The number of units that are demanded, X, follows a known distribution with density $f(x)$, distribution function $F(x)$, and mean μ.

Assume that your unit cost of the product is c (say, $c = \$3$) and that you can sell the product for $c + u$ (say, $\$5$), realizing the profit $u > 0$ (here, $u = \$2$). However, if you cannot sell the product, its salvage price falls to $s < c$ (say, $s = \$2.50$). The profit u is called the "underage" cost per unit, because that is what you lose when your order is "under" (below) the demand. The difference between the cost and the salvage price ($e = c - s > 0$) is called the "overage" cost per unit and represents your unit loss when ordering more than your demand.

(a) Determine the optimal order quantity that maximizes the expected profit. Show that it is given by $Q_{opt} = F^{-1}[u/(u + e)]$.

*(b) Assume that the profit is $u = 2$, the difference between the cost and the salvage price is $e = 0.5$, and the demand has a normal distribution with a mean of 50 units and a standard deviation of 10 units. Determine the optimal order quantity.

*(c) Assume that the demand follows a uniform distribution on the interval from 0 to 100. Determine the optimal order quantity and calculate the expected profit.

Hints for (a): Assume that you order Q units. It is useful to write out in detail the profit for various realizations of the uncertain demand:

$$
\begin{aligned}
x = 0 \quad &: \quad -eQ \qquad\qquad = \quad -eQ \\
x = 1 \quad &: u1 - e(Q - 1) = u1 - e(Q - 1) \\
x = 2 \quad &: u2 - e(Q - 2) = u2 - e(Q - 2) \\
&\quad\vdots \qquad\qquad\qquad \vdots \\
x = Q - 2 &: u(Q - 2) - e2 = u(Q - 2) - e2 \\
x = Q - 1 &: u(Q - 1) - e\ = u(Q - 1) - e \\
x = Q \quad &: uQ \qquad\qquad = uQ \\
x = Q + 1 &: uQ \qquad\qquad = u(Q + 1) - u \\
x = Q + 2 &: uQ \qquad\qquad = u(Q + 2) - u2 \\
x = Q + 3 &: uQ \qquad\qquad = u(Q + 3) - u3 \\
&\quad\vdots \qquad\qquad\qquad \vdots
\end{aligned}
$$

The expected profit, for given order quantity Q, is

$$
\mu(Q) = u\mu - u\int_Q^\infty (x - Q)f(x)dx - e\int_0^Q (Q - x)f(x)dx
$$

$$
= u\mu - u\int_Q^\infty xf(x)dx + uQ[1 - F(Q)] - eQF(Q) + e\int_0^Q xf(x)dx.
$$

Take the derivative with respect to Q. Then use the fundamental theorem of calculus, which implies that $\dfrac{d}{dq}\int_0^q g(x)dx = g(q)$ and $\dfrac{d}{dp}\int_p^\infty g(x)dx = -g(p)$, to show that

$$\frac{d}{dQ}\mu(Q) = u - uF(Q) - eF(Q).$$

Set the derivative equal to zero and show the required result.

Hints for (c): You need to substitute the optimal value of Q (which you found in (a)) into the equation $\mu(Q) = u\mu - u\int_Q^\infty (x - Q)f(x)dx - e\int_0^Q (Q - x)f(x)dx$. The integrals are easy to evaluate for the uniform distribution.

*Project 2: Information theory and entropy

Entropy describes the information that is contained in a signal or an event. C. E. Shannon introduced the idea of information entropy in his 1948 paper, "A Mathematical Theory of Communication," *Bell System Technical Journal,* 27, pp. 379–423 and 623–656, July and October. Information entropy relates to the amount of uncertainty of an event that is inherent in its probability distribution. As an example, consider a box containing many colored balls. If the balls are all of different colors and no color dominates, then the uncertainty about the color of a randomly drawn ball is maximal. By contrast, if the box contains more blue balls than any other color, then there is less uncertainty about the result, as there is a higher chance of drawing a blue ball. Finally, if we know that all of the balls are of one known color, then there is no uncertainty at all about the color of a randomly selected ball. Telling someone the color of a ball that is drawn provides more information in the first case than it does in the second case, because there is more uncertainty about what might happen in the first case. As a result, the entropy of the "signal" (the type of ball drawn, as calculated from the probability distribution) is higher in the first case.

Shannon defined entropy as a measure of the average information content that is associated with a random outcome. In terms of a discrete random variable X with possible outcomes (states) $x_1, x_2, \ldots, x_n$ and probabilities $p(x_1) = P(X = x_1), \ldots, p(x_n) = P(X = x_n)$, entropy is defined as

$$H(X) = \sum_{i=1}^n p(x_i)\log\left(\frac{1}{p(x_i)}\right) = -\sum_{i=1}^n p(x_i)\log p(x_i).$$

Entropy makes use of the product of the probability of the outcome x_i and the logarithm of the inverse of the probability $p(x_i)$. The inverse probability expresses the event's *surprisal.* You are bound to be surprised by the occurrence of an event with low probability (i.e., large inverse probability). If the information is measured in units of *bits,* then the logarithm in this equation is taken to the base 2.

(a) Write the entropy as $H(X) = H_n(p_1, p_2, \ldots, p_n)$, where $p_i = p(x_i)$. Prove the following statements:

    i. For any n, $H(X) = H_n(p_1, p_2, \ldots, p_n)$ is a continuous and symmetric function of the variables $p_1, p_2, \ldots, p_n$.

    ii. An event of probability 0 does not contribute to entropy; that is, $H_{n+1}(p_1, p_2, \ldots, p_n, 0) = H_n(p_1, p_2, \ldots, p_n)$.

    iii. Entropy is maximized when the probability distribution is uniform; that is, for all n, $H_n(p_1, p_2, \ldots, p_n) \le H_n\left(\frac{1}{n}, \frac{1}{n}, \ldots, \frac{1}{n}\right)$.

Note: Prove this statement for the *binary entropy function* with only two outcomes and probabilities p and $1 - p$, $H_2(p) = -p\log p - (1 - p)\log(1 - p)$.

The proof of the general situation assumes properties of probability (e.g., Jensen's inequality) that we have not studied in this text.

(b) Discuss important applications of entropy. Use a Web search.

Note: Information theory is important for studying the communication over a channel (such as a telephone line or an Ethernet wire). As anyone who's ever used a telephone knows, such channels often fail to produce an exact reconstruction of the signal. The quality of the communication is frequently degraded by noise, periods of silence, and various forms of corruption of the signal. It is important to study how much of the information can be communicated over a noisy (or otherwise imperfect) channel.

Information-theoretic concepts are being used in making and breaking cryptographic systems. Intelligence agencies use information theory to keep classified information secret and to discover as much information as possible about an adversary.

*(c) Entropy for a continuous random variable X with probability density $p(x)$ is defined as

$$H(X) = -\int_{-\infty}^{\infty} p(x)\log p(x)dx.$$

Apply this formula to the following distributions:

i.   Uniform distribution on (a, b): $H(X) = \log(b - a)$.

ii. Triangular distribution on $(0, 1)$, with $p(x) = 4x$ for $0 \le x < 0.5$ and $p(x) = 4(1 - x)$ for $0.5 \le x < 1$: $H(X) = (1/2) + \log(1/2)$.

iii. Exponential distribution with $p(x) = (1/\beta)\exp(-x/\beta)$ for $0 \le x < \infty$: $H(X) = 1 + \log(\beta)$.

*(d) A measure of the entropy of a distribution $p(x)$, *relative* to another (reference) distribution $q(x)$, is expressed by the *Kullback–Leibler distance*

$$D_{KL}(p|q) = \int_{-\infty}^{\infty} p(x)\log\frac{p(x)}{q(x)}dx \quad \text{in the continuous case,}$$

$$= \sum p_i \log\frac{p_i}{q_i} \quad \text{in the discrete case.}$$

D_{KL} is a natural distance from a "true" probability distribution (density $p(x)$ in the continuous case and probabilities p_i in the discrete case) to an arbitrary probability distribution $q(x)$. Typically, $p(x)$ represents data, observations, or a calculated probability distribution, and $q(x)$ represents a theory, a model, or an approximation of $p(x)$. The Kullback–Leibler distance is always nonnegative, and is zero if and only if $p(x) = q(x)$.

Obtain the Kullback–Leibler distance when $p(x)$ is the uniform distribution on $(0, 1)$ and $q(x)$ is the triangular distribution on $(0, 1)$.

Statistical Inference: Sampling Distributions, Confidence Intervals, and Tests of Hypotheses

4.1 Sampling Distributions

4.1-1 Introduction and Motivation

We have estimated parameters such as the fraction of defectives, p, of the binomial distribution; the mean μ and the standard deviation σ of the normal distribution; and the parameters α and β of the gamma distribution. We did so by using the observations $x_1, x_2, \ldots, x_n$ that were obtained by sampling from the distribution of interest. However, we recognize that usually these estimates do not equal the true values of the parameters of the underlying distribution. For example, if the sampling were repeated and we obtained another random sample from the very same distribution, we would no doubt get different values for the estimates. Since the estimates vary from sample to sample, it becomes important to investigate their reliability.

We can answer the question of reliability by studying *sampling distribution theory*. In this endeavor, we recognize that the estimates vary from sample to sample; that is, the *estimator* (the function used to estimate the parameter before the sample is taken) is in fact a random variable, and we must know something about its distribution before we can place a reliability factor on the observed estimate.

To explain the preceding remarks in more detail, as before, we let $X_1, X_2, \ldots, X_n$ represent the observations before the random sample is taken from a distribution with p.d.f. $f(x)$. Each X_i has the same distribution with p.d.f. $f(x)$. Events such as $(a_1 < X_1 < b_1), (a_2 < X_2 < b_2), \ldots, (a_n < X_n < b_n)$ are mutually independent, so we say that $X_1, X_2, \ldots, X_n$ are mutually independent random variables—of course, with the same p.d.f. $f(x)$.

Estimators such as $\overline{X} = (X_1 + X_2 + \cdots + X_n)/n$ are functions of $X_1, X_2, \ldots, X_n$, and in general, we denote such functions by $u(X_1, X_2, \ldots, X_n)$ or $v(X_1, X_2, \ldots, X_n)$. For example, the sample variance

$$u(X_1, X_2, \ldots, X_n) = \frac{\sum_{i=1}^{n}(X_i - \overline{X})^2}{n - 1} = S^2$$

is an estimator of σ^2. Once the sample is observed to be $x_1, x_2, \ldots, x_n$, the *computed* value, $u(x_1, x_2, \ldots, x_n) = \sum_{i=1}^{n}(x_i - \overline{x})^2/(n - 1) = s^2$, is said to be an estimate of σ^2, but s^2 will usually not equal σ^2.

Since $X_1, X_2, \ldots, X_n$ are random variables, any function of them, such as the estimator $u(X_1, X_2, \ldots, X_n)$, is also a random variable. So the estimator has a distribution. If, on the one hand, the variance of this distribution is small, we can place a great deal of reliance on a particular estimate $u(x_1, x_2, \ldots, x_n)$. If, on the other hand, the variance of the distribution of the estimator is large, we do not have as much confidence in the observed value of $u(X_1, X_2, \ldots, X_n)$ as an estimate of the parameter.

Figure 4.1-1 makes this point graphically. The first estimator of our parameter— say, μ—has a sampling distribution that is centered at μ with relatively tight variation around it. The second estimator, while also centered at μ, has considerably wider variation. Which estimator would you prefer? Obviously, the first one; because, in practice, you get to look at only one sample, you prefer the estimator that leads to realizations close to the unknown parameter μ. With the second estimator, you may end up with

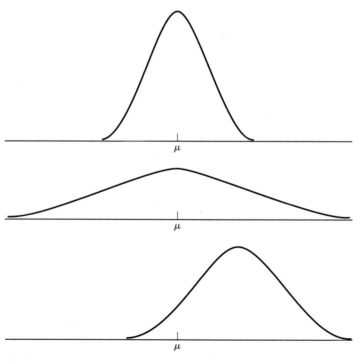

Figure 4.1-1 Sampling distributions of three estimators of μ

an estimate far from μ. The estimator at the bottom of Figure 4.1-1 also needs to be discussed, as it is not centered at the unknown parameter μ. In general, we want to avoid such estimators, because they "overestimate" the parameter, on average. There is more than a 50 percent chance (here it is about 90 percent) that a realization of such an estimator is larger than μ.

In these introductory remarks, we would like to emphasize that the concept of a sampling distribution is important when we cannot look at each element of a large fixed population. The variability of the measurement of interest, X, on an item in the population can be described by a distribution with density $f(x)$; hence, the mean of the population is the same as the mean of the distribution of X. Taking a randomly selected part (i.e., a random sample) from a population allows us to make probability statements about a parameter that applies to the whole population. For example, it is often sufficient to investigate a sample of our daily production if we are interested in the daily proportion of defectives. As another example, it may be sufficient to take a small sample of four or five items from our production to determine whether the dimensions of a certain product have gone out of control. In yet a third example, the results of a representative sample of engineers may give us a good estimate of the mean income of all engineers.

Sampling introduces variability in the estimator $u(X_1, X_2, \ldots, X_n)$. We call this source of variability the *sampling variability* or the *variability due to sampling*. Before we can attach reliability factors to our estimate, we must study the sampling distribution of the estimator, or at least characteristics such as its mean and variance.

4.1-2 Distribution of the Sample Mean $\overline{X}$

Let $X_1, X_2, \ldots, X_n$ be a *random sample* of size n from a distribution with mean μ and variance σ^2. Recall that the term "random sample" implies that (1) $X_1, X_2, \ldots, X_n$ are drawn from the *same* distribution and (2) the drawings are *mutually independent*.

Let us consider the sample mean

$$\overline{X} = \frac{1}{n}\sum_{i=1}^{n} X_1 = \left(\frac{1}{n}\right)X_1 + \left(\frac{1}{n}\right)X_2 + \cdots + \left(\frac{1}{n}\right)X_n,$$

and let us study its sampling distribution. We first determine the mean and the variance of the sampling distribution of $\overline{X}$. The sample mean is a linear combination of $X_1, X_2, \ldots, X_n$ with weights $a_i = 1/n$. Also, $X_1, X_2, \ldots, X_n$ are mutually independent and from the same distribution; thus, $E(X_i) = \mu_i = \mu$ and $\text{var}(X_i) = \sigma_i^2 = \sigma^2$. From our earlier result about the mean and variance of a linear combination of independent random variables (Sections 2.5 and 3.5), it follows that

$$E(\overline{X}) = \sum_{i=1}^{n} \left(\frac{1}{n}\right)\mu = \mu \quad \text{and} \quad \text{var}(\overline{X}) = \sum_{i=1}^{n}\left(\frac{1}{n}\right)^2 \sigma^2 = \frac{\sigma^2}{n}.$$

That is, the mean of $\overline{X}$ is the same as that of the underlying distribution; the variance is that of the underlying distribution, divided by the sample size n.

The sample mean $\overline{X}$ is the usual estimator of the mean of the underlying distribution (population) μ. Now, almost certainly, the mean of one particular sample will not equal μ, unless the investigator is extremely lucky. Moreover, the sample means will vary with repeated samples. The result $E(\overline{X}) = \mu$ is important, showing that the mean of the sampling distribution of $\overline{X}$ is the same as the mean of the underlying distribution that we want to estimate. We call such an estimator an *unbiased estimator.*

The second result, $\text{var}(\overline{X}) = \sigma^2/n$, shows that the variability of the estimator $\overline{X}$ around the population mean μ decreases as the number of observations in the sample increases. In the limit, as n becomes larger and larger, we essentially look at all elements in the population, and we can estimate μ precisely. The standard deviation of the estimator $\overline{X}$,

$$\sigma_{\overline{X}} = \sqrt{\text{var}(\overline{X})} = \frac{\sigma}{\sqrt{n}},$$

is also called the *standard error* of the estimator $\overline{X}$. It represents the standard deviation of the estimation error that one makes when using $\overline{X}$ as an estimator of the unknown μ. The standard error of the estimator $\overline{X}$ shrinks with the square root of the sample size.

The sample mean is an unbiased estimator of μ. But one can think of many other unbiased estimators of the population mean. For example, the average of the first three observations in the sample, $(X_1 + X_2 + X_3)/3$, or the sixth observation that gets recorded, X_6, are also unbiased estimators. However, since $\overline{X}$ uses more information than the other two estimators, our intuition tells us to prefer $\overline{X}$. This intuition is correct, because, for $n > 3$, $\text{var}(\overline{X}) = \sigma^2/n$ is smaller than both

$$\text{var}\left(\frac{X_1 + X_2 + X_3}{3}\right) = \frac{\sigma^2}{3} \quad \text{and} \quad \text{var}(X_6) = \sigma^2.$$

The sample mean $\overline{X}$, calculated from all n observations, is less variable than the other two estimators. We certainly prefer an estimator that varies around the population parameter as little as possible. This is the reason we say that $\overline{X}$ is *more efficient* or *more reliable* than the other two estimators.

We have already determined the mean and the variance of the sampling distribution of $\overline{X}$ and found that $E(\overline{X}) = \mu$ and $\text{var}(\overline{X}) = \sigma^2/n$. This implies that the standardized random variable

$$Z = \frac{\overline{X} - \mu}{\sqrt{\text{var}(\overline{X})}} = \frac{\overline{X} - \mu}{\sigma/\sqrt{n}}$$

has mean zero and variance 1, a property that follows from the fact that, for any random variable W,

$$E(a + bW) = a + bE(W) \quad \text{and} \quad \text{var}(a + bW) = b^2 \text{var}(W).$$

That is,

$$E(Z) = \frac{\mu - \mu}{\sigma/\sqrt{n}} = 0 \quad \text{and} \quad \text{var}(Z) = \left(\frac{\sqrt{n}}{\sigma}\right)^2 \text{var}(\overline{X}) = \frac{n\,\sigma^2}{\sigma^2\,n} = 1.$$

We say that $Z = (\overline{X} - \mu)/(\sigma/\sqrt{n})$ is a *standardized* random variable having mean zero and variance 1.

What more can we say about the distribution of $\overline{X}$ or, if we prefer, the distribution of Z? There is a theorem in mathematical statistics which says that a linear combination of *normal* random variables (which, in general, need not be independent) is again normally distributed. This theorem implies the following result:

Theorem: If we sample from a normal distribution with mean μ and variance σ^2, then the distribution of $\overline{X}$ is $N(\mu, \sigma^2/n)$ and the distribution of Z is $N(0, 1)$.

Example 4.1-1 Let $\overline{X}$ be the mean of a random sample of size $n = 25$ taken from the normal population $N(\mu = 75, \sigma^2 = 100)$. Thus, the distribution of $\overline{X}$ is normal with mean 75 and standard deviation $10/\sqrt{25} = 2$. We can calculate probabilities, such as

$$P(71 < \overline{X} < 79) = P\left(\frac{71 - 75}{2} < \frac{\overline{X} - 75}{2} < \frac{79 - 75}{2}\right) = P(-2 < Z < 2)$$

$$= \Phi(2) - \Phi(-2) = 0.9544.$$

This result implies that the *mean of 25 observations* has a 0.9544 chance of falling between 71 and 79. By contrast, the probability that an *individual observation* X falls within these limits is

$$P(71 < X < 79) = P\left(\frac{71 - 75}{10} < \frac{X - 75}{10} < \frac{79 - 75}{10}\right) = P(-0.4 < Z < 0.4)$$

$$= \Phi(0.4) - \Phi(-0.4) = 0.6554 - (1 - 0.6554) = 0.3108. \quad \blacksquare$$

4.1-3 The Central Limit Theorem

The result that $\overline{X}$ is $N(\mu, \sigma^2/n)$ applies if each X arises from the normal distribution $N(\mu, \sigma^2)$. But what can we say if the underlying distribution is not normal—for example, if we sample from a uniform, an exponential, or some other distribution? The central limit theorem provides the answer to this question.

Central Limit Theorem: If $\overline{X}$ is the mean of a random sample $X_1, X_2, \ldots, X_n$ from a distribution with mean μ and finite variance $\sigma^2 > 0$, then the distribution of

$$Z = \frac{\overline{X} - \mu}{\sigma/\sqrt{n}} = \frac{\sum X_i - n\mu}{\sigma\sqrt{n}}$$

approaches a distribution that is $N(0, 1)$ as n becomes large.

This theorem says that, for a sufficiently large sample size n, the distribution of $\overline{X}$ can be approximated by a distribution that is $N(\mu, \sigma^2/n)$. Similarly, the distribution of the sum $\sum_{i=1}^{n} X_i$ can be approximated by the $N(n\mu, n\sigma^2)$ distribution.

Example 4.1-2 Because we have not given a formal proof of the central limit theorem (doing so would require more mathematics), we include a simulation experiment which should convince you that the theorem is, in fact, true.

Let us sample from the uniform distribution on the interval $(0, 1)$ having p.d.f. $f(x) = 1$, for $0 \le x < 1$. In Example 3.1-5, we showed that $E(X) = 0.5$ and

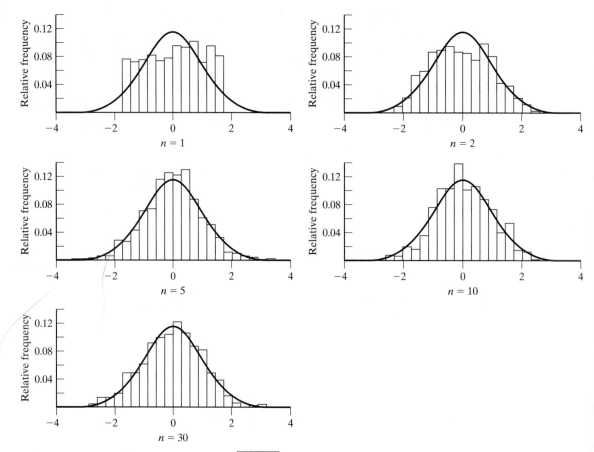

Figure 4.1-2 Histograms of $(\bar{x} - 0.5)/\sqrt{1/(12n)}$ for sample sizes $n = 1, 2, 5, 10$, and 30; 500 simulations each. The p.d.f. of the approximating $N(0, 1)$ distribution is given for comparison

var$(X) = 1/12$. Let us generate 30 columns of 500 numbers each. We explained in Section 3.5 how to do this. You may want to use Minitab's random-number-generation routine. From these numbers, calculate averages of size $n = 1, 2, 5, 10$, and 30 by averaging the numbers across the first 1 (2, 5, 10, 30) columns.

In Figure 4.1-2, we summarize the results of the 500 samples and we plot, for each n separately, the histogram of $(\bar{x} - 0.5)/\sqrt{1/(12n)}$. The smooth curve on these figures corresponds to the p.d.f. of the $N(0, 1)$ distribution. The histograms are centered at zero, and their variance is about 1. For $n = 1$, the distribution is uniform, because the "averages" correspond to individual observations from a uniform distribution. The distribution of the means from samples of size $n = 5$ is already starting to look normal, despite the fact that we are sampling from a uniform distribution. As we expect from the central limit theorem, the approximation to the $N(0, 1)$ distribution becomes better and better as we increase the sample size n. For $n = 30$, the approximation is already extremely good. This experiment shows that, for sufficiently large sample size, the distribution of the sample mean is approximately normal, even if we sample from a distribution that is quite different from a normal distribution. ∎

Example 4.1-3   One can also use a theoretical example to illustrate the approximation. Consider $Y = X_1 + X_2$, the sum of two independent observations from the uniform distribution $f(x) = 1$, for $0 \le x < 1$. Because $\mu = 1/2$ and $\sigma^2 = 1/12$, the mean and the variance of Y are, respectively, $E(Y) = 1$ and $\text{var}(Y) = (1/12) + (1/12) = 1/6$.

Even though $n = 2$ is extremely small, let us use the normal approximation to find

$$P(0.5 < Y < 1.5) = P\left(\frac{-0.5}{1/\sqrt{6}} < \frac{Y-1}{1/\sqrt{6}} < \frac{0.5}{1/\sqrt{6}}\right)$$

$$\approx \Phi(0.5\sqrt{6}) - \Phi(-0.5\sqrt{6})$$

$$= \Phi(1.22) - \Phi(-1.22) = 0.8888 - 0.1112 = 0.7776.$$

The exact p.d.f. of Y is determined in Exercise 4.1-10. Using this result, we find that the exact probability is

$$P(0.5 < Y < 1.5) = \int_{0.5}^{1.5} g(y)\, dy$$

$$= 1 - 2\int_{0}^{0.5} y\, dy = 1 - 0.25 = 0.75.$$

Even for such a small n, the normal approximation is not too bad; it will improve for larger n.   ■

4.1-4 Normal Approximation of the Binomial Distribution

The central limit theorem implies that the binomial distribution with parameters n and p can be approximated by the $N[np, np(1-p)]$ distribution. The reason for this is that a binomial random variable Y can be thought of as the sum $\sum_{i=1}^{n} X_i$, where $X_1, X_2, \ldots, X_n$ is a random sample from a Bernoulli distribution with mean p and variance $p(1-p)$. However, note that Y can take on only integer values, $0, 1, 2, \ldots, n$; thus, we approximate a discrete probability distribution by a continuous p.d.f. $f(y)$, which is $N[np, np(1-p)]$. Hence, it is natural to approximate the true probability $P(Y = k)$, where k is one of those integers, by the area under the normal p.d.f. between $k - 0.5$ and $k + 0.5$. That is,

$$P(Y = k) \approx \int_{k-0.5}^{k+0.5} f(y)\, dy.$$

Since $f(y)$ is the p.d.f. of an $N[np, np(1-p)]$ distribution, this probability is

$$P(Y = k) \approx \Phi\left(\frac{k + 0.5 - np}{\sqrt{np(1-p)}}\right) - \Phi\left(\frac{k - 0.5 - np}{\sqrt{np(1-p)}}\right).$$

The next example illustrates this approximation.

Example 4.1-4   Let Y be binomial with $n = 16$ and $p = 1/2$. Although n is not very large here, we have selected a symmetric binomial distribution; this improves the approximation. If p is different from $1/2$, we would need a larger value of n to attain the same quality of approximation.

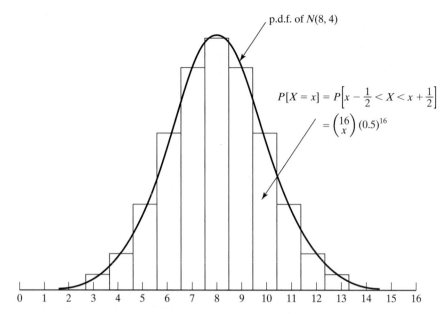

Figure 4.1-3 The p.d.f. of the $b(n = 16, p = 0.5)$ distribution and the approximating $N(8, 4)$ distribution

With $n = 16$, we can compare the approximation with the exact binomial probabilities. In Figure 4.1-3, we have drawn a probability histogram, $b(n = 16, p = 0.5)$, together with the normal p.d.f. of $N[np = 8, np(1 - p) = 4]$. The normal p.d.f. approximates the shape of the probability histogram quite well. For example, the approximation of $P(Y = 7)$ is

$$P(6.5 < Y < 7.5) \approx \Phi\left(\frac{7.5 - 8}{2}\right) - \Phi\left(\frac{6.5 - 8}{2}\right)$$

$$= \Phi(-0.25) - \Phi(-0.75) = 0.7734 - 0.5987 = 0.1747.$$

From the cumulative probabilities of the binomial distribution with $n = 16$ and $p = 0.5$, the true probability is

$$P(Y = 7) = P(Y \le 7) - P(Y \le 6) = 0.4018 - 0.2272 = 0.1746;$$

thus, the approximation and the exact probability are in excellent agreement. As another illustration with the same distribution, consider the probability

$$P(Y = 6, 7, 8, 9, 10) = P(5.5 < Y < 10.5)$$

$$\approx \Phi\left(\frac{10.5 - 8}{2}\right) - \Phi\left(\frac{5.5 - 8}{2}\right) = 0.8944 - 0.1056 = 0.7888.$$

This result agrees well with the exact probability

$$P(Y \le 10) - P(Y \le 5) = 0.8949 - 0.1051 = 0.7898.$$ ■

Exercises 4.1

*4.1-1 Let $\overline{X}$ be the mean of a random sample of size $n = 48$ from the uniform distribution on the interval $(0, 2)$; that is, $f(x) = 1/2$ for $0 < x < 2$. Approximate the probability $P(0.9 < \overline{X} < 1.1)$.

4.1-2 Let $\overline{X}$ be the mean of a random sample of size $n = 15$ from a distribution with p.d.f. $f(x) = (3/2)x^2$, for $-1 < x < 1$. Approximate $P(-0.02 < \overline{X} < 0.1)$. Do you expect this to be a very good approximation?

4.1-3 Let $\overline{X}$ be the mean of a random sample of size $n = 36$ from an exponential distribution with mean 4. Approximate $P(3.1 < \overline{X} < 4.6)$.

4.1-4 Let $\overline{X}$ be the mean of a random sample of size $n = 12$ from a chi-square distribution with six degrees of freedom. Approximate $P(5.1 < \overline{X} < 7.2)$.

*4.1-5 Let Y be $b(n = 16, p = 0.2)$. This particular binomial distribution is quite skewed.

*(a) From the binomial tables or from your statistical software, find $P(Y = 1, 2, 3, 4)$.

*(b) Use the normal distribution to approximate the probability in part (a).

   Hint: Write it as $P(0.5 < Y < 4.5)$.

*(c) Is the approximation as good in this skewed case as in the symmetric case with $p = 0.5$?

*4.1-6 Let Y have the Poisson distribution with mean $\lambda = 9$.

*(a) Find $P(7 \le Y \le 11)$ from the Poisson table or statistical software.

*(b) Approximate the probability in part (a), using the p.d.f. of the $N(9, 9)$ distribution.

   Hint: Y has a discrete distribution, so first write that probability as $P(6.5 < Y < 11.5)$. Also, Y has an approximate normal distribution, as it can be thought of as the sum of $n = \lambda = 9$ independent Poisson variables, each with mean and variance equal to 1.

*4.1-7 Let $\overline{X}$ be the mean of a random sample of size $n = 10$ from a distribution with p.d.f. $f(x) = 6x(1 - x)$, for $0 < x < 1$. Find the mean and the variance of $\overline{X}$.

*4.1-8 Let W have the triangular p.d.f. $f(w) = 2w$ for $0 < w < 1$.

(a) Show that the mean and the variance of W are $2/3$ and $1/18$, respectively.

*(b) What are the mean and the variance of $U = \sqrt{18}(W - 2/3)$?

*(c) Find the distribution function $G(u) = P(U \le u)$ and the p.d.f. $g(u) = G'(u)$.

(d) What type of distribution does U have?

*4.1-9 A cereal manufacturer packages cereal in boxes with a 12-ounce label weight. Suppose that the actual distribution of weights is $N(\mu = 12.2, \sigma^2 = 0.04)$.

*(a) What percentage of the boxes have cereal weighing under 12 ounces?

*(b) Suppose $\overline{X}$ is the mean weight of the cereals in $n = 4$ boxes selected at random. Compute $P(\overline{X} < 12)$.

*4.1-10 Let X_1 and X_2 be two independent observations of the uniform distribution on the interval $(0, 1)$ with p.d.f. $f(x) = 1$, for $0 \le x < 1$. Let $Y = X_1 + X_2$. Draw the space of (X_1, X_2) and, using a geometric argument, determine the probability

$$G(y) = P(Y \le y) = P(X_1 + X_2 \le y).$$

Distinguish between the two cases of $0 \le y \le 1$ and $1 < y < 2$. Show that

$$g(y) = G'(y) = \begin{cases} y, & 0 \le y \le 1 \\ 2 - y, & 1 < y < 2. \end{cases}$$

4.1-11 The profits from investments in individual stocks follow a normal distribution with mean 1 and standard deviation 5.

(a) If you buy a *single* randomly selected stock, what is the probability that you are making money (i.e., profit greater than zero)?

(b) If you buy a *portfolio* of 25 stocks, what is the probability that you are making money (i.e., an average profit greater than zero)?

4.1-12 You are interested in the *proportion of smokers* in a certain population. Consider a population consisting of $N = 4$ subjects: the first two persons smoke, while the remaining two do not. That is, the population space is {S, S, N, N}.

(a) In this population, what is the proportion of smokers?

(b) You take a random sample of size $n = 2$, sampling the subjects without replacement. How many different random samples are possible?

(c) Write down the sampling distribution of $\hat{p}$, the sample proportion of smokers in a random sample of size $n = 2$ from this population. That is, list the possible outcomes (sample proportions of smokers) and the associated probabilities.

(d) Calculate the mean of the sampling distribution, and interpret the results. Could you have guessed the result from the central limit theorem?

(e) Repeat steps (a) through (d) with two smokers in a population of size $N = 5$ and with a random sample of size $n = 3$.

4.1-13 Train cars for the Crandic Railway Company carry 25 standard-sized containers. The weight of each container is described by a distribution with a mean of 380 pounds and a standard deviation of 50 pounds. Steve, the train engineer, becomes upset if a train car contains more than 10,000 pounds. What fraction of train cars would make him angry?

4.1-14 Several applets for illustrating the central limit effect are available on the Web, and you can find them by searching for "central limit effect applet." One of them is listed under http://www.ruf.rice.edu/~lane/stat_sim/. Experiment with this applet.

4.2 Confidence Intervals for Means

If a random sample $X_1, X_2, \ldots, X_n$ arises from a distribution with unknown mean μ and variance $\sigma^2 > 0$, then the sample mean $\overline{X}$ is an unbiased estimator of μ; that is, $E(\overline{X}) = \mu$. We now study the reliability of $\overline{X}$ as an estimator and develop an interval that covers the unknown mean μ with high probability. We use the distribution of $\overline{X}$, which, according to the central limit theorem, is approximately $N(\mu, \sigma^2/n)$, provided that n is large. Of course, $\overline{X}$ is exactly $N(\mu, \sigma^2/n)$ if the underlying distribution is $N(\mu, \sigma^2)$. Thus, the probability statements that follow should be read as approximate or exact, depending on whether $\overline{X}$ has an approximate or exact normal distribution. In either case, the distribution of $\overline{X}$ is said to be a sampling distribution—here $N(\mu, \sigma^2/n)$, because it describes how $\overline{X}$ varies from sample to sample.

For a given probability $1 - \alpha$ (usually, α is small, e.g., $0.01, 0.05$, or 0.10), we can find a value, say, $z(\alpha/2)$, from the normal table such that

$$P\left[-z(\alpha/2) \le \frac{\overline{X} - \mu}{\sigma/\sqrt{n}} \le z(\alpha/2)\right] = 1 - \alpha.$$

Make a graph of the normal p.d.f., and convince yourself that $z(\alpha/2)$ is the $100(1 - \alpha/2)$th percentile of the $N(0, 1)$ distribution; that is, $P[Z \le z(\alpha/2)] = 1 - \alpha/2$. Because $P[Z \ge z(\alpha/2)] = \alpha/2$, $z(\alpha/2)$ is also called the *upper $100(\alpha/2)$ percentage point* of the $N(0, 1)$ distribution. Also, because the $N(0, 1)$ distribution is symmetric around zero, it follows that

$$P[Z \le -z(\alpha/2)] = P[Z \ge z(\alpha/2)] = \alpha/2.$$

For example, if $1 - \alpha = 0.95$, then $z(\alpha/2) = z(0.025) = 1.96$. If $1 - \alpha = 0.90$, then $z(\alpha/2) = z(0.05) = 1.645$.

The inequalities

$$-z(\alpha/2) \le \frac{\overline{X} - \mu}{\sigma/\sqrt{n}} \le z(\alpha/2)$$

are equivalent to the statements

$$-z(\alpha/2)\frac{\sigma}{\sqrt{n}} \le \overline{X} - \mu \le z(\alpha/2)\frac{\sigma}{\sqrt{n}},$$

$$-\overline{X} - z(\alpha/2)\frac{\sigma}{\sqrt{n}} \le -\mu \le -\overline{X} + z(\alpha/2)\frac{\sigma}{\sqrt{n}},$$

and

$$\overline{X} - z(\alpha/2)\frac{\sigma}{\sqrt{n}} \le \mu \le \overline{X} + z(\alpha/2)\frac{\sigma}{\sqrt{n}}.$$

Thus, the probabilities of each set of inequalities are exactly the same, $1 - \alpha$. In particular,

$$P\left[\overline{X} - z(\alpha/2)\frac{\sigma}{\sqrt{n}} \le \mu \le \overline{X} + z(\alpha/2)\frac{\sigma}{\sqrt{n}}\right] = 1 - \alpha.$$

Note that the random variable $\overline{X}$ is in the extremes of these inequalities and the constant, but unknown, parameter μ is in the middle. Hence, the probability that the *random interval*

$$\left[\overline{X} - z(\alpha/2)\frac{\sigma}{\sqrt{n}}, \overline{X} + z(\alpha/2)\frac{\sigma}{\sqrt{n}}\right]$$

includes the unknown mean μ is $1 - \alpha$. For simplicity, we sometimes write the random interval as $\overline{X} \pm z(\alpha/2)\sigma/\sqrt{n}$.

Suppose that an experiment is carried out and we observe the sample $x_1, x_2, \ldots, x_n$ and compute the sample mean $\bar{x}$. Assuming that σ is known, we can then compute the interval

$$\left[\bar{x} - z(\alpha/2)\frac{\sigma}{\sqrt{n}}, \bar{x} + z(\alpha/2)\frac{\sigma}{\sqrt{n}}\right],$$

or, more simply, $\bar{x} \pm z(\alpha/2)\sigma/\sqrt{n}$. Now, either this computed interval includes μ, or it does not. However, because, prior to sampling, the random interval includes μ with probability $1 - \alpha$, we continue to associate that number $1 - \alpha$ with the computed interval. We say that we are $100(1 - \alpha)$ percent *confident* that the computed interval $\bar{x} \pm z(\alpha/2)\sigma/\sqrt{n}$ includes μ. More formally, $\bar{x} \pm z(\alpha/2)\sigma/\sqrt{n}$ is referred to as a $100(1 - \alpha)$ percent *confidence interval* for μ, and $1 - \alpha$ is called the *confidence coefficient*. The confidence interval is centered at the estimate $\bar{x}$; the limits of the confidence intervals are obtained by adding and subtracting a multiple of the standard error of the estimate, $\sigma_{\bar{X}} = \sqrt{\text{var}(\overline{X})} = \sigma/\sqrt{n}$.

Example 4.2-1   A random sample of size $n = 25$ from the distribution $N(\mu, \sigma^2 = 4)$ results in the sample mean $\bar{x} = 13.2$. Then $\bar{x} \pm 2\sigma/\sqrt{n}$ provides a 95.44 percent confidence interval for μ. That is, $13.2 \pm 2(2/5)$ or, equivalently, the interval from 12.4 to 14.0 is a 95.44 percent confidence interval for μ. Note that it is easy to remember how to form a 95.44 percent confidence interval for μ: Take the sample mean and add to it and subtract from it two standard deviations (standard errors) of the sample mean. ∎

Example 4.2-2   The mean breaking strength $\bar{x}$ of a sample of $n = 32$ steel beams equals 42,196 pounds per square inch (psi). From past experience, we know that the standard deviation of an individual measurement is about $\sigma = 500$ psi. Thus, $\bar{x} \pm 1.645\sigma/\sqrt{32}$, which equals $42{,}196 \pm 1.645(500/\sqrt{32})$, or $(42{,}051; 42{,}341)$, provides an approximate 90 percent confidence interval for the mean μ of the population from which the 32 beams were sampled. In other words, we are 90 percent confident that the unknown mean μ is within the stated interval.

In the context of this example, we would like to remark that it is always important to identify clearly from which population we have sampled. For example, taking just beams from the scrap pile and testing their breaking strengths may tell us nothing about the overall breaking strength of our production. We would only learn about the breaking strength of the rejected beams, which most likely would be different from that of the rest of the production. If we want to get information on today's production, we have to sample from today's production and make certain that we give each beam the same chance of being selected. ∎

Of course, if the population mean μ is unknown, then in many situations the population standard deviation σ is also unknown, and we cannot compute the interval $\bar{x} \pm z(\alpha/2)\sigma/\sqrt{n}$. If the sample size is large (at least 30), most statisticians simply approximate σ by the sample standard deviation s, and use $\bar{x} \pm z(\alpha/2)s/\sqrt{n}$ as an approximate $100(1 - \alpha)$ percent confidence interval for μ. This result is usually quite satisfactory, particularly if we recognize that it involves an additional approximation, so that a quoted 95 percent confidence interval might really be one of 93 percent, 96.5 percent, or some other percentage number reasonably close to 95. The quantity $s/\sqrt{n}$ is also known as the *estimated standard error* of the estimator $\overline{X}$.

Example 4.2-3   In Example 4.2-2, we assumed that the standard deviation of individual measurements is known. Let us now take a more realistic position and suppose that σ is estimated by the sample standard deviation s. From the $n = 32$ observations, we compute not only $\bar{x} = 42{,}196$, but also the sample standard deviation $s = 614$. Hence, an approximate 95 percent confidence interval for μ is given by $\bar{x} \pm (1.96)s/\sqrt{n}$. Since this is an approximation anyway, we may as well replace $z(0.025) = 1.96$ by 2 and use $\bar{x} \pm 2s/\sqrt{n}$ as an approximate 95 percent confidence interval. In our example, the approximate 95 percent confidence interval for μ is given by $42{,}196 \pm 2(614)/\sqrt{32}$, or $(41{,}979; 42{,}413)$. ∎

4.2-1 Determination of the Sample Size

Suppose we are planning an experiment and want to be $100(1 - \alpha)$ percent confident that our estimate is within h units of the unknown parameter μ. For example, we may want to be 95 percent confident that our estimate is within, say, $h = 2$ units of μ. How large must we choose our sample size n in order to estimate μ with such accuracy?

We know that

$$P\left[\bar{X} - z(\alpha/2)\frac{\sigma}{\sqrt{n}} \le \mu \le \bar{X} + z(\alpha/2)\frac{\sigma}{\sqrt{n}}\right] = 1 - \alpha,$$

so we must choose n such that $h = z(\alpha/2)(\sigma/\sqrt{n})$, or, equivalently,

$$n = \frac{\sigma^2[z(\alpha/2)]^2}{h^2}.$$

Choosing n thus requires knowledge of σ^2. In many situations, the experimenter may have a reasonable approximation for σ^2. That is, results from pilot studies or from experiments with other related variables can suggest an appropriate planning value for σ^2.

The preceding expression for n shows that the required sample size increases with increasing confidence coefficient $1 - \alpha$ and decreases with increasing half-width h. Sometimes we may require too much accuracy, and as a result, we find that n is very large. If we cannot afford to collect that many observations, we may want to reduce the confidence coefficient or, alternatively, increase the half-width h.

Example 4.2-4 A process used to manufacture paint yields, on the average, 70 tons of paint each day. Yields, however, vary from day to day due to changes in raw materials and plant conditions.

Suppose, however, it is fairly well established that the daily yields are normally distributed, that the variability from one day to the other is more or less independent, and that the standard deviation is $\sigma = 3$ tons.

Because of increasing demand for this type of paint, certain modifications are suggested, and we are interested in estimating the mean yield of this modified process. How many sample observations do we have to obtain if we want to be 95 percent certain that our estimate is within 1 ton of the true, but unknown, mean yield μ? Here, $\sigma = 3$ and $z(0.025) = 1.96$. Thus, $n = (9)(1.96)^2/1 = 34.6$; that is, we must sample at least 35 days' worth of paint. ∎

4.2-2 Confidence Intervals for $\mu_1 - \mu_2$

Let $X_1, X_2, \ldots, X_{n_1}$ and $Y_1, Y_2, \ldots, Y_{n_2}$ be two *independent* random samples of sizes n_1 and n_2 from two distributions with respective parameters μ_1, σ_1^2 and μ_2, σ_2^2. For example, the two distributions may describe the breaking strengths of two different fabrics (also referred to as populations): fabric A, an "all-synthetic" material, and fabric B, a material with 15 percent cotton content. From fabric A, the investigator selects at random n_1 small pieces of cloth and determines their breaking strengths, $X_1, X_2, \ldots, X_{n_1}$. Similarly, $Y_1, Y_2, \ldots, Y_{n_2}$ are the breaking strengths of n_2 fabric specimens that are taken at random from the production of fabric B.

The sample means $\overline{X}$ and $\overline{Y}$ are estimators of μ_1 and μ_2. If the samples are taken from normal distributions, the respective distributions of $\overline{X}$ and $\overline{Y}$ are exactly $N(\mu_1, \sigma_1^2/n_1)$ and $N(\mu_2, \sigma_2^2/n_2)$. However, even if the samples arise from nonnormal distributions, the distributions of $\overline{X}$ and $\overline{Y}$ are *approximately* normal according to the central limit theorem, provided that the sample sizes n_1 and n_2 are large enough.

The appropriate estimator for $\mu_1 - \mu_2$ is the difference of the sample means $\overline{X} - \overline{Y}$. Because we assumed that the two random samples are selected independently, it follows that $\overline{X}$ and $\overline{Y}$ are also independent. Hence,

$$\text{var}(\overline{X} - \overline{Y}) = \text{var}(\overline{X}) + \text{var}(\overline{Y}) = \frac{\sigma_1^2}{n_1} + \frac{\sigma_2^2}{n_2};$$

the square root of this expression, $\sqrt{\text{var}(\overline{X} - \overline{Y})} = \sqrt{\dfrac{\sigma_1^2}{n_1} + \dfrac{\sigma_2^2}{n_2}}$, is known as the standard error of the estimator $\overline{X} - \overline{Y}$. Thus, the sampling distribution of $\overline{X} - \overline{Y}$ is

$$N\left(\mu_1 - \mu_2, \frac{\sigma_1^2}{n_1} + \frac{\sigma_2^2}{n_2}\right).$$

Accordingly,

$$P\left[-z(\alpha/2) \le \frac{(\overline{X} - \overline{Y}) - (\mu_1 - \mu_2)}{\sqrt{\sigma_1^2/n_1 + \sigma_2^2/n_2}} \le z(\alpha/2)\right] = 1 - \alpha.$$

Rewriting these inequalities so that $\mu_1 - \mu_2$ is in the middle expression, we find that the probability that the random interval

$$\left[(\overline{X} - \overline{Y}) - z(\alpha/2)\sqrt{\frac{\sigma_1^2}{n_1} + \frac{\sigma_2^2}{n_2}}, (\overline{X} - \overline{Y}) + z(\alpha/2)\sqrt{\frac{\sigma_1^2}{n_1} + \frac{\sigma_2^2}{n_2}}\right]$$

includes $\mu_1 - \mu_2$ is $1 - \alpha$. Once the samples are observed, so that we can compute $\bar{x}$ and $\bar{y}$, we obtain the computed interval:

$$\left[(\bar{x} - \bar{y}) - z(\alpha/2)\sqrt{\frac{\sigma_1^2}{n_1} + \frac{\sigma_2^2}{n_2}}, (\bar{x} - \bar{y}) + z(\alpha/2)\sqrt{\frac{\sigma_1^2}{n_1} + \frac{\sigma_2^2}{n_2}}\right].$$

This is a $100(1 - \alpha)$ percent confidence interval for $\mu_1 - \mu_2$.

Here, we have assumed that σ_1^2 and σ_2^2 are known. If they are unknown, but the sample sizes are large, we can replace σ_1^2 by the sample variance s_x^2 and σ_2^2 by s_y^2 and use

$$\bar{x} - \bar{y} \pm z(\alpha/2)\sqrt{\frac{s_x^2}{n_1} + \frac{s_y^2}{n_2}}$$

as an approximate $100(1 - \alpha)$ percent confidence interval for $\mu_1 - \mu_2$.

Example 4.2-5 The analysis of the breaking strength (in pounds per square inch) of $n_1 = 35$ specimens selected from the all-synthetic fabric A led to the mean $\bar{x} = 25.2$ and standard deviation $s_x = 5.2$. The $n_2 = 30$ specimens from fiber B, which includes some cotton, resulted in the mean $\bar{y} = 28.5$ and standard deviation $s_y = 5.9$. Since the sample sizes are quite large, we can use the sample variances in place of the unknown population variances. Using 2 in place of 1.96, we find that an approximate 95 percent confidence interval for the unknown difference of the population means is given by

$$25.2 - 28.5 \pm 2\sqrt{\frac{(5.2)^2}{35} + \frac{(5.9)^2}{30}},$$

or, equivalently, $(-6.0, -0.6)$. In other words, we are approximately 95 percent confident that the interval from -6.0 to -0.6 covers the unknown difference $\mu_1 - \mu_2$. Because the interval does not include zero, the data provide some evidence that $\mu_1 - \mu_2$ is negative and that the mean breaking strength of the all-synthetic fabric is less than the one that includes some cotton. ∎

Remark: The derivation of this confidence interval for $\mu_1 - \mu_2$ depends heavily on the independence of the two random samples, as the variance of $\bar{X} - \bar{Y}$ is calculated under the assumption of independence. But how does one know whether it is safe to make this assumption? It depends on the way the experiment is conducted! An experiment that assigns two treatments (say, paints A and B) *randomly* to *different* experimental units (say, wooden boards in a durability comparison) guarantees, in general, independence among the two samples. However, the situation is different if two treatments are assigned to the same experimental unit. Assume, for example, that both paints are applied to the same board and that n such boards are used in the comparison. Then observations made on the same board are most likely similar. (If one paint scores higher than average, so will the other.) In this case, the assumption of independence is no longer valid, and confidence intervals, as calculated in this section, are no longer appropriate. The analysis of the paired (blocked) experiment, in which the measurements X and Y are correlated, will be discussed in Section 4.6-2, where it is shown that the appropriate blocking of an experiment can greatly increase the precision of the comparison.

Exercises 4.2

*4.2-1 A random sample of size $n = 36$ from the $N(\mu, \sigma^2 = 25)$ distribution has mean $\bar{x} = 49.2$. Find a 90 percent confidence interval for μ.

4.2-2 The mean and standard deviation of $n = 42$ mathematics SAT test scores (selected at random from the entering freshman class of a large private university) are $\bar{x} = 680$ and $s = 35$. Find an approximate 99 percent confidence interval for the population mean μ.

*4.2-3 Let a population have mean μ and standard deviation $\sigma = 5$. Find the sample size n such that we are 95 percent confident that the estimate $\bar{x}$ is within ± 1.5 units of the true mean μ.

4.2-4 The average biological oxygen demand (BOD) at a particular station has to be estimated. From measurements at other similar stations, we know that the variance of BOD measurements is about $8.0(\text{mg/liter})^2$. How many observations should we sample if we want to be 90 percent confident that our sample average is within 1 mg/liter of the true mean?

*4.2-5 In comparing the times until failure (in hours) of two different types of light bulbs, we obtain the sample characteristics $n_1 = 45$, $\bar{x} = 984$, $s_x^2 = 8{,}742$, and $n_2 = 52$, $\bar{y} = 1{,}121$, $s_y^2 = 9{,}411$. Find an approximate 90 percent confidence interval for the difference of the two population means. Interpret the result, and explain why we can use the normal table here despite the fact that the distribution of individual failure times is probably Weibull.

4.2-6 We are interested in comparing the biological oxygen demand (BOD) at a river station before and after a certain chemical plant has made "improvements" in treating its waste. We are planning to take n observations before and n observations after the improvements are made. From past experience, we know that the variance of individual BOD measurements, both before and after, is about $8.0(\text{mg/liter})^2$. How large must we choose n such that we are 95 percent confident that the estimated before–after difference in BOD is within 2 mg/liter of the true change?

*4.2-7 Consider the melting points of the $n = 50$ alloy filaments in Exercise 1.3-2. Calculate a 95 percent confidence interval for the mean melting point μ.

4.2-8 Consider the compressive strengths of the $n = 90$ concrete cylinders in Table 1.3-1. Construct a 95 percent (99 percent) confidence interval for the mean of the population from which this sample was taken.

4.2-9 Applets for illustrating confidence intervals for a population mean are available on the Web, and you can find them by searching for "confidence interval applet." A useful one is listed under http://www.ruf.rice.edu/~lane/stat_sim/.

4.3 Inferences from Small Samples and with Unknown Variances

In Section 4.2, we examined confidence intervals for means and differences of means when either the populations had known variances or the sample sizes were large enough that the sample variances could be used to approximate those of the populations. However, what should we do if the sample size is small, such as $n = 13$? Then we do not have as much reliance on s as a good approximation to σ. Accordingly, we must replace the value $z(\alpha/2)$ in the confidence interval by a larger value and thus lengthen the interval to take this additional uncertainty into account. The amount of the increase from $z(\alpha/2)$ raises another interesting sampling distribution problem that we now consider.

Let us assume that the random sample $X_1, X_2, \ldots, X_n$ arises from the normal distribution $N(\mu, \sigma^2)$. Normality is important here, because the discussion that follows

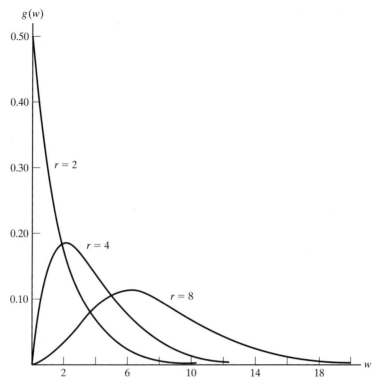

Figure 4.3-1 Probability densities of three different chi-square distributions

depends on it. The estimators $\overline{X}$ and $S^2 = [1/(n-1)]\sum_{i=1}^{n}(X_i - \overline{X})^2$ of μ and σ^2 are random variables, and in our usual notation, they are denoted by capital letters. We already know that the distribution of the sample mean $\overline{X}$ is $N(\mu, \sigma^2/n)$. Although we cannot prove it at this level, it can be shown that $\overline{X}$ and the sample variance S^2 are independent random variables. Moreover, $W = (n-1)(S^2/\sigma^2) = \sum_{i=1}^{n}(X_i - \overline{X})^2/\sigma^2$ has that very special distribution called *chi-square with $n-1$ degrees of freedom* that was first considered in Section 3.3. For brevity, we say that W is $\chi^2(r = n-1)$. Chi-square densities for three different values of the parameter r are given in Figure 4.3-1. Upper percentage points $\chi^2(\alpha; r)$ for selected tail probabilities such that $P[W > \chi^2(\alpha; r)] = \alpha$ are given in Table C.5 of Appendix C. Also, you can determine them from one of the many readily available computer software packages. In Minitab, one would use "Calc > Probability Distributions > Chi-square," specify inverse cumulative probability, and obtain the percentile of order $1 - \alpha$.

Example 4.3-1 Let W be $\chi^2(8)$. Then the value d for which $P(W > d) = 0.05$ is $d = \chi^2(0.05;8) = 15.507$. This is the upper fifth percentage point, or equivalently, the 95th percentile of the distribution of W. We also find that $P(W > 2.180) = 0.975$ and $P(W > 20.090) = 0.01$.   ■

Following are several other properties of chi-square random variables:

1. The mean of a $\chi^2(r)$ random variable is r, while the variance is $2r$.
2. Z^2, the square of an $N(0, 1)$ random variable, follows a $\chi^2(1)$ distribution. (See Exercise 4.3-5 for a proof.)
3. The sum of m independent chi-square random variables with $r_1, r_2, \ldots, r_m$ respective degrees of freedom, has a chi-square distribution with $r = r_1 + r_2 + \cdots + r_m$ degrees of freedom.

The first property and the fact that $(n - 1)(S^2/\sigma^2)$ follows a chi-square distribution with $n - 1$ degrees of freedom together imply that

$$E\left[(n - 1)\frac{S^2}{\sigma^2}\right] = n - 1.$$

Hence, $E(S^2) = \sigma^2$. This shows that the sample variance is an unbiased estimator of the variance σ^2, which in turn explains why the division of $\Sigma(X_i - \overline{X})^2$ by $(n - 1)$ instead of n is preferable in defining the sample variance. Because $E[\Sigma(X_i - \overline{X})^2/n] = (n - 1)\sigma^2/n \neq \sigma^2$, we find that the variance of the empirical distribution, $v = \Sigma(X_i - \overline{X})^2/n$, is a slightly *biased* estimator of σ^2.

This brings us to the study of the ratio we want to consider, namely,

$$\frac{\overline{X} - \mu}{S/\sqrt{n}},$$

where σ in $(\overline{X} - \mu)/(\sigma/\sqrt{n})$ has been replaced by S. The ratio can be written as

$$T = \frac{\overline{X} - \mu}{S/\sqrt{n}} = \frac{(\overline{X} - \mu)/(\sigma/\sqrt{n})}{\sqrt{\dfrac{(n - 1)S^2}{\sigma^2} \bigg/ (n - 1)}} = \frac{Z}{\sqrt{W/r}},$$

where $r = n - 1$ and where the random variables $Z = (\overline{X} - \mu)/(\sigma/\sqrt{n})$ and $W = (n - 1)(S^2/\sigma^2)$ are independent $N(0, 1)$ and $\chi^2(r)$ variables, respectively.

It was exactly this ratio that W. S. Gosset considered in the earlier part of the 20th century. He found that it had a special distribution. However, his employer (an Irish brewery) did not want the other breweries to know that just the one brewery was using statistical methods, so Gosset was asked to publish his result under a pseudonym. He selected "A Student," and since that day, this special distribution has been called *Student's t-distribution with $r = n - 1$ degrees of freedom*.

The p.d.f. of Student's t-distribution with r degrees of freedom is

$$h(t) = \frac{c}{(1 + t^2/r)^{(r+1)/2}}, \quad -\infty < t < \infty,$$

where c is selected so that the area under $h(t)$ is 1. For convenience, we say that T is $t(r)$. It can be shown that $E(T) = 0$ and $\text{var}(T) = r/(r - 2)$, for $r > 2$. As we might suspect, this density looks very much like that of the $N(0, 1)$ distribution, particularly when r is large. It is symmetric about zero, but has thicker tails than the standard normal distribution. (See Figure 4.3-2.) Upper percentage points for selected tail probabilities α such that $P[T > t(\alpha; r)] = \alpha$ are recorded in Table C.6. You can also

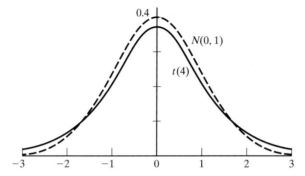

Figure 4.3-2 Probability densities of the $N(0, 1)$ and $t(4)$ distributions. $N(0, 1)$: broken line; $t(4)$: solid line

use the Minitab command "Calc > Probability Distribution > t-distribution" to obtain the cumulative and inverse cumulative probabilities (percentiles). Because the distribution of T is symmetric about zero, we have

$$P[T < -t(\alpha; r)] = P[T > t(\alpha; r)].$$

Example 4.3-2 For a t-distribution with $r = 8$ degrees of freedom, we find that

$$P[T > 2.306] = 0.025.$$

As r increases, say, to $r = 25$, we obtain

$$P[T > 2.060] = 0.025,$$

and we note that with larger r values, $t(0.025; r)$ approaches the value $z(0.025) = 1.96$ associated with the $N(0, 1)$ distribution. ∎

Once we accept the fact that $T = (\overline{X} - \mu)/(S/\sqrt{n})$ is $t(n - 1)$, we can write

$$P\left[-t(\alpha/2; n - 1) \le \frac{\overline{X} - \mu}{S/\sqrt{n}} \le t(\alpha/2; n - 1)\right] = 1 - \alpha.$$

However, as before, this formula can be rewritten as

$$P\left[\overline{X} - t(\alpha/2; n - 1)\frac{S}{\sqrt{n}} \le \mu \le \overline{X} + t(\alpha/2; n - 1)\frac{S}{\sqrt{n}}\right] = 1 - \alpha.$$

Once the sample is observed and the characteristics $\bar{x}$ and s are computed,

$$\left[\bar{x} - t(\alpha/2; n - 1)\frac{s}{\sqrt{n}}, \bar{x} + t(\alpha/2; n - 1)\frac{s}{\sqrt{n}}\right]$$

provides a $100(1 - \alpha)$ percent confidence interval for μ. Again, for ease of presentation, we frequently write this as $\bar{x} \pm t(\alpha/2; n - 1)s/\sqrt{n}$. The width of the interval depends on the chosen confidence coefficient and on $s/\sqrt{n}$, the estimated standard error of the estimator $\overline{X}$.

Recall that this result about the sampling distribution of $T = (\overline{X} - \mu)/(S/\sqrt{n})$ required that the random sample be taken from $N(\mu, \sigma^2)$. What if the underlying distribution is not normal? We still tend to use $\overline{x} \pm t(\alpha/2; n - 1)s/\sqrt{n}$ for the confidence interval for μ, unless we believe that the sample has come from an underlying distribution with very long and thick tails or a distribution that is highly skewed. One or two extreme observations, or outliers, in the sample are usually an indication of the latter situations. If this is the case, *robust methods* must be used to form confidence intervals for μ. These procedures are discussed in more advanced courses.

Example 4.3-3   The mean μ of the times it takes a lab technician to perform a certain task is of interest. The lab technician was observed on $n = 16$ different occasions, and the mean and standard deviation of these 16 times were, respectively, $\overline{x} = 4.3$ minutes and $s = 0.6$ minute. Thus, $r = n - 1 = 15$, and with $\alpha = 0.05$, so that $\alpha/2 = 0.025$ and $t(0.025; 15) = 2.131$, a 95 percent confidence interval for μ is $4.3 \pm (2.131)(0.6/4)$. Our best estimate for the mean time is thus 4.3 minutes. Furthermore, we are 95 percent confident that the interval (3.98 minutes, 4.62 minutes) covers the unknown mean time μ.   ∎

×4.3-1 Tolerance Limits

Confidence intervals assess the reliability of point estimates of unknown population parameters. The fact that, prior to sampling, a certain interval covers the population parameter μ in $100(1 - \alpha)$ percent of the cases allows us to judge the quality of the estimate that we calculate from our sample. Although a confidence interval for an unknown mean μ of a distribution is extremely important, the production supervisor frequently wants a different interval—namely, an interval that contains most of the *individual* measurements. It is hoped that the latter interval is within the specifications of the process, because otherwise some measurements will fall outside the specification limits. That is, defectives will be produced. For instance, suppose that the inside diameter of a washer is normally distributed with $\mu = 0.503$ and $\sigma = 0.005$. Then we know, for example, that 99 percent of the washers have diameter $0.503 \pm 2.576(0.005)$ or, equivalently, are between 0.490 and 0.516. If the specifications for these diameters are given by 0.500 ± 0.005 (i.e., the diameter of a useful washer should be between 0.495 and 0.505), this observation causes some concern, because the interval (0.490, 0.516) is not within specifications and some washers will be outside the specification limits. Possibly, the specifications are unrealistic, and could be taken wider without harming the usefulness of the washers. However, if the specifications are correct, we must make certain changes to the process in order to bring the washers within specifications. Certainly, in this illustration, the mean μ should be shifted to 0.500 and the standard deviation should be reduced.

Unfortunately, we do not always know μ and σ; hence, they must be estimated by $\overline{x}$ and s, respectively. Once this is done, however, the interval $\overline{x} \pm (2.576)s$ does not necessarily contain 99 percent of the measurements, as μ and σ have only been approximated by $\overline{x}$ and s. If we restrict ourselves to intervals such as $\overline{x} \pm ks$, what value of k is needed so that the probability that the random interval $\overline{X} \pm kS$ covers at least $100p$ percent of the population is large, say, $1 - \alpha$? For example, p and α might be 0.99 and 0.05, respectively. The theoretical problem to be solved is this: Find the k such that

$$P[F(\overline{X} + kS) - F(\overline{X} - kS) \geq p] = 1 - \alpha,$$

where F is the cumulative distribution function of a distribution that is $N(\mu, \sigma^2)$. The solution is beyond the scope of this course, but statisticians have prepared tables of k values for various values of p and α, and one adaptation is Table C.9. Once we select k and determine the observed interval $\bar{x} \pm ks$, we say that $(\bar{x} - ks, \bar{x} + ks)$ is a $100(1 - \alpha)$ percent *tolerance interval* for at least $100p$ percent of the population (distribution). Because we wish to cover at least $100p$ percent of the population rather than simply the mean μ, $\bar{x} \pm ks$ must be much wider than the confidence interval for μ, $\bar{x} \pm t(\alpha/2; n - 1)s/\sqrt{n}$.

In Example 4.3-3, the interval from 3.98 to 4.62 is a 95 percent confidence interval for the mean μ. If, however, we want a 95 percent tolerance interval for at least $100p = 90$ percent of the population of the lab technician's times, we find from Table C.9 that, for $n = 16, k = 2.437$. Thus,

$$4.3 \pm (2.437)(0.6)$$

is the desired tolerance interval, given that $\bar{x} = 4.3$ and $s = 0.6$. That is, we are 95 percent confident that the interval from 2.84 to 5.76 minutes covers at least 90 percent of the individual times. This interval is much wider than the 95 percent confidence interval for μ, as it should be. Note that, for large n, the constant k approaches 1.645, nothing else than the upper fifth percentage point of the $N(0, 1)$ distribution. For large n, the estimates $\bar{x}$ and s approximate the population parameters and we can use the normal table to construct the tolerance interval.

Example 4.3-4   In a personal communication, George S. Kalemkarian of Deere & Company reported that an engineer in that company's Corporate Engineering Standards Group was interested in the hardness of certain castings. The engineer mentioned casually that he wanted to construct "a reasonably accurate interval that would describe the hardness of most castings." After some discussion, it became apparent that he was not interested in an interval for the *mean* hardness of a casting, but that he instead wanted an interval for the *individual hardness* values. That is, he wanted a tolerance interval for individual observations, rather than a confidence interval for the mean. These two types of intervals are quite different, but they are often confused in practice. The difference is that, whereas the length of the confidence interval for the mean becomes very small with large sample sizes, the bounds in the tolerance interval approach, with large n, the corresponding percentiles of the population.

Table 4.3-1 lists the Brinell hardness numbers (BHN) of 105 tested castings. The table is in the form of an ordered stem-and-leaf display that serves as a histogram for the data and describes the variability that we can expect among the different castings.

The product specifications for the hardness of such castings call for a target value of 214 BHN, with a lower specification of 173 BHN and an upper specification of 255 BHN. Table 4.3-1 shows that all 105 castings are within the specification limits. An interesting feature of this data set is that it consists of only seven different values—197, 207, 212, 217, 223, 229, and 241—even though, in theory, Brinell hardness is measured

Table 4.3-1 Hardness Values (BHN) of 105 Engine Castings

Stem	Leaf
19	7 7 7 7 7 7 7 7 7 7 7 7 7 7 7 7
20	7 7 7 7 7 7 7 7 7 7 7 7 7 7 7 7 7 7
21	2 2 7 7 7 7 7 7 7 7 7 7 7 7 7 7 7 7 7 7 7 7 7 7 7 7 7 7 7 7 7 7 7 7 7 7 7 7 7 7 7 7 7 7 7 7 7 7 7
22	3 3 3 3 9 9 9 9 9 9 9 9 9 9 9 9 9 9 9 9 9 9 9 9 9 9 9 9 9 9
23	
24	1

on a continuous scale. This feature is due to the particular measurement procedure that was used. A discussion with the engineer who was taking these measurements revealed that hardness is determined by impressing a 3,000-kilogram load into a surface of the casting and measuring the indentation diameter of the ball to the nearest 0.1 millimeter (mm). For example, a diameter of 4.1 mm converts to 217 BHN. Due to rounding, the data set includes only the aforesaid seven discrete points. Incidentally, the measurements were taken by the same person, so person-to-person variation is eliminated.

If we take the rounding into account, it appears that, apart from the one unusual value at 241, the distribution of the data is fairly symmetric. It also seems that the variability in the measurements can be approximated, at least roughly, by a normal distribution. Once we accept the fact that the normal distribution provides an adequate description of the variability, we can use subsequent sample information to construct tolerance intervals and check whether the manufactured items are within specifications. For example, suppose that the Brinell hardness values of a sample of $n = 10$ recently produced items are as follows:

$$217 \quad 217 \quad 197 \quad 207 \quad 217 \quad 223 \quad 229 \quad 229 \quad 217 \quad 217$$

Here, $\bar{x} = 217$ and $s = 9.57$. From Table C.9, with $n = 10$, we find that $k = 2.839$ if we want to be 95 percent confident that the resulting tolerance interval $(\bar{x} - ks, \bar{x} + ks)$ includes at least 90 percent of *all* casting hardness values. That is, we have

$$\text{lower tolerance limit} = 217 - (2.839)(9.57) = 189.8,$$

$$\text{upper tolerance limit} = 217 + (2.839)(9.57) = 244.2.$$

This result implies that we are highly confident that at least 90 percent of our castings are within the specification limits of 173 and 255 BHN. Similarly, 95 percent tolerance limits for at least 95 percent of the values are 184.7 and 249.3, again within specifications. Because 95 percent tolerance limits for at least 99 percent of the values extend beyond the specification limits, a very small percentage of the production may be outside the specifications. In sum, it appears that, at the moment, we are producing castings that are within specifications. However, we know that processes vary; thus, we should check at regular intervals whether we are continuing to produce at a satisfactory level.

Incidentally, it is interesting to observe that the 95 percent confidence interval for the mean μ of the population is

$$\bar{x} \pm t(0.025; 9)\frac{s}{\sqrt{n}} = 217 \pm 2.262\frac{9.57}{\sqrt{10}},$$

or, equivalently, the interval from 210.2 to 223.8. Of course, this interval is much shorter than the tolerance interval. ∎

4.3-2 Confidence Intervals for $\mu_1 - \mu_2$

Let us consider independent samples from two populations in which σ_1^2 and σ_2^2 are unknown, but the sample sizes, n_1 and n_2, are small. The confidence intervals for $\mu_1 - \mu_2$ in Section 4.2-2 may be poor approximations, because sample variances calculated from small samples are imprecise estimates of variances. Accordingly, we must make additional assumptions to obtain a confidence interval for $\mu_1 - \mu_2$. The additional assumptions are that the underlying populations are normal and the unknown variances are equal (or nearly so); that is, $\sigma_1^2 = \sigma_2^2 = \sigma^2$. With these assumptions, we know that

$$\frac{(n_1 - 1)S_X^2 + (n_2 - 1)S_Y^2}{\sigma^2}$$

is $\chi^2(r)$, where $r = n_1 + n_2 - 2$, because $(n_1 - 1)S_X^2/\sigma^2$ and $(n_2 - 1)S_Y^2/\sigma^2$ are independent $\chi^2(n_1 - 1)$ and $\chi^2(n_2 - 1)$. Here, we used the fact that the sum of independent chi-square random variables has a chi-square distribution. Thus, the ratio

$$T = \frac{\overline{X} - \overline{Y} - (\mu_1 - \mu_2)}{\sqrt{\sigma^2/n_1 + \sigma^2/n_2}} \bigg/ \sqrt{\frac{(n_1 - 1)S_X^2 + (n_2 - 1)S_Y^2}{\sigma^2}} \bigg/ (n_1 + n_2 - 2)$$

has a Student's t-distribution with $r = n_1 + n_2 - 2$ degrees of freedom, because the numerator is $N(0, 1)$, the denominator is the square root of a chi-square random variable divided by the number of its degrees of freedom, and the numerator and denominator are independent. The preceding equation can be rewritten as

$$T = \frac{\overline{X} - \overline{Y} - (\mu_1 - \mu_2)}{\sqrt{\dfrac{(n_1 - 1)S_X^2 + (n_2 - 1)S_Y^2}{n_1 + n_2 - 2}\left(\dfrac{1}{n_1} + \dfrac{1}{n_2}\right)}} = \frac{\overline{X} - \overline{Y} - (\mu_1 - \mu_2)}{\sqrt{S_p^2\left(\dfrac{1}{n_1} + \dfrac{1}{n_2}\right)}}.$$

The expression

$$S_p^2 = \frac{(n_1 - 1)S_X^2 + (n_2 - 1)S_Y^2}{n_1 + n_2 - 2}$$

is called the *pooled variance,* because it "pools" or combines the variance estimates from the two samples in proportion to their degrees of freedom. Now, S_X^2 and S_Y^2 are unbiased estimators of the common variance, so $E(S_p^2) = \sigma^2$. The inequalities, with $r = n_1 + n_2 - 2$, in

$$P[-t(\alpha/2; r) \le T \le t(\alpha/2; r)] = 1 - \alpha$$

can be rewritten as

$$\overline{X} - \overline{Y} - t(\alpha/2; r)S_p\sqrt{\frac{1}{n_1} + \frac{1}{n_2}} \le \mu_1 - \mu_2 \le \overline{X} - \overline{Y} + t(\alpha/2; r)S_p\sqrt{\frac{1}{n_1} + \frac{1}{n_2}}.$$

Once the samples are observed, the expression

$$\bar{x} - \bar{y} \pm t(\alpha/2; r = n_1 + n_2 - 2)s_p\sqrt{\frac{1}{n_1} + \frac{1}{n_2}}$$

provides a $100(1 - \alpha)$ percent confidence interval for $\mu_1 - \mu_2$.

Remark: The use of a pooled estimate of σ^2 is not recommended if there is suspicion that the variances of the two distributions are not the same. Pooling is especially dangerous in this circumstance if the sample sizes n_1 and n_2 are quite different. In such a case, it is safer to use the approach in Section 4.2-2 and calculate a confidence interval that uses $s_x^2/n_1 + s_y^2/n_2$ as the estimate of $\text{var}(\bar{X} - \bar{Y})$.

Example 4.3-5 An experiment compared the durability of two types of exterior house paint; type A was an oil-based paint, while type B was a latex paint. A total of $n_1 = 10$ and $n_2 = 12$ wood panels were covered with paints A and B, respectively. The panels were then exposed to a sequence of tests that involved extreme temperature, light, and moisture conditions. At the end of the experiment, the quality of the painted surface was rated on several characteristics (such as paint loss, luster of painted surface, cracking, and peeling). Scores between 0 (extremely poor) and 100 (excellent) were assigned to the 22 panels. For paint A (with $n_1 = 10$), it was found that $\bar{x} = 52$ and $s_x^2 = 400$; for paint B (with $n_2 = 12$), it was found that $\bar{y} = 46$ and $s_y^2 = 340$. Because individual measurements are influenced by numerous random factors, it is reasonable to assume that they are normally distributed. We could check this assumption by constructing normal probability plots, one for each group. (See Section 3.6.) A comparison of $s_x^2 = 400$ and $s_y^2 = 340$ shows that the sample variances are quite similar. This provides evidence that the population variances are about equal and justifies the pooling of the individual sample variances. A formal procedure for assessing whether two population variances are the same is given in Section 4.4-1.

Using the preceding theory, we can construct a 90 percent confidence interval for $\mu_1 - \mu_2$. Because $t(\alpha/2 = 0.05; n_1 + n_2 - 2 = 20) = 1.725$, this interval is given by

$$(52 - 45) \pm (1.725)\sqrt{367}\sqrt{\frac{1}{10} + \frac{1}{12}},$$

where the pooled variance is

$$s_p^2 = \frac{(9)(400) + (11)(340)}{10 + 12 - 2} = 367.$$

The 95 percent confidence interval 6 ± 14.1, or $(-8.1, 20.1)$ includes zero, so there is no evidence in these data that the mean durabilities of the two paints are different. ∎

Exercises 4.3

*4.3-1 The mean μ of the tear strength of a certain paper is under consideration. The $n = 22$ determinations (taken at random) yielded $\bar{x} = 2.4$ pounds.

    *(a) If the standard deviation of an individual measurement is known to be $\sigma = 0.2$, find an approximate 95 percent confidence interval for μ.

*(b) If the standard deviation σ is unknown, but the sample standard deviation is $s = 0.2$, determine a 95 percent confidence interval for μ.

*4.3-2 Let W be $\chi^2(12)$.

*(a) Determine the mean and the variance of W.

*(b) Find d_1, d_2, d_3, and d_4 so that $P(W > d_1) = 0.05$, $P(W > d_2) = 0.99$, $P(W > d_3) = 0.005$, and $P(W < d_4) = 0.025$. Use tables or computer software.

*4.3-3 Let T be $t(r = 11)$.

*(a) Determine the mean and the variance of T.

*(b) Determine d_1 so that $P(T > d_1) = 0.05$.

*(c) Determine d_2 so that $P(-d_2 < T < d_2) = 0.95$.

4.3-4 Let μ be the mean mileage of a certain brand of tire. A sample of $n = 14$ tires was taken at random, resulting in $\bar{x} = 32{,}132$ and $s = 2{,}596$ miles. Find a 99 percent confidence interval for μ.

4.3-5 Let X be $N(\mu, \sigma^2)$. Then $Z = (X - \mu)/\sigma$ is $N(0, 1)$. Find the distribution function of $W = Z^2$; that is, find $G(w) = P(W \leq w) = P(-\sqrt{w} \leq Z \leq \sqrt{w})$, for $w \geq 0$, such that

$$G(w) = 2 \int_0^{\sqrt{w}} \frac{1}{\sqrt{2\pi}} \exp\left(\frac{-z^2}{2}\right) dz.$$

Show that the p.d.f. $g(w) = G'(w)$ of W is $\chi^2(1)$.

Hint: From the fundamental theorem of calculus, we know that

$$\frac{d}{dx}\left[\int_a^{u(x)} g(y)\,dy\right] = g[u(x)]\frac{d}{dx}u(x).$$

4.3-6 We stated without proof that the quantity $(n - 1)S^2/\sigma^2$ follows a chi-square distribution with $n - 1$ degrees of freedom if our random sample is taken from the $N(\mu, \sigma^2)$ distribution. Convince yourself that this is true by performing the following experiment: Use a computer program that generates values from an $N(\mu = 10, \sigma^2 = 4)$ distribution. From samples of size $n = 4$, calculate $\bar{x}$, s^2, and $(n - 1)s^2/\sigma^2 = 3s^2/4$. Repeat the experiment 500 times and construct a normalized relative frequency histogram of $3s^2/4$. [*Note:* A normalized frequency histogram divides relative frequencies by the widths of the intervals; Minitab calls this a density histogram, and you can get it by clicking on density in the Scale, Y-Scale Type window.] Compare the relative frequencies from the simulations with the p.d.f. of the $\chi^2(3)$ distribution, which is $f(w) = cw^{1/2}e^{-w/2}$, where $c = 0.4$.

*4.3-7 The effectiveness of two methods of teaching statistics is compared. A class of 24 students is randomly divided into two groups and each group is taught according to a different method. The test scores of the two groups at the end of the semester show the following characteristics:

$$n_1 = 13, \quad \bar{x} = 74.5, \quad s_x^2 = 82.6$$

and

$$n_2 = 11, \quad \bar{y} = 71.8, \quad s_y^2 = 112.6.$$

Assuming underlying normal distributions with $\sigma_1^2 = \sigma_2^2$, find a 95 percent confidence interval for $\mu_1 - \mu_2$.

4.3-8 Two rubber compounds were tested for tensile strength. Rectangular materials were prepared and pulled in a longitudinal direction. A sample of 14 specimens, 7 from compound A and 7 from compound B, was prepared, but it was later found that 2 B specimens were defective, and they had to be removed from the test. The tensile strengths measured (in units of 100 pounds per square inch) are as follows:

A: 32   30   33   32   29   34   32

B: 33   35   36   37   35

Calculate a 95 percent confidence interval for the difference in the mean tensile strengths of the two rubber compounds. State your assumptions.

4.3-9 Consider the ratio (diesel use of test truck)/(diesel use of control truck) for the data in Project 3 of Chapter 1. Compare the baseline and the test runs, and construct a 95 percent confidence interval for the mean difference.

*4.3-10 Refer to Exercises 4.3-1 and 4.3-4, and in each case compute a 95 percent tolerance interval for 95 percent of the individual values in the population.

4.4 Other Confidence Intervals

In Sections 4.2 and 4.3, we discussed confidence intervals for population means. Now we extend the discussion to other population parameters: the variance σ^2 and the proportion p.

4.4-1 Confidence Intervals for Variances

A critical assumption in the ensuing development of confidence intervals for σ^2 is that we sample from a normal distribution. Let $S^2 = \sum_{i=1}^{n}(X_i - \overline{X})^2/(n - 1)$ be the sample variance from a random sample of size n taken from a $N(\mu, \sigma^2)$ distribution. We know from the preceding section that S^2 is an unbiased estimate of σ^2 and that the sampling distribution of $(n - 1)S^2/\sigma^2$ is $\chi^2(n - 1)$. Thus,

$$P\left[\chi^2(1 - \alpha/2; n - 1) \leq \frac{(n - 1)S^2}{\sigma^2} \leq \chi^2(\alpha/2; n - 1)\right] = 1 - \alpha,$$

where $\chi^2(\alpha/2; n - 1)$ is the upper $100(\alpha/2)$ percentage point [in other words, the $100(1 - \alpha/2)$ percentile] of the $\chi^2(n - 1)$ distribution. Rearranging terms within the probability statement so that σ^2 is in the middle of the interval leads to

$$P\left[\frac{(n - 1)S^2}{\chi^2(\alpha/2; n - 1)} \leq \sigma^2 \leq \frac{(n - 1)S^2}{\chi^2(1 - \alpha/2; n - 1)}\right] = 1 - \alpha.$$

This expression shows that under these particular assumptions (i.e., sampling from a normal distribution), the interval

$$\left[\frac{(n - 1)S^2}{\chi^2(\alpha/2; n - 1)}, \frac{(n - 1)S^2}{\chi^2(1 - \alpha/2; n - 1)}\right]$$

represents a $100(1 - \alpha)$ percent confidence interval for the unknown population variance σ^2. Furthermore, since the last probability statement remains unchanged when the square root of each of the three terms is taken, we find that a $100(1 - \alpha)$ percent confidence interval for the standard deviation σ is given by

$$\left[\sqrt{\frac{(n-1)S^2}{\chi^2(\alpha/2; n-1)}}, \sqrt{\frac{(n-1)S^2}{\chi^2(1-\alpha/2; n-1)}} \right].$$

Example 4.4-1   In Example 4.3-3, we studied the times it takes a lab technician to complete a certain task. A sample of $n = 16$ times led to the sample mean $\bar{x} = 4.3$ minutes and standard deviation $s = 0.6$ minute. With $r = n - 1 = 15$, we have $\chi^2(0.025; 15) = 27.488$ and $\chi^2(0.975; 15) = 6.262$. We then find that a 95 percent confidence interval for σ^2 is given by $[(15)(0.6)^2/27.488, (15)(0.6)^2/6.262]$, or $(0.20, 0.86)$. Taking the square root of each of these limits, we are 95 percent confident that the interval from 0.44 minute to 0.93 minute will cover the unknown σ. (Note that this interval is quite large.)   ∎

Remark: Confidence intervals for σ^2 and σ are usually quite wide, unless the sample size is very large. An estimate of a variance will always be less precise than an estimate of a mean, because an average of squared observations is more variable than an average of observations.   ∎

Now suppose that we have two independent random samples from normal distributions and we want to compare their variances σ_1^2 and σ_2^2 by finding a confidence interval for their ratio, σ_1^2/σ_2^2. Before we can do this, we have to introduce another distribution: the F-distribution. As we will see later, this is a very important distribution in statistics.

Let U and V be independent chi-square random variables with r_1 and r_2 degrees of freedom, respectively. Then we say that

$$F = \frac{U/r_1}{V/r_2}$$

has an *F-distribution with r_1 and r_2 degrees of freedom;* for brevity, we say that F is $F(r_1, r_2)$. Clearly, the reciprocal of F,

$$\frac{1}{F} = \frac{V/r_2}{U/r_1},$$

must be $F(r_2, r_1)$, because U and V (and the associated degrees of freedom, r_1 and r_2) have exchanged roles. The probability density functions of F-distributions look very much like the right-skewed densities of chi-square distributions in Figure 4.3-1, except that the F-distributions are centered around 1 and the two parameters r_1 and r_2 allow greater flexibility. The upper percentage points $F(\alpha; r_1, r_2)$, such that $P[F > F(\alpha; r_1, r_2)] = \alpha$, for tail probabilities $\alpha = 0.05$ and $\alpha = 0.01$, are given in Table C.7. Because $\alpha = P[F > F(\alpha; r_1, r_2)] = P[1/F < 1/F(\alpha; r_1, r_2)]$ and $1/F$ has an $F(r_2, r_1)$ distribution, it follows that $F(1 - \alpha; r_2, r_1) = 1/F(\alpha; r_1, r_2)$.

Example 4.4-2 Let F be $F(5, 8)$. Use statistical software to get cumulative probabilities and inverse cumulative probabilities (i.e., percentiles). You can use the Minitab command "Calc > Probability Distributions > F." The 95th percentile of $F(5, 8)$ is 3.69, and the 99th percentile is 6.63. Hence, the upper percentage points are $F(0.05; 5, 8) = 3.69$ and $F(0.01; 5, 8) = 6.63$. Similarly, $F(0.05; 8, 5) = 4.82$. ∎

We can use the F-distribution to find confidence intervals for σ_1^2/σ_2^2. With independent random samples of sizes n_1 and n_2 from the respective $N(\mu_1, \sigma_1^2)$ and $N(\mu_2, \sigma_2^2)$ distributions, we know that

$$\frac{(n_1 - 1)S_1^2}{\sigma_1^2} \text{ is } \chi^2(n_1 - 1)$$

and

$$\frac{(n_2 - 1)S_2^2}{\sigma_2^2} \text{ is } \chi^2(n_2 - 1).$$

Because the samples are independent random samples, these two random variables are independent. Hence,

$$F = \frac{\dfrac{(n_2 - 1)S_2^2}{\sigma_2^2} \Big/ (n_2 - 1)}{\dfrac{(n_1 - 1)S_1^2}{\sigma_1^2} \Big/ (n_1 - 1)} = \frac{\sigma_1^2 S_2^2}{\sigma_2^2 S_1^2}$$

is $F(r_1 = n_2 - 1, r_2 = n_1 - 1)$, and therefore,

$$P\left[F(1 - \alpha/2; r_1, r_2) \leq \frac{\sigma_1^2 S_2^2}{\sigma_2^2 S_1^2} \leq F(\alpha/2; r_1, r_2)\right] = 1 - \alpha.$$

Because $F(1 - \alpha/2; r_1, r_2) = 1/F(\alpha/2; r_2, r_1)$, the inequality can be written as

$$P\left[\frac{1}{F(\alpha/2; r_2, r_1)}\frac{S_1^2}{S_2^2} \leq \frac{\sigma_1^2}{\sigma_2^2} \leq F(\alpha/2; r_1, r_2)\frac{S_1^2}{S_2^2}\right] = 1 - \alpha.$$

That is, the probability that the random interval

$$\left[\frac{1}{F(\alpha/2; n_1 - 1, n_2 - 1)}\frac{S_1^2}{S_2^2}, F(\alpha/2; n_2 - 1, n_1 - 1)\frac{S_1^2}{S_2^2}\right]$$

includes σ_1^2/σ_2^2 is $1 - \alpha$. Once the samples are observed and s_1^2 and s_2^2 are computed, the corresponding computed interval provides a $100(1 - \alpha)$ percent confidence interval for σ_1^2/σ_2^2.

Example 4.4-3 Consider again Example 4.3-5, about the accelerated life testing of paints. There, $n_1 = 10$, $s_1^2 = 400$, $n_2 = 12$, and $s_2^2 = 340$. Thus, $F(0.05; 9, 11) = 2.90$ and $F(0.05; 11, 9) = 3.10$, and the interval

$$\left[\left(\frac{1}{2.90}\right)\left(\frac{400}{340}\right), \ (3.10)\left(\frac{400}{340}\right)\right]$$

or, equivalently, $(0.41, 3.65)$ is a 90 percent confidence interval for σ_1^2/σ_2^2. This interval contains the number 1; hence, we would not doubt the statement that the variances, σ_1^2 and σ_2^2, are equal. Note, however, that the interval is very wide. This is a common, and objectionable, feature of confidence intervals for σ_1^2/σ_2^2. ∎

4.4-2 Confidence Intervals for Proportions

Let us now discuss confidence intervals for an unknown population *proportion p*—for example, the proportion p of manufactured items that are acceptable (or flawed) or the proportion p of companies that use statistical techniques for process control. Each sampled item (product, company, etc.) can be classified according to one of two categories—say, "success" and "failure," so the outcomes in a random sample correspond to realizations of n independent Bernoulli trials with unknown probability of success p. Let Y be the number of successes in our random sample, and let Y/n be the sample proportion (i.e., relative frequency of success). Then Y follows a binomial distribution with mean np and variance $np(1-p)$, and the sample proportion $\hat{p} = Y/n$ is an unbiased estimator of p with standard deviation (standard error) $\sqrt{\mathrm{var}(\hat{p})} = \sqrt{\dfrac{p(1-p)}{n}}$. Furthermore, the central limit theorem implies that

$$\frac{Y - np}{\sqrt{np(1-p)}} = \frac{(Y/n) - p}{\sqrt{p(1-p)/n}}$$

is approximately $N(0, 1)$. This means that

$$P\left[-z(a/2) \le \frac{(Y/n) - p}{\sqrt{p(1-p)/n}} \le z(\alpha/2)\right] \approx 1 - \alpha.$$

The inequalities in the probability statement can be written as

$$\frac{Y}{n} - z(\alpha/2)\sqrt{\frac{p(1-p)}{n}} \le p \le \frac{Y}{n} + z(\alpha/2)\sqrt{\frac{p(1-p)}{n}}.$$

However, once Y is observed to be y, these limits cannot be evaluated, because they still contain the unknown p. But if p in those limits is approximated by the sample proportion of successes, y/n, then we obtain the approximate $100(1-\alpha)$ percent confidence interval

$$\frac{y}{n} \pm z(\alpha/2)\sqrt{\frac{(y/n)(1 - y/n)}{n}}$$

for p.

Example 4.4-4   Suppose we interview $n = 200$ voters, of which $y = 104$ say that they plan to vote for a certain candidate. Then, with $\alpha = 0.10$ and $z(\alpha/2) = 1.645$, we find that

$$\frac{104}{200} \pm 1.645\sqrt{\frac{(0.52)(0.48)}{200}}$$

or, equivalently, $(0.462, 0.578)$ is an approximate 90 percent confidence interval for p, the fraction of all voters favoring this candidate.   ■

We emphasize at this point that it is very important that the subjects be selected *randomly* from the underlying population. If they are not selected at random, the binomial distribution may no longer be appropriate, and we may not be able to attach confidence intervals to our estimates.

Example 4.4-5   Let Y be $b(n, p)$. How large should the sample size n be such that we are $100(1 - \alpha)$ percent confident that our estimate is within h units of the unknown population proportion—for example, 99.74 percent certain that it is within $h = 0.03$, or three percentage points, of p? From

$$P\left[\left|\frac{Y}{n} - p\right| \leq z(\alpha/2)\sqrt{\frac{p(1 - p)}{n}}\right] = 1 - \alpha,$$

it follows that

$$h = z(\alpha/2)\sqrt{\frac{p(1 - p)}{n}} \quad \text{and} \quad n = p(1 - p)\left[\frac{z(\alpha/2)}{h}\right]^2.$$

Now, p is unknown; but $p(1 - p)$ reaches a maximum when $p = 1/2$. Hence, with $p = 1/2$ and $z(\alpha/2) = z(0.0013) = 3.0$, we find

$$n = \frac{1}{4}\left[\frac{z(\alpha/2)}{h}\right]^2 = \frac{1}{4}\left(\frac{3.0}{0.03}\right)^2 = 2500.$$

The sample size $n = 2,500$ is more than adequate for us to be 99.74 percent confident that the estimate is within three percentage points of p. This example explains why many of the well-known polls interview from 2,000 to 3,000 persons to obtain a reliable estimate of the percentage of voters favoring a certain candidate.   ■

If there are two different methods of performing a certain task, we may want to compare their probabilities of success—say, p_1 and p_2. Let Y_1 and Y_2 be the respective numbers of successes from the two independent random samples of sizes n_1 and n_2. Then the random variables Y_1 and Y_2 have binomial distributions $b(n_1, p_1)$ and $b(n_2, p_2)$, respectively. From the central limit theorem, we know that the distributions of Y_1 and Y_2, or of Y_1/n_1 and Y_2/n_2, can be approximated by normal distributions. Hence, their difference $Y_1/n_1 - Y_2/n_2$ has an approximate normal distribution with mean $p_1 - p_2$ and variance $[p_1(1 - p_1)/n_1] + [p_2(1 - p_2)/n_2]$. The standardized random variable

$$Z = \frac{(Y_1/n_1 - Y_2/n_2) - (p_1 - p_2)}{\sqrt{p_1(1 - p_1)/n_1 + p_2(1 - p_2)/n_2}}$$

has an approximate $N(0, 1)$ distribution. Rewriting the inequalities in

$$P[-z(\alpha/2) \leq Z \leq z(\alpha/2)] \approx 1 - \alpha,$$

we find that the probability that the random interval

$$\frac{Y_1}{n_1} - \frac{Y_2}{n_2} \pm z(\alpha/2)\sqrt{\frac{p_1(1-p_1)}{n_1} + \frac{p_2(1-p_2)}{n_2}}$$

includes $p_1 - p_2$ is about $1 - \alpha$. Performing the experiments, observing y_1 and y_2, and approximating p_1 and p_2 under the radical sign by y_1/n_1 and y_2/n_2, respectively, we obtain the approximate $100(1 - \alpha)$ percent confidence interval for $p_1 - p_2$:

$$\frac{y_1}{n_1} - \frac{y_2}{n_2} \pm z(\alpha/2)\sqrt{\frac{(y_1/n_1)(1-y_1/n_1)}{n_1} + \frac{(y_2/n_2)(1-y_2/n_2)}{n_2}}.$$

Example 4.4-6   A company suspects that its two major plants produce different proportions of "grade A" items. Samples of sizes $n_1 = n_2 = 300$ were selected from a week's production of the two factories, and $y_1 = 213$ and $y_2 = 189$ items were classified as grade A. Thus, an approximate 95.44 percent confidence interval for $p_1 - p_2$ is

$$0.71 - 0.63 \pm 2\sqrt{\frac{(0.71)(0.29)}{300} + \frac{(0.63)(0.37)}{300}}$$

or, equivalently, $(0.004, 0.156)$. Because the confidence interval does not include zero, we conclude that the first factory produces, on average, a higher percentage of grade A items than the second.   ∎

Exercises 4.4

4.4-1   Refer to Exercise 4.2-2. Calculate 95 percent confidence intervals for σ^2 and σ.

4.4-2   Refer to Exercises 4.3-7 and 4.3-8, and in each case, compute 90 percent confidence intervals for σ_1^2/σ_2^2, the ratio of the population variances.

4.4-3   One tire manufacturer found that after 5,000 miles, $y = 32$ of $n = 200$ steel-belted tires selected at random were defective. Find an approximate 99 percent confidence interval for p, the proportion of defective tires in the total production.

*4.4-4   To test two different training methods, 200 workers were divided at random into two groups of 100 each. At the end of the training program, there were $y_1 = 62$ and $y_2 = 74$ successes. Find an approximate 90 percent confidence interval for $p_1 - p_2$, the difference of the true proportions of success.

*4.4-5   A sample of $n = 21$ observations from a $N(\mu, \sigma^2)$ distribution leads to $\bar{x} = 74.2$ and $s^2 = 562.8$. Determine a 90 percent confidence interval for σ^2.

*4.4-6   Let Y be $b(n, p)$. When Y is observed to be y, we want $y/n \pm 0.05$ to be an approximate 95 percent confidence interval for p.

(a) If we know that p is around $1/4$, how should we choose the sample size n?

(b) If we did not have the prior information that p was about $1/4$, what size sample should we take?

4.4-7   We want to be 90 percent confident that the difference of two sample proportions is within 0.06 of the difference of the population proportions, $p_1 - p_2$.

If we take n observations from each population, how should we select this value of n?

4.4-8   In a *New York Times*/CBS poll, 56 percent of 2,000 randomly selected voters in New York City said that they would vote for the incumbent in a certain two-person race. Calculate a 95 percent confidence interval for the population proportion. Discuss its implication. Discuss carefully what is meant by the population, how you would carry out the random sampling, and what other factors could lead to differences between the responses in the survey and the actual votes on the day of the election.

4.4-9   Using your statistical software, plot the densities of the $F(2, 10)$ and $F(10, 2)$ distributions. In Minitab, use "Calc > Probability Distributions > F" and the "probability density" tab.

4.5   Tests of Characteristics of a Single Distribution

4.5-1   Introduction

The collection and the analysis of data are important components of the *scientific method* of learning. Data are used to confirm or refute existing theories, as well as to revise them and to formulate new ones. Theories and hypotheses should always be tested against data from carefully planned experiments. If the data contradict existing theories, one must search for better theories. Data often provide the information required for revisions of an inadequate theory. In cases where there is no existing theory, an exploratory analysis of the data usually provides a good starting point for formulating theories.

These introductory remarks illustrate the fact that sample data are routinely used to decide between competing hypotheses. *Statistical testing of hypotheses* is an area of statistics that deals with procedures for confirming and refuting hypotheses about distributions of random variables. The hypotheses are usually framed in terms of distribution (population) parameters. For example, two competing hypotheses may specify that the mean breaking strength of a certain alloy is greater (not greater) than 2,000 pounds, that the mean yield of product A is smaller (not smaller) than that of product B, that the mean tensile strengths of five different alloys are the same (not the same), and so on. In the discussion that follows, we motivate this decision-theoretic concept with an example.

Suppose that we want to determine whether certain changes in a production setup reduce the amount of time it takes to complete a certain task. Suppose also that, under the standard setup, it takes a worker 30 minutes, on average, to assemble a product. We already know that individual completion times vary and usually do not equal 30 minutes exactly. The variation may be due to slight changes in work conditions due to variability among the individual components (not all parts fit perfectly) or to varying abilities of workers (the training may not be uniform). That is, the completion time is a random variable and we denote it by X. Furthermore, suppose that we have enough past data to know that completion times follow a normal distribution with mean $\mu = 30$ and standard deviation $\sigma = 1$. Then, about 95 percent of the completion times are between 28 and 32 minutes.

   In this example, the hypotheses to be tested are in terms of the mean parameter μ. That is, we assume that the proposed changes do not affect the standard deviation $\sigma = 1$. The "no change" hypothesis specifies that the revised setup has had no effect and that the mean completion time is still $\mu = 30$. Let us now collect some data, and, on the basis of the information collected, let us decide between the two hypotheses $\mu = 30$ and $\mu < 30$. Initially, let us select just a single worker and assume that his completion time under the revised setup is 29 minutes. Could we conclude that there has been a reduction in the mean completion time? It should be fairly obvious that such a conclusion would stand on very shaky grounds: Just one completion time that is only one standard deviation below the hypothesized value is certainly not convincing evidence of a reduction in the mean. We know that for a random variable having the normal distribution $N(30, 1)$, the probability of a realization that is 29 or smaller is about 0.16. Thus, it is certainly not very unusual to obtain a realization as small as $x = 29$. However, let us now assume that we have $n = 5$ observations under the revised setup, and suppose that the average of these observations is 29. Then the evidence that the mean completion time has been reduced is much stronger. Or, even better, assume that the average of $n = 25$ completion times is 29. It is fairly obvious from Figure 4.5-1 that there is now little doubt which of the two hypotheses is more appropriate; we would conclude that $\mu < 30$. The central limit theorem in Section 4.1 tells us that the distribution of an average of a random sample of size n from an $N(30, 1)$ distribution is again normal with mean 30 and standard deviation $1/\sqrt{n}$. With $n = 25$, the probability of obtaining an average of 29 or smaller is

$$P(\overline{X} \leq 29) = P\left(\frac{\overline{X} - 30}{1/5} \leq \frac{29 - 30}{1/5}\right) = P(Z \leq -5) = 0.0000003,$$

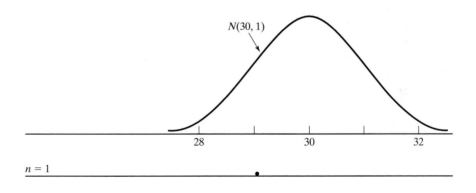

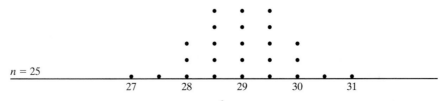

Figure 4.5-1   A plot of the $N(\mu = 30, \sigma^2 = 1)$ distribution, together with hypothetical sample realizations from samples of size $n = 1$ and $n = 25$

which makes an average that is 29 or smaller an extremely rare event among samples of size $n = 25$. This event certainly is not consistent with the hypothesis that $\mu = 30$, and we are on very safe ground to conclude that the mean completion time has been reduced. We say that there is a *statistically significant difference* between the observed mean, 29, and the hypothesized mean, 30. In other words, the difference from the hypothesized value (i.e., the difference $29 - 30 = -1$) is *statistically significant*. Figure 4.5-1 shows this graphically.

And now, a word of caution about a very common misunderstanding: *A difference that is statistically significant is not always of practical importance*. A saving of 1 minute per completion may not lead to savings for the company and may be practically irrelevant. For example, workers may be on the job for 8 continuous hours, and 16 minutes saved per shift may not be enough to start a new piece. Modifications in the revised setup may also be expensive and not worth this improvement in time. It may be, however, that a difference of 1 minute will save a lot of money, especially if the company produces many items, if tasks can be shared, and if the time savings exceed the cost of setup changes. We emphasize that the question of practical relevance can be answered only in the context of the cost structure; it cannot be answered by a statistical significance approach, which just quantifies how likely (or rare) it is that a particular sample result differs from a hypothesized value.

4.5-2 Possible Errors and Operating Characteristic Curves

We begin by introducing some terminology through an illustration. A manufacturer of printed circuits had observed that, on average, about $p = 0.03$ of the circuits failed. One of the engineers suggested certain changes in the design that she thought would improve the product. It was decided to try her suggestion. To check on this new design, $n = 100$ circuits were to be made in the new way. It was agreed that if only zero or one failed, the company would adopt the new procedure. Another engineer, who had taken a statistics course, immediately recognized that the new p could still equal 0.03 and yet only zero or one failure in $n = 100$ circuits might result; thus, the company would mistakenly accept the new design. This is an error that is referred to as a *type I error*. Still, the other engineer also recognized that the new design might truly be an improvement and not be accepted; for example, p could be reduced to 0.01, yet two or more failures could be observed. This second kind of error is called a *type II error*. To study the probabilities of these two types of errors with the rule that had been agreed upon, the engineer decided to model the situation as follows:

Let Y be the number of failures in $n = 100$ independent trials having the binomial distribution $b(100, p)$. The "no change" theory, called the *null hypothesis*, is $H_0: p = 0.03$. The engineer's theory of improvement by the new design, called the *alternative hypothesis*, is $H_1: p < 0.03$. The decision rule is that we reject H_0 and accept H_1 if $Y \le 1$. Or, equivalently, we accept $H_0: p = 0.03$ if $Y \ge 2$. Let us compute the probability of accepting H_0 for various values of p. We denote that probability by

$$OC(p) = P(Y \ge 2; p) = 1 - P(Y \le 1; p) = 1 - \sum_{y=0}^{1} \binom{100}{y} p^y (1 - p)^{100-y}$$

We call this function of p the *operating characteristic curve*. Because p is small and n is large here, the binomial distribution can be approximated by the Poisson distribution with $\lambda = np$. For example, if $p = 0.03$ and $np = 3$, then

$$OC(0.03) = 1 - P(Y \leq 1; p = 0.03) = 1 - 0.199 = 0.801$$

where we are using the cumulative probability of the Poisson distribution. (The Minitab command "Calc > Probability Distributions" can be used.) Other selected values of the operating characteristic curve are

$$OC(0.005) = 1 - P(Y \leq 1; p = 0.005) = 1 - 0.910 = 0.090,$$

$$OC(0.01) = 1 - P(Y \leq 1; p = 0.01) = 1 - 0.736 = 0.264,$$

and

$$OC(0.02) = 1 - P(Y \leq 1; p = 0.02) = 1 - 0.406 = 0.594.$$

Clearly, we want the probability of accepting H_0: $p = 0.03$ to be large if, in fact, p is actually equal to 0.03. Is 0.801 high enough? Maybe not; many persons like this probability to be at least 0.95. By contrast, if the new design has improved the process so that p is reduced, say, to $p = 0.01$, we want the probability of accepting H_0: $p = 0.03$ to be low. Is the probability 0.264 low enough? We doubt it. So here, at $p = 0.03$ and $p = 0.01$, the probabilities of the two types or errors are

$$P(\text{type I error}) = P(\text{rejecting } H_0 \text{ when } H_0 \text{ is true})$$

$$= 1 - OC(0.03) = 1 - 0.801 = 0.199 = \alpha$$

when $p = 0.03$, and

$$P(\text{type II error}) = P(\text{accepting } H_0 \text{ when } H_1 \text{ is true}) = OC(0.01) = 0.264 = \beta$$

when $p = 0.01$. These two probabilities are usually denoted by α and β, respectively. $\alpha = P(\text{type I error})$ is also called the *significance level of the test*. The operating characteristic curve and the preceding two probabilities are depicted in Figure 4.5-2.

These probabilities of the two types of errors might be too large and can be reduced by increasing the sample size. For example, in Exercise 4.5-1 we show that,

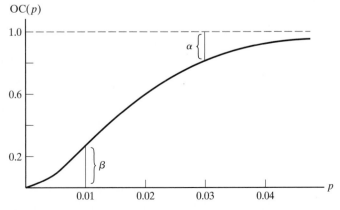

Figure 4.5-2 Operating characteristic curve $OC(p)$

with $n = 300$, if we reject H_0 when $Y \leq 5$, then $\alpha = 0.116$ and $\beta = 0.084$ (when $p = 0.01$). These probabilities are more satisfactory, but they still may be too large.

Let us now look more formally at the terms associated with tests of statistical hypotheses. A statement about a parameter of a distribution is called a *statistical hypothesis*. Hence, in our example, $p = 0.03$ is a statistical hypothesis, and we denote it by H_0: $p = 0.03$; we usually refer to it as the *null hypothesis*. The "research worker's hypothesis," which usually suggests that a change in the parameter has taken place, is called the *alternative hypothesis*. In this example, it is H_1: $p < 0.03$.

The *test of a statistical hypothesis* is a procedure by which we accept or reject H_0. In our example, the test of the null hypothesis H_0: $p = 0.03$ against the alternative hypothesis H_1: $p < 0.03$ was to take a sample of $n = 100$ observations and reject H_0 if the number of defectives Y is less than or equal to 1; that is, we reject H_0 and accept H_1 if $Y \leq 1$. The region of the statistic defined by the rejection rule — here, $Y \leq 1$ — is called the *critical region of the test*.

The two errors that are made in tests of hypotheses are

Type I error: rejecting H_0 when it is true.

Type II error: accepting H_0 when it is false (i.e., when H_1 is true).

The probability of a type I error, denoted by $\alpha = 0.199$ in our example, is called the *significance level of the test*. We found that the probability of a type II error when $p = 0.01$ is $\beta = 0.264$. Information about the probabilities of a type I and type II error is summarized in the *operating characteristic curve*, which represents the probability of accepting H_0 for various values of the parameter. In our example, we want the OC(p) curve to be high when $p = 0.03$, but low when $p < 0.03$, particularly when p is as small as 0.01.

4.5-3 Tests of Hypotheses When the Sample Size Can Be Selected

Example 4.5-1   Let us return to one of the illustrations in the introduction to this section: We want to determine whether certain changes in a production setup reduce the amount of time it takes to complete a certain task. We assume that in the past it has taken a worker 30 minutes, on average, to assemble a certain product and that the completion time X follows a distribution that is $N(30, \sigma^2 = 1)$.

At a recent meeting, engineers, supervisors, and workers discussed possible ways of reducing the required assembly time. One of the workers suggested certain changes in the production sequence that, in his opinion, should reduce the time. Let us assume that these changes affect only the mean; they do not alter the type of the distribution or the variance. That is, under the new procedure, X is $N(\mu, \sigma^2 = 1)$. To determine whether the worker's suggestion leads to an improvement ($\mu < 30$), it was decided to make the changes, obtain information on completion times, and test the null hypothesis H_0: $\mu = 30$ (no decrease in assembly time) against the alternative (here the worker's) hypothesis H_1: $\mu < 30$.

After several practice runs under the new procedure, the plant manager plans to (secretly) time the n workers as they assemble an item. Their completion times are denoted as $X_1, X_2, \ldots, X_n$. We are interested in the mean μ of the distribution of assembly times, so we plan to calculate the sample mean $\overline{X} = (1/n)\sum_{i=1}^{n} X_i$ of the n observations.

On the basis of the value of $\overline{X}$, we decide whether to accept H_0: $\mu = 30$ or H_1: $\mu < 30$. Clearly, $\overline{X} \leq c$, where c is some appropriate constant, is a good critical (rejection) region. That is, if $\overline{X} \leq c$, we accept H_1: $\mu < 30$ (and reject H_0: $\mu = 30$) and, accordingly, conclude that the new method has led to a reduction in assembly time.

The problem that remains is the selection of the sample size n and the constant c. To illustrate this selection, let us first use $n = 4$ and $c = 29.5$, even though these may be poor choices. That is, we take $n = 4$ observations and reject H_0: $\mu = 30$ if the sample mean $\overline{X} \leq 29.5$. Let us find the operating characteristic curve, OC(μ), of this test.

The operating characteristic curve is the probability of accepting H_0: $\mu = 30$. Thus, given that $\overline{X}$ is $N(\mu, \sigma^2/n)$ and $\sigma/\sqrt{n} = 1/\sqrt{4} = 0.5$, we have

$$\text{OC}(\mu) = P(\overline{X} > 29.5; \mu) = P\left(\frac{\overline{X} - \mu}{0.5} > \frac{29.5 - \mu}{0.5}; \mu\right) = 1 - \Phi\left(\frac{29.5 - \mu}{0.5}\right).$$

Substituting various values of μ, we obtain

μ	28.5	29	29.5	30
$\text{OC}(\mu)$	0.0228	0.1587	0.500	0.8413

The OC(μ) curve is plotted in Figure 4.5-3.

We note that if the new method has *not* improved the situation (i.e., if μ is still equal to 30), then the probability of accepting H_0: $\mu = 30$ is 0.8413. That is, when $\mu = 30$, the probability of falsely rejecting H_0: $\mu = 30$ is $\alpha = 0.1587$, the significance level of the test. This value of α is probably a little larger than we would like. Moreover, if μ has actually decreased to 29.5 under the new procedure, the probability of falsely accepting $\mu = 30$ is $\beta = 0.5$, probably much larger than desirable.

The reason the probabilities for these two types of errors are so large is that the sample size is too small. Thus, we must increase n. But rather than guess at n and c until we obtain a reasonable OC(μ) curve, let us first decide on a realistic OC(μ) curve and then find the necessary n and c. Suppose that the situation under consideration suggests that OC(30) = 0.95 and OC(29.5) = 0.10. The significance level of the test should be $\alpha = 0.05$; we want the chance of correctly accepting H_0 to be 0.95 if, in fact, $\mu = 30$. Furthermore, we want the chance of falsely accepting H_0 to be

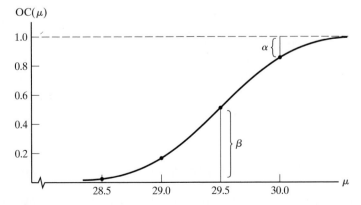

Figure 4.5-3 Operating characteristic curve, OC(μ), for $n = 4$ and $c = 29.5$

$\beta = 0.10$ if $\mu = 29.5$. The company produces a large number of items each month, so a decrease of 30 seconds may result in significant savings. Thus, the company wants to keep the probability of such an error at around 10 percent. The information on α and β provides the two equations

$$P(\overline{X} > c; \mu = 30) = 0.95$$

and

$$P(\overline{X} > c; \mu = 29.5) = 0.10.$$

Because $\overline{X}$ is $N(\mu, \sigma^2/n = 1/n)$, we have

$$0.95 = P\left(\frac{\overline{X} - 30}{1/\sqrt{n}} > \frac{c - 30}{1/\sqrt{n}}; \mu = 30\right) = 1 - \Phi\left(\frac{c - 30}{1/\sqrt{n}}\right),$$

$$0.10 = P\left(\frac{\overline{X} - 29.5}{1/\sqrt{n}} > \frac{c - 29.5}{1/\sqrt{n}}; \mu = 29.5\right) = 1 - \Phi\left(\frac{c - 29.5}{1/\sqrt{n}}\right);$$

thus,

$$\frac{c - 30}{1/\sqrt{n}} = -1.645, \quad \frac{c - 29.5}{1/\sqrt{n}} = 1.282.$$

Solving these two linear equations for c and $1/\sqrt{n}$ leads to

$$0.5 = 2.927\left(\frac{1}{\sqrt{n}}\right), \quad c = 29.5 + 1.282\left(\frac{0.5}{2.927}\right)$$

and

$$\sqrt{n} = 5.854, \quad c = 29.719.$$

So $n = 34$ and $c = 29.72$ together provide an approximate solution to this problem. That is, we should obtain 34 measurements. Then, if the observed average time $\overline{x}$ is less than or equal to 29.72, we reject the null hypothesis $H_0: \mu = 30$ and accept the alternative hypothesis $H_1: \mu < 30$, concluding that the new method is better than the old one. ∎

So far, we have discussed how to test whether the mean μ is smaller than a certain specified value μ_0. The research hypothesis that we want to prove or disprove becomes the alternative hypothesis $H_1: \mu < \mu_0$. The null hypothesis H_0 is the complement of the alternative hypothesis—that is, $H_0: \mu \geq \mu_0$—which says that the mean is *not* smaller than μ_0. Sometimes, if values larger than μ_0 are not possible or undesirable, we simply write the null hypothesis as $H_0: \mu = \mu_0$, because even a mean of μ_0 would make the alternative hypothesis false. We also assumed that σ^2, the variance of individual observations X_i, is known.

4.5-4 Tests of Hypotheses When the Sample Size Is Fixed

In Example 4.5-1, we selected both the critical value c and the sample size n so that our test led to specified probabilities for type I and type II errors. However, quite often the sample size n is already given. A researcher may already have collected data from a certain sample. All that is left to do in this situation is to select the critical value c. This selection is carried out such that the corresponding test achieves a specified significance level α (i.e., a specified probability of making a type I error). To achieve a test with significance level α, we choose the *critical region* (i.e., the *rejection region* of H_0) as $\overline{X} \leq c$ such that

$$\alpha = P(\overline{X} \leq c; \mu = \mu_0).$$

Now, under H_0, $\overline{X}$ is $N(\mu_0, \sigma^2/n)$; thus, this critical region is given by

$$\frac{\overline{X} - \mu_0}{\sigma/\sqrt{n}} \leq -z(\alpha) \quad \text{or} \quad \overline{X} \leq \mu_0 - z(\alpha)\frac{\sigma}{\sqrt{n}},$$

where $z(\alpha)$ is an upper percentage point from the $N(0, 1)$ distribution. The quantity $(\overline{X} - \mu_0)/(\sigma/\sqrt{n})$, called the *standardized test statistic*, calculates the difference between the sample mean and the hypothesized value and "standardizes" this difference by dividing by the standard deviation (i.e., the standard error) of $\overline{X}$. If the observed standardized test statistic is a large negative value—even smaller than $-z(\alpha)$—we reject H_0 in favor of H_1: $\mu < \mu_0$. Otherwise, if the standardized test statistic is larger than $-z(\alpha)$, we say that there is not enough evidence to reject H_0. Note that, alternatively, we can base our decision on the (nonstandardized) test statistic $\overline{X}$, which we then compare with $\mu_0 - z(\alpha)\sigma/\sqrt{n}$.

Assume that, in Example 4.5-1, $n = 34$ and $\overline{x} = 29.68$. Then the standardized test statistic is $(29.68 - 30)/(1/\sqrt{34}) = -1.87$. Because this number is smaller than $-z(0.05) = -1.645$, we reject H_0 in favor of H_1 at the $\alpha = 0.05$ significance level. We conclude that there has been a significant decrease in the average time it takes to complete the task in question. Note that one could also compare $\overline{x} = 29.68$ with the critical value $c = \mu_0 - z(\alpha)\sigma/\sqrt{n} = 30 - (1.645)(1)/\sqrt{34} = 29.72$ and reject H_0, because $\overline{x} \leq 29.72$.

Students confronted with statistical tests of hypothesis often have difficulty setting up the hypotheses and selecting the correct orientation of the critical region. The discussion that follows should help avoid these problems. First, remember that the research hypothesis is taken as the alternative hypothesis H_1. So far, we have dealt only with tests measuring reductions in the mean level μ (or proportions p); that is, H_1: $\mu < \mu_0$. The null hypothesis H_0 simply becomes the complement of H_1; thus, H_0: $\mu \geq \mu_0$, or simply $\mu = \mu_0$ if increases are irrelevant. It helps to draw this information on a graph such as the one in Figure 4.5-4. Draw in μ_0. The region to the left of μ_0 represents H_1; the region to the right represents H_0. Next, remember that a test for μ should be based on its sample estimate $\overline{X}$. (Alternatively, if the hypotheses are about p, the test should be carried out in terms of the sample proportion.) If $\overline{X}$ is small (much smaller than μ_0), we reject H_0. If $\overline{X}$ is large, we accept H_0.

Thus, the left region on our graph is the rejection region of H_0, and the right one is the acceptance region. The question now becomes where to draw the critical value c. Note that the test should achieve a certain probability of a type I error. If the true mean is indeed μ_0, given by H_0, then $\overline{X}$ follows a sampling distribution that is normal and centered at μ_0; let us draw in that normal distribution. We now want to add a line at c such that the probability of rejecting H_0 is α. That is, the area under the curve

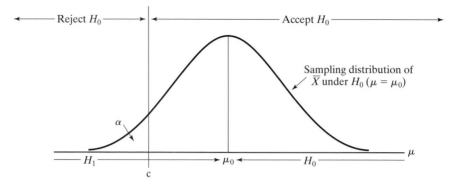

Figure 4.5-4 Graphical illustration of the test of H_0: $\mu \geq \mu_0$ against H_1: $\mu < \mu_0$

that is to the left of the line at c should be α. This simple picture shows that the critical value c must be smaller than μ_0; in fact, it is given by $\mu_0 - z(\alpha)\sigma/\sqrt{n}$.

How should one select the significance level α? The size of α depends on the particular application. A typical value in many applications is $\alpha = 0.05$. If large costs are associated with a type I error, one may want to consider even smaller values for α. However, a very small α implies a large value for $z(\alpha)$; thus, the resulting test is conservative, rejecting H_0 only in extreme cases.

There is yet another way to conduct a hypothesis test: Use the concept of a *probability value,* or a *p value.* When we test for a decrease in μ, we reject H_0 when the observed sample mean is small. In our example, $\bar{x} = 29.68$. One could now determine the probability under H_0 of obtaining a sample mean that small or even smaller:

$$p \text{ value} = P(\bar{X} \leq \bar{x}) = P\left(Z \leq \frac{\bar{x} - \mu_0}{\sigma/\sqrt{n}}\right) = P\left(Z \leq \frac{29.68 - 30}{1/\sqrt{34}}\right)$$

$$= \Phi(-1.87) = 0.031.$$

This probability is depicted in Figure 4.5-5. If the p value is less than or equal to the significance level α, it is highly unlikely that the observed sample average $\bar{x}$ comes from a distribution with mean $\mu_0 = 30$. We reject H_0 in such a case. If the p value

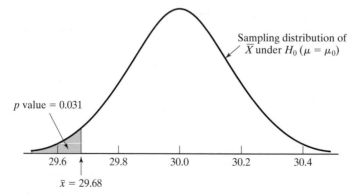

Figure 4.5-5 The probability value

is larger than α, it is quite possible that the sample mean $\bar{x}$ has come from the $N(\mu_0, \sigma^2/n)$ distribution, and we accept H_0. Rejecting H_0 when the p value is smaller than α leads to exactly the same conclusion as the procedure that uses the significance level α to construct the critical region. Most computer programs print the p value, and it is convenient to compare this value with a chosen significance level α. Note that then it is no longer necessary to obtain the critical value from the normal table. In our example, the p value is smaller than $\alpha = 0.05$. Thus, we reject H_0 at the $\alpha = 0.05$ significance level, meaning that there is evidence for a reduction in the mean time.

Remarks: There are two possible outcomes to a statistical hypothesis test: Either we reject the null hypothesis in favor of the alternative, or we accept the null hypothesis. We emphasize at this point that an acceptance of the null hypothesis does not necessarily imply that the null hypothesis is "true." What we mean to say is that, with the data at hand, there is not enough evidence to reject the null hypothesis.

A statistical hypothesis test tells us how likely it is that a particular sample result has come from a certain distribution. It assesses the statistical significance of an observed result. Earlier, we mentioned that a statistically significant reduction in the mean assembly time—say, from 30 to 29.68 minutes—may or may not be of practical significance. In some situations it may not make any difference whether the mean assembly time is reduced by 0.32 minute, but in certain other cases it may result in significant savings. It depends on the particular circumstance, and this has nothing to do with statistics. The point that we are making here is that, in addition to calculating the p value, we should always interpret the magnitude of the observed change and assess whether it is of practical importance.

Example 4.5-2   The tar content of a certain type of cigarette has been averaging 11.5 milligrams per cigarette with a standard deviation of 0.6 milligram. A researcher has discovered a new filter that he claims will reduce the mean tar content from 11.5. That is, he claims that if μ is the mean tar content of cigarettes with the new filter, $\mu < 11.5$. We consider $n = 40$ randomly selected cigarettes having the new filter and find that the average tar content is $\bar{x} = 11.4$. Is this enough evidence to reject $H_0: \mu = 11.5$ at the $\alpha = 0.10$ significance level?

Because we are looking for a decrease, we reject H_0 and accept H_1 whenever $\bar{X} \leq c$, or whenever the standardized test statistic is less than or equal to $-z(\alpha) = -z(0.10) = -1.28$. Here $(11.4 - 11.5)/(0.6/\sqrt{40}) = -1.05$ is the standardized test statistic. This is not less than the critical value, -1.28; hence, there is not enough evidence to reject H_0, which asserts that the mean tar content is 11.5. Equivalently, we could have based the decision on the

$$p \text{ value} = P(\bar{X} \leq 11.4; \mu = 11.5) = P(Z \leq -1.05)$$
$$= \Phi(-1.05) = 0.147 > \alpha = 0.10. \qquad \blacksquare$$

So far, we have discussed how to test whether a mean has decreased. How do we have to change our procedure if we want to test whether there has been an *increase*, in which case we are testing $H_0: \mu = \mu_0$ against $H_1: \mu > \mu_0$? In this case, we reject

H_0 in favor of H_1 when $\overline{X} \geq c$. If you are uncertain about the orientation of the critical region, go through the same steps that led to Figure 4.5-4, but reverse H_0 and H_1. If we want a test with significance level α, we require that the probability of a false rejection of H_0 be α; that is,

$$P(\overline{X} \geq c; \mu = \mu_0) = P\left(Z = \frac{\overline{X} - \mu_0}{\sigma/\sqrt{n}} \geq \frac{c - \mu_0}{\sigma/\sqrt{n}}\right) = \alpha.$$

Thus, we reject H_0 if the standardized test statistic

$$\frac{\overline{X} - \mu_0}{\sigma/\sqrt{n}} \geq z(\alpha), \quad \text{or} \quad \overline{X} \geq \mu_0 + z(\alpha)\frac{\sigma}{\sqrt{n}}.$$

Equivalently, we could base our decision on

$$p \text{ value} = P(\overline{X} \geq \overline{x}; \mu = \mu_0) = 1 - \Phi\left(\frac{\overline{x} - \mu_0}{\sigma/\sqrt{n}}\right).$$

Example 4.5-3   Let p be the average fraction of "grade A" items produced by a certain company. In the past, p has been about 0.62. Of the remaining 38 percent of the items, 3 percent were scrap and 35 percent were "seconds." A new method is proposed that is thought to increase p. Under the new method, $n = 250$ items are produced and $y = 172$ grade A items are found among them. Under the null hypothesis $H_0: p = 0.62$, the distribution of the number of grade A items Y has mean $250(0.62) = 155$, variance $250(0.62)(0.32) = 58.9$, and standard deviation $\sqrt{58.9} = 7.675$. Approximating this distribution by a normal distribution (due to the central limit theorem), we can calculate the approximate

$$\text{probability value} = P(Y \geq 172; p = 0.62) = P\left(\frac{Y - 155}{7.675} \geq \frac{171.5 - 155}{7.675}\right).$$

$$\approx 1 - \Phi(2.15) = 0.0158.$$

Because the probability value is smaller than, for example, 0.05, we reject $H_0: p = 0.62$ and accept $H_1: p > 0.62$ at the $\alpha = 0.05$ significance level.   ∎

Tests of $H_0: \mu = \mu_0$ against $H_1: \mu < \mu_0$ or $H_0: \mu = \mu_0$ against $H_1: \mu > \mu_0$ are called *one-sided* tests because the alternative hypotheses are one sided, not two sided as in $H_1: \mu \neq \mu_0$. There are occasions when two-sided tests are appropriate. Suppose that, in the past, mathematics test scores have had a mean of about 75 points. Assume that there has been an intervention (say, a new type of standardized test is being used) and we really do not know whether scores will increase, decrease, or stay about the same. Thus, we wish to test the null (no change) hypothesis $H_0: \mu = 75$ against the two-sided alternative, $H_1: \mu \neq 75$.

Assume that test scores are normally distributed with mean μ and variance $\sigma^2 = 100$. We reject $H_0: \mu = \mu_0$ if the standardized test statistic $(\overline{X} - \mu_0)/(\sigma/\sqrt{n})$ is large in absolute value. A test with significance level α rejects H_0 in favor of H_1 if

$$\left|\frac{\overline{X} - \mu_0}{\sigma/\sqrt{n}}\right| \geq z(\alpha/2),$$

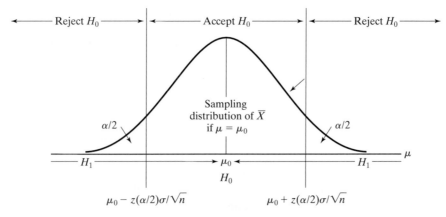

Figure 4.5-6 Graphical illustration of the two-sided test $H_0: \mu = \mu_0$ against $H_1: \mu \neq \mu_0$

because the probability of this event is α when $H_0: \mu = \mu_0$ is true. Alternatively, in terms of the (nonstandardized) test statistic $\overline{X}$, we reject H_0 if either $\overline{X} \leq \mu_0 - z(\alpha/2)\sigma/\sqrt{n}$ or $\overline{X} \geq \mu_0 + z(\alpha/2)\sigma/\sqrt{n}$. Figure 4.5-6 illustrates the latter situation graphically.

   Assume that we find the average to be $\bar{x} = 72$ with a sample of $n = 25$ tests. Then $(\bar{x} - \mu_0)/(\sigma/\sqrt{n}) = (72 - 75)/(10/5) = -1.5$. Because the absolute value of -1.5 is smaller than $z(\alpha/2) = z(0.025) = 1.96$, we accept H_0 at the $\alpha = 0.05$ significance level. The probability value in this two-sided example is

$$p \text{ value} = 2P\left(Z \geq \left| \frac{\bar{x} - \mu_0}{\sigma/\sqrt{n}} \right| \right) = 2P(Z \geq 1.5) = 0.134.$$

We double the probability here because we can reject H_0 with large or small values of the standardized variable Z.

   There is an obvious relationship between confidence intervals and two-sided tests. If a $100(1 - \alpha)$ percent confidence interval $\overline{X} \pm z(\alpha/2)(\sigma/\sqrt{n})$ does not include the value μ_0, then $|(\overline{X} - \mu_0)/(\sigma/\sqrt{n})| > z(\alpha/2)$ and we reject $H_0: \mu = \mu_0$ and accept $H_1: \mu \neq \mu_0$ at significance level α; the converse is also true. For example, if a confidence interval does not include zero, we can reject $H_0: \mu = 0$. If zero is included, we accept H_0.

   Thus far, we have assumed that the variance of the underlying distribution is known. Because the sampling distribution of $(\overline{X} - \mu)/(\sigma/\sqrt{n})$ is $N(0, 1)$ (either exactly if the X values are normally distributed or approximately if X is not normal but the sample size is large enough), we have used the normal tables to determine the critical values. Now let us assume that the underlying distribution of X values is normal, but that the variance σ^2 is estimated by the sample variance S^2. Then the sampling distribution of $(\overline{X} - \mu)/(S/\sqrt{n})$ is a t-distribution with $n - 1$ degrees of freedom and the critical values $z(\alpha)$ must be replaced by $t(\alpha; n - 1)$. Of course, if n is large (at least 30), we can use the normal table.

Example 4.5-4   A company claims that the mean deflection of its 10-foot steel beams is 0.012 inch. A construction contractor who purchases large quantities of steel beams suspects that the manufacturer misleads its customers and that the true deflection is, in fact, larger than the one claimed. To see whether this suspicion is justified, the contractor selects $n = 10$ beams at random from his inventory, determines their deflection, and conducts a test of H_0: $\mu = 0.012$ against H_1: $\mu > 0.012$. The 10 measurements are as follows:

| 0.0132 | 0.0138 | 0.0108 | 0.0126 | 0.0136 |
| 0.0112 | 0.0124 | 0.0116 | 0.0127 | 0.0131 |

From these measurements, the contractor calculates the mean $\bar{x} = 0.0125$ and the standard deviation $s = 0.0010$. The test statistic is $(0.0125 - 0.012)/(0.001/\sqrt{10}) = 1.56$, which is smaller than the critical value $t(0.05; n - 1 = 9) = 1.833$. Thus, this sample alone does not provide enough evidence to reject H_0 at significance level $\alpha = 0.05$. Until more measurements are taken, the contractor cannot claim that the company has made a false claim. ∎

Exercises 4.5

4.5-1 In the introductory illustration of Section 4.5-2, let $n = 300$ (instead of the $n = 100$ used there). Using the Poisson approximation, show that the critical region of $Y \leq 5$ (acceptance region of $Y \geq 6$) has an OC curve such that $OC(0.03) = 0.884$ and $OC(0.01) = 0.084$.

*4.5-2 Let $\bar{X}$ be the mean of a random sample of size $n = 36$ from $N(\mu, \sigma^2 = 9)$. Our decision rule is to reject H_0: $\mu = 50$ and to accept H_1: $\mu > 50$ if $\bar{X} \geq 50.8$. Determine the $OC(\mu)$ curve and evaluate it at $\mu = 50.0, 50.5, 51.0$, and 51.5. What is the significance level of the test?

*4.5-3 Let p be the fraction of engineers who do not understand certain basic statistical concepts. Unfortunately, in the past this number has been high, about $p = 0.73$. A new program to improve the engineers' knowledge of statistical methods has been implemented, and it is expected that under this program p would decrease from 0.73. To test H_0: $p = 0.73$ against H_1: $p < 0.73$, 300 engineers in the new program were tested, and 204 (i.e., 68 percent) did not comprehend certain basic statistical concepts. Compute the probability value required to determine whether these results indicate progress. That is, can we reject H_0 in favor of H_1? Use $\alpha = 0.05$.

4.5-4 In a certain industry, about 15 percent of the workers showed some signs of ill effects due to radiation. After management had claimed that improvements had been made, 140 workers were tested and 19 experienced some ill effects due to radiation. Does this result support management's claim? Use $\alpha = 0.05$.

*4.5-5 The mean time required to repair breakdowns of a certain copying machine is 93 minutes. The company which manufactures the machine claims that breakdowns of its new, improved model are easier to fix. To test this claim, $n = 73$ breakdowns of the new model were observed, resulting in a mean repair time of $\bar{x} = 88.8$ minutes and a standard deviation of $s = 26.6$ minutes. Use the significance level $\alpha = 0.05$. What is your conclusion?

4.5-6 In an industrial training program, students have been averaging about 65 points on a standardized test. The lecture system was replaced by teaching machines with a lab instructor. There was some doubt as to whether the scores would decrease, increase, or stay the same. Hence, $H_0: \mu = 65$ was tested against the two-sided alternative $H_1: \mu \neq 65$. A sample of $n = 50$ students using the teaching machines was tested, resulting in $\bar{x} = 68.2$ and $s = 13.2$. Use a significance level of $\alpha = 0.05$. What is your conclusion?

*4.5-7 Consider a $N(\mu, \sigma^2 = 40)$ distribution. To test $H_0: \mu = 32$ against $H_1: \mu > 32$, we reject H_0 if the sample mean $\bar{X} \geq c$. Find the sample size n and the constant c such that $OC(\mu = 32) = 0.90$ and $OC(\mu = 35) = 0.15$.

*4.5-8 Let Y have a binomial distribution with parameters n and p. In a test of $H_0: p = 0.25$ against $H_1: p < 0.25$, we reject H_0 if $Y/n \leq c$. Find n and c if $OC(p = 0.25) = 0.90$ and $OC(p = 0.20) = 0.05$. State your assumptions.

4.5-9 Let $X_1, X_2, \ldots, X_{10}$ be a random sample from a Poisson distribution with mean λ. In a test of $H_0: \lambda = 1.1$ against $H_1: \lambda < 1.1$, we reject H_0 if the sum of the 10 observations, Y, is less than or equal to 8. Using the Poisson table, find the $OC(\lambda)$ curve at $\lambda = 0.5, 0.7, 0.9,$ and 1.1. What is the significance level of the test?

Hint: Use the fact that a sum of n independent Poisson random variables with parameter λ is again Poisson with parameter $n\lambda$.

4.5-10 Recently, a commuter was told that it takes 60 minutes, on average, to travel by car from Philadelphia to Princeton, New Jersey. Anxious to learn whether the 60-minute figure is correct, the commuter takes measurements on $n = 20$ consecutive weekdays and finds that the average time is 68 minutes and the standard deviation is 6 minutes. Is there enough evidence that the time given to the commuter was too low? State the assumptions that you have made in your analysis.

4.5-11 Let $\bar{X}$ be the mean of random sample of size $n = 16$ from a normal distribution with mean μ and standard deviation $\sigma = 8$. To test $H_0: \mu = 35$ against $H_1: \mu > 35$, we reject H_0 if $\bar{X} \geq 36.5$.

(a) Determine the OC curve at $\mu = 35, 36,$ and 38.5.

(b) What is the probability of a type I error?

(c) What is the probability of a type II error at $\mu = 36$?

4.5-12 Sixty-four randomly selected fuses were subjected to a 20 percent overload, and the time to failure was recorded. It was found that the sample average and standard deviation were $\bar{x} = 8.5$ and $s = 2.4$, respectively.

(a) Compute a 99 percent confidence interval for μ, the mean time to failure, under such conditions.

(b) Test $H_0: \mu = 8$ against $H_1: \mu > 8$; use a significance level of $\alpha = 0.05$. What is your conclusion?

4.5-13 Your null hypothesis asserts that a certain game of chance has even odds (i.e., winning and losing are equally likely). Ten people play the game, and you find that 9 of the 10 lose. Calculate the probability value of this sample result, and indicate whether you can reject your null hypothesis (of even odds) in favor of the two-sided alternative at a significance level of 0.05.

4.5-14 The American Association of University Professors claims that the mean income of tenured professors at public universities is $110,000. Our hypothesis is that the mean salary is lower than $110,000. To test whether the mean salary is lower than $110,000, we take a random sample of $n = 36$ professors. Their salaries led to a sample average of $105,000 and a sample standard deviation

$s = 12,000$. Calculate the standardized test statistic and obtain its probability value. Assuming a significance level of 5 percent, what is your conclusion?

*4.5-15 A student publication claims that undergraduate engineering students spend $400, on average, on textbooks each semester.

*(a) Our hypothesis is that the mean amount is lower than $400. To test whether the mean amount is lower than $400, we take a random sample of $n = 60$ students and ask them how much they spent on textbooks. Their answers led to a sample average of $380 and a sample standard deviation $s = 40$. Calculate the standardized test statistic and obtain its probability value. Assuming a significance level of 5 percent, what is your conclusion?

*(b) Assume that the sample average is $430, with sample standard deviation $s = 40$. Would you conclude that $H_0: \mu = 400$ or that $H_1: \mu < 400$? [Note that, in this case, even the sample average is within H_0.] What is your conclusion as regards to a two-sided alternative ($H_1: \mu \neq 400$)?

*4.5-16 A sample of $n = 50$ loaves of bread is taken from the sizable production that left our bakery this morning. We find that the average weight of the 50 loaves is 1.05 pounds; the standard deviation is $s = 0.06$ pound.

*(a) Obtain a 95 percent confidence interval for the mean weight of this morning's production.

*(b) One of the employees claims that the current process produces loaves that are heavier than 1 pound, on average. Is there enough information in our sample to reject the null hypothesis that $\mu = 1.00$ lb in favor of the alternative that $\mu > 1.00$ lb?

*(c) Assume that the distribution of weights is normal. Using the estimates $\bar{x} = 1.05$ and $s = 0.06$ as the true values, calculate the proportion of loaves that are underweight (i.e., that weigh less than 1.0 pound).

*(d) Predict the weight of the next (single) loaf from this morning's production. Obtain an approximate 95 percent prediction interval.

Hint: The weight of the next loaf is a random variable. Attach bounds to your prediction such that the resulting interval covers 95 percent of the distribution.

*4.5-17 An ESP experiment had 82 people try to guess the suits of 16 randomly chosen cards from a regular 52-card deck, the score being the number of correctly guessed cards. The sample mean and the sample standard deviation of the number of cards guessed correctly were 3.86 and 1.61, respectively. Test the null hypothesis that people are guessing randomly.

Hint: If everyone is guessing randomly, the mean number of cards guessed correctly should be 4.

4.6 Tests of Characteristics of Two Distributions

One of the most important tests made in statistics is that in which two different "methods" are compared. We may compare the effectiveness of two educational systems, the durability of two different paints, the effectiveness of two medical treatments, the gas mileage of two makes of automobile, the measurements made under the "old" method and a "new" one, and so on. Usually, we are interested in comparing the means of two distributions. That is, we want to show that, on average, the responses under the new method are higher than the ones under the old method. Hence, we wish to test the null (no difference) hypothesis H_0: $\mu_1 = \mu_2$ against the alternative hypothesis H_1: $\mu_1 > \mu_2$, where μ_1 and μ_2 are the means of the response distributions under the new and the old methods, respectively. In other applications, we may want to test H_0: $\mu_1 = \mu_2$ against H_1: $\mu_1 < \mu_2$, or H_0: $\mu_1 = \mu_2$ against H_1: $\mu_1 \neq \mu_2$.

4.6-1 Comparing Two Independent Samples

Let us consider *independent* random samples from two distributions with respective means μ_1, μ_2 and variances σ_1^2, σ_2^2. We denote these random samples of respective sizes, n_1 and n_2, by $X_1, X_2, \ldots, X_{n_1}$ and $Y_1, Y_2, \ldots, Y_{n_2}$, respectively. Of course, the sample means, $\overline{X}$ and $\overline{Y}$, are approximately $N(\mu_1, \sigma_1^2/n_1)$ and $N(\mu_2, \sigma_2^2/n_2)$, where the approximation is exact in case the underlying distributions are normal. For the moment, we assume that σ_1^2 and σ_2^2 are known.

Under the null hypothesis H_0: $\mu_1 = \mu_2$ or $\mu_1 - \mu_2 = 0$, the random variable

$$Z = \frac{\overline{X} - \overline{Y}}{\sqrt{\sigma_1^2/n_1 + \sigma_2^2/n_2}}$$

in $N(0, 1)$. Thus, if a sample realization of Z is much larger than zero, we believe that the alternative hypothesis, H_1: $\mu_1 > \mu_2$, is true. For a test with significance level α, we reject H_0 and accept H_1 when $Z \geq z(\alpha)$.

If the variances σ_1^2 and σ_2^2 are unknown, they can be replaced by the respective sample variances S_X^2 and S_Y^2. Then the critical region $Z \geq z(\alpha)$ provides an approximate α-level test, provided that the sample sizes n_1 and n_2 are reasonably large (not smaller than 20).

Example 4.6-1 In this example, we illustrate how to obtain the operating characteristic curve of the test just described. Let $\overline{X}$ and $\overline{Y}$ be the means of the two independent random samples of sizes $n_1 = 22$ and $n_2 = 34$ from the normal distributions $N(\mu_1, \sigma_1^2 = 115)$ and $N(\mu_2, \sigma_2^2 = 96)$. Then the variance of $\overline{X} - \overline{Y}$ is $(115/22) + (96/34) = 8.05$. Accordingly, a test of H_0: $\mu_1 = \mu_2$ against H_1: $\mu_1 > \mu_2$ with significance level $\alpha = 0.025$ rejects H_0 when $Z = (\overline{X} - \overline{Y})/\sqrt{8.05} \geq 1.96$ or $(\overline{X} - \overline{Y}) \geq 1.96\sqrt{8.05}$.

What can we say about the probabilities of this event when $\theta = \mu_1 - \mu_2$ is not equal to zero, and how can we determine the OC(θ) curve? Given that the operating characteristic curve is the probability of accepting H_0, for different values of $\theta = \mu_1 - \mu_2$, we find that

$$\text{OC}(\theta) = P(\overline{X} - \overline{Y} < 1.96\sqrt{8.05}; \theta = \mu_1 - \mu_2)$$

$$= P\left(\frac{\overline{X} - \overline{Y} - \theta}{2.84} < 1.96 - \frac{\theta}{2.84}; \theta\right) = \Phi\left(1.96 - \frac{\theta}{2.84}\right)$$

because $(\overline{X} - \overline{Y} - \theta)/2.84$ is $N(0, 1)$. The OC curve evaluated for a few θ values is as follows:

θ	0	2	4	6	8	10
OC(θ)	0.9750	0.8962	0.7088	0.4404	0.1949	0.0594

The probability of a type I error is $\alpha = 1 - 0.975 = 0.025$, and the probability of a type II error at $\theta = \mu_1 - \mu_2 = 8$, for example, is $\beta = 0.1949$. This says that the probability of accepting H_0 if, in fact, the mean level of group 1 is eight units larger than that of group 2, is 0.1949. If the magnitude of the probability of this type II error is unacceptable, we must increase the sample sizes or increase the probability α of the type I error. ∎

If the sample sizes n_1 and n_2 are fairly small, we must make additional assumptions to carry out the test. We must assume that the underlying distributions are normal and the variances σ_1^2 and σ_2^2 are about equal. If $\sigma_1^2 = \sigma_2^2$, we can obtain an estimate S_p^2 of the common variance by pooling the sample variances S_X^2 and S_Y^2 in proportion to $n_1 - 1$ and $n_2 - 1$, respectively. Under the null hypothesis H_0: $\mu_1 = \mu_2$, the random variable

$$T = \frac{\bar{X} - \bar{Y}}{\sqrt{S_p^2\left(\dfrac{1}{n_1} + \dfrac{1}{n_2}\right)}} = \frac{\bar{X} - \bar{Y}}{\sqrt{\dfrac{(n_1 - 1)S_X^2 + (n_2 - 1)S_Y^2}{n_1 + n_2 - 2}\left(\dfrac{1}{n_1} + \dfrac{1}{n_2}\right)}}$$

has a Student's $t(n_1 + n_2 - 2)$ distribution. (See Section 4.3.) If we reject H_0 and accept H_1: $\mu_1 > \mu_2$ when $T \geq t(\alpha; n_1 + n_2 - 2)$, we obtain a test with significance level α.

Similarly, in a test of H_0: $\mu_1 = \mu_2$ against H_1: $\mu_1 < \mu_2$, we accept H_1 if $T \leq -t(\alpha; n_1 + n_2 - 2)$. In a test of H_0: $\mu_1 = \mu_2$ against H_1: $\mu_1 \neq \mu_2$, we reject H_0 and accept H_1 if $|T| \geq t(\alpha/2; n_1 + n_2 - 2)$.

Example 4.6-2 An experiment is conducted to compare the crash resistance of two different types of car bumpers. Twenty cars of the same model are divided into two groups of $n_1 = 11$ and $n_2 = 9$ cars. Type A bumpers are mounted on the cars in group 1, while type B bumpers are mounted on the cars in the other group. The 20 cars are then driven into a concrete wall at a speed of 10 miles per hour. The amount of damage (in dollars of needed repairs) to each car is measured, resulting in the following means and variances: $\bar{x} = 235$, $s_x^2 = 421$ for the $n_1 = 11$ type A bumpers, and $\bar{y} = 286$, $s_y^2 = 511$ for the $n_2 = 9$ type B bumpers. We wish to test whether the mean repair cost for cars with bumper A is lower than the cost for cars with bumper B. We test H_0: $\mu_1 = \mu_2$ against H_1: $\mu_1 < \mu_2$. With small sample sizes, we check to see if the distributions are normal and if the population variances are equal (or approximately so). We find that the two sample variances are similar; thus, it seems reasonable to assume that the underlying population variances also are equal. (A formal test of $\sigma_1^2 = \sigma_2^2$ is given in Section 4.6-4.) With only $n_1 = 11$ and $n_2 = 9$ observations in each group, it is difficult to assess whether the observations are from normal distributions. However, normal probability plots, in which we plot the normal scores associated with the observations against the observations, can tell us whether the normal assumption is reasonable. (See Section 3.6.)

Assuming that these assumptions are satisfied, we calculate the observed test statistic

$$t = \frac{235 - 286}{\sqrt{\dfrac{(10)(421) + (8)(511)}{18}\left(\dfrac{1}{11} + \dfrac{1}{9}\right)}} = -5.28.$$

This value is much smaller than $-t(0.01; 18) = -2.552$, so we reject H_0: $\mu_1 = \mu_2$ and accept H_1: $\mu_1 < \mu_2$ at the $\alpha = 0.01$ significance level. That is, there is fairly strong evidence that the type A bumper leads to lower repair costs. The average amount saved is \$51. ∎

Comparative dot diagrams and box-and-whisker displays are always helpful in displaying information gleaned from comparative experiments. The construction of box plots was illustrated in Section 1.4. The two boxes in each diagram are constructed from the 25th, 50th (median), and 75th percentiles, and they represent the second and third quarters of the data. The lines from the extremes to the adjacent quartiles are called the whiskers. If two treatments or groups are compared, one can put the two box plots side by side. If the boxes overlap substantially, then there is usually no statistically significant difference between the two population means. However, if the 25 percent line of the higher box is above the 50 percent line of the lower box, there is often reason to conclude that the two population means may, in fact, be different. Of course, we know that the result of a test depends on the number of observations in these groups, but sample sizes do not enter into the construction of the box plots. Thus, we should treat the foregoing statement only as a very rough rule of thumb.

Nevertheless, these simple graphical displays are quite informative, and they should be made whenever possible. In addition to providing us with information on the centers of the two distributions, they tell us about the variability in the two data sets. The 2-sample t-test for small samples assumes that the underlying population variances in the two groups are the same. If we find that the interquartile ranges are very different, we should not pool the two sample variances. In such a case, we may look for a transformation (e.g., a logarithmic or a square-root transformation) that makes the variances more equal.

In cases where the variances remain different even after transformations, one is better off using the earlier test statistic

$$T = \frac{\overline{X} - \overline{Y}}{\sqrt{\dfrac{S_X^2}{n_1} + \dfrac{S_Y^2}{n_2}}}$$

and a t-distribution for the computation of the critical values and the probability values. The appropriate degrees of freedom of the t-distribution are determined from an approximation due to Welch and are calculated by most computer programs (e.g., through Minitab's "Stat > Basic Statistics > 2-Sample t" test procedure for unequal variances). [See Welch, B. L., "The significance of the difference between two means when the population variances are unequal," *Biometrika,* Vol. 29 (1937), 350–62.]

4.6-2 Paired-Sample t-Test

So far, we have assumed that the observations in one treatment group are independent of the observations in the other. In many situations, this is an appropriate assumption. For example, it is appropriate in a comparison of the durabilities of two paints if we apply paints A and B to n_1 and n_2 different randomly selected boards.

There will be variability in the durability measurements because there will be some variability among the boards. Now, the boards (which, in statistical terminology, are called the experimental units) are different, so the durability measurements on paints A and B should be independent. However, this would no longer be the case if we had applied both paints to each board, painting one-half of the board with A and the other half with B. Because we would take two measurements on the same board (experimental unit), we could expect that the measurements are related.

Let us illustrate this point with another example. Suppose that we want to assess the effect of two types of background music on work efficiency. Let us suppose that we have divided our workforce into two different groups of equal size m. Of course, we have done this randomly, so we spread the variability that is due to workers fairly evenly among the two groups. One group performs a sequence of familiar tasks while listening to classical music; the other group listens to the local AM radio station, which specializes in popular music. At the end of the experiment, we record the time that it took each worker to complete the tasks. The resulting observations are $X_1, X_2, \ldots, X_m$ for group 1 and $Y_1, Y_2, \ldots, Y_m$ for group 2. The observations in the two groups come from different workers, so it is appropriate to assume that the X and Y measurements are independent.

Now let us change the experiment and assume that each of the $2m$ workers completes the task under both types of background music. We could flip a coin to decide whether a particular worker listens first to classical or first to popular music. From this experiment, we obtain $2m$ pairs of observations: $(X_1, Y_1), (X_2, Y_2), \ldots, (X_{2m}, Y_{2m})$. As in the preceding case, we observe a sample of X values and a sample of Y values and can calculate the two averages $\overline{X}$ and $\overline{Y}$. However, these two averages are no longer independent: Because each pair (X_i, Y_i) comes from the same person, it is most likely that the measurements are dependent. For example, if worker i performs better than average under the first type of background music, it is very likely that he performs better than average under the other type also. If this is the case, we cannot use the previously discussed test statistic, where the standard deviation of $\overline{X} - \overline{Y}$ is derived under the independence of $\overline{X}$ and $\overline{Y}$.

Suppose now that we are in a situation in which X and Y are no longer independent. For example, suppose that measurements X_i and Y_i are taken on the same person i before and after he or she completes a certain program. The measurements may be pulse rates and the program a certain type of aerobic conditioning training, or the measurements X_i and Y_i may correspond to the efficiencies of worker i before and after completing a training program, or they may correspond to the durabilities of two rubber sole materials that are put on the left and right shoe of individual i, and so on. In all these cases, we expect that the two measurements X_i and Y_i are related.

In such situations, we analyze the differences $W_i = X_i - Y_i$ because these differences "cancel," or adjust for subject differences that make measurements X_i and Y_i both appear unusually high or low. The differences $W_1 = X_1 - Y_1, W_2 = X_2 - Y_2, \ldots, W_n = X_n - Y_n$ are then a random sample of size n (or $2m$ in the previous illustration) from a distribution with mean $\mu_1 - \mu_2$ and variance σ_W^2. The random variables $W_1, W_2, \ldots, W_n$ are independent because they are observations on different subjects. Assuming that the distribution of W is normal, we can use the test statistic

$$T = \frac{\overline{W}}{S_W/\sqrt{n}},$$

where S_W^2 is the sample variance of the differences, to conduct tests of hypotheses. For example, we would reject H_0: $\mu_1 = \mu_2$ in favor of H_1: $\mu_1 > \mu_2$ if $\overline{W}/(S_W/\sqrt{n}) \geq t(\alpha; n - 1)$. This is called a *paired-sample t-test*. Of course, if n is large, we can use the percentage points from the normal distribution.

Example 4.6-3   Ten engineers' knowledge of basic statistical concepts was measured on a scale of 100 before and after a short course in statistical quality control. The engineers were selected at random. Table 4.6-1 shows the results of the tests, where $\overline{w} = 3.9$ and $s_w^2 = 31.21$; thus, the computed test statistic

$$t = \frac{3.9}{\sqrt{31.21/10}} = 2.21 \geq t(0.05; 9) = 1.833.$$

We reject $\mu_1 = \mu_2$ and accept $\mu_1 > \mu_2$ at the $\alpha = 0.05$ significance level. That is, the engineers' knowledge of basic statistical concepts seems to have increased after a course in statistical quality control.

Table 4.6-1	Statistical Knowledge Before and After Completing a Short Course in Statistical Quality Control		
Engineer	Before, y	After, x	$w = x - y$
1	43	51	8
2	82	84	2
3	77	74	-3
4	39	48	9
5	51	53	2
6	66	61	-5
7	55	59	4
8	61	75	14
9	79	82	3
10	43	48	5

Remarks: Pairing up certain X and Y values helps us control or account for variability that arises from differences among the experimental units. For instance, consider Example 4.6-3 again, and suppose that we would have analyzed the data as if X and Y were independent realizations from normal distributions. Then we would obtain more degrees of freedom for the t-statistic in the independent situation, namely, $n_1 + n_2 - 2 = 18$, instead of the $n - 1 = 9$ in the paired t-test, but the variability among engineers would have "swamped" the difference $\overline{x} - \overline{y} = 3.9$. That is, the test statistic calculated under the assumption of independent samples is very small (in fact, it is $3.9/6.7 = 0.58$), and we could not reject H_0. This can be seen clearly from the dot diagram in Figure 4.6-1. Although the program has been successful for 8 out of 10 engineers, the overall summary in the figure, where measurements are viewed as realizations from two independent samples, would show little difference.

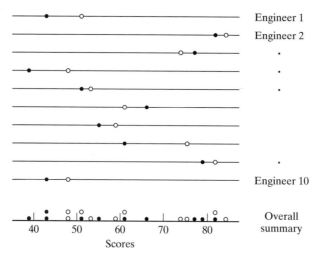

Figure 4.6-1 Graphical display of the data in Example 4.6-3. The symbols • and ○ correspond to the measurements made before and after the training

Thus, more precise comparisons can be made by grouping experimental units into homogeneous blocks, which may be subjects, days, plots of land, temperatures, animals from the same litter, specimens from the same batch, and so on. The precision of the blocked (or paired) comparison is increased, because grouping, or *blocking*, eliminates the differences among the experimental units from the measurements. As in the paired-sample *t*-test, a blocked comparison leads to a smaller denominator in the test statistic and gives us a better chance of detecting a true difference between the two methods. Much more will be said about blocking in Section 6.3.

4.6-3 Test of $p_1 = p_2$

We next consider a test of the equality of means of two independent samples from Bernoulli distributions, namely, p_1 and p_2, which are the probabilities of success resulting from two different processes. For example, a pump manufacturer claims that a larger percentage of his pumps will be operating without repairs in three years than those of a competitor. That is, he claims that $p_1 > p_2$, where p_1 and p_2 are the respective fractions of pumps that operate without repairs after three years.

To test $H_0: p_1 = p_2$ against $H_1: p_1 > p_2$ we select n_1 and n_2 items at random from the two respective populations. The numbers of successes, Y_1 and Y_2, have independent binomial distributions, $b(n_1, p_1)$ and $b(n_2, p_2)$, that can be approximated by normal distributions. Also, the sample proportions $\hat{p}_i = Y_i/n_i$ are approximately $N[p_i, p_i(1 - p_i)/n_i]$, for $i = 1, 2$. Because Y_1 and Y_2 are assumed to be independent, the variance of $\hat{p}_1 - \hat{p}_2 = Y_1/n_1 - Y_2/n_2$ is the sum of the variances, namely,

$$\frac{p_1(1 - p_1)}{n_1} + \frac{p_2(1 - p_2)}{n_2}.$$

If the null hypothesis $H_0: p_1 = p_2 = p$, where p is unknown, is true, then

$$Z = \frac{\hat{p}_1 - \hat{p}_2}{\sqrt{\dfrac{p(1-p)}{n_1} + \dfrac{p(1-p)}{n_2}}}$$

follows an $N(0, 1)$ distribution, at least approximately. Now, p is unknown, so we cannot compute Z from our sample results. To avoid this difficulty, we replace the common p by the pooled estimate $\hat{p} = (Y_1 + Y_2)/(n_1 + n_2)$ and calculate the test statistic

$$Z = \frac{\hat{p}_1 - \hat{p}_2}{\sqrt{\hat{p}(1-\hat{p})[(1/n_1) + (1/n_2)]}}.$$

We reject $H_0: p_1 = p_2$ in favor of $H_1: p_1 > p_2$ if the calculated test statistic is greater than or equal to $z(\alpha)$. An identical test can be performed by finding the probability value associated with the computed Z and then seeing if this probability value is less than or equal to α.

Example 4.6-4 To test the pump manufacturer's claim, $H_1: p_1 > p_2$, we checked on $n_1 = n_2 = 100$ pumps from each of the companies and found $y_1 = 67$ and $y_2 = 62$ working pumps. Here, the pooled estimate of p is $\hat{p} = 129/200 = 0.645$. Because the calculated test statistic

$$z = \frac{0.67 - 0.62}{\sqrt{(0.645)(0.355)[(1/100) + (1/100)]}} = 0.74$$

is rather small and does not exceed $z(0.05) = 1.645$, we have insufficient evidence to reject $H_0: p_1 = p_2$. That is, until more data are observed, it would be difficult to accept the manufacturer's claim. ∎

4.6-4 Test of $\sigma_1^2 = \sigma_2^2$

In concluding this section, we look at a test of the equality of the variances, σ_1^2 and σ_2^2, of two normal distributions. Let S_X^2 and S_Y^2 be the respective variances of independent samples, of sizes n_1 and n_2. We know that

$$F = \frac{S_X^2/\sigma_1^2}{S_Y^2/\sigma_2^2}$$

has an F-distribution with $n_1 - 1$ and $n_2 - 1$ degrees of freedom. Thus, if $H_0: \sigma_1^2 = \sigma_2^2$ is true, then $F = S_X^2/S_Y^2$ is $F(n_1 - 1, n_2 - 1)$. Suppose that the alternative hypothesis is $H_1: \sigma_1^2 > \sigma_2^2$. For example, we may want to test whether the variability in the incoming material from a new supplier (σ_2^2) is smaller than that coming from the current supplier (σ_1^2). We reject $H_0: \sigma_1^2 = \sigma_2^2$ and accept $H_1: \sigma_1^2 > \sigma_2^2$ if $F = S_X^2/S_Y^2 \geq F(\alpha; n_1 - 1, n_2 - 1)$. To illustrate, say that observations on both suppliers resulted in $n_1 = 12, \bar{x} = 26.2, s_x^2 = 29.1$, and $n_2 = 10, \bar{y} = 25.9, s_y^2 = 14.6$. Then $s_x^2/s_y^2 = 29.1/14.6 = 1.99 < F(0.05; 11, 9) = 3.10$. The upper percentage point can be found from statistical tables (e.g., Table C7 in the appendix) or from statistical

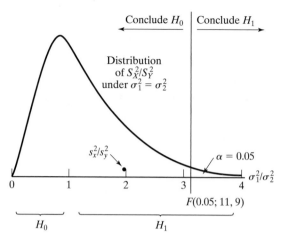

Figure 4.6-2 Graphical description of the test $H_0: \sigma_1^2 = \sigma_2^2$ against $H_1: \sigma_1^2 > \sigma_2^2$

software (such as Minitab's "Calc > Probability Distributions > F distribution," asking for the inverse cumulative probability at 0.95). We find that there is not enough evidence to reject $H_0: \sigma_1^2 = \sigma_2^2$. Figure 4.6-2 shows the situation graphically.

Note that even though the ratio of the two sample variances is almost 2, we still do not reject the equality $(\sigma_1^2 = \sigma_2^2)$. As a matter of fact, that ratio would have to be greater than 3.10 for us to reject $H_0: \sigma_1^2 = \sigma_2^2$. For this reason, most statisticians find this test not to be very useful.

Exercises 4.6

*4.6-1 Two producers of light bulbs each claim that they have the longer lasting bulbs. Accordingly, 100 bulbs were selected at random from each producer and tested. The tests resulted in means of $\bar{x} = 798$ and $\bar{y} = 826$ hours, with respective variances of $s_x^2 = 7{,}982$ and $s_y^2 = 9{,}001$ hours. Test $H_0: \mu_1 = \mu_2$ against the two-sided alternative $H_1: \mu_1 \neq \mu_2$ at the $\alpha = 0.05$ significance level.

4.6-2 Let $\bar{X}$ and $\bar{Y}$ be means of independent random samples of sizes $n_1 = 14$ and $n_2 = 18$ from the respective normal distributions $N(\mu_1, \sigma_1^2 = 26)$ and $N(\mu_2, \sigma_2^2 = 21)$. Let $\theta = \mu_1 - \mu_2$. Construct a test of $H_0: \theta = 0$ against $H_1: \theta < 0$ that has significance level $\alpha = 0.025$. Determine the $OC(\theta)$ curve for $\theta = -5, -4, -3, -2, -1, 0$, and graph this function.

*4.6-3 Consider again Exercise 4.6-2. Take equal sample sizes $n = n_1 = n_2$. Find n so that $\alpha = 0.025$ and $\beta = 0.05$ when $\theta = -3$.

*4.6-4 An engineer wishes to compare the strengths of two types of beams, the first made of steel and the second made of an alloy. Several beams of each type are selected at random, and the deflections (in units of 0.001 inch) are measured when the beams are submitted to a force of 3,000 pounds. The respective sample characteristics are $n_1 = 10, \bar{x} = 82.6, s_x^2 = 6.52,$ and $n_2 = 12, \bar{y} = 78.1, s_y^2 = 7.02$.

Compute the appropriate test statistic, and test whether the mean deflection of steel beams is larger than that made of an alloy.

4.6-5 The golf scores of two competitors, A and B, are recorded over a period of 10 days. Golfer A claims that her game is better than that of golfer B. Use the following data to test this claim:

	Day									
Golfer	1	2	3	4	5	6	7	8	9	10
A	87	86	79	82	78	87	84	81	83	81
B	89	85	83	87	76	90	85	78	85	84

Note: Because the playing conditions on different days were different, it was decided to pair or block the comparison with respect to day.

***4.6-6** A certain washing machine manufacturer claims that the fraction p_1 of his washing machines that need repairs in the first five years of operation is less than the fraction p_2 of another brand. To test this claim, we observe $n_1 = n_2 = 200$ machines of each brand and find that $y_1 = 21$ and $y_2 = 37$ machines need repairs. Do these data support the manufacturer's claim? Use $\alpha = 0.05$.

4.6-7 Two different fabrics are compared on a Martindale wear tester that can compare two materials in a single run. The weight losses (in milligrams) from seven runs are as follows:

	Run						
Fabric	1	2	3	4	5	6	7
A	36	26	31	38	28	37	22
B	39	27	35	42	31	39	21

Analyze the data and discuss whether one fabric is better than the other.

4.6-8 Use the data in Table 1.4-1, and test whether the new process has led to a significant reduction in the average number of lead wires that misfeed during production. State and check your assumptions carefully.

***4.6-9** To find out about the level of air pollution in a home, the amount of suspended particles is measured during a 24-hour period. A sample is taken from homes in which there are no smokers and from homes in which there is at least one smoker. Say that these respective samples yielded $n_1 = 16, \bar{x} = 67.1, s_x^2 = 7.82$, and $n_2 = 13, \bar{y} = 72.3, s_y^2 = 24.12$.

*(a) Test H_0: $\sigma_1^2 = \sigma_2^2$ against H_1: $\sigma_1^2 < \sigma_2^2$ at $\alpha = 0.05$, using $F = s_x^2/s_y^2$.

(b) Investigate H_0: $\mu_1 = \mu_2$ against H_1: $\mu_1 < \mu_2$ at the $\alpha = 0.05$ significance level. If $\sigma_1^2 = \sigma_2^2$ is not rejected in part (a), pool the variance and use the t-test of Section 4.6-1. If $\sigma_1^2 = \sigma_2^2$ is rejected in part (a), test H_0 against H_1 using $(\bar{x} - \bar{y})/\sqrt{s_x^2/n_1 + s_y^2/n_2}$ and the normal distribution, even though the sample sizes are not as large as you might like. Even better, use computer software that determines the number of degrees of freedom in the t-distribution from the Welch approximation.

4.6-10 The depletion of carbon in steel billets is an important indicator of breaking strength and is monitored by steel producers as well as their customers (in the automotive industry). The manager at Northstar Steel suspects that measurements made by the customer show a higher depletion of carbon than measurements taken by the in-house metallurgist. The manager wants to investigate this suspicion by carrying out an experiment. He plans to pick billets at random, remove two small adjacent segments from each billet, give one to the in-house metallurgist, and send the other to the customer's laboratory. From previous experiments, he knows something about the variability of measurement differences (in-house versus laboratory) on the same item; he believes that the standard deviation among measurement differences is 10 units. How many billets does he have to sample such that the probability of concluding incorrectly that the customer's measurements are higher is at most 0.05 and that a difference of 5 units is detected with probability 0.80?

Hint: This is a blocked experiment. Analyze the differences.

4.6-11   We want to determine whether or not a new applied project-oriented way of teaching statistics helps students increase their understanding of probability. The experimenter has developed a set of videotaped instructional materials for both a traditional course and the new applied project-oriented course. He plans to select an equal number of students for each program, have them view the videotapes, and test them on a sequence of new problems. Tests are scored between 0 and 100, and it is known from prior experience that the interquartile range (i.e., the difference between the third and first quartiles) of the test scores is about 15 points. How many students should he select in each of the two groups if (1) the error of concluding incorrectly that the new program leads to higher scores is at most 0.05 when, in fact, the traditional and new programs have the same mean and (2) an increase of 5 points in the new over the traditional method is detected with probability 0.90?

Hint: Assume normality and calculate σ from the interquartile range. The first and third quartiles are given by $Q_1 = \mu - (0.6745)\sigma$ and $Q_3 = \mu + (0.6745)\sigma$, and $\sigma = (Q_3 - Q_1)/1.349$.

*4.6-12   In order to prove or disprove the null hypothesis that there is no salary discrimination, you obtain random samples of male and female employees in your company. From the sample of 30 men, you find an average yearly salary of $60,000 (standard deviation = $4,000). From the sample of 30 women, you find an average salary of $57,500 (standard deviation = $5,000). Your research hypothesis is that the mean salaries for men and women are different. Draw your conclusion at a significance level of 0.05, and discuss whether it answers the question about salary discrimination by gender.

4.7 Certain Chi-Square Tests

Previously, we discussed tests of hypotheses about a single proportion; in particular, in Example 4.5-3 we tested $H_0: p = 0.62$. There, we considered experiments that could result in one of two possible outcomes: a grade of A or a grade different from A. Here, we extend the analysis to experiments in which the outcomes correspond to $k > 2$ mutually exclusive categories, and we discuss tests of hypotheses about the underlying probabilities $p_1, p_2, \ldots, p_k$ of these k categories. For example, a quality inspector may either accept a manufactured item, accept it as a second, or reject it as scrap. Assume that we have sampled $n = 500$ items and have observed the relative frequencies of these three possible states. How can we test whether, for example,

H_0: $p_1 = 0.83$, $p_2 = 0.14$, and $p_3 = 0.03$, where p_1, p_2, and p_3 are the respective probabilities of acceptance, second, and scrap?

Although we cannot develop the theory behind the basic chi-square test, which was first proposed by Karl Pearson in 1900, we can give an intuitive idea of why it works. Let the number of "successes" Y_1 in n independent trials be binomial, or $b(n, p_1)$. Then, when n is large,

$$Z = \frac{Y_1 - np_1}{\sqrt{np_1(1 - p_1)}}$$

has an approximate $N(0, 1)$ distribution. Thus, $Q_1 = Z^2$ is approximate $\chi^2(1)$, due to Exercise 4.3-5, and Q_1 can be written as

$$Q_1 = Z^2 = \frac{(Y_1 - np_1)^2}{np_1(1 - p_1)} = (Y_1 - np_1)^2 \left[\frac{1}{np_1} + \frac{1}{n(1 - p_1)} \right].$$

However, because $(Y_1 - np_1)^2 = [n - Y_1 - n(1 - p_1)]^2$, we can express Q_1 as

$$Q_1 = \frac{(Y_1 - np_1)^2}{np_1} + \frac{(Y_2 - np_2)^2}{np_2},$$

where $Y_2 = n - Y_1$ and $p_2 = 1 - p_1$. That is, using Y_2 as the number of failures, which is $b(n, p_2)$, we see that

$$Q_1 = \sum_{i=1}^{2} \frac{(Y_i - np_i)^2}{np_i}$$

is approximately $\chi^2(1)$.

4.7-1 Testing Hypotheses about Parameters in a Multinomial Distribution

To generalize the preceding formula, let an experiment have k mutually exclusive and exhaustive outcomes, instead of just $k = 2$ as in the binomial case. Let $p_1, p_2, \ldots, p_k$ represent the respective probabilities of these k outcomes. Of course, $\sum_{i=1}^{k} p_i = 1$. Assume that the experiment is performed n independent times, and let $Y_1, Y_2, \ldots, Y_k$ stand for the respective frequencies of the first outcome, the second outcome, . . ., and the kth outcome. Jointly, $Y_1, Y_2, \ldots, Y_k$ are said to have a *multinomial distribution* with multinomial probabilities $p_1, p_2, \ldots, p_k$.

If $y_1, y_2, \ldots, y_k$ are nonnegative integers such that their sum equals n, then the multinomial probability that the first outcome occurs y_1 times, the second outcome y_2 times, . . ., and, finally, the kth outcome y_k times is

$$P(Y_1 = y_1, Y_2 = y_2, \ldots, Y_k = y_k) = \frac{n!}{y_1! y_2! \ldots y_k!} p_1^{y_1} p_2^{y_2} \ldots p_k^{y_k}.$$

Clearly, this equation reduces to a binomial probability if $k = 2$, where $y_2 = n - y_1$ and $p_2 = 1 - p_1$. For example, it is found that, in a certain manufacturing production, about $p_1 = 0.83$ of the items are good, $p_2 = 0.14$ are seconds, and

$p_3 = 0.03$ are defective. An inspector samples $n = 20$ different items at random; the probability of finding 16 good items, 3 seconds, and 1 defective in this sample is

$$P(Y_1 = 16, Y_2 = 3, Y_3 = 1) = \frac{20!}{16!3!1!}(0.83)^{16}(0.14)^3(0.03)^1 = 0.08093.$$

Individually, each Y_i is binomial, $b(n, p_i)$; thus, $E(Y_i) = np_i$. To illustrate, cast a fair die $n = 60$ independent times, and let Y_i be the number of times i spots are on the "up side," $i = 1, 2, \ldots, 6$. Of course, $p_i = 1/6$ and $E(Y_i) = (60)(1/6) = 10$; that is, we would expect 10 of each of $k = 6$ possible outcomes $\{1, 2, 3, 4, 5, 6\}$.

If $Y_1, Y_2, \ldots, Y_k$ have such a multinomial distribution with parameters $n, p_1, p_2, \ldots, p_k$, then

$$Q_{k-1} = \sum_{i-1}^{k} \frac{(Y_i - np_i)^2}{np_i}$$

can be shown to have an approximate $\chi^2(k - 1)$ distribution. This was Karl Pearson's major contribution and extends the earlier result for $k = 2$. In the discussion that follows, we show how it can be used to test hypotheses about the probabilities $p_1, p_2, \ldots, p_k$.

Suppose we hypothesize that the multinomial probabilities equal certain values, say, $H_0: p_i = p_{i0}, i = 1, 2, \ldots, k$, where $\sum_{i=1}^{k} p_{i0} = 1$. For example, in the die illustration we might believe that the die is fair and that $H_0: p_i = 1/6, i = 1, 2, \ldots, 6$, is appropriate. Then, because under H_0, $E(Y_i) = np_{i0}$, we would anticipate Y_i to be about equal to np_{i0}. Thus, if the differences $Y_i - np_{i0}, i = 1, 2, \ldots, k$, are large, we would question the null hypothesis H_0. That is, if the squares $(Y_i - np_{i0})^2$, $i = 1, 2, \ldots, k$, are large, as measured by the *chi-square statistic*

$$Q_{k-1} = \sum_{i=1}^{k} \frac{(Y_i - np_{i0})^2}{np_{i0}},$$

then we reject H_0. In particular, given that Q_{k-1} is approximately $\chi^2(k - 1)$, we reject H_0 if

$$Q_{k-1} = \sum_{i=1}^{k} \frac{(Y_i - np_{i0})^2}{np_{i0}} \geq \chi^2(\alpha; k - 1)$$

for a test having significance level α. The alternative hypothesis H_1 consists of all alternatives to H_0, namely, all cases in which there is at least one inequality $p_i \neq p_{i0}$. Note that with one such inequality, there must be at least one other.

We have introduced the chi-square distribution in Sections 3.3 and 4.3. Cumulative probabilities and inverse cumulative probabilities can be obtained from tables or statistical computer software. For example, Minitab's "Calc > Probability Distributions > Chi-square" with cumulative probability and inverse cumulative probability can be used to obtain $P(\chi^2(3) \leq 7.815) = 0.95$; the upper fifth percentage point of the chi-square distribution with three degrees of freedom is then given by $\chi^2(0.05; 3) = 7.815$.

Example 4.7-1 One famous Mendelian theory concerning the crossing of two types of peas states that the respective probabilities are

$$H_0: p_1 = \frac{9}{16}, p_2 = \frac{3}{16}, p_3 = \frac{3}{16}, p_4 = \frac{1}{16}$$

for the following four mutually exclusive and exhaustive classifications: (1) round and yellow, (2) wrinkled and yellow, (3) round and green, (4) wrinkled and green. It is interesting to note that these probabilities can be obtained by multiplying $(0.75 + 0.25)$ by itself, which might suggest how Mendel arrived at his theory. To test H_0, suppose that we performed $n = 80$ independent experiments, observing the frequencies $y_1 = 42, y_2 = 17, y_3 = 13,$ and $y_4 = 8$. Because the respective expected values are $(80)(9/16) = 45$, $(80)(3/16) = 15$, $(80)(3/16) = 15$, and $(80)(1/16) = 5$, we find that the observed value of Q_3, which is denoted by a lowercase letter, is

$$q_3 = \frac{(42-45)^2}{45} + \frac{(17-15)^2}{15} + \frac{(13-15)^2}{15} + \frac{(8-5)^2}{5} = 2.533.$$

Q_3 is approximately $\chi^2(3)$ and $q_3 = 2.533 < \chi^2(0.05; 3) = 7.815$, so we do not reject H_0 at the $\alpha = 0.05$ level. That is, Mendel's hypothesis is not rejected by these data. ■

In using the chi-square approximation to the distribution of Q_{k-1}, many statisticians suggest that n should be large enough so that $np_{i0} \geq 5, i = 1, 2, \ldots, k$. However, in many cases the approximation also works well for $np_{i0} \geq 1$, provided that the term $(y_i - np_{i0})^2/np_{i0}$ does not contribute too much to q_{k-1} when np_{i0} is small. We do not want a small denominator np_{i0} to inflate $(y_i - np_{i0})^2/np_{i0}$ such that it dominates the other terms in q_{k-1}.

In the basic chi-square test, the probabilities $p_1, p_2, \ldots, p_k$ are completely specified by the null hypothesis. We call such a null hypothesis a *simple* null hypothesis. Sometimes this is not the case, because the probabilities in H_0 may be functions of other parameters that are unknown. We call such a null hypothesis a *composite* null hypothesis.

The two examples that follow illustrate the differences between a simple and a composite null hypothesis. Take Example 4.7-1, in which we considered probabilities of events that are associated with the crossing of two types of peas. There, we tested whether $p_1 = 9/16, p_2 = p_3 = 3/16,$ and $p_4 = 1/16$. We pointed out that these specific values are the terms in the expression $[p + (1 - p)]^2$, where $p = 3/4$. This is an example of a simple null hypothesis. Now, consider the case in which we test whether the probabilities are given by the terms in the expansion $[p + (1 - p)]^2$, but leave p unspecified. Because $H_0: p_1 = p^2, p_2 = p_3 = p(1 - p), p_4 = (1 - p)^2$ includes an unspecified parameter, we call H_0 a composite null hypothesis. As another example, say that we are interested in testing whether the numbers of defectives, X, in samples of size 10 follow a binomial distribution with $p = 0.10$. On the one hand, under $H_0: p = 0.10$, the probability of the event $(X = i)$ is given by $p_i = \binom{10}{i} p^i (1 - p)^{10-i}$, $i = 0, 1, \ldots, 10$. With p specified, we are looking at a simple null hypothesis. On the other hand, if we just want to test whether X follows a binomial distribution $b(10, p)$ and leave p unspecified, we are talking about a composite null hypothesis.

If the p_i are not completely specified under H_0, we cannot compute Q_{k-1} once realizations of $Y_1, Y_2, \ldots, Y_k$ are observed, because we cannot evaluate their expected values $np_i, i = 1, 2, \ldots, k$. However, there is a way out of this difficulty. The probabilities $p_1, p_2, \ldots, p_k$ in

$$Q_{k-1} = \sum_{i=1}^{k} \frac{(Y_i - np_i)^2}{np_i}$$

are usually functions of a smaller set of parameters $\theta_1, \theta_2, \ldots, \theta_h$, where $h < k - 1$. This was indeed the case in our preceding illustrations. For example, the 11 probabilities $p_i, i = 0, 1, \ldots, 10$ were functions of just one unknown: the parameter p. Thus, Q_{k-1} also is a function of $\theta_1, \theta_2, \ldots, \theta_h$. Let us find estimators of these h parameters, so that Q_{k-1} is minimized. We call those estimators the *minimum chi-square estimators* of $\theta_1, \theta_2, \ldots, \theta_h$. Substituting these estimators for p_i, we obtain the minimum value of Q_{k-1}. The distribution of this minimum value Q_{k-1} can be approximated by a chi-square distribution with $k - 1 - h$ degrees of freedom. The only adjustment that we have to make in going from a simple to a composite null hypothesis is in the degrees of freedom of the chi-square distribution. In the context of composite null hypotheses, we lose one degree of freedom for each estimated parameter. We reject the composite null hypothesis if the calculated minimum value, q_{k-1}, exceeds the critical value $\chi^2(\alpha; k - 1 - h)$.

Remark: In many situations, the parameters are *not* estimated by minimizing Q_{k-1}, because minimizing Q_{k-1} is sometimes a difficult problem. If other estimators are used, the computed q_{k-1} is not the smallest possible value; hence, there is a greater chance of rejecting H_0 when the computed q_{k-1} is compared with $\chi^2(\alpha; k - 1 - h)$. Accordingly, the significance level will be somewhat higher than α. Nevertheless, this approximation is still not too bad, particularly if the parameters are estimated as a function of the frequencies $Y_1, Y_2, \ldots, Y_k$.

4.7-2 Contingency Tables and Tests of Independence

A major application of the result of the preceding subsection appears in the context of *contingency tables*. Let the result of a random experiment be classified by two attributes. For example, a radio receiver may be classified as having low, average, or high fidelity and as having low, average, or high selectivity; or graduating engineering students may be classified according to their starting salary and their grade point average. In general, let us label the mutually exclusive and exhaustive levels of the first attribute by $A_1, A_2, \ldots, A_a$ and those of the second by $B_1, B_2, \ldots, B_b$. Doing so leads to the mutually exclusive and exhaustive classification $A_i \cap B_j, i = 1, 2, \ldots, a$, $j = 1, 2, \ldots, b$. Let us repeat the experiment n independent times, and let Y_{ij} denote the frequencies of the $k = ab$ groups in the classification $A_i \cap B_j$. For example, we determine the fidelity and selectivity of n radio receivers, or we ask n graduating engineers about their starting salary and grade point average. For a test of the simple null hypothesis $H_0: P(A_i \cap B_j) = p_{ij}$, where p_{ij} are $k = ab$ specified probabilities that add to 1, we use

$$Q_{ab-1} = \sum_{i=1}^{a} \sum_{j=1}^{b} \frac{(Y_{ij} - np_{ij})^2}{np_{ij}}.$$

Under this null hypothesis, the distribution of Q_{ab-1} is approximately $\chi^2(ab - 1)$, provided that n is large enough to make each np_{ij} reasonably large.

Often, we wish to test the *independence* of the two attributes. For example, is the level of the fidelity of a radio receiver independent of the level of selectivity, or is the level of the starting salary for an engineer independent of the grade point average of that engineer? Let us denote the individual (or marginal) probabilities $P(A_i)$ by $p_{i\bullet}$ and $P(B_j)$ by $p_{\bullet j}$, so that

$$p_{i\bullet} = P(A_i) = \sum_{j=1}^{b} p_{ij} \quad \text{and} \quad p_{\bullet j} = P(B_j) = \sum_{i=1}^{a} p_{ij}.$$

Then the hypothesis of independence,

$$H_0\colon P(A_i \cap B_j) = P(A_i)P(B_j), \quad \text{for all } A_i \text{ and } B_j,$$

can be written as

$$H_0\colon p_{ij} = p_{i\bullet} p_{\bullet j} \begin{cases} i = 1, 2, \dots, a, \\ j = 1, 2, \dots, b. \end{cases}$$

Note that H_0 is a composite null hypothesis with unknown marginal probabilities $p_{1\bullet}, \dots, p_{a-1,\bullet}$, and $p_{\bullet 1}, \dots, p_{\bullet, b-1}$; here, we need just $a - 1$ and $b - 1$ parameters, because

$$p_{a\bullet} = 1 - (p_{1\bullet} + \cdots + p_{a-1,\bullet}) \quad \text{and} \quad p_{\bullet b} = 1 - (p_{\bullet 1} + \cdots + p_{\bullet, b-1}).$$

That is, these $a - 1 + b - 1 = a + b - 2 = h$ parameters are our θ values and can be estimated by

$$\hat{p}_{i\bullet} = \frac{Y_{i\bullet}}{n}, \quad \text{where } Y_{i\bullet} = \sum_{j=1}^{b} Y_{ij}, \quad i = 1, 2, \dots, a$$

and

$$\hat{p}_{\bullet j} = \frac{Y_{\bullet j}}{n}, \quad \text{where } Y_{\bullet j} = \sum_{i=1}^{a} Y_{ij}, \quad j = 1, 2, \dots, b.$$

If we replace $p_{ij} = p_{i\bullet} p_{\bullet j}$ in Q_{ab-1} by the product $\hat{p}_{i\bullet} \hat{p}_{\bullet j}$ of these estimators, then we need to adjust the degrees of freedom of the approximating chi-square distribution by subtracting from $ab - 1$ the number of estimated parameters. Hence Q_{ab-1} has an approximate chi-square distribution with

$$ab - 1 - (a + b - 2) = (a - 1)(b - 1)$$

degrees of freedom. If the computed Q_{ab-1} gets too large (i.e., if it exceeds $\chi^2[\alpha; (a - 1)(b - 1)]$), then we reject the hypothesis $H_0\colon p_{ij} = p_{i\bullet} p_{\bullet j}$ that the two attributes are independent.

Example 4.7-2 Ninety graduating male engineers were classified by two attributes: grade point average (low, average, high) and initial salary (low, high). The following results were obtained:

Grade-Point Average

Salary	Low	Average	High	Total
Low	15	18	7	40
High	5	22	23	50
Total	20	40	30	90

Note that here $y_{1\bullet} = 40, y_{2\bullet} = 50$, and $y_{\bullet1} = 20, y_{\bullet2} = 40, y_{\bullet3} = 30$, and

$$\hat{p}_{1\bullet} = \frac{4}{9}, \hat{p}_{2\bullet} = \frac{5}{9}, \quad \text{and} \quad \hat{p}_{\bullet1} = \frac{2}{9}, \hat{p}_{\bullet2} = \frac{4}{9}, \hat{p}_{\bullet3} = \frac{3}{9}.$$

Assuming independence of the attributes (H_0), we estimate the joint probabilities by $\hat{p}_{ij} = \hat{p}_{i\bullet}\hat{p}_{\bullet j}$. Under H_0, the estimated expected values $n\hat{p}_{i\bullet}\hat{p}_{\bullet j}$ are

$$90\left(\frac{4}{9}\right)\left(\frac{2}{9}\right) = 8.89, \quad 90\left(\frac{4}{9}\right)\left(\frac{4}{9}\right) = 17.78, \quad 90\left(\frac{4}{9}\right)\left(\frac{3}{9}\right) = 13.33,$$

$$90\left(\frac{5}{9}\right)\left(\frac{2}{9}\right) = 11.11, \quad 90\left(\frac{5}{9}\right)\left(\frac{4}{9}\right) = 22.22, \quad 90\left(\frac{5}{9}\right)\left(\frac{3}{9}\right) = 16.67.$$

The computed Q_5 is

$$q_5 = \frac{(15 - 8.89)^2}{8.89} + \frac{(18 - 17.78)^2}{17.78} + \frac{(7 - 13.33)^2}{13.33}$$

$$+ \frac{(5 - 11.11)^2}{11.11} + \frac{(22 - 22.22)^2}{22.22} + \frac{(23 - 16.67)^2}{16.67}$$

$$= 4.20 + 0.00 + 3.01 + 3.36 + 0.00 + 2.40 = 12.97.$$

Because $(a - 1)(b - 1) = (1)(2) = 2$ and $12.97 > \chi^2(0.05; 2) = 5.991$, we reject the hypothesis of independence. That is, on the basis of these data, we conclude that starting salaries and grade point averages are dependent. ∎

4.7-3 Goodness-of-Fit Tests

Sometimes we conjecture that X has a certain kind of distribution which involves one or more parameters. For example, we may think that the number of flaws, X, on a bolt of material has a Poisson distribution with an unknown parameter λ. Or we may conjecture that the compression strengths of concrete cylinders come from a normal distribution with a certain mean and variance. Suppose that we divide the outcome space of the variable X into k mutually exclusive and exhaustive cells. From a sample of size n, we can determine the frequencies of those k cells. Let us denote those frequencies by $Y_1, Y_2, \ldots, Y_k$; of course, $\sum_{i=1}^{k} Y_i = n$. The probabilities $p_1, p_2, \ldots, p_k$, where $\sum_{i=1}^{k} p_i = 1$, and the expected values $E(Y_i) = np_i$ are determined from the distribution of X. If the parameters of the distribution are specified (say, $\lambda = 1$ for Poisson, and $\mu = 45$ and $\sigma^2 = 90$ for the normal), then

$$Q_{k-1} = \sum_{i=1}^{k} \frac{(Y_i - np_i)^2}{np_i}$$

has an approximate chi-square distribution with $k - 1$ degrees of freedom. However, usually those parameters are unknown, and to compute Q_{k-1}, we need to estimate a number of—say, h—parameters of the distribution of X. For example, $h = 1$ in the Poisson distribution and $h = 2$ in the normal distribution. The use of the estimates of these parameters in Q_{k-1} produces a statistic that has an approximate $\chi^2(k-1-h)$ distribution. The null hypothesis H_0 that X has that specified underlying distribution is questioned if the computed q_{k-1} is too large—namely, if it exceeds $\chi^2(\alpha; k-1-h)$.

Example 4.7-3   Suppose we observe $n = 85$ values of a random variable X that is thought to have a Poisson distribution. We obtain the following data:

x	0	1	2	3	4	5
Frequency	41	29	9	4	1	1

The sample average is the appropriate estimate of $\lambda = E(X)$ and is given by

$$\hat{\lambda} = \frac{(41)(0) + (29)(1) + (9)(2) + (4)(3) + (1)(4) + (1)(5)}{85} = 0.8.$$

With this estimated λ value, the expected frequencies for the first three cells are

$$85(0.449) = 38.2, \quad 85(0.360) = 30.6, \quad 85(0.144) = 12.2.$$

We can use tables or statistical software to determine these Poisson probabilities. The expected frequency for the cell $\{3, 4, 5, \ldots\}$ is $85(0.047) = 4.0$. Here, we have combined several of the original cells to obtain an expected value that is not much smaller than 5. The computed Q_3, with $k = 4$ after that combination, is

$$q_3 = \frac{(41 - 38.2)^2}{38.2} + \frac{(29 - 30.6)^2}{30.6} + \frac{(9 - 12.2)^2}{12.2} + \frac{(6 - 4)^2}{4}$$

$$= 0.21 + 0.08 + 0.84 + 1.00 = 2.13.$$

Comparing this value with $\chi^2(0.05; 4 - 1 - 1 = 2) = 5.991$, we find no reason to reject the Poisson distribution. ∎

Exercises 4.7

*4.7-1   We cast a certain die $n = 120$ independent times and find the number of times 1, 2, 3, 4, 5, and 6 spots are on the "up side" to be 16, 21, 22, 14, 19, and 28, respectively. Does it seem as if this die is biased? That is, can we reject H_0: $p_i = 1/6, i = 1, 2, \ldots, 6$, using $\alpha = 0.05$?

*4.7-2   A manufacturer of men's underwear claims that 91 percent of the products are of the "best" quality, 8 percent are "seconds," and only 1 percent are defective. To test this claim, 500 garments selected at random are inspected; the inspection results in the respective frequencies of 434, 48, and 18. Can we reject the manufacturer's claim at the $\alpha = 0.05$ level?

*4.7-3   The starting salaries of $n = 200$ engineers were classified as being in the lower 25 percent (A_1), the second 25 percent (A_2), the third 25 percent (A_3), and the upper 25 percent (A_4). In addition, the 200 engineers were classified according to the color of their eyes. The results are summarized in the following table:

Salary

Eyes	A_1	A_2	A_3	A_4	Total
Blue	22	17	21	20	80
Brown	14	20	20	16	70
Other	14	13	9	14	50
Total	50	50	50	50	200

Are the color of eyes and the starting salary independent attributes? Use $\alpha = 0.05$. If eye color and starting salary are not independent, you may be able to claim discrimination.

4.7-4 A test of the equality of two or more multinomial distributions can be made by using calculations that are associated with a contingency table. For example, $n = 100$ light bulbs were taken at random from each of three brands and were graded as A, B, C, or D as follows:

Grade

Brand	A	B	C	D	Total
1	27	42	21	10	100
2	23	39	25	13	100
3	22	36	23	19	100
Total	72	117	69	42	300

Clearly, we want to test the equality of three multinomial distributions, each with $k = 4$ cells. Because, under H_0, the probability of falling into a particular grade category is independent of brand, we can test this hypothesis by computing Q_{11} and comparing it with $\chi^2(\alpha; (2)(3) = 6)$. Use $\alpha = 0.025$.

*4.7-5 The number X of telephone calls received each minute at a certain switchboard in the middle of a working day is thought to have a Poisson distribution. Data were collected, and the results were as follows:

x	0	1	2	3	4	5	6
Frequency	40	66	41	28	9	3	1

Fit a Poisson distribution. Then find the estimated expected value of each cell after combining {4, 5, 6} to make one cell. Because $k = 5$, compute Q_4 and compare it with $\chi^2(\alpha = 0.05; 3)$. Why do we use three degrees of freedom? Do we accept or reject the Poisson distribution?

4.7-6 Use the compression strength data of Section 1.3. Check to see if a normal curve fits the frequency distribution given in Table 1.3-2. Combine the first two cells and the last three so that $k = 7$. Use $\alpha = 0.05$. Why do we use four degrees of freedom? Note that the goodness-of-fit test provides a formal test for normality. Indeed, it complements the normal probability plots in Section 3.6.

4.7-7 You are planning to publish a new magazine. You want to check whether politically interested people would be inclined to buy the magazine. A survey of $n = 160$ potential customers leads to the following results:

Interested in politics

Would buy	Yes	No
Yes	58	44
No	35	23

Test whether there is an association among these two variables. Use $\alpha = 0.05$.

4.7-8 Two hundred University of Iowa students were selected at random. Of the 120 male students, 60 were against tougher liquor rules. Of the 80 women, 30 were against tougher liquor rules. Test the null hypothesis that there is no association between gender and attitude toward tougher liquor rules. Use $\alpha = 0.05$. Determine the probability value of the appropriate chi-square test.

Chapter 4 Additional Remarks

Why is the normal distribution so important to statisticians? It's because most estimators have approximate normal distributions. As a matter of fact, that is the reason the normal distribution was discovered. Abraham de Moivre, a French Calvinist born in 1667 who had to flee to England because of his religious beliefs, was a brilliant mathematician. But like many other brilliant mathematicians, he was poor throughout his life, supporting himself by tutoring mathematics. His book on probability theory, entitled *The Doctrine of Chances,* was highly prized by gamblers, who realized that probability could help them make money. De Moivre considered the relative frequency Y/n, where Y is binomial $b(n, p)$. Jacob Bernoulli had proved earlier that Y/n converges, with increasing sample size n, to the true proportion p, but it was de Moivre who discovered that Y/n had an interesting approximating distribution: a bell-shaped curve now known as the normal distribution. That is, he had found that the estimator Y/n of p had an approximate normal distribution. Knowing also that its standard deviation was $\sqrt{p(1-p)/n}$, he realized 300 years ago that the interval $p \pm 3\sqrt{p(1-p)/n}$ was almost certain to contain the observed Y/n if n was large.

More than 75 years after de Moivre's discovery, two other mathematicians—Carl Friedrich Gauss and Pierre-Simon Marquis de Laplace—worked independently on a generalization of that earlier result. De Moivre was sampling from a Bernoulli distribution where X_i in the ith trial was either 0 or 1, so $Y/n = \sum_{i=1}^{n} X_i/n = \overline{X}$. Laplace and Gauss were sampling from any distribution with finite mean μ and standard deviation σ, and they found that $\overline{X}$ had an approximate $N(\mu, \sigma^2/n)$ distribution, provided that n was large enough. Gauss was quite secretive about his work, and it is difficult to tell who actually "discovered" the central limit theorem first, but Laplace distributed it a few months before Gauss published his book *Theoria Motus* in 1809. Maybe we should call it a "tie." People seem to forget Laplace's contribution and, worse than that, de Moivre's much earlier result, and today the normal distribution is often called the Gaussian distribution.

Earlier we mentioned that W.S. Gosset published his work on the t-distribution under the pseudonym "A Student," because Guinness did not want other breweries to know that it was using statistical methods. In another story we heard, Gosset used a pseudonym because he did not want Guinness, his employer, to know that he was spending all his extra time on statistics. Whatever the reason for the pseudonym, the distribution will always be known as Student's t. In a certain sense, Gossett was lucky that he found the correct distribution for $T = (\overline{X} - \mu)/(S/\sqrt{n})$, because he computed the first four moments of T [i.e., $E(T)$, $E(T^2)$, $E(T^3)$, $E(T^4)$], and then *assumed* that the distribution of T was in the Pearson's family of distributions. Karl Pearson, another famous statistician of the early 1900s, had proposed this very general family a few years earlier. If the distribution is in that family, only four moments are needed to find its p.d.f. It was a few years later that R.A. Fisher actually proved that T had the p.d.f. Gossett had claimed.

We are also reminded of the conflict between Pearson and Fisher. Fisher was a young, brash man when he told his senior, Pearson, that he should reduce the degrees of freedom of the chi-square distribution by one for every parameter that he estimated in the computation of the chi-square. (See the discussion in Section 4.7.) Pearson never believed this (and, yes, Fisher was right), and as editor of the very prestigious journal *Biometrika,* he would block Fisher from publishing any articles in that journal. Fisher thought that in the long run this was to his advantage, because it made him consider more applied journals, enhancing his development into a well-rounded scientist.

The comments that follow relate to two of the projects. In Project 11, we explain a permutation test that also was an idea of Fisher. The bootstrapping idea explained in Project 10 at the end of this chapter goes back to Brad Efron of Stanford University. He uses the empirical distribution as the best guess for the underlying distribution; thus, he avoids having to make specific parametric assumptions (such as normality) about this distribution. So, to learn about the distribution of a statistic Z, a sample of the same size of the original sample is taken (simulated) from the empirical distribution, and Z is computed from that sample. This is done many times (10,000 times is not unusual), and from the histogram of all the computed Z values, the investigator gets a good idea about the distribution of Z.

There is an interesting story about the name "bootstrap." It seems that the expression to "pull oneself up by his or her bootstraps" comes from the novel *The Surprising Adventures of Baron Munchhausen,* by Rudolf Erich Raspe. The baron had fallen from the sky and hit the earth so hard that he found himself in a hole nine fathoms deep. He had no way to get out, but "Looking down I observed that I had on a pair of boots with exceptionally sturdy straps. Grasping them firmly, I pulled with all my might. Soon I hoisted myself to the top and stepped out on terra firma without further ado." In bootstrapping, statisticians pull themselves up by the empirical distribution; and by replacing theoretical derivations with the results from computer-intensive simulations, they find the approximate distribution of their statistics of interest. This is a flexible approach, because no assumptions about the underlying distribution ever need to be made. It also works quite well. We hope that the baron's technique was equally successful.

Our final comments in these endnotes relate to computer software. Available software packages include programs for the calculation of confidence intervals and the testing of hypotheses. Here we restrict our brief remarks to Minitab and its commands within the "Stat > Basic Statistics" folder. "1-Sample Z" is used for the inference about a population mean from a single random sample, assuming that the standard deviation is known; "1-Sample Z" uses the normal distribution. "1-Sample t" is used to make an inference about a population mean from a single random sample with an estimated standard deviation; "1-Sample t" uses the t-distribution.

"2-Sample t" is used to compare two means when the samples are independent. "Paired t" is used in the paired or blocked situation. "1 Proportion" is used for the inference about a single proportion, while "2 Proportions" is used for the comparison of two proportions. "1 Variance" is used for the inference about a single variance, while "2 Variances" covers the comparison of two variances. The data can be read in as summary statistics, or the program will create these summary statistics from the supplied observations. Available options allow the user to specify one- or two-sided alternatives and to select specific significance levels. The output includes confidence intervals, standard errors, test statistics, and probability values. Furthermore, the numeric analysis can be supplemented with appropriate graphs.

Minitab also includes programs for the calculation of the required sample size such that tests achieve a specified significance level and a specified probability of making a type II error at a certain value under the alternative. These programs can be found under the "Stat > Power and Sample Size" folder.

The commands "Stat > Table > Chi-Square Test (Two-way Table in Worksheet)" and "Stat > Table > Cross Tabulation and Chi-Square" can be used for the chi-square test of independence. The first command is used when the data are already summarized in a contingency table, while the second command extracts the frequency table from the raw (not summarized) data.

Further detailed instructions on how to use these commands are available in the Instructor's Manual. Introductions to Minitab and R, another useful software for statistical analysis, are also available from the Instructor Resource Center.

Projects

Project 1

(a) Consider a normal distribution with mean 5 and standard deviation 2. Generate 10,000 random samples of size $n = 3$. You can do this by generating three columns of normal random variables of length 10,000 and considering the three entries in a row as the elements of a sample. Calculate averages across the three columns to obtain 10,000 averages from samples of size 3. Obtain the histogram of the averages and construct a normal probability plot. Compute the mean and the standard deviation of these 10,000 averages. Convince yourself that the distribution of the averages is normal and centered at mean 5, with standard deviation $2/\sqrt{3}$.

Repeat this exercise for samples of $n = 25$ and $n = 50$. For these sample sizes, you expect the standard deviation of averages to be $2/\sqrt{25}$ and $2/\sqrt{50}$.

Note: Use your statistical software package to generate the random numbers.

(b) Repeat this exercise with samples from a uniform distribution on the interval from 0 to 10. This distribution has mean 5 and standard deviation $\sqrt{100/12}$. What can you say about the distribution of the averages from samples of size $n = 3, 25$, and 50? Do you notice the central limit effect at work? Comment.

Project 2

(a) Consider the normal random variables that you generated in Project 1. Calculate the 10,000 *sample variances* from $n = 3$ (25, 50) observations. Obtain the histogram of the sample variances, as well as a normal probability plot. Comment on your results.

Note: Notice that the distribution of the sample variance is not normal. This should not surprise you, because the central limit effect applies to averages of observations (and not to averages of squares).

(b) Obtain a histogram of $\dfrac{(n-1)s^2}{\sigma^2} = \dfrac{(n-1)s^2}{4}$, and a probability plot with the $\chi^2(n-1)$ distribution as the target distribution. Comment on the results.

Note: Remember the theoretical result in Section 4.3.

(c) Repeat this exercise for samples from a uniform distribution on the interval from 0 to 10 (with variance $\sigma^2 = 100/12$), and obtain the histograms and probability plots in (a) and (b).

Project 3

(a) Consider the normal random variables that you generated in Project 1. Obtain the largest observation among the $n = 3$ (25, 50) observations. Obtain the histogram of the maximum, as well as a normal probability plot. Comment on your results.

(b) Repeat this exercise for samples from a uniform distribution on the interval from 0 to 10.

Project 4

Generate the number of successes in 100 independent trials from a binomial distribution with $p = 0.40$, and store the results of 10,000 such simulations in a column of your work sheet. Obtain the 10,000 sample proportions (dividing the number of successes by the sample size $n = 100$) and construct a histogram and a normal probability plot. Interpret your findings. What is at work here? Obtain the standard deviation of the 10,000 proportions and discuss whether it comes close to the theoretically expected value. (See Section 4.4.)

Repeat the exercise for the binomial distribution with $n = 100$ and $p = 0.02$. Interpret your findings, and discuss why in this case the normal distribution does not provide a close approximation. How would things change if you had generated the successes among $n = 2,000$ (or $n = 10,000$) trials from a binomial distribution with $p = 0.02$?

Note: In Minitab, you would use "Calc > RandomData > Binomial"; in Excel, you would use "Tools > Data Analysis > Random Number Generation."

*Project 5: Issues Involving the Sample Size

You are interested in testing whether the mean starting salary of female engineers is *greater* than that of their male counterparts. From previous studies, you know that the variability among starting salaries for men and women engineers is about the same. Furthermore, you know that the standard deviation among salaries is about $5,000.

You are planning to sample the same number of males as females. How large must your samples be if you want to limit (1) your error to 5 percent of claiming a higher mean starting salary for females when the mean salaries of men and women are actually the same and (2) your error to about 10 percent of missing a $2,000 higher mean salary for women?

(a) Solve this problem from first principles.

Note: Using the result for the standard deviation of a difference of two sample averages, set up two equations—one controlling the probability of a type I error at 5 percent, and the other controlling the probability of a type II error at 10 percent. Solve these equations for the common sample size. (See Sections 4.5 and 4.6.)

(b) Use computer software for the calculations.

Note: Fortunately, many statistical software packages include functions that allow you to calculate the sample size without going through the details in (a). In Minitab, use "Stat > Power and Sample Size > 2-sample t." You need to specify two of three entries: sample size, difference, and power. Here, we skip sample size (this is what we want to determine) and enter 2000 (for the difference), and 0.90 (for the power, which is 1 minus the probability of the type II error). Under options, we check significance level 0.05 and the fact that we are dealing with a one-sided alternative where the mean starting salary for women is assumed greater than the mean salary for men. You will find that $n = 108$ graduates are needed in each group. Alternatively, you can specify a sample size (say, 50 in each group) and determine the power of detecting a difference as large as $2,000. Check that the resulting power is 0.63, implying a type-II error risk of 37 percent.

Project 6: More Issues Involving the Sample Size

In the past, the sign-up rate for a certain credit card has been around 6 percent. Your marketing team wants to decide between two different sets of promotional materials that it plans to send to potential customers: a traditional set that is similar to the one that has been used in the past, and a new bolder set that is expected to increase the sign-up rate. Before switching to the new materials, your company wants to run a comparative experiment that evaluates the two sign-up rates. Assuming a significance level of 0.05, determine the common sample size for the two groups that can detect a 1 percent increase in the sign-up rate with a power of 0.95. How does the sample size change if you require less power (say 0.90 or 0.80)? How does the sample size change if you want to detect a difference of one-half of a percent?

Note: You may want to use computer software to carry out the calculations. In Minitab, use "Stat > Power and Sample Size > 2 Proportions."

Project 7

The data for this project are from Ziegler, E., Nelson, S.E., and Jeter, J.M., "Early Iron Supplementation of Breastfed Infants" (Iowa City, Iowa: Department of Pediatrics, University of Iowa, 2007).

The transfer of iron from mother to infant occurs mostly prenatally. At birth, the infant is endowed with ample amounts of stored iron, which the infant uses to meet the needs of growth during the first four to six months of life. Once the iron endowment is exhausted, the infant depends on exogenous iron to meet its iron needs for growth and for replacement of inevitable losses. Because of its low iron content, breast milk alone meets only a fraction of the infant's needs for iron. Foods that the infant receives upon being weaned tend to be low in iron unless they are fortified with it. This is the reason breastfed infants are at risk of iron deficiency.

The present study was designed to assess the effect of early iron supplementation on body iron stores during, as well as after, the period of supplementation. Iron supplements were provided from 1 month to 5 1/2 months of age — the age at which weaning foods, the breastfed infant's main source of dietary iron, are usually introduced.

The study was designed as a randomized, double-blind, placebo-controlled trial. The subjects, normal breastfed infants, were studied from 1 to 18 months of age. Random assignment to one of the two study groups (called Fe and Placebo) was performed with the use of sealed envelopes at the time of enrollment at 28 days (d) of age. Infants visited the study center every 28 d until they reached 12 months (mo) of age and again at 15 and 18 mo. Study visits occurred within 2 d of the designated age in the first year of life and within 4 d at 15 and 18 mo of age. From 1 to 5.5 mo (168 d), infants in the Fe Group received 7 mg of iron daily. Infants in the Placebo Group received drops of similar appearance and taste that contained no iron. Blood was obtained at enrollment and at 4 mo (112 d), 5.5 mo (168 d), 7 mo (224 d), 9 mo (280 d), 12 mo, 15 mo, and 18 mo.

The subjects were term infants of either gender with birth weight greater than 2,500 grams who were considered normal by their physicians and the investigators. Infants visited the Department of Pediatrics every 28 days. At each visit, weight and length were measured by established methods. At selected ages, blood was drawn by heel stick. The blood was then centrifuged, and plasma was used to determine ferritin, the main indicator of iron level.

The file Project7ZieglerStudy1 includes data on weight, ferritin, age (in days), age (coded, 1 through 8 for the eight age groups), age (grouped by average age in the group), feeding (coded, 1 for placebo and 2 for iron supplement) for 32 infants receiving the placebo and 34 infants receiving the supplement.

Analyze the data. Assess the effect of the iron supplement on ferritin, both graphically and numerically. Do this for each age group separately.

Project 8

In Project 8 of Chapter 1, you surveyed your fellow students about issues surrounding the engineering statistics course. Reanalyze the data, supplementing your earlier analysis with appropriate confidence intervals for means and proportions. For example, add confidence intervals to your computed averages and proportions such as the average number of hours studied and the proportion of students who dislike the statistics course. Apply appropriate hypothesis tests to compare men and women with respect to the mean number of hours they study and the proportion of students who dislike the statistics course. Furthermore, test whether or not dislike of the statistics course is independent of the area of specialization.

Note: Use the chi-square test in Section 4.7.2 to test the independence of two categorical variables.

Project 9

Select three pages at random from your favorite literary book, and for each word on these pages, determine the length of the word. In addition, determine the length (number of words) of each sentence. Analyze the information, both graphically as well as numerically.

Obtain the same information for newspaper or magazine articles. Try to get information on roughly the same number of words and the same number of sentences that you have collected from your literary book.

Compare the information from the book and from the newspaper. Are there differences? Apply the appropriate statistical procedures (confidence intervals, hypothesis tests) to assess any differences you find.

Project 10: Bootstrap Confidence Intervals

The weights (in grams) of 61 one-month old infants are listed as follows:

4960	5130	4260	5160	4050	5240	4350	4360	3930	4410	4610
4550	4460	2940	4160	4110	4410	4800	5130	3670	4550	4290
4950	5210	3210	4030	3580	4360	4360	3920	4050	4630	3756
4586	5336	2828	4172	4256	4594	4866	4784	4520	5238	4320
5330	3836	5916	5010	4344	3496	4148	4044	5192	4368	4180
4102	5210	4382	5070	5044	3530					

The data are taken from the study by Ziegler et al. described in Project 7 of this chapter.

Use a *resampling/bootstrap* approach to obtain 95 percent confidence intervals for the population mean, median, standard deviation, and interquartile range. Compare the bootstrap confidence interval of the mean with the confidence interval that we discussed in Section 4.3, and interpret the difference.

The resampling approach is flexible and can be used to obtain confidence intervals for any characteristic of the population, not just for those which statisticians have studied theoretically. Furthermore, it avoids having to make assumptions about the underlying distribution (such as the assumption of normality) that are so important for statisticians as they seek explicit solutions. The reason the resampling approach is useful is that it is not always obvious whether these assumptions are true.

Although special computer software is available to carry out the resampling approach, we would like you to go through its steps from first principles. Afterwards, you may want to learn about resampling/bootstrap software by searching the Web; some software is actually free.

Use the following procedure to obtain the bootstrap confidence interval: Assume that the population is distributed exactly as in the sample. Construct a discrete probability distribution with the 61 recorded observations as outcomes and their associated probabilities given by 1/61. You will notice that some observations come up more than once; for those outcomes, the associated probability is $j/61$, where j is the multiple of that observation. Draw a sample of size 61 from this discrete distribution (with replacement), determine an estimate of the parameter of interest (mean, median, standard deviation, interquartile range, etc.), and repeat this procedure $B = 10,000$ times. In Minitab, this is easy to do: Just store the distribution (possible values, probabilities) in two columns and generate 10,000 samples (rows) of 61 columns from the discrete distribution in those two columns, using the "Calc > Random Data" command. A histogram of the 10,000 replications tells you about the variability of the estimates. Order the realizations, and obtain the 2.5th and the 97.5th percentiles. (Call them c and d, respectively; they are the realizations with ranks 250 and 9,750). These are the limits of your *naïve* bootstrap interval $[c, d]$. Alternatively, one can calculate the *percentile-method* bootstrap confidence interval $[e - (d - e), e - (c - e)]$, where c and d are the percentiles and e is the estimate that you obtained from your sample.

*Project 11: Permutation Tests

Following are the ferritin levels of 59 infants at age 5 1/2 months:

No Iron Supplement

86.0	48.0	72.0	28.0	142.0	223.0	53.0	14.0	100.0
62.0	17.0	28.0	17.0	16.0	138.0	62.0	21.3	60.7
135.0	35.2	93.8	6.6	43.2	86.1	46.0	37.2	96.3
64.2	94.0	70.7						

Iron Supplement

42.0	60.0	199.0	38.0	124.0	44.0	147.0	243.0	112.0
86.0	124.0	40.0	56.0	11.0	90.0	48.0	75.8	55.8
109.0	75.9	278.0	434.0	146.0	101.0	267.0	24.8	166.0
271.0	48.6							

The first group, with 30 infants, received the placebo; the second group, with 29 infants, received dietary iron supplements. The data are taken from the study by Ziegler et al. described in Project 7 of this chapter.

(a) Compare the two groups and test whether their mean ferritin levels are different (a two-sided alternative). Calculate the two-sample test statistic of Section 4.6, calculate its probability value, and assess its significance.

(b) Compare the result in (a) with that of a permutation/randomization test.

Permutation/randomization tests provide an alternative way of assessing the significance of observed differences. Following this approach, we put all $30 + 29 = 59$ (ferritin) scores into a single pot. Then we randomly select 30 scores from the pot without replacement, and these 30 scores become the observations of the first group. The 29 remaining items make up the second group. Next, we calculate the means of both groups and record the difference of the means. We, repeat the process, obtaining and recording a second difference in means for the resulting sample of 30 scores assigned to one group and the remaining 29 to the other group. We repeat this procedure a large number of (say 10,000) times, and from these replications, we obtain the sampling distribution of the statistic of interest (here, the difference of means). From the sampling distribution, we determine the proportion of the differences that had absolute values as large as or larger than the one observed in our actual data. If this proportion is less than the significance level (typically, 0.05), we reject the null hypothesis that there is no difference among the population means.

An advantage of this approach is that it is general: Not only can you test a hypothesis about the difference of means (as we have done here), but also you can investigate hypotheses about the difference of two medians, the ratio of two standard deviations, and so on.

Note: You may not have access to software that carries out randomization/permutation tests. If you don't, then carry out the procedure we have described from first principles, using the random-sampling features of any available statistical software, such as Minitab.

*Project 12

Foresters study whether trees are spatially random or whether they exhibit some clustering. In the absence of clustering, trees are arranged according to a *Poisson forest*, which is defined as follows:

Divide an area (plot of land) A into nonoverlapping lots of very small areas δA. Let $N(\delta A)$ be the number of trees within the area δA. Assume that the probability of finding a single tree in such a small area is proportional to the area, and suppose that the probability of finding more than one tree is negligible. That is,

$$P[N(\delta A) = 1] \cong \lambda(\delta A), P[N(\delta A) = 0] \cong 1 - \lambda(\delta A) \text{ and } P[N(\delta A) > 1] \cong 0.$$

Furthermore, assume that there is independence from one nonoverlapping area to the next: The occurrence of a tree in one area does not change the probability of occurrence in adjacent areas; in other words, there is no clustering. Then the number of trees in area A is given by a Poisson distribution with parameter δA.

Foresters are interested whether there is clustering or whether the location of the trees within area A is uniform. Several methods have been developed, including the *quadrat method,* which places small nonoverlapping quadrats (areas that can be either square or circular) randomly onto the much larger area A, counts the number of trees in each quadrat, and assesses whether the Poisson distribution describes the distribution of the number of trees per quadrat. Following are the results obtained with 100 circular quadrats (with radius 6 meters) placed onto an area of 16 hectares [see N. A. Cressie, *Statistics for Spatial Data* (New York: Wiley, 1993), Table 8.2]:

Trees per quadrat	Observed frequency
0	34
1	33
2	17
3	7
4	3
5	1
6	1
7	2
8	1
9	0
10	1

Test whether the distribution of the number of trees per quadrat can be modeled with a Poisson distribution indicating the absence of clustering. The mean number of trees per quadrat is 1.43. Use this number as an estimate of the parameter in the Poisson distribution, and carry out the goodness-of-fit test discussed in Section 4.7.3. You may have to combine groups with low frequencies.

Project 13

The studies that follow are described in Rose Martinez-Dawson, "Incorporating Laboratory Experiments in an Introductory Statistics Course," *Journal of Statistics Education,* Vol. 11, No. 1 (2003), www.amstat.org/publications/jse/v11n1/martinez-dawson.html. Replicate some of the studies, obtain your own data, and summarize your findings.

(a) Ecologists are concerned about the effect of pollution on the environment. To illustrate hypothesis testing about a single mean, rainwater samples are obtained and the pH of the rainwater is determined with the use of a pH meter. Carry out a hypothesis test to determine whether the mean rainwater pH is less than 5.6, an indication of "acid rain." Support your analysis with appropriate graphs.

(b) Water samples from different lakes are obtained before and after it rains. The pH of each sample is determined by means of a pH meter. Carry out the appropriate hypothesis test to determine whether the mean pH of lake water after a rain is lower than the mean pH of lake water before a rain. Support your analysis with appropriate graphs. Explain why water samples are obtained from the same lake both before and after it rains.

(c) A spectrophotometer, which measures the transmittance of light, was used to obtain absorbance measurements on two types of solutions. A higher absorbance reading is associated with a darker, more concentrated solution. The mean absorbencies of two brands of red food color, one more expensive than the other, were compared. One drop of red food color from each brand was mixed in 20 milliliters (ml) of distilled water, and three solutions of each type were made. The goal was to determine whether the mean absorbances of these two solutions were different at a wavelength of 570 nanometers (nm).

Several modifications of this experiment were tried. In one version, absorbencies from solutions made with one drop of red food dye were compared with absorbencies from solutions made with two drops of red food color into 20 ml of distilled water. Several solutions of each type were made. In another version, the mean absorbencies at 570 nm for old versus new food color were compared. The old food color was at least one year old.

Comment on the sources that can contribute to variation in the results. For example, comment on the following issues: Were the drops of food dye placed into each beaker of water of uniform size? Were exactly 20 ml of water in each of the beakers? Could errors have been introduced due to reading absorbencies from the scale on the spectrophotometer?

Project 14

Three tests as described in Project 3 of Chapter 1 are conducted with trucks A, B, and C. The first test uses truck A as the control truck and truck B as the test truck. This was the data set that we considered in Project 3 of Chapter 1. Truck A is the control and truck C is the test truck in test 2, while truck C is the control and truck A is the test truck in test 3. Fuel consumption (in pounds) for baseline and test laps of these three tests is as follows:

Test 1

Baseline Laps			Test Laps		
Lap No.	Control	Test	Lap No.	Control	Test
11	50.1488	51.8912	1	50.0354	49.4885
12	49.2472	50.6840	2	50.3589	48.7868
13	50.7082	51.9655	3	49.8449	48.3581
14	49.6399	50.5968	4	50.8973	49.3284
15	50.4372	51.8736	5	51.2444	50.2403
16	50.2273	51.6166	6	55.9998	54.1003
17	50.0609	50.8950	7	55.0668	53.7085
18	50.2046	50.9899	8	52.0332	50.0265
19	50.4621	51.6013	9	51.3607	50.7168
20	50.1678	51.3997	10	51.0776	50.1817
21	49.6908	50.9830	11	50.8610	49.3742
22	49.6244	50.5118	12	50.5268	49.1016
23	47.5206	49.7935	13	49.1427	48.6476
24	49.7272	50.1476			
25	46.9486	48.8342			
26	47.9234	48.9613			
27	48.7489	49.1496			

Test 2

	Baseline Laps			Test Laps	
Lap No.	Control	Test	Lap No.	Control	Test
1	55.3113	55.7334	1	49.8147	50.5716
2	53.6337	54.1822	2	48.9365	50.1891
3	51.7796	51.8036	3	48.8112	49.7914
4	50.5123	50.2009	4	55.7907	54.8131
5	50.4743	51.1479	5	53.0183	52.7244
6	50.0646	50.3411	6	52.7069	52.4025
7	53.2571	52.9607	7	50.8575	50.6479
8	51.4892	50.3907	8	50.9064	51.2019
9	50.7879	50.7934	9	52.2317	51.8950
10	52.4796	51.9957	10	53.4323	52.3988
11	50.2225	51.3380	11	54.7146	55.0461
12	51.9174	53.2850	12	49.7150	49.9162
13	51.1354	52.3657	13	49.4545	49.5722
14	51.5383	51.7294	14	49.8251	48.1274
15	50.9109	52.4779	15	50.0902	51.0376
			16	49.4245	49.6365
			17	49.4206	49.1149

Test 3

	Baseline Laps			Test Laps	
Lap No.	Control	Test	Lap No.	Control	Test
1	53.1619	53.0108	1	52.3952	51.7054
2	51.7278	51.8493	2	52.7375	51.2310
3	51.7645	51.9223	3	54.1094	53.1208
4	51.3215	50.8417	4	53.9236	52.8810
5	52.8214	51.7060	5	50.3270	49.2826
6	51.2873	49.9220	6	51.0103	48.9047
7	51.7809	50.6570	7	52.0884	51.4040
8	54.7923	54.0798	8	55.7383	55.1146
9	53.6278	53.3307	9	53.2446	52.3117
10	52.9175	52.5211	10	51.7210	51.1680
11	53.5807	52.1636	11	52.2943	51.3923
12	53.4555	54.1245	12	51.5441	51.2498
13	53.9870	53.7951	13	52.3488	52.0088
14	52.6708	51.3502	14	53.3372	52.3147

Test 3

Baseline Laps			Test Laps		
Lap No.	Control	Test	Lap No.	Control	Test
15	52.4845	53.5616	15	52.7943	52.4803
16	50.4069	50.4483	16	55.2126	54.7778
17	51.7267	50.8487			
18	50.6869	50.8670			
19	51.3805	51.3316			

Analyze the data and discuss whether the company can make the claim that its additive improves fuel economy. Discuss the strengths and weaknesses of the design.

Hint: For each test, consider the ratio (diesel use of test truck)/(diesel use of control truck).

Chapter

5

STATISTICAL PROCESS CONTROL

5.1 Shewhart Control Charts

All processes have some variation. When we manufacture a product, measurements on the final product inevitably show variation from unintentional changes to the process as well as random variation. Many different factors enter into a production process, and a change in each will cause some variation in the final product. This variation may come from differences among machines, lot-to-lot variation, differences in suppliers and incoming raw materials, changes in the plant environment, and so on. Despite the fact that considerable effort is expended in attempting to control the variability in each of these factors, there will still be variability in the final product. In the end, this variability has to be controlled.

Statistical *control charts* or, more generally, statistical process control methods are procedures for monitoring process variation and for generating information on the stability of a process. It is important to check the stability of processes, because unstable processes will result in lost production, defective products, poor quality, and, in general, loss of consumer confidence. For example, in the production of integrated-circuit boards, which involves several welding procedures, it may be the weld strength that is of importance. Selecting a small sample of such boards at regular intervals and measuring the weld strength by a certain pull test to destruction will provide valuable information on the stability of the welding process. In the production of concrete blocks, it is the compressive strength that is of importance and that needs to be controlled. Measurements on a small number of concrete blocks—say, twice during each production shift—can give us valuable information on the stability of the production process. In the production of thin wafers for integrated-circuit devices by high-temperature furnace oxide growth processes, it is the thickness of the wafers that needs to be controlled. Measurements on the thickness of a few selected wafers from every other furnace run can indicate whether the thickness of the product is stable. Here we have given only three examples. Many others can be found, and we encourage the reader to think of still others.

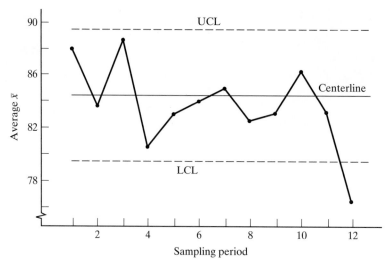

Figure 5.1-1 $\bar{x}$-chart for the sample means in Table 5.1-1 (see p. 297)

5.1-1 $\bar{x}$-Charts and R-Charts

A control chart is a plot of a summary statistic from samples that are taken *sequentially*. Usually, the sample mean and a measure of the sample variability, such as the standard deviation or the range, are plotted on control charts. Figures 5.1-1 and 5.1-2 are two examples. Figure 5.1-1 shows the average compressive strengths of concrete blocks from samples of size $n = 5$. Twice during each shift, five concrete blocks are taken from the production line, their compressive strengths are determined, and the average is entered on the chart. Since we plot *averages*, we call this an *$\bar{x}$-chart* (where "x-bar" stands for average). In Figure 5.1-2, we display the variability within the samples over time and plot the *ranges* from successive samples; we call such a plot an *R-chart* (where R stands for range).

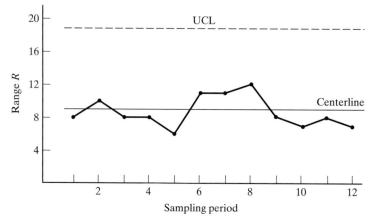

Figure 5.1-2 R-chart for the sample ranges in Table 5.1-1 (see p. 297)

Control charts also include bands, or *control limits,* that help us determine whether a particular average (or range) is "within acceptable limits" of random variation. Through these limits, control charts try to distinguish between variation that can normally be expected (variation due to *common causes*) and variation that is caused by unexpected changes (variation due to *special causes*). One should not tamper with the process if the measurements on these charts fall within the control limits. However, if one notices shifts in the process level on the $\bar{x}$-chart, and if plotted averages are outside the control limits, one must conclude that something has happened to the process. Similarly, if the process variation, as measured on the R-chart, changes by more than what could be expected under usual circumstances, one must conclude that the process variability is no longer stable. In these circumstances, steps have to be taken to uncover the special causes and keep them from recurring.

The theoretical explanation of the control limits in the $\bar{x}$-chart is as follows: Let us suppose that our observations are from a stable distribution with mean μ and variance σ^2. Then the mean $\overline{X}$ of a random sample of size n from this distribution has an approximate normal distribution with mean μ and variance σ^2/n. If averages are plotted on a chart, the probability that a sample average $\overline{X}$ falls within the bounds $\mu - 3\sigma/\sqrt{n}$ and $\mu + 3\sigma/\sqrt{n}$ is 0.9974. If μ and σ^2 are known, we call $\mu - 3\sigma/\sqrt{n}$ and $\mu + 3\sigma/\sqrt{n}$ the *lower control limit* (LCL) and the *upper control limit* (UCL), respectively. Thus, with a stable process, the probability of $\overline{X}$ falling between the LCL and the UCL is very large. Consequently, it is rare that a sample average $\overline{X}$ from such a stable process would fall outside the control limits.

Unfortunately, we usually do not know whether our process is stable, nor do we know the values of μ and σ^2. Thus, we begin by taking several samples, each of size n. Let us say that there are k such samples. Let $\bar{x}_1, \bar{x}_2, \ldots, \bar{x}_k$ and $s_1, s_2, \ldots, s_k$ be the means and standard deviations of these samples. Then we can estimate σ by the average of the standard deviations, namely, $\bar{s} = (s_1 + s_2 + \cdots + s_k)/k$. Moreover, if we are reasonably satisfied with the overall average of the nk observations in these k samples, we can use $\bar{\bar{x}} = (\bar{x}_1 + \bar{x}_2 + \cdots + \bar{x}_k)/k$ as the centerline in our control chart; that is, we use $\bar{\bar{x}}$ as an estimate of μ.

With these estimates, the control limits around the centerline $\bar{\bar{x}}$ are given by $\bar{\bar{x}} - A_3\bar{s}$ and $\bar{\bar{x}} + A_3\bar{s}$, where the constant A_3 is chosen in such a way that $A_3\bar{s}$ is a good estimate of $3\sigma/\sqrt{n}$. The constant A_3 depends on the sample size n and is given in Table C.1. The development behind the selection of A_3 is beyond the scope of this book. Note, however, that A_3 in Table C.1 is always slightly larger than $3/\sqrt{n}$; this is because we use an estimate $\bar{s}$ instead of the unknown standard deviation σ. Note also that, for moderately large sample sizes, A_3 and $3/\sqrt{n}$ are quite similar; for example, for $n = 5$, $A_3 = 1.43$, and $3/\sqrt{n} = 1.34$; for $n = 10$, $A_3 = 0.98$, and $3/\sqrt{n} = 0.95$; for larger n, the difference disappears almost completely.

However, the calculation of the standard deviations involves a considerable amount of computation. An alternative, and simpler, procedure for estimating $3\sigma/\sqrt{n}$ proceeds as follows: Let $R_1, R_2, \ldots, R_k$ be the ranges of the k samples. Calculate the average range $\overline{R} = (R_1 + R_2 + \cdots + R_k)/k$. It can be shown that $A_2\overline{R}$ is a good estimate of $3\sigma/\sqrt{n}$ when one is sampling from a normal distribution. The constant A_2 depends on the sample size n and is also given in Table C.1. With this modification, the $\bar{x}$-chart consists of a centerline at $\bar{\bar{x}}$, which is an estimate of μ, and control limits $\bar{\bar{x}} - A_2\overline{R}$ and $\bar{\bar{x}} + A_2\overline{R}$.

There are several control charts for measures of variability. Either the range R or the standard deviation s of samples of size n can be used to measure variability, and R- and s-charts can be constructed. In this book, we concentrate on R-charts, on which we plot ranges of successive samples.

We approximate the centerline of the R-chart with the average of the k values of R, namely, $\overline{R}$. The lower and upper control limits are taken to be LCL $= D_3\overline{R}$ and UCL $= D_4\overline{R}$, respectively. The constants D_3 and D_4, and thus the limits, are chosen so that, for a stable process, the probability of an individual R falling between the LCL and the UCL is extremely large. The constants D_3 and D_4 are also given in Table C.1.

In sum, the construction of the control charts is very simple. We take samples of a few observations (usually, the sample size n is 4 or 5) at various times. It is often recommended that $k = 10$ to 20 such samples be obtained before constructing the control limits. Depending on the application, these samples can be taken every four hours (see the weld strength example), twice a shift (compressive strength), from every other furnace run (wafer example), every hour, from every tenth batch, and so on. The frequency of the sampling depends on the stability of the process; the more stable the process, the longer is the time between samples. The frequency also depends on the potential loss that is caused when deteriorations of the process are not recognized on time and, of course, on the cost of the sampling inspection. From each sample, we calculate the average $\overline{x} = \sum_{i=1}^{n}x_i/n$ and the range $R = \max(x_1,\ldots,x_n) - \min(x_1,\ldots,x_n)$, and we enter these quantities on the $\overline{x}$-chart and R-chart. From the k sample averages and ranges, we compute the grand average (the average of the averages),

$$\overline{\overline{x}} = \frac{1}{k}\sum_{j=1}^{k}\overline{x}_j,$$

and the average of the ranges,

$$\overline{R} = \frac{1}{k}\sum_{j=1}^{k}R_j.$$

These quantities form the respective centerlines in the $\overline{x}$-chart and the R-chart. The control limits in the $\overline{x}$-chart are given by

$$\text{LCL} = \overline{\overline{x}} - A_2\overline{R} \quad \text{and} \quad \text{UCL} = \overline{\overline{x}} + A_2\overline{R}.$$

The control limits in the R-chart are given by

$$\text{LCL} = D_3\overline{R} \quad \text{and} \quad \text{UCL} = D_4\overline{R}.$$

The constants A_2, D_3, and D_4 can be found in Table C.1 in Appendix C; they depend on the sample size n. These constants are chosen such that almost all future averages $\overline{x}$ and ranges R will fall within the respective control limits, provided that the process has remained under control (which means that the level has not shifted and the variability has not changed). If the process is stable, it is rare that a sample average or sample range will fall outside the control limits. However, if there are shifts and drifts in the process, then the averages or the ranges (or both) are likely to exceed the limits and generate an alarm.

We should also point to a limitation of the R-chart. Table C.1 shows that the constant D_3, and thus also the lower control limit $D_3\overline{R}$, are zero for sample sizes smaller

Table 5.1-1 Compressive Strength of Concrete (kg/cm^2)

Sample		Compressive Strength				$\bar{x}$	R
1	91	88	88	90	83	88.0	8
2	84	89	80	79	87	83.8	10
3	93	90	87	89	85	88.8	8
4	76	84	82	79	82	80.6	8
5	83	85	81	80	86	83.0	6
6	84	84	90	79	83	84.0	11
7	83	89	80	82	91	85.0	11
8	78	79	90	81	85	82.6	12
9	82	81	87	86	79	83.0	8
10	88	90	83	84	87	86.4	7
Mean						$\bar{\bar{x}} = 84.52$	$\bar{R} = 8.9$
11	79	87	82	85	83	83.2	8
12	72	79	76	77	78	76.4	7

Samples used to determine the control limits: (braced for samples 1–10)

than 7. This implies that, for small *n,* the *R*-chart can warn about increases in variability, but not about reductions. That is unfortunate, because, in quality improvement applications, one would also like to know whether certain actions have led to a reduction in variability.

Consider the data in Table 5.1-1, which lists the compressive strength measurements on concrete blocks from $k = 10$ samples of size $n = 5$. The process was sampled twice during each production shift, and the observations were taken while the process was under control, or at least thought to be under control. With $n = 5$ observations in each sample, we find from Table C.1 that the constants are $A_2 = 0.577, D_3 = 0$, and $D_4 = 2.115$. Thus, the control limits for the $\bar{x}$-chart are

$$\text{LCL} = \bar{\bar{x}} - A_2\bar{R} = 84.52 - (0.577)(8.9) = 79.38$$

and

$$\text{UCL} = \bar{\bar{x}} + A_2\bar{R} = 84.52 + (0.577)(8.9) = 89.66.$$

The limits on the *R*-chart are LCL $= D_3\bar{R} = (0)(8.9) = 0$ and UCL $= D_4\bar{R} = (2.115)(8.9) = 18.82$, which are the limits shown in Figures 5.1-1 and 5.1-2. We see that the averages and ranges of all 10 samples are within these limits. We could have expected this, because we were told that the process was in control when the observations were taken. But let us plot the results of the next two samples, also given in Table 5.1-1, on these charts. In that table, we find that the twelfth average $\bar{x} = 76.4$ is smaller than the lower control limit on the $\bar{x}$-chart. This fact should alert the user to the possibility that this particular sample represents the result of an unusual event, called a special cause. Such a finding should lead to an investigation (i.e., discussions with workers on the production line, checking whether there were changes in raw materials, looking for any other unusual condition) that will identify a cause that can be assigned to this event. Finally, any causes found should be eliminated.

Control charts are useful methods that help us assess whether a process is stable. They alert the user to situations in which something has shifted. A point outside the control limits forces us to find an assignable cause of this unusual event and, more important, to make certain changes in the process that prevent such conditions from happening again. Control charts will uncover many external sources that lead to shifts in the mean level and in the variability of the process. Their graphical simplicity makes them a very valuable instrument for process control. The requirement to identify assignable causes and eliminate them forces management and workers to take an aggressive attitude toward maintaining the quality of the work.

Remarks: The use of control charts and a strategy of investigating and eliminating special causes will lead to stable processes. However, we want to make it clear at this point that control limits and specification limits are *not* the same. That we have a stable process (or, to say it differently, a process that is under control) implies that we have been successful in eliminating special, unusual causes. The variability that is due to common causes, however, is still present and may lead to products that are outside the specification limits.

The first step in improving processes is to bring them under control. Once we have eliminated special causes and have made a process stable, we can check whether the process also satisfies the required specification limits. This can be done in the following way:

Because $A_2\bar{R}$ is an estimate of $3\sigma/\sqrt{n}$, where n is the size of each sample, it follows that $\sqrt{n}A_2\bar{R}$ is an estimate of 3σ. Thus, $\bar{\bar{x}} \pm \sqrt{n}A_2\bar{R}$ is an estimate of $\mu \pm 3\sigma$. If the underlying distribution is approximately normal (unimodal and fairly symmetric without long tails), almost all items should be between $\bar{\bar{x}} - \sqrt{n}A_2\bar{R}$ and $\bar{\bar{x}} + \sqrt{n}A_2\bar{R}$. If these two bounds are within the specifications, most items must be within specifications too. However, if one or both of the bounds $\bar{\bar{x}} \pm \sqrt{n}A_2\bar{R}$ are outside the specification limits, it is highly likely that some of the items will be outside the specification limits. The situation should be reviewed carefully, with questions such as "How many items are outside the specifications?" and "Were the specifications determined correctly?" addressed.

If a stable process is not capable of producing items within the specification limits, we must think about making changes to our process. In later chapters on the design of experiments, we will learn how to decide which changes are most promising.

5.1-2 *p*-Charts and *c*-Charts

Control charts are useful not only for averages and ranges, but also for proportions, such as proportions of defectives. Control charts are also useful not just in manufacturing applications, but in other areas as well, such as the service industry. In fact, they can be applied to virtually all situations in which data are taken sequentially in time.

Suppose, for example, that we simply need to judge whether a manufactured item is satisfactory. That is, although we prefer to take more accurate measurements, here we just check an item on a pass–fail basis: that it is within or outside the specifications. Assume that an inspector on the production line checks a sample of n items at certain stated periods (every hour, half-day, day, etc., depending on the numbers of items produced each day) and observes the number of defectives, say, *d,* among the n items. If this is done for k periods, we obtain the number of defectives $d_1, d_2, \ldots, d_k$. The average fraction of defectives is

$$\bar{p} = \frac{(d_1 + d_2 + \cdots + d_k)}{nk}.$$

Statistical theory implies that, in a stable process (i.e., a process that produces defectives at the rate $\bar{p}$), almost all of the future fractions of defectives, d/n, will be between the lower and upper control limits:

$$\text{LCL} = \bar{p} - 3\sqrt{\frac{\bar{p}(1 - \bar{p})}{n}} \quad \text{and} \quad \text{UCL} = \bar{p} + 3\sqrt{\frac{\bar{p}(1 - \bar{p})}{n}}.$$

These 3σ limits are obtained from the sampling distribution of a proportion which has variance $p(1 - p)/n$. The control limits LCL and UCL, together with the centerline at $\bar{p}$, are plotted on a chart; because we are plotting fractions of defectives, or percentages, we call it a p-chart. Fractions outside these limits suggest that the process has gone out of control and that the fraction of defectives has changed. In particular, a point exceeding the upper control limit indicates that the process has deteriorated. In such a situation, we should look for possible reasons for the sudden increase in the number of defectives.

Example 5.1-1   Each hour, $n = 50$ fuses are tested. For the first $k = 20$ hours, we find the following number of defectives:

$$1 \quad 1 \quad 3 \quad 0 \quad 2 \quad 4 \quad 0 \quad 0 \quad 1 \quad 2 \quad 3 \quad 2 \quad 0 \quad 1 \quad 1 \quad 1 \quad 3 \quad 0 \quad 0 \quad 2.$$

Thus, given that $nk = 1,000$, it follows that

$$\bar{p} = 27/1000 = 0.027$$

is the average fraction of defectives. We must first decide whether this fraction is representative for our particular process. If it is, then

$$\text{LCL} = 0.027 - 3\sqrt{\frac{(0.027)(0.973)}{50}} = -0.042$$

and

$$\text{UCL} = 0.027 + 3\sqrt{\frac{(0.027)(0.973)}{50}} = 0.096.$$

Because LCL < 0, and because the fraction defective, d/n, can never be less than zero, we plot the LCL at zero or omit it entirely. In Figure 5.1-3, we have plotted the preceding 20 values of the fraction defective, together with 6 more recent ones (those with d values of 1, 2, 2, 2, 4, and 5, and fraction of defectives $1/50 = 0.02$, $2/50 = 0.04$, $2/50 = 0.04$, $2/50 = 0.04$, $4/50 = 0.08$, and $5/50 = 0.10$). Additional values would also be plotted as long as the process is under control. However, we find that the sixth additional fraction defective is above the UCL. This suggests that the process has become unstable and that corrective action should be taken. In this example, we have assumed that 2.7 percent defective is acceptable and that we are willing to produce at that level; this may not be the case for other items.

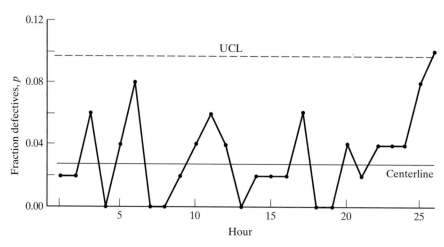

Figure 5.1-3 p-chart for the data in Example 5.1-1   ■

The c-*chart* is similar to the p-chart, except that now we count the number of flaws or defectives on a certain unit (a bolt of fabric, a length of wire, and so on), rather than the number of defectives among n items. Suppose that we determine the number c of blemishes in 50 feet of a continuous strip of tin plate. This is done each hour for k hours, resulting in $c_1, c_2, \ldots, c_k$, with an average of

$$\bar{c} = (c_1 + c_2 + \cdots + c_k)/k.$$

The c-chart is a time-sequence plot of the number of defectives, $c_j, j = 1, 2, \ldots, k$. Its centerline is given by $\bar{c}$, and the respective lower and upper control limits for the c-chart are

$$LCL = \bar{c} - 3\sqrt{\bar{c}}$$

and

$$UCL = \bar{c} + 3\sqrt{\bar{c}}.$$

These are approximations to the 3σ limits $\lambda \pm 3\sqrt{\lambda}$ of a Poisson distribution. The Poisson distribution with parameter λ is appropriate in this context, because it approximates the distribution of the number of defectives. The control limits are obtained after replacing the parameter λ (which is the mean, as well as the variance, of the Poisson distribution) with the sample mean.

Example 5.1-2   We observe $k = 15$ 50-foot tin strips and obtain the following numbers of blemishes:

$$2 \quad 1 \quad 1 \quad 0 \quad 5 \quad 2 \quad 3 \quad 1 \quad 1 \quad 2 \quad 0 \quad 0 \quad 4 \quad 3 \quad 1.$$

The average is $\bar{c} = 26/15 = 1.73$, and we also have

$$LCL = 1.73 - 3\sqrt{1.73} = -2.22 \, (=0) \quad \text{and} \quad UCL = 1.73 + 3\sqrt{1.73} = 5.68.$$

These 15 points, together with the 10 additional observations

$$3 \quad 1 \quad 1 \quad 0 \quad 2 \quad 2 \quad 5 \quad 0 \quad 1 \quad 2,$$

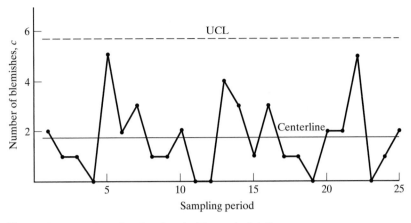

Figure 5.1-4 c-chart for the data in Example 5.1-2

are plotted on the c-chart in Figure 5.1-4. Of course, as long as the process is under control, as it is with these 10 additional points, future points are plotted on this c-chart. Occasionally, new control limits are calculated if the points continue to fall within the control limits; thus, the control limits may change slightly. Points outside the control limits, however, indicate that the process has become unstable. Causes of these unusual events must be found and eliminated. ■

Remarks: The control charts that we have been discussing in this section were developed in the late 1920s by Shewhart in the United States and by Dudding and Jennet in Great Britain. In the United States, they are usually referred to as Shewhart control charts.

Through the commands in "Stat > Control Charts > Variables Charts for Subgroups," Minitab constructs control charts for measurement variables, such as $\bar{x}$- and R-charts, or $\bar{x}$- and s-charts. The commands in "Stat > Control Charts > Attribute Charts" construct control charts for attribute data, such as p- and c-charts.

When viewing time-sequence plots such as Shewhart charts, you should guard against reading too much into short sequences of points. For example, in the desire to improve, a supervisor may believe that three successive points below the centerline of a p-chart are an indication that the process has improved. Of course, we recognize that the probability of getting three points below the centerline is fairly large, even if the process has stayed unchanged. Indeed, this probability is 1/8, as there is only one arrangement with all three points below the current level among $2^3 = 8$ equally likely ones. Sales managers, especially, are known to misinterpret their graphs. Two consecutive sales periods with sales above the stable level are often taken as evidence of improvement. Is that enough, however, to claim improvement when, with no actual changes, the probability of such an arrangement is 1/4? Certainly not. Perhaps five observations above the stable level in succession would be more reason for celebration. The probability of such an event is 1/32, as there is only one arrangement with all five points above the stable level among the $2^5 = 32$ equally likely ones. Because this probability is rather small, it seems more likely that some significant improvement has taken place.

5.1-3 Other Control Charts

Shewhart control charts provide a useful display of the data and give us a simple rule for making decisions as to whether a process has started to become unstable. Such charts require that measurements on the process be taken on a regular basis and that the results be prominently displayed. They create an atmosphere in which the quality of the process is checked on a regular basis. They enhance our awareness of the present state of the process and make us "listen to" the process.

A disadvantage of these charts, however, is their relative insensitivity to small or moderate changes in the mean value. *Cumulative sum (cusum) charts,* by contrast, are more responsive to small changes in the mean level. In cusum charts, we consider deviations $x_i - g$ of the observations (or sample averages) from a reference value g and calculate the cumulative sums

$$S_r = \sum_{i=1}^{r}(x_i - g) = (x_r - g) + S_{r-1}.$$

The mean of the in-control process is usually taken as the reference value g; that is, $g = \bar{\bar{x}}$. The *cumulative sums,* or *cusums* as they are often abbreviated, are then plotted against r. A rising cusum path is an indication that the level of the process may have increased. Statisticians have developed rules that help us decide whether a trend in the cusum path comes from a change in the level or whether it is due to random fluctuations in the process. [Interested readers may consult books on process control, such as D. Montgomery, *Introduction to Statistical Quality Control,* 5th ed. (New York: Wiley, 2004)].

Exercises 5.1

5.1-1 E. L. Grant notes that a particular dimension determines the fit of a molded plastic rheostat knob in its assembly. The dimension was specified by the engineering department as 0.140 ± 0.003 inch. A special gauge was designed to permit quick measurement of the actual value of this dimension. Five knobs from each hour's production were selected and measured. The averages and the ranges (in units of 0.001 inch) for the first 20 hours were as follows:

Hour	Average, $\bar{x}$	Range, R	Hour	Average, $\bar{x}$	Range, R
1	137.8	9	11	139.6	12
2	143.0	8	12	141.4	9
3	141.2	15	13	141.2	3
4	139.8	6	14	140.6	8
5	140.0	10	15	141.6	9
6	139.2	8	16	140.4	13
7	141.2	10	17	140.0	6
8	140.0	8	18	141.8	8
9	142.0	7	19	140.4	8
10	139.2	13	20	138.8	6

Construct an $\bar{x}$-chart and an R-chart. Is this process under control? If so, does it satisfy the specification limits?

[E. L. Grant, *Statistical Quality Control,* 2d ed. (New York: McGraw-Hill, 1952), p. 23.]

*5.1-2   Astro Electronics, an RCA division, uses statistical quality control tools to ensure proper weld strength in several of their welding procedures. Weld strength is measured by a pull test to destruction. A sample of a small number of items (in this case, five) is taken periodically throughout the production process, usually at the beginning, the middle, and the end of each shift. The averages and the ranges of 22 such samples, taken from an article by Shecter, are listed as follows:

Date	Time	Average	Range
1/3	7:30	5.48	1.4
	11:25	5.42	1.6
	16:20	5.42	1.4
	20:00	5.40	0.5
	23:30	5.52	1.7
1/4	7:20	5.32	0.7
	11:25	5.34	1.6
	16:20	5.58	1.2
	20:20	4.54	0.6
	23:00	5.42	1.6
1/7	7:40	5.58	0.5
	10:20	5.06	1.4
	14:00	4.82	1.9
	20:20	4.86	1.3
	23:00	4.68	0.9
1/8	8:00	5.28	1.6
	11:20	4.68	1.1
	14:00	4.94	0.6
1/9	9:00	4.90	1.0
1/10	7:40	4.96	0.7
	11:00	5.06	1.8
	14:00	5.22	0.8

Construct the $\bar{x}$- and R-charts. Interpret your results and comment on the stability of the process. Note that the frequency of sampling should depend on the stability of the process. Actually, Shecter reports that initially the samples were taken every hour, but the data showed a relatively stable process; thus, sampling about every 4 hours (or less) was thought appropriate.

[E. Shecter, "Process Control for High Yields," *RCA Engineer,* 30(3): 38–43 (1985).]

*5.1-3   A one-month record of daily 100 percent inspection of a single critical quality characteristic of a part of an electrical device led to an average fraction defective of $\bar{p} = 0.0145$. After this one month of 100 percent inspection, the company switched to a sampling plan under which a sample of 500 units was selected each day. During the first 10 days, the inspector found 8, 10, 7, 20, 13, 15, 8, 12, 45, and 30 defectives. Using $\bar{p} = 0.0145$ as a centerline, plot this information on a p-chart and interpret your findings.

*5.1-4   A company that produces certain bolts considers their quality adequate as long as the proportion of defectives is not larger than 2.5 percent. To monitor the quality, the company takes a random sample of 100 bolts each hour and counts the number of defective bolts. With average proportion $\bar{p} = 0.025$, calculate the lower and upper control limits of the appropriate control chart. Now suppose that the numbers of defectives in the last six samples were 3, 0, 2, 1, 7, and 8. What conclusions would you draw from this information?

*5.1-5 Past experience has shown that the number of defects per yard of material follows a Poisson distribution with $\lambda = 1.2$. This information was used to establish the control limits of the associated c-chart. If the average number of defects shifts to 2.0, what is the probability that it will be detected by the c-chart on the first observation following the shift? What is the probability that this shift is not recognized for the next 10 (20) observations?

5.1-6 In the production of stainless steel pipes the number of defects per 100 feet should be controlled. From 15 randomly selected pipes of length 100 feet, we obtained the following data on the number of defects: 6, 10, 8, 1, 7, 9, 7, 4, 5, 10, 3, 4, 9, 8, 5. Construct the appropriate control chart. Is this process under control?

*5.1-7 The following data give the results of an inspection of 100-yard pieces of woolen textile: The numbers of defects among the last 12 samples are 3, 6, 3, 0, 5, 2, 4, 0, 1, 0, 3, and 4. Calculate the control limits of the c-chart. Is this process under control?

5.1-8 A company that produces electronic components considers their quality adequate as long as the proportion of defectives is not larger than 2 percent. To monitor the quality, the company takes a random sample of 80 components each hour and counts the number of defective items.

(a) Using $\bar{p} = 0.02$, calculate the centerline and the lower and upper control limits of the appropriate control chart.

(b) Suppose that the numbers of defectives in the last six samples were 2, 0, 4, 1, 3, and 7. What conclusions would you draw from this information?

5.1-9 We are concerned that the level of a process might increase from a specified acceptable level μ_0. Assume that successive observations of X are independent with standard deviation σ. Suppose that one uses the Shewhart chart for *individual* observations ($n = 1$) with centerline at μ_0 and upper control limit of $h = \mu_0 + k\sigma$.

Define the run length T as the time at which the process exceeds the control limit for the first time. T is a random variable that can take on integer values $1, 2, \ldots$. The expectation of this random variable, $E(T)$, is called the *average run length* (ARL).

(a) Show that the ARL for the one-sided Shewhart chart (i.e., our only concern is whether we exceed the upper control limit) is given by

$$ARL = \frac{1}{P(X > h)}.$$

(b) Assume that the observations X come from a normal distribution. Calculate the average run length for $k = 2.0, 2.5, 3.0$.

(c) Calculate the average run lengths in part (b) for a two-sided chart with lower control limit $\mu_0 - k\sigma$ and upper control limit $\mu_0 + k\sigma$, where one is concerned about increases as well as decreases in the level.

Hint: The average run length in (a) is given by

$$ARL = (1)P(T = 1) + (2)P(T = 2) + (3)P(T = 3) + \cdots$$
$$= (1)P(X_1 > h) + (2)P(X_2 > h \text{ and } X_1 < h)$$
$$+ (3)P(X_3 > h \text{ and } X_2 < h \text{ and } X_1 < h) + \cdots$$

You can show the result in (a) by using the independence assumption (which implies that the probability of an intersection of events is the product of the individual probabilities) and properties of geometric sums.

5.1-10 A company manufactures paper containers for a detergent, and the dimension of the containers (in centimeters, from front to back) is of interest. Each hour, four cartons are selected from the production run and their dimensions are measured. Measurements for the last 25 hours are listed as follows:

Time	Measurements			
6:30 A.M.	25.1	25.5	25.0	25.1
7:30	24.8	25.2	25.1	24.9
8:30	25.1	25.2	25.2	25.2
9:30	25.1	25.4	24.8	25.0
10:30	25.2	24.7	24.9	25.3
11:30	25.2	25.2	25.0	25.1
12:30 P.M.	25.2	25.2	25.2	25.3
1:30	25.2	25.1	25.3	25.0
2:30	24.9	25.1	25.2	24.8
3:30	25.1	25.1	25.3	25.4
4:30	25.4	25.0	25.1	24.9
5:30	25.3	25.2	25.1	25.5
6:30	25.2	25.1	25.5	25.2
7:30	25.0	24.9	25.6	25.2
8:30	25.1	25.2	25.1	25.1
9:30	25.0	25.0	24.9	25.0
10:30	25.3	25.1	25.3	24.9
11:30	25.2	25.1	25.2	25.1
12:30 A.M.	25.1	25.1	25.4	24.8
1:30	25.4	25.0	25.2	25.0
2:30	24.8	25.2	25.0	25.0
3:30	25.3	25.4	25.2	25.3
4:30	25.1	24.8	25.2	25.1
5:30	25.0	25.4	25.1	25.1
6:30	25.1	25.3	25.3	25.2

Construct $\bar{x}$- and R-charts, and check whether the level and the variability of the process are under statistical control.

5.2 Process Capability Indices

5.2-1 Introduction

One must check whether processes are capable of producing products that satisfy required specifications. Typically, the customer requires that certain product specifications be met. Specifications are usually given in terms of a target value (T_g), a lower specification limit (LSL), and an upper specification limit (USL); they are also called the tolerances of the product. The specifications are determined by translating customer requirements into suitable product requirements. Engineering considerations and the intended use of the product play important roles in setting the specifications. Once specifications are set, the production process must be monitored to ensure that products meet the specifications. If they do, then we say that the process is *capable* of producing to the required specifications.

In this section, we introduce several process capability indices, examine their importance as well as their shortcomings, and discuss their implementation. The capability measures we examine are expressed in terms of the specifications (the target value and the lower and upper specification limits) and the process characteristics (the process mean μ and the process standard deviation σ). Estimates of capability indices are obtained by taking samples from the process under study and replacing the process characteristics by their sample estimates.

A simple approach in checking conformance is to construct a dot diagram of the measurements (or a histogram if the data set is large), adding the target value and the specification limits to this graph and calculating the proportion of values that are outside these limits. Of course, no (or very few) values should be outside the limits.

We illustrate this approach with data on the width and gauge (i.e., thickness) of steel flats. For the width, the target is 4 inches, with specification limits LSL = 3.97 in. and USL = 4.03 in. For the gauge, it is 0.25 inch, with lower and upper specification limits LSL = 0.235 in. and USL = 0.265 in., respectively. Dot diagrams for the 95 width and gauge measurements in Table 5.2-1 are shown in Figure 5.2-1. We notice 2 of the 95 width measurements (or 2.1 percent) outside the specification limits, while 1 of the 95 gauge measurements (or 1 percent) is outside the specification limits. The figures also show that the process is slightly off target, with process means for both width and gauge below their target values.

Table 5.2-1 Width and Gauge Measurements on 95 Steel Flats

Time	Width	Gauge	Date	Time	Width	Gauge	Date
16.10	3.990	.256	May 19, 1990	12.00	3.988	.242	
16.21	3.993	.252		1.15	4.000	.262	
16.27	3.968	.257		1.20	4.004	.252	
16.32	3.993	.250		1.25	3.998	.247	
17.00	3.998	.248		1.35	3.992	.248	
17.30	4.002	.247		2.00	3.992	.250	
18.00	3.994	.247		2.35	3.989	.248	
18.51	3.990	.273		15.00	3.992	.244	
18.57	3.989	.257		15.15	3.995	.247	
19.00	3.990	.252		15.30	3.992	.249	
19.35	3.988	.257		16.00	3.992	.247	
20.00	3.985	.254		16.30	3.989	.247	

Time	Value	Value		Time	Value	Value	
20.30	3.996	.253		17.00	3.998	.246	
21.00	3.994	.245		17.30	3.997	.246	
21.30	3.988	.250		18.00	3.991	.246	
22.00	3.987	.249		18.30	3.993	.246	
22.30	3.988	.249		19.12	4.002	.251	
23.00	3.988	.249		19.50	3.994	.248	
23.30	3.986	.251		20.00	3.997	.245	
24.00	3.984	.250		20.30	3.994	.246	
0.30	3.984	.239	May 20, 1990	21.00	3.991	.248	
1.00	4.000	.246		21.30	3.988	.250	
1.15	4.012	.249		22.00	3.987	.248	
1.30	4.012	.246		22.30	3.989	.245	
2.00	4.003	.248		23.00	3.997	.245	
2.30	3.994	.252		23.40	3.990	.250	
3.00	3.994	.250		24.00	3.991	.248	
3.30	3.990	.247		0.45	4.006	.248	May 21, 1990
4.00	3.994	.249		1.00	4.006	.249	
4.30	3.989	.249		1.40	4.000	.251	
5.00	4.000	.249		2.00	4.021	.246	
5.30	3.994	.246		2.30	3.998	.250	
7.05	3.969	.253		3.10	3.990	.254	
7.10	3.997	.250		3.30	3.990	.246	
7.15	3.996	.249		4.00	3.990	.245	
7.20	3.992	.250		4.30	3.994	.250	
7.30	4.002	.250		5.00	3.993	.249	
8.00	3.999	.250		5.30	3.990	.246	
8.05	4.000	.249		6.00	4.006	.249	
8.20	4.005	.248		6.30	4.009	.249	
8.30	4.003	.251		7.00	4.009	.249	
9.20	4.009	.250		7.30	4.005	.250	
9.30	3.995	.244		8.30	4.006	.252	
9.50	3.989	.249		8.35	4.000	.249	
10.00	3.990	.244		8.40	3.998	.247	
10.30	3.990	.243		9.00	3.996	.247	
11.00	3.991	.245		9.30	3.995	.246	
11.30	3.987	.245					

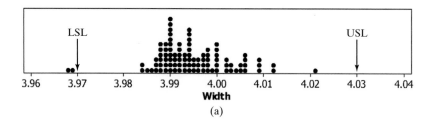

(a)

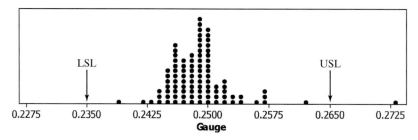

Figure 5.2-1 Dot diagrams of width and gauge measurements

5.2-2 Process Capability Indices

Dot diagrams and histograms are effective graphical summaries of process capability, and we recommend their use. However, it is common practice to calculate capability *indices*, and many companies require their suppliers to document the capability of their processes through such calculations. Capability indices quantify the capability of a process—in other words, the conformance of the process to the required specifications.

The C_p Capability Index A commonly used capability index is

$$C_p = \frac{\text{USL} - \text{LSL}}{6\sigma} = \frac{\text{Allowable Spread}}{\text{Process Spread}},$$

where LSL and USL are, respectively, the lower and upper specification limits and σ is the process standard deviation.

For many distributions, the interval $(\mu - 3\sigma, \mu + 3\sigma)$ covers virtually all of the distribution—in fact, 99.73 percent if the distribution is normal. An interval of length 6σ measures the extent of the process variability and it expresses the process spread. C_p relates the allowable spread, USL – LSL, to the process spread. For capable processes, we expect that the process spread is smaller than the allowable spread and $C_p > 1$. A large C_p indicates small process variability compared with the width of the specification interval. The larger this index is, the better. For normal distributions centered at the target, $C_p = 1$ corresponds to 0.27 percent defectives, or 2,700 defective parts per million. (The specification limits are three standard deviations from the target value, and we learned in Section 3.2 that, for a normal distribution, the probability beyond three sigma limits is 0.0027.) Many companies require that

$C_p > 1.33$ (which implies no more than 63 defective parts per million, as calculations with the normal distribution show) or $C_p > 1.5$ (no more than 7 defective parts per million). Some companies (for example, Motorola) require that C_p be at least 2.0 (implying no more than 0.1 defective part per million).

We estimate C_p by replacing the process standard deviation σ by its estimate s, which we obtain from past process data. The estimated C_p is given by

$$\hat{C}_p = \frac{\text{USL} - \text{LSL}}{6s}.$$

As an illustration, we use the width and gauge measurements in Table 5.2-1. The mean and standard deviation for the width are 3.9947 and 0.0080, respectively. For the gauge, they are 0.24894 and 0.00421. Hence,

$$\hat{C}_p(\text{Width}) = \frac{4.03 - 3.97}{6(0.0080)} = 1.25 \quad \text{and} \quad \hat{C}_p(\text{Gauge}) = \frac{0.265 - 0.235}{6(0.00421)} = 1.19.$$

These values are somewhat smaller than what we would like to see. We would have preferred values at least as large as 1.33.

Caution: C_p makes no reference to the target value. It provides a good description of capability only when the process is on target and the process mean and the target value are the same. However, it is misleading when the process is off target, as illustrated in Figure 5.2-2. In the second illustration, the process is off target, causing a considerable fraction of defectives, despite the fact that the actual process spread, 6σ, is small in comparison to the allowable spread, USL − LSL. C_p is deceptively large in this case, even though the process is not capable of meeting the specifications. Because C_p makes no reference to the target value, we do not recommend its use.

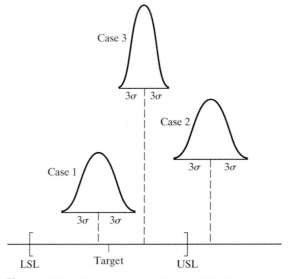

Figure 5.2-2 Illustrating problems with C_p

The C_{pk} Capability Index Processes with small variability, but poor proximity to the target, have sparked the development of several indices similar to C_p. These indices take into account both the process variability and the deviation of the process mean from the specification limits (or the target). One of them is the C_{pk} capability index, given by

$$C_{pk} = \min\left\{ \frac{USL - \mu}{3\sigma}, \frac{\mu - LSL}{3\sigma} \right\}.$$

C_{pk} relates the distances between each specification limit and the process mean to three standard deviations. For capable processes, we expect that the difference between the upper specification limit and the process mean is at least three standard deviations. We expect the same for the difference between the process mean and the lower specification limit. Therefore, we expect that the smaller of these two standardized differences is larger—hopefully, much larger—than 1. A C_{pk} that is smaller than the usual cutoff of 1.33 is taken as an indication that the process fails to satisfy the required specifications.

C_{pk} is a much better measure of capability than C_p, and we recommend its use. For example, C_{pk} for illustration 2 in Figure 5.2-2 is unacceptable; in fact, it is negative, because the process mean is larger than the upper specification limit.

In practice, we need to replace the process characteristics μ and σ by their estimates $\bar{x}$ and s. The estimated C_{pk} is given by

$$\hat{C}_{pk} = \min\left\{ \frac{USL - \bar{x}}{3s}, \frac{\bar{x} - LSL}{3s} \right\}.$$

For the width and gauge measurements in Table 5.2-1, we have

$$\hat{C}_{pk}(\text{Width}) = \min\left\{ \frac{4.03 - 3.9947}{3(0.0080)}, \frac{3.9947 - 3.97}{3(0.0080)} \right\} = \min\{1.47, 1.03\} = 1.03$$

and

$$\hat{C}_{pk}(\text{Gauge}) = \min\left\{ \frac{0.265 - 0.24894}{3(0.00421)}, \frac{0.24894 - 0.235}{3(0.00421)} \right\} = \min\{1.27, 1.10\} = 1.10.$$

These capability indices are also somewhat smaller than what one would have hoped for.

The C_{pm} Capability Index The C_{pm} capability index is defined as

$$C_{pm} = \frac{USL - LSL}{6\sigma*}.$$

C_{pm} is similar to C_p, except that $\sigma* = \sqrt{E(X - T_g)^2}$ measures the variability of process measurements around the target T_g, not around the process mean μ. The distance of a measurement X from the target T_g can be written as $X - T_g = (X - \mu) + (\mu - T_g)$, the sum of the distance of the observation from the process mean and the distance of the process mean from the target. The second component expresses the bias. With

$$\sigma^* = \sqrt{\sigma^2 + (\mu - T_g)^2},$$

we find that

$$C_{pm} = \frac{\text{USL} - \text{LSL}}{6\sqrt{\sigma^2 + (\mu - T_g)^2}} = \frac{\text{USL} - \text{LSL}}{6\sigma\sqrt{1 + \dfrac{(\mu - T_g)^2}{\sigma^2}}} = \frac{C_p}{\sqrt{1 + \dfrac{(\mu - T_g)^2}{\sigma^2}}}.$$

For processes that are on target with $\mu = T_g$, $C_{pm} = C_p$. If the process mean is different from the target, then the denominator in the expression for C_{pm} increases and C_{pm} itself decreases.

We can estimate C_{pm} from

$$\hat{C}_{pm} = \frac{\text{USL} - \text{LSL}}{6\sqrt{s^2 + (\bar{x} - T_g)^2}}.$$

As an example, we use the width and gauge measurements in Table 5.2-1:

$$\hat{C}_{pm}(\text{Width}) = \frac{4.03 - 3.97}{6\sqrt{(0.0080)^2 + (3.9947 - 4.00)^2}} = 1.04,$$

$$\hat{C}_{pm}(\text{Gauge}) = \frac{0.265 - 0.235}{6\sqrt{(0.00421)^2 + (0.24894 - 0.25)^2}} = 1.15.$$

Capability Ratio and Target-Z Several other indices are in use. Some companies use the reciprocal of C_p,

$$\text{CR} = \frac{1}{C_p} = \frac{\text{Process Spread}}{\text{Allowable Spread}}.$$

This measure is known as the *capability ratio*. Multiplied by 100, it represents the proportion of the allowable spread that is taken up by the process spread. Small values of CR are desirable; many companies require that the capability ratio of most of their processes be less than 0.75. Note that this is equivalent to requiring that $C_p > 1.33$.

The capability ratio has the same drawbacks as C_p. It must be supplemented by a measure that assesses whether the process mean is close to the target, namely,

$$\text{Target-}Z = \frac{T_g - \mu}{\sigma}.$$

Target-Z is the standardized difference between the target value T_g and the process mean μ. Many companies require that Target-Z lie between -0.5 and 0.5, which means that the process mean must be within one-half of a standard deviation from the target. Target-Z is estimated from sample data by replacing μ and σ with the sample statistics $\bar{x}$ and s.

In sum, capable processes satisfy *both*

$$CR < 0.75 \quad \text{and} \quad -0.5 < \text{Target-}Z < 0.5.$$

Case 3 in Figure 5.2-2 illustrates a process that leads to acceptable C_p and C_{pk}, but fails the requirements on Target-Z. One could argue that the process depicted is a good one, as the process variability is well within the acceptable range. Nevertheless, the process is clearly off target, and a level adjustment would make things even better.

For the width and gauge measurements, we find the following estimates

$$CR(\text{Width}) = 0.80 \quad \text{and} \quad \text{Target-}Z(\text{Width}) = \frac{4.00 - 3.9947}{0.0080} = 0.66,$$

$$CR(\text{Gauge}) = 0.84 \quad \text{and} \quad \text{Target-}Z(\text{Gauge}) = \frac{0.25 - 0.24894}{0.00421} = 0.25.$$

Neither the width nor the gauge would satisfy the capability standards.

Motorola's Six Sigma Concept   In announcing the achievement of Total Customer Satisfaction as the corporation's fundamental objective, Motorola introduced the concept of Six Sigma as a statistical way of measuring quality. Motorola views its failure rates in terms of parts per million (ppm). Motorola's Six Sigma goal is a 3.4-ppm defect level!

How does the Six Sigma requirement translate into a defect level of 3.4 ppm? Obviously, Motorola requires highly capable processes. In fact, the company requires that processes be on target and that the specification limits be at least 6σ away from the target—hence the name "Six Sigma." Such a requirement on the process translates into a capability index $C_p = 2.0$, better than the standard usually adopted.

Motorola, however, also realizes that the process is not always exactly on target. In practice, mean levels shift dynamically over time. Motorola allows drifts away from the target of at most 1.5σ. Under the Six Sigma goal, and assuming that the observations are from a normal distribution with (shifted) mean $T_g + (1.5)\sigma$ and standard deviation σ, one can calculate the probability that an individual observation falls outside the interval $(T_g - 6\sigma, T_g + 6\sigma)$. This probability is given by

$$P(\text{observation outside specification limits}) = 1 - P(T_g - 6\sigma < X < T_g + 6\sigma)$$

$$= 1 - P\left(\frac{T_g - 6\sigma - (T_g + 1.5\sigma)}{\sigma} < Z < \frac{T_g + 6\sigma - (T_g + 1.5\sigma)}{\sigma}\right)$$

$$= 1 - P(-7.5 < Z < 4.5) \approx 1 - P(Z < 4.5) = 0.0000034,$$

or 3.4 parts per million. Note that the shifted process mean is 4.5 standard deviations below the upper specification limit; hence, $C_{pk} = 1.5$.

One could argue that Motorola's tolerated 1.5-sigma drift of the process mean from the target is somewhat on the large side—certainly larger than the one-half standard deviation that is allowed under Target-Z. However, tolerating drifts in the process mean is quite reasonable, as quite often it is difficult to keep the mean exactly on target. Allowing for drifts also serves as a proxy for the long-term variation in a process (i.e., variation due to slowly changing factors such as operators, materials, and operating instructions), which is probably larger than the standard deviation used in the denominator of the capability index, which reflects the short-term variation in the process.

Process Capability Indices for One-Sided Specifications Consider a situation in which the tolerances on a product are one sided. For example, the product requirements may specify that the concentration of a certain ingredient be at least 50 mg/liter (so that LSL = 50) or that the weight of an item be at most 80 grams (so that USL = 80). In a situation where one-sided specification (or tolerance) is prescribed, we calculate one-sided C_{pk} indices:

$$C_{pk}(\text{lower}) = \frac{\mu - \text{LSL}}{3\sigma} \quad \text{and} \quad C_{pk}(\text{upper}) = \frac{\text{USL} - \mu}{3\sigma}.$$

For example, assume that the process characteristics are given by $\mu = 53$ mg/liter and $\sigma = 0.5$ mg/liter. Then the capability index for the process, relative to a one-sided lower tolerance LSL = 50 mg/liter, is $C_{pk}(\text{lower}) = (53 - 50)/3(0.5) = 2.0$, indicating a highly capable process.

5.2-3 Discussion of Process Capability Indices

Capability indices summarize process information in a succinct manner. They establish a common language that is dimensionless (i.e., it does not depend on the particular units of the observations) and that compares the desired and the actual performance of production processes. Engineering and manufacturing can communicate through these measures, and processes with high (or low) capability can be identified. Capability measures are also useful tools for monitoring capability over time; they indicate improvements, as well as deteriorations, in the process.

However, one must keep in mind that a capability index is just a single summary statistic and can never bring out all features of a distribution. As with any other single summary statistic, there are potential problems. One needs to be aware of these problems in interpreting such statistics.

Specification Limits Capability indices depend on the specifications, which usually are determined by customers and engineers. It is important that careful thought go into the selection of the specifications. It would be wrong to select them too wide, because that may allow inferior products to be manufactured. Usually, products are part of a complicated assembly involving many components. If there is too much variability in one component (because specification limits or tolerances are set too wide), the parts may not fit together as they should. Take, for example, car doors and car frames. If there is too much variability in the door and/or the frame, the door will not close correctly. We refer to this as tolerance "stack-up." However, specifications should not be taken too narrow either. Unnecessary expenses are incurred with narrow specification limits if tight tolerances are not needed.

Assumption of Normality Capability indices are designed for normal distributions; we assumed a normal distribution when we related the C_p index to the proportion of unacceptable parts. It is questionable whether capability indices are meaningful for distributions that are very different from the normal distribution. To illustrate this point, assume that the variability of a process is described by a uniform distribution on the interval from -1 to $+1$. The mean of this distribution is given by 0, and the standard deviation is $\sqrt{4/12} = 0.577$. (See Chapter 3.) Suppose that the specification limits are given by $-a$ and $+a$; consider $a = 1.5$, for example. The probability that this process exceeds the specification limits is zero, because, for the uniform distribution, there is no probability beyond -1 and $+1$. However, $C_{pm} = C_{pk} = C_p = (3)/[(6)(0.5770] = 0.87$, which is rather poor. In this situation, the capability index raises unnecessary concerns. (Keep in mind that it is always useful to supplement these indices by a histogram of the measurements.)

Instead of a uniform distribution with light tails, now consider a symmetric distribution with heavy tails (i.e., a distribution that makes it more probable for observations to fall far from the mean). For such distributions, a result opposite to the one implied by a uniform distribution will occur. The capability index will look quite good, whereas the probability of exceeding the specifications is large (certainly larger than the probabilities implied by the normal distribution).

The Difference Between Process Stability and Process Capability It is worthwhile to repeat earlier comments about the difference between control and specification limits. Specification limits and control limits are *not* the same. The former reflect the requirements of the customer, while the latter provide bounds on the common-cause variability of the process. These two sets of limits must not be confused!

A process should be under control before one assesses its capability. A capable process that is not in control is not very reassuring. Indeed, it is capable of surprising you! The fact that the process is not under control means that some outside factors are causing it to be unstable. Because you don't know what these factors are, there is no guarantee that the unexplained variation will stay in a range such that the products are still satisfactory. If the process is not under control, you don't know how well you will produce in the future.

Confidence Limits for Capability Indices When calculating capability coefficients, we replace the process parameters (μ and σ) by sample estimates. Hence, the calculated capability index is just an estimate of the unknown process capability index, and it is important to evaluate its margin of error, especially if the sample size n from which these statistics are calculated is not very large. Assuming that the process is stable, and assuming that the measurements are normally distributed, it is possible to obtain approximate confidence intervals for the various capability indices. A useful, and also rather simple, approximate 95 percent confidence interval for C_{pk} is given by

$$\hat{C}_{pk}\left\{1 \pm 2\sqrt{\frac{1}{9n(\hat{C}_{pk})^2} + \frac{1}{2(n-1)}}\right\},$$

where n is the number of observations that are used to estimate the capability index. For the $n = 95$ width measurements of the steel flats with $\hat{C}_{pk} = 1.04$, this interval extends from 0.87 to 1.21. [Kushler, R.H., and Hurley , P., "Confidence Bounds for Capability Indices," *Journal of Quality Technology*, 1992 (Vol. 24), 188–195, gives an in-depth review of this and several other approaches to approximating confidence intervals for capability indices.]

Exercises 5.2

*5.2-1 Consider the 50 observations on the compressive strength of concrete listed in Table 5.1-1 (samples 1 through 10), with sample mean 84.52 and standard deviation 4.132. Assume that the target is 84 (kg/cm^2), with lower and upper specification limits given by 77 kg/cm^2 and 91 kg/cm^2, respectively.

    *(a) Confirm the sample mean and standard deviation, and estimate the capability indices C_p, C_{pk}, C_{pm}, CR, and Target-Z. Interpret your findings.

    *(b) Obtain a 95 percent confidence interval for C_{pk}.

*5.2-2 The target value for the pH content of a certain shampoo is 6.15. The lower specification limit is LSL = 5.85 and the upper specification limit is USL = 6.45. A sample of 200 bottles was taken and the pH value for each bottle was determined. It was found that the minimum value was 5.66, the maximum 6.91, the sample average 6.212, and the sample standard deviation 0.123. In addition, the distribution of pH was well approximated by a normal distribution.

    *(a) Calculate and interpret C_p and C_{pk}. Are you satisfied with the capability of your process? What could you do to increase the capability? Is there a problem with not being on target, or is there a problem with excessive variability?

    *(b) Assuming that the distribution is normal, obtain an estimate of the probability of getting an observation (i) above the upper specification limit, (ii) below the lower specification limit, and (iii) outside the specification limits.

5.2-3 A sample ($n = 150$) from another production run of shampoo resulted in a minimum pH value of 5.89, a maximum of 7.12, an average of 6.44, and a standard deviation of 0.15. Repeat Exercise 5.2-2 on this run of shampoo.

5.2-4 Show that $C_{pk} = (1 - k)C_p$, where $k = 2|T_g - \mu|/(USL - LSL)$, with T_g the specified target and LSL and USL the specification limits.

*5.2-5 The requirements on the dimensions of steel flats specify a target of 100 mm, with lower and upper specification limits of 97 mm and 103 mm, respectively. The stability of the process is well established and is being monitored with $\bar{x}$- and R-charts that use subgroups of size $n = 5$. It turns out that the process mean is 0.2 mm below the target; the lower and upper control limits on the $\bar{x}$-chart are 98.8 mm and 100.8 mm. Calculate and interpret C_p, C_{pk}, and C_{pm}.

    Hint: Use the fact that $A_2\bar{R}$ in the control limits of the $\bar{x}$-chart is an estimate of $3\sigma/\sqrt{n}$. Use the latter expression to solve for σ.

5.2-6 Discuss the following statements and calculate the capability indices:

    (a) The $\bar{x}$-chart is used to check whether process variability is stable.

    (b) A stable process (i.e., a process that is under statistical control) is also capable.

    (c) The lower and upper specification limits on a bakery good are 95 and 105 (decagrams), with a target of 100. A sample of $n = 200$ goods resulted in a sample mean $\bar{x} = 100.165$ and standard deviation $s = 1.7419$. Obtain C_p, C_{pk}, C_{pm}, CR, and Target-Z, and interpret your findings.

5.2-7 Consider the data shown in Exercise 5.1-10. There, we studied the stability of the dimensions of cartons and found that both the level and the variability of the process were under statistical control.

The specification limits for this process are 24 to 26 cm, with a target of 25 cm. Calculate C_p and C_{pk}, and check whether the process is capable. Use the mean and the standard deviation of the 100 observations in your calculation.

*5.2-8   Consider a situation with symmetric specification limits around the target T_g; that is, LSL $= T_g - c$ and USL $= T_g + c$, for some $c > 0$. The measurements have a normal distribution with mean μ and standard deviation σ. Assume a capability index of C_p, and suppose that the mean misses the target by a multiple $k > 0$ of the standard deviation; that is, $\mu = T_g + k\sigma$. Relate the proportion of defectives to the capability index and the multiple $k > 0$. In particular, calculate the proportion of defectives if

*(a) $C_p = 2$ and $k = 1.5$ (the process mean is 1.5 standard deviations off the target);

*(b) $C_p = 1.5$ and $k = 0.5$ (the process mean is 0.5 standard deviation off the target);

*(c) $C_p = 1.5$ and $k = 1.0$ (the process mean is 1.0 standard deviation off the target).

Note: This problem has been considered in Tadikamalla, P.R., "The Confusion Over Six-Sigma Quality," in *Quality Progress,* November 1994 (Vol. 27), 83–85.

5.3 Acceptance Sampling

Manufactured parts (items) are often shipped from a supplier (the producer) to another company (the consumer) that uses the parts in the construction of some final product. Items could be sent in boxes—or even boxcars—and a standard grouping of these items is often called a *lot*. Let each lot consist of N items, where N can range from 10 or so to several thousands, depending on the nature of the item. Unfortunately, some of these items are probably defective; here, we let p represent the fraction that is defective in the lot. That is, there are Np defective items among the N parts. If p is small, the supplier is providing acceptable quality; but a large p suggests that the quality is unacceptable and that the lot should be rejected and not be shipped. However, we do not know p or the number Np of defectives in the lot. We must find an estimate of p to help us decide whether to accept or to reject the lot. Of course, we could inspect all of the items in the lot. However, 100 percent inspection is usually extremely expensive (sometimes more costly than the production of the item itself) and sometimes impossible—for example, in the case of destructive inspection, in which an item (e.g., a fuse) is destroyed in the testing. Thus, in most instances, we resort to sampling to determine whether the lot should be accepted or rejected. By observing just a sample (only a fraction of the items in the lot), we may accept certain lots that we would reject if we looked at every item, and we may reject others that we would accept in a 100 percent inspection. To decide whether the acceptance sampling plan is a desirable one, we should consider the probabilities of making these two types of errors.

Example 5.3-1   Let a lot of $N = 1{,}000$ items contain Np defectives, where $0 \leq p \leq 1$. We take $n = 10$ items at random and without replacement and test them. If all are satisfactory, we accept the lot; otherwise, we reject it. By the multiplication rule of probabilities, the probability of obtaining n good items and thus accepting the lot is

$$\left(\frac{N - Np}{N}\right)\left(\frac{N - Np - 1}{N - 1}\right)\left(\frac{N - Np - 2}{N - 2}\right)\cdots\left(\frac{N - Np - (n - 1)}{N - (n - 1)}\right)$$
$$= (1 - p)\left(1 - \frac{Np}{N - 1}\right)\left(1 - \frac{Np}{N - 2}\right)\cdots\left(1 - \frac{Np}{N - (n - 1)}\right).$$

With $n = 10$ and $N = 1{,}000$, it is easy to see that

$$\frac{N}{N - 1} \approx \frac{N}{N - 2} \approx \cdots \approx \frac{N}{N - (n - 1)} \approx 1.$$

Hence, the probability of accepting the lot is approximately equal to the binomial probability $(1 - p)^{10}$. This is the probability that we would have obtained if we had taken the $n = 10$ items at random and *with* replacement—that is, replacing each selected item before selecting the next. The procedure of sampling with replacement is not too practical, however. Nevertheless, if N is considerably larger than n, this binomial probability provides an excellent approximation to the true probability, and we frequently use it.

The probability of accepting the lot is called the *operating characteristic (or OC) curve*. (See the related discussion in Section 4.5.) In this example, the OC curve is approximated by

$$OC(p) \approx (1 - p)^{10}.$$

The OC curve is a function of p, and its graph is given in Figure 5.3-1. The values $OC(0.05) = 0.60$, $OC(0.10) = 0.35$, $OC(0.15) = 0.20$, $OC(0.20) = 0.11$, $OC(0.25) = 0.06$, and $OC(0.30) = 0.03$ can be found from the binomial c.d.f under $P(X \le 0)$ with $n = 10$. The reader may find this OC curve highly undesirable because there is a probability of 0.35 of accepting the lot, even though there are $1{,}000(0.10) = 100$ defective items among the $N = 1{,}000$ items in the lot. That is, there is a 35 percent chance of accepting a seemingly undesirable lot that has 10 percent defectives. In addition, the probability of accepting a desirable lot, say, with $p = 0.02$, is

$$(1 - 0.02)^{10} = (0.98)^{10} = 0.82.$$

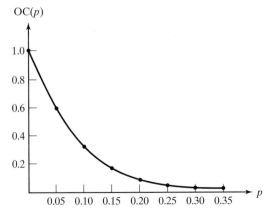

Figure 5.3-1 OC curve. Probability of accepting a lot as a function of the proportion of defectives, p

This means that the probability of rejecting a lot with only 2 percent defectives is 0.18, and that is undesirable, too. In other words, in this example, the probabilities of the two types of error, the *producer's risk* of rejecting a satisfactory lot and the *consumer's risk* of accepting an undesirable lot, are too high: 0.18 and 0.35, respectively. The only way we can correct the situation is by taking a larger sample size and redesigning the acceptance sampling procedure.

The two values of p that were used in this example, $p_0 = 0.02$ and $p_1 = 0.10$, are frequently called *acceptable quality level* (AQL) and *lot tolerance fraction defective* (LTFD). The probabilities of errors at AQL $= 0.02$ and LTFD $= 0.10$, which here are $\alpha = 0.18$ and $\beta = 0.35$, have special names. The probability $\alpha = 0.18$ of rejecting a good lot of AQL $= 0.02$ is called the *producer's risk,* because the producer is hurt if a good lot is rejected. The probability $\beta = 0.35$ of accepting a bad lot of LTFD $= 0.10$ is called the *consumer's risk,* because the consumer loses if a bad lot is accepted. ∎

Example 5.3-2 Suppose that we wish to design an acceptance sampling plan for lots of $N = 5{,}000$ items. Suppose also that, for the particular item that we are manufacturing, $p_0 = $ AQL $= 0.02$ and $p_1 = $ LTFD $= 0.06$. Furthermore, we want the probabilities of errors at those two values of p to be about $\alpha = 0.05$ and $\beta = 0.10$. That is, we desire an OC curve such as that in Figure 5.3-2. To achieve such an OC curve, it seems as if the sample size must be fairly large, so we begin by using the Poisson approximation. The scheme is to take a sample of size n and accept the lot if the number of defectives is less than or equal to an acceptance number, say, Ac. Let us begin with an initial guess of $n = 100$ as the sample size. With $\lambda = (100)(0.02) = 2$, we see from the Poisson distribution that if Ac $= 4$, then OC$(0.02) = P(Y \leq 4) = 0.947$. Now, $1 - 0.947 = 0.053$, so the probability of rejecting a lot of acceptable quality is quite close to the desired α. However, $(100)(0.06) = 6$ and Ac $= 4$ together imply, according to the Poisson distribution with $\lambda = 6$, that OC$(0.06) = P(Y \leq 4) = 0.285$, which is much larger than the desired β value of 0.10. To lower this probability,

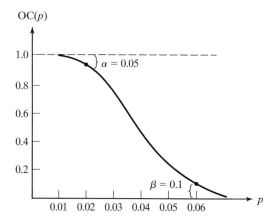

Figure 5.3-2 OC curve for Example 5.3-2. Probability of accepting a lot as a function of the proportion of defectives, p

we must increase the sample size. Let us try $n = 150$, which gives both $\lambda = 150(0.02) = 3$ and $\lambda = 150(0.06) = 9$. Choosing $Ac = 6$ so that

$$OC(0.02) = P(Y \le 6) = 0.966$$

means that $1 - 0.966 = 0.034$ is the probability of rejecting a lot of acceptable quality with $AQL = 0.02$. This probability of 0.034 is close to the desired $\alpha = 0.05$. However,

$$OC(0.06) = P(Y \le 6) = 0.207$$

is still too high and not desirable. Trying various λ values so that the ratio between them is $0.06/0.02 = 3$, we see that $\lambda = 4$ and $\lambda = 12$ with $Ac = 7$ provide the respective OC values of 0.949 and 0.090, probabilities that are close to the desired $1 - \alpha = 0.95$ and $\beta = 0.10$. Also, $(n)(0.02) = 4$ and $(n)(0.06) = 12$ imply that $n = 200$. That is, a highly desirable sampling plan is described by $n = 200$ and $Ac = 7$: Take a sample of $n = 200$ items from each lot. If there are no more than 7 defectives, accept the lot; otherwise, reject it. The approximate operating characteristic curve of this sampling plan is

$$OC(p) \approx \sum_{y=0}^{7} \frac{(np)^y e^{-np}}{y!},$$

where $n = 200$. ∎

   Usually, engineers do not need to construct acceptance sampling plans from first principles because the appropriate plans can be looked up in manuals provided by organizations such as the American National Standards Institute (ANSI) and the American Society for Quality (ASQ). For many years, the federal government had required the use of certain Military Standard plans (MIL-STD-105D and its 1989 revision, MIL-STD-105E); recently, however, the government replaced the mandate with acceptable non governmental standards on sampling procedures.

Example 5.3-3 Suppose that we have a lot of $N = 1,000$ items and we desire an $AQL = 0.025$, or 2.5 percent. From a MIL-STD-105D table that is found in older books on quality control, we find that we should use $n = 80$, $Ac = 5$, and the rejection number $Re = 6$, which means that we should reject the lot if six or more defectives are found in the sample of $n = 80$ items. We believe that the engineer should understand the implications of such a plan. We can help him or her in this regard by constructing an OC curve. Using the Poisson approximation, we calculate the probability of accepting the lot, namely,

$$OC(p) = P(Y \le 5) \approx \sum_{y=0}^{5} \frac{(80p)^y e^{-80p}}{y!}.$$

From the Poisson distribution, we obtain $OC(0.02$ or $\lambda = 1.6) = 0.994$, $OC(0.025$ or $\lambda = 2.0) = 0.983$, $OC(0.03$ or $\lambda = 2.4) = 0.964$, $OC(0.05$ or $\lambda = 4.0) = 0.785$, $OC(0.10$ or $\lambda = 8.0) = 0.191$, and $OC(0.15$ or $\lambda = 12.0) = 0.020$. The OC curve

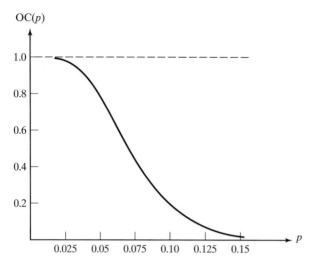

Figure 5.3-3 OC curve for MIL-STD-105D example

through these points is given in Figure 5.3-3. Although the probability $\alpha = 1 - 0.983 = 0.017$ is desirably small at AQL $= 0.025$, we may be somewhat concerned about the high probability of accepting the lot (given by $OC(0.10) = 0.191$) if the lot is 10 percent defective. Thus, we may want to reduce (if possible) the acceptance number to Ac $= 4$, which, when $n = 80$, yields

$$OC(0.025 \text{ or } \lambda = 2.0) = 0.947$$

and

$$OC(0.10 \text{ or } \lambda = 8.0) = 0.100.$$

This seems to us a better plan.   ■

Let as now consider one final concept associated with an acceptance sampling plan. If a lot with fraction defective p is accepted, it is allowed to continue on into the production process. If it is rejected, an agreement is usually made with the supplier that the lot is to be 100 percent inspected and that bad items are to be replaced with good ones before the lot is sent on into the production process. In the first case, the fraction defective entering the process is p; in the second case, it is zero because, after replacing the bad items, all of the items are good. To get an *average outgoing quality* (AOQ), we must average the values p and zero, with weights $OC(p)$ and $1 - OC(p)$, which are their respective probabilities of occurring. That is, AOQ is the expected value

$$AOQ(p) = (p)[OC(p)] + (0)[1 - OC(p)] = p[OC(p)].$$

The maximum of the AOQ curve is called the *average outgoing quality limit* (AOQL). It tells us about the worst possible average of outgoing quality and is usually determined by calculus or empirical means. The AOQ curve associated with the MIL-STD-105D plan of Example 5.3-3 is plotted in Figure 5.3-4; the AOQL is about 4 percent.

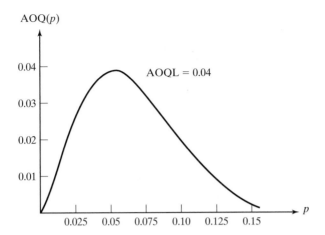

Figure 5.3-4 AOQ curve for MIL-STD-105D example

Example 5.3-4 In Example 5.3-1, we found that $OC(p) \approx (1 - p)^{10}$; thus,

$$AOQ(p) \approx p(1 - p)^{10},$$

which has the maximizing solution $p = 1/11$. Hence,

$$AOQL = AOQ(p = 1/11) = \left(\frac{1}{11}\right)\left(\frac{10}{11}\right)^{10} = 0.035. \quad \blacksquare$$

Acceptance sampling plans are important because we want to avoid letting too many defectives into the consumer's production process. However, we must realize that *quality cannot be inspected into products.* Products have to be built properly in the first place! Therefore, if a supplier has a good quality control program, it is possible to eliminate the acceptance sampling procedures altogether. Reliable suppliers can be trusted and there is no need to inspect their products. But until all suppliers can be classified in this manner, acceptance sampling plans will definitely be needed.

Exercises 5.3

5.3-1 A sampling plan has a sample size of $n = 15$ and an acceptance number $Ac = 1$.

    (a) Using hypergeometric probabilities, calculate the respective probabilities of accepting lots of $N = 50$ items that are 2, 6, 10, and 20 percent defective. Sketch the OC curve.

    (b) The sampling plan is used for large lots. Use binomial probabilities to compute an approximation to the probabilities in part (a).

*5.3-2 A sampling plan uses $n = 100$ and $Ac = 3$. The lot size N is large in comparison to the sample size n.

    (a) Sketch the OC curve.

    *(b) Calculate the producer's risk at $AQL = 0.02$.

    (c) Calculate the consumer's risk at $LTFD = 0.10$.

*(d) Assume that a rejected lot is 100 percent inspected and that defectives are replaced by good items. Plot the AOQ curve and find (approximately) the average outgoing quality limit AOQL.

5.3-3   A sampling plan for the fraction of defectives consists of sampling $n = 50$ items and rejecting the lot if there are more than two defectives. Suppose that the lot size N is large enough to permit us to use the binomial distribution (and, furthermore, because p is small and n relatively large, the Poisson approximation). Calculate the producer's risk at AQL = 0.05 and the consumer's risk at LTFD = 0.10.

*5.3-4   A sampling plan consists of sampling $n = 300$ items from a lot of $N = 5,000$ items. The acceptance number is Ac = 1. The plan stipulates that rejected lots are 100 percent inspected and that defective items are replaced by good ones.

*(a) Plot the OC curve, and observe the producer's risk α at AQL = 0.002 and the consumer's risk β at LTFD = 0.010.

*(b) Plot the AOQ curve and determine the approximate value of AOQL.

*5.3-5   Design an acceptance sampling plan for large lots with AQL = 0.02 and LTFD = 0.08. Control the probabilities of errors at those two values of p at about $\alpha = 0.05$ and $\beta = 0.10$.

*5.3-6   The following sampling plan is used: (1) Select a sample of size 2 from a lot of 20. If both items are good, accept the lot. If both are defective, reject it. If one is good and one defective, take a second sample of one item. (2) If the item in the second sample is good, accept the lot; otherwise, reject it. If the lot is $p = 0.2$ fraction defective, what is the probability that we accept the lot? Repeat your calculations for several other values of the fraction defective p, and plot the OC curve.

Note: This sampling scheme is called a *double-sampling plan,* because it allows the possibility of delaying the decision on the lot until a second sample is chosen. The sampling schemes that we have discussed in this section are *single-sampling plans*—plans according to which we accept or reject a lot on the basis of a single sample.

5.3-7   Specifications require that a product have certain quality characteristics, which can be determined only by a destructive test. The product is made in batches of $N = 1,000$. The current inspection scheme is to select $n = 4$ items from each batch. If all four articles meet the quality specification, we accept the lot. If two or more fail, we reject the lot. If one fails, we take another sample of two. If both pass the inspection, we accept the lot; otherwise, we reject it. What is the probability that we accept a batch that contains 5 percent defectives? Sketch the OC curve of this double-sampling plan.

5.4 Problem Solving

5.4-1 Introduction

The statistical methods that we have discussed so far in this chapter help us monitor processes. They indicate whether something has gone wrong and needs to be fixed. However, it is often quite difficult to determine exactly what has gone wrong and, once the reason is found, how to fix it. Engineers face such problems constantly, and university education should prepare them for these kinds of situations. However, "problem solving" is a difficult subject to teach, and we ourselves are not sure how best to do it. Experience, common sense, and knowledge of the subject are needed as much as anything. It helps to gain this experience by observing and working with experienced problem solvers.

When an engineer is confronted with difficult problems, we do advise that a team of experts familiar with the process be gathered to "brainstorm" the situation. Naturally, the workers with much knowledge about processes and machines must be included. Other members of the team may be supervisors, managers, and, of course, engineers. It is beneficial to include all members who are associated, directly or indirectly, with the project. The size of the team is often determined by the magnitude of the problem: Frequently, two or three persons can handle a minor problem; more may be needed for a major one.

Of course, reliable data are needed to help the team make wise decisions. Information must be collected to reveal facts that help solve the problem at hand. Here we give several examples. To identify the most prevalent defect in a manufacturing process, we collect information on the frequencies of the various defects. To be able to eliminate the most important defects, we have to collect, summarize, and analyze information on their causes. We collect and display information on the percentage of defectives in successive lots to assess whether a production process is stable and to identify sources of its instability—for example, variation in the raw materials, weekday–weekend differences, and failure to maintain machines. We might measure the width, length, weight, or diameter of certain items to check whether they are within specifications. We take measurements on various types of products to determine which type is the best. Processes are run at various settings to determine which factors are important and to learn how we can reach the optimum. Water samples from different locations in a lake are analyzed to discover whether the levels of a particular pollutant depend on location. The amounts of hydrocarbons that are emitted from car engines with and without catalytic converters are obtained to assess the effect of the catalytic converter on those emissions. The fuel efficiencies of various types of automobiles are measured to see how they differ. Also, we may want to check whether fuel efficiency is related to the weight of the car, the design of its engine, the wind resistance of its chassis, and so on. Information on the durability of a certain consumer product is collected to help estimate the amount of money a company will have to spend on warranty repairs.

Reliable data provide the information that is needed to make decisions. Properly collected and analyzed, data can help us understand and solve problems. Data are used routinely to assess the variability of measurements, decide whether a process is under control, compare the effectiveness of various methods, suggest various ways of improving a process, assess the relationships among variables, find optimum conditions, and much more. Proper ways of obtaining and presenting information are extremely important.

5.4-2 Pareto Diagram

Let us start our discussion with a simple *Pareto diagram* that displays the frequencies of various defects. Because this diagram identifies the main sources of defects, it has become a valuable tool in industry.

Defects can arise from a number of sources. For example, in a lens-polishing process, an item can be defective because the lenses are too thick or thin, scarred, unfinished, or poorly coated. In a rubber-molding process, the items may show surface scars, have cracks, be misshapen, or be incomplete. The major contributors to the defects have to be identified, because only then can an appropriate strategy be pursued to eliminate the most important defect(s). The various sources can be identified through a Pareto diagram.

As an example, consider a lens-polishing company which has found that the number of defectives has increased recently. The company has classified the total number of defectives from a day's production ($N = 110$) according to several major causes. The data are given in the first column of Table 5.4-1.

A Pareto diagram is a bar graph that shows the frequencies of the various defects. In Figure 5.4-1, we have arranged the bars in decreasing order of frequency: The most frequent cause is on the left and the least frequent on the right. Note that with qualitative variables such as type of defect, type of engineer (civil, industrial, electrical, etc.), or type of machine, the ordering of types is really arbitrary. However, in the particular context in which we are looking for the most common type of defect, it makes sense to order them in decreasing order of occurrence. In our example, the most important defect arises from poorly coated lenses; there are 45 defectives, or $100(45/110) = 40.9$ percent, in this category. The second most important cause is scarred lenses, contributing another 27.3 percent.

The line graph on the Pareto diagram connects the cumulative percentages of the $k(k = 1, 2, \ldots, 5)$ most frequent defects. For example, the two most frequent defects—poorly coated lenses and scarred lenses—represent $(45 + 30)/110 = 0.68$, or 68 percent of all defective lenses. The three most frequent defects (poorly coated lenses, scarred lenses, and unfinished lenses) represent $(45 + 30 + 15)/110 = 0.82$, or 82 percent, and so on.

Pareto diagrams show that usually only two or three defects account for over 75 percent of the losses. A Pareto diagram is an important component of any quality improvement program because it focuses everyone's attention on the one or two categories that lead to the most defects. It is usually easier to reduce the occurrence of a frequent defect by half than it is to reduce the one occurrence of a rare defect to zero. Rare defects are more or less inevitable, occurring every now and then.

After viewing this information, the company started an effort to improve its lens-coating operation. It turned out that a switch to a cheaper, yet less reliable, supplier

Table 5.4-1 Major Causes of Lens-Coating Defects

	Before	After
Lenses too thick or thin	10	8
Scarred	30	32
Cracked	6	8
Unfinished	15	12
Poorly coated	45	16
Others	4	4
	110	80

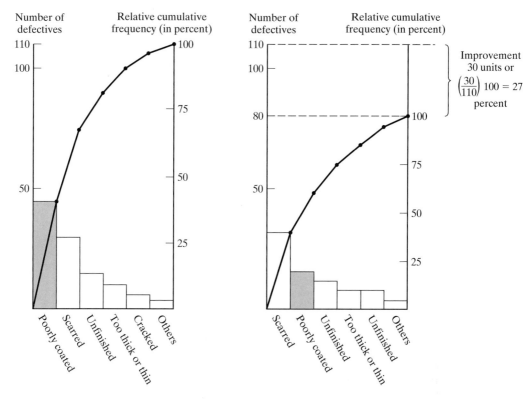

Figure 5.4-1   Pareto diagrams before and after corrective action was taken

of the lens-coating solution had caused an increase in poorly coated lenses. The company decided to return to the original supplier, and in the process it reduced the number of defects due to poorly coated lenses by about two-thirds. A summary of the distribution of defects after corrective action had been taken is given in the second column of Table 5.4-1. A comparison of the two Pareto displays in Figure 5.4-1 also gives a nice summary of the improvements that can be attributed to the corrective action that was taken. Furthermore, these diagrams show that scarring is the next most important defect. Possible causes of scarring should now be investigated. We could start such an investigation by constructing Ishikawa cause-and-effect diagrams, which are discussed in the next section.

5.4-3 Diagnosis of Causes of Defects

A prime objective of many investigations is to *improve* quality—that is, to take actions which lead to better products. Measures must be taken to correct the causes of low quality. Finding the dominant cause of a defect, such as poor coating of lenses or excessive wobble during machine rotation, can be a lengthy project, because we can usually think of many factors that may have contributed to low quality. Kaoru Ishikawa, a Japanese control engineer, has developed certain *cause-and-effect diagrams* which depict the variables that may have affected the response. These diagrams are also called *fishbone* diagrams, because they resemble the skeleton of a fish. There are several different methods for constructing cause-and-effect diagrams, depending on how the information is organized and presented. We reproduce two such diagrams in Figure 5.4-2. Figure 5.4-2(a) lists the main factors that affect wobble during

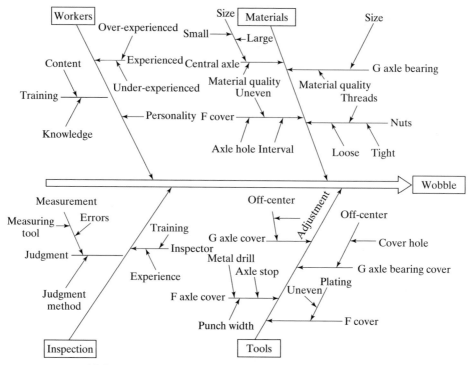

(a) Cause-and-effect diagram for wobbling during machine rotation

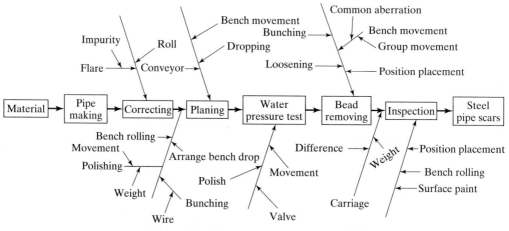

(b) Cause-and-effect diagram for steel pipe scarring.

Figure 5.4-2 Two cause-and-effect diagrams [From Ishikawa, K., *Guide to Quality Control,* second revised edition (Tokyo: Asian Productivity Organization, 1982).]

machine rotation, as well as the factors that influence those main factors. Machine wobble is thought to be a major cause of production defects. So, why does machine wobble occur? One possible factor is the variability among material (components). Thus, "materials" is written on the diagram as a branch. Why does dispersion in the materials occur? It could be because of the variability in the G axle bearing. Hence, "G axle bearing" becomes a twig on the branch. Why does dispersion in the G axle bearing occur? It could be due to variability in the size of the G axle bearing. Therefore, "size" becomes a twig on the twig, and so on. Such a diagram shows why excess wobble may occur, and it helps organize and relate the factors that influence the wobble. The second diagram lists the causes that may affect the scarring of steel pipes. Here, the diagram follows the production process and lists the components that may affect the scarring at the various stages of production.

Cause-and-effect diagrams are useful first steps for recognizing the factors that may be of importance. Successfully composed diagrams usually enhance communication among those who attempt to improve the process. However, after a set of possible factors has been identified, we have to study and quantify their effects in more detail. Well-designed experiments have to be conducted to quantify the cause-and-effect relationships. Strategies for efficient experimentation will be discussed in the next two chapters.

5.4-4 Six Sigma Initiatives

In Section 5.2-2, we introduced the Six Sigma concept and showed how it relates to a 3.4-parts-per-million defect level. The origin of Six Sigma as a statistically based method for reducing variation in electronic manufacturing processes started at Motorola around the mid-1980s. Today, almost 25 years later, Six Sigma has developed into a comprehensive business performance methodology. It is used all over the world, in diverse organizations such as large and small corporations, local governments, hospitals, banks, and the military.

Today, Motorola views Six Sigma as a metric, as a methodology, and as a management system. To Motorola, Six Sigma is all three at the same time. Six Sigma as a *metric* is discussed in Section 5.2-2; it establishes a scale for levels of goodness and quality. Using this scale, Six Sigma equates to 3.4 defects per 1 million opportunities. While Six Sigma started as a defect reduction effort in manufacturing, Motorola realized very quickly that it could be applied to other business processes for the same purpose.

As Six Sigma has evolved, there has been less emphasis on the literal definition of 3.4 defects per 1 million opportunities and on counting defects in products and processes. Six Sigma has become a *business improvement methodology* that guides an organization in understanding and managing customer requirements, aligning key business processes to achieve those requirements, utilizing rigorous data analysis to minimize variation in those processes, and driving rapid and sustainable improvement of many business processes. At the heart of the methodology is the *DMAIC* model for process improvement. *DMAIC* is commonly used by Six Sigma project teams and is an acronym for *D*efine opportunity, *M*easure performance, *A*nalyze opportunity, *I*mprove performance, and *C*ontrol performance. Some authors talk about *DMAICT*, where *T* stands for *T*ransferring best practice to other areas of the organization. The *DMAIC* model is a refinement of the Deming–Shewhart *PDCA* wheel (*Plan*, *Do*, *Check*, *Act*), which was a key paradigm during the Total Quality Management (TQM) movement of the late seventies and early eighties.

Motorola has learned that the disciplined use of metrics and the application of the methodology are still not enough to drive desired breakthroughs and results that are sustainable over time. For the greatest impact, Motorola ensures that process metrics and Six Sigma's structured methodology are applied to opportunities for improvement that are directly linked to the company's organizational strategy. Practiced as a *management system,* Six Sigma becomes a performance-oriented system for executing business strategy, helping organizations align their business strategy with critical improvement efforts, mobilizing teams to attack high-impact projects, accelerating improved business results, and governing efforts to ensure that improvements are sustained.

Six Sigma starts at the top. When an organization decides to implement the method, the executive team has to decide on a strategy (i.e., an improvement initiative) that is focused on all processes necessary to meet customer expectations. Six Sigma processes are executed by Six Sigma Green Belts and Six Sigma Black Belts and are overseen by Six Sigma Master Black Belts. The Sigma Belts are commonly seen as a parallel to the belts that exist in the martial arts.

Brief Outline of Six Sigma's History   In the mid-1980s, Motorola engineers used Six Sigma as an informal name for an in-house initiative for reducing defects in production processes. In the late 1980s, following the success of the original initiative, Motorola extended the Six Sigma methods to its critical business processes, and Six Sigma became an in-house "branded" name for a performance improvement methodology. In 1991, Motorola certified its first "Black Belt" Six Sigma experts, initiating an accredited training of Six Sigma methods. General Electric was one of the first large-scale adopters and advocates of Six Sigma and is considered by most experts to have been responsible for Six Sigma's rapidly achieved high profile. By the mid-1990s, Six Sigma had developed into a transferable "branded" corporate management initiative and methodology, notably in General Electric and other large manufacturing corporations, but also in organizations outside the manufacturing sector. By 2000, Six Sigma was effectively established as an industry in its own right, involving the training, consultancy, and implementation of Six Sigma methodology in organizations around the world.

Importance of Training Activities   Although Six Sigma is capable of leveraging huge performance improvements and cost savings, none of this happens on its own. People need to be trained in Six Sigma methods — especially the measurement and improvement tools (i.e., statistics and data-based problem solving) and the communications/relationships skills that affect the information flow among the various stakeholders of the organization.

Many excellent firms specialize in Six Sigma training. Most training programs use an approach something like this: Say a company wants 15 employees to acquire a good knowledge of a sound quality improvement program. Then the company hires a Six Sigma trainer (usually a statistician with experience in quality improvement methods who also understands the objectives of the firm). The trainer goes to the company for an initial week, teaching the basic philosophy of quality improvement and statistical methods such as Shewhart control charts and principles of good experimental design (discussed in the next two chapters). However, it is the company that selects the projects (usually 10 or so) that need improvement. Some of these projects may be studied by teams of employees, while others may be tackled individually. The trainer is aware of all of the projects and offers helpful suggestions. At the end of the first week, the trainer leaves for about four weeks, during which time the employees keep working on their projects.

The trainer then returns for a second week, giving further instruction on statistical methods and quality improvement, and offering additional suggestions on the various projects. Then the trainer leaves for another four weeks, returning for a third week with some more training on statistics and quality improvement, and further helpful suggestions on the selected improvement projects. The training cycle may be repeated one more time, with the trainer leaving for another four weeks before returning for a final round of instruction.

Clearly, this is an expensive training approach, because good Six Sigma trainers are not cheap and 15 employees have given up at least four weeks of their normal work time. How can companies afford to do this? In almost all cases (General Electric is a good example), the savings produced by improvements on the various projects far exceed the costs of the Six Sigma training. Sometimes an employee who is participating earns a "Black Belt" by saving the company more than a certain amount per project (1 million dollars, for example, but each company has its own figure). Most CEOs are delighted with the program because trainer and employees, through their work on projects in need of improvement, respond to the corporate motto "Show me the money." (This motto was made popular by the movie *Jerry Maguire,* in which it was repeatedly used in a phone exchange between the athlete Rod Tidwell and his agent, Jerry Maguire, emphasizing the importance of the ultimate payback of various initiatives.)

Criticism of Six Sigma Advocates of Six Sigma believe that it can produce quick, noteworthy financial results. However, there is also criticism: (1) Results are often exaggerated, and some companies that have embraced Six Sigma have done so poorly. (2) Six Sigma is based on arbitrary standards; although 3.4 defects per million might work well for certain products and processes, it may not be ideal for others. (3) Many organizations and consulting firms of all sizes deliver Six Sigma training. Seemingly, anyone can start up as a Six Sigma consultant, providing lots of opportunities for charlatans.

Software for Six Sigma Generally, two classes of software support Six Sigma: analysis tools, used to perform statistical or process analysis, and program management tools, used to manage and track a corporation's entire Six Sigma program. Minitab (which is used in this book) has become an important analysis tool for Six Sigma.

Lean Six Sigma New management technologies vie for the attention (and the money) of companies. Even successful and established systems such as Six Sigma must market themselves to business. A recent development is the combination of Six Sigma with Lean production methods, creating "Lean Six Sigma." Lean and Six Sigma are both process improvement methodologies. Six Sigma evolved as a quality initiative to reduce variance in the semiconductor industry and is about precision and accuracy. Lean production methods, which are based on the world-renowned Toyota Production System, arose as methods for optimizing auto manufacturing and are about speed and flow.

Six Sigma will eliminate defects, but it will not address the question of how to optimize the process flow. It is in this area that Lean manufacturing is most useful. Lean manufacturing is a proven approach to streamline operations and reduce waste. It embraces a philosophy of continually increasing value added through the ongoing elimination of waste (such as reducing inventory and floor space requirements). Lean tools include *Value Stream Mapping* (method of visually mapping a product's production path from the producer to the consumer), *Kaizen* (the continual improvement philosophy), and *Elimination of Waste by Focusing on the 7W's* (with the seven wastes being defects, overproduction, waiting, transportation, movement, inappropriate processing, and inventory).

Because of the relative sophistication of most Six Sigma tools, only a comparatively small number of a typical company's employees will be capable of learning and applying them. This limitation leads to Six Sigma's emphasis on green/black belt training programs and its reliance on "expert teams" for problem solving, with relatively little involvement from factory or office workers. In contrast, Lean manufacturing tools are simple enough so that the average worker can understand and apply them. Because Lean can get everyone involved, it increases the likelihood of changing the entire culture of an organization.

Exercises 5.4

5.4-1 Improvement projects and cause-and-effect diagrams:

(a) A large supermarket receives complaints about the quality of its baked goods. Suppose that you are in charge of the bakery. Develop an Ishikawa fishbone diagram that could help you locate the causes of poor quality.

(b) The quality of health care that is provided by a hospital to its patients is the variable of interest. Discuss how one could measure quality of health care, and develop an Ishikawa fishbone diagram that could be a guide to improving quality.

(c) Assume that you are the owner of a business that specializes in residential house painting. You are concerned about the quality of your work, because customers will not pay you for repairs if the paint peels prematurely. Construct a fishbone diagram to locate some of the factors that may be responsible for early peeling.

(d) You notice that a number of color computer display terminals are returned by customers because of misconvergence errors among the three primary colors: red, green, and blue. How would you initiate a process that ultimately leads to the construction of a cause-and-effect diagram?

(e) Universities depend on the quality of their instruction. Discuss ways of measuring the quality of classroom instruction, and construct a diagram which lists the factors that affect the quality.

(f) Think of a problem associated with your engineering education. Form groups, which might include several of your instructors and the dean of the College of Engineering, to brainstorm about this problem. Develop an Ishikawa fishbone diagram.

(g) Consider the registration process at your university. You are concerned because it always seems to take too long to register for next semester's courses. How would you go about improving the process? Identify the main factors that affect registration time and develop appropriate cause-and-effect diagrams.

5.4-2 One hundred twenty students were asked about their views on an engineering statistics course: fifty-five students thought that the presentation of the material was too theoretical, 35 wished that they had a better handout on the use of computer software for statistical analysis, 15 thought that the tests were too difficult, 10 felt that the instructor was not always prepared, and 5 students had difficulty locating the instructor during office hours. Prepare a Pareto chart and discuss ways of improving the course.

Chapter 5 Additional Remarks

This chapter discusses important tools for *quality improvement*. There has been a great deal of development in this area, starting with the Shewhart control charts in the 1920 and 1930s, Statistical Quality Control (during World War II and beyond), Total Quality Management (1980), and Six Sigma (1990 and later). Although W. Edwards Deming convinced the Japanese in the early 1950s to use statistical methods and the quality improvement philosophy discussed in his book *Out of the Crisis,* his message did not catch on in the United States until the early 1980s, when Deming appeared on the NBC television program "If Japan Can, Why Can't We?" After this very important exposure, Deming's short courses were in great demand. He taught four-day short courses, week after week, to hundreds of companies and thousands of employees. He also gained access to CEOs of some of the largest U.S. companies and told them, in no uncertain terms, what they needed to fix if they wanted to stay in business. He continued his activities until shortly before his death in 1993. Deming's courses included all important elements of quality improvement, such as delighting customers, valuing feedback, reducing waste, encouraging team efforts, and using the scientific method to solve problems, with an emphasis on the collection of relevant data and their appropriate statistical analysis.

In the fall of 1991, one of us (Hogg) took a quality fact-finding tour to more than 20 companies, including Ford, GM, Kodak, AT&T, Saturn, Motorola, IBM, and the Mayo Clinics. Arrangements had been made, through statisticians, for Hogg to talk to their boss and their boss' boss about quality, allowing for frank interaction with upper-level management of these companies. One vice president at Kodak mentioned that there was probably 25 percent waste in the company's operations, amounting to a lot of activities with no value added. That percentage seemed high to Hogg, and he mentioned it at AT&T, getting the reply "It's worse than that here—maybe 50%." This response suggested that tremendous amounts of waste were present in every organization, but so were great opportunities for quality improvement efforts. You may argue that this was 15 years ago and, since that time, all problems were fixed. Certainly, improvement has taken place, and at the better companies some of the "low-hanging fruit" (processes or products for which improvements are easy) has "been picked." But much work still needs to be done.

During his fact-finding tour, Hogg had the opportunity to interact with quality gurus such as Ed Deming, Joe Juran, Don Berwick (in the medical sciences), Blan

Godfrey, and Brian Joiner. The latter stated "Complaints are blessings from the sky." Well-articulated complaints from customers and employees alike provide incentives for improvement. Your teachers believe in this principle also, and we encourage students to tell us about areas in which they need help and any bad habits in their teachers that distract from their learning. Your feedback helps us improve. Of course, for this idea to work, we must establish some trust, so that you don't fear that such feedback would lead us (the teachers) to punish you (the students) by handing out bad grades.

One of the highlights of Hogg's road trip was a 90-minute session with Bob Galvin, the former CEO and then chairman of the board of Motorola. He told Hogg that in grade school he was really mad at one of his teachers because she made him retake an entire test, and he had missed only one out of 20 questions! His dad said that the teacher was right, because 5 percent defective was not satisfactory. This was a lesson that Galvin remembered very well. When Hogg asked him why Motorola used the term Six Sigma for its quality improvement program, he implied that, although engineers think they know everything, they certainly don't know much about sigma and how to handle variation. We hope that student engineers taking our course will understand the importance of variability and of the standard deviation as its measure. It is important to reduce variability, so that doors of cars will fit better and the service you receive is at a consistent and high level.

Another important point that Galvin made was that "quality improvement is not just an institutional assignment; it is a daily personal priority obligation." He truly believed this and would practice it every day. For Galvin, quality came first. Harry Roberts, of the University of Chicago, and his coauthor Bernie Sergesketter, of AT&T, picked up on this point and wrote the book *Quality Is Personal* (1993). It shows how an obsession with quality can change your personal life. A simple, but very effective, strategy for personally getting involved in quality improvement is to carry a quality checklist of six to eight activities that you wish to avoid and make a mark on the list every time you engage in one of these undesirable activities. For example, if you want to lose weight, your list may include "eating unhealthy, fattening foods." Now, if you just had a piece of cream pie, take out your list and make a mark. At the end of the day, review your list and perhaps plot your daily scores on a simple time-sequence graph. Try it; construct your own list and carry it with you at all times. Although you will probably find 20 or 30 marks during the first week, you will notice decreasing numbers as time goes on. And the number of infractions may become quite small because you are aware that you carry the list and you do not want to make a mark. Hogg had a list that had the item "not flossing teeth." He tried to avoid a mark at all cost, and his dentist loved the results. There is an added benefit of having people practice personal quality improvement: Such personal efforts will also help companywide quality improvement initiatives. Knowing how to improve your own personal processes will spill over to company-wide quality improvement activities.

The Six Sigma approach to quality, which we discuss in the last section of this chapter, has been embraced by numerous organizations. Six Sigma has many similarities to total quality management (TQM) and other programs of the past, but it also has some distinctive characteristics. One is its focus on defining and responding to customer needs. In doing so, it takes a broader view of quality management compared with some more narrowly focused programs of the past, better integrating quality activities into all areas of the organization and aligning those activities with the strategic goals of the firm. In addition, Six Sigma programs have been more widely applied to service processes, including many implementations in hospitals and other health care organizations.

Projects

*Project 1

Pork producers keep detailed records on their operations. An Iowa company is concerned about its breeding operation. In large-scale breeding operations, the female pigs (sows) of the breeding herd are artificially inseminated at carefully monitored times. In this particular company, about 200 inseminations, called "services," are carried out each week. Weekly records on the breeding herd are kept, including the average number of pigs born alive per litter and the preweaning mortality. Weekly data for the period from July 1995 through June 1996 are listed as follows:

Week	Pigs Born per Litter	Percent Mortality	Heating[1] Degree Days	Cooling[1] Degree Days
7/1/1995	9.8	10.2	3	26
7/8/1995	9.1	8.2	0	121
7/15/1995	9.4	9.4	0	80
7/22/1995	8.8	9.0	0	88
7/29/1995	8.3	12.5	0	84
8/5/1995	9.3	14.6	0	114
8/12/1995	9.4	15.1	0	119
8/19/1995	10.1	14.6	0	71
8/26/1995	9.6	9.2	0	105
9/2/1995	9.4	9.3	9	49
9/9/1995	9.4	7.7	10	21
9/16/1995	8.7	12.0	72	5
9/23/1995	9.1	1.4	45	5
9/30/1995	9.0	4.8	41	0
10/7/1995	9.4	6.2	48	9
10/14/1995	8.9	13.9	75	2
10/21/1995	9.3	10.3	118	0
10/28/1995	9.6	6.3	196	0
11/4/1995	10.0	5.1	201	0
11/11/1995	9.5	9.1	241	0
11/18/1995	8.7	9.7	203	0
11/25/1995	9.3	9.7	226	0
12/2/1995	9.6	3.2	234	0
12/9/1995	8.6	11.6	336	0

Week	Pigs Born per Litter	Percent Mortality	Heating[1] Degree Days	Cooling[1] Degree Days
12/16/1995	9.3	12.9	261	0
12/23/1995	9.1	9.0	280	0
12/30/1995	9.3	10.8	301	0
1/6/1996	9.5	7.3	309	0
1/13/1996	9.3	11.0	270	0
1/20/1996	9.9	3.3	346	0
1/27/1996	9.1	9.6	466	0
2/3/1996	9.0	11.3	323	0
2/10/1996	9.3	10.4	255	0
2/17/1996	9.1	3.3	215	0
2/24/1996	9.5	5.2	247	0
3/2/1996	9.0	10.8	330	0
3/9/1996	9.4	10.0	171	0
3/16/1996	9.4	9.8	209	0
3/23/1996	9.7	6.1	199	0
3/30/1996	9.9	8.3	160	0
4/6/1996	9.6	8.3	130	4
4/13/1996	9.5	12.6	107	0
4/20/1996	9.3	8.4	96	0
4/27/1996	9.0	11.3	125	0
5/4/1996	9.7	12.1	61	1
5/11/1996	9.2	9.8	74	17
5/18/1996	8.7	14.4	15	26
5/25/1996	9.5	4.0	64	0
6/1/1996	9.1	11.9	29	1
6/8/1996	9.4	2.8	2	43
6/15/1996	9.5	7.9	0	61
6/22/1996	9.2	8.9	0	71

[1]A "heating degree day" for a given day is defined as the difference between the "balance point" temperature of 65°F (above which a building is assumed not to need any heating) and the mean temperature on that day. It is zero for days with mean temperature above 65°F. Similarly, a "cooling degree day" is the difference between the mean temperature and 65°F; it is zero for days with mean temperature below 65°F. Weekly heating and cooling degree days are obtained by summing the daily measures over the seven days of the week.

*(a) Use appropriate control charts (in this case, individual observations charts and moving range charts) to check whether the number of pigs born alive per litter and the preweaning mortality are under statistical control. Furthermore, use dot diagrams or histograms to display the variability of these two measures, and calculate the relevant summary statistics.

Note: The subgroup size is $n = 1$, and we graph individual observations against time. The stability of the *level* of the series can be monitored by checking for trends in the time-sequence plot and by making sure that there are no unusual observations. The stability of the *variability* of individual observations can be monitored with a moving range chart. (A "moving" range of individual observations is the absolute value of the difference between adjacent observations.) Both types of chart can be obtained with the Minitab command "Stat > Control Charts > Variables Charts for Individuals (I-MR charts)." (Consult books on statistical process control or search the Web to learn more about moving range (MR) charts.)

*(b) The data include information on weekly heating and cooling degree days for Des Moines, Iowa. Breeding barns are not air-conditioned in the summer and they are poorly heated during the winter. Discuss whether the mortality rate is affected by the weather.

Project 2

Capability indices were discussed in Section 5.2, among them C_p and C_{pk}. Assume that the lower and upper specification limits are -10 and $+10$, respectively. Consider the following processes:

   i. normal with mean 0 and standard deviation 5;

  ii. normal with mean 0 and standard deviation 2.5;

 iii. normal with mean 2.5 and standard deviation 2.5;

 iv. uniform on the interval from -10 to $+10$;

  v. uniform on the interval from -12.5 to $+12.5$.

(a) For each process, calculate the proportion of nonconforming items and determine the capability indices C_p and C_{pk}.

(b) Study the behavior of these indices in samples of sizes $n = 50$ and $n = 100$. Use simulation. Draw samples of sizes $n = 50$ and $n = 100$ from the processes studied and calculate C_p and C_{pk}. Repeat this 1,000 times and display the sampling distribution of the capability indices. Interpret your findings.

6

EXPERIMENTS WITH ONE FACTOR

Frequently, experimenters want to compare more than two treatments. For example, plant breeders compare the yields of different corn hybrids, computer scientists investigate the times between failures of various computer systems, and company managers compare the productivities of several plants within their company. Similarly, engineers study the effect of different gas additives on pollutant emissions, the effectiveness of several rust inhibitors on corrosion, the effects of various catalysts on process yields, and the durability of different types of materials. Experiments have to be designed to investigate possible treatment effects. These experiments have to be designed properly and the resulting data have to be analyzed correctly.

In Section 4.6, we discussed how to compare two treatments. We emphasized the principles of *randomization* and *blocking*. Randomization, which allocates treatments randomly to experimental units, guarantees the validity of the inference we make in the face of unspecified disturbances by making certain that the risk of such disturbances is spread evenly among the treatments. Without randomization, treatment differences might be confounded with other variables that are not controlled by the experimenter. The following example emphasizes this important concept: Suppose that we want to study how two different drying methods affect the compressive strength of concrete cylinders. Suppose also that the plant has produced a total of 30 such cylinders. Because they were made from separate batches of concrete, we can expect some variability among the cylinders. That is, we should expect variability even though the supervisor claims that these batches were mixed exactly the same way. Still, whatever differences we have among the experimental units should not bias our comparison. Thus, we use randomization, assigning the two treatments (the two drying methods) to the experimental units (concrete blocks) at random. We could carry out this randomization by numbering the blocks from 1 through 30 and then selecting blocks at random and without replacement until we have filled our first group of 15 blocks. Then we could flip a coin to determine which of the two drying methods should be applied to that group.

Blocking is the other important concept in the design of comparative experiments. In Section 4.6-2, we discussed how blocking, or running the experiment in pairs if there are only two treatments, can eliminate unwanted sources of variation and can improve the precision of the comparison. Let us assume that in our illustration the batches of concrete were just large enough to produce two concrete cylinders from each batch. Obviously, two cylinders from the same batch would then show less variability than cylinders that come from different batches. Accordingly, we would be better off by assigning the two treatments to the two cylinders in each batch (block) and then analyzing their differences in compressive strength. This approach would eliminate the variability that is introduced by the differences among batches. Randomization would still play a role in the assignment of treatments to the experimental units within each batch. We would flip a coin to determine whether the first cylinder in each batch should be assigned to drying method A or drying method B.

6.1 Completely Randomized One-Factor Experiments

In this and the next two sections, we extend the analysis from two to k levels (treatments) of a single factor. Here and in Section 6.2, we analyze data from the *completely randomized experiment,* in which k treatments are assigned to the experimental units at random. In the analysis that corresponds to this design, we assume that the samples we observe from the k treatment groups are independent. In Section 6.3, we analyze data from the *randomized complete block experiment.* In this design, the experimental units are grouped into homogeneous blocks and all k treatments are randomly assigned to every block.

Let us assume that there are k different treatments (factor levels) under study. For the ith treatment, the response Y is a random variable that varies around an unknown treatment mean $\mu_i, i = 1, 2, \ldots, k$. We assume that, for each treatment group, the distribution of the response Y around its group mean is normal and that the variances are the same for all treatment groups. The latter assumption implies that the precision of the observations is the same in each group. An example with $k = 3$ treatment distributions is shown in Figure 6.1-1.

Now, suppose that we observe n_i independent realizations from the normal distribution that is associated with the ith treatment. Suppose further that these random samples from the different treatment groups are drawn independently. This is an extremely important assumption in the analysis of data from a completely randomized design; it means that the observations in one treatment group are independent of the observations in the other groups. As an example, Figure 6.1-1 plots hypothetical results of samples of respective sizes $n_1 = 5, n_2 = 6$, and $n_3 = 4$.

The preceding assumptions specify that the deviations from the respective group means are independent and normally distributed, with mean zero and constant variance—say, σ^2. Thus, we can write

$$Y_{ij} = \mu_i + \varepsilon_{ij}, \quad i = 1, 2, \ldots, k \quad \text{and} \quad j = 1, 2, \ldots, n_i,$$

where Y_{ij} is the value of the response variable in the jth trial for the ith treatment, μ_i is the ith treatment mean, and ε_{ij} are independent $N(0, \sigma^2)$ random variables. This equation describes a *statistical model* that links the observations and the parameters of the underlying populations.

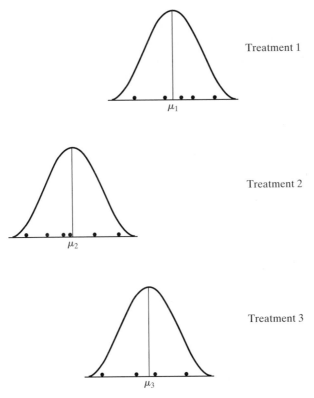

Figure 6.1-1 Graph of three normal distributions with equal variances and a set of possible realizations

The data analysis described in this section depends on these assumptions. Hence, we should always check whether the assumptions in this model are consistent with the nature of the experiment. The model is appropriate, for example, if we are studying, and have obtained several independent measurements of, the breaking strength of three different alloys. Moreover, the measurements must be approximately normally distributed with common variance. The model also is appropriate for a study of the effectiveness of four different dietary supplements for which we have obtained the percentage weight gains of N subjects randomly divided into four dietary groups. We must also assume that the gains in each group are at least approximately normally distributed with common variance. The model is *not* appropriate for an experiment in which each subject receives all four treatments, because then the assumption of independence among observations on the same subject would probably be violated. In the latter case, an analysis that uses the subject as a block would be more appropriate; such an analysis is considered in Section 6.3.

The experimenter wants to determine whether the means $\mu_1, \mu_2, \ldots, \mu_k$ of the k levels (treatments) of this one factor are equal. If they are, we say that there is no difference due to the different levels of that factor; that is, we say that there are no factor effects. If we find that the μ_i are different, we want to determine how they differ. For example, are they different because μ_1 is larger than $\mu_2 = \mu_3$, or is it because all three means are different? Those two issues—testing whether the means are the

Table 6.1-1 One-Factor Experiment

Treatment	Observations	Sample Mean
1	$Y_{11}, Y_{12}, \ldots, Y_{1n_1}$	$\overline{Y}_1 = \dfrac{1}{n_1} \sum\limits_{j=1}^{n_1} Y_{1j}$
.	.	.
.	.	.
.	.	.
i	$Y_{i1}, Y_{i2}, \ldots, Y_{in_i}$	$\overline{Y}_i = \dfrac{1}{n_i} \sum\limits_{j=1}^{n_i} Y_{ij}$
.	.	.
.	.	.
.	.	.
k	$Y_{k1}, Y_{k2}, \ldots, Y_{kn_k}$	$\overline{Y}_k = \dfrac{1}{n_k} \sum\limits_{j=1}^{n_k} Y_{kj}$

same and, if they are not, follow-up testing to determine how they are different—are discussed in this and the next section.

The data from comparative one-factor experiments can be arranged as in Table 6.1-1. The unknown treatment means $\mu_1, \mu_2, \ldots, \mu_k$ are estimated by the sample means

$$\overline{Y}_i = \frac{1}{n_i} \sum_{j=1}^{n_i} Y_{ij}, \quad i = 1, 2, \ldots, k.$$

Example 6.1-1   A civil engineer wishes to compare the strength properties of three different types of beams. Type A is made of steel, while types B and C are made of two different and more expensive alloys. The engineer measures the strength of a beam by setting it in a horizontal position, supported only on each end, applying a force of 3,000 pounds at the center, and then measuring the deflection. Measured in units of $1/1{,}000$ inch, for respective samples of sizes $n_1 = 8$, $n_2 = 6$, and $n_3 = 6$, the deflection is given in Table 6.1-2. The engineer was assured that these beams were representative random samples from the ones that were kept in inventory, not just the first six or eight beams that could be reached most easily. The engineer also was concerned that the measurement procedures might influence the comparison. To make sure that the order in which the experiments were carried out did not affect the comparison, she randomized the order by

Table 6.1-2 Deflections of Three Types of Beams

Type	Observations	$\bar{y}_i = \dfrac{1}{n_i}\sum\limits_{j=1}^{n_i} y_{ij}$	$\sum\limits_{j=1}^{n_i}(y_{ij} - \bar{y}_i)^2$
A (steel)	82 86 79 83 85 84 86 87	84	48
B	74 82 78 75 76 77	77	40
C	79 79 77 78 82 79	79	14

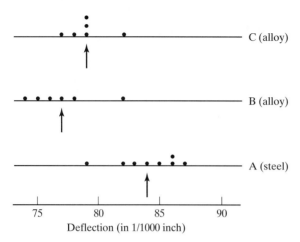

Figure 6.1-2 Dot diagrams of the observations in three groups. Group means are indicated by arrows

assigning a different number from 1 to 20 to each beam, drawing numbers without replacement, and performing the measurements in that particular (random) order. The sample means $\overline{Y}_i$ and the sums of squared deviations of the observations from their respective sample means, $\sum_{j=1}^{n_i}(Y_{ij} - \overline{Y}_i)^2$, $i = 1, 2, 3$, also are given in Table 6.1-2. Dot diagrams of the observations in each treatment group are shown in Figure 6.1-2.

It is always useful to display data graphically, either by separate dot diagrams or by box plots if there are many observations in each treatment group. (See the graphs in Section 1.4.) Dot diagrams of the observations by treatment groups show (1) the variability of the observations from their group average (the *within-treatment* variability) and (2) the variability between the group averages (the *between-treatment* variability). If the between-treatment variability is larger than what could be expected from the variation that occurs within the treatments, we would question whether the population means $\mu_1, \mu_2, \ldots, \mu_k$ are the same. Note that this graphical display of the data also gives us a visual method for checking whether the variabilities in the various treatment groups are roughly the same.

In the case of a single random sample, we used the sum of squares $\sum_{j=1}^{n_i}(Y_{ij} - \overline{Y}_i)^2$ and the sample variance

$$S_i^2 = \frac{1}{n_i - 1}\sum_{j=1}^{n_i}(Y_{ij} - \overline{Y}_i)^2$$

to describe the internal variability in the data. For a single random sample from a normal distribution, we know that $\sum_{j=1}^{n_i}(Y_{ij} - \overline{Y}_i)^2/\sigma^2$ has a chi-square distribution with $n_i - 1$ degrees of freedom. Since we have k independent samples, we know from Section 4.3 that the sum of the k independent chi-square random variables,

$$\frac{\sum_{i=1}^{k}\sum_{j=1}^{n_i}(Y_{ij} - \overline{Y}_i)^2}{\sigma^2},$$

is chi-square with degrees of freedom equal to $\sum_{i=1}^{k}(n_i - 1) = N - k$, where $N = \sum_{i=1}^{k}n_i$ is the total number of observations. The numerator of this chi-square variable is the pooled sum of squares

$$\mathrm{SS}_{\mathrm{Error}} = \sum_{i=1}^{k} \sum_{j=1}^{n_i} (Y_{ij} - \overline{Y}_i)^2$$

and is called the *within-treatment sum of squares* or the *error sum of squares*. This quantity, divided by the degrees of freedom, $N - k$, is known as the *mean square error* (or *mean square due to error*):

$$\mathrm{MS}_{\mathrm{Error}} = \frac{\mathrm{SS}_{\mathrm{Error}}}{N - k} = \frac{\sum_{i=1}^{k} \sum_{j=1}^{n_i} (Y_{ij} - \overline{Y}_i)^2}{N - k}.$$

$\mathrm{MS}_{\mathrm{Error}}$ is a weighted average of the sample variances S_i^2; that is, $\mathrm{MS}_{\mathrm{Error}} = \sum_{i=1}^{k} w_i S_i^2$, where $w_i = (n_i - 1)/(N - k)$. It is an unbiased estimator of the common variance σ^2, which measures the internal variability in the data. Note that $\mathrm{MS}_{\mathrm{Error}}$ is an unbiased estimate of σ^2 regardless of whether the treatment means $\mu_1, \mu_2, \ldots, \mu_k$ are or are not the same.

In this example, we find that the within-treatment sum of squares is

$$\mathrm{SS}_{\mathrm{Error}} = 48 + 40 + 14 = 102$$

and its degrees of freedom are given by $(8 - 1) + (6 - 1) + (6 - 1) = 17$. A pooled estimate of σ^2 is

$$\mathrm{MS}_{\mathrm{Error}} = \frac{102}{17} = 6.0.$$

There is another type of variability that can be calculated, namely, the variation of the treatment averages around the grand mean. The grand mean

$$\overline{Y} = \frac{1}{N} \sum_{i=1}^{k} \sum_{j=1}^{n_i} Y_{ij}$$

is the sum of all observations, divided by the total number of observations. If each mean $\overline{Y}_i$ is given its sample size n_i as its weight in computing a sum of squares, that sum of squares,

$$\mathrm{SS}_{\mathrm{Treatment}} = n_1(\overline{Y}_1 - \overline{Y})^2 + n_2(\overline{Y}_2 - \overline{Y})^2 + \cdots + n_k(\overline{Y}_k - \overline{Y})^2$$

$$= \sum_{i=1}^{k} n_i(\overline{Y}_i - \overline{Y})^2,$$

is called the *between-treatment sum of squares* or simply the *treatment sum of squares*. $\mathrm{SS}_{\mathrm{Treatment}}$ measures the variability among the treatment means and is, of course, zero if the treatment means are equal. There are k components in this sum of squares; however, there is also one restriction, namely, that

$$\sum_{i=1}^{k} n_i(\overline{Y}_i - \overline{Y}) = \sum_{i=1}^{k} n_i \overline{Y}_i - \overline{Y} \sum_{i=1}^{k} n_i = 0.$$

Thus, intuitively, this sum of squares has $k - 1$ degrees of freedom, and we call the ratio

$$\mathrm{MS}_{\mathrm{Treatment}} = \frac{\mathrm{SS}_{\mathrm{Treatment}}}{k - 1}$$

the *mean square due to treatment*.

In this example, we find that the treatment sum of squares is

$$SS_{\text{Treatment}} = 8(84 - 80.4)^2 + 6(77 - 80.4)^2 + 6(79 - 80.4)^2 = 184.8.$$

Because there are $k - 1 = 3 - 1 = 2$ degrees of freedom, the mean square due to treatment is

$$MS_{\text{Treatment}} = \frac{184.8}{2} = 92.4. \qquad \blacksquare$$

6.1-1 Analysis-of-Variance Table

We have defined two types of variation: within- and between-treatment sums of squares. We now show that we can write the total sum of squares around the grand mean—abbreviated SSTO—as the sum of these two sums of squares:

$$
\begin{aligned}
SSTO &= \sum_{i=1}^{k}\sum_{j=1}^{n_i}(Y_{ij} - \overline{Y})^2 = \sum_{i=1}^{k}\sum_{j=1}^{n_i}[(Y_{ij} - \overline{Y}_i) + (\overline{Y}_i - \overline{Y})]^2 \\
&= \sum_{i=1}^{k}\sum_{j=1}^{n_i}(\overline{Y}_i - \overline{Y})^2 + \sum_{i=1}^{k}\sum_{j=1}^{n_i}(Y_{ij} - \overline{Y}_i)^2 + 2\sum_{i=1}^{k}\sum_{j=1}^{n_i}(\overline{Y}_i - \overline{Y})(Y_{ij} - \overline{Y}_i) \\
&= \sum_{i=1}^{k}n_i(\overline{Y}_i - \overline{Y})^2 + \sum_{i=1}^{k}\sum_{j=1}^{n_i}(Y_{ij} - \overline{Y}_i)^2 \\
&= SS_{\text{Treatment}} + SS_{\text{Error}}.
\end{aligned}
$$

Note that the right-most term in line 2 reduces to zero because

$$\sum_{i=1}^{k}\sum_{j=1}^{n_i}(\overline{Y}_i - \overline{Y})(Y_{ij} - \overline{Y}_i) = \sum_{i=1}^{k}(\overline{Y}_i - \overline{Y})\sum_{j=1}^{n_i}(Y_{ij} - \overline{Y}_i) = 0,$$

given that

$$\sum_{j=1}^{n_i}(Y_{ij} - \overline{Y}_i) = 0, \quad i = 1, 2, \ldots, k.$$

This sum-of-squares decomposition is frequently recorded in a table called an *analysis-of-variance table,* or *ANOVA table,* because it concerns various sums of squares or estimates of variances. (See Table 6.1-3.) The first column identifies the

---

Table 6.1-3 ANOVA Table for the Completely Randomized Design

Source	SS	df	MS	F
Treatment	$\sum_{i=1}^{k}n_i(\overline{Y}_i - \overline{Y})^2$	$k - 1$	$MS_{\text{Treatment}} = \dfrac{SS_{\text{Treatment}}}{k-1}$	$F = \dfrac{MS_{\text{Treatment}}}{MS_{\text{Error}}}$
Error	$\sum_{i=1}^{k}\sum_{j=1}^{n_i}(Y_{ij} - \overline{Y}_i)^2$	$N - k$	$MS_{\text{Error}} = \dfrac{SS_{\text{Error}}}{N-k}$	
Total	$\sum_{i=1}^{k}\sum_{j=1}^{n_i}(Y_{ij} - \overline{Y})^2$	$N - 1$		

sources of variation, the second and third columns the corresponding sums of squares and degrees of freedom; the fourth column gives the mean squares, and the last column contains the F-ratio, $F = \text{MS}_{\text{Treatment}}/\text{MS}_{\text{Error}}$ which will be explained later.

The sum-of-squares contributions are usually calculated by computer programs, and we explain how to do this in Section 6.2-6. It is not very likely that you will ever have to carry out the calculations from scratch. But if you have to calculate them by hand, a more convenient and also more accurate way (involving less rounding) is to calculate them from the formulas

$$\text{SSTO} = \sum_{i=1}^{k}\sum_{j=1}^{n_i}(Y_{ij} - \overline{Y})^2 = \sum_{i=1}^{k}\sum_{j=1}^{n_i}Y_{ij}^2 - \frac{\left(\sum_{i=1}^{k}\sum_{j=1}^{n_i}Y_{ij}\right)^2}{N}$$

and

$$\text{SS}_{\text{Treatment}} = \sum_{i=1}^{k}n_i(\overline{Y}_i - \overline{Y})^2 = \sum_{i=1}^{k}\frac{\left(\sum_{j=1}^{n_i}Y_{ij}\right)^2}{n_i} - \frac{\left(\sum_{i=1}^{k}\sum_{j=1}^{n_i}Y_{ij}\right)^2}{N}.$$

Note that one of the three sums of squares in Table 6.1-3 can always be calculated from the other two.

Example 6.1-2 We use the data on the deflection of steel bars. We find that $k = 3, n_1 = 8, n_2 = 6,$ and $n_3 = 6$; thus, $N = 20$. In addition,

$$\sum_{i=1}^{k}\sum_{j=1}^{n_i}Y_{ij} = 82 + 86 + \cdots + 82 + 79 = 1{,}608,$$

$$\sum_{i=1}^{k}\sum_{j=1}^{n_i}Y_{ij}^2 = 82^2 + 86^2 + \cdots + 82^2 + 79^2 = 129{,}570,$$

$$\sum_{j=1}^{n_1}Y_{1j} = 82 + 86 + \cdots + 87 = 672,$$

$$\sum_{j=1}^{n_2}Y_{2j} = 74 + 82 + \cdots + 77 = 462,$$

and

$$\sum_{j=1}^{n_3}Y_{3j} = 79 + 79 + \cdots + 79 = 474.$$

Hence,

$$\text{SSTO} = 129{,}570 - \frac{(1{,}608)^2}{20} = 286.8,$$

$$\text{SS}_{\text{Treatment}} = \frac{(672)^2}{8} + \frac{(462)^2}{6} + \frac{(474)^2}{6} - \frac{(1608)^2}{20} = 184.8,$$

and

$$\text{SS}_{\text{Error}} = 286.8 - 184.8 = 102.0.$$

The ANOVA table is as follows:

Source	SS	df	MS	F
Treatment	184.8	2	92.4	15.4
Error	102.0	17	6.0	
Total	286.8	19		

Check that your computer software will give you this table. Although it may be instructive to calculate the sum of squares once by hand (so that you understand where the entries in the analysis-of-variance table come from), we expect you to use computer software for such calculations. ∎

6.1-2 F-Test for Treatment Effects

We want to test the null hypothesis that all treatment means are the same, $H_0: \mu_1 = \mu_2 = \ldots = \mu_k$, against the alternative hypothesis that there is at least one mean that is different, H_1 : not all μ_i are the same. To do so, we need to construct a test statistic and obtain its distribution under the null hypothesis. The F-ratio, $F = \text{MS}_{\text{Treatment}}/\text{MS}_{\text{Error}}$, in the last column of the ANOVA table is the appropriate statistic. On the one hand, the numerator, and thus the ratio itself, is "small" if H_0 is true, because in that case the sample treatment means are about the same. On the other hand, the numerator and the ratio are "large" if H_1 is true, because in that case the sample means are different. A more formal justification follows.

Earlier, we saw that the within-treatment mean square MS_{Error} is an estimate of σ^2. This is true regardless of whether the treatment means $\mu_1, \mu_2, \ldots, \mu_k$ are or are not the same; that is, $E(\text{MS}_{\text{Error}}) = \sigma^2$. Also, the between-treatment mean square $\text{MS}_{\text{Treatment}}$ provides an estimate of σ^2, but only if the treatment means are the same. If they are different, $\text{MS}_{\text{Treatment}}$ tends to be larger than σ^2. In general, it can be shown (see Exercises 6.1-7 and 6.1-8) that

$$E(\text{MS}_{\text{Treatment}}) = \sigma^2 + \frac{\sum_{i=1}^{k} n_i \tau_i^2}{k-1},$$

where $\tau_i = \mu_i - \mu$ is the deviation of the treatment mean μ_i from the overall (weighted) mean $\mu = (\sum_{i=1}^{k} n_i \mu_i)/N$. The deviations τ_i are also called the treatment effects, and it is true that $\sum_{i=1}^{k} n_i \tau_i = 0$.

If $\mu_1 = \mu_2 = \ldots = \mu_k$, then $\tau_i = 0$ for all i, and $E(\text{MS}_{\text{Treatment}}) = \sigma^2$. In this case, the numerator and the denominator of the F-ratio estimate the same quantity. Moreover, it can be shown that, under H_0, $\text{SS}_{\text{Treatment}}/\sigma^2$ and $\text{SS}_{\text{Error}}/\sigma^2$ are independent chi-square random variables with $k - 1$ and $N - k$ degrees of freedom and thus that

$$F = \frac{(\text{SS}_{\text{Treatment}}/\sigma^2)/(k-1)}{(\text{SS}_{\text{Error}}/\sigma^2)/(N-k)} = \frac{\text{MS}_{\text{Treatment}}}{\text{MS}_{\text{Error}}}$$

has an F distribution with $k - 1$ and $N - k$ degrees of freedom. It is easy to remember the degrees of freedom, as they correspond to the MS entries in the ANOVA table. If some of the means $\mu_1, \mu_2, \ldots, \mu_k$ are different, the numerator in the F-ratio—and thus the F-ratio itself—tends to be larger. This fact leads to the following test procedure: If $F \geq F(\alpha; k - 1, N - k)$, accept H_1. In such a case, we say that the F-statistic is significant and that there are statistically significant differences among the group means. If, however, the F-ratio is smaller than the upper 100α percentage point $F(\alpha; k - 1, N - k)$, then there is not enough evidence in the data to reject $H_0: \mu_1 = \mu_2 = \ldots = \mu_k$.

An identical test procedure uses the *probability value* (or *p value*), which is calculated by most computer programs. Here, we define it as

$$p \text{ value } = P[F(k - 1, N - k) \geq F],$$

where F is the observed F-ratio $MS_{Treatment}/MS_{Error}$; that is, the p value is the probability of obtaining a realization from the $F(k - 1, N - k)$ distribution that is at least as large as the observed F-ratio. If this p value is less than or equal to α, we accept H_1; otherwise, we accept H_0 until further data are obtained.

Example 6.1-3   In Example 6.1-2, the F-ratio is $92.4/6.0 = 15.4$. The critical value with $\alpha = 0.05$ is $F(0.05; 2, 17) = 3.59$; for $\alpha = 0.01$, it is $F(0.01; 2, 17) = 6.11$. Thus, there is very strong evidence that the treatment means are different, and we reject $H_0: \mu_1 = \mu_2 = \mu_3$ at both significance levels $\alpha = 0.05$ and $\alpha = 0.01$. In the next section, we indicate how to follow up on this test. ∎

6.1-3 Graphical Comparison of k Samples

The numerical analysis provided in the ANOVA table should always be supplemented by a graphical display of the data. If there are only a few observations in each group, one can construct k separate dot diagrams and put them side by side or, as we have done in Figure 6.1-2, put one below the other. If there are many observations in each group, one should construct box plots for each group and arrange them side by side on the same scale. This was done in the graphs of Section 1.4. If there is considerable overlap among the boxes of the various groups, the F-statistic from the ANOVA table tends to be insignificant. The box plots also give us information on the variability of the observations within each group, tell us whether the observations need to be transformed, alert us to outliers, and give us a quick summary of the main features of the data. Hence, box plots should be part of any data analysis.

Exercises 6.1

6.1-1   In a one-factor experiment with four treatments, the following results were obtained:

Treatment	n_i	$\bar{y}_i$	s_i^2
1	20	40.2	900
2	20	38.6	800
3	18	43.5	960
4	18	50.0	720

Construct an ANOVA table and test whether there are differences among the treatment means. (Use $\alpha = 0.05$.)

*6.1-2   The female cuckoo lays her eggs in the nests of other species of birds. The "foster parents" are usually deceived, probably because of the similarity in the sizes of the eggs. O. H. Latter investigated this possible explanation and measured the lengths (in millimeters) of cuckoo eggs that were found in the nests of the following three species:

Hedge sparrow:	22.0	23.9	20.9	23.8	25.0
	24.0	21.7	23.8	22.8	23.1
	23.1	23.5	23.0	23.0	
Robin:	21.8	23.0	23.3	22.4	23.0
	23.0	23.0	22.4	23.9	22.3
	22.0	22.6	22.0	22.1	21.1
	23.0				
Wren:	19.8	22.1	21.5	20.9	22.0
	21.0	22.3	21.0	20.3	20.9
	22.0	20.0	20.8	21.2	21.0

Display the information in appropriate graphs. Construct an ANOVA table. Is there a difference in the mean lengths? Investigate.

[O. H. Latter, "The Cuckoo's Egg," *Biometrika*, 1: 164–176 (1901).]

*6.1-3 D. C. Montgomery discusses a case in which the tensile strength of a synthetic fiber used to make cloth for men's shirts is of interest to a manufacturer. It is suspected that the strength is affected by the percentage of cotton in the fiber. Five levels of cotton percentage are considered, and five observations are taken at each level. The 25 experiments are run in random order. The results are as follows:

% Cotton	Tensile Strength (pounds per square inch)				
15	7	7	15	11	9
20	12	17	12	18	18
25	14	18	18	19	19
30	19	25	22	19	23
35	7	10	11	15	11

*(a) Are there differences in the mean breaking strengths due to the percentage of cotton used? Make comparative dot diagrams and construct an ANOVA table.

(b) Plot the treatment averages against the percentage of cotton and interpret your findings.

[D. C. Montgomery, *Design and Analysis of Experiments*, 2d ed. (New York: Wiley, 1984), p. 51.]

*6.1-4 An operator of a feedlot wants to compare the effectiveness of three different cattle feed supplements. He selects a random sample of 15 one-year old heifers from his lot of over 1,000 and divides them into three groups at random. Each group gets a different feed supplement. Upon noting that one heifer in group *A* was lost due to an accident, the operator records the gains in weight over a 6-month period as follows:

Group	Weight Gain (pounds)				
A	500	650	530	680	
B	700	620	780	830	860
C	500	520	400	580	410

*(a) Are there any differences in the mean weight gains due to the three different feed supplements?

(b) Suppose you could start the experiment over. What improvements in its design might help make the conclusions more precise? For example, would a blocking arrangement on initial weight be helpful? [No computation is required here; the analysis will be discussed in Section 6.3.]

*6.1-5 Three workers with different experience manufacture wheels for a magnet brake. Worker A has four years of experience, worker B has seven years, and worker C has one year. The company is concerned about its product's quality, which is measured by the difference between the specified diameter and the actual diameter of the brake wheel. On a given day, the supervisor selects nine brake wheels at random from the output of each worker. The following table lists the deviations between the specified and actual diameters:

Worker				Precision (1/100 inch)					
A	2	3	2.3	3.5	3	2	4	4.5	3
B	1.5	3	4.5	3	3	2	2.5	1	2
C	2.5	3	2	2.5	1.5	2.5	2.5	3	3.5

Are there statistically significant differences in the quality of the wheels produced by the three different workers?

6.1-6 In an experiment to compare different methods of teaching arithmetic, 45 students were randomly divided into five equal-sized groups. Two groups, A and B, were taught by the current method, while the other three were taught by one of three new methods. At the end, each student took a standardized test, with the following results:

Group A:	17	14	24	20	24	23	16	15	14
Group B:	21	23	13	19	13	19	20	21	16
Group C:	28	30	29	24	27	30	28	28	23
Group D:	19	28	26	26	19	24	24	23	22
Group E:	21	14	13	19	15	15	10	18	20

Analyze the data. Note that the first step of any data analysis should be a summary and a graphical representation of the information. Calculate the group means, medians, quartiles, interquartile ranges, and standard deviations, and then construct comparative box-and-whisker displays. What can you conclude from these displays?

The students are allocated to the five groups at random. Thus, it is appropriate to conduct statistical inferences, such as tests of hypotheses. Go ahead and make your test. Discuss whether the conclusion from the ANOVA table gives you more, less, or different information than the comparative box-and-whisker display affords.

Is it always true that a statistically significant difference is of practical importance? Discuss.

[G. B. Wetherill, *Elementary Statistical Methods,* 3d ed. (London: Methuen, 1982).]

6.1-7 Show that the model $Y_{ij} = \mu_i + \varepsilon_{ij}$, where $i = 1, 2, \ldots, k$ and $j = 1, 2, \ldots, n_i$, can also be written as $Y_{ij} = \mu + \tau_i + \varepsilon_{ij}$, where $\mu = (\sum_{i=1}^{k} n_i \mu_i)/N$ is a weighted average of the treatment means, $\tau_i = \mu_i - \mu$ are the treatment effects, and $\sum_{i=1}^{k} n_i \tau_i = 0$.

6.1-8 Using the representation in Exercise 6.1-7, show that

$$E(MS_{\text{Treatment}}) = \sigma^2 + \frac{\sum_{i=1}^{k} n_i \tau_i^2}{k-1}.$$

The following hints should help the mathematically interested reader derive this result:

(a) Use the definition $SS_{\text{Treatment}}$ to show that

$$E(SS_{\text{Treatment}}) = \sum_{i=1}^{k} n_i E[(\overline{Y}_i - \overline{Y})^2].$$

(b) Show that

$$\overline{Y}_i - \overline{Y} = \sum_{j=1}^{n_i}\left(\frac{1}{n_i} - \frac{1}{N}\right)Y_{ij} + \sum_{h \neq i}\sum_{j=1}^{n_h}\left(-\frac{1}{N}\right)Y_{hj}.$$

(c) Show that $E(\overline{Y}_i - \overline{Y}) = \tau_i$ and

$$\text{var}(\overline{Y}_i - \overline{Y}) = \sigma^2\left[\frac{1}{N^2}(N - n_i) + n_i\left(\frac{1}{n_i} - \frac{1}{N}\right)^2\right] = \frac{(N - n_i)\sigma^2}{Nn}.$$

(d) Use the fact that, for any random variable W with finite variance, $E(W^2) = \text{var}(W) + [E(W)]^2$ to obtain $E[(\overline{Y}_i - \overline{Y})^2]$. Substitute this result into part (a) to obtain

$$E(MS_{\text{Treatment}}) = \frac{E(SS_{\text{Treatment}})}{k-1}.$$

6.1-9 Prove that $SS_{\text{Treatment}} = \sum_{i=1}^{k} \frac{\left(\sum_{j=1}^{n_i} Y_{ij}\right)^2}{n_i} - \frac{\left(\sum_{i=1}^{k}\sum_{j=1}^{n_i} Y_{ij}\right)^2}{N}.$

6.1-10 Suppose the following table lists the GPAs of random samples of freshmen, sophomores, and juniors at your university:

Freshmen	Sophomores	Juniors
3.58	2.50	3.64
1.94	2.95	2.01
3.90	2.16	1.47
3.75	3.34	3.16
3.57	1.53	3.65
3.44	3.18	3.78
1.76	3.23	3.98

Test the null hypothesis that freshmen, sophomores, and juniors all have the same mean GPAs.

6.1-11 Test whether there are significant differences in the monthly salaries (in $100) of managers in three different industries, as follows:

Industry A:	92.4,	87.7,	86.6,	92.2,	98.2,	87.2
Industry B:	72.4,	84.3,	89.0,	85.5,	95.3	
Industry C:	96.1,	91.5,	91.0,	100.8,	95.2,	107.4

Interpret your results, both graphically and numerically, carefully discussing the assumptions you have made.

6.2 Other Inferences in One-Factor Experiments

Once we have concluded that the treatment means are not the same by rejecting the hypothesis that they are equal, we must investigate how they differ. For instance, in Example 6.1-1, are the means different because the deflection of steel beams (group A) is larger than the deflection of beams made of alloys (B, C) that produce a similar deflection, or are there differences among all three means?

6.2-1 Reference Distribution for Treatment Averages

In *Statistics for Experimenters* (New York: Wiley, 1978, 2005), G. E. P. Box, W. G. Hunter, and J. S. Hunter describe a graphical procedure that is useful in visually assessing the differences among k treatments. Their procedure involves plots of the treatment means together with that of a certain reference distribution. Let us assume that the sample sizes are the same; that is, $n_1 = n_2 = \ldots = n_k = n$. (If the sample sizes are almost the same, we can set $n = \sum_{i=1}^{k} n_i / k$.) Now, if the k treatments all had the same mean μ, then the sampling distribution of the treatment means $\overline{Y}_i$ would be normal with mean μ and variance σ^2/n. Replacing σ^2 by its estimate, MS_{Error}, from the ANOVA table, we find that $(\overline{Y}_i - \mu)/\sqrt{\text{MS}_{\text{Error}}/n}$ has a t-distribution with $N - k$ degrees of freedom. Thus, we can approximate the distribution of the treatment means by a scaled t-distribution that is centered around μ. The latter distribution is called "scaled" because we have to stretch the axes by the estimate of the standard deviation, $\sqrt{\text{MS}_{\text{Error}}/n}$. Box, Hunter, and Hunter refer to this distribution as the approximate *reference distribution* of the treatment averages $\overline{Y}_i$. Given the appropriate tables for the probability density function of a t-distribution, we could graph this density directly. However, even in the absence of such tables (note that Table C.6 gives only selected percentage points, not the p.d.f.), we can proceed as follows:

First we determine the upper $\alpha = 0.05$ percentage point of the t-distribution with $N - k$ degrees of freedom. Then the 5th and 95th percentiles of the reference distribution for $\overline{Y}_i$ are given respectively by $\mu - t(0.05; N - k)\sqrt{\text{MS}_{\text{Error}}/n}$ and $\mu + t(0.05; N - k)\sqrt{\text{MS}_{\text{Error}}/n}$. (For the moment, assume that μ is an arbitrary constant.) In Example 6.1-1, $N - k = 17$, $t(0.05; 17) = 1.74$, $n = 20/3 = 6.67$, and $\text{MS}_{\text{Error}} = 6.0$; thus, the percentiles are $\mu \pm 1.74\sqrt{6.0/6.67}$, or $\mu \pm 1.65$, and the probability that the treatment average $\overline{Y}_i$ falls outside these limits should be about 10 percent. Of course, the density of the t-distribution reaches its maximum at μ, and the ordinate of the density at μ is approximately four to five times larger than the ordinate at the 5th and 95th percentiles. For example, the ratio is 4.8 for a t-distribution with 10 degrees of freedom, 4.3 for 20 degrees of freedom, 4.1 for 30 degrees of freedom, and 3.9 for the normal distribution. We can use this information to construct a reference distribution. In our example in Figure 6.2-1, we have arbitrarily centered the distribution at $\mu = 78$. If the ordinates at the 5th and 95th percentiles are one

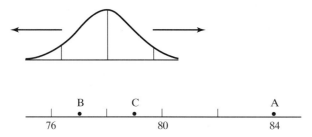

Figure 6.2-1 Three treatment averages and their reference distribution

unit, the ordinate of the density at $\mu = 78$ is approximately 4.3, because $N - k = 17$ is reasonably close to 20. Given that the shape of the p.d.f. of the t-distribution is similar to that of the normal distribution, but with thicker tails, it is reasonable to draw an approximate bell-shaped curve through these three points to approximate that t-density.

Imagine sliding the reference distribution along the horizontal axis. Note that in our example there is no place to center the distribution so that the three treatment averages appear to be typical randomly selected realizations from this reference distribution. Figure 6.2-1 provides evidence that the three treatment means are not the same. This figure is the graphical equivalent of what was formally shown by the significant F-statistic. In addition, the reference distribution shows that although μ_A (the mean deflection of steel beams) is probably larger than μ_B and μ_C, the data do not permit us to distinguish between the deflection means μ_B and μ_C of the beams made of alloys.

6.2-2 Confidence Intervals for a Particular Difference

Let us assume that we want to obtain a confidence interval for the difference of the means of two particular preselected treatments—say, treatments r and s. An estimate of $\mu_r - \mu_s$ is given by the difference of the respective sample means $\overline{Y}_r - \overline{Y}_s$. The variance of this difference of independent averages is

$$\text{var}(\overline{Y}_r - \overline{Y}_s) = \text{var}(\overline{Y}_r) + \text{var}(\overline{Y}_s) = \sigma^2\left(\frac{1}{n_r} + \frac{1}{n_s}\right),$$

which is estimated by

$$\text{MS}_{\text{Error}}\left(\frac{1}{n_r} + \frac{1}{n_s}\right).$$

Thus, a $100(1 - \alpha)$ percent confidence interval for $\mu_r - \mu_s$ is given by

$$\overline{Y}_r - \overline{Y}_s \pm t(\alpha/2; N - k)\sqrt{\text{MS}_{\text{Error}}}\sqrt{\frac{1}{n_r} + \frac{1}{n_s}},$$

where $\nu = N - k$ are the degrees of freedom that are associated with MS_{Error}.

Example 6.2-1   Assume that, prior to running the experiment, the engineer has decided to compare the deflection means of the two alloy beams (B, C). A 95 percent confidence interval for $\mu_C - \mu_B$ is given by

$$(79 - 77) \pm (2.11)\sqrt{6.0}\sqrt{\frac{1}{6} + \frac{1}{6}},$$

or $(-0.98 \le \mu_C - \mu_B \le 4.98)$. Because this interval includes zero, we cannot conclude from the limited amount of data we have that the deflection means of the two alloy beams are different.   ∎

6.2-3 Tukey's Multiple-Comparison Procedure

The confidence interval derived in Section 6.2-2 controls the significance level of one particular preselected comparison. However, if there are k treatments, we can make many paired comparisons; as a matter of fact, there are $\binom{k}{2} = k(k - 1)/2$ of them. If α is the probability of error for one particular comparison, then the probability of making at least one error in $k(k - 1)/2$ comparisons will be much larger than α. For example, assume that there are only two comparisons, each at level α, and that the error in the first comparison is independent of the one in the second. Then the probability of making at least one error in the two comparisons combined (which is the complement of making no error in either comparison) is given by $1 - (1 - \alpha)^2 = \alpha(2 - \alpha)$, which, for small α, is almost twice as large as α. Therefore, if we want a $100(1 - \alpha)$ percent confidence statement for *all possible paired comparisons*, we must make the intervals wider than the ones for individual differences of two means.

That is exactly what is done in Tukey's multiple-comparison procedure. There, the confidence intervals for $\mu_r - \mu_s$ are calculated from

$$\overline{Y}_r - \overline{Y}_s \pm \frac{q(\alpha; k, \nu)}{\sqrt{2}}\sqrt{MS_{Error}}\sqrt{\frac{1}{n_r} + \frac{1}{n_s}},$$

where $q(\alpha; k, \nu)$ is the upper 100α percentage point [i.e., the $100(1 - \alpha)$ percentile] of the *Studentized range* distribution for comparing k means with $\nu = N - k$ degrees of freedom for the mean square error. (A table of upper percentage points of the Studentized range distribution is given in Table C.8.) Because, for $k > 2$, $q(\alpha; k, \nu)/\sqrt{2} > t(\alpha/2; \nu)$, these intervals will always be wider than the confidence intervals that use the t-distribution. (Note that, strictly speaking, this procedure is valid only for equal sample sizes. However, it is a good approximation if the sample sizes n_i are not too different.)

Example 6.2-2   In Example 6.1-1, we studied the differences among the deflection means of three types of beams. In this case, there are a total of $(3)(2)/2 = 3$ pairwise comparisons: those associated with $\mu_A - \mu_B$, those with $\mu_A - \mu_C$, and those with $\mu_C - \mu_B$. Given that $q(0.05; 3, 17) = 3.62$ and $[q(0.05; 3, 17)/\sqrt{2}]\sqrt{MS_{Error}} = 2.56\sqrt{6.0} = 6.27$, Tukey's multiple-comparison intervals are

$$(84 - 77) \pm 6.27\sqrt{\frac{1}{8} + \frac{1}{6}} \quad \text{or} \quad (3.61 \le \mu_A - \mu_B \le 10.39),$$

$$(84 - 79) \pm 6.27\sqrt{\frac{1}{8} + \frac{1}{6}} \quad \text{or} \quad (1.61 \le \mu_A - \mu_C \le 8.39),$$

and

$$(79 - 77) \pm 6.27\sqrt{\frac{1}{6} + \frac{1}{6}} \quad \text{or} \quad (-1.62 \le \mu_C - \mu_B \le 5.62).$$

From these multiple comparisons, we conclude that μ_A is probably larger than μ_B and μ_C, but that μ_B and μ_C are not much different from each other. This is the same conclusion that we reached when we used the technique involving the reference distribution.   ∎

6.2-4 Model Checking

The single-factor ANOVA model assumes that the observations are independent and normally distributed with the same variance in each treatment group. The diagnostic checking procedures focus on the residuals, $e_{ij} = Y_{ij} - \overline{Y}_i$; these are differences between the observations and their respective group averages. We usually construct a histogram (or a dot diagram) of the residuals to see if the normal assumption is realistic. [See Figure 6.2-2(a).] One could also make a normal probability plot of the residuals to check whether the plotted observations fall more or less on a straight line. (See Section 3.6.) To check whether the variances in the treatment groups are about the same, we compare the dot diagrams of the residuals from each group in Figure 6.2-2(b). Also, we plot the residuals against their corresponding group averages $\overline{Y}_i$ in Figure 6.2-2(c). The variability should not depend on the level of $\overline{Y}_i$. If it does, this is often an indication that we should consider a transformation of the data. For example, if the variability grows with the level, a transformation such as $\log Y$ or $\sqrt{Y}$ might be appropriate. To check whether the observations are independent, we can plot the residuals against the time (run) order in which the experiment was performed. Patterns in the residuals, such as runs of positive or negative residuals or patterns with alternating signs, indicate that errors from measurements taken closely together in time are not independent. Such patterns, which occur if there is a carryover effect from one experiment to the next, if there is deterioration in the measurement instrument, or if the operator becomes fatigued, suggest that time (order) is an important factor and that it should be included as a factor (blocking variable) in the experiment. Blocking of experiments is considered in Section 6.3, where it is illustrated that such blocking arrangements will increase the precision of the comparison.

6.2-5 The Random-Effects Model

The one-factor analysis-of-variance model can also be written as

$$Y_{ij} = \mu + \tau_i + \varepsilon_{ij}, \quad i = 1, 2, \ldots, k; \quad j = 1, 2, \ldots, n_i;$$

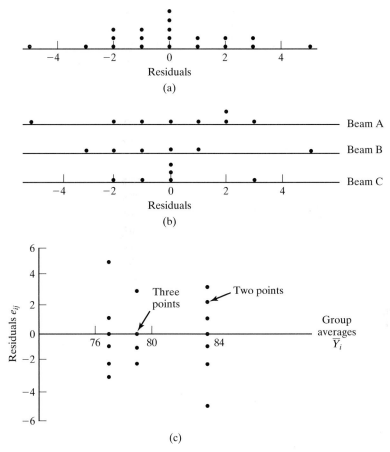

Figure 6.2-2 Various residual plots of the data in Example 6.1-1: (a) overall dot diagram of the residuals; (b) plot of residuals for each group separately; (c) plot of residuals against group averages

where $\mu = (\sum_{i=1}^{k} n_i \mu_i)/N$ is the overall mean and where $\tau_i = \mu_i - \mu$, for $i = 1, 2, \ldots, k$, are the treatment effects. (See Exercise 6.1-7.)

We can distinguish two different situations with respect to the treatment effects. In the first, the treatments are specifically chosen by the experimenter (e.g., the three different types of beams in Example 6.1-1) and the tests of the hypotheses apply only to these particular factor levels. The conclusions that are reached cannot be extended to other treatments that were not explicitly considered. In this particular model, the treatment effects are unknown constants, and the model is called the *fixed-effects model*; it is the one that we have considered so far in this chapter. We have estimated the unknown fixed treatment means μ_i by $\overline{Y}_i$ and the treatment effects τ_i by $\overline{Y}_i - \overline{Y}$. Also, in the derivation of the formula $E(\text{MS}_{\text{Treatment}}) = \sigma^2 + (\sum_{i=1}^{k} n_i \tau_i^2)/(k-1)$ in Exercise 6.1-8, it was assumed that the τ_i are fixed nonrandom constants.

In the second situation, the population treatment means μ_i, $i = 1, 2, \ldots, k$, can be thought of as a random sample of size k from a larger population of treatment means. As an example, suppose we investigate differences in yield that are due to various batches of raw material. In this situation, we would like to extend the conclusions from our sample of batches (treatments) to all batches in the population. Here, if the k batches are selected at random from many possible batches, the treatment effects τ_i are random variables; thus, this model is called the *random-effects model*.

In the random-effects model, the observation

$$Y_{ij} = \mu + \tau_i + \varepsilon_{ij}$$

is the sum of two random variables τ_i and ε_{ij} and the unknown μ. Let us assume that the τ_i and ε_{ij} are independently distributed random variables with respective normal distributions $N(0, \sigma_\tau^2)$ and $N(0, \sigma^2)$. This implies that $\text{var}(Y_{ij}) = \sigma_\tau^2 + \sigma^2$ is the sum of two variance components, one due to the treatment and the other due to the error.

In this situation, we are not interested in testing the equality of the k randomly chosen treatment means. Instead, we are interested in the variability of the treatment effects in general; that is, we want to test $H_0: \sigma_\tau^2 = 0$ against $H_1: \sigma_\tau^2 > 0$. Despite the fact that the random-effects model differs from the fixed-effects model, the analysis of variance for the single-factor experiment is conducted in identical fashion. The F-test discussed earlier is also appropriate in this case.

The only difference in the two models appears in the expected mean square due to treatment. In the random-effects model,

$$E(\text{MS}_{\text{Treatment}}) = \sigma^2 + \widetilde{n}\sigma_\tau^2, \quad \text{where } \widetilde{n} = \frac{1}{k-1}\left(N - \frac{\sum_{i=1}^{k} n_i^2}{\sum_{i=1}^{k} n_i}\right).$$

Here, $\widetilde{n} = n$ if each sample is of the same size n. The entries in the ANOVA table can be used to estimate the variance components. Since $E(\text{MS}_{\text{Error}}) = \sigma^2$, it follows that estimates of the variance components are given by

$$\hat{\sigma}^2 = \text{MS}_{\text{Error}}$$

and

$$\hat{\sigma}_\tau^2 = \frac{1}{\widetilde{n}}(\text{MS}_{\text{Treatment}} - \text{MS}_{\text{Error}}).$$

Occasionally, the latter expression may lead to a negative estimate, $\hat{\sigma}_\tau^2$. We take this as evidence that the true variance, σ_τ^2, is zero, because a variance must be non-negative.

Example 6.2-3 A certain dairy has received several complaints from its customers that its milk tends to spoil prematurely. Milk spoilage is measured by the number of bacteria that are found in milk after it has been stored for 12 days. The dairy receives the raw milk for processing in large trucked-in shipments from farmers in the surrounding area. Thus, there is a concern that the milk supply may have an important effect on the spoilage of the processed milk. To study a possible shipment (batch) effect, the company selects five shipments at random. After processing each batch, six cartons of milk are selected at random and are stored for 12 days. After that time, the bacteria counts (SPC, standard plate counts in units of 100) are made. From considerations

Table 6.2-1 Milk Shipment Bacteria Data							
Batch	Observations (Square Roots)						Sample Mean
1	24	15	21	27	33	23	23.83
2	14	7	12	17	14	16	13.33
3	11	9	7	13	12	18	11.67
4	7	7	4	7	12	18	9.17
5	19	24	19	15	10	20	17.83

that are explained in a later comment, a square-root transformation is made; the square roots of the bacteria counts are given in Table 6.2-1.

This is clearly an example of a random-effects model, because the batches are taken from a larger population. The analysis, however, is identical to the one in the fixed-effects model, and the ANOVA table is given in Table 6.2-2. We have used a computer program to calculate the entries in the table. If such a program is not available, one can easily calculate them with a calculator. All sample sizes are the same ($\tilde{n} = n = 6$), so the calculation for the treatment sum of squares is simply

$$\text{SS}_{\text{Treatment}} = \frac{1}{n}\sum_{i=1}^{k}\left(\sum_{j=1}^{n}Y_{ij}\right)^2 - \frac{\left(\sum_{i=1}^{k}\sum_{j=1}^{n}Y_{ij}\right)^2}{kn}.$$

The F-statistic, $F = 9.01$, is significant at the $\alpha = 0.01$ significance level, because it exceeds $F(0.01; 4,25) = 4.18$. Thus, the company is justified in concluding that there is considerable variation from shipment to shipment.

We can estimate the variance components, σ_τ^2 and σ^2, from the mean squares in the ANOVA table. The estimates are as follows:

$$\hat{\sigma}^2 = \text{MS}_{\text{Error}} = 22.3;$$

$$\hat{\sigma}_\tau^2 = \frac{1}{6}(200.8 - 22.3) = 29.75.$$

The adequacy of the fitted model should always be checked before the conclusions from the estimated model are accepted. The residuals $e_{ij} = Y_{ij} - \overline{Y}_i$ are analyzed in Figure 6.2-3. It appears from these plots that the assumptions of the one-factor ANOVA model (normality and equal variances) are reasonably satisfied.

Table 6.2-2 ANOVA Table for Bacteria in Milk Shipments				
Source	SS	df	MS	F
Treatment	803.0	4	200.8	9.01
Error	557.2	25	22.3	
Total	1,360.2	29		

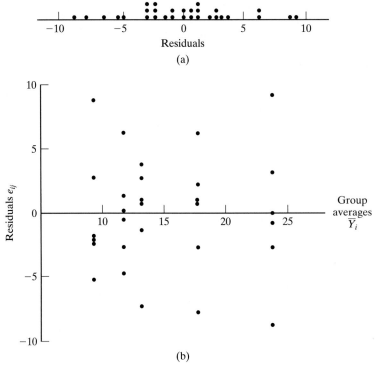

Figure 6.2-3 Various residual plots for the data in Example 6.2-3: (a) overall dot diagram of the residuals; (b) plot of residuals against group averages   ■

Remark: Note that we have analyzed transformed data; that is, $Y = \sqrt{\text{counts}}$. Bacteria counts often follow a Poisson distribution, which implies that groups with higher average counts also have higher variability, because the mean and the variance in the Poisson distribution are the same. This violates two of our assumptions: normality and constant variance. It can be shown that the square root of a Poisson random variable stabilizes the variance and also provides a variable that looks more like a normal random variable.

6.2-6 Computer Software

Programs for calculating an ANOVA table and for testing the hypothesis that the population means are the same are part of most statistical software packages. Minitab's command "Stat > ANOVA > One-way" and "Stat > ANOVA > One-way (Unstacked)" can be used. For the unstacked command, the response data are entered into k ($k = 3$, in Example 6.1-1) different columns. For the first (stacked) command, the data are stacked into a single column, with a second column containing the group labels for the observations ($1, 2, \ldots, k$, or a list of k different numbers). The output contains the ANOVA table, summary statistics for each group, and confidence intervals for the group means. The user can store the fitted values and the residuals (to be used for diagnostic graphs, which are readily available by clicking options under "Graphs"), and calculate Tukey's multiple-comparison confidence intervals.

The analysis of the random-effects model can be obtained through the command Minitab "Stat > ANOVA > Balanced ANOVA." The responses are entered into a stacked column, with group identifiers collected in a second column. The column of group identifiers is entered as the model and also as the random factor. Under "Results", we click the tab for "Display expected mean squares and variance components" and enter the random factors.

Exercises 6.2

6.2-1 In Exercises 6.1-3 to 6.1-6, we asked you to construct ANOVA tables for the one-factor experiments described and to test whether the treatment means are different.

(a) Here, we ask you to perform the appropriate diagnostic checks. In particular, you should construct overall dot diagrams of the residuals, compare dot diagrams of the residuals for the various treatment groups, and plot residuals against treatment averages.

(b) Also, we ask you to conduct the follow-up tests that are discussed in this section. In particular, plot the treatment averages and their reference distribution and calculate 95 percent confidence intervals that control for all possible paired comparisons.

Remark: There are $k(k-1)/2$ possible comparisons of k treatments. It is often convenient to order the treatment averages from smallest to largest and display the $k(k-1)/2$ treatment differences in the upper triangle of a $k \times k$ table. For example, pairwise differences of the $k = 4$ ordered averages $\overline{Y}_A, \overline{Y}_B, \overline{Y}_C$, and $\overline{Y}_D$ can be displayed as

$$\overline{Y}_B - \overline{Y}_A \qquad \overline{Y}_C - \overline{Y}_A \qquad \overline{Y}_D - \overline{Y}_A$$
$$\overline{Y}_C - \overline{Y}_B \qquad \overline{Y}_D - \overline{Y}_B$$
$$\overline{Y}_D - \overline{Y}_C$$

Differences that are significantly different from zero [i.e., greater than $q(\alpha; k, v)\sqrt{\text{MS}_{\text{Error}}/n}$ if all sample sizes $n_i = n$ are the same] are usually circled.

*6.2-2 We are interested in the effect of nozzle type on the rate of fluid flow. Three specific nozzle types are under study. Five runs through each of the three nozzle types led to the following results:

Nozzle Type	Rate of Flow (cubic centimeters)				
A	96.6	97.2	96.4	97.4	97.8
B	97.5	96.4	97.0	96.2	96.8
C	97.0	96.0	95.6	95.8	97.0

*(a) Test whether nozzle type has an effect on fluid flow. Use significance level $\alpha = 0.05$.

(b) Plot the treatment averages, together with their reference distribution.

(c) Perform and interpret the results of the appropriate diagnostic checks.

*6.2-3 A company is concerned about the level of impurity in its manufactured product. One possible source of impurity is the raw material used in the production. To check this hypothesis, the company selects six batches of raw material at random. Three products are selected at random from those produced from each batch, and the impurities are counted, resulting in the following table:

Batch	Impurities		
1	15	20	17
2	12	16	15
3	20	18	14
4	9	10	13
5	12	14	8
6	21	16	19

*(a) Is this a random-effects or a fixed-effects model?

*(b) Test whether there are differences from batch to batch.

(c) Estimate the variance components.

6.2-4 A large bakery receives its flour in batches of 50-pound bags. The company suspects that the batches furnished by the supplier differ significantly in weight. To check this suspicion, the company selects four batches at random. Five bags from each batch are weighed, and their weights are recorded as follows:

Batch	Weight $(y - 50)$				
1	0.09	−0.04	−0.08	−0.01	0.00
2	0.01	0.14	−0.04	0.02	−0.01
3	−0.22	−0.10	−0.13	−0.10	−0.11
4	−0.04	−0.02	−0.11	0.06	−0.10

Is the company's suspicion justified? Discuss.

6.2-5 Consider the aldrin concentrations (or their logarithms if you think that the data should be transformed) in Exercise 1.4-3. Calculate an ANOVA table and test whether the vertical distribution of aldrin in the river is homogeneous. What assumptions did you have to make here? For your analysis to be appropriate, how should the sampling procedure be carried out? Would your analysis be appropriate if the experiment were conducted over a period of 10 days and the measurements in the 10 rows of the table in Exercise 1.4-3 came from samples taken on different days?

6.2-6 A company employs three workers who manufacture brake disks. Being randomly assigned to machines and material, they each make 10 disks. The absolute differences between the actual and the specified diameters are recorded as follows:

Worker, i	n_i	$\bar{y}_i$	s_i^2
1	10	1.2	3.2
2	10	3.1	2.4
3	10	0.8	2.1

Construct an ANOVA table and answer the following questions:

(a) Are there differences among the mean precisions of the three workers $(\alpha = 0.05)$?

(b) Construct a reference distribution for the three averages.

(c) Determine the width of the 95 percent confidence interval obtained by Tukey's multiple-comparison procedure.

6.2-7 Your job in the quality assurance department at Northstar Steel is to analyze the metallurgic consistency of the production of 1/4-inch steel rods. An item of particular interest is the depletion of carbon, as it is related to the strength of these rods.

Every 30 minutes, a rod is picked from the production line and the depletion of carbon is determined. The observations are displayed in a time-sequence plot, and at the end of each week a histogram of the $48 \times 7 = 336$ observations is displayed. The histogram tells us about the variability in our production. Specification limits drawn on this histogram tell us whether we are in compliance with the customer specifications.

There are many factors that influence the depletion of carbon in the final product. Rods are extruded from billets, and billets are produced in batches (called heats). During each heat, scrap metal is melted in huge buckets and then poured into forms to produce several billets. The billets from a single heat are sufficient for a six-hour production of rods. Assume that the first 12 observations in our time-sequence plot come from heat 1, observations 13 to 24 from heat 2, and so on. How can you analyze the data (both graphically as well as analytically) to study whether the production from the same heat is more uniform than the one from different heats?

6.2-8 Suppose you are studying the monthly amounts that seventh- and eighth-grade boys and girls spend on entertainment such as movies, music CDs, and candy. Representative samples of children within a certain school district were selected, and children were asked about their spending habits. The following results were obtained:

	7th-grade boys	8th-grade boys	7th-grade girls	8th-grade girls
Sample size	30.0	25.0	30.0	25.0
Mean ($)	20.1	23.2	19.6	25.0
Standard Deviation ($)	6.0	5.6	5.3	7.0

(a) Test whether the four groups differ with respect to their mean spending amounts.

(b) Follow up on your analysis in (a) if you find differences. In particular, assess whether there are differences in the mean spending amounts of seventh- and eighth-grade boys and in the mean spending amounts of seventh- and eighth-grade girls.

6.2-9 **(Youden plot for interlaboratory experiments)** In interlaboratory studies, nearly identical samples of a certain product are prepared, divided, and sent to each of the participating laboratories. The resulting data from such studies are used to compare laboratory alignment and to assess whether there are differences among the labs.

Here, we consider a situation in which two samples are sent to each of m laboratories. The following results, with numbers scaled from an actual interlaboratory study of vitamin A, are taken from an article by Lynne Hare in the August 2007 issue of *Quality Progress*:*

*Hare, Lynne B. "It's not always what you say, but how you say it." *Quality Progress,* Vol. 40, August 2007, 64–66.

Laboratory	Replicate 1	Replicate 2
A	57.9	70.9
B	57.3	71.1
C	67.9	77.9
D	84.0	78.9
E	84.0	55.3
F	58.1	57.6
G	61.9	55.3
H	39.2	33.9
I	56.4	58.8
J	45.3	51.6
K	63.0	62.7
L	60.9	75.6
M	35.4	112.7
N	81.7	77.6
O	58.9	60.5
P	51.9	55.8

(a) Consider the random-effects model discussed in Section 6.2-5. The factor of interest is the laboratory participating, and the laboratory effects are random. Estimate the components of the variance, and test whether there is significant variability due to the laboratory. Interpret the results.

(b) In a very early paper in *Industrial Quality Control,* W. J. Youden recommends a simple, but very useful, graphical analysis of interlaboratory data.[‡] He recommends using a scatter plot with the two results from each laboratory plotted against each other. With data from m laboratories, the scatter plot contains m points. A diagonal line of slope 1 and intercept 0 is added to the graph. The proximity of a laboratory point to this diagonal line shows agreement of the two replicates.

Plot this graph with the given data, and confirm that laboratory M (and, to a lesser degree, laboratory E) show high variation among their replicates. This lack of agreement of its measurements should be communicated (perhaps in private) to the laboratory, whereupon, hopefully, it will lead the laboratory to standardize its testing procedure.

Youden also proposed a graphical method that can highlight unusually large laboratory deviations from a central value. The coordinates of the center point of the scatter plot can be obtained by calculating the means (or medians) of the observations on the x-axis (replicate 1) and the y-axis (replicate 2), perhaps after omitting obvious outliers. Omitting laboratory M, for example, we find that the coordinates of the center are given by (61.9, 62.9). Youden recommends drawing a circle around this center point, with a radius that is a certain multiple of the interlaboratory standard deviation. [In the hint that follows this exercise, we explain how to estimate the interlaboratory standard deviation, and we show that the correct multiple is 2.45 for 95 percent confidence.] If laboratories follow the same

---

[‡]Youden, W. J. "Graphical Diagnosis of Interlaboratory Test Results." *Industrial Quality Control,* Vol. 15 (May 1959).

procedures (so that there will be no laboratory effects), 95 percent of the plotted points should lie within this circle. If much more than 5 percent of the points lie outside the circle, then there is evidence that these laboratories use different procedures in analyzing the samples. (Actually, it is possible that the laboratories are using the same procedure, but that some other kind of systematic errors have crept into the study. For example, if not all samples were analyzed at the same age, deterioration of samples could influence the measurements.)

Youden claims that these graphical procedures will tell us more than a simple random-effects ANOVA (with laboratories being the random factor) can. The graph he recommends shows two things: (1) Points that are far from the 45-degree line point to laboratories that are not internally consistent in their analyses. These laboratories should be contacted so that they can improve their consistency. (2) Points that are near the 45-degree line, but outside the calculated circle, point to laboratories that are consistent, but are struggling with systematic errors. These laboratories should be contacted also, and one should investigate whether they follow the established protocol, whether samples were handled uniformly, and whether other factors might be causing systematic errors among laboratories.

Construct the scatter plot suggested by Youden. Show that, for our data (omitting laboratory M), $s_{\text{Interlab}} = 7.48$. Draw the circle with radius $(7.48)(2.45) = 18.33$ and interpret the results. (1) Are there laboratories with inconsistent measurements (i.e., with plotted points that lie far from the 45-degree line)? (2) Are there laboratories which use procedures that are different from the procedures the rest of the laboratories use (i.e., with plotted points that fall outside the circle)?

Hint: The following discussion gives some technical details to consider in answering this question:

Interlaboratory standard deviation: An estimate of the interlaboratory standard deviation can be obtained by pooling the m laboratory standard deviations, each obtained from two measurements. That is,

$$s_{\text{Interlab}} = \sqrt{\frac{1}{m}\sum_{i=1}^{m}\{[y_{i1} - (y_{i1} + y_{i2})/2]^2 + [y_{i2} - (y_{i1} + y_{i2})/2]^2\}}$$

$$= \sqrt{\frac{1}{2m}\sum_{i=1}^{m}(y_{i1} - y_{i2})^2},$$

where y_{i1}, y_{i2} are the two replicates of laboratory $i = 1, 2, \ldots, m$. s_{Interlab} is an estimate of the (common) measurement variability σ.

Determination of the multiple: We make the usual assumptions of the random effects model in Section 6.2. We assume a normal distribution and independence among the measurements from the same laboratory. We assume that there is no laboratory effect and that the variability among the data is due solely to (common) measurement variability. In essence, we assume that the variance component due to the laboratory is zero. Under these assumptions, we find that $\left[\dfrac{X - \mu_X}{\sigma}\right]^2 + \left[\dfrac{Y - \mu_Y}{\sigma}\right]^2$ has a chi-square distribution with two degrees of freedom and that

$$P\left\{\left[\frac{X - \mu_X}{\sigma}\right]^2 + \left[\frac{Y - \mu_Y}{\sigma}\right]^2 \le \chi^2(0.05; df = 2) = 5.99\right\} = 0.95$$

and

$$P\{(X - \mu_X)^2 + (Y - \mu_Y)^2 \le \sigma^2\chi^2(0.05; df = 2)\} = 0.95.$$

If laboratory effects are absent, a circle with radius $(s_{Interlab})\sqrt{\chi^2(0.05; df = 2)} = s_{Interlab}(2.45)$ should contain 95 percent of the plotted points.

6.2-10 In his 1959 paper in *Industrial Quality Control,* Youden reports on another interlaboratory study that compares the measurements of the percent insoluble residue in cement from 29 different laboratories, with the following results:

Laboratory	A	B
1	0.31	0.22
2	0.08	0.12
3	0.24	0.14
4	0.14	0.07
5	0.52	0.37
6	0.38	0.19
7	0.22	0.14
8	0.46	0.23
9	0.26	0.05
10	0.28	0.14
11	0.10	0.18
12	0.20	0.09
13	0.26	0.10
14	0.28	0.14
15	0.25	0.13
16	0.25	0.11
17	0.26	0.17
18	0.26	0.18
19	0.12	0.05
20	0.29	0.14
21	0.22	0.11
22	0.13	0.10
23	0.56	0.42
24	0.30	0.30
25	0.24	0.06
26	0.25	0.35
27	0.24	0.09
28	0.28	0.23
29	0.14	0.10

Analyze the data, both with an analysis of variance and with the scatter plot suggested by Youden. What do you find? (1) Are there laboratories with inconsistent measurements (i.e., with plotted points that lie far from the 45-degree line)? (2) Are there laboratories which use procedures that are different from the rest (i.e., with plotted points that fall outside the circle)? Discuss why the Youden plot may give you more information than an analysis of variance can.

6.3 Randomized Complete Block Designs

Suppose that we want to compare a number of treatments—say, k—on certain experimental units—for example, k plant varieties on certain plots of land, k drugs on certain animals, or k test procedures on certain specimens. In designing efficient experiments, we attempt to control the variability that arises from extraneous sources, and some of this variability may come from differences among the experimental units. For example, not all plots of land, especially if they are far apart, may be of the same quality, or not all specimens, especially if they are from different batches, may be exactly the same. In a completely randomized experiment, we allocate the experimental units to the k treatments at random. Such an arrangement is successful in spreading the differences of the experimental units evenly among the k treatments. However, these differences do create a larger mean square error in the corresponding one-factor ANOVA table, making it more difficult to detect differences among the treatments.

More precise comparisons can be made by grouping the experimental units into homogeneous blocks such that the units within a given block resemble each other more than they resemble the units in the other blocks. Then the experimental units within each block are assigned to the k treatments at random. For example, a block may consist of k adjacent plots of land, k animals from the same litter, or k specimens from the same batch. Because all k treatments are used within the same block, we can compare the treatments on the same block and thus adjust for the differences among blocks.

If every one of the k treatments is represented in each block, we call the arrangement a *complete block design*. In *incomplete* block designs, by contrast, we are unable to include all k treatments in each block. Such a situation arises, for example, if we compare $k = 4$ different fabrics on a Martindale wear tester possessing the feature that only three different pieces of cloth can be compared in a single run. Here, the block size (3) is smaller than the number of treatments (4). As another example, in a comparison of k test procedures, it may be impossible to run all k tests on the same day, which could be considered a block. The analysis of data from incomplete block designs is more difficult and is not considered in this text.

We call the complete block design a *randomized* complete block design because, within each block, we assign the k treatments to the experimental units at random. For example, within each block of k adjacent plots of land, we use a random mechanism (drawing numbers from a hat or using random numbers) to decide which plant variety is planted on which plot. Note that this is a *restricted* randomization, as the blocks impose a constraint on how to allocate the experimental units.

Let us begin with an example taken from a study of the compressive strength of concrete. The $k = 3$ treatments correspond to three different drying methods of molded concrete cylinders that are 100 mm in diameter and 200 mm long. Concrete is mixed in batches that are just large enough to produce three cylinders. Although great care is taken to achieve uniformity among the batches, one can expect variability from batch to batch.

The experimental units in this example are the molded concrete cylinders. In a completely randomized experiment, with $n = b$ replications for each treatment, we would divide the $N = kn = kb$ cylinders at random into k groups and then assign the k treatments to these groups at random. However, a more precise comparison

Table 6.3-1 Compressive Strength of Concrete (100 pounds per square inch)

			Batch			
Treatment	1	2	3	4	5	Treatment Mean
A	52	47	44	51	42	47.2
B	60	55	49	52	43	51.8
C	56	48	45	44	38	46.2
Batch Mean	56	50	46	49	41	

can be achieved by blocking the experiment and assigning the $k = 3$ cylinders from each batch at random to the three drying methods. This procedure was followed, and the results of a strength test on $b = 5$ batches are given in Table 6.3-1. For example, the number 49 in row 2 and column 3 represents the compressive strength of a cylinder from batch 3 that was dried according to method B.

In general, the data from a randomized complete block experiment can be represented as in Table 6.3-2. Here Y_{ij}, for $i = 1, 2, \ldots, k$ and $j = 1, 2, \ldots, b$, is the observation on the ith treatment in the jth block. The treatment means are given by

$$\overline{Y}_{i\bullet} = \frac{1}{b} Y_{i\bullet} = \frac{1}{b} \sum_{j=1}^{b} Y_{ij},$$

the block means are

$$\overline{Y}_{\bullet j} = \frac{1}{k} Y_{\bullet j} = \frac{1}{k} \sum_{i=1}^{k} Y_{ij},$$

and the grand mean is

$$\overline{Y}_{\bullet\bullet} = \frac{1}{kb} Y_{\bullet\bullet} = \frac{1}{kb} \sum_{i=1}^{k} \sum_{j=1}^{b} Y_{ij}.$$

The dot in the notation for the sums $(Y_{i\bullet}, Y_{\bullet j}, Y_{\bullet\bullet})$ and the means $(\overline{Y}_{i\bullet}, \overline{Y}_{\bullet j}, \overline{Y}_{\bullet\bullet})$ indicates the index over which the sum or average is taken.

Table 6.3-2 Data from a Randomized Complete Block Experiment

				Block			
Treatment	1	2	$\cdots$	j	$\cdots$	b	Treatment Mean
1	Y_{11}	Y_{12}	$\cdots$	Y_{1j}	$\cdots$	Y_{1b}	$\overline{Y}_{1\bullet}$
2	Y_{21}	Y_{22}	$\cdots$	Y_{2j}	$\cdots$	Y_{2b}	$\overline{Y}_{2\bullet}$
$\vdots$	$\vdots$	$\vdots$		$\vdots$		$\vdots$	$\vdots$
i	Y_{i1}	Y_{i2}	$\cdots$	Y_{ij}	$\cdots$	Y_{ib}	$\overline{Y}_{i\bullet}$
$\vdots$	$\vdots$	$\vdots$		$\vdots$		$\vdots$	$\vdots$
k	Y_{k1}	Y_{k2}	$\cdots$	Y_{kj}	$\cdots$	Y_{kb}	$\overline{Y}_{k\bullet}$
Block Mean	$\overline{Y}_{\bullet 1}$	$\overline{Y}_{\bullet 2}$	$\cdots$	$\overline{Y}_{\bullet j}$	$\cdots$	$\overline{Y}_{\bullet b}$	$\overline{Y}_{\bullet\bullet}$

In the completely randomized one-factor experiment in Sections 6.1 and 6.2, we have written the model that generates the data as $Y_{ij} = \mu_i + \varepsilon_{ij} = \mu + \tau_i + \varepsilon_{ij}$, where $\tau_1, \tau_2, \ldots, \tau_k$ are the treatment effects. Each treatment changes the mean by a certain amount τ_i. In the randomized complete block design, we have two factors that may affect the response: (1) the treatments and (2) the blocks. The model

$$Y_{ij} = \mu + \tau_i + \beta_j + \varepsilon_{ij}$$

incorporates both factors: μ is an overall mean; $\tau_1, \tau_2, \ldots, \tau_k$ are the treatment effects; $\beta_1, \beta_2, \ldots, \beta_b$ are the block effects; and the errors ε_{ij} are independent $N(0, \sigma^2)$ random variables. The treatment and block effects may be either fixed or random. In the fixed-effects model, the effects are nonrandom deviations from an overall mean that satisfy $\sum_{i=1}^{k} \tau_i = 0$ and $\sum_{j=1}^{b} \beta_j = 0$.

We could also take treatment or block effects (or both) as random. In fact, in many cases in which we block with respect to batches, it is more reasonable to consider the block effects as random. That is, the β_j are taken to be independent random variables from a $N(0, \sigma_\beta^2)$ distribution. It turns out that our analysis is unchanged if blocks or treatments are random; only the interpretation is different. If the blocks are random, we expect the inferences about the treatments to be the same throughout the population of blocks from which those used in the experiment were randomly selected.

Note that the model assumes that the treatment and block effects are additive. This means that if block j leads to an increase of $\beta_j = 5$ units in the response, and treatment i to an increase of $\tau_i = 2$ units, then the combined effect is $5 + 2 = 7$. Also, the model implies that the increase of $\beta_j = 5$ units in block j is the same for all treatments and, similarly, the increase $\tau_i = 2$ for treatment i is the same for all blocks. Although this additive model is often adequate, there may be cases for which it should not be used. The effect of treatment i may, in fact, depend on block j; the specific conditions in a specific block (batch) may make a certain treatment completely ineffective. Then we say that there are interactions between blocks and treatments. Later, in our discussion of model diagnostics, we will check whether an additive model structure is appropriate.

6.3-1 Estimation of Parameters and ANOVA

The estimate of the overall mean μ is given by the grand mean

$$\overline{Y}_{\bullet\bullet} = \frac{1}{kb} Y_{\bullet\bullet} = \frac{1}{kb} \sum_{i=1}^{k} \sum_{j=1}^{b} Y_{ij}.$$

The estimate of the treatment effect τ_i is given by $\overline{Y}_{i\bullet} - \overline{Y}_{\bullet\bullet}$, the difference of the ith treatment mean from the grand mean. The estimate of the block effect β_j is $\overline{Y}_{\bullet j} - \overline{Y}_{\bullet\bullet}$, the difference of the jth block mean from the grand mean. Accordingly, an estimate of the error $\varepsilon_{ij} = Y_{ij} - (\mu + \tau_i + \beta_j)$ is given by the residual

$$e_{ij} = Y_{ij} - [\overline{Y}_{\bullet\bullet} + (\overline{Y}_{i\bullet} - \overline{Y}_{\bullet\bullet}) + (\overline{Y}_{\bullet j} - \overline{Y}_{\bullet\bullet})] = Y_{ij} - \overline{Y}_{i\bullet} - \overline{Y}_{\bullet j} + \overline{Y}_{\bullet\bullet}.$$

The residual represents the component of Y_{ij} that remains after allowing for the overall mean and the treatment and block effects.

Associated with the model decomposition

$$Y_{ij} = \mu + \tau_i + \beta_j + \varepsilon_{ij}$$

we find a corresponding data decomposition

$$Y_{ij} = \overline{Y}_{\bullet\bullet} + (\overline{Y}_{i\bullet} - \overline{Y}_{\bullet\bullet}) + (\overline{Y}_{\bullet j} - \overline{Y}_{\bullet\bullet}) + (Y_{ij} - \overline{Y}_{i\bullet} - \overline{Y}_{\bullet j} + \overline{Y}_{\bullet\bullet}).$$

A corresponding decomposition of the total sum of squares is

$$
\begin{aligned}
\text{SSTO} &= \sum_{i=1}^{k}\sum_{j=1}^{b}(Y_{ij} - \overline{Y}_{\bullet\bullet})^2 \\
&= \sum_{i=1}^{k}\sum_{j=1}^{b}[(\overline{Y}_{i\bullet} - \overline{Y}_{\bullet\bullet}) + (\overline{Y}_{\bullet j} - \overline{Y}_{\bullet\bullet}) + (Y_{ij} - \overline{Y}_{i\bullet} - \overline{Y}_{\bullet j} + \overline{Y}_{\bullet\bullet})]^2 \\
&= b\sum_{i=1}^{k}(\overline{Y}_{i\bullet} - \overline{Y}_{\bullet\bullet})^2 + k\sum_{j=1}^{b}(\overline{Y}_{\bullet j} - \overline{Y}_{\bullet\bullet})^2 + \sum_{i=1}^{k}\sum_{j=1}^{b}(Y_{ij} - \overline{Y}_{i\bullet} - \overline{Y}_{\bullet j} + \overline{Y}_{\bullet\bullet})^2 \\
&= \quad\;\; \text{SS}_{\text{Treatment}} \quad + \quad \text{SS}_{\text{Block}} \quad + \quad\quad\quad \text{SS}_{\text{Error}}.
\end{aligned}
$$

The cross-product terms resulting from the squaring of the trinomial, namely,

$$2\sum_{i=1}^{k}\sum_{j=1}^{b}(\overline{Y}_{i\bullet} - \overline{Y}_{\bullet\bullet})(\overline{Y}_{\bullet j} - \overline{Y}_{\bullet\bullet}), \quad 2\sum_{i=1}^{k}\sum_{j=1}^{b}(\overline{Y}_{i\bullet} - \overline{Y}_{\bullet\bullet})(Y_{ij} - \overline{Y}_{i\bullet} - \overline{Y}_{\bullet j} + \overline{Y}_{\bullet\bullet}),$$

and

$$2\sum_{i=1}^{k}\sum_{j=1}^{b}(\overline{Y}_{\bullet j} - \overline{Y}_{\bullet\bullet})(Y_{ij} - \overline{Y}_{i\bullet} - \overline{Y}_{\bullet j} + \overline{Y}_{\bullet\bullet}),$$

are all equal to zero. (See Exercise 6.3-7.)

The sum of squares due to treatment, $\text{SS}_{\text{Treatment}}$, measures the variation among the k treatment means. The sum of squares due to block, SS_{Block}, measures the variation among the b block means. The sum of squares due to error, SS_{Error}, measures the variability in the residuals.

Equivalent computational formulas (which also avoid rounding) for the total, the treatment, the block, and the error sum of squares are, respectively,

$$\text{SSTO} = \sum_{i=1}^{k}\sum_{j=1}^{b}(Y_{ij} - \overline{Y}_{\bullet\bullet})^2 = \sum_{i=1}^{k}\sum_{j=1}^{b}Y_{ij}^2 - \frac{(Y_{\bullet\bullet})^2}{kb},$$

$$\text{SS}_{\text{Treatment}} = b\sum_{i=1}^{k}(\overline{Y}_{i\bullet} - \overline{Y}_{\bullet\bullet})^2 = \frac{1}{b}\sum_{i=1}^{k}(Y_{i\bullet})^2 - \frac{(Y_{\bullet\bullet})^2}{kb},$$

$$\text{SS}_{\text{Block}} = k\sum_{j=1}^{b}(\overline{Y}_{\bullet j} - \overline{Y}_{\bullet\bullet})^2 = \frac{1}{k}\sum_{j=1}^{b}(Y_{\bullet j})^2 - \frac{(Y_{\bullet\bullet})^2}{kb},$$

and

$$\text{SS}_{\text{Error}} = \text{SSTO} - \text{SS}_{\text{Treatment}} - \text{SS}_{\text{Block}}.$$

One of these components can always be found from the other three. These expressions are not that important, as you will use computer software for the calculations. (See the discussion in Section 6.3-6.)

The $N = kb$ components in SSTO satisfy one restriction, namely, that $\sum_i \sum_j (Y_{ij} - \overline{Y}_{\bullet\bullet}) = 0$; hence, the total sum of squares has $kb - 1$ degrees of freedom. The k components in $SS_{\text{Treatment}}$ add to zero; that is, $\sum_i (\overline{Y}_{i\bullet} - \overline{Y}) = 0$; thus, $SS_{\text{Treatment}}$ has $k - 1$ degrees of freedom. Similarly, SS_{Block} has $b - 1$ degrees of freedom. SS_{Error} is the sum of squares of kb residuals. It is easy to show [see Exercise 6.3-7(a)] that

$$\sum_{i=1}^{k} e_{ij} = 0, \; j = 1, 2, \ldots, b \quad \text{and} \quad \sum_{j=1}^{b} e_{ij} = 0 \text{ for } i = 1, 2, \ldots, k.$$

Consequently, we can regard the residuals as entries in a kb table, where the elements in each row and column add to zero. With these restrictions, we need only $(k - 1)(b - 1)$ entries to specify the table completely, and we see that the error sum of squares has $(k - 1)(b - 1)$ degrees of freedom.

Corresponding to the sum-of-squares decomposition

$$\text{SSTO} = SS_{\text{Treatment}} + SS_{\text{Block}} + SS_{\text{Error}}$$

is the decomposition of the degrees of freedom:

$$kb - 1 = (k - 1) + (b - 1) + (k - 1)(b - 1).$$

This information is written in the form of an ANOVA table, Table 6.3-3. The column headed "MS" in the table lists the mean squares: the sums of squares, divided by their corresponding degrees of freedom. The column headed "F" is explained in the next section.

6.3-2 Expected Mean Squares and Tests of Hypotheses

Before the observations are actually made, $Y_{ij} = \mu + \tau_i + \beta_j + \varepsilon_{ij}$; hence, the mean squares are random variables. Therefore, it is appropriate to investigate their distribution and expected values.

The residuals $e_{ij} = Y_{ij} - \overline{Y}_{i\bullet} - \overline{Y}_{\bullet j} + \overline{Y}_{\bullet\bullet}$ estimate the errors ε_{ij}. The mean square error is an unbiased estimator of σ^2; that is, $E(MS_{\text{Error}}) = \sigma^2$. Because the residuals are adjusted for possible treatment and block effects, this is true regardless of whether treatment and block effects are or are not present. One can also show, in the fixed-effects model, that

$$E(MS_{\text{Treatment}}) = \sigma^2 + b \frac{\sum_{i=1}^{k} \tau_i^2}{k - 1}$$

Table 6.3-3 ANOVA for Data from a Randomized Complete Block Design

Source	SS	df	MS	F
Treatment	$SS_{\text{Treatment}}$	$k - 1$	$MS_{\text{Treatment}} = SS_{\text{Treatment}}/(k - 1)$	$MS_{\text{Treatment}}/MS_{\text{Error}}$
Block	SS_{Block}	$b - 1$	$MS_{\text{Block}} = SS_{\text{Block}}/(b - 1)$	$MS_{\text{Block}}/MS_{\text{Error}}$
Error	SS_{Error}	$(k - 1)(b - 1)$	$MS_{\text{Error}} = SS_{\text{Error}}/(k - 1)(b - 1)$	
Total	SSTO	$kb - 1$		

and

$$E(MS_{Block}) = \sigma^2 + k\frac{\sum_{j=1}^{b}\beta_j^2}{b - 1}.$$

Our main interest is in testing whether there are *treatment effects*. The objective is to test the null hypothesis $H_0: \tau_1 = \tau_2 = \ldots = \tau_k = 0$ against the alternative H_1: at least one τ_i, for $i = 1, \ldots, k$, is different from zero. Under H_0, both $MS_{Treatment}$ and MS_{Error} are unbiased estimates of σ^2 and their ratio should vary around 1. Now, we can show that $SS_{Treatment}$ and SS_{Error}, divided by σ^2, are independent chi-square random variables with $k - 1$ and $(k - 1)(b - 1)$ degrees of freedom; hence, under H_0, the sampling distribution of $F_{Treatment} = MS_{Treatment}/MS_{Error}$ is an F-distribution with $k - 1$ and $(k - 1)(b - 1)$ degrees of freedom. Under H_1, this F-statistic tends to be larger than what can be expected under H_0. If $F_{Treatment}$ is an observed value of this F-statistic, then the probability value

$$p \text{ value} = P[F(k - 1, (k - 1)(b - 1)) \geq F_{Treatment}]$$

expresses how likely it is that an F value of at least the size of $F_{Treatment}$ occurs by chance. If the p value is smaller than a chosen significance level α, we conclude that there are differences among treatments. Many computer packages calculate the probability value. If it is not given, we choose a significance level α and use the upper percentage point $F[\alpha; k - 1, (k - 1)(b - 1)]$ as the critical value. If $F_{Treatment}$ is greater than or equal to $F[\alpha; k - 1, (k - 1)(b - 1)]$, then we conclude, at significance level α, that the treatment means are different.

Usually, the treatment comparisons are of primary interest. Blocks are chiefly the means of reducing experimental error, and often we are not interested in testing for their effects. In some instances, however, it may be of interest to test for block effects, because if there are no differences, blocking may not be necessary in future experiments. From the expected mean squares, we find that the hypothesis $H_0: \beta_1 = \beta_2 = \ldots = \beta_b = 0$ may be tested by comparing $F_{Block} = MS_{Block}/MS_{Error}$ with $F[\alpha; b - 1, (k - 1)(b - 1)]$. Note that the degrees of freedom in the appropriate F-distributions are easy to remember because they correspond to those in the numerator and denominator in these F-ratios.

Example 6.3-1 For the data listed in Table 6.3-1, we find that $\sum_i\sum_j Y_{ij}^2 = 35{,}638$ and $\sum_i\sum_j Y_{ij} = 726$; thus, $SSTO = 35{,}638 - (726)^2/15 = 499.6$. Also, $Y_{1\bullet} = 236$, $Y_{2\bullet} = 259$, $Y_{3\bullet} = 231$, and

$$SS_{Treatment} = \frac{1}{5}(236^2 + 259^2 + 231^2) - \frac{(726)^2}{15} = 89.2.$$

In addition, $Y_{\bullet 1} = 168$, $Y_{\bullet 2} = 150$, $Y_{\bullet 3} = 138$, $Y_{\bullet 4} = 147$, $Y_{\bullet 5} = 123$, and

$$SS_{Block} = \frac{1}{3}(168^2 + 150^2 + 138^2 + 147^2 + 123^2) - \frac{(726)^2}{15} = 363.6.$$

Of course, $SS_{Error} = 499.6 - 89.2 - 363.6 = 46.8$. All of these calculations lead to the following ANOVA table:

Source	SS	df	MS	F
Treatment	89.2	2	44.60	7.62
Block	363.6	4	90.90	15.54
Error	46.8	8	5.85	
Total	499.6	14		

Check that your computer software gives this table. The fact that $F_{\text{Treatment}} = 7.62$ indicates that there are differences among the treatment means. The probability value $P[F(2, 8) \geq 7.62] \cong 0.02$ is quite small, and the observed differences among the treatment averages cannot easily be explained as a chance result.   ∎

6.3-3 Increased Efficiency by Blocking

Of the sum of squares that is not accounted for by the treatments, namely, $499.6 - 89.2 = 410.4$, the major part (363.6, or 89 percent) is due to the differences among the blocks. Thus, if we had run a completely randomized experiment, obtained the same observations, and analyzed the data according to a one-factor analysis of variance, then the batch variability would have increased the mean square error. That is, that design would not have been as sensitive as the randomized complete block design. In fact, MS_{Error} from the one-factor ANOVA would have been $(363.6 + 46.8)/(4 + 8) = 34.2$, associated with $4 + 8 = 12$ degrees of freedom. The resulting F-statistic for testing treatment effects, $44.6/34.2 = 1.30$, would have been insignificant compared with $F(0.05; 2, 12) = 3.89$.

On the one hand, this example demonstrates the advantage of blocking if we suspect variability among the experimental units and are able to group them into homogeneous blocks. On the other hand, if blocking is not needed, and if we had run the experiment according to a completely randomized design with $n = b$ replicates, then the degrees of freedom for the error sum of squares would have been $k(b - 1)$. This is larger than $(k - 1)(b - 1)$ in the randomized complete block design. Thus, blocking when not needed has cost us $b - 1$ degrees of freedom for error and has made the analysis less sensitive to detecting differences. Usually, however, the gains from blocking outweigh the small loss in degrees of freedom, and one should, as a rule, block whatever is reasonably possible to block.

6.3-4 Follow-Up Tests

A large value of $F_{\text{Treatment}}$ provides evidence that the treatment means are different. To study how they differ, it is useful to display the treatment means, which are averages of b observations, on their reference distribution. This is done in Figure 6.3-1. The reference distribution for the treatment means $\overline{Y}_{i\bullet}$ is given by a scaled t-distribution with $v = (k - 1)(b - 1) = 8$ degrees of freedom, with scale factor $\sqrt{\text{MS}_{\text{Error}}/b} = \sqrt{5.85/5} = 1.08$. In Figure 6.3-1, we have arbitrarily centered the reference distribution at 49. Because the 95th percentile of the $t(8)$ distribution is $t(0.05; 8) = 1.86$, 10 percent of the probability of the reference distribution is outside $49 \pm (1.86)(1.08) = 49 \pm 2.01$. We have marked a unit ordinate at these points. The ordinate at the mean, 49, is approximately five times higher. The graph of

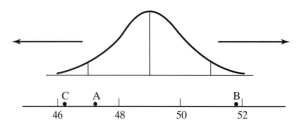

Figure 6.3-1 Plot of the treatment averages and their reference distribution

a "*t*-like" distribution through these three points yields the reference distribution, shown in the figure. Although making such a plot is certainly not an "exact" approach, it provides a good approximation. The reference distribution confirms the conclusion of the *F*-test; it is difficult to center the reference distribution to make the treatment means seem to be a typical sample from that distribution. The figure also illustrates that μ_A and μ_C are most likely the same, but that μ_B is different from both.

A more formal multiple comparison can be made with Tukey's paired-comparison procedure. The $100(1 - \alpha)$ percent confidence intervals for differences in treatment means that take account of all possible pairwise comparisons are given by

$$\overline{Y}_{r\bullet} - \overline{Y}_{s\bullet} \pm \frac{q(\alpha; k, v)}{\sqrt{2}} \sqrt{MS_{Error}} \sqrt{\frac{2}{b}},$$

where $q(\alpha; k, v)$ is the upper 100α percentage point of the Studentized range distribution. In this case, $q(0.05; 3, 8) = 4.04$ and $(4.04/\sqrt{2})\sqrt{5.85}\sqrt{2/5} = 4.37$. Thus, 95 percent confidence intervals for the treatment differences are

$$\mu_A - \mu_C : 1.0 \pm 4.37;$$

$$\mu_B - \mu_A : 4.6 \pm 4.37;$$

$$\mu_B - \mu_C : 5.6 \pm 4.37.$$

These results confirm the conclusions that we have drawn from the reference distribution in Figure 6.3-1; the intervals for $\mu_B - \mu_A$ and $\mu_B - \mu_C$ do not include zero.

We could also construct a reference distribution for the block means, which is a scaled $t[(k - 1)(b - 1)]$ distribution with scale factor $\sqrt{MS_{Error}/k}$. However, inferences on block differences are usually not of major importance, implying that μ_B is different from μ_A and μ_C.

6.3-5 Diagnostic Checking

Before we accept the results of statistical tests, we should always convince ourselves that the assumptions on which those tests are based are actually satisfied. Violations of model assumptions will often invalidate test results.

The diagnostic checks in the randomized complete block arrangement follow closely the ones for the completely randomized experiment. We check the residuals $e_{ij} = Y_{ij} - \overline{Y}_{i\bullet} - \overline{Y}_{\bullet j} + \overline{Y}_{\bullet\bullet}$, which are approximations of the errors, to see if they are about normally distributed. We must also check whether the variability in the various treatment and block groups is roughly the same and whether a plot of residuals against fitted values $\hat{Y}_{ij} = \overline{Y}_{i\bullet} + \overline{Y}_{\bullet j} - \overline{Y}_{\bullet\bullet}$ shows any patterns. We call $\hat{Y}_{ij} = \overline{Y}_{i\bullet} + \overline{Y}_{\bullet j} - \overline{Y}_{\bullet\bullet} = \overline{Y}_{\bullet\bullet} + (\overline{Y}_{i\bullet} - \overline{Y}_{\bullet\bullet}) + (\overline{Y}_{\bullet j} - \overline{Y}_{\bullet\bullet})$ the *fitted value* because it represents the estimate of the mean $\mu_{ij} = \mu + \tau_i + \beta_j$ in the ith treatment group and jth block. With $kb = 15$ observations, as in our example in Figure 6.3-2, it is often difficult to check these assumptions completely. For example, it may well be that the variability for the second treatment or the third block is somewhat smaller than that for the other groups. However, we want to point out that the visual impression can change quickly if we add or remove a single point. At any rate, in small data sets, we can only hope to spot gross violations of the assumptions. In this example, none can be found.

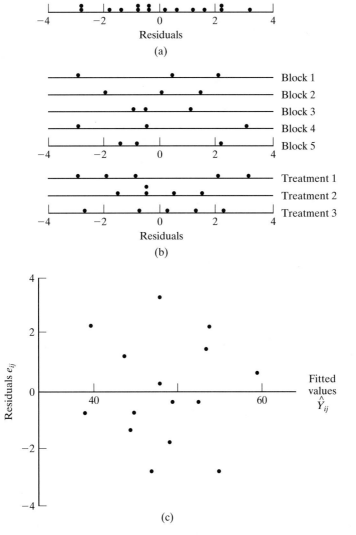

Figure 6.3-2 Various residual plots: (a) overall dot diagram of the residuals; (b) dot diagrams of the residuals by block and treatment; (c) plot of residuals against fitted values

The plot of residuals against fitted values is of particular interest here because a nonrandom pattern would suggest that the block and treatment effects are not additive. Recall that, in our analysis, we have assumed that treatment and block effects in the model $Y_{ij} = \mu_{ij} + \varepsilon_{ij} = \mu + \tau_i + \beta_j + \varepsilon_{ij}$ are additive. However, interactions may be present, and the magnitude of treatment effects may depend on the particular block. (A detailed discussion of interactions will be given in Chapter 7.) In some cases, the model may be nonadditive in the original observations, but additive under a certain transformation of the data. For example, the logarithmic transformation changes the multiplicative model $\mu_{ij} = \mu\tau_i\beta_j$ into an additive one, $\mu_{ij}^* = \mu^* + \tau_i^* + \beta_j^*$. A nonrandom pattern in the plot of e_{ij} against the fitted values $\hat{Y}_{ij}$, like a curvilinear one, is an indication that the model is not additive and that a transformation should be considered.

6.3-6 Computer Software

Minitab and its command "Stat > ANOVA > Two-Way" can be used. The spreadsheet containing the data in Table 6.3-1 consists of three columns: Column 1 contains the response (compressive strength), column 2 the treatment (three different numbers for A, B, and C), and column 3 the blocking groups (1 through 5). There are 15 rows to each column. Row 1 contains 52 in column 1, 1 in column 2, and 1 in column 3; row 2 contains 47 in column 1, 1 in column 2, and 2 in column 3; ...; the 15th row contains 38 in column 1, 3 in column 2, and 5 in column 3. The order in which the rows are entered is arbitrary. The output from this particular Minitab command includes the ANOVA table in Section 6.3-2. Residuals and fitted values can be stored, and several residual diagnostic checks are available.

Alternatively, one can use the "Stat > ANOVA > Balanced ANOVA" command. The two columns for treatments and blocks are entered at the model line. This command has the advantage of being able to specify treatment or block effects (or both) as random.

Exercises 6.3

*6.3-1   Four different fabrics are examined on a Martindale wear tester that can compare four materials in a single run (block). The weight loss (in milligrams) from five runs is measured and the following results are obtained:

	Run (Block)				
Fabric	**1**	**2**	**3**	**4**	**5**
A	36	17	30	30	25
B	38	18	39	40	25
C	36	26	41	38	28
D	30	17	34	33	21

*(a) Our main interest is in comparing the four fabrics. Obtain the appropriate ANOVA table and test whether there are differences among treatment means. Conduct appropriate follow-up tests.

*(b) Has the blocking arrangement been useful? Investigate whether there are differences among the block means.

(c) Check the adequacy of the model.

6.3-2 A certain type of film composition resistor used in electronic equipment is mounted on ceramic plates. An investigation was designed to determine the effects of four different geometric shapes on the current noise of these resistors. Four resistors could be mounted on each ceramic plate, and their location was randomized. The results, measured in the logarithm of the noise measurement, for six ceramic plates are as follows:

Plate (Block)

Shape	1	2	3	4	5	6
A	1.11	1.70	1.60	1.20	1.85	1.50
B	0.95	1.42	1.28	1.05	1.45	1.10
C	1.00	1.45	1.25	1.15	1.55	1.15
D	0.82	1.30	1.10	0.90	1.40	1.10

(a) Estimate the overall mean, the treatment effects, and the block effects.

(b) Formulate an ANOVA table and investigate whether there are treatment differences. Plot the treatment means and construct a reference distribution. Calculate Tukey's 95 percent confidence intervals for all possible pairwise comparisons of treatments.

(c) Check the model assumptions.

(d) Has blocking been helpful in making the treatment comparisons more precise?

*6.3-3 A plant manager wants to investigate the productivity of three groups of workers: those with little, those with average, and those with considerable experience. Because productivity depends to some degree on the day-to-day variability of the available raw materials, which affects all groups in a similar fashion, the manager suspects that the comparison should be blocked with respect to day. The results from five production days are as follows:

Day (Block)

Experience Level	1	2	3	4	5
A	53	58	49	52	60
B	55	57	53	57	64
C	60	62	55	64	69

*(a) Are there differences in productivity among the three groups? Investigate how the groups differ.

*(b) Has blocking made a difference? Test whether there are block effects.

(c) Check the model assumptions.

6.3-4 A psychologist studies the effect of background noise on reaction time. Each subject in this study is exposed to three levels of background noise: no noise, classical music, and rock music. At each level, the subject has to perform a variety of tasks, and an overall estimate of the subject's reaction time is determined.

The particular order of the experiments is randomized for each subject. The results for seven randomly selected subjects are as follows:

Subject (Block)

Noise Level	1	2	3	4	5	6	7
A	10	7	14	10	12	15	8
B	10	8	13	8	11	12	9
C	14	10	17	12	14	18	6

(a) Are there differences among the three noise levels? Discuss.

(b) Are there differences among the subjects? Discuss.

(c) Is the additive model $Y_{ij} = \mu + \tau_i + \beta_j + \varepsilon_{ij}$ appropriate? Plot residuals against fitted values.

6.3-5 Consider the aldrin concentrations (or their logarithms if you think that the data should be transformed) in Exercise 1.4-3. Assume that the experiment was conducted over a period of 10 days, and suppose that the measurements in each row of the table in Exercise 1.4-3 come from samples that were taken on the same day. Analyze the data and discuss whether the vertical distribution of aldrin in the river is homogeneous.

6.3-6 Grab samples were collected every 10 minutes for a period of 2 hours at six different sampling locations in the lateral transect of a large river. Separated by approximately 120 feet in the lateral direction, the sampling points were 2 miles downstream from a sewage treatment plant. Sodium, which is easily detected by spectrophotometry, was chosen as the water-quality variable to assess spatial variability in the river's cross section. The sampling points (the boats) were on a line that was perpendicular to the west bank; each person in the boat took samples approximately 1.5 feet from the water surface. The absorbances, which are linearly related to the sodium concentrations, are given in the following table:

			Location			
Time	1	2	3	4	5	6
1	0.266	0.239	0.227	0.230	0.226	0.231
2	0.299	0.277	0.279	0.274	0.273	0.273
3	0.264	0.247	0.238	0.238	0.231	0.237
4	0.273	0.244	0.244	0.246	0.250	0.244
5	0.263	0.237	0.247	0.238	0.243	0.243
6	0.269	0.246	0.241	0.241	0.247	0.252
7	0.256	0.241	0.249	0.253	0.244	0.247
8	0.260	0.241	0.238	0.246	0.246	0.240
9	0.264	0.244	0.237	0.235	0.235	0.224
10	0.253	0.234	0.222	0.231	0.234	0.222
11	0.271	0.246	0.229	0.229	0.236	0.234
12	0.246	0.226	0.234	0.225	0.235	0.241
Distance from west bank (feet)	155	247	347	474	595	820

Analyze the data and discuss whether the absorbances vary across the lateral transect of the river.

[T. G. Sanders, *Principles of Network Design for Water Quality Monitoring* (Fort Collins, CO: Colorado State University, 1980).]

6.3-7 The set of residuals is given by $e_{ij} = Y_{ij} - \overline{Y}_{i\bullet} - \overline{Y}_{\bullet j} + \overline{Y}_{\bullet\bullet}$. Show that

(a) $\sum\limits_{i=1}^{k} e_{ij} = 0, j = 1, 2, \ldots, b$ and $\sum\limits_{j=1}^{b} e_{ij} = 0$ for $i = 1, 2, \ldots, k$.

(b) $\sum_i \sum_j (\overline{Y}_{i\bullet} - \overline{Y}_{\bullet\bullet}) e_{ij} = \sum_i \sum_j (\overline{Y}_{\bullet j} - \overline{Y}_{\bullet\bullet}) e_{ij} = 0$.

(c) $\sum_i \sum_j (\overline{Y}_{i\bullet} - \overline{Y}_{\bullet\bullet})(\overline{Y}_{\bullet j} - \overline{Y}_{\bullet\bullet}) = 0$.

6.4 Designs with Two Blocking Variables: Latin Squares

In Section 6.3, we discussed how to design a comparison of k treatments in a randomized complete block design that adjusts for the variability that is introduced by a single blocking variable. Our approach was to assign each of the k treatments at random to each level of the blocking variable. Here, we extend this principle to two blocking variables. We design the experiment such that each of the k treatments is assigned to each level of each of two blocking variables in a special way.

Let us explain our intent with a simple illustration. Suppose we want to compare the yields of $k = 4$ different corn hybrids (*A, B, C,* and *D*) resulting from an experiment carried out on a plot of land that consists of $4 \times 4 = 16$ subplots. (See Table 6.4-1.) Because the soil composition, and thus the "fertility," might vary from

---

Table 6.4-1 Various Blocking Arrangements: Agricultural Experiment

C	*A*	*C*	*B*
D	*A*	*B*	*C*
A	*C*	*A*	*D*
B	*D*	*B*	*D*

(a) Completely randomized

A	*C*	*D*	*B*
B	*C*	*A*	*D*
A	*D*	*C*	*B*
D	*C*	*A*	*B*

(b) One blocking variable (row)

B	*C*	*A*	*D*
D	*A*	*C*	*B*
A	*D*	*B*	*C*
C	*B*	*D*	*A*

(c) Two blocking variables; Latin square

plot to plot, we can expect that the variability in fertility might introduce differences among the treatments. To spread this risk evenly among the treatments, one can assign each treatment to 4 of the 16 subplots at random; this set of assignments corresponds to a completely randomized arrangement. [See Table 6.4-1(a).] However, the precision of the comparisons can probably be increased if we block the experiment. In particular, it is reasonable to assume that fertility might vary from row to row. Treating the rows as the levels of our first blocking variable, we can assign the four treatments to the 4 subplots in each row at random, producing the randomized complete block design in Table 6.4-1(b). However, it is just as reasonable to assume that fertility might vary from column to column. Thus, we want an arrangement in which each column also contains all k treatments. Because there are now two blocking variables—rows and columns—we want a design in which each row and each column contains each of the k treatments exactly once. Such arrangements are called *Latin squares*. A Latin square for $k = 4$ is given in Table 6.4-1(c).

6.4-1 Construction and Randomization of Latin Squares

There exist many different Latin square arrangements for a given number k of treatments. For $k = 3$, there are 12 different tables, in each of which each row and each column receives each treatment exactly once. Table 6.4-2 shows three such tables. The first arrangement is called a *standard Latin square*. It is called "standard" because the treatments in the first row and the first column are arranged in standard (alphabetical) order. For $k = 4$ treatments, there are 576 different Latin squares; for $k = 5$ treatments, there are 161,280. Clearly, the number of different Latin squares increases rapidly with k.

Under randomization, one should select one of these tables at random. Each table has the same chance of being chosen, so the risk of uncontrolled factors is spread evenly among the k treatments. Clearly, it is often not feasible to list all possible Latin squares such that one can be selected at random. A different strategy has to be adopted.

It turns out that standard Latin squares can be utilized to achieve the desired randomization. From each standard Latin square, we can obtain all other Latin squares by permuting rows as well as columns. In Table 6.4-3, we have listed the standard Latin squares for various values of k. For $k = 3$, there is only one standard Latin square; for $k = 4$, there are four. For $k = 5$, there are 56; however, we have shown only one. It is easy to remember the construction of at least one standard Latin square. From the first row, given in standard order, we generate the entries in

Table 6.4-2 Three Possible Latin Squares for $k = 3$

A	B	C
B	C	A
C	A	B

C	B	A
A	C	B
B	A	C

A	C	B
C	B	A
B	A	C

the other rows by rearranging the letters; the first letter of a line becomes the last letter in the subsequent line, while all other letters are shifted to the left by one position. Such an arrangement is also shown for $k = 6$; however, note that there are many other standard Latin squares which are not shown.

In the randomization for $k = 3$, we start with the single Latin square in Table 6.4-3 and rearrange (permute) the order of the rows and columns. For example, the second square in Table 6.4-2 is obtained after we rearrange the rows of the standard Latin square in the order (312) (i.e., the third row becomes the first, the first becomes the second, and the second row becomes the third) and then rearrange the columns of the resulting table in the order (132) (i.e., the first column stays the first, the third column becomes the second, and the second column becomes the third).

For $k = 4$ treatments, there are four different standard Latin squares. Under our randomization procedure, we first select a standard Latin square at random and then randomize the rows and columns. For example, the arrangement in Table 6.4-1(c) is obtained from the fourth standard Latin square in Table 6.4-3 after we arrange the rows in the order (1324) and the columns in the order (2314).

Table 6.4-3 Selected Standard Latin Squares

$k = 3$:

A	B	C
B	C	A
C	A	B

$k = 4$:

A	B	C	D
B	C	D	A
C	D	A	B
D	A	B	C

A	B	C	D
B	A	D	C
C	D	B	A
D	C	A	B

A	B	C	D
B	D	A	C
C	A	D	B
D	C	B	A

A	B	C	D
B	A	D	C
C	D	A	B
D	C	B	A

$k = 5$:

A	B	C	D	E
B	C	D	E	A
C	D	E	A	B
D	E	A	B	C
E	A	B	C	D

$k = 6$:

A	B	C	D	E	F
B	C	D	E	F	A
C	D	E	F	A	B
D	E	F	A	B	C
E	F	A	B	C	D
F	A	B	C	D	E

For $k = 5$ treatments, there are 56 standard Latin squares, and it is not practical to list them and select one at random from these 56 tables. An alternative approach is to start with one particular standard Latin square, such as the one in Table 6.4-3, and then randomize the rows, the columns, and, in addition, the letters that characterize the treatments. The randomization of rows and columns was discussed before. The randomization of the treatments tries to make up for the fact that we did not start with a randomly selected standard Latin square. For example, the permutation (31524) means that the previous letter C replaces A, the previous letter A replaces B, and so on. Even though this particular randomization procedure is not based on all possible Latin squares, it is based on a very large subset that is sufficient for our purposes.

6.4-2 Analysis of Data from a Latin Square

In total, there are k^2 observations Y_{ij}, $i = 1, 2, \ldots, k; j = 1, 2, \ldots, k$. We can calculate the grand total G, the row totals R_i, the column totals C_j, and the treatment totals T_h, $h = 1, \ldots, k$. Furthermore, we can calculate the corresponding averages, say, $\overline{G}, \overline{R}_i, \overline{C}_j$, and $\overline{T}_h$. An analysis-of-variance table (Table 6.4-4) partitions the total variation,

$$\text{SSTO} = \sum_{i=1}^{k} \sum_{j=1}^{k} (Y_{ij} - \overline{G})^2 = \sum_{i=1}^{k} \sum_{j=1}^{k} Y_{ij}^2 - \frac{G^2}{k^2},$$

into four components: sums of squares due to treatment, due to rows, due to columns, and due to error.

The four sums of squares, divided by σ^2, are again independent chi-square random variables with the appropriate number of degrees of freedom. The main interest is usually in testing whether there are significant treatment effects. If $\text{MS}_{\text{Treatment}}/\text{MS}_{\text{Error}}$ exceeds the critical value from an F-distribution with $k - 1$ and $k^2 - 1 - 3(k - 1) = k^2 - 3k + 2$ degrees of freedom, then there is significant evidence against the null hypothesis that there are no treatment differences.

Table 6.4-4 ANOVA Table for the $k \times k$ Latin Square

Source	SS	df	MS
Treatment	$k \sum_{h=1}^{k} (\overline{T}_h - \overline{G})^2 = \frac{1}{k} \sum_{h=1}^{k} T_h^2 - \frac{1}{k^2} G^2$	$k - 1$	$\text{MS}_{\text{Treatment}}$
Row	$k \sum_{i=1}^{k} (\overline{R}_i - \overline{G})^2 = \frac{1}{k} \sum_{i=1}^{k} R_i^2 - \frac{1}{k^2} G^2$	$k - 1$	MS_{Row}
Column	$k \sum_{j=1}^{k} (\overline{C}_j - \overline{G})^2 = \frac{1}{k} \sum_{j=1}^{k} C_j^2 - \frac{1}{k^2} G^2$	$k - 1$	$\text{MS}_{\text{Column}}$
Error	$\text{SSTO} - \text{SS}_{\text{Treatment}} - \text{SS}_{\text{Row}} - \text{SS}_{\text{Column}}$	$k^2 - 3k + 2$	MS_{Error}
Total	SSTO	$k^2 - 1$	

**Example
6.4-1**

The yields (in bushels per tenth of an acre) of the $k = 4$ corn hybrids from our earlier example are given in Table 6.4-5. The row, column, and treatment totals are also given in this table. Calculate the entries of the ANOVA table as an exercise. Or use a computer program such as Minitab and its "Stat > ANOVA > General Linear Model" command, with the responses entered into one column and identifiers for rows, columns, and treatments into the other three. For example, the first row of the Minitab worksheet would contain 10 (for the response), 1 (for the row), 1 (for the column), and 1 (for A), while the last row would contain 14, 4, 4, and 1 (for A). The worksheet column information for rows, columns, and treatments is entered at the model line of the "General Linear Model" command. The output should yield the test statistic $MS_{Treatment}/MS_{Error} = (72.5/3)/(10.5/6) = 13.8$. Because this is larger than $F(0.01; 3, 6) = 9.78$, we conclude that there are differences among the mean yields of the four corn hybrids.

Table 6.4-5 Latin Square for $k = 4$ Corn Hybrids

Rows	Columns				Row Total	Treatment Total
1	10 (A)	14 (B)	7 (C)	8 (D)	39	$T_A = 53$
2	7 (D)	18 (A)	11 (B)	8 (C)	44	$T_B = 44$
3	5 (C)	10 (D)	11 (A)	9 (B)	35	$T_C = 30$
4	10 (B)	10 (C)	12 (D)	14 (A)	46	$T_D = 37$
Column Total	32	52	41	39	$G = 164$	

■

Exercises 6.4

6.4-1 Carry out the computations associated with the Latin square design in Example 6.4-1.

*6.4-2 R. A. Fisher discusses an experiment in which he compares the root weights of five different varieties of marigolds. The experiment was conducted on a plot of land that was divided into 25 subplots: five rows and five columns. The treatments (varieties A through E) were applied according to a Latin square arrangement, with the following results:

	Columns				
Rows	376 (D)	371 (E)	355 (C)	356 (B)	335 (A)
	316 (B)	338 (D)	336 (E)	356 (A)	332 (C)
	326 (C)	326 (A)	335 (B)	343 (D)	330 (E)
	317 (E)	343 (B)	330 (A)	327 (C)	336 (D)
	321 (A)	332 (C)	317 (D)	318 (E)	306 (B)

*(a) Construct the appropriate ANOVA table and test whether there are differences in mean root weight among the five varieties of marigolds.

*(b) Discuss whether this particular blocking arrangement has increased the precision of the comparison. That is, suppose we obtained these numbers in a corresponding one-factor randomized design. Would the inference about the treatments have changed?

[R. A. Fisher, *Statistical Methods for Research Workers* (Edinburgh: Oliver & Boyd, 1925).]

6.4-3 An experiment was conducted to assess the differences in wear of four different types of automobile tires. Because tire wear may vary from one automobile to the next, and because it may also be affected by the position of the wheel on which the tire is mounted, it was decided to block the experiment with respect to these two factors. The four types of tires were assigned to the four wheel positions of four selected automobiles in the following Latin square arrangement, with results shown (tire wear is measured in 1/100 mm per 1,000 miles traveled):

Wheel Position		Car		
1	22 (*B*)	21 (*C*)	25 (*D*)	25 (*A*)
2	29 (*D*)	35 (*A*)	16 (*B*)	23 (*C*)
3	17 (*C*)	37 (*D*)	26 (*A*)	20 (*B*)
4	23 (*A*)	19 (*B*)	18 (*C*)	24 (*D*)

(a) Construct the appropriate ANOVA table and test whether there are significant differences in mean wear among the four types of automobile tires.

(b) Discuss whether this blocking arrangement has been useful.

6.4-4 Four test markets, selected from different geographic regions, were used in a study that investigated the impact of four different advertising strategies on the sale of cheese. Each of the four strategies (*A, B, C,* and *D*) was implemented within each test market during a different three-month period. The sequence in which the advertising levels were tested was selected so that each advertising strategy was used in only one test market during any one period. Approximately 30 supermarkets were recruited within each market. The average cheese sales (in pounds per store, for the three-month period) are as follows:

	Binghamton (NY)	Rockford (IL)	Albuquerque (NM)	Chattanooga (TN)
May–Jul 1972	7,360 (*A*)	11,258 (*B*)	11,800 (*C*)	7,776 (*D*)
Aug–Oct 1972	7,364 (*B*)	13,147 (*D*)	11,852 (*A*)	8,501(*C*)
Nov–Jan 1973	8,049 (*C*)	13,153 (*A*)	11,450 (*D*)	7,900(*B*)
Feb–Apr 1973	9,010 (*D*)	13,880 (*C*)	12,089 (*B*)	7,557(*A*)

Analyze the data. Test the effect of advertising. What can you say about the location and the time effects?

Note: Chapters 6 and 7 discuss the design of experiments and the analysis of the resulting data. We have combined the Additional Remarks and the Projects for these two chapters so that they appear at the end of Chapter 7.

Chapter

7

EXPERIMENTS WITH TWO OR MORE FACTORS

7.1 Two-Factor Factorial Designs

In Chapter 6, we investigated the differences among k levels (or treatments) of a *single factor.* Here, we study the response to changes in *two factors;* factor A, observed at factor levels $1, 2, \ldots, a$; and factor B, observed at factor levels $1, 2, \ldots, b$. For example, an engineer in a textile mill may be interested in the effects of temperature and cycle time on the brightness of a fabric in a process involving dye. He or she may study the process at three levels of temperature (350, 375, and 400°F) and two levels of cycle time (40 and 50 cycles), for a total of six factor–level combinations.

Suppose that we have conducted exactly n experiments at every possible factor–level combination. Suppose also that (i, j) represents the combination of the ith level of factor A with the jth level of factor B, where $i = 1, 2, \ldots, a$ and $j = 1, 2, \ldots, b$. Moreover, let us denote the kth observation in the (i, j) factor–level combination, or cell, as Y_{ijk}. Since $i = 1, \ldots, a; j = 1, \ldots, b;$ and $k = 1, \ldots, n$, we have a total of abn observations that can be arranged as in Table 7.1-1. We call such an arrangement an $a \times b$ *factorial design.* The factors A and B are said to be *completely crossed,* because every level of A is combined with every level of B. Here, we consider only the *balanced* case, in which the numbers of observations in the cells are the same.

We assume that the observations in cell (i, j) constitute a random sample from an $N(\mu_{ij}, \sigma^2)$ distribution. The mean levels μ_{ij} may change from cell to cell, but the variances are assumed to be the same. Furthermore, we suppose that the ab random samples, each of size n, are independent. In other words, we assume a completely randomized arrangement.

More formally, we can write the probability model that is generating our observations as

$$Y_{ijk} = \mu_{ij} + \varepsilon_{ijk}, \qquad i = 1, \ldots, a \text{ (factor } A),$$

$$j = 1, \ldots, b \text{ (factor } B),$$

$$k = 1, \ldots, n \text{ (replications)},$$

Table 7.1-1 Observations from a Two-Factor Factorial Design

Factor A	Factor B			
	1	2	$\cdots$	b
1	$Y_{111}, Y_{112}, \ldots, Y_{11n}$	$Y_{121}, Y_{122}, \ldots, Y_{12n}$	$\cdots$	$Y_{1b1}, Y_{1b2}, \ldots, Y_{1bn}$
2	$Y_{211}, Y_{212}, \ldots, Y_{21n}$	$Y_{221}, Y_{222}, \ldots, Y_{22n}$	$\cdots$	$Y_{2b1}, Y_{2b2}, \ldots, Y_{2bn}$
$\vdots$	$\vdots$	$\vdots$		$\vdots$
a	$Y_{a11}, Y_{a12}, \ldots, Y_{a1n}$	$Y_{a21}, Y_{a22}, \ldots, Y_{a2n}$	$\cdots$	$Y_{ab1}, Y_{ab2}, \ldots, Y_{abn}$

where the ε_{ijk} are independent random variables from an $N(0, \sigma^2)$ distribution. This model contains ab different mean levels μ_{ij}, which can also be expressed in the following equivalent, but more informative, representation:

Define the overall mean of the ab mean levels in Table 7.1-2 as

$$\bar{\mu}_{\bullet\bullet} = \frac{1}{ab}\sum_{i=1}^{a}\sum_{j=1}^{b}\mu_{ij},$$

the row (factor A) means as

$$\bar{\mu}_{i\bullet} = \frac{1}{b}\sum_{j=1}^{b}\mu_{ij} \qquad i = 1, 2, \ldots, a,$$

and the column (factor B) means as

$$\bar{\mu}_{\bullet j} = \frac{1}{a}\sum_{i=1}^{a}\mu_{ij} \qquad j = 1, 2, \ldots, b.$$

The respective differences $\alpha_i = \bar{\mu}_{i\bullet} - \bar{\mu}_{\bullet\bullet}, i = 1, 2, \ldots, a$, and $\beta_j = \bar{\mu}_{\bullet j} - \bar{\mu}_{\bullet\bullet}$, $j = 1, \ldots, b$, are called the row (factor A) and column (factor B) effects. They are also called *main effects*, because they express the effects of one factor after averaging over all levels of the other factor. There is one restriction among the a row effects: Deviations from a mean must add to zero; thus, $\sum_{i=1}^{a}\alpha_i = 0$. Similarly, there is one restriction among the b column effects: $\sum_{j=1}^{b}\beta_j = 0$.

A model in which the cell means can be written as $\mu_{ij} = \bar{\mu}_{\bullet\bullet} + \alpha_i + \beta_j$ is called a *main-effects* or *additive* model. In such a model, the effect of factor A does not depend on the level of factor B, and vice versa. In some instances, however, we may find that the effect of one factor depends on the level of the second; in that case, μ_{ij} is

Table 7.1-2 Mean Levels in an $a \times b$ Factorial Arrangement

Factor A	Factor B				
	1	2	$\cdots$	b	Row Mean
1	μ_{11}	μ_{12}	$\cdots$	μ_{1b}	$\bar{\mu}_{1\bullet}$
2	μ_{21}	μ_{22}	$\cdots$	μ_{2b}	$\bar{\mu}_{2\bullet}$
$\vdots$	$\vdots$	$\vdots$		$\vdots$	$\vdots$
a	μ_{a1}	μ_{a2}	$\cdots$	μ_{ab}	$\bar{\mu}_{a\bullet}$
Column Mean	$\bar{\mu}_{\bullet 1}$	$\bar{\mu}_{\bullet 2}$	$\cdots$	$\bar{\mu}_{\bullet b}$	$\bar{\mu}_{\bullet\bullet}$

different from $\bar{\mu}_{\bullet\bullet} + \alpha_i + \beta_j$. Then we say that there exists an *interaction effect*, which is defined as the difference

$$(\alpha\beta)_{ij} = \mu_{ij} - (\bar{\mu}_{\bullet\bullet} + \alpha_i + \beta_j)$$

$$= \mu_{ij} - \bar{\mu}_{i\bullet} - \bar{\mu}_{\bullet j} + \bar{\mu}_{\bullet\bullet}, \quad i = 1,\ldots,a; \quad j = 1,\ldots,b.$$

There are ab interaction effects; it is easy to show that $\sum_{j=1}^{b}(\alpha\beta)_{ij} = 0$, for $i = 1,\ldots,a$, and $\sum_{i=1}^{a}(\alpha\beta)_{ij} = 0$, for $j = 1,\ldots,b$. That is, the row and column sums of the interactions $(\alpha\beta)_{ij}$, arranged in a table similar to Table 7.1-2, are zero. Because of these restrictions, it is sufficient to specify only $(a-1)(b-1)$ of these interactions.

The two illustrations in Table 7.1-3, with two levels for factor A and three levels for factor B, illustrate these concepts. The first is a 2×3 factorial arrangement without interaction, whereas the second includes interaction terms. In Figure 7.1-1, we display the means separately for levels 1 and 2 of factor A. Alternatively, we could have displayed the means separately for the three levels of B. In illustration 1, with no interaction, the effects of moving from level 1 to level 2 of factor A are the same at all levels of factor B, and the two corresponding graphs in Figure 7.1-1 are parallel. In illustration 2, this is not the case. For example, the effect is positive at level 1 of B, but is negative at level 3 of B. The two factors interact, so the corresponding graphs in Figure 7.1-1 are not parallel.

Our discussion and the definition of row, column, and interaction effects have shown that we can always reparameterize the ab cell means as

$$\mu_{ij} = \bar{\mu}_{\bullet\bullet} + \alpha_i + \beta_j + (\alpha\beta)_{ij}.$$

Table 7.1-3 Two Examples of Cell Means

	Illustration 1					Illustration 2			
	Factor *B*					Factor *B*			
Factor *A*	1	2	3	$\bar{\mu}_{i\bullet}$	Factor *A*	1	2	3	$\bar{\mu}_{i\bullet}$
1	1	2	6	3	1	1	2	6	3
2	3	4	8	5	2	5	8	2	5
$\bar{\mu}_{\bullet j}$	2	3	7	4	$\bar{\mu}_{\bullet j}$	3	5	4	4

$\alpha_1 = 3 - 4 = -1$ $\beta_1 = 2 - 4 = -2$
$\alpha_2 = 5 - 4 = 1$ $\beta_2 = 3 - 4 = -1$
$\beta_3 = 7 - 4 = 3$

$\alpha_1 = 3 - 4 = -1$ $\beta_1 = 3 - 4 = -1$
$\alpha_2 = 5 - 4 = 1$ $\beta_2 = 5 - 4 = 1$
$\beta_3 = 4 - 4 = 0$

$(\alpha\beta)_{11} = 1 - 3 - 2 + 4 = 0$
$(\alpha\beta)_{12} = 2 - 3 - 3 + 4 = 0$
$(\alpha\beta)_{13} = 6 - 3 - 7 + 4 = 0$

$(\alpha\beta)_{21} = 3 - 5 - 2 + 4 = 0$
$(\alpha\beta)_{22} = 4 - 5 - 3 + 4 = 0$
$(\alpha\beta)_{23} = 8 - 5 - 7 + 4 = 0$

$(\alpha\beta)_{11} = 1 - 3 - 3 + 4 = -1$
$(\alpha\beta)_{12} = 2 - 3 - 5 + 4 = -2$
$(\alpha\beta)_{13} = 6 - 3 - 4 + 4 = 3$

$(\alpha\beta)_{21} = 5 - 5 - 3 + 4 = 1$
$(\alpha\beta)_{22} = 8 - 5 - 5 + 4 = 2$
$(\alpha\beta)_{23} = 2 - 5 - 4 + 4 = -3$

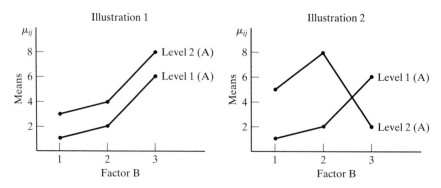

Figure 7.1-1 Population means for the two illustrations in Table 7.1-3

Substituting this expression into the data-generating model

$$Y_{ijk} = \mu_{ij} + \varepsilon_{ijk}$$

leads to

$$Y_{ijk} = \overline{\mu}_{\bullet\bullet} + \alpha_i + \beta_j + (\alpha\beta)_{ij} + \varepsilon_{ijk}$$
$$Y_{ijk} = \overline{\mu}_{\bullet\bullet} + (\overline{\mu}_{i\bullet} - \overline{\mu}_{\bullet\bullet}) + (\overline{\mu}_{\bullet j} - \overline{\mu}_{\bullet\bullet}) + (\mu_{ij} - \overline{\mu}_{i\bullet} - \overline{\mu}_{\bullet j} + \overline{\mu}_{\bullet\bullet}) + \varepsilon_{ijk}.$$

The parameters μ_{ij} and, consequently, the implied effects α_i, β_j, and $(\alpha\beta)_{ij}$ are unknown and must be estimated from sample data. From the observations in Table 7.1-1, we estimate the means, $\mu_{ij}, \overline{\mu}_{i\bullet}, \overline{\mu}_{\bullet j}$, and $\overline{\mu}_{\bullet\bullet}$, by the respective sample averages,

$$\overline{Y}_{ij\bullet} = \frac{1}{n}Y_{ij\bullet} = \frac{1}{n}\sum_{k=1}^{n} Y_{ijk},$$

$$\overline{Y}_{i\bullet\bullet} = \frac{1}{bn}Y_{i\bullet\bullet} = \frac{1}{bn}\sum_{j=1}^{b}\sum_{k=1}^{n} Y_{ijk},$$

$$\overline{Y}_{\bullet j\bullet} = \frac{1}{an}Y_{\bullet j\bullet} = \frac{1}{an}\sum_{i=1}^{a}\sum_{k=1}^{n} Y_{ijk},$$

$$\overline{Y}_{\bullet\bullet\bullet} = \frac{1}{abn}Y_{\bullet\bullet\bullet} = \frac{1}{abn}\sum_{i=1}^{a}\sum_{j=1}^{b}\sum_{k=1}^{n} Y_{ijk}.$$

Thus, the row effects α_i are estimated by $\overline{Y}_{i\bullet\bullet} - \overline{Y}_{\bullet\bullet\bullet}$, the column effects β_j by $\overline{Y}_{\bullet j\bullet} - \overline{Y}_{\bullet\bullet\bullet}$, and the interaction effects $(\alpha\beta)_{ij}$ by $\overline{Y}_{ij\bullet} - \overline{Y}_{i\bullet\bullet} - \overline{Y}_{\bullet j\bullet} + \overline{Y}_{\bullet\bullet\bullet}$. Approximations of the errors ε_{ijk} are given by the residuals $e_{ijk} = Y_{ijk} - \overline{Y}_{ij\bullet}$.

Corresponding to the data-generating model, we observe the data decomposition

$$Y_{ijk} = \overline{Y}_{\bullet\bullet\bullet} + (\overline{Y}_{i\bullet\bullet} - \overline{Y}_{\bullet\bullet\bullet}) + (\overline{Y}_{\bullet j\bullet} - \overline{Y}_{\bullet\bullet\bullet})$$
$$+ (\overline{Y}_{ij\bullet} - \overline{Y}_{i\bullet\bullet} - \overline{Y}_{\bullet j\bullet} + \overline{Y}_{\bullet\bullet\bullet}) + (Y_{ijk} - \overline{Y}_{ij\bullet}).$$

The corresponding sum-of-squares decomposition is

$$\text{SSTO} = \text{SS}_A + \text{SS}_B + \text{SS}_{AB} + \text{SS}_{\text{Error}}$$

where

$$\text{SSTO} = \sum_i \sum_j \sum_k (Y_{ijk} - \bar{Y}_{...})^2 = \sum_i \sum_j \sum_k (Y_{ijk})^2 - \frac{(Y_{...})^2}{abn},$$

$$\text{SS}_A = bn \sum_i (\bar{Y}_{i..} - \bar{Y}_{...})^2 = \frac{1}{bn} \sum_i (Y_{i..})^2 - \frac{(Y_{...})^2}{abn},$$

$$\text{SS}_B = an \sum_j (\bar{Y}_{\cdot j \cdot} - \bar{Y}_{...})^2 = \frac{1}{an} \sum_j (Y_{\cdot j \cdot})^2 - \frac{(Y_{...})^2}{abn},$$

$$\text{SS}_{AB} = n \sum_i \sum_j (\bar{Y}_{ij\cdot} - \bar{Y}_{i..} - \bar{Y}_{\cdot j \cdot} + \bar{Y}_{...})^2$$

$$= \frac{1}{n} \sum_i \sum_j (Y_{ij\cdot})^2 - \frac{(Y_{...})^2}{abn} - \text{SS}_A - \text{SS}_B,$$

and

$$\text{SS}_{\text{Error}} = \text{SSTO} - \text{SS}_A - \text{SS}_B - \text{SS}_{AB}$$

are the respective sums of squares associated with the total, factor A, factor B, interaction, and error components. It is easy to show this decomposition: We write $(Y_{ijk} - \bar{Y}_{...})$ as the sum of the four terms in the data decomposition, and then we square the expression $(Y_{ijk} - \bar{Y}_{...})$ and sum the resulting components over all replications $k(k = 1, \ldots, n)$, levels of factor $A(i = 1, \ldots, a)$, and levels of factor $B(j = 1, \ldots, b)$. Note that all cross-product terms sum to zero. The sums of squares are displayed in the ANOVA table in Table 7.1-4.

A short note on the computation of these sums of squares is in order. Computer programs are usually available to obtain the entries in the ANOVA table, so there is no need for hand calculations. We explain how to utilize such software at the end of Section 7.1. However, knowledge of what these sums of squares measure also makes it straightforward to calculate them by hand if no computer is available. Note that it is necessary to calculate only four sums of squares, as the fifth one can be obtained from the identity. It is probably easiest to calculate SSTO first; this is the sum of the squared deviations from the overall mean. Next, we calculate SS_A, the sum of the squared distances from the row means to the overall mean, multiplied by nb, the number of observations that go into calculating each row mean. SS_B is calculated in a similar manner. The error sum of squares, SS_{Error}, is also easy to calculate: We

Table 7.1-4 Analysis-of-Variance Table in the Two-Factor Factorial Experiment: Fixed Effects

Source	SS	df	MS	F
A	SS_A	$a - 1$	MS_A	$F_A = \text{MS}_A/\text{MS}_{\text{Error}}$
B	SS_B	$b - 1$	MS_B	$F_B = \text{MS}_B/\text{MS}_{\text{Error}}$
AB	SS_{AB}	$(a - 1)(b - 1)$	MS_{AB}	$F_{AB} = \text{MS}_{AB}/\text{MS}_{\text{Error}}$
Error	SS_{Error}	$ab(n - 1)$	MS_{Error}	
Total	SSTO	$abn - 1$		

simply add up the ab sums of squares that are associated with the ab cells; each of these sums is computed by summing the squared deviations from the respective cell mean $\overline{Y}_{ij\bullet}$. The interaction sum of squares can then be obtained from the identity $\text{SS}_{AB} = \text{SSTO} - \text{SS}_A - \text{SS}_B - \text{SS}_{\text{Error}}$. Of course, one can also calculate these sums from the foregoing computation equations. The computations are a little more difficult, but, in a sense, better, as they are not subject to so many rounding errors.

The third column in the ANOVA table lists the degrees of freedom for the various sums of squares. Because of the restrictions among the main and interaction effects, there are $(a - 1)$ degrees of freedom for SS_A, $(b - 1)$ for SS_B, and $(a - 1)(b - 1)$ for SS_{AB}. The degrees of freedom for SS_{Error} are $ab(n - 1)$ as we combine ab error sums of squares, each with $(n - 1)$ degrees of freedom. The sum of these degrees of freedom equals $abn - 1$, which is the number of degrees of freedom for SSTO. The fourth column lists the mean squares and the fifth gives the respective F-statistics, which are explained later.

The expected values of these mean squares suggest how to conduct tests of certain null hypotheses—that is, hypotheses about interactions in which $(\alpha\beta)_{ij} = 0$ for all $i = 1, \ldots, a$ and $j = 1, \ldots, b$; about main effects of factor A, in which $\alpha_1 = \ldots = \alpha_a = 0$; and about main effects of factor B, in which $\beta_1 = \ldots = \beta_b = 0$. It is quite useful to know whether interactions are present, as their absence implies that the effects are additive and that the effects of changes in factor A do not depend on the particular level of factor B. Furthermore, it is important to know about main effects, because they tell us about the effects of factor A (or B) on the mean response.

In the derivation of the expected mean squares, we must distinguish between *fixed* and *random effects*. In the *fixed-effects model*, we assume that the α_i, β_j, and $(\alpha\beta)_{ij}$ are nonrandom unknown quantities and that our comparisons are restricted to the specifically chosen factor levels. The only random components are the errors ε_{ijk}. After some algebra, which is similar to that in the one-factor comparison, it can be shown that

$$E(\text{MS}_A) = \sigma^2 + bn\frac{\sum_{i=1}^{a}\alpha_i^2}{a - 1},$$

$$E(\text{MS}_B) = \sigma^2 + an\frac{\sum_{j=1}^{b}\beta_j^2}{b - 1},$$

$$E(\text{MS}_{AB}) = \sigma^2 + n\frac{\sum_{i=1}^{a}\sum_{j=1}^{b}[(\alpha\beta)_{ij}]^2}{(a - 1)(b - 1)},$$

and

$$E(\text{MS}_{\text{Error}}) = \sigma^2.$$

Because $\text{SS}_A, \text{SS}_B, \text{SS}_{AB}$, and SS_{Error}, once divided by σ^2, can be shown to be independent chi-square random variables, the F-ratios in Table 7.1-4 are the appropriate statistics for testing the significance of main and interaction effects. The statistic $F_{AB} = \text{MS}_{AB} / \text{MS}_{\text{Error}}$ tests the significance of the interactions through a comparison with the critical value $F[\alpha; (a - 1)(b - 1), ab(n - 1)]$ for a test with significance level α. The statistic $F_A = \text{MS}_A / \text{MS}_{\text{Error}}$ tests the significance of the factor A main effects by a comparison with the upper percentage point $F[\alpha; a - 1, ab(n - 1)]$. Finally, $F_B = \text{MS}_B / \text{MS}_{\text{Error}}$ provides a test of the significance of the factor B main effects via a comparison with $F[\alpha; b - 1, ab(n - 1)]$.

The significance of the main effects of a factor is determined by comparing the averages of the observations at the various levels of that factor. Because the averaging is over all levels of the other factor, it is meaningful to talk about and test for the significance of main effects only if there are fairly small and unimportant interactions. If there are large interaction effects, we should ignore tests of main effects and display the nature of the interactions by plotting and connecting the sample averages, as we did with the population means μ_{ij} in Figure 7.1-1. We refer to such plots as *interaction plots*.

Frequently, transformations of the response, such as $\log Y$, $1/Y$, or $\sqrt{Y}$, take a model with interactions into one without interactions. A simple illustration in which the logarithmic transformation leads to a simplification is the multiplicative model $\mu_{ij}^* = \mu^* \alpha_i^* \beta_j^*$, which becomes

$$\log \mu_{ij}^* = \log \mu^* + \log \alpha_i^* + \log \beta_j^*, \quad \text{or} \quad \mu_{ij} = \mu + \alpha_i + \beta_j$$

in a notation that is obvious.

The reader should be aware that there can be situations in which we observe zero main effects (namely, $\alpha_i = 0$ for all $i = 1, \ldots, a$), but some interactions are nonzero. Such a situation does not imply that changes in factor A have no effect on the mean response. They do have an effect; however, because of the nonzero interactions, the effects of changes in A depend on the level of factor B. It would be a mistake to gloss over the important interactions and conclude that A is not important just because the effects of changes in A "average out" to zero when averaging is done over all levels of B. To emphasize this point, take $a = 2, b = 3, \mu_{11} = 1, \mu_{12} = -2, \mu_{13} = 1, \mu_{21} = -1, \mu_{22} = 2$, and $\mu_{23} = -1$. Then all main effects of A (as well as of B) are zero, but it is obvious that changes in factor A affect the mean response: The change in A from level 1 to level 2 leads to a reduction of 2 units if B is at level 1, an increase of 4 units if B is at level 2, and a reduction of 2 units if B is at level 3.

If no interaction effects are present, we should conduct *follow-up tests* on the row (factor A) and column (factor B) means. We can display the row averages $\overline{Y}_{i \bullet \bullet}$, together with their reference distribution, which in this case is the scaled $t[v = ab(n - 1)]$ distribution, with scale factor $\sqrt{\text{MS}_{\text{Error}}/bn}$. Similarly, we can display the column averages $\overline{Y}_{\bullet j \bullet}$, together with their reference distribution, which is the scaled $t[v = ab(n - 1)]$ with scale factor $\sqrt{\text{MS}_{\text{Error}}/an}$.

Diagnostic checks should always be conducted before the conclusions drawn from an analysis are adopted. As before, the primary tool is residual analysis, where the residuals $e_{ijk} = Y_{ijk} - \overline{Y}_{ij \bullet}$ are the deviations of the observations from their respective cell averages. A dot diagram, a histogram, or a normal probability plot of these residuals allows us to check whether the error distribution is approximately normal. Dot diagrams of the residuals, drawn separately for each of the ab cells, but displayed on the same scale, provide information as to whether the variances in the ab cells are approximately the same. Also, a plot of the residuals e_{ijk} against the fitted values $\overline{Y}_{ij \bullet}$ should show no apparent nonrandom patterns. A pattern in which the variability is related to the level of the cell would violate the equal-variance assumption.

Example 7.1-1   Suppose the yield (in cups of popped corn from 1/4 cup of popcorn) is the variable of interest. We wish to test the effects of two factors. Factor A is the type of popcorn maker; we compare an "oil-based" popper (level 1) with a certain "air-based" popper (level 2). Factor B is popcorn brand; level 1 corresponds to a gourmet-type popcorn

Table 7.1-5 Results for Popcorn Example

Popper	Gourmet	National Brand	Generic	Row Sum	Row Average
Oil	5.5, 5.5, 6	4.5, 4.5, 4	3.5, 4, 3	40.5	4.50
Air	6.5, 7, 7	5, 5.5, 5	4, 5, 4.5	49.5	5.50
Column Sum	37.5	28.5	24.0	$y_{\cdots} = 90.0$	
Column Average	6.25	4.75	4.00	$\bar{y}_{\cdots} = 5.00$	
				$\sum\sum\sum y_{ijk}^2 = 472$	

(its cost is \$2.20 per pound), level 2 is a national brand (cost \$1.00 per pound), and level 3 is a generic brand (cost \$0.76 per pound). A 2×3 factorial experiment with $n = 3$ replications was performed, where the order of the 18 experiments was randomized to spread the variability that results from other uncontrolled factors evenly among all treatment groups. Table 7.1-5 lists the outcomes from these experiments. From the data, we calculate the following sums of squares:

$$\text{SSTO} = 472 - \frac{(90)^2}{18} = 22;$$

$$\text{SS}_A = \frac{1}{9}\left[(40.5)^2 + (49.5)^2\right] - \frac{(90)^2}{18} = 4.5;$$

$$\text{SS}_B = \frac{1}{6}\left[(37.5)^2 + (28.5)^2 + (24)^2\right] - \frac{(90)^2}{18} = 15.75;$$

$$\text{SS}_{AB} = \frac{1}{3}\left[(17)^2 + (20.5)^2 + (13)^2 + (15.5)^2 + (10.5)^2 + (13.5)^2\right] - \frac{(90)^2}{18} - 4.5 - 15.75$$

$$= 0.0833;$$

$$\text{SS}_{\text{Error}} = 22 - 4.5 - 15.75 - 0.0833 = 1.6667.$$

Alternatively, we could have calculated these sums of squares from their definitions. Either way, we obtain the following ANOVA table:

Source	SS	df	MS	F
A (Popper)	4.5	1	4.5	32.4
B (Brand)	15.75	2	7.8750	56.7
AB	0.0833	2	0.0417	0.3
Error	1.6667	12	0.1389	
Total		17		

You can confirm this table with your available computer software. There is no indication of an interaction: $F_{AB} = 0.3$ is much smaller than the critical value $F(0.01; 2, 12) = 6.93$. Also, the interaction plot of the cell averages in Figure 7.1-2(a) shows the absence of interaction: The two graphs are almost parallel. Thus, it is appropriate to investigate the factor main effects. Both factors (popper and brand) are highly significant: 32.4 compared with $F(0.01; 1, 12) = 9.33$, and 56.7 compared

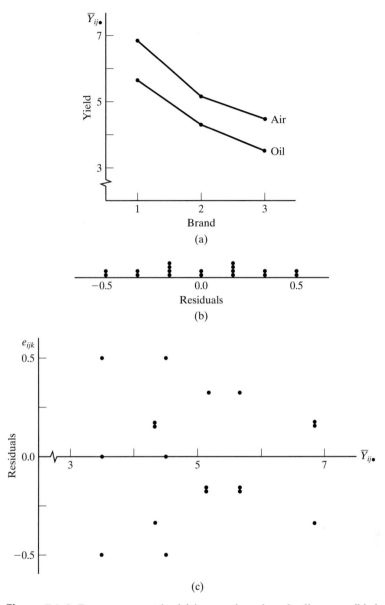

Figure 7.1-2 Popcorn example: (a) interaction plot of cell means; (b) dot diagram of the residuals (c) plot of residuals against fitted values

with $F(0.01; 2, 12) = 6.93$. That is, there is very strong evidence that these differences cannot be explained as being chance results. The dot diagram of the residuals [Figure 7.1-2(b)] and the plot of the residuals against fitted values [Figure 7.1-2(c)] do not point to any major inadequacies of the fitted model. The variability of the residuals that correspond to low fitted values does appear to be a bit larger; however, due to the small sample size, it is difficult to tell whether this is a serious violation of the assumption of constant variance.

Our analysis shows that the gourmet brand leads to a much higher yield than that of the other two brands. However, if we take the differences in price into account, the conclusion changes. Using the cost of the generic popcorn as a base and recalling that it yields 4 cups, on average, we find that the national brand yields $(4.75)(0.76)/(1.00) = 3.61$ cups for the same cost and the gourmet brand only $(6.25)(0.76)/(2.20) = 2.16$ cups. Thus, judging on volume alone, it does not pay to buy the gourmet brand. Also, the air-based popper is much better than the oil-based popper, using volume yield as a measure. Clearly, taste will play some role in the determination of the type of popcorn and popper a consumer will purchase. ∎

7.1-1 Graphics in the Analysis of Two-Factor Experiments

It is always an excellent strategy to display the response averages for the various levels of each factor. Furthermore, in order to recognize the joint effects of factors A and B, we recommend that, for each level of one factor, you plot and connect the cell means that correspond to the various levels of the other factor. You can overlay these graphs, as we have done in the interaction graphs in Figures 7.1-1 and 7.1-2. Roughly parallel curves suggest a model in which the effects of A and B are additive; that is, the interaction is zero.

If the number n of observations in each cell is fairly large, we can also calculate the quartiles, the interquartile range, and the standard deviation for each of the ab cells. Box-and-whisker diagrams can be prepared for each cell, and the ab diagrams can be put side by side. For example, we can display the b box-and-whisker diagrams for level 1 of factor A; next to them, on the same scale, the b diagrams that correspond to level 2 of A; and so on. Also, instead of displaying these a groups of diagrams (each of size b) side by side, we can try to overlay the box plots, as we have done with the cell averages in Figures 7.1-1 and 7.1-2. Looking across the interquartile ranges in these diagrams, we can assess whether the variability in these cells is roughly the same. It is important to check this property, as the F-test in the analysis of variance assumes equal variances. If the variability is not constant, we should check whether it is related to the level of the observations. A plot of the ab interquartile ranges or the standard deviations against the cell averages is often helpful in selecting an appropriate transformation. If the interquartile ranges (or standard deviations) grow in proportion to the cell averages, then the logarithm of the response is the appropriate transformation. If they grow in proportion to the square root of the cell averages, then the square root of the response is a better transformation.

If we look across the box-and-whisker diagrams, we can also check whether the patterns among the first b diagrams [those for cells $(1, 1), (1, 2), \ldots, (1, b)$] are different from the patterns in the second group of b diagrams [those for cells $(2, 1)$, $(2, 2), \ldots, (2, b)$], in the third group, and so on. If they are just shifted by constant amounts, we can adopt an additive model. In such a situation, we can average over the other factor. For example, we can ignore factor B and, for each level of factor A, construct a box-and-whisker diagram from the bn observations; these diagrams can then be compared side by side. A similar plot can be shown for the b levels of factor B.

Usually, there is no single best way of displaying a data set. It depends on the particular problem and, of course, the imagination of the investigator.

7.1-2 Special Case: $n = 1$

If there is only a single observation Y_{ij} in each cell, then the model with $\mu_{ij} = \overline{\mu}_{\bullet\bullet} + \alpha_i + \beta_j + (\alpha\beta)_{ij}$ includes the same number of parameters as observations. Thus, the error sum of squares is zero, and it is not possible to make inferences about the parameters. Accordingly, simplifying assumptions must be made about the model before we can make inferences. One such simplification assumes that the model is additive: $\mu_{ij} = \overline{\mu}_{\bullet\bullet} + \alpha_i + \beta_j$. This assumption implies that the interaction sum of squares can be used as the error sum of squares. It then follows that the analysis is the same as the one in the randomized complete block arrangement (Section 6.3), with the second factor replacing the blocking variable.

7.1-3 Random Effects

So far, we have assumed that the factor levels of A and B are fixed. Now let us assume that the factor levels are chosen at random from larger populations; thus, our inference is about all possible levels in these populations, not just the ones we have selected. The terms, except $\overline{\mu}_{\bullet\bullet}$, in the model

$$Y_{ijk} = \overline{\mu}_{\bullet\bullet} + \alpha_i + \beta_j + (\alpha\beta)_{ij} + \varepsilon_{ijk}$$

are now random variables: The α_i values are from $N(0, \sigma_\alpha^2)$, the β_j values from $N(0, \sigma_\beta^2)$, the $(\alpha\beta)_{ij}$ values from $N(0, \sigma_{\alpha\beta}^2)$, and the ε_{ijk} from $N(0, \sigma^2)$. In addition, we assume that these random variables are mutually independent.

The basic analysis-of-variance table in the random-effects model is the same as that for the fixed-effects model; the sums of squares, degrees of freedom, and mean squares are calculated the same way. However, the expected mean squares are now

$$E(MS_A) = \sigma^2 + n\sigma_{\alpha\beta}^2 + bn\sigma_\alpha^2,$$
$$E(MS_B) = \sigma^2 + n\sigma_{\alpha\beta}^2 + an\sigma_\beta^2,$$
$$E(MS_{AB}) = \sigma^2 + n\sigma_{\alpha\beta}^2,$$

and

$$E(MS_{Error}) = \sigma^2.$$

Thus, in the test of main effects, the correct denominator in the F-ratios is the interaction mean square MS_{AB}, not MS_{Error} as in the fixed-effects model. For example, to test whether there is variability due to factor A ($H_0: \sigma_\alpha^2 = 0$), we compare $F = MS_A / MS_{AB}$ against the critical value $F[\alpha; (a-1), (a-1)(b-1)]$. The entries in the mean-square column of the ANOVA table can also be used to estimate the variance components. It is straightforward to show that these estimates are as follows:

$$\hat{\sigma}^2 = MS_{Error}; \qquad \hat{\sigma}_{\alpha\beta}^2 = \frac{MS_{AB} - MS_{Error}}{n};$$

$$\hat{\sigma}_\beta^2 = \frac{MS_B - MS_{AB}}{an}; \qquad \hat{\sigma}_\alpha^2 = \frac{MS_A - MS_{AB}}{bn}.$$

Consider the example in Exercise 7.1-5, in which a manufacturer of 13-inch brake wheels is concerned about the quality of its production. Here, quality is measured by the absolute values of the differences between specified and actual diameters. Because operators and machines were thought to be important factors, the quality control engineer selected three operators and three machines at random and took samples of size $n = 5$ at each of the nine operator–machine combinations.

7.1-4 Computer Software

Minitab's command "Stat > ANOVA > Two-Way" can be used. The spreadsheet containing the data consists of three columns: column 1, which contains the response (the yield of popcorn); column 2, which lists the popper (1 or 2); and column 3, which contains the brand (1, 2, and 3). There are 18 rows to each column. Row 1 contains 5.5 in column 1, 1 in column 2, and 1 in column 3. Row 2 contains 5.5 in column 1, 1 in column 2, and 1 in column 3; ...; and the 18th row contains 4.5 in column 1, 2 in column 2, and 3 in column 3. The order in which the rows are entered is arbitrary. If you want to fit an additive model without the interaction term, click the "Fit Additive Model" box; otherwise do not check that box. The output lists the ANOVA table. Residuals and fitted values can be stored, and various residual diagnostic checks are available. The commands "Stat > ANOVA > Main Effects Plots" and "Stat > ANOVA > Interactions Plots" provide plots of the main effects and the interactions, respectively.

Alternatively, you can use Minitab's "Stat > ANOVA > Balanced ANOVA" command. In the model line, you need to enter column 2, column 3, and the product column1* column2 (for the interaction). Leaving off this last term will give you the analysis of the additive model.

Note that the "Stat > ANOVA > Two-Way" and "Stat > ANOVA > Balanced ANOVA" commands will not work if you have an unbalanced situation. Suppose that you planned for three replications in your popcorn example, but that one experiment failed and that there are only two responses in one of the six cells of Table 7.1-5. Then the program will give you an error message telling you that you have an unbalanced situation. The problem is that the additive sums-of-squares decomposition into unconditioned sums-of-squares components (factor1, factor2, interaction, and error) works only in the balanced case. In the unbalanced case, the sums of the various cross-product terms are no longer zero. You can use the "Stat > ANOVA > General Linear Model" command, and you will get an ANOVA table. But we will need to learn more about linear models and regression before we can interpret this table. We will come back to the analysis of the unbalanced situation after our discussion of regression. (See Section 8.5-1.)

Exercises 7.1

7.1-1 Consider the following 2 × 4 table of cell means, μ_{ij}, $i = 1, 2; j = 1, 2, 3, 4$:

	Factor B			
Factor A	1	2	3	4
1	2	5	6	6
2	4	5	7	9

Calculate the row, column, and interaction effects.

7.1-2 For the popcorn example, plot the three brand averages and their reference distribution. Plot the two popper averages and their reference distribution.

*7.1-3 A company wants to evaluate the effectiveness of three different training methods (factor B) designed to improve the job performance of its entering sales workforce. New workers are classified on the basis of their educational background (factor A): high school and college graduates. A random sample of 12 high school graduates is randomly divided into three groups of four. Each subgroup is then randomly assigned to one of the three job training programs. An analogous procedure is followed for a random sample of 12 college graduates. The sales (in $10,000) for the first quarter after the training of the new employees are as follows:

		Training	
Education	1	2	3
High School	8, 4, 2, 3	10, 8, 6, 7	8, 6, 4, 7
College	14, 10, 6, 9	8, 9, 7, 10	15, 13, 9, 12

Use graphs to display the information presented in the table. Calculate the ANOVA table and conduct the appropriate tests. Summarize your findings. Check the assumptions of your model.

*7.1-4 An engineer in a textile mill studies the effect of temperature and time on the brightness of a synthetic fabric in a process involving dye. Several small randomly selected fabric specimens were dyed under each temperature and time combination. The brightness of the dyed fabric was measured on a 50-point scale, and the results of the investigation are as follows:

	Temperature (degrees Fahrenheit)		
Time (cycles)	350	375	400
40	38, 32, 30	37, 35, 40	36, 39, 43
50	40, 45, 36	39, 42, 46	39, 48, 47

Display the information presented in the table through well-chosen graphs. Estimate the row, column, and interaction effects. Construct the ANOVA table, and conduct tests for interaction and main effects if appropriate. Check the adequacy of your model.

*7.1-5 A company is concerned about the quality of its production of 13-inch brake wheels. The engineer who is in charge of quality control suspects that machines and operators may influence the quality of the final product. She conducts a small experiment to investigate the validity of her suspicion by selecting three machines and three operators at random. Each operator uses each machine. At each of the nine machine–operator combinations, she takes a sample of $n = 5$ brake wheels and determines the absolute difference in actual and specified diameter (in units of 1/100 inch). The results are as follows:

	Operator		
Machine	1	2	3
1	2.0, 3.0, 2.5, 3.5, 3.0	3.5, 3.0, 4.5, 3.0, 3.0	2.5, 3.0, 2.0, 2.5, 1.5
2	3.0, 3.0, 3.5, 3.0, 4.0	4.0, 5.0, 3.0, 2.5, 2.5	2.0, 2.0, 2.5, 3.0, 3.0
3	3.5, 2.5, 3.0, 3.0, 4.0	5.0, 4.0, 2.5, 3.5, 4.0	2.0, 2.5, 3.0, 3.0, 2.0

*(a) Are the operator and machine effects fixed or random effects? Discuss.

*(b) Calculate the ANOVA table.

*(c) Test whether there is an interaction effect.

*(d) Test for operator and machine main effects.

Remark: The expected mean squares in the random-effects model in Section 7.1-3 show that the appropriate denominator in the test for main effects, say, of factor A, is the interaction mean square MS_{AB}. Note, however, that it makes sense to interpret and to test for main effects only if the interactions are negligible. But if there are no interactions, then the mean square MS_{AB} is also an estimate of σ^2. In this situation, we can pool SS_{AB} and SS_{Error} to obtain a new and better estimate of σ^2:

$$MS_{Error(pooled)} = \frac{SS_{AB} + SS_{Error}}{(a-1)(b-1) + ab(n-1)}.$$

Given that $\sigma_{\alpha\beta}^2 = 0$ in $E(MS_A)$, we can use this estimate as the denominator in the test for the main effects of factor A and compare $MS_A / MS_{Error(pooled)}$ with the percentage points of the $F[(a-1), (a-1)(b-1) + ab(n-1)]$ distribution. This test will be more reliable because pooling increases the number of degrees of freedom in the denominator.

7.1-6 Show that

$$n \sum_i \sum_j (\overline{Y}_{ij\bullet} - \overline{Y}_{\bullet\bullet\bullet})^2 = SS_A + SS_B + SS_{AB}.$$

This result says that the variability of the cell averages around the overall average can be written as the sum of the main effects and interaction sums of squares. Use this fact to show that

$$SS_{AB} = n \sum_i \sum_j (\overline{Y}_{ij\bullet} - \overline{Y}_{i\bullet\bullet} - \overline{Y}_{\bullet j\bullet} + \overline{Y}_{\bullet\bullet\bullet})^2$$

$$= \frac{1}{n} \sum_i \sum_j (Y_{ij\bullet})^2 - \frac{(Y_{\bullet\bullet\bullet})^2}{abn} - SS_A - SS_B.$$

7.1-7 Following is the partially completed ANOVA table from a two-factor 3×4 factorial experiment with $n = 3$ observations in each cell:

Source	SS	df	MS	F
A	20.12			
B	15.21			
AB	15.52			
Error				
Total	102.00			

(a) Complete the ANOVA table, and test (at significance level $\alpha = 0.05$) whether an interaction of the two factors is present. What about the significance of the main effects?

(b) What assumptions can be checked on the basis of the residuals and by what means?

7.1-8 Consider a factorial design in which a response is measured at four temperatures and three treatments, with four independent observations at each combination of temperature and treatment. Following is a summary table of means:

Table of Means (Four Observations per Cell)

	Temperature (degrees C)			
Treatment	0	5	10	15
A	6	8	12	17
B	10	14	18	23
C	12	8	7	6

Furthermore, we are told that $SS_{Error} = 550$. Construct an ANOVA table and
carry out appropriate tests. Interpret your findings. Use graphical methods to
support your conclusions.

7.1-9 Volunteers who had a smoking history classified as heavy, moderate, and non-
smoker were accepted until nine men were in each category. Three men in each
category were randomly assigned to each of three stress tests: bicycle ergome-
ter, treadmill, and step tests. The time until maximum oxygen uptake was
recorded in minutes. The data are as follows:

	Test		
Smoking History	Bicycle	Treadmill	Step
Nonsmoker	12.8	16.2	22.6
	13.5	18.1	19.3
	11.2	17.8	18.9
Moderate	10.9	15.5	20.1
	11.1	13.8	21.0
	9.8	16.2	15.9
Heavy	8.7	14.7	16.2
	9.7	13.2	6.1
	7.5	8.1	17.8

(a) Analyze the results of this experiment. Obtain the ANOVA table and test
for main effects and interactions.

(b) Construct reference distributions for the row (smoking history) means and
the column (test) means.

7.2 Nested Factors and Hierarchical Designs

Excessive variability among manufactured items should be a major concern of companies that are committed to high-quality production. To increase the quality of the
production, we have to identify factors that affect the variable of interest and that
contribute the most to its variability. The $\bar{x}$- and R-charts in Chapter 5 are useful first
steps in identifying those factors. An $\bar{x}$-chart, for example, may show segments of
time during which the plotted averages are particularly low (or high). If these periods coincide with different operators or shifts (day–night, weekday–weekend), or
with changed input variables or changes in manufacturing conditions, we can tenta-
tively identify the factors in question as contributors to the variation.

The analysis of nested or hierarchical designs gives us a formal method of identifying and estimating the sources of the variability. Let us begin with an illustration. An engineer working for a company believes that the variability in a measured property of manufactured forgings is excessive. The raw material for the forgings consists of rods that the company purchases from a nearby foundry. The forgings are stamped from the rods, and measurements on the property of interest are made on each forging. The engineer wants to identify the sources of the excessive variability. The variability may be due to that among the purchased rods, to variability among the stamped forgings, or to variability in measurement. It is important to know about these components of the variance. If the company finds that the variability is due mostly to the rods, it would insist that the supplier of the rods strengthen its quality control program. If the variability is due to the forgings, the company would look for improved stamping methods. If measurement variability is the major factor, the company would carefully review its measurement procedures.

To estimate the variance components, the engineer conducts the following experiment: First, a rods are selected at random. Then, b forgings are taken at random from each of the selected rods. Finally, n independent measurements are made on each resulting forging. Figure 7.2-1 illustrates the design for $a = 5$ rods, $b = 3$ forgings, and $n = 2$ measurements. Such an arrangement is called a *nested,* or *hierarchical, design.*

It is important to recognize the difference between the arrangement in Figure 7.2-1 and the $a \times b$ factorial experiment. In the factorial, or *crossed,* arrangement, we combine every level of factor B with every level of factor A. In the arrangement in Figure 7.2-1, the "levels" of factor B (forgings) are not the same for different levels of factor A (rods). In fact, there are 15 different forgings, three from each of the five rods. Because, for each rod (i.e., each level of factor A), we are talking about different forgings (i.e., different levels of factor B), we say that the levels of factor B are *nested* within the levels of factor A. In short, we say that B is nested within A and write $B(A)$. Applying this terminology to the error or measurement component, we can also say that the error is nested within the rod-and-forging factor-combination subgroups.

Let us define the kth measurement on the jth forging (factor B) from the ith rod (factor A) as Y_{ijk}. Then the model for the hierarchical $a \times b \times n$ design can be written as

$$Y_{ijk} = \mu + \alpha_i + \beta_{j(i)} + \varepsilon_{k(i,j)}, \quad \begin{cases} i = 1, 2, \ldots, a, \\ j = 1, 2, \ldots, b, \\ k = 1, 2, \ldots, n. \end{cases}$$

where μ is an overall process level; the α_i values are the effects of factor A; the $\beta_{j(i)}$ values are the effects of factor B, which are nested within the ith level of factor A;

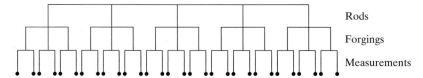

Rods

Forgings

Measurements

Figure 7.2-1 Nested design

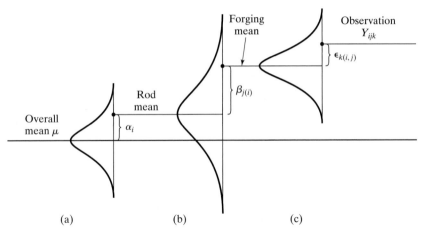

Figure 7.2-2 Three components of variance: (a) distribution of rod means about the process mean; (b) distribution of forging means about the ith rod mean; (c) distribution of individual measurements about the mean of the jth forging, taken from the ith rod

and the random variables $\varepsilon_{ijk} = \varepsilon_{k(i,j)}$ are the error components in the (i, j)th subgroup. Note that we cannot include interaction terms between factors A and B, as not every level of factor B appears with every level of factor A.

We are usually interested in the variance components and assume that the factor effects are random. In other words, we assume that α_i, $\beta_{j(i)}$, and $\varepsilon_{k(i,j)}$ are mutually independent random variables with the respective normal distributions $N(0, \sigma_\alpha^2)$, $N(0, \sigma_\beta^2)$, and $N(0, \sigma^2)$. In our example, the variation in the variable of interest reflects a combination of three variabilities: those inherent in measurements, those due to forgings, and those caused by rods. Figure 7.2-2 illustrates these components.

Table 7.2-1 lists the results of the $5 \times 3 \times 2$ hierarchical design associated with Figure 7.2-1. From these data, we can calculate the overall average $\overline{Y}_{\bullet\bullet\bullet}$ as an estimate of μ. The differences between the factor A averages $\overline{Y}_{i\bullet\bullet}$ and the overall average $\overline{Y}_{\bullet\bullet\bullet}$ are estimates of α_i. Furthermore, within each level of factor A, we can calculate the deviations of the factor B averages $\overline{Y}_{ij\bullet}$ from their respective group average $\overline{Y}_{i\bullet\bullet}$; these are estimates of $\beta_{j(i)}$. Finally, the deviations of the individual measurements Y_{ijk} from their subgroup average $\overline{Y}_{ij\bullet}$ are estimates of $\varepsilon_{k(i,j)}$.

Corresponding to the data-generating model, we can write the data decomposition as

$$Y_{ijk} = \overline{Y}_{\bullet\bullet\bullet} + (\overline{Y}_{i\bullet\bullet} - \overline{Y}_{\bullet\bullet\bullet}) + (\overline{Y}_{ij\bullet} - \overline{Y}_{i\bullet\bullet}) + (Y_{ijk} - \overline{Y}_{ij\bullet}).$$

Table 7.2-1 Individual Measurements, Forging, and Rod Averages

Rod, $\bar{y}_{i\bullet\bullet}$:	76.2	80.9	84.3	80.7	81.8
Forgings, $\bar{y}_{ij\bullet}$:	76.5, 78.2, 73.9	81.1, 78.7, 82.9	85.8, 81.2, 85.9	79.0, 79.4, 81.9	81.9, 80.7, 82.8
Measurements,	76.0, 77.6, 74.1	80.0, 79.0, 82.0	85.1, 81.5, 85.9	78.3, 79.1, 81.7	81.9, 80.1, 82.8
y_{ijk}:	77.0, 78.8, 73.7	82.2, 78.4, 83.8	86.5, 80.9, 85.9	79.7, 79.7, 82.1	81.9, 81.3, 82.8
		$y_{\bullet\bullet\bullet} = 2419.8$,	$\bar{y}_{\bullet\bullet\bullet} = 80.66$,	$\sum\sum\sum y_{ijk}^2 = 195{,}478.26$	

This formula leads to the sum-of-squares decomposition

$$\text{SSTO} = \text{SS}_A + \text{SS}_{B(A)} + \text{SS}_{\text{Error}},$$

where

$$\text{SSTO} = \sum_i \sum_j \sum_k (Y_{ijk} - \overline{Y}_{\bullet\bullet\bullet})^2 = \sum_i \sum_j \sum_k (Y_{ijk})^2 - \frac{(Y_{\bullet\bullet\bullet})^2}{abn},$$

$$\text{SS}_A = bn \sum_i (\overline{Y}_{i\bullet\bullet} - \overline{Y}_{\bullet\bullet\bullet})^2 = \frac{1}{bn} \sum_i (Y_{i\bullet\bullet})^2 - \frac{(Y_{\bullet\bullet\bullet})^2}{abn},$$

$$\text{SS}_{B(A)} = n \sum_{i=1}^{a} \sum_{j=1}^{b} (\overline{Y}_{ij\bullet} - \overline{Y}_{i\bullet\bullet})^2 = \frac{1}{n} \sum_i \sum_j (Y_{ij\bullet})^2 - \frac{1}{bn} \sum_i (Y_{i\bullet\bullet})^2$$

$$= \frac{1}{n} \sum_i \sum_j (Y_{ij\bullet})^2 - \frac{(Y_{\bullet\bullet\bullet})^2}{abn} - \text{SS}_A,$$

and

$$\text{SS}_{\text{Error}} = \sum_i \sum_j \sum_k (Y_{ijk} - \overline{Y}_{ij\bullet})^2 = \text{SSTO} - \text{SS}_A - \text{SS}_{B(A)}.$$

SS_A measures the variability among the averages at the a levels of factor A and has $a - 1$ degrees of freedom. $\text{SS}_{B(A)}$, the sum of squares of factor B nested within A, combines the sums of squares of $\overline{Y}_{ij\bullet}$ around the average $\overline{Y}_{i\bullet\bullet}$ over the a factor levels of A. Because each component has $b - 1$ degrees of freedom, and because there are a levels, $\text{SS}_{B(A)}$ has $a(b - 1)$ degrees of freedom. The error sum of squares is the sum of ab sums of squares of individual observations around their respective subgroup averages. Thus, SS_{Error} has $ab(n - 1)$ degrees of freedom. The various sums of squares, the degrees of freedom, and the mean squares are summarized in Table 7.2-2. Most computer programs inquire whether the factors are crossed (factorial design) or nested (hierarchical design) and will then calculate the appropriate ANOVA table. If the computer package can handle only the factorial arrangement, we can calculate $\text{SS}_{B(A)}$ from the identity $\text{SS}_{B(A)} = \text{SS}_B + \text{SS}_{AB}$. (See Exercise 7.2-5.)

The expected values of the mean squares for the nested random-effects model are

$$E(\text{MS}_A) = \sigma^2 + n\sigma_\beta^2 + bn\sigma_\alpha^2,$$

$$E(\text{MS}_{B(A)}) = \sigma^2 + n\sigma_\beta^2,$$

and

$$E(\text{MS}_{\text{Error}}) = \sigma^2.$$

Table 7.2-2 ANOVA of a Nested Design

Source	SS	df	MS	F
A	SS_A	$a - 1$	MS_A	$\text{MS}_A / \text{MS}_{B(A)}$
$B(A)$	$\text{SS}_{B(A)}$	$a(b - 1)$	$\text{MS}_{B(A)}$	$\text{MS}_{B(A)} / \text{MS}_{\text{Error}}$
Error	SS_{Error}	$ab(n - 1)$	MS_{Error}	
Total	SSTO	$abn - 1$		

If the levels of factor A are fixed instead of random, then σ_α^2 in $E(\text{MS}_A)$ is replaced by $\sum_{i=1}^a \alpha_i^2/(a-1)$.

The expected values suggest (1) how to construct the appropriate statistics for testing the factor effects and (2) how to estimate the variance components σ_α^2, σ_β^2, and σ^2.

The statistic for testing the effects of factor A (H_0: $\sigma_\alpha^2 = 0$) is $F_A = \text{MS}_A/\text{MS}_{B(A)}$. To assess its significance, we compare it with the upper percentage points from the $F[(a-1), a(b-1)]$ distribution. The statistic for testing the effects of factor B (H_0: $\sigma_\beta^2 = 0$) is $F_{B(A)} = \text{MS}_{B(A)}/\text{MS}_{\text{Error}}$. There, the appropriate reference distribution is $F[a(b-1), ab(n-1)]$.

Estimates of the variance components can be found after replacing the expected values of the mean squares by the observed mean squares and solving the equations for σ^2, σ_α^2, and σ_β^2. We thus obtain

$$\hat{\sigma}^2 = \text{MS}_{\text{Error}},$$

$$\hat{\sigma}_\beta^2 = \frac{\text{MS}_{B(A)} - \text{MS}_{\text{Error}}}{n},$$

$$\hat{\sigma}_\alpha^2 = \frac{\text{MS}_A - \text{MS}_{B(A)}}{bn}.$$

In some instances, one (or even both) of the estimates $\hat{\sigma}_\beta^2$ and $\hat{\sigma}_\alpha^2$ is negative. We take this as an indication that the corresponding variance component is zero, because a variance must be nonnegative.

Example 7.2-1   Consider the data in Table 7.2-1. The various sums of squares in the ANOVA table are as follows:

$$\text{SSTO} = 195{,}478.26 - \frac{(2419.8)^2}{30} = 297.192,$$

$$\text{SS}_A = (3)(2)[(76.2)^2 + (80.9)^2 + (84.3)^2 + (80.1)^2 + (81.8)^2] - \frac{(2419.8)^2}{30}$$

$$= 208.872,$$

$$\text{SS}_{B(A)} = (2)[(76.5)^2 + (78.2)^2 + \cdots + (80.7)^2 + (82.8)^2] - \frac{(2419.8)^2}{30} - 208.872$$

$$= 79.68,$$

$$\text{SS}_{\text{Error}} = 297.192 - 208.872 - 79.68 = 8.64.$$

Note that in this computation we used the facts that

$$bn(\overline{Y}_{i\bullet\bullet})^2 = \frac{(Y_{i\bullet\bullet})^2}{bn} \quad \text{and} \quad n(\overline{Y}_{ij\bullet})^2 = \frac{(Y_{ij\bullet})^2}{n}.$$

As an exercise, obtain the sums of squares and the ANOVA table from your computer software. Minitab's command "Stat > ANOVA > Fully Nested ANOVA" can be used. The spreadsheet containing the data from Table 7.2-1 consists of three columns: a column of responses (say, C1), a column for the rods (C2), and a

Table 7.2-3 ANOVA for Forging Example

Source	SS	df	MS	F
Rods	208.872	4	52.218	$52.218/7.968 = 6.55$
Forgings (rods)	79.680	10	7.968	$7.968/0.576 = 13.83$
Error	8.640	15	0.576	
Total	297.192	29		

column for forgings (C3). There are 30 rows to each column. Row 1 contains 76.0 in column 1, 1 in column 2 (for rod), and 1 in column 3 (for forging). Row 2 contains 77.0 in column 1, 1 in column 2, and 1 in column 3. Row 3 contains 77.6 in column 1, 1 in column 2, and 2 in column 3. The last row contains 82.8 in column 1, 5 in column 2 (rod), and 3 in column 3 (forging). You obtain the ANOVA table by entering the columns C2 (rod) and C3 (forging) as the factors.

The ANOVA table is summarized in Table 7.2-3. There is significant variability due to factor A because $F_A = 6.55$ is larger than $F(0.01; 4, 10) = 5.99$. In addition, $F_{B(A)} = 13.83$ is larger than $F(0.01; 10, 15) = 3.80$, so factor B is significant.

The variance components are estimated as

$$\hat{\sigma}^2 = 0.576,$$

$$\hat{\sigma}_\beta^2 = \frac{7.968 - 0.576}{2} = 3.696,$$

$$\hat{\sigma}_\alpha^2 = \frac{52.218 - 7.968}{(3)(2)} = 7.375.$$

This analysis shows that the two major contributors to the estimated variability are the variability among rods and the variability due to forgings. This simple experiment has quantified the major sources of variability: The variability due to rods is twice as large as that due to forgings. Action in these areas is likely to be most rewarding. ∎

Exercises 7.2

*7.2-1 A chemical company is concerned about excessive variability in the moisture content of its pigment paste, produced in large batches of approximately 80 drums. The moisture content is measured on samples from randomly selected batches. The variation of a single test measurement reflects the analytical test variation, the variation due to sampling, and the batch-to-batch variability. To isolate the variabilities, the company conducted a $15 \times 2 \times 2$ hierarchical design. Selecting 15 batches at random, the company took two samples of paste at random from each batch, giving a total of 30 samples, each of which was divided in two. Measurements were made on the resulting 60 subsamples. To make certain that these subsamples were not treated differently by the lab, they were randomly introduced into the normal work process. The results of the experiment were as follows:

Batch	Sample	Measurements
1	1	40, 39
	2	30, 32
2	1	26, 28
	2	25, 26
3	1	29, 28
	2	14, 15
4	1	30, 31
	2	24, 24
5	1	19, 20
	2	17, 17
6	1	33, 32
	2	26, 24
7	1	23, 24
	2	32, 33
8	1	34, 34
	2	29, 29
9	1	27, 27
	2	31, 31
10	1	13, 16
	2	27, 24
11	1	25, 23
	2	25, 27
12	1	29, 29
	2	31, 32
13	1	19, 20
	2	29, 30
14	1	23, 24
	2	25, 25
15	1	39, 27
	2	26, 28

*(a) For the given data, construct the appropriate ANOVA table that resulted from this design.

*(b) Estimate the variance components. Interpret your findings. Identify the area(s) in which a reduction in variance might be most rewarding.

(c) Check the adequacy of the model. In particular, construct a dot diagram of the residuals $e_{ijk} = y_{ijk} - \bar{y}_{ij\bullet}$ and a scatter plot of residuals against fitted values $\bar{y}_{ij\bullet}$.

[G. E. P. Box, W. G. Hunter, and J. S. Hunter, *Statistics for Experimenters* (New York: Wiley, 1978, 2005), Chapter 18.]

7.2-2 (a) The Environmental Protection Agency (EPA) monitors the water quality of a certain medium-sized lake by collecting and measuring three water samples each at 15 randomly selected test sites. Measurements are made at the same sites over a period of 10 randomly selected days, resulting in a total of $3 \times 15 \times 10 = 450$ observations. Discuss how you would analyze this experiment. Specify the appropriate ANOVA table and discuss how to test the various hypotheses.

(b) How would your analysis change if the 15 locations were not the same each day?

*7.2-3 A company producing plastic wrap is concerned about excessive variability in the rupture strength of its product. In a study designed to isolate the sources of variability, the company selected six batches at random. From each of these six batches, it took two samples and made three independent measurements on each of the 12 samples. The partial ANOVA table is as follows:

Source	SS	df	MS	F
Batch	109.6			
Sample (Batch)	124.2			
Error				
Total	340.8			

Complete the ANOVA table, estimate the variance components, and interpret the results.

7.2-4 A company is concerned about the variability in the tensile strength of the steel beams it produces. The company produces steel beams at three different plants and suspects that some of the differences may be due to location. Furthermore, it suspects that some of the variability could come from the variability in the ingots from which the beams are extruded.

To study this problem, the company selected four ingots at random at each plant. From each ingot's production, it selected three beams at random and measured their tensile strengths. Note that in this example factor A (plant) is fixed. Construct the appropriate analysis-of-variance table and test for plant and ingot effects.

Plant	Ingot	Beam Measurements
1	1	45, 46, 43
	2	44, 47, 45
	3	55, 60, 56
	4	49, 58, 50
2	1	44, 45, 50
	2	55, 51, 54
	3	68, 74, 76
	4	55, 59, 63
3	1	49, 47, 44
	2	53, 54, 51
	3	48, 51, 47
	4	65, 57, 55

7.2-5   Show that

$$\mathrm{SS}_{B(A)} = n\sum_{i=1}^{a}\sum_{j=1}^{b}(\overline{Y}_{ij\bullet} - \overline{Y}_{i\bullet\bullet})^2 = an\sum_{j=1}^{a}(\overline{Y}_{\bullet j\bullet} - \overline{Y}_{\bullet\bullet\bullet})^2$$

$$+ n\sum_{i=1}^{a}\sum_{j=1}^{b}(\overline{Y}_{ij\bullet} - \overline{Y}_{i\bullet\bullet} - \overline{Y}_{\bullet j\bullet} + \overline{Y}_{\bullet\bullet\bullet})^2.$$

Your proof demonstrates formally that $\mathrm{SS}_{B(A)}$ is the sum of SS_{B} and SS_{AB}. However, because, in the nested arrangement, not every level of factor A is combined with every level of factor B, it is incorrect to interpret the second component as an interaction sum of squares. That is, in hierarchical designs, $\mathrm{SS}_{B(A)}$ should not be partitioned; it is done in this exercise only to prove that $\mathrm{SS}_{B(A)} = \mathrm{SS}_{B} + \mathrm{SS}_{AB}$ in case the computer program provides the latter two terms and not $\mathrm{SS}_{B(A)}$.

7.2-6   A plant needed instrumentation to obtain measurements of very small parts. The required precision was in the 5-micron range (in which the wavelength of light is an inherent limitation), so the measurement system developed was complex. Nevertheless, it was found that the variability of the measurements on supposedly similar parts was quite large and unacceptable. At this stage, it was not clear whether this variability was due to genuine differences in size, whether it was due to the measurement process, or whether it was due to the measurement instrument.

The following experiment was performed: (1) Five representative small parts were selected. (2) The first part was placed in the measurement fixture, following the usual description of the measurement process, and the measurement of interest was taken. Then four additional measurements were taken without disturbing the small part in any way. Note that, from this series of five measurements, we can estimate the "short-term" measurement variability, which is the variability due to the measurement instrument. (3) The first part was removed from the apparatus, the next part was set up in the measurement fixture, and five measurements were taken in rapid succession without disturbing the small part in any way. (4) After each of the small parts were measured (similarly) five times, the experimenter again set up one of the five parts in the equipment and measured its size rapidly five more times. Similarly, measurements were obtained on the four other parts. (5) The process was repeated three more times, until five rounds of measurements were obtained, each involving the same five small parts; in each round, each part was measured five times in rapid succession. Note that the variability of the averages that are calculated from each of the five rapid-succession quintuples measures the "long-term" measurement variability— the variability that is due to the measurement process (i.e., due to the setup procedure, adjustment of the measurement instrument, etc.).

The experiment leads to a total of 125 observations; Y_{ijk} corresponds to the kth rapid measurement in setup j of part i. Explain how you would set up the ANOVA table, and discuss what information you could extract from this experiment.

[Coleman, D. E., "Measuring Measurements," *RCA Engineer*, 1985, 30(3): 16–23.]

Remark: The tolerance limits for measurement errors listed by the manufacturer of a measurement instrument usually address only short-term measurement variability. They do not incorporate long-term variability, which is caused by factors that enter into the measurement process.

7.2-7 A paper company is concerned about excessive variability in the rupture strength of its tissue. To study the sources of variability, the company selected five batches at random, from each of which it took three samples and made four independent measurements of strength. Complete the following ANOVA table, estimate the variance components, and interpret the results.

Source	SS	df	MS	F
Batch	119.6			
Sample (Batch)				
Error	77.0			
Total	320.4			

7.3 General Factorial and 2^k Factorial Experiments

In Section 7.1, we studied the factorial experiment with two factors: A and B. Each of the a levels of factor A is combined with each of the b levels of factor B, so there is a total of ab level combinations of the two factors. The cell means, denoted by μ_{ij}, are modeled as $\mu_{ij} = \bar{\mu}_{\bullet\bullet} + \alpha_i + \beta_j + (\alpha\beta)_{ij}$. Recall that the ith main effect of factor A, $\alpha_i = \bar{\mu}_{i\bullet} - \bar{\mu}_{\bullet\bullet}$, is the difference between the average response at the ith level of factor A and the overall average. Similarly, the jth main effect of factor B equals $\beta_j = \bar{\mu}_{\bullet j} - \bar{\mu}_{\bullet\bullet}$. The two-factor interaction,

$$(\alpha\beta)_{ij} = \mu_{ij} - (\bar{\mu}_{\bullet\bullet} + \alpha_i + \beta_j) = \mu_{ij} - \bar{\mu}_{i\bullet} - \bar{\mu}_{\bullet j} + \bar{\mu}_{\bullet\bullet},$$

is that part of μ_{ij} that cannot be explained by the main effects alone. A nonzero interaction implies that the effect of one factor depends on the level of the other.

Now, if we consider three factors—say, factors A, B, and C—with number of levels a, b, and c, respectively, there is a total of abc level combinations of the three factors and thus abc cell means μ_{ijh}, $i = 1, \ldots, a; j = 1, \ldots, b; h = 1, \ldots, c$. Because there are three factors, we can define the main effects, the two-factor interactions (between A and B, A and C, and B and C), and now three-factor interactions. Main effects and two-factor interactions need little additional explanation, having been defined earlier. The only difference from the two-factor experiment is that now we have to average over the levels of yet another factor. For example, the main effects of factor C are given by $\gamma_h = \bar{\mu}_{\bullet\bullet h} - \bar{\mu}_{\bullet\bullet\bullet}$, and the two-factor interactions between factors A and C are $(\alpha\gamma)_{ih} = \bar{\mu}_{i\bullet h} - \bar{\mu}_{i\bullet\bullet} - \bar{\mu}_{\bullet\bullet h} + \bar{\mu}_{\bullet\bullet\bullet}$. The three-factor interactions between factors A, B, and C represent that part of μ_{ijh} which is not explained by a model that includes only main effects and two-factor interactions:

$$(\alpha\beta\gamma)_{ijh} = \mu_{ijh} - [\bar{\mu}_{\bullet\bullet\bullet} + \alpha_i + \beta_j + \gamma_h + (\alpha\beta)_{ij} + (\alpha\gamma)_{ih} + (\beta\gamma)_{jh}].$$

A nonzero three-factor interaction implies that the two-factor interactions depend on the levels of the third factor. The "dot notation" should be fairly obvious: a dot in place of an index means that we have averaged the means over that particular index.

For the three-factor factorial experiment with n replications in each cell, the ANOVA table is given in Table 7.3-1. Computer programs are usually available to calculate the entries in the table, and there is little reason to illustrate the computation

Table 7.3-1 ANOVA Table for the Factorial Experiment with Three Factors: Fixed Effects

Source	SS	df	MS	F
A	SS_A	$a - 1$	MS_A	MS_A / MS_{Error}
B	SS_B	$b - 1$	MS_B	MS_B / MS_{Error}
C	SS_C	$c - 1$	MS_C	MS_C / MS_{Error}
AB	SS_{AB}	$(a - 1)(b - 1)$	MS_{AB}	MS_{AB} / MS_{Error}
AC	SS_{AC}	$(a - 1)(c - 1)$	MS_{AC}	MS_{AC} / MS_{Error}
BC	SS_{BC}	$(b - 1)(c - 1)$	MS_{BC}	MS_{BC} / MS_{Error}
ABC	SS_{ABC}	$(a - 1)(b - 1)(c - 1)$	MS_{ABC}	MS_{ABC} / MS_{Error}
Error	SS_{Error}	$abc(n - 1)$	MS_{Error}	
Total	SSTO	$abcn - 1$		

of the various sums of squares. The Minitab command "Stat > ANOVA > Balanced ANOVA" can be used. The spreadsheet containing the data consists of four columns: three columns (say, C1, C2, and C3) for the levels of factors A, B, and C, and a column for the response. If you want to obtain the ANOVA table in Table 7.3-1, you specify the model as C1, C2, C3, C1*C2, C1*C3, C2*C3, C1*C2*C3. The command "Stat > ANOVA > Interactions Plot" provides the interaction plots.

If we assume that the three factors are fixed and all observations are independent and normally distributed, the test statistics for main and interaction effects are the ratios of the corresponding mean squares and the mean square due to error. Because two-factor interactions and main effects are meaningful only if there are no three-factor interactions, we should first test for the presence of a three-factor interaction. To do so, we compare MS_{ABC} / MS_{Error} with the critical value $F[\alpha;(a - 1)(b - 1)(c - 1), ab(n - 1)]$. Tests for two-factor interactions and main effects are constructed in a similar fashion.

It should be obvious how to extend these models to more than three factors. Note that it is not necessary to include all interactions; for example, you may want to omit certain higher order interactions (such as the three-factor interactions). Again, the command "Stat > ANOVA > Balanced ANOVA" will not work in the unbalanced situation; in that case, one must use "Stat > ANOVA > General Linear Model."

Factorial experiments with three or more factors require many experimental runs, especially if each factor is studied at several levels and if the experiment is replicated at each combination ($n > 1$). The number of runs can be reduced if we study the factors at only two levels. This leads us to factorial experiments with k factors at two levels each.

7.3-1 2^k Factorial Experiments

Here we assume that each of the k factors occurs at two levels: a "high" level, coded as $+1$ or just as $+$, and a "low" level, coded as -1 or as $-$. A complete factorial arrangement leads to a total of 2^k runs.

Table 7.3-2 The 2^2 Factorial Experiment

Run	Design x_1	x_2	x_1x_2	Observation
1	−	−	+	Y_1
2	+	−	−	Y_2
3	−	+	−	Y_3
4	+	+	+	Y_4

The 2^2 Factorial Let us start with $k = 2$ factors and the $2^2 = 4$ factor combinations (low, low), (high, low), (low, high), and (high, high). In coded units, the four runs are $(-1, -1)$, $(1, -1)$, $(-1, 1)$, and $(1, 1)$. We have arranged these runs in Table 7.3-2 in what is called the *standard order*. We start the levels of factor 1 with one minus sign and alternate the signs: $-, +, -, +$. The levels of factor 2 start with two minus signs, and the signs alternate in blocks of two: $-, -, +, +$. In Table 7.3-2, we display the levels of factor 1 in the column headed x_1 and those of factor 2 in the column headed x_2. The arrangement of the factor levels in these two columns is called the *design matrix*. The standard order is a convenient systematic list of all possible factor–level combinations. The actual order in which the experiments are carried out, however, should be randomized. A particular order may be as follows: run 3, run 1, run 2, run 4.

There are four independent and normally distributed observations in this unreplicated 2^2 factorial. Only a single observation is taken at each factor–level combination; hence, we can estimate at most four quantities: the overall mean, the two main effects, and the interaction between factors 1 and 2. The overall mean is estimated by

$$\text{average} = \frac{Y_1 + Y_2 + Y_3 + Y_4}{4}.$$

The main effect of factor 1, denoted by (1), is one-half of the difference between the averages of the responses at the high and low levels of factor 1; that is,

$$(1) = \frac{1}{2}\left(\frac{Y_2 + Y_4}{2} - \frac{Y_1 + Y_3}{2}\right) = \frac{-Y_1 + Y_2 - Y_3 + Y_4}{4}.$$

Similarly, the main effect of factor 2 is estimated by

$$(2) = \frac{1}{2}\left(\frac{Y_3 + Y_4}{2} - \frac{Y_1 + Y_2}{2}\right) = \frac{-Y_1 - Y_2 + Y_3 + Y_4}{4}.$$

The interaction between the two factors is estimated as one-half the difference between the main effect of factor 1 at the high level of factor 2, namely, $(Y_4 - Y_3)/2$, and that at the low level of factor 2, namely, $(Y_2 - Y_1)/2$. That is, the interaction is

$$(12) = \frac{1}{2}\left(\frac{Y_4 - Y_3}{2} - \frac{Y_2 - Y_1}{2}\right) = \frac{Y_1 - Y_2 - Y_3 + Y_4}{4}.$$

Note that this is the same as one-half the difference between the main effects of factor 2 at the high and low levels of factor 1. If the main effects of one factor are the same at the high and low levels of the other, there is no interaction.

There is an easy way to remember how to calculate these effects: The sequences of + and − signs in columns x_1 and x_2 tell us how to combine the observations to get the main effects; the column headed x_1x_2 of the products of x_1 and x_2 in Table 7.3-2 contains the appropriate coefficients for the interaction. All linear combinations are then divided by $2^2 = 4$.

Remarks This definition of main and interaction effects is consistent with our definition in the general $a \times b$ factorial experiment in Section 7.1. (See Exercise 7.3-2.) Some other books and computer programs define main effects as the difference between the averages at the high and low levels of a factor, not as one-half of this difference as we do here. If this alternative definition is adopted, we must divide the linear combinations by $2^{2-1} = 2$, the number of plus signs (or minus signs) in the columns for the coefficients. The effects are then twice as large as the ones that result from our definition, and they measure the effect of moving from the low (−1) to the high (+1) level of a factor, which corresponds to a change of 2 units; the effects in our definition express the effect of a one-unit change.

Example 7.3-1
The effects of temperature (factor 1) and reaction time (factor 2) on the percent yield of a certain chemical reaction (response Y) are studied. There are two levels of temperature, namely, 110°C (coded as −) and 130°C (coded as +), and two levels of reaction time, namely, 50 minutes (−) and 70 minutes (+). The 2^2 factorial experiment was replicated ($n = 2$) and the order of the eight runs was randomized. The results are given in Table 7.3-3. The average yields from the two replications are used to estimate the effects, and the individual observations are used later to calculate standard errors of the estimated effects. We find that the interaction is quite small; thus, we can proceed to interpret the main effects. The temperature main effect is 2.4, which means that, on average, a temperature increase of 10°C leads to a 2.4 percent increase in yield. Stating this differently, we find that a temperature increase from the low level of 110°C to the high level of 130°C leads to a $(2)(2.4) = 4.8$ percent increase in yield. The main effect of time is estimated as 4.2, which implies that a 10-minute increase in reaction time increases the yield by 4.2 percent. That is, changing the reaction time from 50 minutes to 70 minutes increases the yield by $(2)(4.2) = 8.4$ percent, on average. This information is shown graphically in the interaction plot on the bottom of Table 7.3-3. Because the interaction is negligible, the two lines on the graph are essentially parallel.

These results show that the yield can be increased by increasing the temperature and reaction time. The coded time (x_2) effect is $4.2/2.4 = 1.75$ times greater than the coded temperature (x_1) effect, so we should increase these factors along the line $x_2 = 1.75x_1$. This line is called the path of *steepest ascent*. For example, on the basis of the information from our initial 2^2 factorial experiment, the experiments with coded points $(x_1 = 0.5, x_2 = 0.875)$, $(x_1 = 1.0, x_2 = 1.75)$, and $(x_1 = 1.5, x_2 = 2.625)$ would move us quickly toward higher yields. These points are equivalent to (125°C, 68.75 minutes), (130°C, 77.5 minutes), and (135°C, 86.25 minutes), respectively. We would then conduct further experiments along this path of steepest ascent. The determination of optimal conditions is discussed in detail in Section 8.6.

Table 7.3-3 Example of a 2^2 Factorial Experiment

Run	Design x_1	x_2	x_1x_2	Average Yield	Individual Observations
1	−	−	+	55.0	55.5, 54.5
2	+	−	−	60.6	60.2, 61.0
3	−	+	−	64.2	64.5, 63.9
4	+	+	+	68.2	67.7, 68.7

$$\text{Average} = (55.0 + 60.6 + 64.2 + 68.2)/4 = 62.0$$
$$(1) = (-55.0 + 60.6 - 64.2 + 68.2)/4 = 2.4$$
$$(2) = (-55.0 - 60.6 + 64.2 + 68.2)/4 = 4.2$$
$$(12) = (55.0 - 60.6 - 64.2 + 68.2)/4 = -0.4$$

The 2^3 Factorial In the 2^3 factorial, we have a total of eight runs. The levels of the three factors are listed in standard order in the columns headed x_1, x_2, and x_3 in Table 7.3-4. For example, the first run puts all three factors at their low levels, the second run sets factor 1 at the high level and factors 2 and 3 at low levels, and so on. From this list, it is easy to become familiar with the standard order of the eight runs in the 2^3 factorial experiment. Start the x_1 column with one minus sign, and then alternate the signs until the eighth run is reached. Start the x_2 column with two minus signs, and alternate the signs in blocks of two. The x_3 column starts with four minus signs and we alternate the signs in blocks of four.

From the eight observations in the 2^3 factorial, we can estimate a mean, three main effects, three two-factor interactions, and one three-factor interaction. The mean is estimated by

$$\text{average} = \frac{Y_1 + Y_2 + Y_3 + Y_4 + Y_5 + Y_6 + Y_7 + Y_8}{8}.$$

Table 7.3-4 The 2^3 Factorial Experiment

Run	x_1	x_2	x_3	x_1x_2	x_1x_3	x_2x_3	$x_1x_2x_3$	Observation
1	−	−	−	+	+	+	−	Y_1
2	+	−	−	−	−	+	+	Y_2
3	−	+	−	−	+	−	+	Y_3
4	+	+	−	+	−	−	−	Y_4
5	−	−	+	+	−	−	+	Y_5
6	+	−	+	−	+	−	−	Y_6
7	−	+	+	−	−	+	−	Y_7
8	+	+	+	+	+	+	+	Y_8

The main effect of factor i ($i = 1, 2, 3$) is one-half the difference between the averages of the responses at the high and low levels of factor i. Thus, from the cube in Table 7.3-4, we find that

$$(1) = \frac{1}{2}\left(\frac{Y_2 + Y_4 + Y_6 + Y_8}{4} - \frac{Y_1 + Y_3 + Y_5 + Y_7}{4} \right)$$

$$= \frac{-Y_1 + Y_2 - Y_3 + Y_4 - Y_5 + Y_6 - Y_7 + Y_8}{8},$$

$$(2) = \frac{1}{2}\left(\frac{Y_3 + Y_4 + Y_7 + Y_8}{4} - \frac{Y_1 + Y_2 + Y_5 + Y_6}{4} \right)$$

$$= \frac{-Y_1 - Y_2 + Y_3 + Y_4 - Y_5 - Y_6 + Y_7 + Y_8}{8},$$

$$(3) = \frac{1}{2}\left(\frac{Y_5 + Y_6 + Y_7 + Y_8}{4} - \frac{Y_1 + Y_2 + Y_3 + Y_4}{4} \right)$$

$$= \frac{-Y_1 - Y_2 - Y_3 - Y_4 + Y_5 + Y_6 + Y_7 + Y_8}{8}.$$

The interaction between factors i and $j \neq i$ is one-half the difference between the main effect of factor i at the high level of factor j and that at the low level of factor j. Thus,

$$(12) = \frac{1}{2}\left[\frac{1}{2}\left(\frac{Y_4 + Y_8}{2} - \frac{Y_3 + Y_7}{2}\right) - \frac{1}{2}\left(\frac{Y_2 + Y_6}{2} - \frac{Y_1 + Y_5}{2}\right)\right]$$

$$= \frac{Y_1 - Y_2 - Y_3 + Y_4 + Y_5 - Y_6 - Y_7 + Y_8}{8},$$

$$(13) = \frac{1}{2}\left[\frac{1}{2}\left(\frac{Y_6 + Y_8}{2} - \frac{Y_5 + Y_7}{2}\right) - \frac{1}{2}\left(\frac{Y_2 + Y_4}{2} - \frac{Y_1 + Y_3}{2}\right)\right]$$

$$= \frac{Y_1 - Y_2 + Y_3 - Y_4 - Y_5 + Y_6 - Y_7 + Y_8}{8},$$

$$(23) = \frac{1}{2}\left[\frac{1}{2}\left(\frac{Y_7 + Y_8}{2} - \frac{Y_5 + Y_6}{2}\right) - \frac{1}{2}\left(\frac{Y_3 + Y_4}{2} - \frac{Y_1 + Y_2}{2}\right)\right]$$

$$= \frac{Y_1 + Y_2 - Y_3 - Y_4 - Y_5 - Y_6 + Y_7 + Y_8}{8}.$$

The three-factor interaction is one-half the difference between the two-factor interactions between factors 1 and 2 at the high and low levels of factor 3. Note that, due to symmetry, the three-factor interaction is the same as one-half the difference between the (13) interactions at high and low levels of factor 2 or one-half of the difference between the (23) interactions at high and low levels of factor 1:

$$(123) = \frac{1}{2}\left[\frac{1}{2}\left(\frac{Y_8 - Y_7}{2} - \frac{Y_6 - Y_5}{2}\right) - \frac{1}{2}\left(\frac{Y_4 - Y_3}{2} - \frac{Y_2 - Y_1}{2}\right)\right]$$

$$= \frac{-Y_1 + Y_2 + Y_3 - Y_4 + Y_5 - Y_6 - Y_7 + Y_8}{8}.$$

Again, it is easy to remember the calculation of these effects from the $+$ and $-$ signs in Table 7.3-4. The signs in $x_1, x_2,$ and x_3 tell us how to combine the observations when calculating the main effects. The signs in the remaining product columns, $x_1 x_2$, $x_1 x_3,$ $x_2 x_3,$ and $x_1 x_2 x_3,$ which are also called the calculation columns, tell us how to combine the observations when estimating the interaction effects. These linear combinations are then divided by $2^k = 2^3 = 8.$ Also, it is instructive to draw a cube, as we have done in the table, and display the responses at its vertices.

The 2^k Factorial With k factors at two levels each, we have a total of 2^k independent and normally distributed observations. It is easy to write down the runs in standard order: Start column x_1 with one minus sign, and alternate the signs until you reach row (run) 2^k. Start x_2 with two minus signs, and alternate signs in blocks of 2 until row 2^k is reached. Start x_3 with $2^2 = 4$ minus signs, and alternate in blocks of 4. Start x_4 with $2^3 = 8$ minus signs, and alternate signs in blocks of 8. Continue until you reach column x_k, which now consists of 2^{k-1} minus signs followed by 2^{k-1} plus signs.

Example 7.3-2 As an illustration of a 2^4 factorial, we use the data from an experiment designed to evaluate the effect of laundering on certain fire-retardant treatments for fabrics. [M. G. Natrella, *Experimental Statistics*, National Bureau of Standards Handbook 91 (Washington, D.C.: U.S. Government Printing Office, 1963)]. Factor 1 is the type of fabric (sateen or monk's cloth), factor 2 corresponds to two different fire-retardant

Table 7.3-5 A 2^4 Factorial Experiment

x_1	x_2	x_3	x_4	Y	Effect
$-$	$-$	$-$	$-$	42	Average $= 575/16 = \quad 35.94$
$+$	$-$	$-$	$-$	31	$(1) = -129/16 = -8.06$
$-$	$+$	$-$	$-$	45	$(2) = \quad 1.56$
$+$	$+$	$-$	$-$	29	$(3) = -0.56$
$-$	$-$	$+$	$-$	39	$(4) = -0.56$
$+$	$-$	$+$	$-$	28	$(12) = -2.19$
$-$	$+$	$+$	$-$	46	$(13) = -0.31$
$+$	$+$	$+$	$-$	32	$(14) = -1.56$
$-$	$-$	$-$	$+$	40	$(23) = \quad 0.81$
$+$	$-$	$-$	$+$	30	$(24) = \quad 0.06$
$-$	$+$	$-$	$+$	50	$(34) = -0.31$
$+$	$+$	$-$	$+$	25	$(123) = \quad 0.31$
$-$	$-$	$+$	$+$	40	$(124) = -1.19$
$+$	$-$	$+$	$+$	25	$(134) = -0.56$
$-$	$+$	$+$	$+$	50	$(234) = -0.44$
$+$	$+$	$+$	$+$	23	$(1234) = \quad 0.06$

For example, the $x_1 x_2 x_3$ column reads (written as a row)

$$- + + - + - - + - + + - + - - +.$$

Thus

$$(123) = \frac{\begin{array}{c} -42 + 31 + 45 - 29 + 39 - 28 - 46 + 32 \\ -40 + 30 + 50 - 25 + 40 - 25 - 50 + 23 \end{array}}{16} = 0.31.$$

treatments, factor 3 describes the laundering condition (no laundering, after one laundering), and factor 4 corresponds to two different methods of conducting the flame test. The observations listed in Table 7.3-5 are in inches burned, measured on a standard-size sample fabric after a flame test.

The estimated effects are also listed in the table. Each is a linear combination of the observations, where the weights are the elements of the appropriate calculation column. For example, to get the (123) interaction, you multiply the x_1, x_2, and x_3 columns to obtain $x_1 x_2 x_3$; the signs in this column tell you how to combine the observations. Finally, you divide this combination by $2^4 = 16$. The expression (1234) corresponds to the four-factor interaction and is one-half the difference between the three-factor interactions at the high and low levels of the fourth factor. If these three-factor interactions are the same, the four-factor interaction is zero. To estimate this four-factor interaction, we combine the observations according to the coefficients in the product column $x_1 x_2 x_3 x_4$ and divide by $2^4 = 16$.

After you have completed and checked these computations, you will agree that the calculation of the average and the 15 effects is quite cumbersome. In particular, you had to form 16 sequences of $+$ and $-$ signs and then use them to form 16 linear combinations. Fortunately, good software is available, and we will discuss computation in Section 7.4-3. ∎

7.3-2 Significance of Estimated Effects

We distinguish two situations. In the first, we calculate the standard errors of the effects from several independent observations at each of the 2^k different factor–level combinations. That is, we repeat the 2^k design at least once. In the second, unreplicated, situation, we use *normal probability plots* to assess the importance of the estimated effects.

Let us assume that there are n independent observations $Y_{i1}, Y_{i2}, \ldots, Y_{in}$ with variance estimate $S_i^2 = [\sum_{j=1}^{n}(Y_{ij} - \overline{Y}_i)^2]/(n - 1)$ at each of the 2^k level combinations, $i = 1, 2, \ldots, 2^k$. The 2^k variance estimates can be pooled to obtain the overall variance estimate

$$S^2 = \frac{1}{2^k}\sum_{i=1}^{2^k}S_i^2 = \frac{1}{(n - 1)2^k}\sum_{i=1}^{2^k}\sum_{j=1}^{n}(Y_{ij} - \overline{Y}_i)^2.$$

Because the estimate of the variance of an average $\overline{Y}_i$ at a particular level combination is S^2/n, and because the overall average and each estimated effect can be written as $(1/2^k)\sum_{i=1}^{2^k}c_i\overline{Y}_i$, where the coefficients c_i are either $+1$ or -1, we find that the estimate of the variance of an effect is

$$\text{var(effect)} = \text{var(average)} = \frac{1}{(2^k)^2}\sum_{i=1}^{2^k}\text{var}(\overline{Y}_i) = \frac{S^2}{n2^k}.$$

The estimated effects, together with estimates of their standard deviations which are also known as standard errors, indicate the statistical significance of the various effects.

Remark: That an effect is insignificant does not necessarily imply that the associated factor is unimportant; it just says that the response is unaffected if the factor is varied over a certain range (from -1 to $+1$ in coded units). For example, it could be that a factor is quite important, but that a change over a certain small range has no effect on the response.

Example 7.3-3   Using the result derived in Exercise 7.3-1, we find for the data in Table 7.3-3 that

$$s^2 = \frac{1}{4}\left[\frac{(55.5 - 54.5)^2}{2} + \frac{(60.2 - 61.0)^2}{2} + \frac{(64.5 - 63.9)^2}{2} + \frac{(67.7 - 68.7)^2}{2}\right]$$
$$= 0.375,$$

and

$$\text{var(effect)} = \text{var(average)} = \frac{1}{(2)(4)}(0.375) = 0.0469.$$

Hence, the standard error of an estimated effect, as well as of the average, is

$$[\text{var(effect)}]^{1/2} = [\text{var(average)}]^{1/2} = 0.22.$$

Thus, the two-sigma limits around the estimates are 62.0 ± 0.44 for the mean, 2.4 ± 0.44 for the main effect of factor 1, 4.2 ± 0.44 for the main effect of factor 2, and -0.40 ± 0.44 for the two-factor interaction. These intervals are approximate 95 percent confidence intervals and indicate large main effects, but negligible interaction. ∎

In the second situation, we have only one observation at each factor–level combination. Consequently, we cannot estimate the standard error of the effects from replications. However, a normal probability plot (see the discussion in Section 3.6) of the estimates can shed some light on the importance of the various effects. If all of the effects are zero, then the observations are like a random sample drawn from a normal distribution with a fixed mean. It then follows that the $2^k - 1$ main and interaction effects, which are linear combinations of the observations, are normally distributed about zero. The effects would therefore plot on normal probability paper approximately as a straight line. Those effects which do not fit reasonably well on a straight line are not easily explained as chance occurrences and therefore are treated as significant effects.

Example 7.3-4 Consider again the 2^4 factorial in Example 7.3-2. The estimated effects, ordered from smallest to largest, are given in the second column of Table 7.3-6. The ranks are given in the third column; in case of ties, we assign the average rank. The ordered effects, $effect_{(i)}, i = 1, 2, \ldots, m = 2^k - 1$, are the quantiles of order $P_i = (i - 0.5)/m$. The values of P_i are listed in the fourth column of the table.

Now, if all of the effects are zero and the estimated effects are just a sample from a normal distribution with mean zero, then a scatter plot of P_i against $effect_{(i)}$ on ordinary graph paper should resemble the "S-shaped" c.d.f. of a normal distribution. Furthermore, because the mean of the normal distribution is zero, this curve should pass through the point $(0, 0.5)$. We could also plot the points $(effect_{(i)}, P_i)$ on normal probability paper; if we do, we should find that they fall more or less on a straight line that goes through the point $(0, 0.5)$.

Table 7.3-6 Normal Scores of the $m = 2^{4-1} = 15$ Estimated Effects from Example 7.3-2

Identity of Effect	Effect by Magnitude	Rank	$P_i = \dfrac{i - 0.5}{m}$	z_i
(1)	−8.06	1	0.033	−1.84
(12)	−2.19	2	0.100	−1.28
(14)	−1.56	3	0.167	−0.97
(124)	−1.19	4	0.233	−0.73
(3)	−0.56	6	0.367	−0.34
(4)	−0.56	6	0.367	−0.34
(134)	−0.56	6	0.367	−0.34
(234)	−0.44	8	0.500	0.00
(13)	−0.31	9.5	0.600	0.25
(34)	−0.31	9.5	0.600	0.25
(24)	0.06	11.5	0.733	0.62
(1234)	0.06	11.5	0.733	0.62
(123)	0.31	13	0.833	0.97
(23)	0.81	14	0.900	1.28
(2)	1.56	15	0.967	1.84

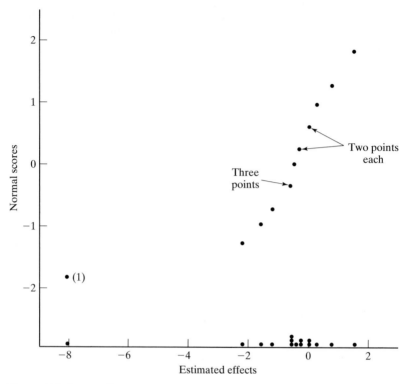

Figure 7.3-1 Dot diagram and normal probability plot of the estimated effects from Example 7.3-2

Alternatively, with today's computers, we can easily calculate z_i, the normal scores that correspond to the estimated effects. The normal scores are the quantiles of order P_i from the standard normal distribution; that is, they satisfy $P_i = P(Z \le z_i) = \Phi(z_i)$. The normal scores are listed in the fifth column of Table 7.3-6. If all of the true effects are zero, the points (effect$_{(i)}$, z_i), when plotted on ordinary graph paper, should lie on a straight line. Furthermore, given that the mean of the distribution is zero, the line should go through the point $(0, 0)$. In the case of Example 7.3-2, this plot is shown in Figure 7.3-1. Note that the point associated with effect (1) is far from a straight line going through the other points. Hence, the main effect of factor 1, the type of fabric, appears to be nonzero.   ∎

Exercises 7.3

7.3-1 Show that the sample variance from a sample of $n = 2$ observations is one-half of the squared difference of the two observations. Thus, for $n = 2$, the variance estimate in the replicated 2^k factorial in Section 7.3-2 is

$$s^2 = \frac{1}{2^{k+1}} \sum_{i=1}^{2^k} (y_{i1} - y_{i2})^2.$$

7.3-2 The main and interaction effects in the general two-factor factorial experiment in Section 7.1 were defined as $\alpha_i = \bar{\mu}_{i\bullet} - \bar{\mu}_{\bullet\bullet}$, $\beta_j = \bar{\mu}_{\bullet j} - \bar{\mu}_{\bullet\bullet}$, and

$(\alpha\beta)_{ij} = \mu_{ij} - \bar{\mu}_{i\bullet} - \bar{\mu}_{\bullet j} + \bar{\mu}_{\bullet\bullet}$, for $i = 1,\dots,a$ and $j = 1,\dots,b$. Using the obvious notation $\mu_{--}, \mu_{+-}, \mu_{-+}, \mu_{++}$ for the four means, show that for $a = b = 2$,

(a) $\alpha_+ = -\alpha_- = (-\mu_{--} + \mu_{+-} - \mu_{-+} + \mu_{++})/4$.

(b) $\beta_+ = -\beta_- = (-\mu_{--} - \mu_{+-} + \mu_{-+} + \mu_{++})/4$.

(c) $(\alpha\beta)_{++} = -(\alpha\beta)_{-+} = -(\alpha\beta)_{+-} = (\alpha\beta)_{--}$
$= (\mu_{--} - \mu_{+-} - \mu_{-+} + \mu_{++})/4$.

*7.3-3 Percent yields from a certain chemical reaction for changing temperature (factor 1), reaction time (factor 2), and concentration (factor 3) are listed as follows, where each $\bar{y}$ is computed from $n = 5$ replications:

x_1	x_2	x_3	$\bar{y}$
$-$	$-$	$-$	79.3
$+$	$-$	$-$	74.3
$-$	$+$	$-$	76.7
$+$	$+$	$-$	70.0
$-$	$-$	$+$	84.0
$+$	$-$	$+$	81.3
$-$	$+$	$+$	87.3
$+$	$+$	$+$	73.7

*(a) Estimate the main effects, the three two-factor interactions, and the three-factor interaction.

*(b) The pooled variance estimate obtained from the $n = 5$ replications in each cell is $s^2 = 40$. Calculate standard errors for the estimated effects and interpret your findings.

*7.3-4 Box, Hunter, and Hunter studied the effects of catalyst charge (10 pounds, 20 pounds), temperature ($220°C, 240°C$), pressure (50 psi, 80 psi) and concentration (10 percent, 12 percent) on percent conversion of a certain chemical (Y). The results of a 2^4 factorial experiment, in standard order, are given in the following table (the actual order of the 16 experiments was randomized):

x_1	x_2	x_3	x_4	y	Run Order
$-$	$-$	$-$	$-$	71	8
$+$	$-$	$-$	$-$	61	2
$-$	$+$	$-$	$-$	90	10
$+$	$+$	$-$	$-$	82	4
$-$	$-$	$+$	$-$	68	15
$+$	$-$	$+$	$-$	61	9
$-$	$+$	$+$	$-$	87	1
$+$	$+$	$+$	$-$	80	13
$-$	$-$	$-$	$+$	61	16
$+$	$-$	$-$	$+$	50	5
$-$	$+$	$-$	$+$	89	11
$+$	$+$	$-$	$+$	83	14
$-$	$-$	$+$	$+$	59	3
$+$	$-$	$+$	$+$	51	12
$-$	$+$	$+$	$+$	85	6
$+$	$+$	$+$	$+$	78	7

*(a) Estimate the main effects and the two-, three-, and four-factor interactions.

*(b) Make a normal probability plot of the estimated effects and assess the significance of the various effects.

[G. E. P. Box, W. G. Hunter, and J. S. Hunter, *Statistics for Experimenters* (New York: Wiley, 1978), p. 325.]

7.3-5 Anderson and McLean discuss a 2^4 factorial experiment that arose in the finishing of metal strips in a metallurgical process. The response that was measured was a score for smoothness of the surface finish, where a small reading was desirable and a large number indicated roughness. The factors (two levels each) that were included in this experiment were solution temperature (x_1), solution concentration (x_2), roll size (x_3), and roll tension (x_4). Results are listed as follows:

Factors					**Factors**				
x_1	x_2	x_3	x_4	y	x_1	x_2	x_3	x_4	y
−	−	−	−	17	−	−	−	+	11
+	−	−	−	14	+	−	−	+	18
−	+	−	−	8	−	+	−	+	8
+	+	−	−	16	+	+	−	+	21
−	−	+	−	11	−	−	+	+	14
+	−	+	−	15	+	−	+	+	21
−	+	+	−	9	−	+	+	+	12
+	+	+	−	4	+	+	+	+	13

(a) Estimate the main effects and all possible interactions.

(b) Make a normal probability plot of the estimated effects and discuss the importance of the various effects.

[V. L. Anderson and R. A. McLean, *Design of Experiments: A Realistic Approach* (New York: Marcel Dekker, 1974), p. 232.]

*7.3-6 Gunter, Tadder, and Hemak use factorial designs to improve the picture-tube phosphor screening process at RCA's picture-tube manufacturing facility. The phosphor screening process involves applying, exposing, and developing a photosensitive phosphor slurry to the inside of the picture-tube faceplate. The operation requires precise control of the slurry's chemical and physical properties. (A slurry is a mixture of numerous chemicals, including three surfactant additives labeled *A, B,* and *C* that help make the slurry smoother and more able to wet the surface of the faceplate.)

When Gunter, Tadder, and Hemak started their investigation, the process was already in a state of good statistical control, meaning that the results were consistent over time. However, to reduce costs, it was essential to reduce the losses, which were measured by screen-room process rejects. Company personnel had a great deal of knowledge about the process and had found satisfactory levels for the three surfactant additives, but systematic statistical studies had not been made before, either in the laboratory or online in the plant.

Gunter, Tadder, and Hemak used an evolutionary operation (or EVOP, for short) approach in which small changes in process levels are made without interfering significantly with the ongoing process. The changes were made according to a predetermined pattern (i.e., a design), and the resulting changes in the response were recorded, analyzed, and interpreted. [For a more detailed description of EVOP, refer to G. E. P. Box and N. R. Draper, *Evolutionary Operation* (New York: Wiley, 1969).]

Here, the three surfactant additives were varied in a 2^3 factorial experiment. The existing settings of the three additives were varied slightly, and low $(-)$ and high $(+)$ levels were determined. In addition to the eight combinations of high and low settings, it was decided to run two experiments at the standard settings. (Each coded variable equals zero.) The 10 runs required about 14 production days. The order of the runs was randomized, except that a run with standard settings was conducted each week. The losses from the $2^3 = 8$ factorial runs were as follows:

Surfactant Additive

A	B	C	Loss
−	−	−	1.28
+	−	−	1.93
−	+	−	0.94
+	+	−	1.31
−	−	+	0.36
+	−	+	1.11
−	+	+	1.16
+	+	+	1.58

*(a) Calculate the main and interaction effects.

*(b) The replications at the center point (i.e., experiments at the standard settings) led to losses of 0.83 and 1.40 units. Use these two values to calculate a crude estimate of the variance of an individual observation, and make a rough estimate of the standard error of the estimated effects.

*(c) Use the standard error in part (b) to interpret the results of your experiment. How would you change the surfactant additives to reduce your losses? Why would it be reasonable to use the $(-, -, +)$ point of the 2^3 cube as the center point of a new EVOP cube and to start a new cycle around this point?

[B. H. Gunter, A. L. Tadder, and C. M. Hemak, "Improving the Picture Tube Phosphor Screening Process with EVOP," *RCA Engineer,* 30(3): 54–59 (1985).]

7.3-7 In a personal communication, Ronald Snee describes the application of a factorial design in a product impurity study at du Pont. The experiment was carried out by an engineer who had taken a du Pont short course on the strategy of experimentation. The objective of this study was to find ways of reducing the level of impurity found in a chemical product. Three factors were thought to be important: the type of polymer that was used (the engineer compared the standard polymer *B* with a new, but more expensive, one, *A*), the polymer concentration (which was varied between the low level, 0.01 percent, and the high level, 0.04 percent), and the amount of a certain additive (which was varied from a low level of 2 pounds to a high level of 12 pounds). The additive was known to have an effect at 2 pounds, and it was of interest to determine whether higher levels would be of benefit. A factorial experiment with factors at these low and high levels was conducted, and the percentage impurities from the runs that were conducted are given in the following table (note that there are replications at some (but not all) design points):

Polymer	Polymer Concentration (percent)	Additive (pounds)	Impurity (percent)
A	0.01	2	1.0
B	0.01	2	1.0, 1.2
A	0.04	2	0.2
B	0.04	2	0.5
A	0.01	12	0.9, 0.7
B	0.01	12	1.1
A	0.04	12	0.2, 0.3
B	0.04	12	0.5

(a) Give a pictorial representation of the information in the table. You may want to draw a cube and display the impurities at its vertices. Interpret the results.

(b) Estimate the main and interaction effects. (Use the averages from the replications in your calculations.) Calculate an estimate of the process variance σ^2 and an estimate of the variance of the effects.

Hint: Use the three cases with replications to get an estimate of σ^2. Then use σ^2 to calculate the variances of the responses (which are either individual observations or averages of replicates). For example, var$(Y_1) = \sigma^2$ and var$(\bar{Y}_2) = \sigma^2/2$, because we took the average of 1.0 and 1.2 when we calculated the effects, and so on. Adding these variances and dividing by $8^2 = 64$ will give you the variance of an effect.

(c) You will find that a switch from the standard polymer B to the more expensive polymer A will decrease the impurities. Is this sufficient reason to switch to polymer A, or should additional analyses be conducted?

Hint: Recall cost–benefit analysis.

(d) Is it appropriate to conclude that the additive has no effect on impurity? What recommendations would you make concerning the level of the additive?

(e) What level of polymer concentration would you recommend? What considerations would you take into account to reach such a decision?

7.3-8 An experiment is performed to compare the cleaning action of two detergents, detergent A and detergent B. Of 32 swatches of cloth soiled with grease, 16 are washed with detergent A and 16 with detergent B in an agitator machine and then measured for "whiteness." Criticize the following aspects of the experiment:

(a) The entire experiment is performed with soft water.

(b) To accelerate the testing procedure, the experimenter used very hot water and 10-minute washing times.

(c) All experiments with detergent A were done first.

(d) Discuss briefly how the experiment should be done. Assume that softness of the water, water temperature, and washing time might be important.

7.3-9 The yields (in percent) of a certain chemical reaction for changing levels of temperature (factor 1), reaction time (factor 2), and concentration (factor 3) are as follows (each $\bar{y}$ is computed from $n = 3$ observations; note that the runs are not in standard order):

Temp (°F)	Time (min)	Concentration (percent)	$\bar{y}$
250	50	3	71.8
200	50	5	81.5
250	60	5	71.2
250	50	5	78.8
200	50	3	77.2
200	60	3	74.2
200	60	5	84.8
250	60	3	67.5

(a) Estimate the main effects and the two- and three-factor interactions.

(b) The three replications in each cell are used to obtain the variance estimate of an individual measurement, $s^2 = 24$. Find standard errors of the estimated effects, calculate approximate confidence intervals for the effects, and interpret your findings.

7.3-10 A chemical engineer is trying to improve the efficiency of a reaction that converts a raw material into a product. The response variable Y is called the *degree of conversion*. The engineer varies three factors: catalyst type (C), reactant concentration (R), and reaction temperature (T); two replications were made in each cell. The data, in coded form ($Y - 59$), are as follows:

	Catalyst, C			
	Type 1		Type 2	
	Reactant Concentration R			
Temperature	0.1%	0.5%	0.1%	0.5%
120	25	26	2	8
	16	33	1	15
160	29	27	4	27
	23	31	0	35

(a) Estimate main and interaction effects, obtain their standard errors, and interpret your findings.

(b) Display the results graphically. For example, display the cell means of the other two factors separately for each level of catalyst (as in Table 7.3-3). Use this plot to describe the nature of the three-factor interaction.

(c) Which factor–level combination produces the greatest degree of conversion?

7.3-11 Meredith Corporation, the publisher of *Ladies' Home Journal*, sends more than a million letters each year to potential subscribers, hoping to secure as many subscriptions as possible. The marketing team looks for the right mix of promotional materials, and it experiments constantly with various aspects of the brochure, the order card, enclosed testimonials, and offers. The June 2005 campaign, for example, tested different versions of the front page of the brochure and different messages on the front and the back of the order card.

Front of brochure: One version (level −1) shows a radiant-looking Kelly Ripa (the star of the ABC show *Live with Regis and Kelly*), while the other version (level +1) features Dr. Phil (known from his nationally syndicated TV show and publications on life strategies and relationships).

Front of the order card: Level 1 (−1) highlights the message "Double our Best Offer," while level 2 (+1) draws attention to the message "We never had a bigger sale."

Back of the order card: Level 1 (−1) emphasizes "Two extra years free," while level 2 (+1) features magazine covers of previous issues.

The results (number of letters sent and number of orders that were received) are as follows:

Order Card, Front	Order Card, Back	Brochure	Letters Sent	Orders	Proportion
−1	−1	−1	15,042	573	0.0380933
1	−1	−1	15,042	644	0.0428135
−1	1	−1	15,042	563	0.0374285
1	1	−1	15,042	616	0.0409520
−1	−1	1	15,042	564	0.0374950
1	−1	1	15,042	550	0.0365643
−1	1	1	15,042	575	0.0382263
1	1	1	15,042	553	0.0367637

Analyze the data. Estimate main and interaction effects. Display the effects graphically through main effects and interaction plots. Assess the significance of the effects. Summarize your conclusions.

Hint: Obtain the variance of the estimated effect. Each estimated effect is a linear combination of the eight sample proportions, with weights that are either −1/8 or 1/8. Use the fact that the variance of each sample proportion in the last column of the table is given by $\dfrac{p(1-p)}{n}$, where n is the sample size (here, $n = 15{,}042$) and $p = (573 + 644 + \cdots + 553)/(15{,}042)(8)$ is the pooled proportion of orders.

7.4 2^{k-p} Fractional Factorial Experiments

Investigators frequently have an extensive list of factors that may have an effect on a response. However, it is often not feasible to conduct a full 2^k factorial experiment when the number of factors k is large, because that would require experiments (runs) at all combinations of two levels of each of k factors. The number of runs increases rapidly with k; for example, a full factorial with seven factors requires $2^7 = 128$ runs. In this section, we show that certain well-chosen parts of the 2^k factorial design allow us to estimate the main relationships without too much loss of information. If a part is only $1/2^p$ of the full 2^k factorial design, it is called a 2^{k-p} *fractional factorial design*. If $p = 1$, we talk about a half fraction, as we perform one-half of the full factorial; if $p = 2$, a quarter fraction; and so on.

7.4-1 Half Fractions of 2^k Factorial Experiments

A 2^{3-1} Fractional Factorial Let us consider an example with $k = 3$ factors. Instead of studying the independent and normally distributed responses at all eight factor–level combinations in the 2^3 factorial, we consider a half fraction ($p = 1$). The question now becomes, Which four runs from the cube in Figure 7.4-1 should be selected in the 2^{3-1} fractional factorial design? One possible and, in fact, very good half fraction uses the factor–level combinations listed in Table 7.4-1. These points correspond to the bulleted corners of the cube in Figure 7.4-1.

Figure 7.4-1 illustrates that this particular fraction supplies a complete factorial in each of the three pairs of factors, because any projection of the 2^{3-1} design onto a two-dimensional plane produces a complete 2^2 factorial experiment. Such a design is therefore good in screening situations in which it can reasonably be assumed that, out of three factors, no more than two are of importance.

Obviously, since there are only four runs, we cannot estimate three main effects, three two-factor interactions, and the three-factor interaction separately. In fact, we note from Table 7.4-1 that the levels of the third factor in x_3 are selected so that $x_3 = x_1 x_2$. Thus, the linear combination that corresponds to the third column, denoted by $L_3 = (Y_1 - Y_2 - Y_3 + Y_4)/4$, *confounds* the main effect (3) and the two-factor interaction (12). Our notation $L_3 \rightarrow (3) + (12)$ in Table 7.4-1 expresses the fact that L_3 estimates the sum of (3) and (12).

The formal multiplication of one column by itself gives a column of $+$ signs, which is denoted by I; that is, $I = x_1 x_1 = x_2 x_2 = x_3 x_3$ and $Ix_1 = x_1$, $Ix_2 = x_2$, and

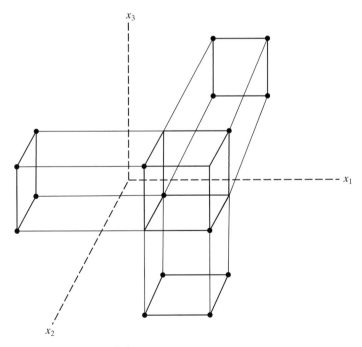

Figure 7.4-1 The 2^{3-1} fractional factorial design and its projection into three 2^2 factorials

Table 7.4-1 The 2^{3-1} Fractional Factorial Design with $x_3 = x_1 x_2$

Run	Design x_1	x_2	x_3	Y
1	$-$	$-$	$+$	Y_1
2	$+$	$-$	$-$	Y_2
3	$-$	$+$	$-$	Y_3
4	$+$	$+$	$+$	Y_4

$$L_0 = (Y_1 + Y_2 + Y_3 + Y_4)/4 \rightarrow \mu + (123)$$
$$L_1 = (-Y_1 + Y_2 - Y_3 + Y_4)/4 \rightarrow (1) + (23)$$
$$L_2 = (-Y_1 - Y_2 + Y_3 + Y_4)/4 \rightarrow (2) + (13)$$
$$L_3 = (Y_1 - Y_2 - Y_3 + Y_4)/4 \rightarrow (3) + (12)$$

$Ix_3 = x_3$. Hence, $x_3 = x_1 x_2$ is equivalent to $I = x_1 x_2 x_3$, as can be seen from the multiplication with x_3. Also, $x_1 = x_1 x_1 x_2 x_3 = x_2 x_3$. Accordingly, we find that $L_1 = (-Y_1 + Y_2 - Y_3 + Y_4)/4$ confounds (1) and (23). In addition, the fact that $x_2 = x_2 x_1 x_2 x_3 = x_1 x_3$ means that $L_2 = (-Y_1 - Y_2 + Y_3 + Y_4)/4$ confounds (2) and (13). Finally, because the $x_1 x_2 x_3$ product column is equal to a column of $+$ signs, which we denote by I, the average $L_0 = (Y_1 + Y_2 + Y_3 + Y_4)/4$ confounds the overall mean level μ and the three-factor interaction (123).

The relationship that generates the levels of the third factor, $x_3 = x_1 x_2$, is called the *generator* of the 2^{3-1} fractional factorial and is equivalent to $x_3 x_3 = x_1 x_2 x_3$. Thus, $I = x_1 x_2 x_3$. As seen earlier, the confounding patterns can be obtained from the equation $I = x_1 x_2 x_3$:

$$Ix_1 = x_1(x_1 x_2 x_3), \quad \text{or} \quad x_1 = x_2 x_3, \quad \text{because } x_1 x_1 = I;$$
$$Ix_2 = x_2(x_1 x_2 x_3), \quad \text{or} \quad x_2 = x_1 x_3, \quad \text{because } x_2 x_2 = I;$$
$$Ix_3 = x_3(x_1 x_2 x_3), \quad \text{or} \quad x_3 = x_1 x_2, \quad \text{because } x_3 x_3 = I.$$

The factor–level combinations and the confounding patterns in the 2^{3-1} fractional factorial are summarized in Table 7.4-1.

Because pairs of effects have been confounded, we cannot tell which member of a pair might cause a significant measurement. Thus, we say that one member of such a pair is the *alias* of the other. In the preceding example, we have alias pairs (1) and (23), (2) and (13), and (3) and (12).

A 2^{4-1} Fractional Factorial Now let us study $k = 4$ factors in only eight runs. This calls for a half fraction ($p = 1$) of the 2^4 factorial, or a 2^{4-1} fractional factorial. We have $2^3 = 8$ runs, so we construct our fractional factorial by first writing down a complete factorial in any three factors and then associating one of the interaction terms with the levels of the fourth factor. We can use any of the two- or three-factor interactions; however, it is usually best to confound the levels of the fourth factor with the highest possible interaction. Thus, we select the generator $x_4 = x_1 x_2 x_3$ or, equivalently, $I = x_1 x_2 x_3 x_4$ because $x_4 x_4 = I$.

Table 7.4-2 The 2_{IV}^{4-1} Fractional Factorial Design with $x_4 = x_1x_2x_3$

Run	x_1	x_2	x_3	x_1x_2	x_1x_3	x_2x_3	$x_4 = x_1x_2x_3$	Y
1	$-$	$-$	$-$	$+$	$+$	$+$	$-$	$y_1 = 21$
2	$+$	$-$	$-$	$-$	$-$	$+$	$+$	$y_2 = 23$
3	$-$	$+$	$-$	$-$	$+$	$-$	$+$	$y_3 = 27$
4	$+$	$+$	$-$	$+$	$-$	$-$	$-$	$y_4 = 24$
5	$-$	$-$	$+$	$+$	$-$	$-$	$+$	$y_5 = 23$
6	$+$	$-$	$+$	$-$	$+$	$-$	$-$	$y_6 = 26$
7	$-$	$+$	$+$	$-$	$-$	$+$	$-$	$y_7 = 33$
8	$+$	$+$	$+$	$+$	$+$	$+$	$+$	$y_8 = 37$

Design   Design

$$26.75 = L_0 = (y_1 + y_2 + y_3 + y_4 + y_5 + y_6 + y_7 + y_8)/8 \to \mu + (1234)$$
$$0.75 = L_1 = (-y_1 + y_2 - y_3 + y_4 - y_5 + y_6 - y_7 + y_8)/8 \to (1) + (234)$$
$$3.50 = L_2 \to (2) + (134)$$
$$3.00 = L_3 \to (3) + (124)$$
$$-0.50 = L_4 = (y_1 - y_2 - y_3 + y_4 + y_5 - y_6 - y_7 + y_8)/8 \to (12) + (34)$$
$$1.00 = L_5 \to (13) + (24)$$
$$1.75 = L_6 \to (23) + (14)$$
$$0.75 = L_7 = (-y_1 + y_2 + y_3 - y_4 + y_5 - y_6 - y_7 + y_8)/8 \to (4) + (123)$$

Table 7.4-2 lists the factor–level combinations for the eight runs. Run 1 sets all four factors at their low levels. Run 2 sets factors 1 and 4 at their high levels and factors 2 and 3 at their low ones. Run 3 has factors 1 and 3 at their low levels and factors 2 and 4 at their high ones, and so on. Table 7.4-2 also shows the confounding patterns that result from this particular fraction. For example, because $x_1 = x_2x_3x_4$, the linear combination $L_1 = (-Y_1 + Y_2 - Y_3 + Y_4 - Y_5 + Y_6 - Y_7 + Y_8)/8$ estimates the sum of (1) and (234).

We say that this fractional factorial design is of *resolution* IV. The reason for this terminology is that main effects are confounded with three-factor interactions $(1 + 3 = 4)$ and two-factor interactions are confounded with two-factor interactions $(2 + 2 = 4)$, but main effects are not confounded with other main effects or with two-factor interactions. We denote the design by 2_{IV}^{4-1} and call it a 2^{4-1} *fractional factorial of resolution IV*.

If one assumes that three- and four-factor interactions are negligible, which is a very reasonable assumption in most cases, then the linear combinations L_1, L_2, L_3, and L_7 in this fractional factorial arrangement provide unconfounded estimates of the main effects. The two-factor interactions, however, are confounded with other two-factor interactions.

Note that the 2^{4-1} design, with $I = x_1x_2x_3x_4$, is not the only possible fractional design of four factors in eight runs. For example, we could have selected the levels of the fourth factor, x_4, such that $x_4 = x_1x_2$. However, this design confounds the main effect of factor 4 with the (12) interaction and thus is of resolution III. It is not as desirable as the resolution IV design, as the two-factor interaction may not be negligible. We have already studied one resolution III design: the 2^{3-1} design with generator $I = x_1x_2x_3$.

Example 7.4-1 This example of a 2^{4-1} design with $I = x_1x_2x_3x_4$ comes from a producer of cake mixes concerned about the texture of its cake frostings. Four different constituents that go into the manufacture of frosting were thought to have an effect on texture. A 2^{4-1} fractional factorial was conducted; the texture readings and the estimated effects are given in Table 7.4-2. The results show that there are three large estimates: $L_2 = 3.50$, $L_3 = 3.00$, and $L_6 = 1.75$. The most reasonable interpretation of these results is that the main effects of factors 2 and 3, together with their interaction (23), are important. Another interpretation would be that main effects (2) and (3) and the (14) interaction are important. However, this interpretation is less plausible because models that include only interactions of factors, such as (14), and not their main effects, (1) and (4) here, are not as common. We treat the estimates L_2, L_3, and L_6 as important effects, because they are large in comparison to the others. Ideally, however, we should assess their significance by comparing them with a standard error estimate from replications. If replications are not available, we should make a normal probability plot of the effects. (Recall our discussion in Section 7.3.) ∎

A 2^{5-1} Fractional Factorial As another illustration, consider a 2^{5-1} fractional factorial to study five factors in $2^4 = 16$ runs. We can generate the runs of the design by first writing down the 4 columns associated with a full 2^4 factorial in four factors and then adding the 11 columns that contain all possible products of the initial 4 columns (that is, $x_1x_2, x_1x_3, \ldots, x_1x_2x_3, \ldots, x_1x_2x_3x_4$). We associate the levels of the fifth factor with the four-factor interaction; that is, we choose $x_5 = x_1x_2x_3x_4$, or $I = x_1x_2x_3x_4x_5$, as the generator, because $x_5x_5 = I$. In Exercise 7.4-2, we ask you to write down the level combinations of this design and determine the confounding patterns. For example, we find that the linear combination which uses the column of weights (± 1) in x_1, say, $L(x_1)$, estimates (1) + (2345), because $x_1 = x_2x_3x_4x_5$. The combination that uses the weights in x_1x_2, $L(x_1x_2)$, estimates (12) + (345), because $x_1x_2 = x_3x_4x_5$. [*Note:* The argument in parentheses after L reminds us of the column of weights that is being used in the linear combination.] This fractional factorial design confounds main effects with four-factor interactions and two-factor interactions with three-factor interactions. Thus, it is said to be of resolution V, because $1 + 4 = 5$ and $2 + 3 = 5$. We denote it by 2^{5-1}_V. If we can assume that interactions of three or more factors are zero, a resolution V design leads to unconfounded estimates of all main effects and two-factor interactions.

7.4-2 Higher Fractions of 2^k Factorial Experiments

So far, we have considered only half fractions of 2^k factorials—that is, 2^{k-1} fractional factorials. Now let us consider higher fractions. For example, let us consider a design to study seven factors in only $2^3 = 8$ runs. Since $2^7 = 128$, this is a 16th fraction, or a 2^{7-4} fractional factorial.

One can generate the eight design points by first writing down a full 2^3 factorial in three factors and then associating the two- and three-factor interaction columns with the four additional factors. Thus, there are four generators: $x_4 = x_1 x_2$, $x_5 = x_1 x_3$, $x_6 = x_2 x_3$, and $x_7 = x_1 x_2 x_3$. Multiplying these by x_4, x_5, x_6, and x_7, respectively, and noting that $x_4 x_4 = x_5 x_5 = x_6 x_6 = x_7 x_7 = I$, we obtain the *generating relation*

$$I = x_1 x_2 x_4 = x_1 x_3 x_5 = x_2 x_3 x_6 = x_1 x_2 x_3 x_7.$$

The factor–level combinations for the 2^{7-4} fractional factorial are given in Table 7.4-3. The linear combinations $L_0, L_1, \ldots, L_7$ will confound main effects and interactions. How can one determine these relationships? A straightforward approach, which, however, is often laborious if there are several generators, is as follows. Note that the products of any number of the four expressions in the generating relationship will again lead to I. Considering all possible products of two, three, and four of these expressions, we find that

$$I = x_1 x_2 x_4 = x_1 x_3 x_5 = x_2 x_3 x_6 = x_1 x_2 x_3 x_7$$

$$= x_2 x_3 x_4 x_5 = x_1 x_3 x_4 x_6 = x_3 x_4 x_7 = x_1 x_2 x_5 x_6$$

$$= x_2 x_5 x_7 = x_1 x_6 x_7 = x_4 x_5 x_6 = x_1 x_4 x_5 x_7$$

$$= x_2 x_4 x_6 x_7 = x_3 x_5 x_6 x_7 = x_1 x_2 x_3 x_4 x_5 x_6 x_7.$$

We call this equation the *defining relation*. There are four generators in the 2^{7-4} fractional factorial, so there are $2^4 = 16$ expressions (including I) in the defining relation.

Table 7.4-3 The 2_{III}^{7-4} Fractional Factorial Design

Run	x_1	x_2	x_3	$x_4 = x_1 x_2$	$x_5 = x_1 x_3$	$x_6 = x_2 x_3$	$x_7 = x_1 x_2 x_3$	Y
1	−	−	−	+	+	+	−	$y_1 = 68.4$
2	+	−	−	−	−	+	+	$y_2 = 77.7$
3	−	+	−	−	+	−	+	$y_3 = 66.4$
4	+	+	−	+	−	−	−	$y_4 = 81.0$
5	−	−	+	+	−	−	+	$y_5 = 78.6$
6	+	−	+	−	+	−	−	$y_6 = 41.2$
7	−	+	+	−	−	+	−	$y_7 = 68.7$
8	+	+	+	+	+	+	+	$y_8 = 38.7$

Design

$$65.1 = L_0 \rightarrow \mu$$
$$-5.4 = L_1 \rightarrow (1) + (24) + (35) + (67)$$
$$-1.4 = L_2 \rightarrow (2) + (14) + (36) + (57)$$
$$-8.3 = L_3 \rightarrow (3) + (15) + (26) + (47)$$
$$1.6 = L_4 \rightarrow (4) + (12) + (56) + (37)$$
$$-11.4 = L_5 \rightarrow (5) + (13) + (46) + (27)$$
$$-1.7 = L_6 \rightarrow (6) + (23) + (45) + (17)$$
$$0.3 = L_7 \rightarrow (7) + (34) + (25) + (16)$$

The precise confounding patterns can be obtained from the defining relation. For example, multiplying that relation by x_1, we find that the linear combination L_1 estimates

$$(1) + (24) + (35) + (1236) + (237) + (12345) + \cdots + (234567).$$

The resolution of this design is III, since it confounds main effects with two-factor interactions. We refer to it as the 2_{III}^{7-4} design. The confounding relations in Table 7.4-3 are obtained after we assume that interactions of three or more factors can be disregarded (i.e., assuming that these interactions are essentially zero).

Due to the very small fraction (1/16 of a full 2^7 factorial), we confound main effects with two-factor interactions even after assuming that higher order interactions are equal to zero. Thus, at first there may be considerable ambiguity due to the confounding of particular estimates. However, fractional factorial experiments are of value in situations where experiments are performed in sequence. Having performed one experiment, we can review the results, and if there is ambiguity, a further group of experiments can be selected to resolve the uncertainty. The basic ideas behind this sequential approach to fractional factorial experimentation are outlined briefly in Exercises 7.4-6 and 7.4-7. A detailed discussion of the strategy can be found in books on the design of experiments, such as G. E. P. Box, W. G. Hunter, and J. S. Hunter, *Statistics for Experimenters* (New York: Wiley, 1978, 2005), and J. Ledolter and A. J. Swersey, *Testing 1–2–3: Experimental Design with Applications in Marketing and Service Operations* (Stanford, CA: Stanford University Press, 2007).

Example 7.4-2   Box and Hunter [in "The 2^{k-p} Fractional Factorial Designs," *Technometrics*, 3: 311–351 (1961)] describe an application of a 2_{III}^{7-4} fractional factorial design in which a company experienced considerable difficulties at the filtration stage in the start-up of a new manufacturing unit. Various explanations for the much longer filtration times were discussed, and the following seven variables were proposed as being possibly responsible: water supply (town, local well), raw material (two suppliers), temperature at filtration (low, high), recycling (included, omitted), rate of addition of caustic soda (slow, fast), type of filter cloth (old, new), and holdup time (two different storage times prior to filtration). The results listed in Table 7.4-3 show three large estimates: $L_1 = -5.4$, $L_3 = -8.3$, and $L_5 = -11.4$. The simplest interpretation would be that the main effects of factors 1, 3, and 5 are important. However, other interpretations are also possible. It could be that factors 3 and 5 and their interaction (35), or factors 1 and 5 and (15), or 1 and 3 and (13) are important. At this stage, we simply do not know. However, following up the design with another fraction could resolve the ambiguities that are due to the confounding of the estimates. For example, Exercise 7.4-6 shows that a "fold-over" of the original design would leave all main effects unconfounded. Changing the levels of a particular factor—say, the fifth one—would leave the main effect of factor 5 and all two-factor interactions with factor 5 unconfounded (Exercise 7.4-7). ∎

The 2_{III}^{7-4} design just considered is also called a *saturated* design, because every calculation column in the original 2^3 factorial generates the levels of an additional four factors. Another example of a saturated design is the 2_{III}^{15-11} design. This design allows us to study 15 variables in 16 runs. Every interaction column from the original

2^4 factorial generates the levels of an additional factor ($x_5 = x_1x_2, x_6 = x_1x_3, \ldots,$ $x_{11} = x_1x_2x_3, \ldots, x_{15} = x_1x_2x_3x_4$). Saturated designs are useful in screening situations in which a large number of factors could have an effect on the response.

7.4-3 Computer Software

Minitab (http://www.minitab.com) and *JMP* (The Statistical Discovery Software, http://www.JMPdiscovery.com) are two general statistical software packages with strong process control and design components. Excellent software packages specifically targeted to the construction of experimental designs and the analysis of the resulting data are also available. *Design-Ease* and *Design-Expert,* distributed by Stat-Ease (htpp://www.statease.com), share many of the features found in JMP and Minitab.

These software packages construct the designs (i.e., they tell us about the factor levels of the runs and the order in which the runs should be carried out) and help analyze the resulting data once the experiment has been performed. In this section, we discuss Minitab, even though JMP is equally versatile and useful.

Minitab's features are included under the "Stat > DOE > Factorial" tab. Minitab makes it easy for the user to construct two-level factorial and fractional factorial designs. The user enters the number of factors and is then presented a list of full and fractional designs with their respective run sizes. After deciding on the number of runs, the user can construct the design either through default generators that optimize the resolution of the design or by writing out specific generators. In either situation, Minitab indicates the design columns, displays a randomized arrangement of the runs (if such display is desired), and lists the confounding patterns of the particular fraction that is selected. If default generators are used, Minitab displays the generators adopted. The user can add center points, replicate the design, block the experiment by specifying blocking generators, and modify the design by considering fold-overs (which can be complete fold-overs where the signs of all factors are changed or fold-overs of individual factors).

Once the design has been carried out, Minitab facilitates an efficient analysis of the data. The analysis of two-level factorial and fractional factorial designs includes estimates of the effects. Two different kinds of estimates are listed: estimates labeled "Coef," which are the estimates discussed in this book, and estimates called "Effects," which are twice the size of the coefficients expressing the differences in the response averages at high and low levels. Standard errors of the estimated coefficients are calculated if replications are available. Normal probability plots and Pareto plots for assessing the importance of the effects, as well as main-effects and interaction plots for assessing the nature of the relationships, are also available.

Minitab's normal probability plot has a feature that makes it particularly easy to spot significant (i.e., active) effects. Minitab uses the pseudo standard error (PSE) that has been developed by R.L. Lenth in his paper "Quick and Easy Analysis of Unreplicated Factorials" (*Technometrics,* Vol. 31, 1989, 469–474). If none of the factors are active, then the standard deviation of the m estimated effects (say, $f_1, f_2, \ldots, f_m$, where $m = 7$ in the 2^3 factorial and $m = 15$ in the 2^4 factorial) serves as the standard error of the estimated effects. However, if some effects are active, this estimate is too large, as it incorporates not only random variability, but also the effects of active factors. Hence, one needs to omit the estimates of all active factors from the calculation of the standard deviation. The normal probability plot discussed previously does this informally when it determines the "best fitting" straight line from just the estimates in the linear portion of the middle part of the graph and not from the estimates on the extreme left and right sides that do not appear to fit the line through the middle.

Lenth uses the fact that the median of the absolute values of the estimated non-active effects, suitably normalized, provides an estimate of the standard deviation. He starts by calculating

$$s = (1.5)\text{Median}\big(|f_1|, |f_2|, \dots, |f_m|\big).$$

The factor 1.5 in the normalization arises from the relationship between the standard deviation and the median of the absolute value of a mean-zero normal random variable. In the next step, Lenth omits all estimates with absolute values larger than $2.5s$ from the calculation, and he calculates a revised standard deviation

$$\text{PSE} = (1.5)\underset{|f_i| < 2.5s}{\text{Median}}\big(|f_1|, |f_2|, \dots, |f_m|\big).$$

He calls this quantity the *pseudo standard error* (PSE), and he uses it for the confidence intervals of the effects. His 95% confidence interval for an effect, Estimated Effect $\pm (t)(\text{PSE})$, uses the 97.5th percentile (upper 2.5th percentage point) of a t-distribution with $m/3$ degrees of freedom. For a standard error that is estimated with reasonable confidence and that comes from many observations, one would use

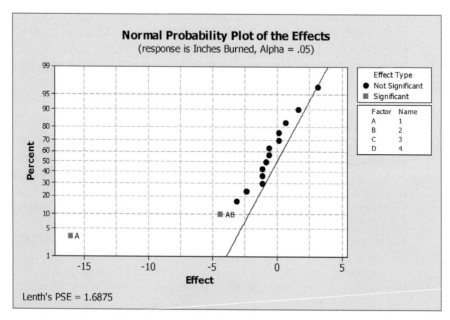

Figure 7.4-2 Minitab normal probability plot of the estimated effects from Example 7.3-2

Note: Minitab defines the estimated effects as twice the effects that are used in this book. Minitab's effects express the change in the response when a factor changes from its low to its high level. The normal probability plot shows that the main effect of factor 1 is quite significant. The question whether the interaction (12) is significant is up for debate. (It becomes insignificant for alpha = 0.045, for example.)

the 97.5th percentile of the standard normal distribution, or simply $t = 2$. However, in the unreplicated situation, the PSE comes from very few observations, and Lenth found through simulations that the t-distribution with $m/3$ degrees of freedom works best. For $m = 7$ effects, $t = 3.76$; for $m = 15$, $t = 2.57$, and for $m = 31$, $t = 2.22$. Lenth compares the estimated effects with $\pm(t)(\mathrm{PSE})$. If an estimate exceeds the limits $\pm(t)(\mathrm{PSE})$, it is likely that the associated factor is active (i.e., significant).

Let us illustrate this procedure with the effects in Example 7.3-2 of Section 7.3. There, we analyzed the results of a 2^4 experiment, finding the estimated effects shown in Table 7.3-5. The probability plot of the estimated effects, carried out through the Minitab command "Stat > DOE > Factorial > Analyze Factorial Design (checking Graph: Effects Plot)," is shown in Figure 7.4-2. The probability plot and Lenth's PSE suggest that the main effect of factor 1 and, to a much lesser extent, the (12) interaction are significant. [These significant effects are indicated in Minitab by (red) squares.]

Exercises 7.4

7.4-1 Consider the 2^{4-1}_{IV} fractional factorial design with generator $I = x_1x_2x_3x_4$. Show that the confounding patterns in Table 7.4-2 are correct.

*7.4-2 Consider the 2^{5-1}_{V} fractional factorial design with generator $I = x_1x_2x_3x_4x_5$. Specify the factor–level combinations for this design and obtain the confounding patterns.

Hint: Start with the 2^4 factorial design and write down the two-, three-, and four-factor interaction columns. Associate the levels of the fifth factor with $x_1x_2x_3x_4$, the four-factor interaction.

*7.4-3 Consider the 2^{5-2} fractional factorial with generators $x_4 = x_1x_2$ and $x_5 = x_1x_2x_3$. Specify the factor–level combinations for this design and determine the confounding patterns. What is the resolution of the design?

Hint: Start with a 2^3 factorial design and specify all interaction columns. Associate x_4 with x_1x_2 and x_5 with $x_1x_2x_3$. Because there are two generators, there are $2^2 = 4$ terms in the defining relation $I = x_1x_2x_4 = x_1x_2x_3x_5 = x_3x_4x_5$.

7.4-4 Consider the 2^{6-2} fractional factorial with $x_5 = x_1x_2x_3$ and $x_6 = x_2x_3x_4$. Repeat Exercise 7.4-3, but start with a 2^4 factorial design.

7.4-5 Consider the 2^{7-4}_{III} design given in this section. Assuming that interactions among three or more factors are zero, show that we obtain the confounding patterns that are given in Table 7.4-3.

*7.4-6 The 2^{7-4}_{III} design in Table 7.4-3 was one particular fraction of the 2^7 factorial with generators $x_4 = x_1x_2$, $x_5 = x_1x_3$, $x_6 = x_2x_3$, and $x_7 = x_1x_2x_3$. Consider the fraction you obtain by changing *all* the signs in the $x_1, x_2, x_3, x_4, x_5, x_6, x_7$ columns in Table 7.4-3. This is called the *fold-over* of the original fraction.

(a) Convince yourself that these eight runs are the same as the ones that you get by writing down a 2^3 and selecting the levels of the other four factors as

$$x_4 = -x_1x_2, x_5 = -x_1x_3, x_6 = -x_2x_3, \text{ and } x_7 = x_1x_2x_3.$$

(b) Using the generators in part (a), write down the defining relation for this fractional factorial design.

*(c) Assuming that interactions of three or more factors are zero, determine the confounding relationships for the estimates in this fold-over fractional design. Call the estimates $L_0', L_1', \ldots, L_7'$.

(d) Combine the linear combinations from the original (L_i) and the second (L_i') fraction, and form $(L_i + L_i')/2$ and $(L_i - L_i')/2$. Show that these linear combinations lead to unconfounded estimates of all seven main effects, while two-factor interactions are still confounded. For example, we find that $(L_1 + L_1')/2 \rightarrow (1)$ and $(L_1 - L_1')/2 \rightarrow (24) + (35) + (67)$.

7.4-7   Use the same original 2_{III}^{7-4} design as in Exercise 7.4-6, but now consider another fraction: the one we obtain by changing the signs of a single factor, say, factor 5, in the original design.

(a) Convince yourself that these eight new runs are the same as the ones you obtain by writing down a 2^3 and selecting the levels of the other four factors as

$$x_4 = x_1 x_2, \quad x_5 = -x_1 x_3, \quad x_6 = x_2 x_3, \quad \text{and } x_7 = x_1 x_2 x_3.$$

(b) Using these generators, write down the defining relation for this fractional design.

(c) Assuming that interactions of three or more factors are zero, determine the confounding relationships for, say, the estimates $L_0'', L_1'', \ldots, L_7''$.

(d) Combine these estimates with the ones from the original 2_{III}^{7-4}, and form $(L_i + L_i'')/2$ and $(L_i - L_i'')/2$. Show that this approach will lead to unconfounded estimates of the main effect of factor 5 and of all two-factor interactions with factor 5.

*7.4-8   Snee reports on a study designed to determine the factors that affect the color of a certain chemical product. The study identified five factors that were thought to be important. It was decided to examine these variables over the following ranges:

Process Variable	Low ($-$)	High ($+$)
$x_1 =$ solvent/reactant	Low	High
$x_2 =$ catalyst/reactant	0.025	0.035
$x_3 =$ temperature (°C)	150	160
$x_4 =$ reactant purity (percent)	92	96
$x_5 =$ pH of reactant	8.0	8.7

The variables were studied with a 2^{5-1} fractional factorial design with generator $x_5 = x_1 x_2 x_3 x_4$ or, equivalently, $I = x_1 x_2 x_3 x_4 x_5$. Determine the confounding patterns that are implied by this particular fraction.

The color of the product (in coded units) produced in each of the 16 runs (which were carried out in random order) is given in the following table:

Run	x_1	x_2	x_3	x_4	x_5	Color
1	$-$	$-$	$-$	$-$	$+$	-0.63
2	$+$	$-$	$-$	$-$	$-$	2.51
3	$-$	$+$	$-$	$-$	$-$	-2.68
4	$+$	$+$	$-$	$-$	$+$	-1.66
5	$-$	$-$	$+$	$-$	$-$	2.06
6	$+$	$-$	$+$	$-$	$+$	1.22

Run	x_1	x_2	x_3	x_4	x_5	Color
7	−	+	+	−	+	−2.09
8	+	+	+	−	−	1.93
9	−	−	−	+	−	6.79
10	+	−	−	+	+	6.47
11	−	+	−	+	+	3.45
12	+	+	−	+	−	5.68
13	−	−	+	+	+	5.22
14	+	−	+	+	−	9.38
15	−	+	+	+	−	4.30
16	+	+	+	+	+	4.05

Analyze the data. Plot the estimated effects on normal probability paper, or equivalently, use the computer to plot the normal scores of the effects against the effects. In your analysis, you may assume that interactions of order three or greater are zero.

[R. D. Snee, "Experimenting with a Large Number of Variables," in R. D. Snee, L.B. Hare, and J. R. Trout, eds., *Experiments in Industry: Design, Analysis and Interpretation* (Milwaukee, WI: American Society for Quality Control, 1985).]

*7.4-9 A study (also described by Snee in the book listed in Exercise 7.4-8) was initiated because of the perceived large variation in viscosity measurements obtained by an analytical laboratory. It was decided to conduct a "ruggedness test" of the measurement process to determine which variables, if any, were influencing the measurement process. The following seven variables were thought to be important:

Process Variable	Low (−)	High (+)
x_1 = sample preparation	M1	M2
x_2 = moisture measurement	Volume	Weight
x_3 = mixing speed (rpm)	800	1,600
x_4 = mixing time (hours)	0.5	3
x_5 = healing time (hours)	1	2
x_6 = spindle	S1	S2
x_7 = protective lid	Absent	Present

The sample was prepared by one of two methods (M1, M2), using moisture measurements made on a volume or weight basis. The sample was then put into a machine, mixed at a given speed for a specified period, and allowed to "heal" for 1 to 2 hours. There were two different spindles used in the mixer; the apparatus had a protective lid, and it was decided to run tests with and without the lid.

A 2^{7-3} fractional factorial design with generators $x_5 = x_2x_3x_4$, $x_6 = x_1x_3x_4$, and $x_7 = x_1x_2x_3$ was used. The viscosity measurements of the 16 runs, all made on samples from a common source product, are listed in the following table:

Run	x_1	x_2	x_3	x_4	x_5	x_6	x_7	Viscosity
1	−	−	−	−	−	−	−	27.9
2	+	−	−	−	−	+	+	24.6

Run	x_1	x_2	x_3	x_4	x_5	x_6	x_7	Viscosity
3	−	+	−	−	+	−	+	29.0
4	+	+	−	−	+	+	−	23.2
5	−	−	+	−	+	+	+	27.5
6	+	−	+	−	+	−	−	37.7
7	−	+	+	−	−	+	−	24.2
8	+	+	+	−	−	−	+	33.8
9	−	−	−	+	+	+	−	22.2
10	+	−	−	+	+	−	+	25.5
11	−	+	−	+	−	+	+	20.8
12	+	+	−	+	−	−	−	24.1
13	−	−	+	+	−	−	+	32.2
14	+	−	+	+	−	+	−	23.8
15	−	+	+	+	+	−	−	32.0
16	+	+	+	+	+	+	+	23.6

To guard against effects from other, unspecified sources, the experiments were conducted in random order. Determine the confounding patterns that are implied by this particular fraction. Construct a normal probability plot of the estimated effects. Interpret the results, assuming that interactions of order 3 or greater are zero.

Hint: Use a computer program to estimate the effects. The defining relation of this fractional factorial tells you about the confounding patterns. For example, the (13) interaction is confounded with (46) and (27).

7.4-10 As manager of a large collection agency, you are looking for ways to increase success in collecting overdue bills. The initial step in your collection process is to send a letter to the debtor. You are not certain whether a nice reminder or a threatening letter works best, nor do you know whether the letter should be sent by the collection agency or by a lawyer.

(a) Describe how you would set up a simple experiment to settle some of these open questions. You have available a large number of overdue accounts to select from, and for each account, you know its size as well as the length of time it has been overdue. This is important information, as the success of your strategy may depend on these characteristics. To simplify the situation, you may assume that you have small and large, and recent and long-overdue, accounts.

You have to give some thought on how to measure success. One possibility is to count the number of debts that are collected.

Can you determine the economic feasibility of your letter-writing strategy? Make the appropriate assumptions that allow you to carry out such an analysis. Assume that it costs $\$a_1$ to send a letter (and $\$a_2 > \$a_1$ to send it through the lawyer) and that you are paid $\$c_1$ for bringing in an overdue small account and $\$c_2$ for bringing in a large one. (Also, the length of time that the account has been overdue may enter into your calculations.)

(b) There are usually many delinquent clients who do not respond to a letter. Follow up on your nonrespondents with a phone call that can be either threatening or nicely reminding. How would you design this follow-up experiment, and how would you analyze the resulting information? Does a threatening phone call work best in all situations? Is it always worthwhile to make a follow-up phone call? (Assume that it costs $\$a_3$ to make a call.) Can you identify situations in which it would be beneficial to call?

Chapters 6 and 7 Additional Remarks

Designed experiments are an important component of the scientific learning process. The goals of this process are to confirm or refute prior knowledge and to suggest new hypotheses for future study. Clearly, it is important that the experimental approach be efficient and lead to the right answers.

A strategy of carrying out one single, all-encompassing experiment is usually ill conceived, as it leaves no room for subsequent experimentation and it shortcuts the accumulation of knowledge. Learning is sequential, and as R.A. Fisher has said so well, the best time to plan an experiment is after you have done it. It is a reasonable strategy to spend only a portion of the available resources on the initial experiment and save the remainder for follow-up runs.

At the outset of a study, one often encounters a large number of conflicting theories and numerous factors that are thought to have an effect on the response. At that stage, two-level fractional factorial designs like those discussed in Section 7.4 are especially useful for screening purposes, to separate the "vital few" factors from the "trivial many."

The design approaches in the last two chapters are powerful, but not foolproof. The key is to experiment. Missing something occasionally will be of small consequence compared with the accumulation of insights over time.

Much can be learned by studying a textbook, reading case studies, and solving end-of-chapter exercises. However, it has been our experience that, to really master the material, you must apply the methods in the real world. Go out and experiment! Discovering the unexpected is more important than confirming what you know.

In the randomized block experiment, the term "block" comes from the origins of this design in agricultural studies. Blocks were created by aggregating contiguous parcels that were homogeneous in terms of soil composition and hence fertility. R. A. Fisher describes these kinds of experiments in his book *Statistical Methods for Research Workers,* published in 1925.

The term *analysis of variance (ANOVA)* gives no indication that the procedure is about comparing means. But as we have seen, we test whether several means differ by comparing between-group and within-group variances. The F-distribution plays a key role in the analysis of the data, and perhaps not surprisingly, F stands for *Fisher,* the most important statistician of the 20th century. But Fisher did not invent this distribution; it was derived by George Snedecor, who named it F to honor Fisher.

R.A. Fisher, The Life of a Scientist, is an interesting biography of Sir Ronald Fisher written by his daughter, Joan Fisher Box. Fisher is one of the statisticians included in *The Lady Tasting Tea: How Statistics Revolutionized Science in the Twentieth Century,* by David Salsburg. The title of the book comes from a paper that Fisher wrote and that is included in Fisher's book *The Design of Experiments.* As the story goes, a lady claimed that, by tasting it, she could tell whether milk or tea was put into the cup first. Fisher designed an experiment to test her claim.

Computer software has simplified the construction of suitable designs and the efficient analysis of the resulting data. Computer software avoids tedious hand (or even calculator) computations and simplifies the construction of graphical displays. Throughout this book, we have emphasized the value of good statistical software. Use the computer to your advantage, but do not trust its outcomes blindly. Check the reasonableness of the results, as results depend on assumptions that may be violated. There is no substitute for common sense.

Chapters 6 and 7 Projects

Project 1

The website http://www.misd.net/Mathematics/hsmathscience/syllabus.htm is designed to make science and math exciting and relevant to students. The underlying pedagogical theme is the importance of *doing* real science through experimentation. The site describes many simple experiments that can be carried out with basic supplies that are readily available to most students.

The article "Some Ideas about Teaching Design of Experiments with 2^5 Examples of Experiments Conducted by Students," by W.G. Hunter in *The American Statistician* (Vol. 31, 1977, pp. 12–17), is another useful reference for simple experiments. In one experiment, popcorn yield was measured for combinations of two brands of popcorn, two batch sizes, and two popcorn-to-oil ratios. In another experiment, differing amounts of nickel, manganese, and carbon were used to affect the strength of a certain steel alloy. The extent of iron corrosion was the response in an experiment that considered pH, dissolved oxygen content in water, and temperature. Yet another project dealt with the number of days it took mail to reach its destination; the factors studied included the type of stamp (first class or airmail), zip code (used or not), and the time of day the letter was mailed.

Use the suggestions in these two references to come up with your own experiment. Conduct the experiment, analyze the resulting data, and write a brief report summarizing your findings. Discuss what you have learned.

Project 2

StatEase, Inc., of Minneapolis (http://www.statease.com/) is an innovative company that offers design-of-experiments (DOE) software, books, training, and consulting services. The company's mission is to help firms improve the quality of their products, develop efficient processes, solve manufacturing problems, and make breakthrough discoveries by applying powerful statistical methods.

One of StatEase's publications (http://www.statease.com/pubs/doe-self.pdf) compiles a list of simple experiments, including studying the "bounciness" of play putty, measuring the strength of two types of paper clips, and measuring the diameter of a crater formed in sand by dropping balls of different sizes. Carry out some of the suggested experiments and analyze the resulting data. George Box's helicopter experiment is a good place to start.

Also, you may want to look at StatEase's publication on case studies and articles (http://www.statease.com/articles.html). Select one or more of these case studies and analyze the data reported.

Project 3

The NIST/SEMATECH *Engineering Statistics Handbook* (2003, continuously updated; http://www.itl.nist.gov/div898/handbook/) discusses many interesting projects and gives a detailed description of the experiments and a list of the resulting data. The handbook serves as an excellent resource for industry and university training alike. Look at the two case studies "Eddy Current Probe Sensitivity Case Study" and "Sonoluminescent Light Intensity Case Study" in Section 5.6 of the handbook. Discuss the designs and reanalyze the data.

Project 4

Investigate the effects of the following five factors on the expansion of pinto beans:

Soaking fluid: water ($-$) or beer ($+$)

Salinity: no salt ($-$) or salt ($+$)

Acidity: no vinegar ($-$) or vinegar ($+$)

Soaking temperature: refrigerator temperature ($-$) or room temperature ($+$)

Soaking time: 2 hours ($-$) or 6 hours ($+$)

Carry out the following experiment: Select a pinto bean, measure its "size," put the bean into a soaking fluid, and—after a certain amount of time—measure its size again. Use five tablespoons of soaking fluid to soak each bean. Make sure that the liquid covers the bean. For salt, add 1/4 teaspoon to the soaking fluid. For vinegar, add 1 teaspoon to the soaking fluid.

(a) Discuss how you measure the "size of a pinto bean" and its "expansion." Give a detailed description of your measurement procedure, so that it can be carried out by other people; that is, give an operational definition.

(b) Design, set up, and execute the experiment as a 2^5 factorial experiment. Conduct two replications, which may be run concurrently. Analyze the effects of the five factors. Write a short report summarizing your findings. Support your results with appropriate graphs and calculations. What did you learn? What was the most difficult part of your experiment? If you had to do it over again, what would you change?

(c) Instead of carrying out a full factorial with 32 runs and two replications, economize on the number of runs and conduct a half fraction (or a quarter fraction) of the 2^5 factorial design. Explain how you would carry out such fractional factorial designs. What are their advantages and what is the price you pay compared with the full 2^5 factorial experiment?

Project 5

Treat your friends to some waffles. Experiment with three factors: two types of pancake mix (the cheapest budget mix and the most expensive product that you can find in the store); the water-to-mix ratio (the amount suggested on the package and one that has 20 percent more water); and the type of oil (corn oil versus olive oil). The tastiness of the waffle is the response.

Design an appropriate experiment. Document your procedure and explain how you measure the tastiness of a waffle. Carry out your experiment and analyze the data. What are your findings?

*Project 6

Jay Harris, publisher of *Mother Jones* magazine, was constantly searching for ways to increase circulation. He realized that the makeup of the promotional mailing was essential in getting people to subscribe to his magazine.

A typical mailing was sent to about 400,000 potential subscribers. Under the experimental design approaches, numerous changes are being tested simultaneously. The following seven factors were identified for testing:

Factor	Factor Levels	
	−	+
A	no act-now insert	act-now insert
B	no credit card	credit card
C	hard offer	harder offer
D	strong guarantee	stronger guarantee
E	no testimonials	testimonials
F	no bumper sticker	bumper sticker
G	gutsy	ballsy

Harris wanted a design that would not only show main effects, but if possible, identify two-factor interactions as well. He anticipated that a typical response rate for a mailing was about 2 percent, and he thought that a 1/4 percentage increase in response was worth learning about.

(a) Devise an appropriate design for the mailing. Label each of the factors given in this project as $A, B, C, \ldots$. What is the confounding pattern for the design that you have selected (ignoring interactions of order 3 or higher)? Explain briefly why you picked this design.

(b) Recall that a 1/4 percentage increase from the current subscription rate of 2 percent was considered important. If we call each run a package, how many people should be sent each package (i.e., how many mailings are there for each run)? What is the total number of mailings (i.e., the total sample size)?

Note: This exercise involves determining the sample size. Control the risks of type I and type II errors to be around 5–10 percent. (See Project 6 of Chapter 4 for helpful hints, and use the available computer software.)

*(c) The data that resulted from one specific experiment are given in J. Ledolter and A.J. Swersey, "Using a Fractional Factorial Design to Increase Direct Mail Response at *Mother Jones* Magazine," *Quality Engineering,* 18(2006), 469–475. The authors consider a 2^{7-3} fractional factorial design in 16 runs. Each run consists of a particular combination of factor settings, with each combination sent to $n = 2,500$ persons. The response variable is the response rate (in percent), which is the percentage of people who subscribed and paid. Read the paper, analyze the information in the data file Chapter7Project6MotherJones, and summarize your findings.

REGRESSION ANALYSIS

The objective of many scientific investigations is to understand and explain relationships among variables. Frequently, one wants to know how and to what extent a certain response variable is related to a set of explanatory variables. As an example, consider the chemical engineer who is interested in the relationship between the yield and the temperature and reaction time of a chemical process; or the automotive engineer who studies the relationship between fuel efficiency and the weight, engine characteristics, and chassis design of an automobile; or the oncologist who relates the incidence of lung cancer to its many possible causes, such as genetic makeup, diet, cigarette consumption, and environmental factors.

In rare instances, the relationships are known *exactly,* so that the response and explanatory variables are functionally related. We call such relationships *deterministic,* because repeated experiments at a given setting of the explanatory variables will always lead to the same response. Examples of these relationships can be found among some of the traditional laws of physics and chemistry.

In most cases, however, the relationships are not known and, furthermore, are much too complicated to be described by a small set of explanatory variables. In such situations, we have to approximate the relationships and develop models that characterize the main features of the relationships. These models are no longer deterministic, but *statistical* in nature. The response variable, let us call it *Y,* is treated as a random variable that varies around a mean value which depends on the values of the explanatory variables. Repeated trials with identical values of the explanatory variables will no longer lead to the same response every time.

Regression analysis is concerned with developing such statistical "approximating" models. The technique is a very useful and widely employed tool of data analysis that leads to simple, yet often powerful, descriptions of the main features of the relationships among variables.

In Sections 8.1 through 8.3, we describe the *simple linear regression model*. In this type of model, we approximate the relationship between a response variable and a single explanatory variable by a linear function. In Sections 8.4 and 8.5, we extend the analysis to more than one explanatory variable. In Section 8.6, we use regression modeling to find the settings of certain input (or explanatory) variables that lead to the optimal conditions of a process.

8.1 The Simple Linear Regression Model

In this section, we consider only one explanatory variable, x, and assume that the statistical relationship between the response variable Y and this explanatory variable is linear. The model is written as

$$Y_i = \beta_0 + \beta_1 x_i + \varepsilon_i, \qquad i = 1, 2, \ldots, n.$$

The usual assumptions about the parameters and variables in this model are the following:

1. x_i is the ith observation on the explanatory variable. In planned experiments, the known constants $x_1, x_2, \ldots, x_n$ correspond to particular settings of the explanatory variable that are chosen by the investigator.

2. Y_i is the response that corresponds to the setting x_i of the explanatory variable, for $i = 1, 2, \ldots, n$.

3. β_0 and β_1 are the coefficients (parameters) in the linear relationship; β_0 is the intercept and β_1 is the slope. A change of one unit in the explanatory variable x translates into a change of β_1 units in the response variable.

4. The random variables $\varepsilon_1, \varepsilon_2, \ldots, \varepsilon_n$ are errors that create the scatter around the linear relationship $\beta_0 + \beta_1 x_i$, $i = 1, 2, \ldots, n$, respectively. We assume that these errors are mutually independent and normally distributed with mean zero and variance σ^2. That is,

$$E(\varepsilon_i) = 0 \quad \text{and} \quad \text{var}(\varepsilon_i) = \sigma^2, i = 1, 2, \ldots, n.$$

Furthermore, the information obtained from one error does not imply something about another.

From these assumptions, we note that the response $Y_i = \beta_0 + \beta_1 x_i + \varepsilon_i$ is the sum of two components. The first, $\beta_0 + \beta_1 x_i$, is not random, because β_0 and β_1 are fixed parameters and x_i is a known constant; we call this component the *signal*. The second component, ε_i, is a random variable that we call the *noise*. Hence, Y_i is also a random variable, and it follows from our assumptions that Y_i has a normal distribution with mean

$$E(Y_i) = E(\beta_0 + \beta_1 x_i + \varepsilon_i) = \beta_0 + \beta_1 x_i = \mu_i$$

and variance

$$\text{var}(Y_i) = \text{var}(\beta_0 + \beta_1 x_i + \varepsilon_i) = \text{var}(\varepsilon_i) = \sigma^2.$$

Furthermore, $Y_1, Y_2, \ldots, Y_n$ are mutually independent, because we assume the mutual independence of $\varepsilon_1, \varepsilon_2, \ldots, \varepsilon_n$. Thus, we can summarize the assumptions of the simple linear regression model by stating that $Y_1, Y_2, \ldots, Y_n$ are mutually independent random variables that have the respective distributions $N(\beta_0 + \beta_1 x_i, \sigma^2)$, $i = 1, 2, \ldots, n$. $E(Y) = \beta_0 + \beta_1 x$ is called the *regression function*.

Example 8.1-1 Consider the relationship between the weight of an automobile (x) and fuel consumption (Y), where the latter is measured by gpm, the amount of fuel (in gallons) that is needed to drive 100 miles. Fuel consumption is roughly proportional to the

y(GPM)

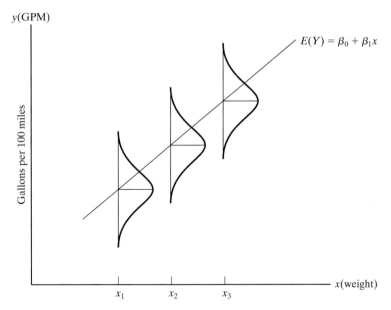

$E(Y) = \beta_0 + \beta_1 x$

Gallons per 100 miles

x_1 x_2 x_3 x(weight)

Figure 8.1-1 Simple linear regression model

effort (force times distance) that is required to move the car. Now, force is proportional to weight, so we expect fuel consumption also to be proportional to weight. Thus, it makes sense to use the model

$$Y = \beta_0 + \beta_1 x + \varepsilon,$$

where ε is a random component. Note that this model specifies a statistical relationship. A completely deterministic relationship would not be appropriate, because different automobiles of the same weight can be expected to have somewhat different fuel efficiencies. That is, weight is not the only factor that determines fuel consumption. There are many others, such as shape of chassis, engine design, type of tires, and so on. In addition, it is difficult to measure fuel efficiency without error, and our model must allow for measurement error. Hence, for fixed x, we assume that the response Y is a random variable with mean $E(Y) = \beta_0 + \beta_1 x$. Furthermore, let us suppose that we can assume that $\text{var}(Y) = \sigma^2$ does not depend on x and that each Y has a normal distribution. The latter assumptions must be checked carefully because they may not be satisfied in practice. However, if they are satisfied, this linear regression model is relatively simple to use. It is depicted in Figure 8.1-1.   ■

Remarks: Note that the explanatory variable x and the settings $x_1, x_2, \ldots, x_n$ are denoted by lowercase letters. This is because these levels are fixed constants and not random variables. The use of lowercase letters is consistent with the notation in this book, whereby we reserve capital letters for random variables. The responses $Y_1, Y_2, \ldots, Y_n$ are denoted by capital letters, because, before the observations are taken, they are random variables. After a particular sample has been taken, they become particular realizations, or numbers, and are denoted by the lowercase letters $y_1, y_2, \ldots, y_n$. You should not be overly concerned about the distinction between Y

and y. The subtle change in notation merely reflects whether you look at the response before or after the experiment has been carried out.

8.1-1 Estimation of Parameters

There are three parameters in the simple linear regression model: the coefficients β_0 and β_1 in the regression function and the variance σ^2, which accounts for the random scatter around the regression line. Ordinarily, these parameters are unknown and must be estimated from sample data.

Let us assume that we have n pairs of observations $(x_1, y_1), (x_2, y_2), \ldots, (x_n, y_n)$. As an example, consider the weight (x) and the fuel consumption gpm (y) of the $n = 10$ cars listed in Table 8.1-1. We have already seen this data set in an earlier chapter. We used it in Section 1.5 to illustrate scatter plots and the sample correlation coefficient. We also used it in Section 1.5 to introduce the basic principles behind regression analysis. A scatter plot of fuel consumption against weight is shown in Figure 8.1-2. It confirms that the relationship is approximately linear.

Remark: You probably have noticed that the data set on fuel efficiencies and weights is somewhat dated. The data are from 1978–79 model-year cars and pertain to vehicles such as the AMC Concord, the Toyota Corona, and the VW Rabbit. The American Motors Company (AMC) was long bought out by another company (Chrysler), and the Toyota Corona (not Corolla) and the VW Rabbit have not been manufactured for many years (Volkswagen revived the Rabbit nameplate in 2006). Nevertheless, we find this very small data set instructive, as it allows us to illustrate most regression issues. Also, not much has changed as regards fuel efficiency since the early 1980s. Cars have become heavier, and there have been some (many people would say rather modest) improvements in fuel efficiency. It is of interest to check

Table 8.1-1 Fuel Consumption Data

Car	Weight (1000 pounds)	Fuel Consumption, gpm (gallons 100 miles)
AMC Concord	3.4	5.5
Chevy Caprice	3.8	5.9
Ford Country Squire Wagon	4.1	6.5
Chevette	2.2	3.3
Toyota Corona	2.6	3.6
Ford Mustang Ghia	2.9	4.6
Mazda GLC	2.0	2.9
AMC Sprint	2.7	3.6
VW Rabbit	1.9	3.1
Buick Century	3.4	4.9

Source: Data are taken from H. V. Henderson and P. F. Velleman, "Building Multiple Regression Models Interactively," *Biometrics,* 37: 391–411 (1981). The full data set is given in Exercise 8.3-6.

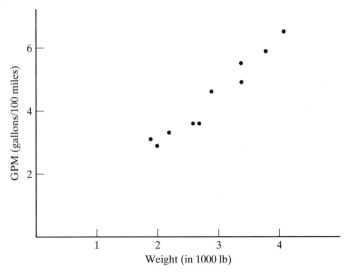

Figure 8.1-2 Scatter plot of gpm against weight

whether the relationship between fuel consumption and weight has changed. For that, in Exercise 8.4-4 we ask you to analyze a new data set consisting of 2007 model-year cars, repeat the analysis, and check whether there have been changes to the intercept and the slope of the linear relationship. A graph of both data sets on the same graph will turn out to be quite informative. ■

Our objective now is to find estimates $\hat{\beta}_0$ and $\hat{\beta}_1$ of the coefficients β_0 and β_1, respectively, such that the fitted line $\hat{\beta}_0 + \hat{\beta}_1 x$ is "close to" the observation points $(x_1, y_1), (x_2, y_2), \ldots, (x_n, y_n)$. Closeness between the observations y_i and a regression line $\beta_0 + \beta_1 x_i$ can be measured in several different ways. One popular measure is to consider the square of each deviation, $[y_i - (\beta_0 + \beta_1 x_i)]^2$, and add all the squared deviations together to obtain $S(\beta_0, \beta_1) = \sum_{i=1}^{n}[y_i - (\beta_0 + \beta_1 x_i)]^2$. This sum of squares is a measure of the closeness between the regression line and the observations. It is obviously equal to zero if the line goes through all the points. Using the sum-of-squares measure, we select the estimates of β_0 and β_1 such that $S(\beta_0, \beta_1)$ is as small as possible. Because we are minimizing a sum of squares, we call the values of β_0 and β_1 at which the minimum is obtained the *least-squares estimates* and denote them by $\hat{\beta}_0$ and $\hat{\beta}_1$, respectively.

To find the minimum of the function $S(\beta_0, \beta_1)$ of the two arguments β_0 and β_1, we set its two first partial derivatives equal to zero. This leads to the two equations

$$\frac{dS(\beta_0, \beta_1)}{d\beta_0} = 2\sum_{i=1}^{n}[y_i - (\beta_0 + \beta_1 x_i)](-1) = 0$$

and

$$\frac{dS(\beta_0, \beta_1)}{d\beta_1} = 2\sum_{i=1}^{n}[y_i - (\beta_0 + \beta_1 x_i)](-x_i) = 0.$$

The least-squares estimates $\hat{\beta}_0$ and $\hat{\beta}_1$ must satisfy these two equations. Rewriting the equations in terms of $\hat{\beta}_0$ and $\hat{\beta}_1$, we obtain

$$\hat{\beta}_0 n + \hat{\beta}_1 \sum x_i = \sum y_i,$$

$$\hat{\beta}_0 \sum x_i + \hat{\beta}_1 \sum x_i^2 = \sum x_i y_i.$$

In this form, the two equations are called the *normal equations*. (This is unfortunate terminology, as they have nothing to do with the normal distribution.) The solution of these two equations provides the *least-squares estimates*

$$\hat{\beta}_1 = \frac{\sum x_i y_i - (\sum x_i)(\sum y_i)/n}{\sum x_i^2 - (\sum x_i)^2/n}$$

and

$$\hat{\beta}_0 = \bar{y} - \hat{\beta}_1 \bar{x}.$$

An equivalent expression for $\hat{\beta}_1$ is

$$\hat{\beta}_1 = \frac{\sum (x_i - \bar{x})(y_i - \bar{y})}{\sum (x_i - \bar{x})^2} = \frac{\sum (x_i - \bar{x}) y_i}{\sum (x_i - \bar{x})^2} = r \frac{s_y}{s_x},$$

where r is the correlation coefficient and s_x and s_y are the respective sample standard deviations. (See Exercise 8.1-5.)

The more mathematically oriented reader surely will note that we should check certain conditions about the second partial derivatives to ensure that $\hat{\beta}_0$ and $\hat{\beta}_1$, as given here, yield the minimum value of $S(\beta_0, \beta)$. We accept the fact that these conditions are actually fulfilled, and $\hat{\beta}_0$ and $\hat{\beta}_1$ are, in fact, the least-squares estimates.

Example 8.1-2   For the gas mileage data in Table 8.1-1, we find that

$$\sum_{i=1}^{10} x_i = 29.0, \qquad \sum_{i=1}^{10} x_i^2 = 89.28, \qquad \sum_{i=1}^{10} y_i = 43.9,$$

$$\sum_{i=1}^{10} y_i^2 = 207.31, \qquad \sum_{i=1}^{10} x_i y_i = 135.80.$$

Substituting these values into our expressions for $\hat{\beta}_1$ and $\hat{\beta}_0$, we obtain

$$\hat{\beta}_1 = \frac{135.80 - (29.0)(43.9)/10}{89.28 - (29.0)^2/10} = \frac{8.49}{5.18} = 1.639,$$

$$\hat{\beta}_0 = 4.39 - (1.639)(2.9) = -0.363.$$

The slope estimate $\hat{\beta}_1 = 1.639$ implies that each additional unit (1,000 pounds) of weight requires an additional 1.639 gallons of fuel for the car to be driven 100 miles. Note that our data set includes cars with weights that range from 1,900 to 4,100 pounds. Over this range, we can approximate the relationship between weight and fuel efficiency by a linear function. However, the fuel efficiencies of cars with weights outside this range may follow a different law, but no data are available to investigate the form of the relationship. The intercept estimate is $\hat{\beta}_0 = -0.363$, but we cannot attach too much meaning to it; certainly, it is not the fuel consumption of cars of zero weight! ∎

Remark: The principle of least squares is convenient mathematically because it leads to explicit expressions for the estimates. One can also adopt other principles for fitting equations. For example, the principle of *least absolute deviations* chooses the estimates that minimize $S^*(\beta_0, \beta_1) = \Sigma |y_i - (\beta_0 + \beta_1 x_i)|$. The calculation of these estimates, however, is much more cumbersome than the calculation of those associated with least squares; moreover, explicit expressions for the least absolute deviations estimates cannot be given. Thus, this and other procedures are not used as often as the method of least squares.

8.1-2 Residuals and Fitted Values

The least-squares estimates $\hat{\beta}_0$ and $\hat{\beta}_1$ lead to the *estimated (fitted) regression line*, which is given by

$$\hat{y} = \hat{\beta}_0 + \hat{\beta}_1 x.$$

The calculation of this expression at the levels $x_1, x_2, \ldots, x_n$ of the explanatory variable provides the *fitted* values

$$\hat{y}_i = \hat{\beta}_0 + \hat{\beta}_1 x_i, \qquad i = 1, 2, \ldots, n.$$

The respective differences between the observations $y_1, y_2, \ldots, y_n$ and the fitted values $\hat{y}_1, \hat{y}_2, \ldots, \hat{y}_n$ are called *residuals* and are given by

$$e_i = y_i - \hat{y}_i = y_i - (\hat{\beta}_0 + \hat{\beta}_1 x_i), \qquad i = 1, 2, \ldots, n.$$

Example 8.1-3 For the mileage data, with $x_1 = 3.4$, the corresponding fitted value is $\hat{y}_1 = -0.363 + (1.639)(3.4) = 5.21$ and the residual is $e_1 = y_1 - \hat{y}_1 = 5.5 - 5.21 = 0.29$. The fitted values and the residuals for all 10 observations are given in Table 8.1-2.

Table 8.1-2 Fitted Values and Residuals

Observation Number, i	Weight, x_i	Observed Value, y_i	Fitted Value, $\hat{y}_i$	Residual, $e_t = y_i - \hat{y}_i$
1	3.4	5.5	5.21	0.29
2	3.8	5.9	5.87	0.03
3	4.1	6.5	6.36	0.14
4	2.2	3.3	3.24	0.06
5	2.6	3.6	3.90	−0.30
6	2.9	4.6	4.39	0.21
7	2.0	2.9	2.91	−0.01
8	2.7	3.6	4.06	−0.46
9	1.9	3.1	2.75	0.35
10	3.4	4.9	5.21	−0.31
				$\Sigma e_i = 0.00$

We notice from the table that the residuals sum to zero. This is true in general and is not just due to this particular data set. In fact, we note that the n residuals satisfy the following two restrictions:

$$\sum_{i=1}^{n} e_i = 0 \quad \text{and} \quad \sum_{i=1}^{n} e_i x_i = 0.$$

The residuals satisfy these two equations because the equations are essentially the two first partial derivatives of $S(\beta_0, \beta_1)$ evaluated at the least-squares estimates; thus, they must equal zero. That is,

$$\sum [y_i - (\hat{\beta}_0 + \hat{\beta}_1 x_i)] = 0 \quad \text{and} \quad \sum [y_i - (\hat{\beta}_0 + \hat{\beta}_1 x_i)] x_i = 0. \quad \blacksquare$$

8.1-3 Sampling Distribution of $\hat{\beta}_0$ and $\hat{\beta}_1$

For a given sample $(x_1, y_1), (x_2, y_2), \ldots, (x_n, y_n)$, the least-squares *estimate* $\hat{\beta}_1$ is a certain numerical value. However, prior to sampling, the responses Y_i at the selected fixed levels x_i, $i = 1, 2, \ldots, n$, are normally distributed random variables with means $\beta_0 + \beta_1 x_i$ and variance σ^2. Thus, the *estimator*

$$\hat{\beta}_1 = \frac{\sum (x_i - \bar{x}) Y_i}{\sum (x_i - \bar{x})^2} = \sum_{i=1}^{n} c_i Y_i, \quad \text{where} \quad c_i = \frac{x_i - \bar{x}}{\sum_{j=1}^{n} (x_j - \bar{x})^2},$$

is a linear function of $Y_1, Y_2, \ldots, Y_n$ and is itself a random variable. In fact, $\hat{\beta}_1$ has a normal distribution with mean

$$E(\hat{\beta}_1) = \sum_{i=1}^{n} c_i E(Y_i) = \sum_{i=1}^{n} c_i (\beta_0 + \beta_1 x_i) = \beta_0 \sum_{i=1}^{n} c_i + \beta_1 \sum_{i=1}^{n} c_i x_i = \beta_1.$$

This is so because $\sum_{i=1}^{n} c_i = 0$ and

$$\sum_{i=1}^{n} c_i x_i = \sum_{i=1}^{n} \frac{(x_i - \bar{x}) x_i}{\sum (x_j - \bar{x})^2} = \frac{\sum (x_i - \bar{x})^2}{\sum (x_j - \bar{x})^2} = 1,$$

given that $\sum (x_i - \bar{x}) x_i = \sum (x_i - \bar{x})(x_i - \bar{x}) = \sum (x_i - \bar{x})^2$.

This shows that $\hat{\beta}_1$ is an unbiased estimator of β_1. In addition, using the results obtained on variances in Section 3.5, we find that the variance of $\hat{\beta}_1$ is

$$\text{var}(\hat{\beta}_1) = \sum_{i=1}^{n} c_i^2 \sigma^2 = \sigma^2 \sum_{i=1}^{n} \frac{(x_i - \bar{x})^2}{\left[\sum (x_j - \bar{x})^2 \right]^2} = \frac{\sigma^2}{\sum (x_i - \bar{x})^2}.$$

That is, $\hat{\beta}_1$ is $N[\beta_1, \sigma^2 / \sum (x_i - \bar{x})^2]$.

If σ^2 is known (at least approximately), we can use the preceding result to find a confidence interval for $\hat{\beta}_1$. The probability statement

$$P\left[-z(\alpha/2) \leq \frac{\hat{\beta}_1 - \beta_1}{\sigma / \sqrt{\sum (x_i - \bar{x})^2}} \leq z(\alpha/2) \right] = 1 - \alpha$$

is equivalent to

$$P\left[\hat{\beta}_1 - z(\alpha/2)\frac{\sigma}{\sqrt{\sum(x_i - \bar{x})^2}} \le \beta_1 \le \hat{\beta}_1 + z(\alpha/2)\frac{\sigma}{\sqrt{\sum(x_i - \bar{x})^2}}\right] = 1 - \alpha.$$

That is, the observed interval

$$\hat{\beta}_1 \pm z(\alpha/2)\frac{\sigma}{\sqrt{\sum(x_i - \bar{x})^2}}$$

is a $100(1 - \alpha)$ percent confidence interval for β_1. Usually, σ^2 is unknown and must be estimated. As we shall see in the next section, this leads to a confidence interval for β_1 that is based on a t-distribution.

Similarly, we can show that the sampling distribution of the estimator $\hat{\beta}_0 = \bar{Y} - \hat{\beta}_1\bar{x}$ is normal with mean β_0 and $\text{var}(\hat{\beta}_0) = \sigma^2[(1/n) + \bar{x}^2/\sum(x_i - \bar{x})^2]$. (See Exercise 8.1-6.) This result allows us to calculate the following confidence interval for β_0:

$$\hat{\beta}_0 \pm z(\alpha/2)\sigma\sqrt{\frac{1}{n} + \frac{\bar{x}^2}{\sum(x_i - \bar{x})^2}}.$$

Exercises 8.1

*8.1-1 The attendance at a racetrack (x) and the amount that was bet (y) on $n = 10$ selected days are given in the following table:

Attendance, x (hundreds)	Amount Bet, y (millions of dollars)
117	2.07
128	2.80
122	3.14
119	2.26
131	3.40
135	3.89
125	2.93
120	2.66
130	3.33
127	3.54

(a) Make a scatter plot of y against x.

*(b) Fit a simple linear regression model to the data, and calculate the least-squares estimates, the fitted values, and the residuals. (It is useful to calculate these estimates once "by hand" to appreciate the computations that are involved. However, we recommend that after you have done this once, to use a computer program.)

*8.1-2 An experiment was conducted to study the relationship between baking temperature x (in units of $10°$ Fahrenheit) and yield y (as a percentage) of a popular cake mix. The results were as follows:

Baking Temperature, x	Percent Yield, y
10	21.2
10	19.9
11	22.5
11	23.7
12	25.0
15	30.3
17	36.1
19	38.6
20	41.5
20	42.7
23	45.0
25	50.0
27	53.9
30	62.1

(a) Make a scatter plot of y against x.

*(b) Estimate the parameters of the simple linear regression model by least squares, and calculate the fitted values and residuals.

*8.1-3 The following data of Snedecor and Cochran show the initial weights and the gains in weight (both in grams) of 15 female rats on a high-protein diet that was administered from the 24th day to the 84th day of their age:

Initial Weight, x	Weight Gain, y
50	128
64	159
76	158
64	119
74	133
60	112
69	96
68	126
56	132
48	118
57	107
59	106
46	82
45	103
65	104

The interest in these data is whether the weight gain depends on the initial weight.

(a) Make a scatter plot of y against x.

*(b) Calculate the least-squares estimates of the parameters in the simple linear regression model. Calculate the fitted values and the residuals.

[G. W. Snedecor and W. G. Cochran, *Statistical Methods,* 6th ed. (Ames, IA: Iowa State University Press, 1967).]

8.1-4 Show that the fitted regression line $\hat{y} = \hat{\beta}_0 + \hat{\beta}_1 x$ goes through the point $(\bar{x}, \bar{y})$. *Hint:* Substitute $\bar{x}$ into $\hat{\beta}_0 + \hat{\beta}_1 x$ and show that the result equals $\bar{y}$.

8.1-5 Show that the least-squares estimate $\hat{\beta}_1$ can be written as

$$\hat{\beta}_1 = \frac{\sum (x_i - \bar{x}) y_i}{\sum (x_i - \bar{x})^2} = r(s_y/s_x),$$

where

$$r = \sum (x_i - \bar{x})(y_i - \bar{y}) / \left[\sum (x_i - \bar{x})^2 \sum (y_i - \bar{y})^2 \right]^{1/2}$$

is the sample correlation coefficient discussed in Section 1.5 and where s_x and s_y are the respective sample standard deviations.

8.1-6 Consider the least-squares estimator $\hat{\beta}_0 = \bar{Y} - \hat{\beta}_1 \bar{x}$ of the intercept β_0.

(a) Show that $\hat{\beta}_0 = \sum k_i Y_i$, where $k_i = (1/n) - \bar{x} c_i$, $c_i = (x_i - \bar{x})/\sum (x_i - \bar{x})^2$. Thus, $\hat{\beta}_0$, a linear function of $Y_1, Y_2, \ldots, Y_n$, has a normal distribution.

(b) Recalling that $\sum c_i = 0$, $\sum c_i x_i = 1$, and $\sum c_i^2 = 1/\sum (x_i - \bar{x})^2$, show that $E(\hat{\beta}_0) = \beta_0$ and $\text{var}(\hat{\beta}_0) = \sigma^2 [(1/n) + \bar{x}^2 / \sum (x_i - \bar{x})^2]$.

8.1-7 In the notation of this section, show that $E[(\hat{\beta}_0 - \beta_0)(\hat{\beta}_1 - \beta_1)] = \sigma^2 \sum k_i c_i$. Substituting $k_i = (1/n) - \bar{x} c_i$, prove that this equals

$$\text{cov}(\hat{\beta}_0, \hat{\beta}_1) = E[(\hat{\beta}_0 - \beta_0)(\hat{\beta}_1 - \beta_1)] = -\bar{x}\sigma^2 / \sum (x_i - \bar{x})^2$$
$$= -\bar{x}\,\text{var}(\hat{\beta}_1).$$

*8.1-8 Consider the simple linear regression model $Y_i = \beta x_i + \varepsilon_i$ through the origin.

*(a) Using principles from calculus, determine the least-squares estimate $\hat{\beta}$ that minimizes

$$\sum_{i=1}^{n} (y_i - \beta x_i)^2.$$

*(b) Prior to sampling, the Y_i's are random variables. Show that the least-squares estimator $\hat{\beta}$ can be written as a linear function of $Y_1, Y_2, \ldots, Y_n$, and derive its mean and variance.

8.1-9 Suppose that you are interested in estimating the effect of a variable x on a certain response variable. You have the opportunity to design the relevant experiment (i.e., to choose the levels of the x variable) before you estimate the simple linear regression model $Y_i = \beta_0 + \beta_1 x_i + \varepsilon_i$. In designing your experiment, you are restricted to the feasible range $[c_1, c_2]$ of your x variable.

Assume that your sample size n is even. Show that the best arrangement of the x levels is to allocate $n/2$ experiments to $x = c_1$ and $n/2$ experiments to $x = c_2$, an arrangement that minimizes the variance of $\hat{\beta}_1$. Discuss why this arrangement is not very good if we want to check the adequacy of the linear model.

8.2 Inferences in the Regression Model

Regression analysis attempts to use the information obtained from the explanatory variable to explain some of the variability in the response variable Y. Ignoring the information that is contained in $x_1, x_2, \ldots, x_n$, we can measure the variation among the responses $Y_1, Y_2, \ldots, Y_n$ by $\Sigma(Y_i - \overline{Y})^2$, which is called the *total sum of squares* and is written as

$$\text{SSTO} = \sum_{i=1}^{n}(Y_i - \overline{Y})^2 = \sum_{i=1}^{n}Y_i^2 - \frac{\left(\sum_{i=1}^{n}Y_i\right)^2}{n}.$$

Some of the variation in $Y_1, Y_2, \ldots, Y_n$ may be due to the different levels $x_1, x_2, \ldots, x_n$ of the explanatory variable. For example, in the mileage data set of Section 8.1, we consider cars that range from 1,900 to 4,100 pounds in weight. Having cars with different weights in our sample is bound to lead to large variation among fuel consumption values (gpm). This was the very reason for considering a regression model that relates the response to the explanatory variable and "explains" the observation Y_i through the fitted value $\hat{Y}_i = \hat{\beta}_0 + \hat{\beta}_1 x_i, i = 1, 2, \ldots, n$. The variation of the fitted values about the mean $\overline{Y}$ measures the variability that is explained by the regression model. The sum of squares,

$$\text{SSR} = \sum_{i=1}^{n}(\hat{Y}_i - \overline{Y})^2,$$

is called the *sum of squares due to regression* or simply the *regression sum of squares*.

Usually, however, not all the variation is explained by the regression model; that is, part of the variation is left unexplained. The residuals $e_i = Y_i - \hat{Y}_i, i = 1, 2, \ldots, n$, express that unexplained component. The unexplained variation is measured by

$$\text{SSE} = \sum_{i=1}^{n}e_i^2,$$

and is called the *sum of squares due to error* or the *error sum of squares*.

We have introduced three sums of squares: the total sum of squares, SSTO; the regression sum of squares, SSR; and the error sum of squares, SSE. In fact, one can show that

$$\text{SSTO} = \text{SSR} + \text{SSE}.$$

The proof of this decomposition of the total sum of squares is as follows:

$$\text{SSTO} = \sum_{i=1}^{n}(Y_i - \overline{Y})^2 = \sum_{i=1}^{n}[(Y_i - \hat{Y}_i) + (\hat{Y}_i - \overline{Y})]^2$$

$$= \sum_{i=1}^{n}(\hat{Y}_i - \overline{Y})^2 + \sum_{i=1}^{n}(Y_i - \hat{Y}_i)^2 + 2\sum_{i=1}^{n}(\hat{Y}_i - \overline{Y})(Y_i - \hat{Y}_i)$$

$$= \text{SSR} + \text{SSE},$$

because

$$\sum_{i=1}^{n}(\hat{Y}_i - \overline{Y})(Y_i - \hat{Y}_i) = \sum_{i=1}^{n}e_i(\hat{Y}_i - \overline{Y}) = \sum_{i=1}^{n}e_i(\hat{\beta}_0 + \hat{\beta}_1 x_i) - \overline{Y}\sum_{i=1}^{n}e_i$$

$$= \hat{\beta}_0 \sum_{i=1}^{n} e_i + \hat{\beta}_1 \sum_{i=1}^{n} e_i x_i - \overline{Y} \sum_{i=1}^{n} e_i = 0.$$

(Recall that $\Sigma e_i = \Sigma e_i x_i = 0$.)

Let us illustrate these three sums of squares in the context of two special cases. First, consider $\hat{\beta}_1 = 0$, which says that the explanatory variable has no linear association with the response. Then all fitted values are $\hat{Y}_i = \overline{Y}$, irrespective of the level x, and it follows that SSR $= 0$ and SSE $=$ SSTO. This says that the regression model and the so-called explanatory variable x explain none of the variation in $Y_1, Y_2, \ldots, Y_n$. In the second situation, assume that the fitted regression line passes through each observation; that is, $\hat{Y}_i = Y_i$. Then SSR $=$ SSTO and SSE $= 0$. This says that the regression model with the variable x explains all the variability in $Y_1, Y_2, \ldots, Y_n$.

Remarks on Computation: The calculation of all three sums of squares is very easy. Fitted values and residuals are needed in the calculation of SSR and SSE. Virtually every statistics computer software, as well as most sophisticated pocket calculators, calculate these sums of squares. Should hand calculation be necessary, because $\overline{Y} = \hat{\beta}_0 + \hat{\beta}_1 \overline{x}$ (see Exercise 8.1-4), one can use the following computation formula for SSR:

$$\text{SSR} = \sum_{i=1}^{n} (\hat{Y}_i - \overline{Y})^2 = \sum_{i=1}^{n} [\hat{\beta}_0 + \hat{\beta}_1 x_i - (\hat{\beta}_0 + \hat{\beta}_1 \overline{x})]^2$$

$$= \hat{\beta}_1^2 \sum_{i=1}^{n} (x_i - \overline{x})^2 = \frac{\left[\sum x_i Y_i - \left(\sum x_i \right)\left(\sum Y_i \right)/n \right]^2}{\sum x_i^2 - \left(\sum x_i \right)^2/n}.$$

This formula avoids some of the rounding that is necessary if fitted values are used. Of course, SSE can be obtained from SSE $=$ SSTO $-$ SSR. Details on how to use computer software are given at the end of Section 8.3.

8.2-1 Coefficient of Determination

The *coefficient of determination,* a summary statistic that measures how well the regression equation fits the data, is given by

$$R^2 = \frac{\text{SSR}}{\text{SSTO}} = 1 - \frac{\text{SSE}}{\text{SSTO}},$$

because SSR $=$ SSTO $-$ SSE. This coefficient expresses the variability that is explained by the regression model as a fraction of the total sum of squares. From $0 \leq \text{SSR} \leq \text{SSTO}$, it follows that

$$0 \leq R^2 \leq 1.$$

$R^2 = 0$ means that SSR $= 0$ and SSE $=$ SSTO. In such a case, the simple linear regression model explains none (i.e., zero percent) of the variation in the Y values. By contrast, $R^2 = 1$ means that SSR $=$ SSTO and SSE $= 0$. In this case, all n

observations lie on the fitted regression line and all (i.e., 100 percent) of the variation in $Y_1, Y_2, \ldots, Y_n$ is explained by the linear relationship with the explanatory variable.

The coefficient of determination in the simple linear regression model is related to the sample correlation coefficient that we studied in Section 1.5. Substitution of SSR, as expressed in the previous computation formula, into $R^2 = \text{SSR}/\text{SSTO}$ leads to

$$R^2 = \frac{\left[\sum x_i Y_i - \left(\sum x_i\right)\left(\sum Y_i\right)/n\right]^2}{\left[\sum x_i^2 - \left(\sum x_i\right)^2/n\right]\left[\sum Y_i^2 - \left(\sum Y_i\right)^2/n\right]}.$$

This equation shows that R^2 is the square of the sample correlation coefficient.

Example 8.2-1 For the mileage data in Table 8.1-1, we find that

$$\text{SSTO} = 207.31 - \frac{(43.9)^2}{10} = 14.589,$$

$$\text{SSR} = \frac{[135.80 - (29.0)(43.9)/10]^2}{89.28 - (29.0)^2/10} = 13.915,$$

and

$$\text{SSE} = 14.589 - 13.915 = 0.674.$$

Alternatively, we could have also calculated SSE by summing the squares of the residuals given in Table 8.1-2. It follows that $R^2 = 13.915/14.589 = 0.954$. This means that the explanatory variable, weight, explains 95.4 percent of the variability in fuel consumption. Or, to put it differently, the simple linear regression model reduces the variability in the response (fuel consumption) by 95.4 percent. ∎

8.2-2 Analysis-of-Variance Table and F-Test

The decomposition of the total sum of squares is usually summarized in an analysis-of-variance (ANOVA) table, similar to the ones we discussed in Chapters 6 and 7 for data from designed experiments. The first column in Table 8.2-1 shows the sources of variation. The second column gives the corresponding sums of squares: The total sum of squares (SSTO) is partitioned into a sum of squares due to regression (SSR) and a sum of squares due to error (SSE).

The third column contains the number of degrees of freedom for the various sums of squares just mentioned. The number of degrees of freedom can be thought

Table 8.2-1 ANOVA Table for the Simple Linear Regression Model

Source	SS	df	MS	F
Regression	SSR	1	MSR = SSR/1	MSR/MSE
Error	SSE	$n - 2$	MSE = SSE/$(n - 2)$	
Total	SSTO	$n - 1$		

of as the number of independent components that are necessary to calculate a sum of squares. For example, the number of degrees of freedom for SSTO $= \Sigma(Y_i - \overline{Y})^2$ is $n - 1$. There are n deviations $Y_i - \overline{Y}$ in this sum of squares; however, because $\Sigma(Y_i - \overline{Y}) = 0$, one needs only $n - 1$ deviations to calculate SSTO, as the remaining one can always be calculated from the others. The number of degrees of freedom for the error sum of squares is $n - 2$, which is n minus the number of estimated coefficients in the regression function. There are n residuals in SSE $= \Sigma e_i^2$; however, there are two restrictions among the residuals ($\Sigma e_i = \Sigma e_i x_i = 0$). Thus, for given x values, we need only $n - 2$ residuals to calculate this sum of squares. Because there is only one explanatory variable, there is only one degree of freedom for the regression sum of squares. The equation SSR $= \hat{\beta}_1^2 \Sigma(x_i - \overline{x})^2$ shows that, apart from the given x values, we need only one quantity—the estimate $\hat{\beta}_1$—to calculate this sum of squares.

The fourth column in the ANOVA table is called the mean-square column and contains the ratios of the various sums of squares and their degrees of freedom. MSR $=$ SSR/1 is the *mean square due to regression,* and MSE $=$ SSE/$(n - 2)$ is the *mean square due to error* or the *mean-square error.*

The F-ratio, $F =$ MSR/MSE, in the fifth column of the table, provides a statistic for testing whether $\beta_1 = 0$. To understand this test, we must investigate the sampling distribution of the F-ratio. Let us first discuss its sampling distribution when $\beta_1 = 0$. In this case, the explanatory variable x has no effect on the response and $Y_i = \beta_0 + \varepsilon_i, i = 1, 2, \ldots, n$, are independent random variables from a normal distribution with mean β_0 and variance σ^2. We know from earlier chapters that, in this situation, SSTO/σ^2 follows a chi-square distribution with $n - 1$ degrees of freedom. Next, let us consider the regression sum of squares and determine the distribution of SSR/σ^2. Using the computational formula for SSR discussed prior to Section 8.2-1, we can write

$$\frac{\text{SSR}}{\sigma^2} = \frac{\hat{\beta}_1^2 \sum_{i=1}^{n}(x_i - \overline{x})^2}{\sigma^2} = \left[\frac{\hat{\beta}_1 - 0}{\sigma/\sqrt{\sum(x_i - \overline{x})^2}}\right]^2.$$

Because $\sigma/\sqrt{\sum(x_i - \overline{x})^2}$ is the standard deviation of $\hat{\beta}_1$, it follows that, for $\beta_1 = 0$, SSR/σ^2 is the square of an $N(0, 1)$ random variable and hence is $\chi^2(1)$.

So far, we have shown that the first two terms in the identity

$$\frac{\text{SSTO}}{\sigma^2} = \frac{\text{SSR}}{\sigma^2} + \frac{\text{SSE}}{\sigma^2}$$

follow chi-square distributions with $n - 1$ and 1 degrees of freedom, respectively. What can we say about the distribution of the last term, SSE/σ^2? Applying a well-known decomposition theorem from mathematical statistics [see R. V. Hogg, J. W. McKean, and A. T. Craig, *Introduction to Mathematical Statistics,* 6th ed. (New York: Prentice Hall, 2004)], we can also show that SSE/σ^2 follows a chi-square distribution with $n - 2$ degrees of freedom and that SSR and SSE are independent. Note that the number of degrees of freedom on the right side of the preceding equation add up to the number of degrees of freedom on the left side: $n - 1 = 1 + (n - 2)$.

The chi-square distributions for SSR/σ^2 and SSE/σ^2 imply that both MSR $=$ *SSR*/1 and MSE $=$ SSE/$(n - 2)$ provide unbiased estimates of σ^2 if $\beta_1 = 0$. This property follows from the fact that the mean of a chi-square random variable is equal to its degrees of freedom. (See Section 4.3.) The independence of SSR and SSE also implies that

$$F = \frac{(SSR/\sigma^2)/1}{(SSE/\sigma^2)/(n-2)} = \frac{SSR/1}{SSE/(n-2)} = \frac{MSR}{MSE}$$

follows an F-distribution with 1 and $n - 2$ degrees of freedom.

Next, let us consider the situation when $\beta_1 \neq 0$. One can show that, in this case also, the standardized error sum of squares, SSE/σ^2, follows a $\chi^2(n-2)$ distribution. Thus, in either case ($\beta_1 = 0$ or $\beta_1 \neq 0$), $MSE = SSE/(n-2)$ is an unbiased estimator of σ^2. This fact is not surprising, as the residuals approximate the unknown error components. However, if $\beta_1 \neq 0$, MSR is no longer an unbiased estimator of σ^2. In fact, we show in Exercise 8.2-3 that $E(MSR) = \sigma^2 + \beta_1^2 \Sigma (x_i - \bar{x})^2$ is always larger than σ^2. Therefore, the F-ratio, $F = MSR/MSE$, will tend to be larger when $\beta_1 \neq 0$ than when $\beta_1 = 0$.

Large F-ratios indicate that β_1 is different from zero. This relationship implies the following decision rule for testing the null hypothesis $H_0: \beta_1 = 0$ against the alternative $H_1: \beta_1 \neq 0$: Calculate the F-ratio from the data. If it exceeds the upper 100α percentage point of the $F(1, n-2)$ distribution—that is, if $F \geq F(\alpha; 1, n-2)$— we reject $H_0: \beta_1 = 0$ in favor of $H_1: \beta_1 \neq 0$ at significance level α. If, however, $F < F(\alpha; 1, n-2)$, we do not reject H_0; that is, we conclude that, with these data, there is insufficient evidence to say that β_1 is different from zero.

Example 8.2-2   Following is the ANOVA table for the mileage data in Table 8.1-1:

Source	SS	df	MS	F
Regression	13.915	1	13.915	165.2
Error	0.674	8	0.084	
Total	14.589	9		

The mean square error $MSE = 0.084$ is the estimate of σ^2. The F-ratio is much larger than the critical value $F(0.01; 1, 8) = 11.26$; thus, at significance level $\alpha = 0.01$, we conclude that $\beta_1 \neq 0$.   ∎

8.2-3 Confidence Intervals and Tests of Hypotheses for Regression Coefficients

In Section 8.1, we obtained the variance of the least-squares estimate $\hat{\beta}_1$, namely, $\text{var}(\hat{\beta}_1) = \sigma^2/\Sigma(x_i - \bar{x})^2$. We also showed that the standardized random variable

$$Z = \frac{\hat{\beta}_1 - \beta_1}{\sqrt{\text{var}(\hat{\beta}_1)}} = \frac{\hat{\beta}_1 - \beta_1}{\sqrt{\sigma^2/\Sigma(x_i - \bar{x})^2}}$$

follows an $N(0, 1)$ distribution, and we used this result to obtain confidence intervals for β_1. This, however, required that σ^2 be known.

In practice, σ^2 is unknown and must be estimated from the data. Replacing σ^2 in $\text{var}(\hat{\beta}_1)$ by the mean square error $\text{MSE} = \text{SSE}/(n-2)$ and taking the square root, we obtain an estimate of the standard deviation of $\hat{\beta}_1$:

$$s(\hat{\beta}_1) = \sqrt{\frac{\text{MSE}}{\sum(x_i - \bar{x})^2}}.$$

We also call this the (estimated) *standard error* of the estimate $\hat{\beta}_1$. Replacing the standard deviation $\sqrt{\text{var}(\hat{\beta}_1)}$ in the denominator of Z by its estimate changes the sampling distribution slightly. We can show that

$$\frac{\hat{\beta}_1 - \beta_1}{s(\hat{\beta}_1)} = \frac{\hat{\beta}_1 - \beta_1}{\sqrt{\text{MSE}/\sum(x_i - \bar{x})^2}}$$

follows a t-distribution with $n-2$ degrees of freedom. The degrees of freedom are easy to remember as they correspond to the degrees of freedom of the error sum of squares in the ANOVA table.

In the proof of this result, we use the fact that SSE/σ^2 is $\chi^2(n-2)$ and independent of $\hat{\beta}_1$. Dividing the standard normal variable Z given earlier by $\sqrt{(\text{SSE}/\sigma^2)/(n-2)}$, the square root of a $\chi^2(n-2)$ variable divided by its degrees of freedom, leads to the result that $(\hat{\beta}_1 - \beta_1)/\sqrt{\text{MSE}/\sum(x_i - \bar{x})^2}$ has a $t(n-2)$ distribution.

We can use this sampling distribution result to construct confidence intervals for $\hat{\beta}_1$ as well as test hypotheses about β_1 following the general discussion in Chapter 4. A $100(1-\alpha)$ percent confidence interval for β_1 is given by

$$\hat{\beta}_1 \pm t(\alpha/2; n-2)s(\hat{\beta}_1).$$

This interval covers the true, but unknown, β_1 in $100(1-\alpha)$ percent of the cases.

To test the null hypothesis $H_0: \beta_1 = \beta_{10}$ against the alternative $H_1: \beta_1 \neq \beta_{10}$, where β_{10} is some given value, we compare the test statistic $(\hat{\beta}_1 - \beta_{10})/s(\hat{\beta}_1)$ with critical values from the $t(n-2)$ distribution. At significance level α, we accept H_1 if

$$\frac{|\hat{\beta}_1 - \beta_{10}|}{s(\hat{\beta}_1)} \geq t(\alpha/2; n-2);$$

otherwise, there is insufficient evidence to reject H_0.

Frequently, we are interested in testing $H_0: \beta_1 = 0$ against $H_1: \beta_1 \neq 0$. On the one hand, if H_0 is not rejected, we conclude that the explanatory variable x is not important in explaining the variable Y. On the other hand, concluding H_1 implies that x does have a significant linear association with Y. In this particular case $(\beta_{10} = 0)$, the test statistic becomes $t(\hat{\beta}_1) = \hat{\beta}_1/s(\hat{\beta}_1)$; $t(\hat{\beta}_1)$ is called the *t-statistic* or *t-ratio*. At significance level α, we accept H_1 if

$$|t(\hat{\beta}_1)| \geq t(\alpha/2; n-2).$$

Otherwise, there is not enough evidence to reject $H_0: \beta_1 = 0$. Applied statisticians call the estimate $\hat{\beta}_1$ statistically *significant* if $H_1(\beta_1 \neq 0)$ is accepted. They call the estimate $\hat{\beta}_1$ *insignificant* if $H_0(\beta_1 = 0)$ is accepted.

The test involving the t-statistic is equivalent to the F-test, which was discussed previously. Because

$$[t(\hat{\beta}_1)]^2 = \frac{\hat{\beta}_1^2}{[s(\hat{\beta}_1)]^2} = \frac{\hat{\beta}_1^2 \sum (x_i - \bar{x})^2}{\text{MSE}} = \frac{\text{MSR}}{\text{MSE}} = F,$$

the inequalities

$$|t(\hat{\beta}_1)| \geq t(\alpha/2; n - 2) \quad \text{and} \quad F \geq F(\alpha; 1, n - 2)$$

provide exactly the same test of $H_0: \beta_1 = 0$ against $H_1: \beta_1 \neq 0$. This follows from the fact that percentiles of the $t(r)$ and $F(1, r)$ distributions are related by

$$[t(\alpha/2; r)]^2 = F(\alpha; 1, r).$$

(See Exercise 8.2-4.)

Alternatively, we can conduct the test of H_0 by using the probability value. The p value is the probability, under H_0, of obtaining a value of the test statistic at least as "extreme" as the one actually observed. We know that, under H_0, the t-statistic, $t(\hat{\beta}_1) = \hat{\beta}_1/s(\hat{\beta}_1)$, follows a t-distribution with $n - 2$ degrees of freedom. Thus, in the case of a two-sided test,

$$p \text{ value } = 2P[t(n - 2) \geq |t(\hat{\beta}_1)|],$$

where $t(\hat{\beta}_1)$ is the observed value of the test statistic. Of course, the p value can be looked up in extensive t-tables, but it is usually part of the computer output. If this p value is less than or equal to the chosen significance level α, we reject H_0 and accept H_1. That is, the probability, under H_0, of getting a sample value $\hat{\beta}_1$ of that size or larger is just too small for us to believe in $H_0: \beta_1 = 0$. If, however, the p value is greater than α, we do not reject H_0; that is, the sample result $\hat{\beta}_1$ is not extreme enough to doubt our null hypothesis H_0.

Example 8.2-3   For the mileage data set presented earlier, the estimate of β_1 is $\hat{\beta}_1 = 1.639$. Because MSE $= 0.084$ and $\sum (x_i - \bar{x})^2 = 5.18$, it follows that the estimated variance of $\hat{\beta}_1$ is $s^2(\hat{\beta}_1) = 0.084/5.18 = 0.0162$; the standard error or $\hat{\beta}_1$ is $s(\hat{\beta}_1) = 0.127$. A 95 percent confidence interval for β_1 is given by $1.639 \pm (2.306)(0.127)$, because $t(0.025; 8) = 2.306$. This equals 1.639 ± 0.293, or equivalently, $1.35 \leq \beta_1 \leq 1.93$. The confidence interval provides strong evidence that $\beta_1 > 0$ and that heavier cars do indeed require more fuel (higher gpm values). More formally, one could test $H_0: \beta_1 = 0$ against the one-sided alternative hypothesis $H_1: \beta_1 > 0$. Note that this requires a one-sided test; in this case, the critical value is given by $t(\alpha; n - 2)$, with $\alpha = 0.05$ and $n = 10$. The t-statistic is given by $t(\hat{\beta}_1) = 1.639/0.127 = 12.91$ and is much larger than the critical value $t(0.05; 8) = 1.86$; thus, there is very strong evidence that $\beta_1 > 0$.

In Section 8.1-3, we obtained the variance of $\hat{\beta}_0$, the least-squares estimate of the intercept of the regression line. Replacing σ^2 in this expression by MSE and taking the square root gives us the (estimated) standard error of $\hat{\beta}_0$:

$$s(\hat{\beta}_0) = \sqrt{\text{MSE}\left[\frac{1}{n} + \frac{\bar{x}^2}{\sum (x_i - \bar{x})^2}\right]}.$$

Now, $(\hat{\beta}_0 - \beta_0)/s(\hat{\beta}_0)$ is $t(n - 2)$, so a $100(1 - \alpha)$ percent confidence interval for β_0 is

$$\hat{\beta}_0 \pm t(\alpha/2; n - 2)s(\hat{\beta}_0).$$ ∎

Exercises 8.2

*8.2-1 In the linear regression situation, where $Y_i = \beta_0 + \beta_1 x_i + \varepsilon_i$, with the usual assumptions (normality, independence, and common variance), we compute, from $n = 16$ points,

$$\hat{\beta}_1 = 0.35, \quad \text{MSE} = 2.3, \quad \sum(x_i - \bar{x})^2 = 100.0.$$

Test $H_0: \beta_1 = 0.20$ against $H_1: \beta_1 > 0.20$ at the $\alpha = 0.05$ significance level. Obtain a 90 percent confidence interval for β_1.

*8.2-2 Consider Exercises 8.1-1 to 8.1-3. For each of the three exercises, obtain the ANOVA table, calculate and interpret R^2, and test whether the slope β_1 is different from zero. In addition, obtain and interpret 95 percent confidence intervals for the slope β_1 and the intercept β_0.

8.2-3 In the notation of this section, show that, for a general β_1,

$$E(\text{MSR}) = E[\hat{\beta}_1^2 \sum(x_i - \bar{x})^2] = \sigma^2 + \beta_1^2 \sum(x_i - \bar{x})^2.$$

Thus, if $\beta_1 = 0$, then $F = \text{MSR/MSE}$ should be near 1, because $E(\text{MSR}) = E(\text{MSE}) = \sigma^2$. If $\beta_1 \neq 0$, then F tends to be larger.

Hint: Recall that $E(\hat{\beta}_1^2) = \text{var}(\hat{\beta}_1) + [E(\hat{\beta}_1)]^2$.

8.2-4 In general, $t(r) = Z/\sqrt{\chi^2(r)/r}$, where Z is a standard normal variable that is independent of $\chi^2(r)$, a chi-square random variable with r degrees of freedom.

(a) Argue that $[t(r)]^2 = (Z^2/1)/[\chi^2(r)/r]$ is $F(1, r)$.

Hint: What is the distribution of Z^2? (See Exercise 4.3-5.)

(b) Take $r = 10$ and $\alpha = 0.05$, and note that $[t(0.025; 10)]^2 = F(0.05; 1, 10)$ by observing these respective values in the t- and F-tables.

8.2-5 *Estimation of the mean response.* Let $\hat{\beta}_0$, $\hat{\beta}_1$, and MSE be the unbiased estimators of β_0, β_1, and σ^2, respectively, as given in this and the previous section. Let us consider the estimation of the *mean response*, $E(Y_k) = \beta_0 + \beta_1 x_k$, at a certain new level $x = x_k$. The obvious estimator is $\tilde{Y}_k = \hat{\beta}_0 + \hat{\beta}_1 x_k$. This is an unbiased estimator of $E(Y_k)$, because $E(\tilde{Y}_k) = \beta_0 + \beta_1 x_k$.

(a) Using the results of Section 8.1 and Exercises 8.1-6 and 8.1-7, show that

$$\text{var}(\tilde{Y}_k) = \text{var}(\hat{\beta}_0) + x_k^2 \text{var}(\hat{\beta}_1) + 2x_k E[(\hat{\beta}_0 - \beta_0)(\hat{\beta}_1 - \beta_1)]$$

$$= \sigma^2 \left[\frac{1}{n} + \frac{(x_k - \bar{x})^2}{\sum(x_i - \bar{x})^2}\right],$$

where the sums $\bar{x} = \sum x_i/n$ and $\sum(x_i - \bar{x})^2$ do not include the new level x_k. Observe that this variance is smallest if $x_k = \bar{x}$—that is, if x_k is at the center of the experimental region. Can you think of an intuitive explanation of this fact?

(b) Argue that

$$\frac{[\tilde{Y}_k - (\beta_0 + \beta_1 x_k)]/\sqrt{\text{var}(\tilde{Y}_k)}}{\sqrt{\dfrac{\text{SSE}}{\sigma^2}/(n-2)}} = \frac{\tilde{Y}_k - (\beta_0 + \beta_1 x_k)}{\sqrt{\text{MSE}\left[\dfrac{1}{n} + \dfrac{(x_k - \bar{x})^2}{\sum(x_i - \bar{x})^2}\right]}}$$

has a $t(n-2)$ distribution.

(c) Hence,

$$\hat{\beta}_0 + \hat{\beta}_1 x_k \pm t(\alpha/2; n-2)\sqrt{\text{MSE}\left[\frac{1}{n} + \frac{(x_k - \bar{x})^2}{\sum(x_i - \bar{x})^2}\right]}$$

provides a $100(1 - \alpha)$ percent confidence interval for the mean response $E(Y_k) = \beta_0 + \beta_1 x_k$.

(d) Use the mileage data set in Table 8.1-1 and show that, for an automobile weighing $x_k = 2,500$ pounds, a 95 percent confidence interval for $E(Y_k)$ is given by $3.73 \pm (2.306)(0.105)$ or, equivalently, $3.49 \le E(Y_k) = \beta_0 + \beta_1 x_k \le 3.97$.

8.2-6 Consider the *prediction* of a new observation $Y_k = \beta_0 + \beta_1 x_k + \varepsilon_k$ at the level $x = x_k$. Note that this is different from Exercise 8.2-5, in which we estimate the mean response.

(a) Use $\tilde{Y}_k = \hat{\beta}_0 + \hat{\beta}_1 x_k$ as the prediction and define the prediction error as $W = Y_k - \tilde{Y}_k$. From the result in Exercise 8.2-5, show that

$$\text{var}(W) = \sigma^2 + \text{var}(\tilde{Y}_k) = \sigma^2\left[1 + \frac{1}{n} + \frac{(x_k - \bar{x})^2}{\sum(x_i - \bar{x})^2}\right].$$

(b) Argue that

$$\frac{W/\sqrt{\text{var}(W)}}{\sqrt{\dfrac{\text{SSE}}{\sigma^2}/(n-2)}} = \frac{Y_k - (\hat{\beta}_0 + \hat{\beta}_1 x_k)}{\sqrt{\text{MSE}\left[1 + \dfrac{1}{n} + \dfrac{(x_k - \bar{x})^2}{\sum(x_i - \bar{x})^2}\right]}}$$

has a $t(n-2)$ distribution.

(c) Hence,

$$\hat{\beta}_0 + \hat{\beta}_1 x_k \pm t(\alpha/2; n-2)\sqrt{\text{MSE}\left[1 + \frac{1}{n} + \frac{(x_k - \bar{x})^2}{\sum(x_i - \bar{x})^2}\right]}$$

provides a $100(1 - \alpha)$ percent *prediction interval* for Y_k. This interval is always wider than the confidence interval for the mean response. Give an intuitive explanation.

(d) Use the mileage data set in Table 8.1-1 and show that, for an automobile weighing $x_k = 2,500$ pounds, a 95 percent prediction interval for Y_k is $3.73 \pm (2.306)(0.308)$ or, equivalently, $3.02 \le Y_k \le 4.44$.

*8.2-7 Consider the data in Exercise 8.1-3. Use the result in Exercise 8.2-5(c) to construct a 95 percent confidence interval for the mean weight gain of rats with an initial weight of 60 grams. Repeat you calculations for rats with an initial weight of 100 grams. Would you put much trust in this second interval? Discuss.

8.2-8 Consider the data shown in Exercise 8.1-1. Use the result of Exercise 8.2-6(c) to construct a 90 percent prediction interval for the amount bet when the attendance at the track is 12,000.

8.3 The Adequacy of the Fitted Model

8.3-1 Residual Checks

A simple linear model should be fit only to data that exhibit at least a rough linear relationship. A linear fit is not appropriate for data that follow a quadratic (or a more complicated) pattern. Such a pattern would not be consistent with the assumptions associated with a simple linear regression model. Moreover, the regression model should be fit only to data for which the variability in the Y values is approximately constant for all values of x. For example, such a model is not appropriate for data in which the variability in Y increases with increasing or decreasing levels of x or Y, a property that would violate the constant-variance assumption.

A standard regression analysis is also not appropriate if the error terms ε_i are correlated, as that would violate the independence assumption. Correlation among the errors may occur if the observations $(x_i, Y_i), i = 1, 2, \ldots, n$, are collected sequentially in time and the index i stands for time or run order. One such example is observations on the yield Y and a certain input variable x from consecutive batches in a chemical production process. In such a situation, it is highly likely that there is a "carryover" effect from batch to batch. This carryover could introduce correlation among the errors.

It is important to always check whether the assumptions of the regression model are satisfied. Computers and handheld calculators now make it so easy to estimate regression functions that a number of different models are frequently fit to the data. Thus, it becomes especially important that users of regression methods check to see if these models really are appropriate.

Our main tools in model checking are various residual plots. In particular, for $i = 1, 2, \ldots, n$, we check

1. a plot of the residuals e_i against the fitted values $\hat{y}_i$.
2. a plot of the residuals e_i against x_i.
3. plots of the residuals e_i against other explanatory variables that were not included in the original model (e.g., time or run order if the data are collected sequentially).
4. a plot of the residuals e_i against the lagged e_{i-1} if we collected the data sequentially.

If the regression assumptions are satisfied, we should see no patterns in these scatter plots. The residuals should give the impression of varying independently within a 2σ horizontal band around zero. That is, because the residuals are estimates of independent $N(0, \sigma^2)$ errors, most of the residuals (roughly 95 percent) should fall within $\pm 2\sqrt{\text{MSE}}$ of zero. Or, equivalently, we could calculate standardized residuals $e_i/\sqrt{\text{MSE}}$; 95 percent of these should fall within ± 2 of zero.

To illustrate checking residuals, we consider two cases in which some of the assumptions are violated. In case 1, the regression function is specified incorrectly: The data follow a quadratic relationship, but a simple linear regression model is fitted. [See Figure 8.3-1(a).] In case 2, the equal-variance assumption is violated because the variability grows with the level of Y. [See Figure 8.3-1(b).] In each case,

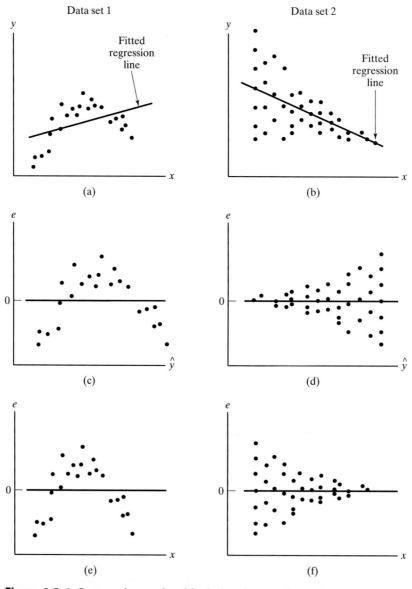

Figure 8.3-1 Scatter plots and residual plots for two data sets

we have calculated the fitted values and the residuals from the simple linear regression model. The respective plots of e_i against $\hat{y}_i$ are given in Figures 8.3-1(c) and (d), and those of e_i against x_i are given in Figures 8.3-1(e) and (f). In both cases, we can recognize patterns in the residual plots. For example, in the first one (c), the residuals are negative for low and high fitted values and positive for x's in the intermediate range. In the second case (d), the variability in the residuals grows with the fitted values. These residuals certainly do not give the impression of varying independently within a horizontal band around zero. The graphs in (e) and (f), in which we plot the residuals against the x values, convey similar information as the graphs in (c) and (d).

Just the recognition that there is a problem with the regression model is not enough: Actions must be taken to fix the inadequacies in the model. In the case of fitting the incorrect model form, one has to revise the regression function. For our first example, we would fit the quadratic model $Y_i = \beta_0 + \beta_1 x_i + \beta_2 x_i^2 + \varepsilon_i$. (See Section 8.4.) In the case of variances that are not constant, we should look for a transformation of the response variable Y that stabilizes the variance and then relate the transformed observations to the explanatory variable. We know that the logarithmic transformation ($\log Y$) is appropriate if the *standard deviation* of Y increases proportionally with the level of Y. If the *variance* of Y increases proportionally with the level of Y, then we should try the square root transformation ($\sqrt{Y}$). In practice, we try various transformations, such as $1/Y$, $\log Y$, $\sqrt{Y}$, and other powers of Y. Plots of these transformed observations against their x values indicate which particular transformation stabilizes the variance the best.

Transformations of the response variable Y, as well as those of the explanatory variable x, may also help us simplify the regression function. The fact that the variables are measured in a particular metric is usually not a good enough reason to analyze them in that metric. For example, fuel efficiency is usually given in miles per gallon (mpg). However, we have illustrated that gpm = 100/mpg = gallons per 100 miles is a more sensible metric to use in analyzing the relationship between fuel efficiency and weight.

Example 8.3-1 In Exercise 8.3-6, we give the weights and fuel efficiencies [Y = mpg (miles per gallon)] of $n = 38$ cars. A scatter plot of mpg against weight is given in Figure 8.3-2(a), together with the linear fit. The residual plots (against fitted values and weights) are given in Figures 8.3-2(b) and (c), respectively. There are patterns in these plots indicating that the simple linear regression model is not an appropriate model. For example, most of the residuals for cars with low and high weights are positive, while the ones for cars with weights in the intermediate range are negative.

One way to correct this problem is to fit a more general model—say, one that allows for a quadratic component, namely, $Y = \beta_0 + \beta_1 x + \beta_2 x^2 + \varepsilon$. Another possibility is to look for transformations that simplify the structure of the model. In fact, we have given some explanation of why the transformation gpm = 100/mpg should be appropriate. A scatter plot of gpm against weight is given in Figure 8.3-3(a), together with the linear fit. The residual plots (against fitted values and weights) are given in Figures 8.3-3(b) and (c), respectively. We notice that this model fits better; in particular, the residuals in Figure 8.3-3 show fewer nonrandom patterns than the ones in Figure 8.3-2.

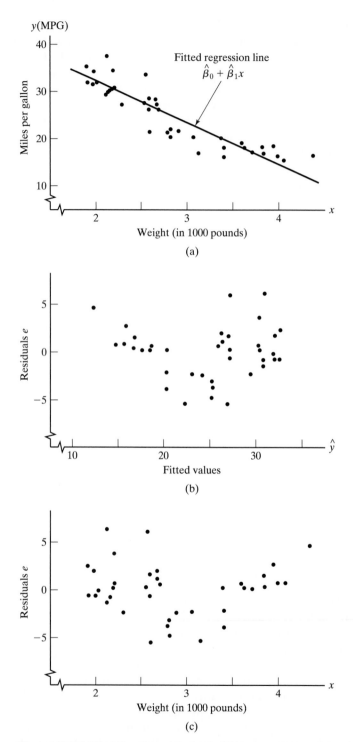

Figure 8.3-2 Plots for Example 8.3-1: (a) mpg against weight; (b) residuals against fitted values; (c) residuals against weight. Fitted values and residuals are from the model $mpg = \beta_0 + \beta_1(weight) + \varepsilon$

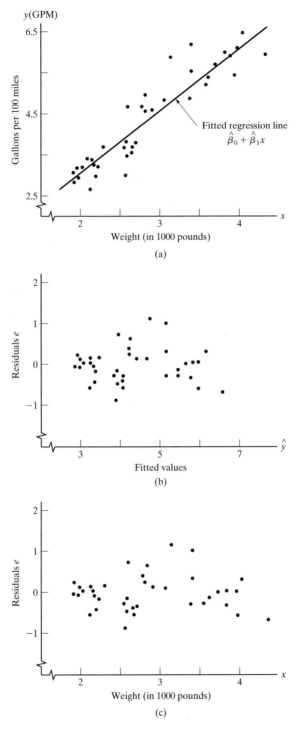

Figure 8.3-3 Plots for Example 8.3-1: (a) gpm against weight; (b) residuals against fitted values; (c) residuals against weight. Fitted values and residuals are from the model gpm $= \beta_0 + \beta_1(\text{weight}) + \varepsilon$

■

Plots of residuals against explanatory variables other than the one already in the model may also be useful. Patterns in these plots may tell us that the unexplained component in the regression model can be "explained" by variables such as the number of cylinders, engine displacement, or horsepower. Associations in these plots indicate that these variables are important and that they should be included in the model. ∎

We also assume that the errors in the regression model are independent. If an experiment is conducted sequentially, it may happen that the error at time i depends on the error at time $i - 1$. Consider, for example, a chemical batch process with a certain amount of carryover from batch to batch. This carryover could be negative if a lower-than-average yield from batch $i - 1$ usually means a higher-than-average yield from the next batch. Of course, the carryover could also be positive. Because the residuals e_i are estimates of the errors ε_i, a plot of e_i against e_{i-1} in these cases of influential carryover would show a negative (or positive) association. An example of a plot of (e_{i-1}, e_i), $i = 2, 3, \ldots, n$, that points to a violation of the independence assumption is given in Figure 8.3-4.

We could carry this analysis one step further and calculate the correlation coefficient between the residuals and the lagged residuals. Substituting $x_i = e_{i-1}$ and $y_i = e_i$ into the equation for the correlation coefficient given in Section 1.5, we can calculate the correlation from the $n - 1$ pairs $(e_1, e_2), (e_2, e_3), \ldots, (e_{n-1}, e_n)$. Of course, the n residuals sum to zero; thus,

$$\bar{x} = \frac{\sum_{i=2}^{n} e_{i-1}}{n - 1} \approx 0, \quad \bar{y} = \frac{\sum_{i=2}^{n} e_i}{n - 1} \approx 0,$$

and if n is reasonably large,

$$\sum_{i=2}^{n} e_{i-1}^2 \approx \sum_{i=2}^{n} e_i^2 \approx \sum_{i=1}^{n} e_i^2.$$

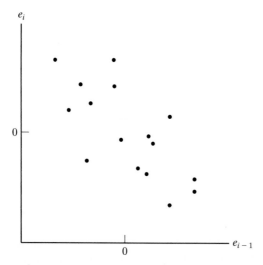

Figure 8.3-4 Plots of the residuals e_i against the lagged residuals e_{i-1}

With these approximations, the correlation coefficient is about equal to

$$r_1 = \frac{\sum_{i=2}^{n} e_{i-1} e_i}{\sum_{i=1}^{n} e_i^2}.$$

This is called the *lag 1 autocorrelation* of the residuals. It is of lag 1 because it calculates the approximate correlation coefficient between adjacent residuals. It is called an *autocorrelation* because there is only one series involved; that is, it is a correlation of a series with itself. Note that we can extend this concept to *lag k autocorrelations*:

$$r_k = \frac{\sum_{i=k+1}^{n} e_{i-k} e_i}{\sum_{i=1}^{n} e_i^2}, \quad k = 1, 2, 3, \ldots.$$

If the lag 1 autocorrelation is close to zero, we can conclude that adjacent residuals (and therefore errors) are uncorrelated. If its absolute value, $|r_1|$, is large (say, larger than $2/\sqrt{n}$), then the independence of the errors would be seriously questioned. Here, we are using the fact that if there is no correlation among the errors, then the sampling distribution of r_1 is approximately $N(0, 1/n)$.

Alternatively, the *Durbin–Watson statistic* could be calculated:

$$DW = 2(1 - r_1).$$

Because r_1 is a correlation coefficient and thus between -1 and $+1$, the Durbin–Watson statistic is between 0 and 4. If DW is close to 2, so that r_1 is close to zero, we can safely assume that there is no lag 1 autocorrelation among the errors. If DW is much different from 2 (toward either zero or 4), we would have to question the independence of the errors.

What should we do if we notice correlation among the errors? Clearly, in that case the techniques we have discussed are compromised. The answer to the question posed is beyond the scope of this introductory book. The interested reader is referred to books that combine regression and time-series models, such as Abraham and Ledolter, *Introduction to Regression Modeling* (Belmont, CA: Thomson Brooks/Cole 2006).

Remark: The plot of e_i against e_{i-1}, the lag 1 autocorrelation r_1, and the Durbin–Watson test statistic DW are meaningful only if the index i refers to time or run order. They are not appropriate, for example, in the mileage example, where i is just an arbitrary index; that is, we could have arranged the order of the cars in many different ways.

8.3-2 Output from Computer Programs

Many computer programs are available for plotting data and for carrying out regression calculations. Minitab, SAS, JMP, SPSS, and R are excellent packages with many regression options. In Minitab, you use the programs within the "Stat > Regression" tab, especially "Regression" and "Fitted Line Plot." The fitted line plot constructs a scatter diagram of the response against a single explanatory variable, adding the fitted least-squares line (which can be a linear, quadratic, or cubic polynomial) to the graph. A graph such as this makes it easy to spot violations of the model. Executing the "Regression" function gives you the least-squares coefficients $\hat{\beta}_0$ and $\hat{\beta}_1$, together with their standard errors, t-ratios, and probability values. The "Regression" function also provides an ANOVA table, with mean squares, the F-test, and coefficient of determination R^2. The square root of the mean square error is labeled "s"; it is the estimate of $\sigma = \sqrt{\text{var}(\varepsilon)}$. Residuals and fitted values can be stored. Graphing options in the "Regression" command generate the residual plots discussed in Section 8.3-1.

Example 8.3-2   The data in Table 8.3-1 give the monthly steam consumption Y (in pounds of steam used) and the average monthly atmospheric temperature x (in degrees Fahrenheit) for 25 consecutive months. The Figures listed are taken from a steam plant of a large industrial company and are part of a larger data set that includes other variables, such as the number of operating days per month and total production. For a detailed

Table 8.3-1 Steam Usage and Average Temperature for 25 Consecutive Months

Month	Steam Usage (pounds)	Average Temperature (°F)
1	10.98	35.3
2	11.13	29.7
3	12.51	30.8
4	8.40	58.8
5	9.27	61.4
6	8.73	71.3
7	6.36	74.4
8	8.50	76.7
9	7.82	70.7
10	9.14	57.5
11	8.24	46.4
12	12.19	28.9
13	11.88	28.1
14	9.57	39.1
15	10.94	46.8
16	9.58	48.5
17	10.09	59.3
18	8.11	70.0
19	6.83	70.0
20	8.88	74.5
21	7.68	72.1
22	8.47	58.1
23	8.86	44.6
24	10.36	33.4
25	11.08	28.6

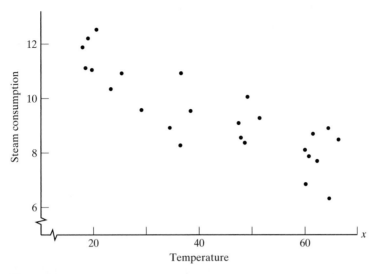

Figure 8.3-5 Scatter plot of monthly steam consumption against average monthly temperature

analysis, see Draper and Smith, *Applied Regression Analysis*, 2d ed. (New York: Wiley, 1981). The objective of the analysis that follows is to find a model that explains the variation in steam consumption. An obvious starting point is to relate steam consumption Y to the average atmospheric temperature x. The scatter plot in Figure 8.3-5 shows an approximate linear relationship that suggests the simple linear regression model

$$Y_i = \beta_0 + \beta_1 x_i + \varepsilon_i, \quad i = 1, 2, \ldots, 25.$$

The output (somewhat edited) from a computer run is given in Table 8.3-2. The parameter of primary interest, β_1, is estimated as $\hat{\beta}_1 = -0.0798$; its standard error is $s(\hat{\beta}_1) = 0.0105$ and the corresponding t-ratio is $t(\hat{\beta}_1) = \hat{\beta}_1/s(\hat{\beta}_1) = -7.59$. This

Table 8.3-2 Edited Computer Output from the Simple Linear Regression of Steam Usage on Average Temperature

The fitted regression equation is $\hat{y} = 13.62 - 0.0798x$

Coefficient	Estimate	Standard Error	t-Ratio
β_0	13.62	0.58	23.43
β_1	−0.0798	0.0105	−7.59

$\sqrt{\text{MSE}} = 0.890$; $R^2 = 0.714$.

ANOVA table:

Source	SS	df	MS	F
Regression	45.592	1	45.592	57.5
Error	18.223	23	0.792	
Total	63.816	24		

Durbin–Watson statistic $= 2.70$.

Table 8.3-3 Observations, Fitted Values, and Residuals for the Steam Usage Data

i	x_i	y_i	$\hat{y}_i$	$e_i = y_i - \hat{y}_i$
1	35.3	10.98	10.81	0.17
2	29.7	11.13	11.25	−0.12
3	30.8	12.51	11.16	1.35
4	58.8	8.40	8.93	−0.53
5	61.4	9.27	8.72	0.55
6	71.3	8.73	7.93	0.80
7	74.4	6.36	7.68	−1.32
8	76.7	8.50	7.50	1.00
9	70.7	7.82	7.98	−0.16
10	57.5	9.14	9.03	0.11
11	46.4	8.24	9.92	−1.68
12	28.9	12.19	11.32	0.87
13	28.1	11.88	11.38	0.50
14	39.1	9.57	10.50	−0.93
15	46.8	10.94	9.89	1.05
16	48.5	9.58	9.75	−0.17
17	59.3	10.09	8.89	1.20
18	70.0	8.11	8.04	0.07
19	70.0	6.83	8.04	−1.21
20	74.5	8.88	7.68	1.20
21	72.1	7.68	7.87	−0.19
22	58.1	8.47	8.98	−0.51
23	44.6	8.86	10.06	−1.20
24	33.4	10.36	10.96	−0.60
25	28.6	11.08	11.34	−0.26

number is much larger in absolute value than the critical value $t(0.025; 23) = 2.069$, so there is strong evidence that $\beta_1 \neq 0$. The estimate of β_1 implies that we can expect, on average, a 0.0798-pound reduction in steam use for each 1-degree increase in temperature. The R^2 from the simple linear regression model is $R^2 = 0.714$, which implies that temperature explains 71.4 percent of the variation in steam use.

The fitted values $\hat{y}_i$ and the residuals e_i are listed in Table 8.3-3. The respective residual plots of e_i against $\hat{y}_i$ and of e_i against x_i in Figures 8.3-6(a) and (b) show no apparent patterns. The residuals fall within two horizontal bands around zero. Moreover, the variance of the errors seems to be unaffected by the level of the response variable or that of the explanatory variable. Thus, the simple linear regression model seems adequate.

Because the data were collected sequentially, it is of interest to check whether adjacent residuals are correlated. The plot of e_i against e_{i-1} in Figure 8.3-6(c) indicates some, but a rather minor, negative correlation. The lag one autocorrelation is

$$r_1 = \frac{(0.17)(-0.12) + (-0.12)(1.35) + \cdots + (-0.60)(-0.26)}{(0.17)^2 + (-0.12)^2 + \cdots + (-0.26)^2} = -0.35,$$

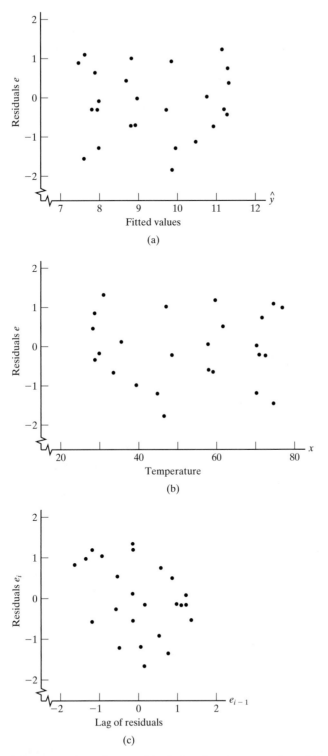

Figure 8.3-6 Various residual plots for the steam data: (a) residuals e against fitted values; (b) residuals e against temperature, x; (c) residuals e_i against lagged residuals e_{i-1}

and the Durbin–Watson statistic is DW $= 2(1 - r_1) = 2.70$. If we used the critical values in tables provided by Durbin and Watson, we would find that the correlation is significant at the $\alpha = 0.05$ level. A simpler, but only approximate, test is to check whether $|r_1| = |-0.35|$ exceeds 2 times the standard error of r_1, which is approximately equal to $1/\sqrt{n} = 1/\sqrt{25} = 0.20$. Here, $|r_1|$ is slightly smaller than 0.40. This is a borderline case of a possible moderate correlation among the residuals.   ∎

8.3-3 The Importance of Scatter Plots in Regression

Scatter plots of the regression variables are absolutely necessary in any regression analysis. Figure 8.3-7 illustrates this point clearly. The figure shows plots of four equal-sized data sets that all yield the same regression summaries (i.e., same estimates, same standard errors, same ANOVA table, same R^2). But in only one of them, namely, in part (a), is it really appropriate to fit a linear model. The pattern in part (b) is quadratic, and it would be a mistake to approximate it with a linear function, at least over the range from $x = 5$ to $x = 15$. The figures in parts (c) and (d) show the effect of a single observation on the regression estimates. Ten observations in part (c) lie on a straight line, but one response, the one at $x = 13$, does not follow their pattern. This one observation, called an *outlier* in the response variable y, changes the slope of the regression line. The plot in part (d) is also interesting because it illustrates the influence of a single

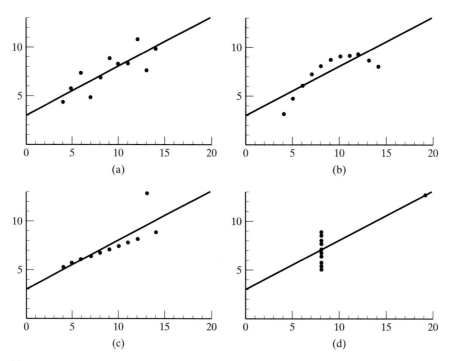

Figure 8.3-7 Anscombe's plots of four equal-sized data sets, all of which lead to the same regression summaries. The line refers to the least-squares fit [F.J. Anscombe, "Graphs in Statistical Analysis," *The American Statistician*, 27: 17–21 (1973)]

observation whose x value is far from all others in the experiment—that is, outside the usual experimental region. Because least-squares minimizes the sum of the squared distances from the observations to the regression line, the fitted line in part (d) is not very far from the response that corresponds to this outlying x value. Different responses at this particular x level would lead to different fitted regression lines.

If data are just "fed into" a regression computer program, and if the data or the residuals from the regression fit are not plotted, certain features of a data set would not be recognized. Scatter plots are important because they alert the investigator to special features, such as outlying observations. Scatter plots should be followed by an investigation of the particular circumstances that were present at the time the observations were taken. Frequently, special experimental conditions or errors in measuring or recording the data are the causes of outliers. Given that we do not want one observation to have a large influence on the fitted regression line, we usually drop such observations from the data set. But we also want to emphasize that very often much useful information can be uncovered by studying how those points were actually obtained.

The scatter plots in Figure 8.3-7—especially in (c) and (d)—raise important issues. Consider the observation (case) with $x = 19$ and $y = 12.5$ in graph (d). This is clearly an outlying observation that is different from the rest of the observations. It is the x-dimension of this case that is so different from the remaining cases. The case exerts high "leverage" on the fitted regression line. Imagine changing the response from $y = 12.5$ to $y = 7$. Then the fitted line becomes parallel to the x-axis. Change the response to $y = 0$ and the slope becomes negative. One single response has "hijacked" the slope. The fact that the x-value of this case is so different from the rest of the x-values gives it its leverage. We call an observation with such a different x-level a *high-leverage case.*

A case with high leverage may be "influential" (i.e., the inclusion of that case changes the slope of the fitted line considerably). Although this is true in graph (d), it does not have to be. Imagine $n - 1$ observations on a straight line and one case with a very different x-value, but a response y that lies on the very same line. In this situation, it does not matter whether the case with high leverage is included in the least-squares fitting: The estimate of the slope will stay the same, whether we include that point or not. We thus have a high-leverage case, but one that is not influential.

Now consider graph (c). You notice that the case ($x = 13, y = 13$) does not follow the pattern that has been established by the other cases. This point is not an outlying observation in the x-dimension and hence not a point with high leverage, but is an outlier in the y-dimension. It pulls up the fitted regression line and it changes the slope somewhat. However, note that the pull on the slope would have been much larger if the x-value of the point had been more different from the rest of the x-values. Consider the situation in which the x of the outlying y-value is right in the middle of the x-range (say, $x = 10$). Such a point would pull up the line (and change the intercept), but leave the slope largely unchanged.

Scatter plots are useful for discovering high-leverage and influential cases, but the plotting strategy works only for a single explanatory variable. In higher dimensions (with several explanatory variables), one needs diagnostic measures for leverage and influence. Such statistics have been developed, but their discussion goes beyond this introductory text.

Exercises 8.3

8.3-1 Previously, you were asked to fit simple linear regression models to the data in Exercises 8.1-1 to 8.1-3. Check the appropriateness of the simple linear regression model for each of these data sets by conducting appropriate residual plots.

8.3-2 Consider the following four fictitious data sets, each consisting of 11 (x, y) pairs (for the first three data sets, the x-values are the same and are listed only once):

x	y	y	y	x	y
10	8.04	9.14	7.46	8	6.58
8	6.95	8.14	6.77	8	5.76
13	7.58	8.74	12.74	8	7.71
9	8.81	8.77	7.11	8	8.84
11	8.33	9.26	7.81	8	8.47
14	9.96	8.10	8.84	8	7.04
6	7.24	6.13	6.08	8	5.25
4	4.26	3.10	5.39	19	12.50
12	10.84	9.13	8.15	8	5.56
7	4.82	7.26	6.42	8	7.91
5	5.68	4.74	5.73	8	6.89

Show that, for each of the four data sets listed, the simple linear regression fit leads to the same output (i.e., same estimates and standard errors, same ANOVA table, same R^2). Carry out appropriate diagnostics, and convince yourself that the simple linear regression model is suitable for only one data set.

[F.J. Anscombe, "Graphs in Statistical Analysis," *The American Statistician*, 27: 17–21 (1973).]

8.3-3 In a regression problem with $n = 10$ pairs of (x_i, y_i), we obtain $\Sigma x_i = 15$, $\Sigma y_i = 20$, $\Sigma x_i y_i = 33$, $\Sigma x_i^2 = 31.5$, and $\Sigma y_i^2 = 49$.

(a) Find the least-squares estimates of the regression coefficient in the simple linear regression model.

(b) Why is it important to look at a scatter plot of y against x?

8.3-4 Show that, in the simple linear regression model, the relationship between the F-statistic and R^2 is given by $F = (n - 2)R^2/(1 - R^2)$.

*8.3-5 Box, Hunter, and Hunter have analyzed the dispersion of an aerosol spray as a function of its age. The dispersion Y is measured as the reciprocal of the number of particles in a unit volume. Age x is measured in minutes. The experiments were performed in random order, but their results are listed here in increasing order of x:

Age, x	Dispersion, y
8	6.16
22	9.88
35	14.35
40	24.06
57	30.34

73	32.17
78	42.18
87	43.23
98	48.76

(a) Make a scatter plot of the observations.

*(b) Calculate the least-squares estimates, the ANOVA table, and R^2.

*(c) Construct a test of $\beta_1 = 0$ against $\beta_1 > 0$.

*(d) Test whether $\beta_0 = 0$. Can you simplify the model? Are dispersion and age proportional?

*(e) Estimate the mean response at age $x_k = 110$. Construct a 95 percent confidence interval. Discuss the possible dangers of such an extrapolation.

   Hint: See Exercise 8.2-5.

[G. E. P. Box, W.G. Hunter, and J. S. Hunter, *Statistics for Experimenters* (New York: Wiley, 1978).]

8.3-6 Consider the regression model that relates gas mileage and weight of automobiles. Thirty-eight cars from the model year 1978–1979 were selected, and their weights (in units of 1,000 pounds) and fuel efficiencies mpg (miles per gallon) were measured. The results are listed in the following table (these data are analyzed in Example 8.3-1; here, we ask you to repeat the analysis with locally available computer software):

Weight	mpg	Weight	mpg
4,360	16.9	3.830	18.2
4.054	15.5	2.585	26.5
3.605	19.2	2.910	21.9
3.940	18.5	1.975	34.1
2.155	30.0	1.915	35.1
2.560	27.5	2.670	27.4
2.300	27.2	1.990	31.5
2.230	30.9	2.135	29.5
2.830	20.3	2.670	28.4
3.140	17.0	2.595	28.8
2.795	21.6	2.700	26.8
3.410	16.2	2.556	33.5
3.380	20.6	2.200	34.2
3.070	20.8	2.020	31.8
3.620	18.6	2.130	37.3
3.410	18.1	2.190	30.5
3.840	17.0	2.815	22.0
3.725	17.6	2.600	21.5
3.955	16.5	1.925	31.9

(a) Make a scatter plot of mpg against weight, and discuss whether the simple linear regression model mpg $= \beta_0 + \beta_1(\text{weight}) + \varepsilon$ is appropriate.

(b) Plot gpm $= 100/$mpg (gallons per 100 miles traveled) against weight. Discuss whether the simple linear regression model gpm $= \beta_0 + \beta_1(\text{weight}) + \varepsilon$ is appropriate.

(c) Consider the regression model in part (b). Compute the least-squares estimates, their standard errors, and the t-ratios. Interpret your findings.

(d) Discuss whether the model fitted in part (c) leads to an adequate representation. Plot the residuals against weight. Can you see patterns in the plots? If so, suggest modifications.

(e) Estimate the mean fuel consumption (gpm) of cars with a weight of 3,500 pounds. Construct a 90 percent confidence interval for the mean response.

[H. V. Henderson and P. F. Velleman, "Building Multiple Regression Models Interactively," *Biometrics*, 37: 391–411 (1981).]

8.3-7 In the notation associated with the simple linear regression model, the *adjusted* R^2 is defined as

$$R_a^2 = 1 - \frac{\text{SSE}/(n-2)}{\text{SSTO}/(n-1)} = \frac{[\text{SSTO}/(n-1)] - \text{MSE}}{\text{SSTO}/(n-1)}.$$

The term "adjusted" refers to the fact that the sums of squares are adjusted by their respective degrees of freedom. Then, because $[\text{SSE}/(n-2)]/[\text{SSTO}/(n-1)]$ represents the fraction of variance remaining after fitting the regression line, R_a^2 represents the fractional reduction from $\text{SSTO}/(n-1)$ to $\text{MSE} = \text{SSE}/(n-2)$; that is, R_a^2 represents the reduction in variance that can be attributed to the regression. The fractional reduction in standard deviation is

$$\frac{\sqrt{\text{SSTO}/(n-1)} - \sqrt{\text{MSE}}}{\sqrt{\text{SSTO}/(n-1)}} = 1 - \sqrt{1 - R_a^2}.$$

(a) Show that $1 \geq R^2 \geq R_a^2 \geq 1 - \sqrt{1 - R_a^2}$.

(b) For the mileage data in Table 8.1-1, compute and interpret R_a^2 and $1 - \sqrt{1 - R_a^2}$.

*8.3-8 Student enrollment at the University of Iowa for the fall and spring semesters combined (given in units of 1,000 students) from 1963–1964 (time $x = 1$) to 1979–1980 ($x = 17$) is given in the following table:

Time, x	Enrollment, y
1	25.3
2	28.3
3	32.2
4	34.6
5	36.4
6	37.9
7	39.2
8	40.1
9	39.6
10	39.1
11	39.8
12	41.6

13	43.6
14	43.5
15	44.2
16	44.6
17	45.6

*(a) Plot enrollment against time. Fit the regression model $Y_i = \beta_0 + \beta_1 x_i + \varepsilon_i$. Calculate the least-squares estimates, fitted values, residuals, and the entries in the ANOVA table.

*(b) Check the appropriateness of the regression model by constructing residual plots. Calculate the lag 1 autocorrelation of the residuals and the Durbin–Watson statistic. Interpret your findings.

Note: It can be shown that the standard error $s(\hat{\beta}_1) = \sqrt{MSE / \sum (x_i - \bar{x})^2}$ underestimates the true standard error of $\hat{\beta}_1$ if there is positive autocorrelation among the residuals. Thus, it is inappropriate in these cases to use the t-ratio $\hat{\beta}_1 / s(\hat{\beta}_1)$ for testing $H_0: \beta_1 = 0$.

(c) You may want to repeat this exercise with data from your academic institution.

8.3-9 Polychlorinated biphenyl (PCB), an industrial pollutant, is thought to have harmful effects on the thickness of eggshells. The amount of PCB (in parts per million) and the thickness of the shell (in millimeters) in 65 Anacapa pelican eggs is given in the following table:

$x = $ **PCB**	$y = $ **Thickness**
452	0.14
139	0.21
166	0.23
175	0.24
260	0.26
204	0.28
138	0.29
316	0.29
396	0.30
46	0.31
218	0.34
173	0.36
220	0.37
147	0.39
216	0.42
216	0.46
206	0.49
184	0.19
177	0.22
246	0.23
296	0.25

x = PCB	y = Thickness
188	0.26
89	0.28
198	0.29
122	0.30
250	0.30
256	0.31
261	0.34
132	0.36
212	0.37
171	0.40
164	0.42
199	0.46
115	0.20
214	0.22
177	0.23
205	0.25
208	0.26
320	0.28
191	0.29
305	0.30
230	0.30
204	0.32
143	0.35
175	0.36
119	0.39
216	0.41
185	0.42
236	0.47
315	0.20
356	0.22
289	0.23
324	0.26
109	0.27
265	0.29
193	0.29
203	0.30
214	0.30
150	0.34
229	0.35
236	0.37

144	0.39
232	0.41
87	0.44
237	0.49

Investigate the relationship between the thickness of the shell and the amount of PCB in pelican eggs. Construct a scatter plot and fit a linear regression model. Calculate a 95 percent confidence interval for the slope. Obtain the ANOVA table and the coefficient of determination, R^2. Interpret the results and comment on the adequacy of the model.

[Data taken from Risebrough, R.W., "Effects of environmental pollutants upon animals other than man." *Proceedings of the 6th Berkeley Symposium on Mathematics and Statistics, VI.* California: University of California Press, 1972, 443–463.]

8.3-10 Search the web for useful *regression applets*. Many such applets are available, and you can find them by searching for "regression applet."

After entering points on a scatter plot, these applets calculate the least-squares estimates, draw in the fitted regression line, and calculate summary statistics such as the correlation coefficient or the coefficient of determination, R^2. Applets allow you to change points, and they illustrate the effect of such changes on the regression results.

Applets also illustrate the standard errors of the estimates, taking repeated samples of a certain size from a given population of points and, for each sample, calculating an estimate of the regression slope. Estimates of the slope coefficients of repeated draws from the population are displayed in form of histograms illustrating the sampling variability of the estimate. [See http://www.kuleuven.ac.be/ucs/java/version2.0/Content_Regression.htm.]

Experiment with these applets and write a brief essay on what you can learn from them. Note that the applets are designed mostly for regression with a single explanatory variable, as it is difficult to display observations in higher dimensional space.

8.3-11 Scott Simms, an engineer at the Ford Transmission Plant in Sharonville, Ohio, collected experimental data to support a reduction in destructive testing of a laser weld that attaches a stamped steel clutch housing to a micro alloy hub. The Product Engineering Department required the production facility to carry out a destructive push test on two items once every two hours of production. For the plant's current operating schedule, it was estimated that the yearly cost of destructive testing (the cost of manufacturing and destroying the parts) was about $150,000. Hence, any reduction in destructive testing represented an opportunity for sizeable improvement.

As part of a larger experiment, Simms measured the actual power (in kilowatts) of the company's laser generator (a 6.0-Kw Rofin Sinar 860HF CO_2 laser generator with a mirror with a 250-mm focal length) and related it to the weld penetration (mm) and the weld strength (push-out force, in pounds) of the item that was manufactured under these conditions. His study showed that the actual power of the laser was an excellent indicator of weld penetration and hence of weld strength. He concluded that expensive destructive testing can be avoided by controlling the actual power of the laser. As a result of his study, a case was made for reducing the destructive testing from two pieces of the laser every two hours to one piece every eight hours under normal production. The cost savings amounted to $115,000 per year. Considering the facility had 23 similar welding systems at the time, this cost savings amounted to millions of dollars of savings.

Analyze the data. In particular, graph weld strength against laser power, and weld penetration against laser power. Fit appropriate regression models and interpret the fit of these models.

Simms's data are as follows:

Laser Power (KW)	Weld Penetration (mm at 90 degrees)	Push-out Force (lb)
5.85	4.55	30,214
5.82	4.45	33,381
5.83	4.62	32,168
5.79	4.54	31,988
5.80	4.62	34,872
5.86	4.65	34,127
5.82	4.58	32,885
5.80	4.47	29,564
5.79	4.54	33,217
5.81	4.53	31,080
5.51	4.23	30,451
5.49	4.12	29,756
5.50	4.07	29,841
5.50	4.01	31,625
5.51	4.03	32,417
5.52	4.13	29,555
5.48	4.14	29,673
5.49	4.10	30,567
5.48	4.12	30,793
5.50	4.19	29,899
5.23	3.78	28,456
5.23	3.85	28,564
5.21	3.91	29,014
5.23	3.76	27,893
5.20	3.85	26,478
5.23	3.82	27,848
5.19	3.76	29,327
5.23	3.62	28,776
5.18	3.67	28,567
5.19	3.71	29,042
4.52	3.34	25,551
4.59	3.28	25,730
4.53	3.21	24,796
4.47	3.39	22,985

4.53	3.31	26,559
4.51	3.24	23,678
4.56	3.26	26,793
4.60	3.33	25,049
4.50	3.18	25,733
4.52	3.16	25,648
4.01	2.56	20,367
4.02	2.75	20,990
4.01	2.61	22,004
3.99	2.59	21,746
4.08	2.74	23,074
4.03	2.48	22,243
4.05	2.60	21,356
3.95	2.51	19,963
3.89	2.59	20,154
4.00	2.66	22,840

8.4 The Multiple Linear Regression Model

In Sections 8.1 through 8.3, we studied the *simple linear regression model* $Y_i = \beta_0 + \beta_1 x_i + \varepsilon_i$. There was just *one* explanatory variable; moreover, the model involved only a linear relationship.

Models that are linear in the explanatory variable x are often good approximations of complicated nonlinear functional relationships, especially if the range of the x-values is not too large. Over a small enough region, many complicated nonlinear functions can be approximated by a linear one; for example, consider a Taylor series expansion through only the linear term(s). However, in certain cases, nonlinear relationships are needed; for example, we could fit the *quadratic* $Y_i = \beta_0 + \beta_1 x_i + \beta_2 x_i^2 + \varepsilon_i$ or, more generally, the *polynomial* $Y_i = \beta_0 + \beta_1 x_i + \cdots + \beta_k x_i^k + \varepsilon_i$. These relationships are nonlinear in the explanatory variable. However, they are still linear in the regression coefficients $\beta_0, \beta_1, \ldots, \beta_k$; for this reason, they are called *linear regression models*.

Polynomials are not the only nonlinear relationships that we can consider. Models of the form $Y_i = (\beta_0 + \beta_1 x_i)e^{\beta_2 x_i} + \varepsilon_i$ or $Y_i = \beta_0 x_i/(1 + \beta_1 x_i) + \varepsilon_i$, are also nonlinear in x. In addition, they are nonlinear in the regression coefficients; for this reason, they are called *nonlinear regression models*. Section 8.5-4 provides an introduction to this topic.

All the models that we have described so far include just *one* explanatory variable. Frequently, more than one variable has an influence on the response. For example, gas mileage depends not only on weight, but may also depend on the engine design and the wind resistance of the car. Also, the yield of a chemical process may be influenced not only by the concentration, but, in addition, by the temperature and the type of catalyst that is used in the reaction.

To study such situations, let us initially consider models with two explanatory variables: x_1 and x_2. We use the following notation: Y_i is the response that corresponds to the level x_{i1} of the first variable x_1 and level x_{i2} of the second variable x_2, $i = 1, 2, \ldots, n$. The linear model

$$Y_i = \beta_0 + \beta_1 x_{i1} + \beta_2 x_{i2} + \varepsilon_i, \quad i = 1, 2, \ldots, n,$$

where $\varepsilon_1, \varepsilon_2, \ldots, \varepsilon_n$ are independent $N(0, \sigma^2)$ random variables, is probably the simplest regression model with two explanatory variables. The expected value $E(Y) = \beta_0 + \beta_1 x_1 + \beta_2 x_2$ defines a plane in three-dimensional space. The coefficient β_j measures the effect on the response of a unit change in the variable $x_j, j = 1, 2$.

A slightly more complicated model is

$$Y_i = \beta_0 + \beta_1 x_{i1} + \beta_{11} x_{i1}^2 + \beta_2 x_{i2} + \beta_{22} x_{i2}^2 + \varepsilon_i.$$

The effect of the variable $x_j, j = 1, 2,$ is now quadratic. The model

$$Y_i = \beta_0 + \beta_1 x_{i1} + \beta_{11} x_{i1}^2 + \beta_2 x_{i2} + \beta_{22} x_{i2}^2 + \beta_{12} x_{i1} x_{i2} + \varepsilon_i$$

is even more general: Due to the term $\beta_{12} x_{i1} x_{i2}$, the effect of a unit change in the variable x_1 depends on the value of the second explanatory variable x_2. (It did not in the previous models.) We say that the variables interact, and the term $\beta_{12} x_{i1} x_{i2}$ is called an *interaction* component.

Here, we have given three different models with two explanatory variables. The first is linear in the two x variables; the other two are quadratic in x_1 and x_2, and in the last one, the variables x_1 and x_2 interact. Note that all these models are linear in the regression coefficients and thus are linear regression models.

Now let us define the *multiple linear regression model* with p explanatory variables (or regressors) $x_1, x_2, \ldots, x_p$ as

$$Y_i = \beta_0 + \beta_1 x_{i1} + \beta_2 x_{i2} + \cdots + \beta_p x_{ip} + \varepsilon_i, \quad i = 1, 2, \ldots, n,$$

where $\varepsilon_1, \varepsilon_2, \ldots, \varepsilon_n$ are independent $N(0, \sigma^2)$ random variables. The random variable Y_i is the ith response that corresponds to the levels $x_{i1}, x_{i2}, \ldots, x_{ip}$ ($i = 1, 2, \ldots, n$) of the p regressors $x_1, x_2, \ldots, x_p$. The p regressors can be p different variables, or functions (such as products) of a smaller set of variables. For example, in the model

$$Y_i = \beta_0 + \beta_1 x_{i1} + \beta_{11} x_{i1}^2 + \beta_2 x_{i2} + \beta_{22} x_{i2}^2 + \beta_{12} x_{i1} x_{i2} + \varepsilon_i,$$

there are only two variables, x_1 and x_2, but there are $p = 5$ variables (regressors) in the regression model: $x_1, x_1^2, x_2, x_2^2,$ and $x_1 x_2$. To use the general linear multiple regression model in this case, we redefine these $p = 5$ regressors to be $x_1, x_2, x_3, x_4,$ and x_5 respectively.

8.4-1 Estimation of the Regression Coefficients

The regression coefficients $\beta_0, \beta_1, \ldots, \beta_p$, as well as the variance of the errors, $\mathrm{var}(\varepsilon_i) = \sigma^2$, are usually unknown and have to be estimated from observations $(y_i, x_{i1}, \ldots, x_{ip}), i = 1, 2, \ldots, n$. As in the simple linear regression model, we can use the *least-squares criterion* to determine the estimates of the regression coefficients. With least squares, we chose the estimates of $\beta_0, \beta_1, \ldots, \beta_p$ such that the sum of the squared deviations between the observations and the regression surface, namely,

$$S(\beta_0, \beta_1, \ldots, \beta_p) = \sum_{i=1}^{n} [y_i - (\beta_0 + \beta_1 x_{i1} + \cdots + \beta_p x_{ip})]^2,$$

is minimized. If we equate the $(p + 1)$ partial derivatives $dS(\beta_0, \beta_1, \ldots, \beta_p)/d\beta_j$, $j = 0, 1, \ldots, p$, to zero, we obtain the $p + 1$ equations

$$\sum_{i=1}^{n} [y_i - (\beta_0 + \beta_1 x_{i1} + \cdots + \beta_p x_{ip})](-1) = 0,$$

$$\sum_{i=1}^{n} [y_i - (\beta_0 + \beta_1 x_{i1} + \cdots + \beta_p x_{ip})](-x_{ij}) = 0, \quad j = 1, 2, \ldots, p.$$

The solutions of these equations are the least-squares estimates, $\hat{\beta}_0, \hat{\beta}_1, \ldots, \hat{\beta}_p$, since it can be shown that the second partial derivatives evaluated at $\hat{\beta}_0, \hat{\beta}_1, \ldots, \hat{\beta}_p$ satisfy the conditions for a minimum of $S(\beta_0, \beta_1, \ldots, \beta_p)$. Rearranging some of the terms in these equations and substituting $\hat{\beta}_j$ for β_j leads to the following $(p + 1)$ equations involving the $(p + 1)$ estimates $\hat{\beta}_0, \hat{\beta}_1, \ldots, \hat{\beta}_p$:

$$n\hat{\beta}_0 + \left(\sum x_{i1}\right)\hat{\beta}_1 + \left(\sum x_{i2}\right)\hat{\beta}_2 + \cdots + \left(\sum x_{ip}\right)\hat{\beta}_p = \sum y_i,$$

$$\left(\sum x_{i1}\right)\hat{\beta}_0 + \left(\sum x_{i1}^2\right)\hat{\beta}_1 + \left(\sum x_{i1}x_{i2}\right)\hat{\beta}_2 + \cdots + \left(\sum x_{i1}x_{ip}\right)\hat{\beta}_p = \sum x_{i1}y_i,$$

$$\vdots \qquad\qquad\qquad\qquad\qquad\qquad\qquad\qquad \vdots$$

$$\left(\sum x_{ip}\right)\hat{\beta}_0 + \left(\sum x_{i1}x_{ip}\right)\hat{\beta}_1 + \left(\sum x_{i2}x_{ip}\right)\hat{\beta}_2 + \cdots + \left(\sum x_{ip}^2\right)\hat{\beta}_p = \sum x_{ip}y_i.$$

These are called the *normal equations*. It is quite easy to solve the normal equations and determine the least-squares estimates. In fact, we can write each estimate as a linear combination of the responses $y_1, \ldots, y_n$; the coefficients in these linear combinations depend only on the x_{ij} values ($i = 1, \ldots, n$ and $j = 1, \ldots, p$). Because computer programs for least-squares estimation are generally available, we do not discuss the numerical solution of the normal equations, nor do we introduce the solution in matrix notation. We expect you to use locally available computer software.

A Further Remark on Estimation: There is yet another, somewhat simpler, method for calculating the least-squares estimates. This is achieved by writing the regression model as

$$Y_i = \beta_0^* + \beta_1(x_{i1} - \bar{x}_1) + \cdots + \beta_p(x_{ip} - \bar{x}_p) + \varepsilon_i,$$

where $\beta_0^* = \beta_0 + \beta_1\bar{x}_1 + \cdots + \beta_p\bar{x}_p$ is a new intercept and $\bar{x}_j = (1/n)\sum_{i=1}^{n} x_{ij}$ is the average level of variable x_j, $j = 1, \ldots, p$. Replacing the x_{ij} in the first normal equation by $x_{ij} - \bar{x}_j$ leads to the least-squares estimate $\hat{\beta}_0^* = \bar{y} = (1/n)\sum_{i=1}^{n} y_i$. This formula shows that the multiple linear regression model can be written in a "mean-corrected" form:

$$(Y_i - \bar{Y}) = \beta_1(x_{i1} - \bar{x}_1) + \cdots + \beta_p(x_{ip} - \bar{x}_p) + \varepsilon_i, \quad i = 1, 2, \ldots, n.$$

Thus, the number of normal equations is reduced from $p + 1$ to p, and $\hat{\beta}_0$ can be calculated from $\hat{\beta}_0 = \bar{y} - \hat{\beta}_1\bar{x}_1 - \cdots - \hat{\beta}_p\bar{x}_p$.

We can use the mean-corrected regression model to obtain explicit expressions for the least-squares estimates in the linear regression model with $p = 2$ explanatory variables. The substitution of $y_i - \bar{y}$ for y_i and $x_{ij} - \bar{x}_j$ for x_{ij} in the $p = 2$ remaining normal equations leads to

$$s_{11}\hat{\beta}_1 + s_{12}\hat{\beta}_2 = s_{1y},$$

$$s_{12}\hat{\beta}_1 + s_{22}\hat{\beta}_2 = s_{2y},$$

where $\quad s_{11} = \Sigma(x_{i1} - \bar{x}_1)^2, \quad s_{22} = \Sigma(x_{i2} - \bar{x}_2)^2, \quad s_{12} = \Sigma(x_{i1} - \bar{x}_1)(x_{i2} - \bar{x}_2),$
$s_{1y} = \Sigma(x_{i1} - \bar{x}_1)(y_i - \bar{y})$, and $s_{2y} = \Sigma(x_{i2} - \bar{x}_2)(y_i - \bar{y})$. The solution of these
two equations is given by

$$\hat{\beta}_1 = \frac{s_{22}s_{1y} - s_{12}s_{2y}}{s_{11}s_{22} - s_{12}^2},$$

$$\hat{\beta}_2 = \frac{-s_{12}s_{1y} + s_{11}s_{2y}}{s_{11}s_{22} - s_{12}^2}.$$

The estimate of the intercept can be calculated from $\hat{\beta}_0 = \bar{y} - \hat{\beta}_1\bar{x}_1 - \hat{\beta}_2\bar{x}_2$.

8.4-2 Residuals, Fitted Values, and the Sum-of-Squares Decomposition

With the least-squares estimates $\hat{\beta}_0, \hat{\beta}_1, \ldots, \hat{\beta}_p$, we can calculate the *fitted values*

$$\hat{y}_i = \hat{\beta}_0 + \hat{\beta}_1 x_{i1} + \cdots + \hat{\beta}_p x_{ip}, \quad i = 1, 2, \ldots, n,$$

and the residuals

$$e_i = y_i - \hat{y}_i = y_i - (\hat{\beta}_0 + \hat{\beta}_1 x_{i1} + \cdots + \hat{\beta}_p x_{ip}), \quad i = 1, 2, \ldots n.$$

Because there are $p + 1$ regression coefficients, there are $p + 1$ restrictions among the n residuals. These restrictions are given by $\Sigma_{i=1}^{n} e_i = 0$ and $\Sigma_{i=1}^{n} e_i x_{ij} = 0$ for $j = 1, \ldots, p$, as can be seen by evaluating the $p + 1$ first partial derivatives of $S(\beta_0, \beta_1 \ldots, \beta_p)$ at the least-squares estimates and setting them equal to zero.

The components in the regression sum-of-squares decomposition, SSTO = SSR + SSE, can be calculated from SSTO = $\Sigma(y_i - \bar{y})^2 = \Sigma y_i^2 - (\Sigma y_i)^2/n$, SSR = $\Sigma(\hat{y}_i - \bar{y})^2$, and SSE = $\Sigma(y_i - \hat{y}_i)^2 = \Sigma e_i^2$ = SSTO − SSR. As in the simple linear regression model, we display the sums of squares in an ANOVA table. (See Table 8.4-1.) The error sum of squares, SSE, now has $n - p - 1$ degrees of freedom, since there are $p + 1$ restrictions among the n residuals. A more formal explanation is that, under the assumption that the errors $\varepsilon_1, \varepsilon_2, \ldots, \varepsilon_n$ are independent $N(0, \sigma^2)$ variables, the sampling distribution of SSE$/\sigma^2 = \Sigma(Y_i - \hat{Y}_i)^2/\sigma^2$ is a chi-square distribution with $n - (p + 1)$ degrees of freedom. The regression sum of squares, SSR, has p degrees of freedom because there are p explanatory variables in the model. The mean square due to error, MSE = SSE$/(n - p - 1)$, and the mean square due to regression, MSR = SSR$/p$, are given in the fourth column of the table. The mean square error, MSE = SSE$/(n - p - 1)$, is an unbiased estimate of σ^2.

Table 8.4-1 ANOVA for the Multiple Linear Regression Model

Source	SS	df	MS	F
Regression	SSR	p	MSR = SSR$/p$	MSR/MSE
Error	SSE	$n - p - 1$	MSE = SSE$/(n - p - 1)$	
Total	SSTO	$n - 1$		

From the ANOVA table, we can calculate the coefficient of determination:

$$R^2 = \frac{\text{SSR}}{\text{SSTO}} = 1 - \frac{\text{SSE}}{\text{SSTO}}, \quad 0 \le R^2 \le 1.$$

R^2 expresses the regression sum of squares as a fraction of the total sum of squares. A high R^2 indicates that the regression model with the p explanatory variables explains a large proportion of the variability among the observations $y_1, y_2, \ldots, y_n$. However, the value of R^2 by itself does not tell us which of the p variables are the most important ones.

A closely related measure of fit of an estimated regression model compares the variance estimates with and without the explanatory variables. That is, the sample variance of the observations $y_1, y_2, \ldots, y_n$, which is $\text{SSTO}/(n-1)$, can be compared with the variance estimate in the regression model, $\text{MSE} = \text{SSE}/(n-p-1)$. The proportionate reduction in variance due to regression is the adjusted R^2, namely,

$$R_a^2 = 1 - \frac{\text{SSE}/(n-p-1)}{\text{SSTO}/(n-1)}.$$

It can be shown that $R_a^2 \le R^2$.

8.4-3 Inference in the Multiple Linear Regression Model

The regression model includes a normally distributed random component ε; thus, the response variable Y is also normally distributed. The least-squares estimates, the sums of squares in the ANOVA table, and R^2 are all functions of $Y_1, \ldots, Y_n$. Hence, prior to sampling, they are random variables and each has a sampling distribution. To make statistical inferences, we must know these sampling distributions.

Frequently, the first question that one wants answered is whether the regressors $x_1, \ldots, x_p$ explain some of the variation in the response variable Y. Accordingly, we are interested in testing $H_0: \beta_1 = \beta_2 = \ldots = \beta_p = 0$ against the alternative H_1: not all $\beta_j = 0$. Under the null hypothesis, because none of the x variables are involved, the regression model reduces to the simple model $Y_i = \beta_0 + \varepsilon_i$—that is, the model in which the normal variables $Y_1, Y_2, \ldots, Y_n$ are independent and identically distributed. Under the alternative hypothesis, there is *at least one* explanatory variable that explains some of the variation in Y. The appropriate test statistic is the F-ratio, $F = \text{MSR}/\text{MSE}$, from the ANOVA table.

If $\varepsilon_1, \varepsilon_2, \ldots, \varepsilon_n$ are independent normal random variables with mean zero and common variance σ^2, then the distribution of SSE/σ^2 is chi-square with $n-p-1$ degrees of freedom; thus, $E(\text{MSE}) = \sigma^2$. Now, if H_0 is correct, it is also true that SSR/σ^2 is $\chi^2(p)$ and $E(\text{MSR}) = \sigma^2$. It can be shown that SSE and SSR are independent; therefore, $F = \text{MSR}/\text{MSE}$ follows an F-distribution with p and $n-p-1$ degrees of freedom. If, however, H_1 is true, then $E(\text{MSR}) > \sigma^2$ and the F-ratio will tend to be larger than what can be expected under H_0. This chain of reasoning leads to the following decision rule: Calculate $F = \text{MSR}/\text{MSE}$ from the ANOVA table, and if $F \ge F(\alpha; p, n-p-1)$, accept H_1 at significance level α; otherwise, do not reject H_0. Alternatively, many computer programs calculate the p value $= P[F(p, n-p-1) \ge F]$, where F is the calculated F-ratio. If this probability value is small (less than α), it is highly unlikely that such a large F-ratio has come from H_0. We then accept H_1 and conclude that the regressor variables explain some of the variation in the response variable Y. Note that the F-ratio and R^2 are related through the simple equation

$$F = \frac{R^2}{1 - R^2}\left(\frac{n - p - 1}{p}\right).$$

Thus, a high R^2 implies a large F-ratio.

The conclusion that there is a significant regression relationship is only the first step in our analysis. Next, we want to assess the significance of each individual regression coefficient $\hat{\beta}_j$. Prior to sampling, $\hat{\beta}_j$ is a random variable and has a sampling distribution. Because it is a linear combination of the normal random variables $Y_1, Y_2, \ldots, Y_n$—say, $\hat{\beta}_j = \sum_{i=1}^{n} c_{ij} Y_i, j = 0, 1, \ldots, p$—$\hat{\beta}_j$ is normally distributed. The c_{ij}'s depend only on the nonrandom levels of the explanatory variables, and it is possible to write down explicit expressions for these coefficients. However, this is not done here because computer programs are available for their calculation. As in the simple linear regression model, we can show that $\hat{\beta}_j$ is an unbiased estimator of β_j; that is, $E(\hat{\beta}_j) = \beta_j$. Moreover, we can find the variance, $\text{var}(\hat{\beta}_j) = \sigma^2 \sum_{i=1}^{n} c_{ij}^2$, of these sampling distributions, $j = 0, 1, \ldots, p$. Consequently, the standardized random variable $(\hat{\beta}_j - \beta_j)/\sqrt{\text{var}(\hat{\beta}_j)}$ follows an $N(0,1)$ distribution. Replacing the unknown variance σ^2 by its estimate MSE leads to the estimated variance of $\hat{\beta}_j$, namely, $s^2(\hat{\beta}_j) = \text{MSE} \sum_{i=1}^{n} c_{ij}^2$. The square roots of these quantities, the estimated standard errors $s(\hat{\beta}_j)$, are displayed in the regression output. With this standard error, the sampling distribution of $(\hat{\beta}_j - \beta_j)/s(\hat{\beta}_j)$ is $t(n - p - 1)$, where the degrees of freedom in the t-distribution, $n - p - 1$, come from the degrees of freedom of the error sum of squares in the ANOVA table.

With this sampling distribution, we can calculate $100(1 - \alpha)$ percent confidence intervals for β_j, namely, $\hat{\beta}_j \pm t(\alpha/2; n - p - 1)s(\hat{\beta}_j)$. Also, to test the significance of an individual regression coefficient, we test $H_0: \beta_j = 0$ against $H_1: \beta_j \neq 0$. The appropriate test statistic is the t-ratio, $t(\hat{\beta}_j) = \hat{\beta}_j/s(\hat{\beta}_j)$, and if $|t(\hat{\beta}_j)| \geq t(\alpha/2; n - p - 1)$, we conclude H_1 at significance level α; otherwise, we do not reject $\beta_j = 0$. Alternatively, one can compare the p value $= 2P[t(n - p - 1) \geq |t(\hat{\beta}_j)|]$, where $t(\hat{\beta}_j)$ is the computed value of the t-statistic, with the significance level α. If the p value is less than or equal to α, we conclude that the regressor variable x_j has a significant impact on the response. These t-tests are also called *partial t-tests*, because they assess the partial (or additional) significance of the variable x_j, over and above the impact of all other variables in the model. More will be said on this point in the next section.

8.4-4 A Further Example: Formaldehyde Concentrations

Data were collected to check whether the presence of urea formaldehyde foam insulation (UFFI) has an effect on the ambient formaldehyde concentration (CH_2O) inside a house. Twelve homes with, and 12 homes without, UFFI were studied, and the average weekly CH_2O concentration (in parts per billion) was measured. It was

Table 8.4-2 CH$_2$O, Airtightness of the House, and the Presence of UFFI

Response: CH$_2$O	Air Tightness	Presence of UFFI
31.33	0	0
28.57	1	0
39.95	1	0
44.98	4	0
39.55	4	0
38.29	5	0
50.58	7	0
48.71	7	0
51.52	8	0
62.52	8	0
60.79	8	0
56.67	9	0
43.58	1	1
43.30	2	1
46.16	2	1
47.66	4	1
55.31	4	1
63.32	5	1
59.65	5	1
62.74	6	1
60.33	6	1
53.13	7	1
56.83	9	1
70.34	10	1

thought that the CH$_2$O concentration was also influenced by the amount of air that can move through the house via windows, cracks, chimneys, etc. A measure of air-tightness, on a scale of 0 to 10 (with 0 representing a "leaky" house and 10 an airtight house), was determined for each home.

The data are shown in Table 8.4-2. CH$_2$O concentration is the response variable (Y) that we try to explain through two explanatory variables: the airtightness of the home (x) and the absence or presence of UFFI (z). The absence or presence of UFFI is expressed through the indicator variable UFFI. If insulation is present, then UFFI $= 1$; if it is absent, then UFFI $= 0$. A scatter plot of CH$_2$O against airtightness, for homes with and without UFFI, is shown in Figure 8.4-1. The points on the scatter plot are labeled with squares and circles, depending whether UFFI is, respectively, present or absent. The plot shows strong evidence that CH$_2$O concentrations increase with increasing airtightness of the home.

The basic objective of this particular observational study is to see whether differences in the CH$_2$O concentrations can be attributed to the presence of a certain

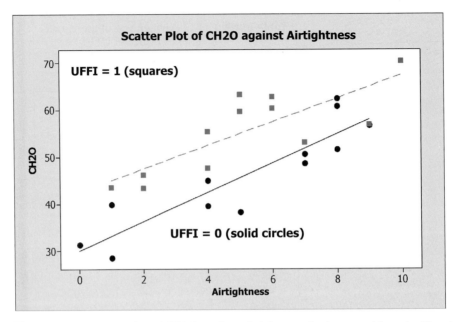

Figure 8.4-1 Scatter plot of CH_2O against airtightness, for homes with and without UFFI

insulation. Let us first consider two separate regressions: one for houses with UFFI (the squares) and one for houses without UFFI (the circles). These two analyses provide the two lines shown on the graph. The output of the two regressions is shown in Table 8.4-3. Note that the slopes are fairly similar (3.12 for UFFI = 0 and 2.50 for UFFI = 1). The ambient formaldehyde concentration in the house increases by about 2.5 to 3.0 units with each unit of airtightness. There is a large difference in the intercepts (30.0 for UFFI = 0 and 42.5 for UFFI = 1), indicating that the presence of UFFI increases the ambient formaldehyde concentration in the house by about $42.5 - 30.0 = 12.5$ units and that this increase appears not to be influenced by the airtightness of the house. The points in Figure 8.4-1 suggest two parallel lines, with the only difference being in their intercepts.

Next, we combine these two models into a single analysis. We estimate the model

$$Y = \beta_0 + \beta_1 x + \beta_2 z + \varepsilon,$$

where x is the airtightness of the house and z is an indicator that is 1 if there is UFFI insulation and 0 if there is not. The deterministic component, $\beta_0 + \beta_1 x + \beta_2 z$, is the sum of three parts. The intercept β_0 measures the average CH_2O concentration for airtight houses ($x = 0$) without UFFI insulation ($z = 0$). The parameter β_2 can be explained as follows: Consider two houses with the same value for airtightness (x), the first house with UFFI ($z = 1$) and the second house without it ($z = 0$); β_2 represents the difference in their average CH_2O concentrations. This is exactly the quantity we are interested in. If $\beta_2 = 0$, we cannot link the presence of UFFI to the formaldehyde concentration.

The parameter β_1 is the expected change in CH_2O concentrations that is due to a unit change in airtightness in homes with (or without) UFFI. Our model assumes

Table 8.4-3 Regression Output for $CH_2O = \beta_0 + \beta_1$ Airtightness $+\ \varepsilon$. Two groups: UFFI = 0 and UFFI = 1

UFFI = 0

The regression equation is

$CH_2O = 30.0 + 3.12$ Tightness

Predictor	Coef	SE Coef	T	P
Constant	29.998	2.879	10.42	0.000
Tightness	3.1208	0.4810	6.49	0.000
S = 5.03726	R − Sq = 80.8%		R − Sq(adj) = 78.9%	

Analysis of Variance

Source	DF	SS	MS	F	P
Regression	1	1068.1	1068.1	42.09	0.000
Residual Error	10	253.7	25.4		
Total	11	1321.8			

UFFI = 1

The regression equation is

$CH_2O = 42.5 + 2.50$ Tightness

Predictor	Coef	SE Coef	T	P
Constant	42.476	3.449	12.32	0.000
Tightness	2.5023	0.6026	4.15	0.002
S = 5.48720	R − Sq = 63.3%		R − Sq(adj) = 59.6%	

Analysis of Variance

Source	DF	SS	MS	F	P
Regression	1	519.19	519.19	17.24	0.002
Residual Error	10	301.09	30.11		
Total	11	820.29			

that this change is the same for homes with and without UFFI. This assumption is a consequence of the additive structure of the model: The contributions of the two explanatory variables, $\beta_1 x$ and $\beta_2 z$, get added. However, additivity does not *have* to be the rule: The more general model that involves the product of x and z, namely,

$$Y = \beta_0 + \beta_1 x + \beta_2 z + \beta_3(xz) + \varepsilon,$$

allows airtightness to affect the two types of homes differently. For a house without UFFI, $E(Y) = \beta_0 + \beta_1 x$ and β_1 expresses the effect on the CH_2O concentrations of a unit change in airtightness. For a house with UFFI, $E(Y) = (\beta_0 + \beta_2) + (\beta_1 + \beta_3)x$ and $(\beta_1 + \beta_3)$ expresses the effect of a unit change in airtightness. The effect is now different by the factor β_3.

The fitting results for the two models are given in Table 8.4-4. We find that the interaction term is insignificant ($\hat{\beta}_3 = -0.6185$, with probability value 0.429, which is

Table 8.4-4 Regression Output for the Models $CH_2O = \beta_0 + \beta_1 x + \beta_2 z + \varepsilon$ and $CH_2O = \beta_0 + \beta_1 x + \beta_2 z + \beta_3 xz + \varepsilon$

$CH_2O = \beta_0 + \beta_1 x + \beta_2 z + \varepsilon$:

The regression equation is

$CH_2O = 31.4 + 2.85\,X + 9.31\,Z$

Predictor	Coef	SE Coef T	P
Constant	31.373	2.461	12.75	0.000
X	2.8545	0.3764	7.58	0.000
Z	9.312	2.133	4.37	0.000
S = 5.22309		R − Sq = 78.3%	R − Sq(adj) = 76.2%

Analysis of Variance

Source	DF	SS	MS	F	P
Regression	2	2063.3	1031.6	37.82	0.000
Residual Error	21	572.9	27.3		
Total	23	2636.1			

$CH_2O = \beta_0 + \beta_1 x + \beta_2 z + \beta_3 xz + \varepsilon$

The regression equation is

$CH_2O = 30.0 + 3.12\,X + 12.5\,Z - 0.618\,XZ$

Predictor	Coef	SE Coef	T	P
Constant	29.998	3.011	9.96	0.000
X	3.1208	0.5030	6.20	0.000
Z	12.478	4.475	2.79	0.011
XZ	−0.6185	0.7665	−0.81	0.429
S = 5.26704		R − Sq = 79.0%	R − Sq(adj) = 75.8%

Analysis of Variance

Source	DF	SS	MS	F	P
Regression	3	2081.32	693.77	25.01	0.000
Residual Error	20	554.83	27.74		
Total	23	2636.15			

much larger than the significance level, 0.05); the two lines are indeed parallel. The two intercepts are different. We conclude that UFFI insulation increases the ambient formaldehyde concentration by 9.312, irrespective of the airtightness of the house.

Exercises 8.4

*8.4-1 The model $Y_i = \beta_0 + \beta_1 x_{i1} + \beta_2 x_{i2} + \varepsilon_i$ is fit to $n = 8$ observations. The calculation results in

$$\sum y_i = 8, \quad \sum x_{i1} = \sum x_{i2} = \sum x_{i1}x_{i2} = 0,$$
$$\sum x_{i1}^2 = 20, \quad \sum x_{i2}^2 = 40, \quad \sum x_{i1}y_i = 10, \quad \sum x_{i2}y_i = 30.$$

Using the normal equations, calculate the least-squares estimates of $\beta_0, \beta_1,$ and β_2.

*8.4-2 Consider data on per-capita beer consumption (y), per-capita real income (x_1), and the relative price of beer (x_2). The linear regression model $Y_i = \beta_0 + \beta_1 x_{i1} + \beta_2 x_{i2} + \varepsilon_i$ is estimated from $n = 17$ observations.

    *(a) Using a regression program, we find that the estimates and their standard errors (given in parentheses) are $\hat{\beta}_0 = 1.37$ (0.35), $\hat{\beta}_1 = 1.14$ (0.16), and $\hat{\beta}_2 = -0.83$ (0.20). Calculate 95 percent confidence intervals for β_1 and β_2. Conduct partial t-tests at the $\alpha = 0.05$ significance level. Interpret your results.

    *(b) A partially completed ANOVA table is as follows:

Source	SS	df	MS	F
Regression				
Error	34			
Total	100			

Complete the table, calculate R^2, and use the F-statistic to test $H_0: \beta_1 = \beta_2 = 0$ at the $\alpha = 0.05$ level.

*8.4-3 A linear regression model $Y_i = \beta_0 + \beta_1 x_{i1} + \beta_2 x_{i2} + \varepsilon_i$ was fit to $n = 20$ data points. It was found that SSTO $= 200$ and SSR $= 66$.

    *(a) Construct the ANOVA table.

    *(b) Calculate and interpret R^2.

    *(c) Test $H_0: \beta_1 = \beta_2 = 0$ at the $\alpha = 0.05$ significance level.

    *(d) The individual t-statistics were $t(\hat{\beta}_0) = 4.30, t(\hat{\beta}_1) = -3.60,$ and $t(\hat{\beta}_2) = 0.80$. Make appropriate tests about the parameters and interpret the results. Does it seem that the regression model can be simplified?

8.4-4 The data set in Table 8.1-1 lists weights and fuel efficiencies of 1978–1979 model-year cars. The fuel efficiency was the result of an actual 123-mile test drive. (*Source:* Henderson and Velleman, *Biometrics,* Vol. 37 (1981), 391–411). The following data set lists weights and EPA fuel efficiencies of forty-one 2007 model-year cars:

Brand	Class	Type	MPG City	MPG Hwy	Curb Weight
Ford	Compact	Focus Sedan SE	27	37	2.636
Ford	Midsize	Fusion V6 SE	20	28	3.280
Ford	Large	500 SEL FWD	21	29	3.643
Ford	Minivan	Freestar Wagon SE	17	23	4.275
Chevrolet	Compact	PT Cruiser Base	22	29	3.152
Chevrolet	Midsize	Sebring Sedan Base	24	32	3.287
Chevrolet	Large	300 Base	21	28	3.712
Chevrolet	Minivan	Town and Country LWB LX	19	26	4.239

Brand	Class	Type	MPG City	MPG Hwy	Curb Weight
Toyota	Compact	Matrix	30	36	2.756
Toyota	Midsize	Camry Solara SE	24	34	3.240
Toyota	Large	Avalon SL	22	31	3.495
Toyota	Minivan	Sienna LE AWD	18	23	4.398
Honda	Compact	Honda Civic Sedan	30	38	2.740
Honda	Midsize	Accord Sedan	20	29	3.344
Honda	Large	Honda Acura RL	18	25	4.014
Honda	Minivan	Honda Odyssey EX	18	25	4.475
Audi	Compact	A3 2.0T	23	32	3.263
Audi	Midsize	A6 3.2	21	29	3.825
Audi	Large	A8 sedan	18	25	4.288
BMW	Compact	3-series 328i Sedan	20	29	3.406
BMW	Midsize	5-series 525i Sedan	20	29	3.450
BMW	Large	7-series 750i Sedan	17	25	4.486
Chrysler	Compact	Pacifica FWD	18	25	4.337
Chrysler	Midsize	PT Cruiser Base	22	29	3.076
Chrysler	Large	300 Base	21	28	3.712
Chrysler	Minivan	Town & Country LWB LX	19	26	4.239
Mercedes	Compact	C-Class C230 Sport Sedan	19	25	3.405
Mercedes	Midsize	E-Class E350 Sedan	19	26	3.740
Mercedes	Large	S-Class S550	16	24	4.465
Mercedes	Minivan	R-Class R350	16	21	4.829
Volkswagen	Compact	Jetta 2.5L Sedan	22	30	3.230
Volkswagen	Midsize	Passat Sedan 2.0T	23	32	3.305
Volkswagen	Large	Phaeton V8	16	22	5.194
Nissan	Compact	Versa 1.8S Sedan	30	34	2.720
Nissan	Midsize	Altima 2.5	26	35	3.055
Nissan	Large	Maxima SE	21	28	3.591
Nissan	Minivan	Quest 3.5	18	25	4.040
Kia	Compact	Rio Base	32	35	2.365
Kia	Midsize	Optima LX	24	34	3.142
Kia	Large	Amanti Base	17	25	4.012
Kia	Minivan	Sedona LX LWB	18	25	4.387

Compare the relationships between GPM = 100/MPG and weight for these two data sets. Consider both city and highway GPM. Compare the scatter plots of GPM against weight on the same graph, together with the fitted regression lines. Mintab's "Graph > Scatterplot > With Regression and Groups" allows you to show all three graphs on the same plot. Comment on the changes that you see between the two data sets and between city and highway GPM.

8.4-5 Consider the regression model Salary($1,000) = 60 + (2)$x$ + (10)z + (1)(xz), where x is the number of years of experience and z is an indicator variable that is 1 if the employee has obtained an MBA degree and 0 otherwise; xz is the product between years of experience and the indicator variable z.

Graph salary (y) against years of experience (x). Do this for both groups (without MBA and with MBA) on the same graph. Does experience matter? Is it worth getting an MBA? Interpret the graph.

8.4-6 To prove or disprove the null hypothesis that there is no salary discrimination in the company for which you work you obtain random samples of male and female employees of the company.

(a) From the sample of 30 men, you find an average yearly salary of $40,000 (standard deviation, $4,000). From the sample of 30 women, you find an average salary of $34,000 (standard deviation, $5,000). Your research hypothesis is that women earn less than men, on average. Test this hypothesis at the 5 percent significance level. (See Chapter 4.)

(b) Proponents of the no-discrimination hypothesis argue that the salary difference arises because of a difference in experience levels and that this factor should be taken account of in drawing a conclusion. They recommend fitting a regression of salary (y) on years of experience (x) and an indicator variable IND that is 1 for male and 0 for female employees.

The estimated regression model (using the 60 employees) is given by

$$\text{Estimated Salary} = 20 + (2)(\text{Experience}) + (4.5)\text{IND}$$

The standard errors of the three regression coefficients are 3.5, 0.5, and 2.0, respectively. Interpret the results.

8.4-7 Wabtec Corporation (http://www.wabtec.com) is a leading supplier of technology-based products and services for rail and transit industries. The company builds locomotives and manufactures a broad range of products for locomotives, freight cars, and passenger transit vehicles. Ann Grimm, an engineer in the Systems Engineering Group at Wabtec Railway Electronics, had collected field data on the stopping distances (in feet) of $n = 83$ trains. She also kept information on the speed of the train at which the brakes were applied (enforcement speed, in mph), whether the train was loaded, and the actual weight of the train (in tons). The data can be found in the file titled Section8.4Exercise8.4-7Train StoppingDistances, which is reproduced below.

Analyze the data. In particular,

(a) Graph stopping distance against speed, separating the data for unloaded and loaded trains (preferably on the same graph). Discuss the relationship. Specify and estimate regression models that describe the relationship.

   Note: You may want to estimate quadratic models separately for loaded and unloaded trains.

(b) Transform the stopping distance, and experiment with transformations such as the logarithm of stopping distance and the square root of stopping distance. Can you find a transformation for which the relationship between the transformed response and enforcement speed is roughly linear? What about a model which assumes that the square root of stopping distance is linearly related to enforcement speed? Estimate the coefficients separately for unloaded and loaded trains. Discuss.

(c) Can you improve your analysis by considering the actual weight of the train, as opposed to the indicator information that you used in (a) and (b)?

Note: The variability among weights of unloaded (and loaded) trains is rather small; hence, the added information on the actual weight may not help.

Stopping Distance (feet)	Enforcement Speed (mph)	Loaded (0 = No; 1 = Yes)	Train Weight (tons)
148	8.0	0	3300
11	0.1	0	3300
111	7.0	0	3470
11	0.8	0	3490
174	8.9	0	3490
63	3.2	0	3810
95	4.5	0	3810
5	0.7	0	4850
42	2.7	0	3473
127	6.9	0	3473
32	2.5	0	3630
5	0.2	0	3630
37	3.5	0	3500
312	10.8	0	3500
63	3.7	0	3459
612	19.5	0	3374
876	25.9	0	3374
21	3.0	0	3367
5	1.5	0	3340
21	1.8	0	3340
216	9.2	0	3340
607	22.4	0	3340
21	2.8	0	3481
69	3.9	0	3481
232	13.5	0	3481
63	2.6	0	3910
37	3.2	0	3910
174	7.2	0	3910
37	3.2	0	3430
359	12.9	0	3810
21	2.2	0	3346
5	0.8	0	3630
121	5.2	0	3630
264	11.1	0	3630

253	9.8	0	3830
195	10.7	0	3730
53	4.5	0	3786
929	23.9	0	3306
11	0.8	0	3399
5	1.2	0	3399
132	6.5	0	3516
301	12.7	0	3516
11	0.4	0	3420
16	1.1	0	4430
16	4.0	1	19297
1415	26.4	1	19297
565	16.6	1	19297
132	5.2	1	19297
655	14.2	1	19248
2033	31.9	1	19248
201	7.4	1	19248
106	4.1	1	19248
327	8.6	1	19067
1362	21.9	1	19067
32	3.1	1	19299
232	7.0	1	18010
1896	29.7	1	17648
327	8.8	1	18368
100	6.8	1	19053
26	1.4	1	19053
153	4.7	1	19191
438	13.4	1	19170
2408	33.8	1	19220
1769	26.1	1	19220
2170	31.2	1	19250
158	5.1	1	19250
90	4.7	1	19171
370	10.8	1	19220
1061	20.0	1	18956
190	6.0	1	18956
26	1.5	1	18706
269	10.5	1	18706
132	4.2	1	18638
32	2.2	1	19098
364	11.7	1	19006

Stopping Distance (feet)	Enforcement Speed(mph)	Loaded (0 = No; 1 = Yes)	Train Weight (tons)
718	16.6	1	18934
301	8.8	1	16504
74	3.1	1	16504
972	18.0	1	16504
935	16.6	1	18891
5	1.2	1	18891
243	10.9	1	18851
597	13.3	1	20297

8.5   More on Multiple Regression

8.5-1   Multicollinearity among the Explanatory Variables

In this section, we discuss how multicollinearity affects the regression analysis. *Multicollinearity* refers to a condition in the explanatory variables and occurs when explanatory variables convey similar information and when there is "near" linear dependence among the variables $x_1, x_2, \ldots, x_p$.

To discuss multicollinearity among the explanatory variables, let us consider the multiple regression model with $p = 2$ regressors. In the *mean-corrected* form, the basic model can be written as

$$Y_i - \overline{Y} = \beta_1(x_{i1} - \overline{x}_1) + \beta_2(x_{i2} - \overline{x}_2) + \varepsilon_i.$$

If the explanatory variables x_1 and x_2 convey the same or very similar information, we say that there is multicollinearity. If they express the same information through an approximate linear relationship between x_1 and x_2, then the absolute value of the correlation coefficient

$$r_{12} = \frac{\sum (x_{i1} - \overline{x}_1)(x_{i2} - \overline{x}_2)}{\left[\sum (x_{i1} - \overline{x}_1)^2 \sum (x_{i2} - \overline{x}_2)^2 \right]^{1/2}}$$

is close to 1. By contrast, if x_1 and x_2 provide different and independent information, r_{12} is close to zero.

As a simple and trivial example, consider the mileage data presented earlier, with gpm as the dependent variable Y and with weight in kilograms as x_1 and weight in pounds as x_2. Obviously, x_1 and x_2 express the same information, because x_1 and x_2 are proportional; thus, $r_{12} = 1$. In many practical and nontrivial applications, r_{12} will not be exactly 1, but may be very large. For example, heavy cars tend to have larger engines; hence, the correlation between weight and engine displacement of cars will be quite high.

If the correlation coefficient between x_1 and x_2 is $r_{12} = 0$, we also say that x_1 and x_2 are *orthogonal*. An example of such a situation arises in the context of factorial experiments. (See Section 7.3.) In a 2^2 factorial experiment, the investigator conducts the experiment at two levels for each of the two variables: the low level is coded as -1 and the high level is given by $+1$. There are $n = 4$ observations if there are no replications, and the data are given by $(y_1, -1, -1), (y_2, +1, -1), (y_3, -1, +1)$, and $(y_4, +1, +1)$. It is easy to check that the correlation coefficient between x_1 and x_2 is $r_{12} = 0$.

If $r_{12} = 0$, the normal equations in Section 8.4 simplify to

$$\left[\sum (x_{i1} - \bar{x}_1)^2 \right] \hat{\beta}_1 = \sum (x_{i1} - \bar{x}_1)(y_i - \bar{y}),$$
$$\left[\sum (x_{i2} - \bar{x}_2)^2 \right] \hat{\beta}_2 = \sum (x_{i2} - \bar{x}_2)(y_i - \bar{y}).$$

The estimates of the regression coefficients are

$$\hat{\beta}_1 = \frac{\sum (x_{i1} - \bar{x}_1)(y_i - \bar{y})}{\sum (x_{i1} - \bar{x}_1)^2} = \frac{s_{1y}}{s_{11}},$$

$$\hat{\beta}_2 = \frac{\sum (x_{i2} - \bar{x}_2)(y_i - \bar{y})}{\sum (x_{i2} - \bar{x}_2)^2} = \frac{s_{2y}}{s_{22}}.$$

These estimates are the same as the ones we get by estimating β_1 and β_2 in the two individual models $Y_i - \bar{Y} = \beta_1(x_{i1} - \bar{x}_1) + \varepsilon_i$ and $Y_i - \bar{Y} = \beta_2(x_{i2} - \bar{x}_2) + \varepsilon_i$. In other words, because $r_{12} = 0$, the addition of the second variable x_2 in the model and the joint estimation of β_1 and β_2 do not change our estimate of β_1.

Of course, this is not true if $r_{12} \neq 0$, because, from the normal equations in Section 8.4, we have

$$\hat{\beta}_1 = \frac{s_{22}}{s_{11}s_{22} - s_{12}^2} s_{1y} - \frac{s_{12}}{s_{11}s_{22} - s_{12}^2} s_{2y}.$$

Now, $r_{12} = s_{12}/\sqrt{s_{11}s_{22}}$, so we can rewrite the preceding estimate as

$$\hat{\beta}_1 = \left(\frac{1}{1 - r_{12}^2} \right) \frac{s_{1y}}{s_{11}} - \left(\frac{r_{12}}{1 - r_{12}^2} \right) \frac{s_{2y}}{\sqrt{s_{11}s_{22}}}.$$

From this equation, it becomes clear that, in general, when $r_{12} \neq 0$, the estimate of β_1 will change if we include another variable x_2 in the model. This is the reason we emphasized the *partial* nature of individual regression coefficients and of the corresponding t-tests; we always have to consider the contribution of an explanatory variable in relation to all other variables in the regression model.

The latter point should be kept in mind when we interpret the multiple-regression output. For example, if individual t-tests show that *both* $\hat{\beta}_1$ and $\hat{\beta}_2$ are insignificant, this does not necessarily imply that *both* x_1 and x_2 are unimportant and can be omitted from the model. Because the t-ratio $t(\hat{\beta}_2) = \hat{\beta}_2/s(\hat{\beta}_2)$ represents a partial test statistic, it tests only the additional importance of the variable x_2, over and above the contribution of x_1. It is like adding an extra explanatory variable to the model and then asking for its contribution beyond that of all other variables in the model. Similarly, $t(\hat{\beta}_1) = \hat{\beta}_1/s(\hat{\beta}_1)$ tests the significance of the regression contribution of x_1, over and above that explained by x_2. If x_1 and x_2 are highly correlated, it could be that both individual t-statistics are insignificant, even though the response Y is related to each of them. In such a case, the F-statistic, which tests H_0: $\beta_1 = \beta_2 = 0$ against the alternative that at least one of the coefficients is nonzero, would be significant. In sum, this discussion implies that we must proceed cautiously when we seek to simplify regression models. We should not omit all insignificant coefficients at once, but should proceed sequentially, omitting one insignificant variable at a time and reestimating the remaining parameters each time.

Example 8.5-1 An engineer in the paint division of a chemical company is studying the effects of latex solid contents (x_1) and drying temperature (x_2) on the stiffness of paint. The engineer varies latex contents between 40 and 50 percent and drying temperature between 75 and 95°F. To study possible interaction effects between latex solid contents and drying temperature, she varies these two variables simultaneously. The experiment is run at the following four conditions of latex solid contents and drying temperature: (40 percent, 75°F), (50 percent, 75°F), (40 percent, 95°F), and (50 percent, 95°F). This is a two-level factorial experiment, because each of the two levels of one variable (factor) is combined with each of the two levels of the other. Following our notation in Sections 7.3 and 7.4, we code the two levels as ± 1. We let $x_1 = -1$ if latex contents is 40 percent and $x_1 = +1$ if it is 50 percent; thus, a unit change in x_1 corresponds to a 5 percent change in latex solid contents. Similarly, $x_2 = -1$ if the temperature is at 75°F, and $x_2 = +1$ if the temperature is at 95°F; a unit change in x_2, then, corresponds to a 10°F change in temperature. Now, suppose that, at each level combination, the engineer conducts two separate and independent experiments. That is, eight experiments are run altogether. To guarantee the validity of the resulting inference, the engineer randomizes the order of the eight experiments and obtains the data listed in Table 8.5-1.

We want to estimate the regression model

$$Y_i = \beta_0 + \beta_1 x_{i1} + \beta_2 x_{i2} + \beta_3 x_{i3} + \varepsilon_i, \quad i = 1, 2, \ldots, 8,$$

where $x_{i3} = x_{i1} x_{i2}$. We observe that $\bar{x}_1 = \bar{x}_2 = \bar{x}_3 = 0$; thus, we know from the first normal equation in Section 8.4 that $\hat{\beta}_0 = \bar{y} = 135$. Furthermore, we note that the correlation coefficients between the three explanatory variables are $r_{12} = r_{13} = r_{23} = 0$; so the three x variables are orthogonal. The estimates, with standard errors in parentheses, are given by $\hat{\beta}_0 = 135 \, (1.98), \hat{\beta}_1 = 8.75 \, (1.98), \hat{\beta}_2 = 2.5 \, (1.98),$ and $\hat{\beta}_3 = -1.25 \, (1.98)$. The fact that the standard errors are the same arises because of the special nature of the x-values in this example. The ANOVA table is given in Table 8.5-2. The coefficient of determination is $R^2 = 675/800 = 0.84$. The overall regression model is significant because $F = 225/31.25 = 7.20$ is larger than the critical value $F(0.05; 3, 4) = 6.59$.

Table 8.5-1 Data from a Replicated 2^2 Factorial Design

x_1	x_2	$x_3 = x_1 x_2$	Y
−1	−1	1	120
−1	−1	1	125
1	−1	−1	140
1	−1	−1	145
−1	1	−1	125
−1	1	−1	135
1	1	1	140
1	1	1	150

Table 8.5-2 ANOVA Table for Example 8.5-1

Source	SS	df	MS	F
Regression	675	3	225	7.20
Error	125	4	31.25	
Total	800	7		

The model can be simplified because the coefficient that corresponds to the interaction $x_3 = x_1 x_2$ is insignificant. Its t-ratio, $t(\hat{\beta}_3) = -1.25/1.98 = -0.63$, is much smaller in absolute value than the critical value $t(0.025; 4) = 2.776$. This leads us to the *main-effects* model, $Y_i = \beta_0 + \beta_1 x_{i1} + \beta_2 x_{i2} + \varepsilon_i$. Such a model is much easier to interpret, as the effect of a unit change in one variable does not depend on the level of the other. In general, we should estimate β_1 and β_2 again, because omitting a variable from the model will lead to different estimates of the remaining coefficients. In this special case, however, the three sets of x variables are orthogonal; thus, the estimates remain unchanged.

Finally, Table 8.5-3 summarizes the results of the estimation of three regression models, where we have eliminated the explanatory variables in a logical order in moving from one model to the next. For this orthogonal example, we note that (1) the regression coefficient estimates do not change as we drop the other explanatory variables from the model. The standard errors change somewhat because the estimates of σ^2 are different. (2) The second variable x_2 (temperature) is insignificant; the t-ratio from the model with x_1 and x_2, $t(\hat{\beta}_2) = 2.5/1.85 = 1.35$, is smaller than the critical value $t(0.025; 8 - 3 = 5) = 2.571$. (3) In sum, it appears that, over the ranges considered, x_1 is the only important variable, because $t(\hat{\beta}_1) = 8.75/1.98 = 4.43$ is larger than $t(0.025; 8 - 2 = 6) = 2.447$. A 5 percent change in latex solid contents leads to an 8.75 unit change (increase) in the stiffness rating. Of course, this statement applies only for changes in the 40 to 50 percent range of latex contents. We have no data outside this region, so we should not extrapolate too far outside of it.

You may wonder whether it is possible for this data set to include quadratic terms x_1^2 and x_2^2 in the regression model. With these data, it is not possible to estimate quadratic effects, because we have only two levels and there is no way to determine a quadratic curve from just two points. If we want to fit a quadratic model, we need

Table 8.5-3 Regression Estimates and Standard Errors (in Parentheses) for Example 8.5-1

Regression on:	β_0	β_1	β_2	β_3	MSE
$x_1, x_2, x_3 = x_1 x_2$	135.00	8.75	2.50	−1.25	31.25
	(1.98)	(1.98)	(1.98)	(1.98)	
x_1, x_2	135.00	8.75	2.50		27.50
	(1.85)	(1.85)	(1.85)		
x_1	135.00	8.75			31.25
	(1.98)	(1.98)			

observations at more than two levels. For example, including a center point where $x_1 = x_2 = 0$, or considering a factorial experiment with three levels for each variable, resulting in the nine (x_1, x_2) design points $(-1, -1), (0, -1), (1, -1), (-1, 0),$ $(0, 0), (1, 0), (-1, 1), (0, 1),$ and $(1, 1)$, would provide the necessary information to fit a quadratic model.   ■

Example 8.5-2   Consider the mileage data shown in Section 8.1. With those data, we found a strong positive linear association between gpm (y) and weight (x_1). Now we add a second variable, horsepower (x_2), and consider the multiple-regression model $Y_i = \beta_0 + \beta_1 x_{i1} + \beta_2 x_{i2} + \varepsilon_i$. Because more powerful cars require more fuel, we expect a positive association between gpm and horsepower and thus a positive estimate for β_2. The weights and horsepowers of the 10 cars in our sample are given in Table 8.5-4.

The plot of horsepower against weight shown in Figure 8.5-1, with a correlation coefficient of $r_{12} = 0.98$, indicates a strong linear relationship between these two explanatory variables. The explanatory variables for this particular data set constitute a "poor design" for estimating their effects, because one is essentially a linear function of the other. To study the partial effects of weight and horsepower on fuel consumption, we would need the gpm of light cars with high horsepower. However, such cars are not part of our sample and may, in fact, not even exist. To illustrate why it is difficult to partition the effects in such a situation, let us assume that the linear relationship between x_2 and x_1 is exact; that is, say that $x_2 = \alpha_0 + \alpha_1 x_1$. For our data, $\alpha_1 \approx 35$. If we substitute this for x_2 into $Y = \beta_0 + \beta_1 x_1 + \beta_2 x_2 + \varepsilon$, we obtain $Y = \beta_0 + \beta_1 x_1 + \beta_2(\alpha_0 + \alpha_1 x_1) + \varepsilon = \beta_0^* + \beta_1^* x_1 + \varepsilon$. Now $\beta_1^* = \beta_1 + \alpha_1 \beta_2$ may be estimated with good precision. For example, in Section 8.1 we found a highly significant regression coefficient for the variable weight x_1; its value was $\hat{\beta}_1^* = 1.64$. But there are many solutions of $1.64 = \beta_1 + 35\beta_2$; in fact, there are an infinite number of solutions. One of them is $\beta_1 = 1.64$ and $\beta_2 = 0.0$, another is $\beta_1 = 0.0$ and $\beta_2 = 0.047$, a third is $\beta_1 = 0.82$ and $\beta_2 = 0.024$, and so on. If the linear relationship between x_2 and x_1 is not exact, but only approximate, there is still considerable uncertainty on how to estimate the coefficients β_1 and β_2. This uncertainty is reflected in the large standard errors of the estimates. In a multicollinear situation,

Table 8.5-4 Gpm, Weight, and Horsepower of 10 Selected Cars

Gpm, Y	Weight (1000 pounds), x_1	Horsepower, x_2
5.5	3.4	120
5.9	3.8	130
6.5	4.1	142
3.3	2.2	68
3.6	2.6	95
4.6	2.9	109
2.9	2.0	65
3.6	2.7	80
3.1	1.9	71
4.9	3.4	105

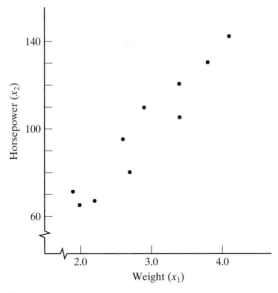

Figure 8.5-1 Scatterplot of horsepower against weight

the standard errors tend to be much larger than the standard errors in the simplified models in which we have omitted one of the collinear explanatory variables—and that is exactly what we observe in this example. The resulting estimates for the three models $Y_i = \beta_0 + \beta_1 x_{i1} + \beta_2 x_{i2} + \varepsilon_i$, $Y_i = \beta_0 + \beta_1 x_{i1} + \varepsilon_i$, and $Y_i = \beta_0 + \beta_2 x_{i2} + \varepsilon_i$ in Table 8.5-5 show that the standard errors of the estimates of β_1 and β_2 are much larger in the full model than in either of the simplified models.

This table leads to the following observations: (1) The estimate of β_1 has changed from 1.64 in the regression on weight only to 0.82 in the more general model that includes both weight and horsepower as explanatory variables. The same observation can be made for the coefficient of horsepower, which changes from 0.046 to 0.024 when weight is introduced as an additional explanatory variable. (2) Weight still has a significant partial effect (at significance level $\alpha = 0.10$) after horsepower has been included in the model. However, its significance is greatly reduced, from $1.64/0.13 = 12.85$ in the regression on weight only to $0.82/0.36 = 2.26$

Table 8.5-5 Regression Estimates and Standard Errors (in Parentheses) for Example 8.5-2				
Regression on	β_0	β_1	β_2	R^2
x_1, x_2	−0.33	0.82	0.024	0.974
	(0.30)	(0.36)	(0.010)	
x_1 (weight)	−0.36	1.64		0.954
	(0.38)	(0.13)		
x_2 (horsepower)	−0.12		0.046	0.955
	(0.36)		(0.003)	

in the regression on both weight and horsepower. The same observation applies to the coefficient of horsepower. (3) The coefficient of determination, R^2, changes very little. Also, the fitted values that are implied by the three models are similar, which means that all three models lead to similar fitted surfaces. However, this is true only as long as one is interested in the mean responses to values from the experimental region, which is this narrow band of values in the scatterplot of x_2 against x_1 in Figure 8.5-1. The extrapolations for x_1 and x_2 values that lie outside this region will be quite different for the three fitted models. Because there are no data points to check the adequacy of the models outside of this experimental region, it is difficult to determine which of the three models is more appropriate. ■

A Note on Orthogonality: The designs in Chapters 6 and 7 are such that, for any two design factors, each factor–level combination has the same number of runs. We call such designs *orthogonal*. The two-factor designs in Section 7.1, the three-factor design in Section 7.3, and the 2^k factorial and 2^{k-p} fractional factorial designs in Sections 7.3 and 7.4 are orthogonal designs. Orthogonality simplifies the analysis of the data. Orthogonality is responsible for the additive sum-of-squares decomposition in the analysis of variance (recall that the sums of cross products were always zero) and for the fact that the estimated effects of factors in 2^k designs are unaffected by the presence or absence of all other factors of the design (see Table 8.5-3).

Orthogonality is no longer a given if the design is unbalanced or if observations for some factor–level combinations are missing. But then, this complicates the analysis. Unbalanced designs can still be analyzed with the Minitab "Stat > ANOVA > General Linear Model" command, but the estimates of the factor effects and their associated sums of squares have to be interpreted as conditional on the presence of other factors in the model. Addressing these issues fully would go beyond this introductory text; we refer the interested reader to more advanced statistics books.

8.5-2 Another Example of Multiple Regression

Data concerning the durability of pavement are given in Table 8.5-6. Specifically, measurements are shown on the change in rut depth (y) of 31 experimental asphalt pavements that were prepared under different conditions specified by the values of five explanatory variables: viscosity of asphalt (x_1), percentage of asphalt in surface course (x_2), percentage of asphalt in base course (x_3), percentage of fines in surface course (x_4), and percentage of voids in surface course (x_5). The sixth variable, x_6, is an *indicator variable* that separates the results for 16 pavements tested in one set of runs from results obtained on 15 pavements tested in the second run. Note that asphalt viscosity is considerably higher in the second set of runs. The data are taken from a study by J. W. Gorman and R. J. Toman, "Selection of Variables for Fitting Equations to Data" [*Technometrics,* 8: 27–51 (1966)], who give a detailed description of the experiment. An analysis of the data is also given in a book by C. Daniel and F. S. Wood, *Fitting Equations to Data,* 2d ed. (New York: Wiley, 1980).

An objective of this experiment is to determine the important variables that affect the change in rut depth. Viscosity of the asphalt is certainly one such variable,

Table 8.5-6 Pavement Durability Data[a]

y	x_1	x_2	x_3	x_4	x_5	x_6
6.75	2.80	4.68	4.87	8.4	4.916	−1
13.00	1.40	5.19	4.50	6.5	4.563	−1
14.75	1.40	4.82	4.73	7.9	5.321	−1
12.60	3.30	4.85	4.76	8.3	4.865	−1
8.25	1.70	4.86	4.95	8.4	3.776	−1
10.67	2.90	5.16	4.45	7.4	4.397	−1
7.28	3.70	4.82	5.05	6.8	4.867	−1
12.67	1.70	4.86	4.70	8.6	4.828	−1
12.58	0.92	4.78	4.84	6.7	4.865	−1
20.60	0.68	5.16	4.76	7.7	4.034	−1
3.58	6.00	4.57	4.82	7.4	5.450	−1
7.00	4.30	4.61	4.65	6.7	4.853	−1
26.20	0.60	5.07	5.10	7.5	4.257	−1
11.67	1.80	4.66	5.09	8.2	5.144	−1
7.67	6.00	5.42	4.41	5.8	3.718	−1
12.25	4.40	5.01	4.74	7.1	4.715	−1
0.76	88.00	4.97	4.66	6.5	4.625	1
1.35	62.00	5.01	4.72	8.0	4.977	1
1.44	50.00	4.96	4.90	6.8	4.322	1
1.60	58.00	5.20	4.70	8.2	5.087	1
1.10	90.00	4.80	4.60	6.6	5.971	1
0.85	66.00	4.98	4.69	6.4	4.647	1
1.20	140.00	5.35	4.76	7.3	5.115	1
0.56	240.00	5.04	4.80	7.8	5.939	1
0.72	420.00	4.80	4.80	7.4	5.916	1
0.47	500.00	4.83	4.60	6.7	5.471	1
0.33	180.00	4.66	4.72	7.2	4.602	1
0.26	270.00	4.67	4.50	6.3	5.043	1
0.76	170.00	4.72	4.70	6.8	5.075	1
0.80	98.00	5.00	5.07	7.2	4.334	1
2.00	35.00	4.70	4.80	7.7	5.705	1

Source: J. W. Gorman and R. J. Toman, "Selection of Variables for Fitting Equations to Data," *Technometrics,* 8: 27–51 (1966).
[a] y, Change in rut depth in inches per million wheel passes; x_1, viscosity of asphalt; x_2, percentage of asphalt in surface course; x_3, percentage of asphalt in base course; x_4, percentage of fines in surface course; x_5, percentage of voids in surface course: x_6, indicator variable to separate two sets of runs.

as we can see from the plot in Figure 8.5-2(a). The relationship between Y (change in rut depth) and x_1 (asphalt viscosity) is highly nonlinear. In many cases, transformations of the response variable, as well as of the explanatory variables, may lead to simplifications of the relationship. Logarithmic transformations are usually the first

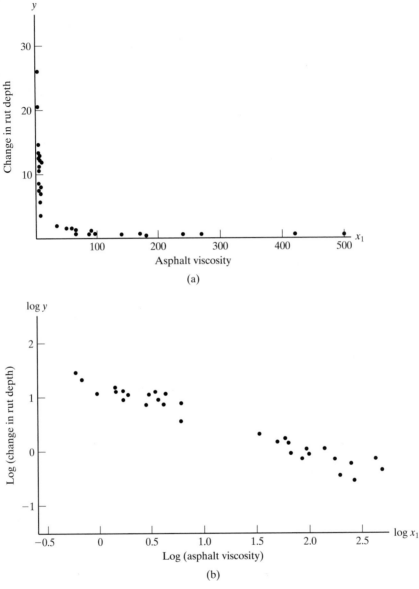

Figure 8.5-2 Pavement durability data: (a) plot of the change in rut depth (y) against viscosity of asphalt, x_1; (b) plot of log y against log x_1

ones tried, especially when variables are bounded from below, as is the case with the change in rut depth. A plot of log y against log x_1 is given in Figure 8.5-2(b), and it shows that these logarithmic (base-10) transformations lead to a relationship that is approximately linear. Thus, we start with a model of the form log $Y_i = \beta_0 + \beta_1 \log x_{i1} + \varepsilon_i$.

We mentioned that the experiment was run in two batches. Accordingly, one may be interested in learning whether there is an effect due to the two different sets of runs. To do so, we include the indicator variable x_6 in the model and consider $\log Y_i = \beta_0 + \beta_1 \log x_{i1} + \beta_6 x_{i6} + \varepsilon_i$. An indicator variable is a variable that expresses the absence or presence of a condition. It is either 0 or 1 or, as in our example, -1 or $+1$. A nonzero value of β_6 would change the intercept in the linear regression model to $\beta_0 - \beta_6$ for the first group and $\beta_0 + \beta_6$ for the second. Thus, β_6 measures the additional effect that is due to the different sets of runs, after we have adjusted for the different levels of asphalt viscosity. Note that the term $\beta_6 x_6$ affects only the intercept of the relationship between $\log y$ and $\log x_1$; the slope (which measures the effect of $\log x_1$) stays the same. In other words, we assume that the effects of $\log x_1$ and x_6 are additive.

The remaining variables $x_2, \ldots, x_5$ also may have a significant impact on the response variable $\log Y$. To study their effects, we consider the model

$$\log Y_i = \beta_0 + \beta_1 \log x_{i1} + \beta_2 x_{i2} + \beta_3 x_{i3} + \beta_4 x_{i4} + \beta_5 x_{i5} + \beta_6 x_{i6} + \varepsilon_i.$$

The reader may ask why we have included these variables in their original metric and not considered a transformation (possibly logarithmic). For one thing, individual scatter plots of $\log y$ against x_2 through x_5 show more or less linear relationships. Furthermore, the range of these variables is rather small. Because most nonlinear functions can be approximated by linear ones over a limited range, it does not matter too much whether we use x_j or $\log x_j$.

The regression output for this general model with the six explanatory variables is given in Table 8.5-7. This regression explains 97.2 percent of the variation in $\log Y$ and is highly significant because $F = 140.1 > F(0.01; 6, 24) = 3.67$. That is, the null hypothesis $H_0: \beta_1 = \beta_2 = \ldots = \beta_6 = 0$ is rejected in favor of H_1: not all $\beta_j = 0, j = 1, \ldots, 6$. Looking at the partial (individual) t-tests and comparing the t-ratios with $t(0.025; 24) = 2.064$, we find that the important variables which affect

Table 8.5-7 Regression Output for the Example

	Estimate	Standard Error	t-Ratio
Constant	-2.645	1.091	-2.43
$\log x_1$	-0.513	0.073	-7.03
x_2	0.498	0.115	4.32
x_3	0.101	0.142	0.71
x_4	0.019	0.034	0.55
x_5	0.138	0.048	2.87
x_6	-0.134	0.064	-2.10

ANOVA table:

Source	SS	df	MS	F
Regression	10.752	6	1.792	140.1
Error	0.307	24	0.0128	
Total	11.059	30		

$R^2 = 10.752/11.059 = 0.972.$

the change in rut depth (as measured by its logarithm) are asphalt viscosity, as measured by log x_1, percent of asphalt in surface course (x_2), and percent of voids in surface course (x_5). In addition, it appears that the runs in the second set lead to smaller changes in rut depth than the runs in the first set. Thus, there are some differences in the rutting rates from the two sets of runs that are not accounted for by the change in asphalt viscosity, the percent of asphalt, and the percent of voids in the surface course. These differences may be due to a difference between the batches of asphalt, but could also be due to changes of tires on test machines, slightly different uncontrolled test conditions, and so on. We simply do not know.

To check the adequacy of the fitted models, we use various residual plots that we discussed in Section 8.3. One of them, a plot of the residuals against fitted values, is given in Figure 8.5-3. It shows no apparent patterns. In addition, we find that the plots of residuals against the explanatory variables show no inadequacies in the model. Thus, it appears that the model in Table 8.5-7 gives an appropriate representation of the data.

Two of the coefficients in this model were found insignificant. One can simplify the model by omitting insignificant variables. However, we cautioned earlier that the variables should be omitted one by one, estimating the coefficients of the new model at each stage. In Table 8.5-7, the coefficient with the smallest t-ratio in absolute value is x_4 (percentage of fines in surface course). Because it is insignificant, we drop it from the model and use

$$\log Y_i = \beta_0 + \beta_1 \log x_{i1} + \beta_2 x_{i2} + \beta_3 x_{i3} + \beta_5 x_{i5} + \beta_6 x_{i6} + \varepsilon_i.$$

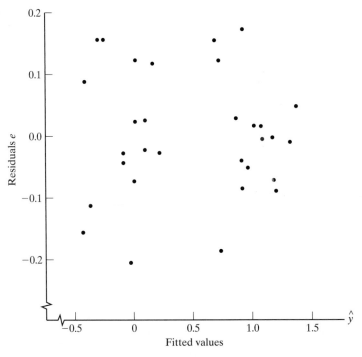

Figure 8.5-3 Pavement durability data. Plot of residuals against fitted values

	Estimate	Standard Error	t-Ratio	Estimate	Standard Error	t-Ratio
Table 8.5-8 Results of Estimation for the Example						
Constant	-2.692	1.072	-2.51	-1.853	0.658	-2.81
$\log x_1$	-0.522	0.070	-7.42	-0.547	0.065	-8.36
x_2	0.502	0.114	4.43	0.465	0.107	4.35
x_3	0.129	0.130	0.99			
x_4						
x_5	0.146	0.045	3.24	0.144	0.045	3.20
x_6	-0.131	0.063	-2.09	-0.111	0.059	-1.87
		$R^2 = 0.972$			$R^2 = 0.971$	

The estimation results for this model are summarized in the first three columns of Table 8.5-8. We note that x_3 is insignificant (t-ratio $= 0.99$); this leads to a further simplification of the model, namely,

$$\log Y_i = \beta_0 + \beta_1 \log x_{i1} + \beta_2 x_{i2} + \beta_5 x_{i5} + \beta_6 x_{i6} + \varepsilon_i.$$

The estimates, standard errors, and t-ratios for this model are given in the last three columns of Table 8.5-8. The smallest t-ratio is the one associated with the indicator variable x_6. It is still significant at the $\alpha = 0.10$ level, because $t(0.05; 31 - 5 = 26) = 1.706$. Thus, we should not simplify this model further and should consider it to be our best description of the relationship we are investigating. In sum, we find that asphalt viscosity, percent of asphalt in surface course, and percent of voids in surface course have an effect on the change in rut depth. In addition, we note some differences in the rutting rates from the two sets of runs.

The selection procedure that we have described here is called *backward elimination*. One starts with the largest possible model and omits insignificant variables, one at a time. Several other "search" procedures to choose the best-fitting model have been developed. For example, we could start with the simplest model with just one explanatory variable and then add variables if they are found significant. Such an approach is known as *forward selection*.

Another approach, which is usually preferable to either backward elimination or forward selection, examines the results of all possible regressions. The number of all possible regressions is large in most cases; for p regressor variables, there are $2^p - 1$ possible regression models: p models with one variable, $p(p - 1)/2$ models with two variables, ..., and one model with all of the variables included. Today's computers, however, can perform these estimations very rapidly. Various criteria for comparing the fitted models are suggested in the literature. For example, one can select the model that leads to the smallest mean-square error. The resulting "optimal" model, as well as every other model that achieves a mean-square error which is close to the minimum and which includes not more variables than the optimal model, should be considered for a more detailed study. Texts on regression analysis, such as Draper and Smith, *Applied Regression Analysis*, 3d ed. (New York: Wiley, 1998); Montgomery, Peck, and Vining, *Introduction to Linear Regression Analysis*, 3d ed. (New York: Wiley, 2001); Weisberg, *Applied Linear Regression*, 3d ed. (New York: Wiley,

2001); and Abraham and Ledolter, *Introduction to Regression Modeling* (Belmont, CA: Thomas Brooks/Cole, 2006), give a detailed description of these procedures.

Our model for the change in rut depth is linear in the transformed variables. We could ask whether it is necessary to include interaction terms. In particular, one may be interested in interactions with the indicator variable x_6. An interaction with an indicator variable amounts to a change in slope in the regression model. Consider, for example, adding the term $\beta_{16}(\log x_1)x_6$ to our model to obtain

$$\log Y = \beta_0 + \beta_1(\log x_1) + \beta_2 x_2 + \beta_5 x_5 + \beta_6 x_6 + \beta_{16}(\log x_1)x_6 + \varepsilon.$$

The coefficient β_6 represents the change in the intercept. The coefficient β_{16} represents the change in the regression coefficient of $\log x_1$. If $x_6 = -1$, as in the first set of runs, the regression coefficient for $\log x_1$ is $\beta_1 - \beta_{16}$; if $x_6 = +1$, it is $\beta_1 + \beta_{16}$. The model with the added interaction component was estimated. However, we found that the estimate of β_{16} was insignificant.

8.5-3 A Note on Computer Software

Regression analysis can be carried out through various programs listed under the Minitab "Stat > Regression" tab, namely, Regression, Fitted Line Plot, Best Subsets, and Stepwise.

The "Stat > Regression > Regression" dialog box requires the specification of the response variable and the explanatory variables. The output provides the coefficient estimates, their standard errors, t-ratios, and probability values, as well as the ANOVA table and R^2. The output (residuals and fitted values, as well as measures of leverage and influence that were mentioned briefly in Section 8.3-3) can be stored in unused columns of the worksheet. Residual plots can be executed through the "Graph" dialog box or by first storing the needed variables and then using standard plotting functions.

The "Stat > Regression > Fitted Line Plot" is used when there is only one explanatory variable. A scatter plot of the data is shown, together with the superimposed fitted model (which can be linear, quadratic, or cubic).

The "Stat > Regression > Best Subsets" feature fits all possible regressions and displays useful summary statistics that help select a parsimonious model. For example, for 10 explanatory variables, the feature will estimate $2^{10} - 1$ different models (the 10 models with a single explanatory variable, the 45 models with two explanatory variables, ..., the model with all 10 explanatory variables) and will organize the fit of these models according to their R^2 (as well as several other criteria).

The "Stat > Regression > Stepwise" feature includes the backward elimination method discussed in Section 8.5-2, where we started with the largest model and then omitted variables, one by one, until no variable could be omitted anymore. The "Stepwise" procedure extends this approach, omitting and adding variables until a satisfactory model is found. The resulting final model gives you a good starting point. Of course, the model needs to be checked to see whether it satisfies all of the assumptions.

Minitab includes a useful command for creating indicator variables. "Calc > Make Indicator Variables" creates m indicator variables from a single variable with m different values.

8.5-4 Nonlinear Regression

So far, we have assumed that our regression models are linear in their parameters, expressed as the equation $Y = \beta_0 + \beta_1 x + \varepsilon$. Least-squares estimation, which minimizes the sum of squared deviations $S(\beta_0, \beta_1) = \sum_{i=1}^{n}[y_i - (\beta_0 + \beta_1 x_i)]^2$, was easy to carry out. The derivatives of $S(\beta_0, \beta_1)$ with respect to the parameters β_0 and β_1 led to linear functions in the parameters, and setting them equal to zero gave us explicit expressions for the least-squares estimates. (See Section 8.1-1.)

Explicit algebraic solutions for least-squares estimates are no longer possible if the regression model is nonlinear in its parameters. Consider, for example, the *logistic* regression function

$$Y_i = \mu_i + \varepsilon_i = \frac{\alpha}{1 + \beta \exp(-\gamma x_i)} + \varepsilon_i \quad \alpha > 0, \beta > 0, \gamma > 0.$$

With increasing x, this model describes an S-shaped, or "sigmoidal," curve described by early rapid growth, followed by a more mature period of slower growth, and concluding with growth that is bounded by some limit that is ultimately attainable. The starting value for μ at $x = 0$ is $\alpha/(1 + \beta)$; the limiting value as x approaches ∞ is α. The parameters α and β determine the initial and limiting values. Once these values are set, the parameter γ determines the shape of the function.

The derivatives of $S(\alpha, \beta, \gamma) = \sum_{i=1}^{n}\left[y_i - \dfrac{\alpha}{1 + \beta \exp(-\gamma x_i)}\right]^2$ with respect to the parameters (α, β, γ) are no longer linear in the parameters; hence, it is not possible to obtain explicit expressions of the least-squares estimates. However, the estimates can always be obtained by evaluating the sum of squares $S(\alpha, \beta, \gamma)$ over a grid of values and locating the values (α, β, γ) for which the sum of squares is a minimum. In practice, this tedious grid-search approach is not needed, as statisticians and numerical mathematicians have developed iterative procedures that lead to a solution quickly, usually in a few iterations. This area of statistics is known as *nonlinear least-squares estimation,* and computer software is available for performing its operations—not in Minitab, but in other programs, such as SPSS, SAS, and R. Nonlinear least-squares estimation goes beyond an introduction to regression; hence, we will not pursue it further. For more information, consult texts on regression, such as Abraham and Ledolter, *Introduction to Regression Modeling* (Belmont, CA: Thomas Brooks/Cole, 2006.)

Exercises 8.5

*8.5-1 Ten randomly selected rats were fed various doses of a dietary supplement (x, in grams). The growth rates (y, in coded units) of the rats are given in the following table:

Observation Number	Amount of Supplement, x	Growth Rate, y
1	10	73
2	10	78
3	15	85

Observation Number	Amount of Supplement, x	Growth Rate, y
4	20	90
5	20	91
6	25	87
7	25	86
8	25	91
9	30	75
10	35	65

*(a) A simple linear regression (i.e., $Y = \beta_0 + \beta_1 x + \varepsilon$) was fitted to the data. Calculate the ANOVA table and test whether $\beta_1 = 0$. Check the adequacy of the fitted model.

*(b) The experimenter also considered a second-order model $Y = \beta_0 + \beta_1 x + \beta_2 x^2 + \varepsilon$. Using available computer software, calculate the least-squares estimates and test $\beta_2 = 0$ at the $\alpha = 0.05$ significance level.

(c) How can a residual analysis of the model in part (a) indicate whether the simple linear regression model is adequate?

*8.5-2 The cloud point of a liquid is a measure of the degree of crystallization in a stock that can be measured by the refractive index. It has been suggested that the percentage of I-8 in the base stock is an excellent predictor of cloud point. The data in the following table were collected on stocks with known percentages of I-8.

Percent I-8, x	Cloud Point, y	Percent I-8, x	Cloud Point, y
0	22.1	0	21.9
1	24.5	2	26.1
2	26.0	4	28.5
3	26.8	6	30.3
4	28.2	8	31.5
5	28.9	10	33.1
6	30.0	0	22.8
7	30.4	3	27.3
8	31.4	6	29.8
		9	31.8

(a) Make a scatter plot of the data.

*(b) Consider the linear model $Y = \beta_0 + \beta_1 x + \varepsilon$. Calculate the least-squares estimates, the ANOVA table, and R^2. Conduct appropriate diagnostic checks and assess the adequacy of this first-order model.

*(c) Consider the second-order model $Y = \beta_0 + \beta_1 x + \beta_2 x^2 + \varepsilon$. Repeat the analysis of part (b). Which model would you prefer?

[N. R. Draper and H. Smith, *Applied Regression Analysis,* 2d ed. (New York: Wiley, 1981), p. 283.]

*8.5-3 An experimenter performed a 2^2 factorial experiment to study the effects of temperature and concentration on the yield of a chemical reaction. The experimenter conducted two runs at each of the four possible combinations of temperature and concentration, for a total of eight runs. The levels for temperature were 120 and 140°F, and the levels for concentration were 15 and 25 percent. The levels are denoted by ± 1. The order of the eight experiments was randomized. The results are as follows:

Temperature, x_1	Concentration, x_2	Yield, y
−1	−1	58
−1	−1	63
+1	−1	75
+1	−1	85
−1	+1	65
−1	+1	59
+1	+1	88
+1	+1	94

Consider the regression model $Y = \beta_0 + \beta_1 x_1 + \beta_2 x_2 + \beta_3 x_1 x_2 + \varepsilon$.

*(a) Calculate the least-squares estimates, the ANOVA table, and R^2.

*(b) Test for overall significance of the regression; that is, test $H_0: \beta_1 = \beta_2 = \beta_3 = 0$ against H_1: at least one regression coefficient is different from zero. Consider partial t-tests and interpret your findings.

8.5-4 A certain response Y is measured at 2-minute intervals. It is established that the response variable increases linearly with time x_1. At time zero, a "treatment" is applied, and it is hypothesized that this causes the response curve to shift by a constant, but positive, amount. Use an indicator variable x_2, and formulate an appropriate regression model. Estimate the parameters from the following data:

x_1	x_2	y
−5	0	5
−3	0	7
−1	0	10
1	1	14
3	1	17
5	1	18

Take $\alpha = 0.10$. Is there significant evidence that such a shift has occurred?

*8.5-5 Snedecor and Cochran have described an investigation of the source from which corn plants in various Iowa soils obtain their phosphorus. The concentrations (in parts per million) of inorganic (x_1) and organic (x_2) phosphorus in the soils were determined chemically. The phosphorus content (y, in ppm) of corn plants grown in these soils was also measured. The results appear in the following table:

Soil Sample	x_1	x_2	y
1	0.4	53	64
2	0.4	23	60
3	3.1	19	71
4	0.6	34	61
5	4.7	24	54
6	1.7	65	77

Soil Sample	x_1	x_2	y
7	9.4	44	81
8	10.1	31	93
9	11.6	29	93
10	12.6	58	51
11	10.9	37	76
12	23.1	46	96
13	23.1	50	77
14	21.6	44	93
15	23.1	56	95
16	1.9	36	54
17	26.8	58	168
18	29.9	51	99

*(a) Estimate the multiple regression model $Y_i = \beta_0 + \beta_1 x_{i1} + \beta_2 x_{i2} + \varepsilon_i$.

*(b) Check the adequacy of the fitted model. Carry out the appropriate residual plots. In particular, check whether there are outliers in the data.

Note: An outlier is a point that does not follow the general patterns exhibited by the other observations in the data set. A residual that is more than $3\sqrt{MSE}$ from zero is usually a good indication that the corresponding observation is different from the rest of the observations. If an outlier is indicated, the experimenter should scrutinize this point and check whether the observation is due to a recording error or to some other special circumstance that is not captured in the data (variables) recorded. In the absence of such specific information, the outlying data point may be omitted and the model reestimated.

*(c) Calculate the ANOVA table and R^2, and test the overall significance of the regression model. Use $\alpha = 0.05$.

*(d) Calculate the standard errors of the estimates and the t-ratios. Test the significance of the individual regression coefficients. Use $\alpha = 0.05$. Can you simplify the model? If so, estimate the new model and interpret your findings.

[G. W. Snedecor and W. G. Cochran, *Statistical Methods,* 6th ed. (Ames, IA: Iowa State University Press, 1967).]

*8.5-6 Business schools are interested in predicting the GPA (y) of MBA students as a function of their GMAT test scores (x_1) and their undergraduate grade point average (x_2). The data for a sample of 12 students are as follows:

x_1	x_2	y
560	3.32	3.20
540	3.23	3.44
520	2.97	3.70
580	2.79	3.10
520	3.35	3.00
620	2.90	4.00
660	3.52	3.38
630	3.54	3.83

x_1	x_2	y
550	2.64	2.67
550	3.06	2.75
600	2.41	2.33
537	3.44	3.75

*(a) Estimate the regression model $Y = \beta_0 + \beta_1 x + \beta_2 x_2 + \varepsilon$, test the significance of the regression coefficients and draw conclusions.

(b) Discuss why GMAT test scores and undergraduate grade point averages may not appear to be such important predictor variables.

Hint: Consider the limited range of the explanatory variables, the fact that the experiment is not designed, the fact that other important predictor variables may be missing, and so on.

8.5-7 T. Kyotani et al. have studied the variation in the traction coefficients of 25 lubricating oils. Traction measurements were obtained on a special two-disk machine. After oil was applied to the cleaned faces of two disks, the disks were pressed into contact by a spring. A certain load was applied and the rotation speed of one disk was increased from 44 centimeters per second (cm/s) to 52 cm/s, while the speed of the other disk was kept constant at 45 cm/s. The effect of the sliding speed on the traction coefficient was studied, and the maximum traction coefficients for the 25 lubricating oils were recorded in the following table (loads of 100 kg and 150 kg were used in the experiment):

	Maximum Traction Coefficient $f_{max} \times 10^3$		
Fluid	**Load (100 kg)**	**Load (150 kg)**	**Flow Activation Volume, FAV (cm³/mol)**
1	66	74	51.8
2	65	68	50.7
3	65	70	51.0
4	59	60	40.1
5	54	56	40.4
6	58	58	39.5
7	63	63	39.5
8	57	57	39.7
9	83	77	76.1
10	74	75	61.5
11	91	93	67.8
12	92	93	92.2
13	90	88	80.5
14	88	87	87.8
15	88	89	88.7
16	91	93	91.5
17	87	88	94.4
18	69	75	57.2
19	67	68	53.0

Fluid	Maximum Traction Coefficient $f_{max} \times 10^3$		Flow Activation Volume, FAV (cm^3/mol)
	Load (100 kg)	Load (150 kg)	
20	69	72	53.9
21	66	66	57.6
22	76	75	58.7
23	76	74	73.5
24	76	78	61.2
25	85	83	71.9

It is thought that the traction coefficients increase with the rigidity of the molecular structure of these different lubricating fluids. The flow activation volume (FAV) is taken as a measure of rigidity. It measures the average size of holes into which the flow segments are required to move. Because flexible molecules can separate into small flow segments with ease, most of these molecules have small values of FAV. Conversely, rigid molecules have large values of FAV.

(a) Consider the traction coefficients at a load of 100 kg. Plot the traction coefficients against the flow activation volumes. Develop a regression model that explains the variability in the traction coefficients.

(b) Repeat this exercise for traction coefficients at a load of 150 kg.

(c) Use the graphs and discuss whether the two fitted regression lines are the same.

[T. Kyotani, H. Yoshitake, T. Ito, and Y. Tamai, "Correlation Between Flow Properties and Traction of Lubricating Oils," *Transactions of the American Society of Lubricating Engineers*, 29: 102–106 (1986).]

8.5-8 The mean rate of oxygenation from the atmospheric reaeration process for a stream depends on the mean velocity of the stream flow and the average depth of the stream bed. The data from 12 experiments are as follows:

Mean Velocity, x_1 (ft/sec)	Mean Depth, x_2 (ft)	Mean Oxygenation Rate, y (ppm per day)
3.69	5.09	1.44
3.07	3.27	2.27
2.10	4.42	0.98
2.68	6.14	0.50
2.78	5.66	0.74
2.64	7.17	1.13
2.92	11.41	0.28
2.47	2.12	3.36
3.44	2.93	2.79
4.65	4.54	1.57
2.94	9.50	0.46
2.51	6.29	0.39

Consider the following two models:

(1) $E(Y) = \beta_0 + \beta_1 x_1 + \beta_2 x_2;$

(2) $E(Y) = \beta_0 x_1^{\beta_1} x_2^{\beta_2}.$

Determine the least-squares estimates of the coefficients and discuss the adequacy of these two models. Which model would you prefer?

Hint: Consider a logarithmic transformation for model (2).

8.5-9 Indicator variables are variables that denote the absence and presence of a certain factor. Indicator variables take on the values 0 and 1 (or -1 and $+1$, as in the example in Section 8.5-2). Assume that the response variable Y is linearly related to an explanatory variable x_*. At a certain known threshold, say, x_*, the slope coefficient changes as shown in the following graph:

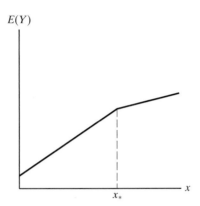

Explain how you would set up the regression model to test the significance of the change in slope.

8.5-10 William J. Hill and Robert A. Wiles ["Plant Experimentation (PLEX)," *Journal of Quality Technology*, 7: 115–122 (1975)] describe an experiment on a large continuous chemical operation consisting of several separate reaction steps. The variable of interest in their investigation is percent yield. The factors believed to affect yield are the concentration of reactant A in solvent S (x_1), the ratio of reactant B to reactant A (x_2), and the temperature in the reactor (x_3). The coded variables are given in the following table (note that Hill and Wiles do not tell us the original units; the 15 runs in this experiment were performed in random order over a period of 10 days, with 8 hours at each operating condition):

x_1	x_2	x_3	**Percent yield**
-1	-1	-1	75.4
1	-1	-1	73.9
-1	1	-1	76.8
1	1	-1	72.8
-1	-1	1	75.3 75.3
1	-1	1	71.4
-1	1	1	76.5 77.2
1	1	1	72.3
0	0	0	74.4 74.5
-2	0	0	79.0 78.4
2	0	0	69.2

Note that there are replicates at some of the operating conditions. In reading your data into a computer program, you have to enter them as $(-2, 0, 0, 79.0)$, $(-2, 0, 0, 78.4)$, and so on.

Develop a model that relates yield to these three explanatory variables. Consider quadratic terms and interactions if necessary. Carefully check the adequacy of your fitted model and discuss its implications. For example, what does it mean if the regression coefficients associated with a variable are found insignificant?

8.5-11 S. Weisberg [*Applied Linear Regression*, 2d ed. (New York: Wiley, 1985)] describes the results of an experiment that evaluates the performance of a certain cutting-tool material in cutting steel on a lathe. The aim of the experiment is to understand how tool life (measured in minutes) varies with cutting speed (S, in feet per minute) and feed rate (F, in thousandths of an inch per revolution). The following table lists the results of the experiment:

S	F	Life							
-1	-1	54.5	66.0						
1	-1	11.8	14.0						
-1	1	5.2	3.0						
1	1	0.8	0.5						
0	$-\sqrt{2}$	86.5							
0	$\sqrt{2}$	0.4							
$-\sqrt{2}$	0	20.1							
$\sqrt{2}$	0	2.9							
0	0	3.8	2.2	3.2	4.0	2.8	3.2	4.0	3.5

The levels of the two factors are coded and centered to give $S = $ (speed $-$ 900)/300 and $F = $ (feed rate $- 13)/6$. The order of the runs was randomized.

Analyze the information. Use graphical procedures as well as analytic methods. You may want to consider transformations of the response variable.

Are there certain observations that you would question? If there are, repeat your analysis without those observations. (Of course, the analyst should always discuss any unusual observations with the scientist who carried out the experiment. Sometimes, unusual observations arise because something went wrong in the experiment. But often, these unusual observations help us make our models more realistic, pointing to factors that we previously ignored.)

8.5-12 You are fitting a regression model with two regressors: x_1 and x_2. The resulting R^2 is very high (indicating that x_1 or x_2 (or both) is useful in predicting y), but the two partial t-tests for the regression coefficients show that neither x_1 nor x_2 is significant. Does this mean that both x_1 and x_2 can be omitted from the model? Note that such a decision would leave a model with just an intercept, contradicting the high R^2 you found in the model with both x_1 and x_2. Explain this apparent contradiction.

*8.5-13 Consider a multiple-regression model of price of house (y) on three explanatory variables: taxes paid (x_1), number of bathrooms (x_2), and square feet (x_3). The incomplete Minitab output from a regression on $n = 28$ houses is as follows:

The regression equation is price $= -10.7 + 0.190$ taxes $+ 81.9$ baths $+ 0.101$ sqft

Predictor	Coef	SE Coef	T	P
Constant	−10.65	24.02		
taxes	0.18966	0.05623		
baths	81.87	47.82		
sqft	0.10063	0.03125		

Analysis of Variance

Source	DF	SS	MS	F	P
Regression	3	504541			
Residual Error					
Total	27	541119			

*(a) Calculate the multiple coefficient of determination, R^2.

*(b) Test the null hypothesis that all three regression coefficients are zero $(H_0: \beta_1 = \beta_2 = \beta_3 = 0)$. Use a significance level of 0.05.

*(c) Obtain a 95% confidence interval of the regression coefficient for "taxes." Can you simplify the model by dropping "taxes" from the model? Obtain a 95% confidence interval of the regression coefficient for "baths." Can you simplify the model by dropping "baths" from the model?

*8.5-14 (Continuation of Exercise 8.5-13.) The incomplete Minitab output from a multiple regression of price of house on the two explanatory variables of taxes paid and square feet is as follows:

The regression equation is price $= 4.9 + 0.242$ taxes $+ 0.134$ sqft

Predictor	Coef	SE Coef	T	P
Constant	4.89	23.08		
taxes	0.24237	0.04884		
sqft	0.13397	0.02537		

Analysis of Variance

Source	DF	SS	MS	F	P
Regression	2	500074	250037		
Residual Error					
Total		541119			

*(a) Calculate the multiple coefficient of determination, R^2.

*(b) Test the null hypothesis that both regression coefficients are zero $(H_0: \beta_1 = \beta_2 = 0)$. Use a significance level of 0.05.

*(c) Test whether you can omit the variable "taxes" from the regression model. Use a significance level of 0.05.

*(d) Comment on the fact that the regression coefficients for taxes and square feet are different from those shown in Exercise 8.5-13.

*8.5-15 The data in this exercise are taken from Ryan, T.A., Joiner, B.L., and Ryan, B.F., *The Minitab Student Handbook* (Boston: Duxbury Press, 1985). Measurements of the volume (cubic feet), height (feet), and diameter at breast height (inches, measured at 54 inches above the ground) of 31 black cherry trees in the Allegheny National Forest in Pennsylvania are listed in the following table:

x_1 = Diameter	x_2 = Height	y = Volume
8.3	70	10.3
8.6	65	10.3
8.8	63	10.2
10.5	72	16.4
10.7	81	18.8
10.8	83	19.7
11.0	66	15.6
11.0	75	18.2
11.1	80	22.6
11.2	75	19.9
11.3	79	24.2
11.4	76	21.0
11.4	76	21.4
11.7	69	21.3
12.0	75	19.1
12.9	74	22.2
12.9	85	33.8
13.3	86	27.4
13.7	71	25.7
13.8	64	24.9
14.0	78	34.5
14.2	80	31.7
14.5	74	36.3
16.0	72	38.3
16.3	77	42.6
17.3	81	55.4
17.5	82	55.7
17.9	80	58.3
18.0	80	51.5
18.0	80	51.0
20.6	87	77.0

Develop a model that relates the volume of a tree to its diameter and height. A study of the volume of a (tapered) cylinder will suggest an appropriate model. Construct scatter plots of volume against diameter and height, considering appropriate transformations. Fit appropriate regression models, obtain and interpret the estimates of the coefficients, calculate ANOVA tables, and discuss the adequacy of the fitted models. Consider appropriate residual diagnostics. Use your models to obtain a 95 percent confidence interval for the mean volume of a tree with diameter 11 inches and height 70 feet. Discuss whether it is also reasonable to obtain a confidence interval for the mean volume of a tree with diameter 11 inches and height 95 feet.

8.5-16 The following table gives the residual sums of squares $\sum_{i=1}^{20}(y_i - \hat{y}_i)^2$ for the 16 possible regression models in a regression of monthly sales y (in thousands of cases) on four possible predictor variables ($n = 20$ months are available for the analysis):

Model Number	Variables Included	$SSE = \sum_{i=1}^{20}(y_i - \hat{y}_i)^2$
1	None (mean only)	19.9961
2	X1	19.7956
3	X2	19.8947
4	X3	19.8966
5	X4	19.9961
6	X1,X2	19.6964
7	X1,X3	19.6977
8	X1,X4	19.6986
9	X2,X3	0.2034
10	X2,X4	19.6995
11	X3,X4	19.6957
12	X1,X2,X3	0.2022
13	X1,X2,X4	0.1101
14	X1,X3,X4	19.6079
15	X2,X3,X4	0.2024
16	X1,X2,X3,X4	0.0671

The predictor variables consist of two different economic indicators (X1 and X4), the dollar value of sales promotions (X2), and a measure of sales force activity (X3). All models include an intercept. Note that there is a very high degree of multicollinearity among the explanatory variables. What model would you select?

8.5-17 You are given pairs of observations (x_i, y_i) with $x_i > 0$. Consider the nonlinear regression model $Y = \dfrac{\beta}{1 + \beta x} + \varepsilon$. Assume that $\beta > 0$ and sketch the model.

8.6 Response Surface Methods

Often, experimenters wish to find the conditions under which a certain process attains its optimal results. That is, they want to determine the levels of the design (explanatory) variables at which the response reaches its optimum. This optimum could be either a maximum or a minimum of a function of the design variables. For example, consider a chemical process whose yield is a function of temperature and pressure; it is of interest to determine the levels of temperature and pressure that lead to the largest possible yield. Or consider a biologist who wants to maximize the growth of an organism as a function of the percentage of glucose, the concentration of yeast extract, and the time allowed for the organism to grow. Or a production engineer may be trying to maximize the seal strength of a certain plastic wrap by varying the sealing temperature and the percentage of a polyethylene additive. Methods for determining the optimum conditions and for exploring the response surface in the neighborhood of these "best" conditions are discussed in books on *response surface methods*. Here we give only a brief introduction to the topic.

We start our discussion by emphasizing that the "change one variable at a time" approach is an inefficient and usually unsuccessful method for determining the optimum conditions. A much better strategy results by changing the variables simultaneously in a systematic way.

Experimentation is often started in a region that may be far from optimal. However, strategies for reaching the optimum region through successive iterative experimentation have been developed. A good approach consists of first performing a factorial experiment in the initial region, approximating the unknown response surface by a linear function of the design variables, and determining the path of steepest ascent. Once the neighborhood of the optimum is reached, we design a more elaborate experiment and estimate second-order models that include quadratic and interaction components. We discuss briefly two such designs: the 3^k factorial and the central composite design. We also illustrate how the optimum conditions, as well as the nature of the response surface in their vicinity, can be determined from the estimated second-order model.

8.6-1 The "Change One Variable at a Time" Approach

Suppose that a chemist wants to maximize the yield of a chemical reaction by varying the reaction time (T) and the reaction pressure (P). Suppose also that the contours of the true response function, which is unknown to the chemist, are as given in Figure 8.6-1. The curves on such a *contour plot* connect points that lead to the same yield.

In the "change one variable at a time" approach, the experimenter fixes a certain level for one of the factors. Assume that he starts with a reaction temperature of 220°C. He then conducts experiments for different levels of pressure, say, at 80, 90, 100, 110, and 120 pounds per square inch (psi). Results are obtained at those five experimental points and are plotted in Figure 8.6-2(a). With the temperature fixed at 220°C, the experimenter finds that the maximum yield is obtained at a pressure of about 100 psi. In the next step, the pressure is fixed at this "best" value and the temperature is varied. With the pressure fixed at 100 psi, results are obtained at temperatures of 180, 200, 240, and 260°C, as plotted in Figure 8.6-2(b). The experimenter finds that for this "best" value of pressure (where "best" is to be understood in the sense that this particular pressure has led to the highest response when the temperature is fixed at 220°C), the "best" value of temperature is not far from the temperature of 220°C (possibly a little higher) that was used in the first set of runs. Thus, it might seem justified to conclude that the overall maximum is achieved with a temperature of 220°C (or somewhat higher) and a pressure of 100 psi.

However, this reasoning is incorrect, as is shown by the contour plot in Figure 8.6-1. The maximum yield is obtained when the temperature is about 270°C and the pressure

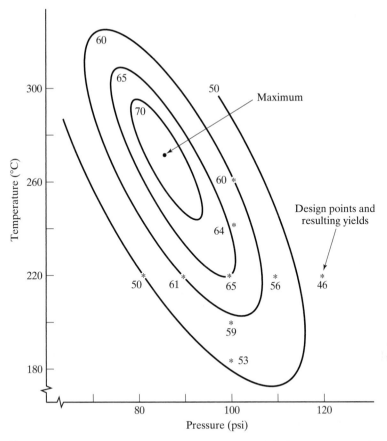

Figure 8.6-1 Contours of a response surface. Minitab's command "Graph > Contour Plot" can be used to create contour plots

is at 85 psi. Why has this approach failed? The graphs in Figure 8.6-2(a) and (b) show that if either temperature or pressure, *individually,* is changed from the operating condition (220°C, 100 psi), the yield will decrease. But Figure 8.6-1 also shows that yield will increase if we change temperature and pressure *together* and move toward the upper left-hand corner of the figure. The "change one variable at a time" approach has failed because it assumes that the effects of changes in one factor are independent of the effects of changes in the other. This, however, is usually not true. An alternative and more successful approach is to change the variables together.

8.6-2 Method of Steepest Ascent

Experimentation is often started in a region that may be far from optimal. However, a sequence of well-designed experiments should lead us toward the optimum. We may start with a 2^k factorial, possibly with several center points added. Experience from previous experiments may help us in selecting the initial levels of the factors. Once those are chosen, we can express the design points in coded form by ($x_1 = \pm 1, x_2 = \pm 1, \ldots, x_k = \pm 1$), with ($x_1 = 0, x_2 = 0, \ldots, x_k = 0$) for the center point.

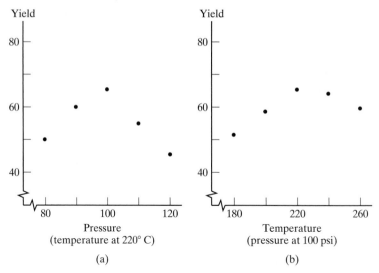

Figure 8.6-2 "Change one variable at a time" approach: (a) plot of yield against pressure, at a fixed temperature of 220°C; (b) plot of yield against temperature, at a fixed pressure of 100 psi

Initially, the response surface is approximated by a first-order (or linear) model, and the results from this initial experiment are used to calculate least-squares estimates of the coefficients in $Y = \beta_0 + \beta_1 x_1 + \cdots + \beta_k x_k + \varepsilon$. Even though the linear model may not be an adequate description of the relationship, it usually provides a good starting point for our analysis.

The objective is now to find the path of *steepest ascent* that leads us to the optimum as quickly as possible. We want to locate the point, among all points that are a fixed distance (say, R units) from the center $(0, 0, \ldots, 0)$, that gives the largest response. In mathematical terms, this involves the maximization of $\hat{\beta}_0 + \hat{\beta}_1 x_1 + \cdots + \hat{\beta}_k x_k$, subject to the constraint $\sum_{i=1}^{k} x_i^2 = R^2$, where $\hat{\beta}_0, \hat{\beta}_1, \ldots, \hat{\beta}_k$ are the estimates of $\beta_0, \beta_1, \ldots, \beta_k$. It can be shown (see Exercise 8.6-8) that the maximum is achieved at the point with coordinates

$$x_i = \frac{R\hat{\beta}_i}{\left(\sum_{j=1}^{k}\hat{\beta}_j^2\right)^{1/2}}, \quad i = 1, \ldots, k.$$

Thus, the path of steepest ascent from the center $(0, 0, \ldots, 0)$ is in the direction of the point

$$(x_1, \ldots, x_k) = \left(\frac{R\hat{\beta}_1}{\left(\sum_{j=1}^{k}\hat{\beta}_j^2\right)^{1/2}}, \ldots, \frac{R\hat{\beta}_k}{\left(\sum_{j=1}^{k}\hat{\beta}_j^2\right)^{1/2}}\right).$$

Because the coded design points of the original 2^k factorial are one unit from the origin, and because we should not trust extrapolations that go much beyond the experimental region, it is reasonable to generate points on this path by taking $R = 1, 2, 3, \ldots$ and by conducting new experiments at these levels. To illustrate, assume that this is done and that we find that the responses along this path reach their maximum for a certain R. Then the design point corresponding to this R is taken as the center point for a new and usually more elaborate experiment.

A Remark on Center Points in 2^k Factorial Experiments: Observations at the center point in a 2^k factorial experiment have no effect on the slope estimates $\hat{\beta}_1, \ldots, \hat{\beta}_k$ in the linear model $Y = \beta_0 + \beta_1 x_1 + \cdots + \beta_k x_k + \varepsilon$, nor do they affect the path of steepest ascent. This is easy to see from the normal equations in Section 8.4-1: Observations at the center point with $x_{ij} = 0$ do not contribute to the sums in those equations.

   Why, then, would we want to run experiments at the center point in a 2^k factorial? There are several reasons for doing so. One is that replications at the center point give us an estimate of the variability among individual observations, and this information can be used to obtain standard errors for the estimates. A second reason is that observations at the center point can tell us about the curvature of the surface. If there is no curvature, and if the linear model is appropriate over the region of our initial experiment, the average of the n_1 observations at the center point, say, $\overline{Y}_1$, and the average of the 2^k observations in the factorial, say, $\overline{Y}_2$, are both estimates of β_0, and they should be roughly equal. A nonzero difference of these two averages indicates that there is curvature. The sample variance s^2, calculated from the n_1 observations at the center point, can be used to obtain a standard error for this difference. We have $\mathrm{var}(\overline{Y}_1) \approx s^2/n_1$ and $\mathrm{var}(\overline{Y}_2) \approx s^2/2^k$, so $\mathrm{var}(\overline{Y}_1 - \overline{Y}_2) \approx s^2[(1/n_1) + (1/2^k)]$. The standardized difference $(\overline{Y}_1 - \overline{Y}_2)/\sqrt{\mathrm{var}(\overline{Y}_1 - \overline{Y}_2)}$ can be used to assess the statistical significance of the observed difference of the two averages.

   Why is it so important to know that, over a certain region, there is curvature in the surface? The reason is that we base our path of steepest ascent on a linear approximation of the surface. If there is sizeable curvature, our path may be poor and we may find that, on this path, responses will soon start to decrease. In such a situation, it is important to perform another simple experiment and recalculate the direction of the path of steepest ascent.

   There is another lesson to be learned from this discussion: Experimentation must be iterative. It is important to incorporate any prior knowledge into designing the experiment at the next stage. It is usually inefficient to perform one big experiment that uses up most of the allocated resources, because such a strategy does not allow for subsequent learning. This is the reason many experimenters recommend spending only 25 or 30 percent of one's budget on the initial stage of an experiment.

8.6-3 Designs for Fitting Second-Order Models: The 3^k Factorial and the Central Composite Design

First-order models do not lead to surfaces that describe maxima or minima. For that, one needs at least a second-order model. The 2^k factorial cannot be used to estimate the coefficients in the second-order model

$$Y = \beta_0 + \sum_{i=1}^{k}\beta_i x_i + \sum_{i=1}^{k}\beta_{ii} x_i^2 + \sum\sum_{i<j}\beta_{ij} x_i x_j + \varepsilon$$

Table 8.6-1 The 3^2 Factorial

x_1	x_2
-1	-1
0	-1
$+1$	-1
-1	0
0	0
$+1$	0
-1	$+1$
0	$+1$
$+1$	$+1$

because we need at least three different levels for each factor in order to estimate a second-order function. One can use a 3^k factorial experiment, where each factor is observed at three levels: a low (-1), an intermediate (0), and a high $(+1)$ level. For $k = 2$ factors, this involves the nine design points given in Table 8.6-1.

One problem with 3^k factorials is that the number of runs increases very rapidly with k. Thus, other arrangements, such as the *central composite design,* have been developed. In the central composite design, we start with the ordinary 2^k factorial and add $2k$ axial or "star" points, $(\pm w, 0, \ldots, 0), (0, \pm w, 0, \ldots, 0), \ldots, (0, \ldots, 0, \pm w)$. In addition, several (say, n_1) observations are taken at the center point $(0, 0, \ldots, 0)$. This leads to a total of $n = 2^k + 2k + n_1$ runs, a number that is usually much smaller than 3^k. Furthermore, if $w \neq 1$, each variable is actually measured at five different levels, namely, $-w, -1, 0, 1$, and w, allowing us to get a better estimate of the curvature.

The central composite design for $k = 2$ factors is given in Table 8.6-2. The value of w is selected by the experimenter. Various criteria can be used in making this

Table 8.6-2 Central Composite Design for $k = 2$ Factors

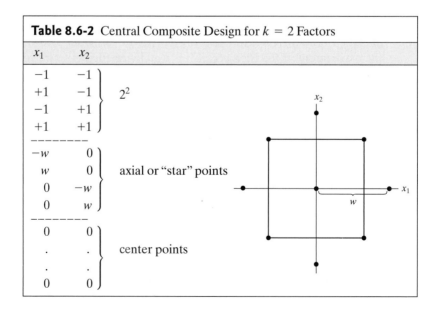

x_1	x_2	
-1	-1	
$+1$	-1	2^2
-1	$+1$	
$+1$	$+1$	
$-w$	0	
w	0	axial or "star" points
0	$-w$	
0	w	
0	0	
$\cdot$	$\cdot$	center points
$\cdot$	$\cdot$	
0	0	

Table 8.6-3 Central Composite Design for $k = 3$ Factors

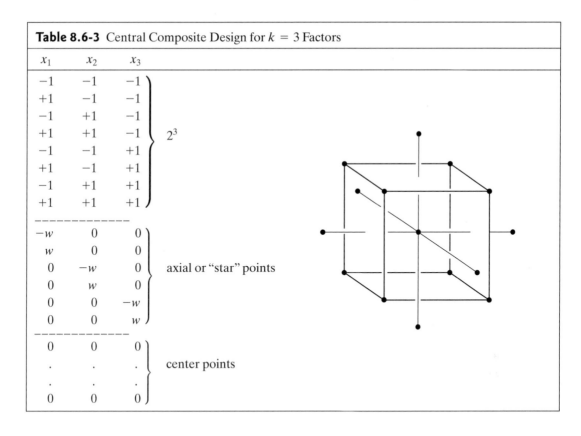

x_1	x_2	x_3	
-1	-1	-1	
$+1$	-1	-1	
-1	$+1$	-1	
$+1$	$+1$	-1	2^3
-1	-1	$+1$	
$+1$	-1	$+1$	
-1	$+1$	$+1$	
$+1$	$+1$	$+1$	
$-w$	0	0	
w	0	0	
0	$-w$	0	axial or "star" points
0	w	0	
0	0	$-w$	
0	0	w	
0	0	0	
$\cdot$	$\cdot$	$\cdot$	center points
$\cdot$	$\cdot$	$\cdot$	
0	0	0	

choice. For example, we can choose $w = \sqrt{2}$. Then the design is *rotatable*, because, geometrically, all design points lie on a circle of radius $\sqrt{2}$.

The central composite design for $k = 3$ factors is given in Table 8.6-3. Frequently, with $k = 3$, w is chosen to be $w = 2^{3/4} = 1.682$. Note that with one center point ($n_1 = 1$) this design consists of only 15 runs, while the 3^3 factorial requires 27. For more than three variables, the advantage, in terms of the number of runs, of the central composite design over the factorial arrangement with three levels increases even more.

8.6-4 Interpretation of the Second-Order Model

A second-order model with $k = 2$ variables is given by

$$E(Y) = \beta_0 + \beta_1 x_1 + \beta_2 x_2 + \beta_{11} x_1^2 + \beta_{22} x_2^2 + \beta_{12} x_1 x_2;$$

depending on the values of its coefficients, this model can describe several different response surfaces. The three most common ones are those with a maximum, those with a minimum, and those with a saddle point. Contour plots of these three types of surfaces are shown in Figure 8.6-3. Exercise 8.6-1 gives several examples of second-order models that lead to such surfaces.

For most readers, surfaces with a maximum (or a minimum) need very little explanation: Leaving the critical point (the point of the optimum) in any direction

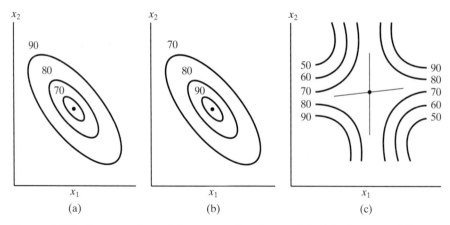

Figure 8.6-3 Contour plots of three second-order models: (a) minimum; (b) maximum; (c) saddle point

results in a decrease (or increase) in the response. However, in the case of a saddle point, the experimenter may get an increase or a decrease in the response when moving away from the critical point, depending on which direction is taken. The surface looks like a saddle—ergo its name.

To determine the critical point, which is sometimes also called the *stationary point,* we set the two first partial derivatives equal to zero:

$$\frac{dE(Y)}{dx_1} = \beta_1 + 2\beta_{11}x_1 + \beta_{12}x_2 = 0,$$

$$\frac{dE(Y)}{dx_2} = \beta_2 + 2\beta_{22}x_2 + \beta_{12}x_1 = 0.$$

This leads to the stationary point

$$x_{1,0} = \frac{\beta_{12}\beta_2 - 2\beta_{22}\beta_1}{4\beta_{11}\beta_{22} - \beta_{12}^2},$$

$$x_{2,0} = \frac{\beta_{12}\beta_1 - 2\beta_{11}\beta_2}{4\beta_{11}\beta_{22} - \beta_{12}^2}.$$

Note that if the denominator in these expressions is zero, the solution corresponds to what is called a *ridge.* (See Exercise 8.6-2.)

To determine the nature of the surface at the stationary point, one has to investigate the second derivatives:

$$\frac{d^2E(Y)}{dx_1^2} = 2\beta_{11}, \qquad \frac{d^2E(Y)}{dx_1dx_2} = \beta_{12}, \qquad \frac{d^2E(Y)}{dx_2^2} = 2\beta_{22}.$$

If the two solutions of the quadratic equation $(2\beta_{11} - \lambda)(2\beta_{22} - \lambda) - \beta_{12}^2 = 0$—say, λ_1 and λ_2—are both negative, the function has a maximum at the stationary point $(x_{1,0}, x_{2,0})$. If they are both positive, the function has a minimum at the stationary point. If the two solutions are of different signs, the function has a saddle point at the stationary point.

Now, various analytic methods can be used to investigate the nature of the response surface. For example, in the case of a saddle point, these methods indicate the direction in which we must move to increase the response. In the case of a maximum, they show in which direction the decrease in the response is slowest. This information is important, because it tells the experimenter about the direction in which the response is least sensitive to changes in the input factors. We do not discuss these methods here, as they are beyond the scope of an introductory book. Also, with only $k = 2$ factors, we can always plot the contours of the estimated second-order surface and make our assessments from the graph. Computer programs for contour plotting are a part of most statistical packages. The Minitab command "Graph > Contour Plot" can be used.

8.6-5 An Illustration

An experimenter attempts to gain insight into the influence of the sealing temperature (T) and the percentage of a polyethylene additive (P) on the seal strength $(Y,$ measured in grams per centimeter) of a certain plastic wrap.

Assume that, *unknown to our experimenter,* the response function is given by

$$E(Y) = -20 + (0.85)T + (1.5)P - (0.0025)T^2 - (0.375)P^2 + (0.025)TP.$$

The contours of this response surface are shown in Figure 8.6-4. Its maximum is at temperature $T = 216°C$ and percentage of polyethylene additive $P = 9.2$ percent. Of course, due to the uncontrolled variability in raw materials, laboratory conditions, measurement errors, and so on, the actual results of the experiment will vary around these expected values.

The experimenter starts his investigation at a temperature of $140°C$ and an additive of 4 percent. This combination was suggested by a colleague who had experimented with a similar additive. The experimenter performs a factorial experiment around this center point. In his experiment, a unit change in temperature corresponds to $20°C$ and a unit change in P corresponds to 2 percentage points.

The coded levels are obtained as

$$x_1 = \frac{T - 140}{20} \quad \text{and} \quad x_2 = \frac{P - 4}{2}.$$

The results of the 2^2 factorial with two center points are given in Table 8.6-4. The observations were generated by adding small random errors to the mean response $E(Y)$. From these results, the experimenter fits the linear model and obtains the least-squares equation $\hat{y} = 62 + 5x_1 + 4x_2$. The results in Section 8.6-2 are used and the path of steepest ascent is obtained; it is described by the points on $(5R/\sqrt{41}, 4R/\sqrt{41})$, or equivalently, $(1.25R_1, R_1)$.

Experiments on this path are conducted when $R_1 = 1, 2, \ldots, 5$, and the results are given in Table 8.6-5. The responses on this path of steepest ascent seem to reach their maximum at about $(x_1 = 3.75$ and $x_2 = 3)$ or, in terms of uncoded variables, at $T = 215°C$ and $P = 10$ percent. This point is then taken as the center point of a central composite design. Choosing the rotatable design with $w = \sqrt{2}$, our experimenter conducts the nine runs listed in Table 8.6-6: The first four points are those of

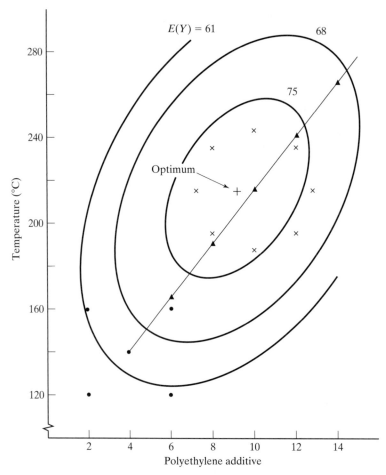

Figure 8.6-4 Response surface and design points for the example in Section 8.6-5. The symbol • denotes the design points of the initial 2^2 factorial design with one center point; ▲ represents the design points on the path of steepest ascent; × designates the design points of the central composite design

Table 8.6-4 Results from the 2^2 Factorial with Two Center Points

T	P	Coded T	Coded P	Y
120	2	−1	−1	52
160	2	+1	−1	62
120	6	−1	+1	60
160	6	+1	+1	70
140	4	0	0	63
140	4	0	0	65

Table 8.6-5 Results of Experiments on
Path of Steepest Ascent

		Coded		
T	P	T	P	y
165	6	1.25	1	72
190	8	2.50	2	77
215	10	3.75	3	79
240	12	5.00	4	76
265	14	6.25	5	70

a 2^2 factorial design, the next four are axial points, and the last is a center point. Note that we kept the initial coding $x_1 = (T - 140)/20$ and $x_2 = (P - 4)/2$; the coded levels in the central composite design are expressed in relation to the original center point at $T = 140°$ C and $P = 4$ percent. A second-order model is fitted to the data from Table 8.6-6.

The estimated second-order equation is

$$\hat{y} = 40.44 + 14.57x_1 + 9.33x_2 - 2.38x_1^2 - 2.63x_2^2 + 1.25x_1x_2.$$

The stationary solution is given by

$$x_{1,0} = \frac{(1.25)(9.33) - 2(-2.63)(14.57)}{4(-2.38)(-2.63) - (1.25)^2} = 3.76,$$

$$x_{2,0} = \frac{(1.25)(14.57) - 2(-2.38)(9.33)}{4(-2.38)(-2.63) - (1.25)^2} = 2.67.$$

These equations correspond to a temperature $T = 215°$ C and percent additive $P = 9.3$. The solutions of

$$[2(-2.38) - \lambda][2(-2.63) - \lambda] - (1.25)^2 = 0$$

Table 8.6-6 Results from the Central Composite Design

		Coded			
T	P	T	P	y	
195	8	2.75	2	78	
235	8	4.75	2	76	2^2
195	12	2.75	4	72	
235	12	4.75	4	75	
187	10	2.34	3	74	
243	10	5.16	3	76	axial points
215	7.2	3.75	1.59	77	
215	12.8	3.75	4.41	72	
215	10	3.75	3	80	center point

are both negative, so the stationary point is indeed a maximum. It is very close to the maximum of $E(Y)$, which was at $T = 216$ and $P = 9.2$, but of course, this fact is unknown to the experimenter. Thus, despite the experiment's having started in a region that was not optimal, a sequence of iterative experiments led to the optimum fairly quickly. Of course, the variability of the error component around the expected response surface will influence the variability in the direction of the path of steepest ascent, which, in turn, may affect how quickly the optimum is reached.

In this example, the variability is rather small. We can see this from the replications at the design point $T = 140°$, $P = 4$ percent (leading to responses 63 and 65) and at the design point $T = 215°$, $P = 10$ percent (leading to responses 79 and 80). The variance of an observation is estimated by pooling the two variances: $s^2 = \{[(63 - 64)^2 + (65 - 64)^2] + [(79 - 79.5)^2 + (80 - 79.5)^2]\}/2 = 1.25$. The standard deviation $s = 1.12$ is rather small, considering that the observations from the initial 2^2 factorial experiment range from 52 to 70 and those from the central composite design from 72 to 80. Hence, it is not surprising that we reach the optimum rather quickly.

Exercises 8.6

*8.6-1   Consider the following second-order models:

*(a) $E(Y) = 3 + 2x_1 + 4x_2 - 0.6x_1^2 - 0.9x_2^2 - 0.4x_1x_2$

*(b) $E(Y) = 13 - 2x_1 - 4x_2 + 0.6x_1^2 + 0.9x_2^2 + 0.4x_1x_2$

*(c) $E(Y) = 13 + 2x_1 + 4x_2 + 0.2x_1^2 - 0.7x_2^2 - 1.2x_1x_2$

*(d) $E(Y) = 6 + 3x_1 + 5x_2 - 4x_1^2 - 3x_2^2 - 2x_1x_2$

Determine the stationary solution and discuss the nature of the response surface. Sketch its contours.

*8.6-2   Consider the second-order model

$$E(Y) = 10 + 2x_1 + 4x_2 - x_1^2 - 4x_2^2 - 4x_1x_2.$$

In this case, $4\beta_{11}\beta_{22} - \beta_{12}^2 = 0$. Sketch the contours of this response surface. Show that its maximum is described by the points that satisfy the equation $x_1 + 2x_2 = 1$. We call such a surface a *stationary ridge*.

*8.6-3   The percentage concentration of a certain input material and the temperature at which the associated reaction takes place are thought to be important factors in determining the yield of a chemical batch process. A 2^2 factorial experiment (replicated once) led to the following results:

Concentration (percent)	Temperature (°C)	Yield (in grams)
12	100	70, 73
14	100	78, 80
12	110	84, 86
14	110	85, 88

Determine the path of steepest ascent. Where should the next observation be taken?

*8.6-4   A 3^2 factorial experiment is conducted to determine the effect of temperature (x_1) and reaction time (x_2) on the yield (y, in grams) of a chemical reaction. The coded factors are given by

$$x_1 = \frac{\text{temperature} - 145°C}{5}$$

and

$$x_2 = \frac{\text{time} - 90 \text{ minutes}}{10}.$$

The results are as follows:

x_1	x_2	y
−1	−1	31.1
0	−1	33.7
+1	−1	30.7
−1	0	31.1
0	0	34.1
+1	0	32.2
−1	+1	28.4
0	+1	33.4
+1	+1	31.9

Estimate a second-order response surface, determine its stationary solution (expressed in coded as well as original units), and discuss its nature.

8.6-5   A central composite design with $k = 2$, $w = \sqrt{2}$, and four center points is used in an experiment on the precipitation of a stoichiometric dehydrate of calcium hydrogen orthophosphate. The response is percentage yield. The two factors are the mole ratio of NH_3 to $CaCl_2$ in the calcium chloride solution and the starting pH of the $NH_4H_2PO_4$ solution. The coded factors are

$$x_1 = \frac{(NH_3/CaCl_2) - 0.85}{0.09}$$

$$x_2 = \frac{pH - 3.5}{0.9}.$$

The data are given in the following table:

x_1	x_2	y
−1	−1	54.1
1	−1	66.0
−1	+1	74.5
1	+1	80.6
$-\sqrt{2}$	0	68.1
$\sqrt{2}$	0	89.2
0	$-\sqrt{2}$	40.8

x_1	x_2	y
0	$\sqrt{2}$	78.0
0	0	72.3
0	0	75.9
0	0	74.6
0	0	75.5

Estimate the second-order response surface and determine its stationary solution. Discuss the nature of the estimated response surface.

8.6-6 Assume that, in a certain experiment, the expected response surface is given by

$$E(Y) = 50 + 2.6x_1 + 1.4x_2 - 0.25x_1^2 - 0.20x_2^2 - 0.05x_1x_2,$$

but, of course, the experimenter does not know this response surface.

(a) The experimenter starts her investigation with a 2^2 factorial with one center point; that is, she takes observations at $(-1, -1)$, $(1, -1)$, $(-1, 1)$, $(1, 1)$, and $(0, 0)$. Generate random responses at these design points as follows: Perform a sequence of coin tosses. If the result is "heads," add 1 to $E(Y)$; otherwise, subtract 1.

(b) Suppose the results are given to the experimenter. Calculate the path of steepest ascent. The experimenter performs additional runs along this path. Use the procedure in part (a) to calculate the response for these design points.

(c) After the experimenter has located the maximum on the path of steepest ascent, she performs a central composite design with $w = \sqrt{2}$. Generate the responses at these design points and estimate a second-order model. Determine the stationary solution, describe the nature of the response surface, and discuss whether the true maximum has been reached.

*8.6-7 Myers conducted an experiment to explore the synthesis of mercaptobenzothiazole (MBT) as a function of reaction time and temperature. A rotatable central composite design with coded factors

$$x_1 = \frac{\text{time(hours)} - 12}{5.6}$$

and

$$x_2 = \frac{\text{temperature}(^\circ\text{C}) - 250}{20}$$

was conducted, and the following results were observed:

x_1	x_2	y
-1	-1	81.3
1	-1	85.3
-1	$+1$	83.1
1	$+1$	72.7
$-\sqrt{2}$	0	82.9
$\sqrt{2}$	0	81.7
0	$-\sqrt{2}$	84.7

0	$\sqrt{2}$	57.9
0	0	82.9
0	0	81.2
0	0	82.4

Estimate the second-order model, determine the stationary point, and discuss the nature of the response surface (in coded as well as original units).

Hint: To facilitate the interpretation, draw the contours for $\hat{y} = 75, 80$, and 85. You should obtain a saddle point.

[R. H. Myers, *Response Surface Methodology* (Boston: Allyn and Bacon, 1971).]

8.6-8 Determine the values of $x_i, i = 1, \ldots, k$, that maximize $\hat{\beta}_0 + \hat{\beta}_1 x_1 + \cdots + \hat{\beta}_k x_k$, subject to the constraint $\sum_{i=1}^{k} x_i^2 = R^2$.

Hint: Use a Lagrange multiplier and maximize

$$Q(x_1, \ldots, x_k) = (\hat{\beta}_0 + \hat{\beta}_1 x_1 + \cdots + \hat{\beta}_k x_k) - \lambda \left(\sum_{i=1}^{k} x_i^2 - R^2 \right).$$

*8.6-9 Consider the 2^k factorial experiment. Discuss the importance of adding center points. In particular, explain how the n_1 observations at the center point can be used to (1) estimate the standard errors of main effects and interactions and (2) assess whether quadratic components, such as x_1^2, x_2^2, and so on, are needed in the model.

Chapter 8 Additional Remarks

Regression modeling is an activity that leads to a mathematical description of a process in terms of a set of associated variables. Regression modeling models the functional relationship between a response variable and one or more explanatory variables. In some instances, one has a fairly good idea about the form of these models. The laws of physics or chemistry may tell us how a response is related to explanatory variables. These laws may involve complicated mathematical equations that contain functions such as logarithms and exponentials. Sometimes the constants in the equations are known, but more often the constants need to be determined empirically by "fitting" the models to data. However, in many applications, theoretical models are absent and one must develop empirical models that describe the main features of the relationship entirely from data.

Models are important tools for understanding complex phenomena, and very often they are the only tools we have. We need models and simplifying assumptions to understand the economy, the onset and progression of disease, the universe and how it began, the climate and issues relating to global warming, and so on. Those who claim to be able to do any of these things without making simplifying assumptions are only fooling themselves. By their very nature, models are only approximations to a much more complicated reality, and by virtue of the many simplifications they make, models are literally false. Still, they can be valuable. George Box, one of the most influential statisticians of the second half of the 20th century, summarized the situation very well when he said, "Essentially, all models are wrong, but some are useful. The practical question is how wrong do models have to be to not be useful." Yes, we need to be skeptical of our models and we should test them as thoroughly as we can. However, too much skepticism shouldn't block us from putting our models to work and using them. They will always be wrong, but they may be good enough approximations to shed light on the issues we are interested in.

Once satisfactory models have been found, they can be used for several different purposes:

- We learn which of the explanatory variables have an effect on the response. This tells us which explanatory variables we have to change in order to affect the response. If a variable does not affect a response, then there may be little reason to measure or control it. Not having to keep track of something that is not needed can lead to significant savings.

- The functional relationship between the response and the explanatory variables allows us to estimate the response for given values of the explanatory variables. It makes it possible to infer the response for values of the explanatory variables that were not studied directly. Also, the model allows us to ask "what-if" type questions and to control the response variable at certain desired levels.

- Prediction of future events is another important application. We may have a good model for sales over time, and want to know the likely sales for the next several future periods.

- A regression analysis may show that a variable which is difficult and expensive to measure can be explained to a large extent by variables that are easy and cheap to obtain. This is important information, as we can substitute the cheaper measurements for the more expensive ones. It may be quite expensive to determine someone's body fat as this requires that the whole body be immersed in water. Or, it may be expensive to obtain a person's bone density. However, variables such as height, weight, thickness of thighs or biceps are easy and cheap to obtain. If there is a good model that can explain the expensively measured variable through the variables that are easy and cheap to obtain, then one can save money and effort by using the latter variables as proxies.

Projects

Project I

Investigate the annual compensation paid to chief executive officers of large U.S. companies. Study whether the compensation paid is a reflection of performance.

(a) Develop operational definitions of the variables. What do you mean by "compensation"? What do you mean by "performance"?

(b) Collect the relevant information for the last available year. You may want to check for this information in libraries (issues of business magazines and publications) or on relevant websites. Depending on the available data, you may have to change your operational definition. How did you select the CEOs and their companies?

(c) Analyze the data graphically and through summary statistics. What can you say about CEO compensation? What does the distribution of compensation look like? Why are the mean and median compensations so different?

(d) What can you say about the relationship between compensation and performance? What about gender and type and size of the company? Discuss.

*Project 2

Traditionally, the quality of the Bordeaux vintage is first evaluated by experts in March of the next year. However, it turns out that these first ratings are rather unreliable. What about using a statistical approach that predicts the price of the vintage as a function of its age, rainfall, and temperature?

Consider data on average temperature during the growing season (April through September, in degrees centigrade), rain during the harvest season (rain in August and September, in total millimeters), rain prior to the growing season (October of the previous year to March of the current year, in total millimeters), age of the vintage, and average price for the Bordeaux vintage relative to the year 1961 (the response variable). Data for the years 1952 through 1980 are listed in the following table (the data for 1954 and 1956 are missing; relative prices for these two vintages could not be established because they were poor vintages and very little wine is now sold from those two years):

Year	Temperature	Rainfall	Previous Rainfall	Age (1983 = 0)	Relative Price
1952	17.12	160	600	31	0.368
1953	16.73	80	690	30	0.635
1955	17.15	130	502	28	0.446
1957	16.13	110	420	26	0.221
1958	16.42	187	582	25	0.180
1959	17.48	187	485	24	0.658
1960	16.42	290	763	23	0.139
1961	17.33	38	830	22	1.000
1962	16.30	52	697	21	0.331
1963	15.72	155	608	20	0.168
1964	17.27	96	402	19	0.306
1965	15.37	267	602	18	0.106
1966	16.53	86	819	17	0.473
1967	16.23	118	714	16	0.191
1968	16.20	292	610	15	0.105
1969	16.55	244	575	14	0.117
1970	16.67	89	622	13	0.404
1971	16.77	112	551	12	0.272
1972	14.98	158	536	11	0.101
1973	17.07	123	376	10	0.156
1974	16.30	184	574	9	0.111
1975	16.95	171	572	8	0.301

Year	Temperature	Rainfall	Previous Rainfall	Age (1983 = 0)	Relative Price
1976	17.65	247	418	7	0.253
1977	15.58	87	821	6	0.107
1978	15.82	51	763	5	0.270
1979	16.17	122	717	4	0.214
1980	16.00	74	578	3	0.136

(a) Construct a regression model that describes the price of the vintage as a function of its age, rainfall, previous rainfall, and temperature. Interpret the data through appropriate scatter plots (such as plots of price against the various explanatory variables), summary statistics, and the output from the regression analysis. Check the adequacy of the model by looking at residual plots.

Note: It may be beneficial to transform price into logarithm(price).

(b) A more appropriate test of a model compares its predictions with actual observations that are not part of the sample used to establish the model. We refer to this as "*out-of-sample*" prediction, as opposed to the "*within-sample*" diagnostic checking that looks at the residuals. Fortunately, new data are available. Ray C. Fair, in Chapter 6 of his book *Predicting Presidential Elections and Other Things* (Stanford, CA: Stanford University Press, 2002), lists the growing conditions for the years 1987 through 1991, shown in the following table:

Year	Temperature	Rainfall	Previous Rainfall	Age (1983 = 0)	Relative Price
1987	16.98	115	452	−4	0.135
1988	17.10	59	808	−5	0.271
1989	18.60	82	443	−6	0.432
1990	18.70	80	468	−7	0.568
1991	17.70	183	570	−8	0.142

In the table of the historic data, the trend variable Age is set to 0 in 1983; hence, its values for 1987 and beyond are negative. Fair obtained prices for the 1987–1991 vintages from an east-coast wine distributor. He found that the average prices for the 1961 and the 1987–1991 vintages were as follows: $258.33 (1961), $35.00 (1987), $70.00 (1988), $111.67 (1989), $146.67 (1990), and $36.67 (1991). He expressed the price of the 1987–1991 vintages relative to 1961 as $35.00/258.33 = 0.135$, $70.00/258.33 = 0.271$, $111.67/258.33 = 0.432$, $146.67/258.33 = 0.568$, and $36.67/258.33 = 0.142$, and these numbers are shown in the last column of the table.

How successful is the regression approach in obtaining out-of-sample predictions? Can we use the model to predict the price of a new vintage? How well does the model do on more recent data?

Note: Obtain predictions for the five vintages from 1987 to 1991 by substituting the values of the regressor variables into the model equation (that you obtained from previous data). Compare the resulting predictions with the actual values. For example, calculate the errors, the absolute errors, and the absolute percent errors. Summarize these deviations by calculating their averages. For good predictions, the mean error should be close to zero, and the mean of the absolute errors (and percent absolute errors) should be small.

[Ashenfelter, O., Ashmore, D., and Lalonde, R., "Bordeaux Wine Vintage Quality and the Weather," *Chance Magazine,* Vol. 8, No. 4, 1995, 7–14 (see also *Barron's,* December 30, 1996, pp. 17–19).]

Project 3

Some people claim that gender has an important effect on salary. Investigate this claim, using professors' salary data that are available at your educational institution. If you are a student at a public university, you should have no problems obtaining information on salary, as salaries are public. If you are a student at a private university, you may not be able to do this project.

(a) Collect information on salary. If you are a student at a large university, we recommend that you obtain the salaries for the College of Engineering. This will keep your project at a realistic level. Stratify the salary data according to gender. Analyze the information and state your preliminary findings.

(b) Discuss the factors that, in your opinion, affect salary. Do you think that the salary comparison in (a) represents a valid and fair assessment of possible salary discrimination on the basis of gender?

(c) Develop measures for the factors that you have identified in (b). These factors are potential causes, but their effects must still be confirmed by subsequent analysis.

Experience may have an effect. Discuss how you would measure this variable.

[You may define experience in terms of the number of years since obtaining a Ph.D, or you may define it by the "rank" of the professor (assistant, associate, or full professor), or possibly both.]

Excellence in research may have an effect, especially at large research institutions. Discuss how you would measure this factor.

[Excellence in research may be difficult to measure; you could obtain and evaluate professors' resumes. The dean's office or the departmental offices should give you access to the resumes. You may also want to try the website of your school.]

Excellence in teaching may have an effect. Develop measures of this factor.

[Excellence in teaching may be even harder to measure than excellence in research; you may want to start by looking for teaching awards on resumes, by asking fellow students to rate instructors, etc.]

Professors' salaries may depend on *administrative duties*. Keep track of whether the professor is also a dean or a departmental officer.

(d) Discuss any difficulties that arise in the measurement of all of the preceding factors.

(e) With the information you have obtained in hand, construct a regression model that helps explain the salary in terms of gender and the explanatory variables that you identified in (b) and measured in (c). How many of the variables that you had identified as potential causes are really necessary for explaining the differences in salary? Did the estimated regression coefficients for the explanatory variables have the anticipated "correct" signs? Do the estimation results make sense? What is your interpretation of the regression coefficient for gender? Does it help you answer the question whether your institution engages in salary discrimination with respect to gender?

Notes: Gender is a binary variable, so it can be coded as 0 or 1. We hope that your college of engineering has a sufficiently large pool of women professors.

Rank (assistant, associate, full professor) is a categorical variable with three possible outcomes. You may want to transform this information into indicator variables. You will need two such variables: The first one, say IND1, is set to 1 for associate professors and to 0 otherwise. The second, IND2, is set to 1 for full professors and to 0 otherwise. This coding convention makes assistant professors the standard for the comparisons: The regression coefficient for IND1 expresses the difference in salary between associate and assistant professors; the coefficient for IND2 expresses the difference between the salaries for full and assistant professors.

In large and departmentalized colleges of engineering you may want to keep track of the department in which each professor teaches. Some departments may pay higher salaries than others. You may want to use additional indicator variables to characterize this information.

(f) What other data, besides the current salary, could shed light on the issue of gender and salary? For example, would the information on the last yearly (percentage) raises be helpful?

(g) Write a report that summarizes your analysis and your findings. What is your conclusion? What are the shortcomings of your analysis?

Project 4

Lake Neusiedl is a large, shallow lake in the eastern part of Austria. Obtain information on this lake by accessing information on the web. The file Chapter8Project4Neusiedl contains daily data from January 2, 1971, through December 31, 2004, on the following variables:

- Daily water level (meters above Adriatic sea level), measured at 7 A.M. at Neusiedl, a town at the northernmost edge of the lake.

- Average daily water level of the lake (determined from several stations across the lake).

- Daily precipitation (mm/day) for the preceding 24 hours, obtained from measurements at eight weather stations in the vicinity of the lake.

- Wind speed (km/h) and wind direction [0: no wind; 8: wind from east; 16: wind from south; 24: wind from west; 32: wind from north], measured at Neusiedl at three different hours of the day [7 A.M., 2 P.M., and 7 P.M.]. Winds coming from the north blow across the plains toward the lake.

(a) Investigate possible trends in the water level of the lake over time. Investigate whether there are seasonal (monthly) components.

(b) Investigate possible time trends and seasonality in precipitation, wind speed, and wind direction.

Hint: For directional data, such as wind direction, it does not make sense to calculate the ordinary summary statistics. Consider two periods with winds from the northeast (here coded as 4) and the northwest (coded as 28). It wouldn't make sense to calculate the mean (which is 16) and imply that the average wind direction is from the south. A simple graph of directional data on a circle shows that the wind direction north (coded here as 32) is the more appropriate average of these two directions. The *circular mean* for directional data is defined in the statistical literature, and we ask you to run a web search to learn how it is defined.

For your trend analysis, you may avoid the difficulties of averaging directional data by considering monthly frequencies of various wind directions and displaying the frequencies in the form of time-sequence plots.

(c) It is widely thought that the water level at the station in Neusiedl is affected by the wind. Strong winds from the north push the water south (lowering the water level), while strong winds from the south push the water toward Neusiedl (increasing the water level). Investigate this hypothesis. Think about graphical displays that could confirm it.

Hint: Keeping the particular geographical situation in mind, calculate the perpendicular wind speed component WS_p = wind speed × cos(wind direction from north). (See Exercise 1.5-6 for an application of this concept in a different setting.)

Project 5

Data on the utilization of nitrate in bush beans as a function of light intensity were obtained by J. R. Elliott and D. R. Peirson of Wilfrid Laurier University (Canada). Portions of leaves from three 16-day-old plants were subjected to eight levels of light intensity (in microeinsteins per square meter per second), and the nitrate utilization (in nanomols per gram per hour) was measured. The experiment yielded the following results:

Light Intensity ($\mu E/m^2 s$)	Nitrate Utilization (nmol/g hr)
2.2	256
	685
	1,537
5.5	2,148
	2,583
	3,376
9.6	3,634
	4,960
	3,814
17.5	6,986
	6,903
	7,636
27.0	9,884
	11,597
	10,221
46.0	17,319
	16,539
	15,047
94.0	19,250
	20,282
	18,357
170.0	19,638
	19,043
	17,475

Plot nitrate utilization against light intensity. Nitrate utilization should be zero at zero light intensity and should approach an asymptote as the light intensity increases. Consider the following nonlinear regression models $Y = \mu + \varepsilon$ with

$$\mu = f(x; \beta_1, \beta_2) = \frac{\beta_1 x}{\beta_2 + x};$$ (Michaelis–Menton model)

$$\mu = f(x; \beta_1, \beta_2) = \beta_1[1 - \exp(-\beta_2 x)].$$ (exponential rise model)

Fit these models to the data, add the fitted regression lines to your scatter plot, and discuss their adequacy.

Unfortunately, at present Minitab does not support the estimation of nonlinear regression models. However, other programs, such as SAS, SPSS, and R, include such routines. For example, in SPSS you can use "Analyze > Regression > Nonlinear." All you need to do is write out the model in functional form and provide initial starting values for the unknown coefficients. Determine the starting values by investigating the behavior of the observations for small and large values of x —for example, $x = 2.2$ (and average response 800) and $x = 170$ (and average response 19,000). This gives you two equations that you can solve for the starting values of β_1 and β_2.

*Project 6

In 1970, Du Pont developed an Automatic Clinical Analyzer that allows the clinician in a hospital to determine the concentrations of substances in a patient's blood or urine. Measurements of many chemical substances can be made with this device and with custom-made analytical test packs. The analyzer assays a sample of blood serum for a certain substance by transferring a mixed volume of the specimen into a method-specific test pack and adding a controlled amount of a certain buffer solution. The pack contains, in several different compartments, all the reagents that are necessary to complete the assay. Reagents are added to the test sample in a predetermined manner. A chemical reaction is initiated, followed by timed measurements of the absorbance of light of known wavelength by the pack. From these absorbance measurements, the concentration of the chemical substance in the sample can be calculated via a calibration curve.

The pack configuration affects the assay's sensitivity, which is defined as the unit absorbance change per unit concentration of the chemical substance that is in the sample. The configuration is, among many other factors, a function of the reagents and their concentrations and of the buffer and its pH. Response surface methodology is used to develop the relationships between sensitivity and these factors and to select the conditions that optimize the sensitivity.

Humphries, Melnychuk, Donegan, and Snee discuss the use of the duPont analyzer in determining the concentration of plasma ammonia. In this case, the chemical reaction is initiated by the enzyme glutamate dehydrogenase (GLDH). Response surface methods were used to determine the optimum values for the GLDH concentration (x_1) and for the pH (x_2) and molarity (x_3) of the buffer, which in this case was a certain sulfonic acid, to be used in the pack.

A central composite design with $w = 1$ was used to study the effects of these $k = 3$ factors on assay sensitivity. Note that, for $w = 1$, the "star points" in the central composite design correspond to the centers of the six sides of the cube. Hence, this design is frequently called a *face-centered-cube design*. The concentration of GLDH was varied among 90 (low), 125 (medium), and 160 (high); pH was varied among 7.25 (low), 7.45 (medium), and 7.65 (high); and the molarity (mol/liter) of the buffer was varied among

0.04 (low), 0.05 (medium), and 0.06 (high). Three assays were run for each experimental test combination, and the results are shown in the following table (note that several responses are missing):

GLDH Concentration, x_1	pH, x_2	Molarity, x_3	Sensitivity, y
−	−	−	148, 143, 147
+	−	−	175, 188
−	+	−	175, 178, 180
+	+	−	167, 166, 173
−	−	+	141, 139
+	−	+	178, 175
−	+	+	187, 170, 181
+	+	+	164, 177, 170
−	0	0	180, 184, 176
+	0	0	178, 179, 175
0	−	0	173, 160
0	+	0	185, 171, 180
0	0	−	187, 182, 175
0	0	+	181, 182, 180
0	0	0	193, 181, 188

Note: $x_1 = \dfrac{\text{concentration} - 125}{35}$, $x_2 = \dfrac{\text{pH} - 7.45}{0.20}$, $x_3 = \dfrac{\text{molarity} - 0.05}{0.01}$.

*(a) Estimate the quadratic model

$$Y = \beta_0 + \beta_1 x_1 + \beta_2 x_2 + \beta_3 x_3 + \beta_{11} x_1^2 + \beta_{22} x_2^2 + \beta_{33} x_3^2 + \beta_{12} x_1 x_2$$
$$+ \beta_{13} x_1 x_3 + \beta_{23} x_2 x_3 + \varepsilon.$$

(b) You should find that the t-ratios for the coefficients β_3, β_{33}, β_{13}, and β_{23} in this model are insignificant. This suggests that molarity (x_3) has no effect on the response over the range from 0.04 to 0.06 mol/liter.

*(c) Omit molarity from the model, and estimate

$$Y = \beta_0 + \beta_1 x_1 + \beta_2 x_2 + \beta_{11} x_1^2 + \beta_{22} x_2^2 + \beta_{12} x_1 x_2 + \varepsilon.$$

Determine the stationary point of the response surface that is implied by this model in coded as well as original units. Determine its nature (maximum, minimum, or saddle point). Plot the contours of the response function over the region $7.25 < \text{pH} < 7.65$ and $100 < \text{concentration (GLDH)} < 180$.

(d) What levels of GLDH, molarity, and pH would you recommend to maximize sensitivity?

*(e) Your recommendations from part (d) are used to manufacture packs. In practice, it is sufficient to be within 3 percent of the highest attainable sensitivity. Will the resulting packs be insensitive (i.e., robust) to variations in the actual levels of GLDH, pH, and molarity? Recall that, because of variation in the manufacturing process, the levels of the ingredients in the pack will not be *identical* to the target

values. What process variations can be tolerated without affecting the performance of the pack (i.e., the sensitivity of the analysis)?

[B. A. Humphries, M. Melbychuk, E. J. Donegan, and R. D. Snee, "Automated Enzymatic Assay for Plasma Ammonia," *Clinical Chemistry,* 25 (1979), 26–30.]

Project 7

Darius, Portier, and Schrevans have developed helpful computer applets for simulating outcomes of virtual experiments. The virtual environment in which to experiment mimics real situations of interest without requiring the resources for carrying out actual experiments and without being as time consuming as actual experiments are. Visit the website http://ucs.kleuven.be/env2exp and go to the Factory Applet. (You may also want to try some other applets provided by these authors.)

The purpose of the simulation applet is to optimize a process running in a production plant through experimentation on a scaled-down version of the process run on a pilot plant. The process involves three parameters (temperature, time, concentration).

The experiment runs over a period of 39 weeks. With 1 week equivalent to three minutes of real time, the experiment lasts about two hours. The time bar on top of the window reminds you of time. The user employs the pilot plant to learn about the process. The user opens up an experiment window at the pilot plant and specifies the settings of the three parameters for one or more runs. The pilot plant (as well as the production plant) works with raw material delivered in batches, with some batch variability and one batch allowing a maximum of 10 experimental runs. There is a setup cost for the pilot plant (15,000 euros), and additional costs arise with each experimental run (500 euros). The results of the experimental runs are shown after a brief delay in the "History" window. The applet generates the response to the user-selected values of the parameters by adding a random variable to a known model result (which, unknown to the user, is determined as a function of the three parameters and a batch effect). The user analyzes the results of the experiments on the pilot plant, trying to find improved levels for the three factors. This is done by copying the factor-level settings and the responses into a worksheet of the available statistical software and by carrying out the appropriate analyses. The user can continue with the experiment or reset the parameters at the production plant. There is a cost for making a change in the production plant (6,000 euros, because the process needs to be stopped), and it takes 6 weeks to learn about the results of the new process settings. After 6 weeks, the user is shown (in the "Net Extra Profit" window) the response of the production plant to the new process settings (in euros per week) and the extra profit from the new settings (compared with the profit from the process settings at the start of the experiment) is summed over all remaining weeks up to week 90 (which is the planning horizon for this exercise). The user is shown the total extra profit for weeks 1 through 90 (calculated as the sum of the weekly extra profits) and the net extra profit (the total extra profit, minus the costs associated with all experiments and process changes). The goal of the project is to maximize the net extra profit over the 90-week horizon.

The main reason for the delay in getting feedback from a process change is to make it difficult to use the production plant to carry out the experiments; people running the process would get upset if the production plant were stopped for changes too many times. Another reason for the delay is the fact that the financial performance of the production is obtained by averaging results for a large number of batches.

Try the various experiments that you have learned in this book. You may want to use factorial and fractional factorial experiments, with and without center points, and central composite designs. For example, you may want to start with a design that lets you estimate a second-order model. Find the optimum conditions of the process (by taking first derivatives and setting them equal to zero, as we have done in Section 8.6 for the case of two factors). Or you may want to construct contour plots with your data and locate promising regions for the parameters that way. Once you have found good settings, apply them to the production plant. Wait for the response of the production plant (which becomes available after a delay of 6 weeks), compare the extra profit you get in the production plant with the results of the experiments on the pilot plant, run further experiments if you feel they are needed, and fine-tune the production process by making additional changes. Keep in mind the costs; it is much cheaper to experiment on the pilot plant (500 euros versus 6,000 euros), and you also get much quicker feedback (no need to wait for 6 weeks). Keep in mind that you are being evaluated on the performance of the production plant and that you will need to find good settings before week 39 (hopefully, earlier), as no experiments and changes are possible after that week. The last process settings made on or before week 39 will greatly influence your total extra profit over the remaining weeks (until week 90).

Try to have some fun!

[Darius, P. L., Portier, K. M., and Schrevans, E., "Virtual Experiments and Their Use in Teaching Experimental Design," *International Statistical Review,* 75 (2007), 281–294).]

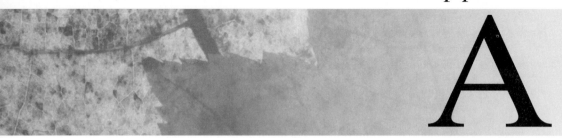

Appendix

A

REFERENCES

Chapter 1

Cleveland, W. S. *Visualizing Data,* rev. ed. Summit, NJ: Hobart Press, 1993.

Cleveland, W. S. *The Elements of Graphing Data.* Summit, NJ: Hobart Press, 1994.

Deming, W. E. *Out of the Crisis.* Cambridge, MA: Cambridge University Press, 1986.

Gaydos, T. M., Stanier, C. O., and Pandis, S. N. "Modeling of in situ Ultrafine Atmospheric Particle Formation in the Eastern United States," *Journal of Geophysical Research,* Vol. 110: D07S12 (2005).

Henderson, H. V., and Velleman, P. F. "Building Multiple Regression Models Interactively." *Biometrics,* Vol. 37 (1981), 391–411.

Jaffe, P. R., Parker, F. L., and Wilson, D. J. "Distribution of Toxic Substances in Rivers." *Journal of the Environmental Engineering Division,* Vol. 108 (1982), 639–649.

Ledolter, J., and Chan, K. S. "Evaluating the Impact of the Increased 65 mph Speed Limit on Iowa Rural Interstates." *The American Statistician,* Vol. 50 (1996), 79–85.

Ledolter, J., and Tiao, G. C. "Statistical Methods for Ambient Air Pollutants with Special Reference to the Los Angeles Catalyst Study (LACS) Data." *Environmental Science and Technology,* Vol. 13 (1979), 1233–1240.

Minitab Statistical Software (http://www.minitab.com).

Moberg, A., Sonechkin, D. M., Holmgren, K., Datsenko, N. M., and Karlén, W. "Highly Variable Northern Hemisphere Temperatures Reconstructed from Low- and High-Resolution Proxy Data." *Nature,* Vol. 433, No. 7026, February 10, 2005, pp. 613–617.

R Project for Statistical Computing (http://cran.us.r-project.org).

Snee, R. D. "Graphical Analysis of Process Variation Studies." *Journal of Quality Technology,* Vol. 15 (April 1983), 76–88.

Snee, R. D., and Pfeifer, C. G. "Graphical Representation of Data," in S. Kotz and N. L. Johnson, eds., *Encyclopedia of Statistical Sciences,* Vol. 3 (New York: Wiley, 1983), 488–511.

Snee, R. D., Hare, L. B., and Trout, J. R. *Experiments in Industry: Design, Analysis, and Interpretation of Results.* Milwaukee: American Society for Quality Control, 1985.

Tufte, Edward R. *The Visual Display of Quantitative Information.* Cheshire, CT: Graphics Press, 1983 (second edition, 2001).

Tufte, Edward R. *Envisioning Information.* Cheshire, CT: Graphics Press, 1990.

Tufte, Edward R. *Visual Explanations: Images and Quantities, Evidence and Narrative.* Cheshire, CT: Graphics Press, 1997.

Tufte, Edward R. *Beautiful Evidence.* Cheshire, CT: Graphics Press, 2006.

Tukey, John W. *Exploratory Data Analysis.* Reading, MA: Addison Wesley, 1977.

Velleman, P., and Hoaglin, D. *Applications, Basics, and Computing of Exploratory Data Analysis.* Boston: Duxbury Press, 1981.

Chapter 2

Bortkiewicz, L. von. *Das Gesetz der Kleinen Zahlen.* Leipzig, Germany: G. Teubner, 1898.

Evans, James R. *Statistics, Data Analysis, and Decision Modeling,* (3d ed.). New York: Prentice Hall, 2007.

Tanis, E. A., and Hogg, R. V. *A Brief Course in Mathematical Statistics.* New York: Prentice Hall, 2006.

Chapter 3

Box, G. E. P, and Muller, M. E. "A Note on the Generation of Random Normal Deviates." *Annals Math. Stat.,* Vol. 29 (1958), 610–611.

Dexter, F., and Ledolter, J. "Bayesian Prediction Bounds and Comparisons of Operating Room Times Even for Procedures with Few or No Historic Data." *Anesthesiology,* Vol. 103 (2005), 1259–1267.

Dexter, F., Epstein, R. H., Marcon, E., and Ledolter, J. "Estimating the Incidence of Prolonged Turnover Times and Delays by Time of Day." *Anesthesiology,* Vol. 102 (2005), 1242–1248.

Hogg, R. V., and Tanis, E. A. *Probability and Statistical Inference,* 7th ed. Upper Saddle River, NJ: Prentice Hall, 2006.

Nelson, W. *Applied Life Data Analysis.* New York: Wiley, 1982.

Shannon, C. E. "A Mathematical Theory of Communication." *Bell System Technical Journal,* Vol. 27 (July/October 1948), 379–423/623–656.

Chapter 4

Cressie, N. A. *Statistics for Spatial Data*. New York: Wiley, 1993.

De Moivre, Abraham. *The Doctrine of Chances*. London: W. Pearson, 1718.

Gauss, Carl Friedrich. *Theoria motus corporum coelestium in sectionibus conicis solem ambientium*. Hamburg, Germany: F. Perthes and I. H. Besser, 1809.

Martinez-Dawson, R. "Incorporating Laboratory Experiments in an Introductory Statistics Course." *Journal of Statistics Education*, Vol. 11, No. 1 (2003), www.amstat.org/publications/jse/v11n1/martinez-dawson.html.

Raspe, Rudolf Erich: *The Surprising Adventures of Baron Munchhausen*. Germany, 1785.

Welch, B. L. "The Significance of the Difference Between Two Means When the Population Variances Are Unequal." *Biometrika*, Vol. 29 (1937), 350–362.

Ziegler, E., Nelson, S. E., and Jeter, J. M. "Early Iron Supplementation of Breastfed Infants." Iowa City, IA: Department of Pediatrics, University of Iowa, 2007.

Chapter 5

Deming, W. E. *Out of the Crisis*. Cambridge, MA: MIT Press, 1986.

Grant, E. L. *Statistical Quality Control*, 2d ed. New York: McGraw-Hill, 1952.

Ishikawa, K. *Guide to Quality Control*, 2d rev. ed. Tokyo: Asian Productivity Organization, 1982.

Kushler, R. H., and Hurley, P. "Confidence Bounds for Capability Indices." *Journal of Quality Technology*, Vol. 24 (1992), 188–195.

Montgomery, D. *Introduction to Statistical Quality Control*, 5th ed. New York: Wiley, 2004.

Roberts, H., and Sergesketter, B. *Quality Is Personal: A Foundation for Total Quality Management*. New York: Free Press, 1993.

Shecter, E. "Process Control for High Yields." *RCA Engineer*, Vol. 30, No. 3 (1985), 38–43.

Tadikamalla, P. R. "The Confusion Over Six-Sigma Quality." *Quality Progress*, Vol. 27 (November 1994), 83–85.

Chapter 6

Box, G. E. P., Hunter, W. G., and Hunter, J. S. *Statistics for Experimenters*. New York: Wiley, 1978 (second edition, 2005).

Fisher, R. A. *Statistical Methods for Research Workers*. Edinburgh, U.K.: Oliver & Boyd, 1925.

Hare, L. B. "It's not always what you say, but how you say it." *Quality Progress,* Vol. 40 (August 2007), 64–66.

Latter, H. O. "The Cuckoo's Egg." *Biometrika,* Vol. 1 (1901), 164–176.

Montgomery, D. C. *Design and Analysis of Experiments,* 2d ed. New York: Wiley, 1984.

Sanders, T. G. *Principles of Network Design for Water Quality Monitoring.* Fort Collins, CO: Colorado State University, 1980.

Wetherill, G. B. *Elementary Statistical Methods,* 3d ed. London: Methuen, 1982.

Youden, W. J. "Graphical Diagnosis of Interlaboratory Test Results." *Industrial Quality Control,* Vol. 15 (May 1959), 133–137.

Chapter 7

Anderson, V. L., and McLean, R. A. *Design of Experiments: A Realistic Approach.* New York: Marcel Dekker, 1974.

Box, G. E. P., and Draper, N. R. *Evolutionary Operation.* New York: Wiley, 1969.

Box, G. E. P., and Hunter, J. S. "The 2^{k-p} Fractional Factorial Designs." *Technometrics,* Vol. 3. (1961), 311–351.

Box, G. E. P., Hunter, W. G., and Hunter, J. S. *Statistics for Experimenters.* New York: Wiley, 1978 (second edition, 2005).

Coleman, D. E. "Measuring Measurements." *RCA Engineer,* Vol. 30, No. 3 (1985), 16–23.

Fisher, R. A. *Statistical Methods for Research Workers.* Edinburgh, U.K.: Oliver and Boyd, 1925.

Fisher, R. A. *The Design of Experiments.* Edinburgh, U.K.: Oliver and Boyd, 1935 (and various later editions).

Fisher Box, J. *R. A. Fisher: The Life of a Scientist.* New York: Wiley, 1978.

Gunter, B. H., Tadder, A. L., and Hemak, C. M. "Improving the Picture Tube Phosphor Screening Process with EVOP." *RCA Engineer,* Vol. 30, No. 3 (1985), 54–59.

Hunter, W. G. "Some Ideas about Teaching Design of Experiments with 2^5 Examples of Experiments Conducted by Students." *The American Statistician,* Vol. 31 (1977), 12–17.

Ledolter, J., and Swersey, A. J. "Using a Fractional Factorial Design to Increase Direct Mail Response at *Mother Jones* Magazine." *Quality Engineering,* Vol. 18 (2006), 469–475.

Ledolter, J., and Swersey, A. J. *Testing 1–2–3: Experimental Design with Applications in Marketing and Service Operations,* Stanford, CA: Stanford University Press, 2007.

Lenth, R. L. "Quick and Easy Analysis of Unreplicated Factorials." *Technometrics,* Vol. 31 (1989), 469–474.

Natrella, M. G. *Experimental Statistics.* National Bureau of Standards Handbook 91. Washington, D.C.: U.S. Government Printing Office, 1963.

NIST/SEMATECH Engineering Statistics Handbook, 2003 (continuously updated, http://www.itl.nist.gov/div898/handbook/).

Salsburg, D. *The Lady Tasting Tea: How Statistics Revolutionized Science in the Twentieth Century.* New York: Owl Books, 2002.

Snee, R. D. "Experimenting with a Large Number of Variables." In Snee, R. D., Hare, L. B., and Trout, J. R., eds., *Experiments in Industry: Design, Analysis, and Interpretation of Results.* Milwaukee, WI: American Society for Quality Control, 1985.

Chapter 8

Abraham, B., and Ledolter, J. *Introduction to Regression Modeling.* Belmont, CA: Duxbury, 2006.

Anscombe, F. J. "Graphs in Statistical Analysis." *The American Statistician,* Vol. 27 (1973), 17–21.

Ashenfelter, O., Ashmore, D., and Lalonde, R. "Bordeaux Wine Vintage Quality and the Weather." *Chance Magazine,* Vol. 8, No. 4 (1995), 7–14; also *Barron's,* December 30, 1996, pp. 17–19.

Box, G. E. P., Hunter, W. G., and Hunter, J. S. *Statistics for Experimenters.* New York: Wiley, 1978 (Second Edition, 2005).

Darius, P. L., Portier, K. M., and Schrevans, E. "Virtual Experiments and Their Use in Teaching Experimental Design." *International Statistical Review,* Vol. 75 (2007), 281–294.

Draper, N. R., and Smith, H. *Applied Regression Analysis,* 3d ed. New York: Wiley, 1998.

Fair, R. C. *Predicting Presidential Elections and Other Things.* Stanford, CA: Stanford University Press, 2002.

Gorman, J. W., and Toman, R. J. "Selection of Variables for Fitting Equations to Data." *Technometrics,* Vol. 8 (1966), 27–51.

Henderson, H. V., and Velleman, O. F. "Building Multiple Regression Models Interactively." *Biometrics,* Vol. 37 (1981), 391–411.

Hill, W. J., and Wiles, R. A. "Plant Experimentation (PLEX)." *Journal of Quality Technology,* Vol. 7 (1975), 115–122.

Hogg, R. V., McKean, J. W., and Craig, A. T. *Introduction to Mathematical Statistics,* 6th ed. Upper Saddle River, NJ: Prentice Hall, 2005.

Humphries, B. A., Melbychuk, M., Donegan, E. J., and Snee, R. D. "Automated Enzymatic Assay for Plasma Ammonia." *Clinical Chemistry,* Vol. 25 (1979), 26–30.

Kyotani, T., Yoshitake, H., Ito, T., and Tamai, Y. "Correlation Between Flow Properties and Traction of Lubricating Oils." *Transactions of the American Society of Lubricating Engineers,* Vol. 29 (1986), 102–106.

Montgomery, D. C., Peck, E. A., and Vining, G. G. *Introduction to Linear Regression Analysis,* 3d ed. New York: Wiley, 2001.

Myers, R. H. *Response Surface Methodology*. Boston: Allyn and Bacon, 1971.

Risebrough, R. W. "Effects of Environmental Pollutants upon Animals Other than Man. *Proceedings of the 6th Berkeley Symposium on Mathematics and Statistics, VI.* Berkeley, CA: University of California Press, 443–463 (1972).

Ryan, T. A., Joiner, B. L., and Ryan, B. F. *The Minitab Student Handbook*. Boston: Duxbury Press, 1985.

Snedecor, G. W., and Cochran, W. G. *Statistical Methods,* 6th ed. Ames, IA: Iowa State University Press, 1967.

Weisberg, S. *Applied Linear Regression,* 3d ed. New York: Wiley, 2001.

ANSWERS TO SELECTED EXERCISES

Chapter 1

1.1-1 Census: population census (every 10 years). Survey: opinion poll, sample inspection. Designed experiment: agricultural field trial; engineering experiment.

1.1-2 Categorical (not ordered): gender, ethnicity, race. Categorical (ordered): size of company, agreement ranging from strongly disagree to strongly agree. Measurement variables: height, weight, profit, sales, yield.

1.1-5 Process variability; measurement variability.

1.1-8 **(a)** Not a random sample, as not every subset of four elements is permitted;

(b) Stratified random sample: good, as it excludes certain "unlucky" samples, such as sets of four managers or four workers.

1.2-1 Evidence of (linear) time trend. Trend changes with time; in beginning, rather flat; then steady increase. "Locally constant" time trends, compared with "globally constant" time trends. Forecasts of future observations should use the most recent observations (perhaps the last 10–20 observations). Or use exponential smoothing forecast methods and weight the importance of the observations according to their age.

1.3-1 **(b)** $\bar{x} = 15.375$ (psi), $s^2 = 3.4107$ (psi^2), $s = 1.85$ (psi).

(c) $Q_1 = (14 + 15)/2 = 14.5$, $Q_2 = (15 + 16)/2 = 15.5$, $Q_3 = (16 + 17)/2 = 16.5$, $Q_3 - Q_1 = 2$.

1.3-2 **(b)** $\bar{x} = 320.1, s^2 = 45.326, s = 6.76$, $Q_1 = 316, Q_2 = 320, Q_3 = 325$.

1.3-3 **(b)** $\bar{x} = 7.29, s^2 = 4.1025, s = 2.03$, $Q_1 = 6.05, Q_2 = 6.95, Q_3 = 8.45$, $Q_3 - Q_1 = 2.40$.

1.3-6 Calculate averages and differences of the replications. The 12 averages describe the variability due to the actual size differences; the 12 differences describe the measurement variability.

1.3-9 **(a)** $Q_1 = 61.12, Q_2 = 87.94, Q_3 = 117.89$, $Q_3 - Q_1 = 56.77$.

(b) $\bar{x} = 94.25$ (grams), $s^2 = 2193.79$, $s = 46.84$ (grams).

1.3-13 **(a)** $\bar{x} = 189.0$;

(b) logarithmic transformation.

1.3-14 **(a)** $\bar{x} = 74.4, s = 2.11, Q_1 = 73.1$, $Q_2 = 74.55, Q_3 = 75.6$.

1.3-18 Minitab output

Variable	N	Mean	St Dev	Median
height (inch)	54	69.889	3.755	70.000
weight (pound)	54	175.54	34.05	177.00
number children	54	0.852	1.156	0.000

		N	Mean	St Dev	Median
height (inch)	f	14	65.143	2.445	65.500
	m	40	71.550	2.501	72.000
weight (pound)	f	14	140.86	20.38	137.50
	m	40	187.68	29.23	185.00
number children	f	14	0.357	0.842	0.000
	m	40	1.025	1.209	1.000

gender	Count	Percent
f	14	25.93
m	40	74.07
N =	54	

Rows: gender   Columns: number children

	0	1	2	3	5	All
f	11	2	0	1	0	14
	78.57	14.29	0.00	7.14	0.00	100.00
m	19	7	10	3	1	40
	47.50	17.50	25.00	7.50	2.50	100.00
All	30	9	10	4	1	54
	55.56	16.67	18.52	7.41	1.85	100.00

Cell Contents: Count
              % of Row

1.4-1 **(a)** $\bar{x} = 9.42$, $s = 2.08$, $x_{min} = 2.9$, $Q_1 = 8.65$, $Q_2 = 9.40$, $Q_3 = 10.25$, $x_{max} = 16.7$; lead concentrations have increased; variability in the two groups is about the same.

1.4-3 **(a)**

	x_{min}	Q_1	Q_2	Q_3	x_{max}
Surface	3.08	3.67	4.33	4.40	5.17
Middepth	3.17	4.26	5.04	6.17	6.57
Bottom	3.76	4.90	5.53	7.30	8.79

Variability increases with level. A logarithmic transformation should be tried.

1.5-2 $\hat{\beta} = \sum x_i y_i / \sum x_i^2 = 1.25$.

1.5-4 $\hat{\alpha} = 3.2761$, $\hat{\beta} = 0.3098$.

1.5-6 **(c)** CO concentrations are related to traffic density. Wind plays the role of transport (under calm conditions or when $WS_p < 0$, only part of the emissions are transported to the measurement site) and diffusion (strong winds diffuse part of the emissions before they reach the measurement site).

Ledolter and Tiao consider models of the form

$$CO = \alpha$$
$$+ \beta_1 (TD) \exp[-\beta_2 (WS_p - \beta_3)^2],$$

where α is a parameter measuring background, β_1 is a parameter proportional to emissions, and β_2 and β_3 are parameters that model the effects of WS_p.

1.5-8 $r = 0.63$.

1.5-9 $r = 0.48$; $\hat{\alpha} = 0.25$ and $\hat{\beta} = 2.53$.

1.5-11 $r = 0.71$ (all data); $r = 0.48$ (male); $r = 0.54$ (female).

1.6-5 $2^3 = 8$. For example, responses 70 for (factor 1 = low, factor 2 = low, factor 3 = low), 90 for (high, low, low), 60 for (low, high, low), 80 for (high, high, low), 80 for (low, low, high), 120 for (high, low, high), 90 for (low, high, high), and 100 for (high, high, high), and starting to change factors with factor 1.

1.7-2 Corruptions in presentations of evidence: effects without causes, cherry picking, overreaching, chart junk, and the rage to conclude. Chart junk in Excel. How PowerPoint can corrupt arguments.

Project 3 (e) Additive increases fuel efficiency. Ratios decrease from an average of 1.0243 (without) to 0.9756 (with additive), an improvement of 4.75 percent.

Chapter 2

2.1-1   0.06, 0.08, 0.12, 0.02.

2.1-3 $\{1, 2, 3, 4, 5, 6\}, 1/6, 1/2$.

2.1-6   0.363, 0.738.

2.1-10 **(a)** 0.6;   **(b)** 0.2;   **(c)** 0.7.

2.2-3   0.0010144

2.2-5 $2^7, 1/128, 7/128$.

2.2-7   0.4095

2.2-8   **(a)** No;   **(b)** $1/2, 2/5, 1/5, 7/10$;
           (c) Yes.

2.2-9   **(a)** 0.06;   **(b)** 0.94;   **(c)** $2/7$;
           (d) $1/93$.

2.2-14 **(a)** $64/105$;   **(b)** $7/32$.

2.2-16 **(a)** 0.000354;   **(b)** 354 cases.

2.2-19 0.471.

2.3-1   **(a)** 0.3;   **(b)** 0.9;   **(c)** 3;   **(d)** 1.

2.3-2   **(a)** $1/325$;   **(b)** $1/36$;   **(c)** 2.

2.3-3   **(b)** $1 - (5/6)^x; x = 1, 2, 3, \ldots$;
           (c) 0.2996;   **(d)** 6.

2.3-5   1.3, 0.81, 0.9.

2.3-6 $34/15$.

2.4-1   **(a)** 0.8171;   **(b)** 2.

2.4-3 $n = 22$.

2.4-4   **(a)** 0.1143;   **(b)** $810, $270.

2.4-10 **(a)** 0.50.

2.4-11 **(a)** 0.5695;   **(b)** 0.813.

2.4-14 0.2180, 0.5834; 13, 2.13.

2.4-15 0.2626.

2.5-1   **(a)** 0.018;   **(b)** 0.196;   **(c)** 0.215.

2.5-3   0.014

2.5-4   **(a)** 0.143;   **(b)** 0.865;   **(c)** $2000.

2.5-5 $30,000.

2.6-1   **(a)** X: 0.4, 0.3, 0.3; Y: 0.6, 0.4;
           (b) X: 1.9, 0.69; Y: 1.4, 0.24;
           (c) 0.14, 0.344;   **(d)** Z: 0.3, 0.3, 0.2, 0.2. Yes

2.6-3   **(a)** $f_1(x)$: 0.3, 0.4, 0.3; $f_2(y)$: 0.4, 0.3, 0.3; No, because $(0.3)(0.4) \neq 0.05$;
           (b) $\mu_X = 2$; $\sigma_X^2 = 0.6$; $\mu_Y = 1.9$; $\sigma_Y^2 = 0.69$; $\text{cov}(X, Y) = -0.30$; $\rho = -0.466$;
           (c) $E(Y|x = 1) = 7/3$; $\text{var}(Y|x = 1) = 5/9$; $E(Y|x = 2) = 2$; $\text{var}(Y|x = 2) = 3/4$; $E(Y|x = 3) = 4/3$; $\text{var}(Y|x = 3) = 2/9$.

2.6-9   **(a)** X: 2, 2/3; Y: 4, 4/3;
           (b) $Y: R = \{2, 3, 4, 5, 6\}$ with probabilities $\{1/9, 2/9, 3/9, 2/9, 1/9\}$;
           (c) $W: R = \{1, 2, 3, 4, 6, 9\}$ with probabilities $\{1/9, 2/9, 2/9, 1/9, 2/9, 1/9\}$; 4, 52/9.

Project 1 Adopt the stepped-up in-house version. Expected savings = $24,767.

Project 3 Select $s = -1$ if $x \leq -1$, and select $s = +1$ if $x > -1$.

Chapter 3

3.1-4   **(a)** 0.604;   **(b)** 0.216;   **(c)** 0.343.

3.1-5   **(a)** $2/3, 1/18, \sqrt{0.9}$;   **(b)** $1/2, 1/20$.

3.1-6   **(a)** 0.841, 0.707, 0.931, 0.974;
           (b) 0.693, 0.288, 1.386, 2.303.

3.1-7   **(a)** $(x - a)/(b - a), a \leq x < b$;
           (b) $(a + b)/2; (b - a)^2/12$.

3.2-1   **(a)** 0.6612;   **(b)** 0.2850;   **(c)** 0.2455.

3.2-3   **(a)** 7.30;   **(b)** 4.28;   **(c)** 8.29;
           (d) 3.92.

3.2-6   **(a)** 0.1057;   **(b)** 0.0062;   **(c)** 0.2113;
           (d) 0.0784;   **(e)** 0.2946.

3.2-7   100.82.

3.2-11 0.8185.

3.2-16 **(a)** 5.2645, 5.3326;   **(b)** 0.16;
           (c) Yes; reduced to about 5 percent.

3.3-2   **(a)** 0.368.

3.3-4   **(a)** Gamma with $\alpha = 2$ and $\beta = 4/3$;
           (b) 0.690.

3.3-5   0.908, 0.017.

3.3-7   0.1490.

3.3-15 **(a)** 0.2865;   **(b)** 0.23(Week) = 1.61 days.

3.4-2   **(a)** $1/2, 1/12, 0.95, 0.75 - 0.25 = 0.50$.

3.5-1   **(a)** Yes;   **(b)** $f_1(x) = 2x$ and $f_2(y) = 2y$; independence.

3.5-2 (a) $c = 1$;   (b) $f_1(x) = (1/2) + x$;
$f_2(y) = (1/2) + y$;
$E(X) = E(Y) = 7/12$.

3.5-3 (a) $\mu_D \approx a_0 = 700/200 = 3.5$,

$$\sigma_D \approx \sqrt{(700)^2 \frac{2(0.2)^2}{(200)^4}} = 0.00495;$$

(b) $\sigma_D \approx \sqrt{(700)^2 \frac{2(0.2)^2}{(200)^4} + \frac{25}{(200)^2}}$
$= 0.0255$.

3.5-7 (a) $11, \sqrt{7.25}$;   (b) $11, \sqrt{17.65}$.

3.6-1 Relatively small data set. Normal distribution is unlikely, as there are 6 observations smaller than 15, while the bulk of the data (16 of 25 observations) is larger than 60.

3.6-2 (a) $G(y) = 1 - \exp(-y/\beta)$.

3.6-3 Normal distribution likely.

3.6-11 (a) Normal distribution likely; (b) Lognormal distribution more appropriate.

3.7-2 $R(x) = [1 + (x/\beta)]e^{-x/\beta}$,

$$\lambda(x) = \frac{f(x)}{R(x)} = \frac{x}{\beta^2 + \beta x}.$$

3.7-3 (a) $\lambda = 0.021$; (b) 47.6; (c) $e^{-1} = 0.368$.

3.7-6 $z_p = [-\ln(1 - p^{1/k})]/\lambda$.

3.7-7 $P(Y > 25) = 1 - P(Y \le 25)$
$= 1 - 0.3840 = 0.6110$;
$P(Y > 35) = 1 - P(Y \le 35)$
$= 1 - 0.6993 = 0.3007$.

Project 1 (b) $Q_{Opt} = 58.42$;

(c) $\mu(Q_{Opt}) = 50 \dfrac{u^2}{u + e}$.

Project 2 (c) i. $\log(b - a)$; ii. $(1/2) + \log(1/2)$;
iii. $\log(\beta) + 1$;

(d) $D_{KL}(p|q) = 1 + \log(1/2)$.

Chapter 4

4.1-1 0.7698.

4.1-5 (a) 0.7701;   (b) 0.7455;   (c) No.

4.1-6 (a) 0.5963;   (b) 0.5934.

4.1-7 $1/2, 1/200$.

4.1-8 (b) $0, 1$;
(c) $(2/3 + u/3\sqrt{2})^2, -2\sqrt{2} \le u < \sqrt{2}$.

4.1-9 (a) 15.87%;   (b) 0.0228.

4.1-10 $G(y) = y^2/2, 0 < y \le 1$;
$G(y) = 1 - (2 - y)^2/2, 1 < y \le 2$.

4.2-1 $(47.83, 50.57)$.

4.2-3 43.

4.2-5 $(-168.9 \text{ to } -105.1)$.

4.2-7 $(318.23 \text{ to } 321.97)$.

4.3-1 (a) $(2.316 \text{ to } 2.484)$;   (b) $(2.311 \text{ to } 2.489)$.

4.3-2 (a) $12, 24$;
(b) $21.026, 3.571, 28.300, 4.404$.

4.3-3 (a) $0, 11/9$;   (b) 1.796;   (c) 2.201.

4.3-7 $(-5.6, 11.0)$.

4.3-10 Exercise 4.3-1: $2.4 \pm (2.7)(0.2)$ or (1.98 to 2.94).

4.4-3 $(0.09, 0.23)$.

4.4-5 $(358.4, 1037.3)$.

4.4-6 (a) 289;   (b) 385.

4.5-2 0.9452, 0.7257, 0.3446, 0.0808; 0.0548.

4.5-3 0.0256; reject H_0: $p = 0.73$.

4.5-5 test statistic $= -1.35$, p value $= 0.0885$, do not reject H_0.

4.5-7 24, 33.66.

4.5-8 589, 0.227.

4.5-15 (a) test statistic $= -3.87$, p value $= 0.00005$; reject H_0, students spend less than 400 on average.
(b) test statistic $= +5.81$, p value $= 1.000$; cannot reject H_0: $\mu = 400$ in favor of the alternative H_1: $\mu < 400$; students do not spend less than \$400 on average. We can reject H_0: $\mu = 400$ in favor of the two-sided alternative; average amount spent on books is different from \$400.

4.5-16 (a) $(1.033, 1.067)$;
(b) test statistic $= 5.89$, p value $= 0.000$; can reject H_0: $\mu = 1$ in favor of H_1: $\mu > 1$;
(c) 0.2023;
(d) $1.05 \pm (2)(0.06)$.

4.5-17 p value $= 0.43$; cannot reject H_0: $\mu = 4$;

4.6-1 $|-2.15| > 1.96$; p value $= 0.033$; reject H_0.

4.6-3 $n = 68$.

4.6-4 $4.03 > 1.725$; reject H_0.

4.6-6 $-2.27 < -1.645$; reject H_0.

4.6-9 (a) $24.12/7.82 = 3.08 > 2.48$; reject H_0.

4.6-12 test statistic $= 2.14, p$ value $= 2P(Z > 2.14)$ $= 0.032$; men and women are paid differently, on average. Need to factor in covariates (equal pay for equal work); see Chapter 8 (regression).

4.7-1 $6.1 < 11.070$; do not reject H_0.

4.7-2 $36.37 > 5.991$; reject H_0.

4.7-3 $3.603 < 12.592$; accept independence.

4.7-5 $1.7 < 7.815$; do not reject H_0.

Project 5 $n = \dfrac{2\sigma^2\{z(\alpha) - [z(1 - \beta)]\}^2}{\delta^2}$
$$= \frac{(2)(25)(1.645 + 1.28)^2}{4} \approx 108.$$

Project 11 (a) p value $= 0.01 < 0.05$; reject H_0: $\mu_{\text{Placebo}} = \mu_{\text{Iron}}$ in favor of H_1: $\mu_{\text{Placebo}} \neq \mu_{\text{Iron}}$.

Project 12 . Reject the null hypothesis of a Poisson forest evidence for clustering.

Chapter 5

5.1-2 $\bar{x} = 5.16$, $\bar{R} = 1.18$; control limits for $\bar{x}$-chart: 4.48, 5.84; control limits for R-chart: $0, 2.49$; process is under control.

5.1-3 Control limits for p-chart: $0, 0.0305$; process has gone out of control.

5.1-4 $LCL = 0, UCL = 0.072 < 0.08$.

5.1-5 $UCL = 4.486, 0.053, 0.58$.

5.1-7 $UCL = 7.40$, yes.

5.2-1 (a) $\hat{C}_p = 0.56$; $\hat{C}_{pk} = 0.52$; $\hat{C}_{pm} = 0.56$; process is on target, but has too much variability.
(b) $(0.38 \text{ to } 0.66)$.

5.2-2 (a) $\hat{C}_p = 0.81, \hat{C}_{pk} = 0.64$; process mean is close to the target, but process variability is too large.
(b) $0.0268; 0.0016; 0.0284$.

5.2-5 $\bar{x} = 99.8$, σ is estimated by $\sqrt{5}/3 = 0.745$, $\hat{C}_p = 1.34, \hat{C}_{pk} = 1.25$.

5.2-8 $P[\text{exceed}] = 1 - [\Phi(3C_p - k) - \Phi(-3C_p - k)]$, where $\Phi(z)$ is the c.d.f. of the standard normal distribution.
(a) 0.00034 (percent);   (b) 0.00320 (percent);   (c) 0.02326 (percent).

5.3-2 (b) 0.143;   (c) 0.010;
(d) $AOQL \approx 0.02$.

5.3-4 (a) $0.122, 0.199$;
(b) $AOQL \approx 0.0028$.

5.3-5 $n = 100, Ac = 4$ gives $\alpha = 0.053$, $\beta = 0.100$.

5.3-6 0.9123.

Project 1 (a) Mortality and pigs/litter in statistical control;   (b) influence of weather not very strong.

Chapter 6

6.1-2

Source	SS	df	MS	F
Treatment	31.112	2	15.56	22.3
Error	29.260	42	0.70	
Total	60.372	44		

$F_{\text{Treatment}} = 22.3 > F(0.05; 2, 42) = 3.22$; mean lengths are different.

6.1-3 (a)

Source	SS	df	MS	F
Treatment	475.76	4	118.94	14.76
Error	161.20	20	8.06	
Total	636.96	24		

$F_{\text{Treatment}} = 14.76 > F(0.05; 4, 20) = 2.87$; there are differences among the mean breaking strengths.

6.1-4 (a)

Source	SS	df	MS	F
Treatment	193,011	2	96,506	12.5
Error	85,160	11	7,742	
Total	278,171	13		

$F = 12.5 > F(0.05; 2, 11) = 3.98$, so we conclude that there are differences among the mean weight gains.

6.1-5

Source	SS	df	MS	F
Treatment	1.5474	2	0.774	1.08
Error	17.2022	24	0.717	
Total	18.7496	26		

$F_{\text{Treatment}} = 1.08 < F(0.05; 2, 24) = 3.40$; no significant differences.

6.2-2 **(a)**

Source	SS	df	MS	F
Treatment	1.6333	2	0.817	2.34
Error	4.1840	12	0.349	
Total	5.8173	14		

$F = 2.34 < F(0.05; 2, 12) = 3.89$; insufficient evidence to conclude that nozzle type affects the flow.

(b) Scaled t-distribution with 12 degrees of freedom and scale factor $(MS_{\text{Error}}/5)^{1/2} = 0.26$.

6.2-3 **(a)** Random-effects model.

(b)

Source	SS	df	MS	F
Batch	170.94	5	34.19	5.13
Error	80.00	12	6.67	
Total	250.94	17		

$F = 5.13 > F(0.05; 5, 12) = 3.11$, indicating significant batch variability.

(c) $\hat{\sigma}^2 = 6.67, \hat{\sigma}_\tau^2 = 9.17$.

6.3-1 **(a)**

Source	SS	df	MS	F
Treatment	165.8	3	55.3	7.5
Block	905.3	4	226.3	30.6
Error	88.7	12	7.4	
Total	1159.8	19		

$F_{\text{Treatment}} = 7.5 > F(0.05; 3, 12) = 3.49$; differences among the fabrics. Use a scaled t-distribution with 12 degrees of freedom and scale factor $(7.4/5)^{1/2} = 1.22$ to compare the treatment averages.

(b) $F_{\text{Block}} = 30.6 > F(0.05; 4, 12) = 3.26$; the blocking arrangement is very useful. If blocks are ignored, the estimate of the error variance is

$$\frac{(905.3 + 88.7)}{4 + 12} = 62.1;$$

then the resulting F-statistic for treatment differences in the completely randomized arrangement,

$$\frac{55.3}{62.1} = 0.89 < F(0.05; 3, 16) = 3.24,$$

is insignificant. Blocking has increased the precision of our comparison.

6.3-3 **(a)**

Source	SS	df	MS	F
Treatment (experience)	147.73	2	73.87	24.3
Block (day)	231.73	4	57.93	19.1
Error	24.27	8	3.03	
Total	403.73	14		

$F_{\text{Treatment}} = 24.3 > F(0.05; 2, 8) = 4.46$; we find that there are differences in productivity among the three groups.

The group averages (already ordered in magnitude) are $\bar{y}_A = 54.4$, $\bar{y}_B = 57.2, \bar{y}_C = 62.0$.

Significant entries in the table of pairwise comparisons

$$\bar{y}_B - \bar{y}_A = 2.8 \quad \bar{y}_C - \bar{y}_A = 7.6^{**}$$
$$\bar{y}_C - \bar{y}_B = 4.8^{**}$$

are denoted by **;
$q(0.05; 3, 8) \sqrt{MS_{\text{Error}}/b} = 3.15$ is used to determine significance.

Plot the treatment averages, and compare them with the scaled $t(v = 8)$ distribution with scale factor $\sqrt{3.03/5} = 0.78$. Group C is very different from A and B.

(b) $F_{\text{Block}} = 19.1 > F(0.05; 4, 8) = 3.84$.

6.4-2 **(a)**

Source	SS	df	MS	F
Treatment	330.24	4	82.56	0.56
Row	4240.24	4	1060.06	7.25
Column	701.84	4	175.46	1.2
Error	1754.32	12	146.19	
Total	7026.64	24		

$F_{\text{Treatment}} = 0.56 < F(0.05; 4, 12) = 3.26$, so the differences among mean root weights are not significant.

(b) Blocking has increased the precision of our comparison. If the same data were obtained from a completely randomized one-factor experiment, we would have obtained $SS_{Error} = 4240.42 + 701.84 +$

$1754.32 = 6696.4$, with $4 + 4 + 12 = 20$ degrees of freedom. The resulting estimate of the error variance, $6696.4/20 = 334.8$, is 2.5 times larger than $MS_{Error} = 146.19$.

Chapter 7

7.1-3

Source	SS	df	MS	F
A (background)	100.04	1	100.04	19.02
B (training)	20.25	2	10.12	1.92
AB	33.58	2	16.79	3.19
Error	94.75	18	5.26	
Total	248.62	23		

Fixed effects.

Interaction is borderline significant, because its probability value $P[F(2, 18) \geq 3.19] = 0.065$ is only slightly larger than the 0.05 level; interaction is significant at the 10 percent significance level.

Main effect A (background) is significant, because $F_A = 19.02 > F(0.05; 1, 18) = 4.41$.

Main effect B (training) is insignificant, because $F_B = 1.92 < F(0.05; 2, 18)$.

However, because of the borderline significant interaction, one should not interpret the effects of factors without looking at the interaction diagram first. Sales by college graduates tend to be higher, on average, than sales by high school graduates. We should investigate the second training program. It works well with high school graduates, but not with college graduates. (This creates the interaction.)

Residual plots show no major model inadequacies.

7.1-4

Source	SS	df	MS	F
A (time)	150.2	1	150.2	9.69
B (temperature)	80.8	2	40.4	2.61
AB	3.4	2	1.7	0.11
Error	186.0	12	15.5	
Total	420.4	17		

Interaction is insignificant, because $F_{AB} < 1$.

Time is an important factor; $F_A = 9.69 > F(0.05; 1, 12) = 4.75$ is significant; probability value $P[F(1, 12) \geq 9.69] = 0.009$.

Temperature main effect is not quite significant, because $F_B = 2.61 < F(0.05; 2, 12) = 3.89$. Its probability value $P[F(2, 12) \geq 2.61] = 0.115$ is slightly larger than the usually required significance level of 0.05.

The interaction plot shows the absence of interaction and the strong main effect of time, with longer times increasing the brightness. The graph shows evidence that increases in temperature increase brightness (linearly over the studied range). However, because of the large variability of individual measurements, this effect is only borderline significant. Additional observations would help strengthen the evidence for a temperature effect.

Residual plots show no major model inadequacies.

7.1-5 **(a)** The machine and operator effects are random.

(b)

Source	SS	df	MS
A (machine)	0.878	2	0.439
B (operator)	9.211	2	4.606
AB	0.489	4	0.122
Error	16.4	36	0.456
Total	26.978	44	

(c) $F_{AB} = MS_{AB}/MS_{Error} = 0.27 < F(0.05; 4, 36) = 2.61$; interaction is insignificant.

(d) $MS_A/MS_{AB} = 3.6 < F(0.05; 2, 4) = 6.94$; machine effect is insignificant.

$MS_B/MS_{AB} = 37.7 > F(0.05; 2, 4)$; operator effect is highly significant.

Conclusions do not change if $MS_{Error(pooled)} = 0.4222$ is used.

7.2-1 **(a)**

Source	SS	df	MS	F
A: batch	1210.93	14	86.495	1.49
B(A): sample (batch)	869.75	15	57.983	63.23
Error	27.50	30	0.917	
Total	2108.18	59		

$MS_{B(A)}/MS_{Error} = 63.23 > F(0.05; 15, 30)$
$= 2.01$; sample variability is significant.
$MS_A/MS_{B(A)} = 1.49 < F(0.05; 14, 15)$
$= 2.43$; batch variability is insignificant.

(b) Measurement: $\hat{\sigma}^2 = 0.92$
Sample: $\hat{\sigma}_\beta^2 = 28.53$
Batch: $\hat{\sigma}_\alpha^2 = 7.13$
Sampling procedure should be improved.

7.2-3

Source	SS	df	MS	F
A: batch	109.6	5	21.92	1.06
B(A): sample (batch)	124.2	6	20.70	4.64
Error	107.0	24	4.46	
Total	340.8	35		

$MS_{B(A)}/MS_{Error} = 4.64 > F(0.05; 6, 24) = 2.51$; sample variability is significant.
$MS_A/MS_{B(A)} = 1.06 < F(0.05; 5, 6) = 4.39$; batch variability is insignificant.
$\hat{\sigma}^2 = 4.46; \hat{\sigma}_\beta^2 = 5.41; \hat{\sigma}_\alpha^2 = 0.20$.

7.3-3 **(a)** Average = 78.375 $(12) = -1.525$
$(1) = -3.55$ $(13) = -0.525$
$(2) = -1.45$ $(23) = 0.375$
$(3) = 3.20$ $(123) = -1.20$.

(b) Main effects of factors 1 and 3 are significant.

7.3-4 **(a)** Average = 72.25 $(4) = -2.75$
$(1) = -4.00$ $(14) = 0.00$
$(2) = 12.00$ $(24) = 2.25$
$(12) = 0.50$ $(124) = 0.25$
$(3) = -1.13$ $(34) = -0.13$
$(13) = 0.38$ $(134) = -0.13$
$(23) = -0.63$ $(234) = -0.38$
$(123) = -0.38$ $(1234) = -0.13$.

(b) Normal probability plot of the 15 effects shows that $(1), (2), (4)$, and (24) are large.

7.3-6 **(a)** Average = 1.209 $(AB) = -0.076$
$(A) = 0.274$ $(AC) = 0.019$
$(B) = 0.039$ $(BC) = 0.279$
$(C) = -0.156$ $(ABC) = -0.006$

(b) $s^2 = (0.83 - 1.40)^2/2 = 0.1625$ is a very crude variance estimate from the two replications at the center point; var(effect) $= s^2/8 = 0.0203$; standard error (effect) $= 0.14$.

(c) (A) and (BC) are large (two standard errors from zero).

Factor A does not interact with the other factors. Its main effect is positive; we should set additive A at its low level in order to reduce losses.

The results at the low level of A show that we should set B at the low level and C at the high one.

7.4-2 Start with the 2^4 factorial, specify the 11 interaction columns, and associate x_5 with $x_1x_2x_3x_4$. Write out the levels of the 16 runs.

There is only a single generator; hence, the defining relation is given by $I = x_1x_2x_3x_4x_5$. In addition to the mean, we can form 15 linear combinations that use the weights (± 1) in the design columns of the original 2^4 [i.e., x_1, x_2, x_3, x_4] and the 11 interaction columns [i.e., $x_1x_2, x_1x_3, x_1x_4,$ $x_2x_3, x_2x_4, x_3x_4,$ and $x_1x_2x_3, x_1x_2x_4, x_1x_3x_4,$ $x_2x_3x_4, x_1x_2x_3x_4$]. The weighted sum is then divided by 16, the number of runs.

Let us specify the column of weights in parentheses, as this makes it clear which column of weights is being used by the linear combination. Then these 15 linear combinations estimate

$$L(x_1) \rightarrow (1) + (2345);$$
$$L(x_2) \rightarrow (2) + (1345);$$
$$L(x_3) \rightarrow (3) + (1245);$$
$$L(x_4) \rightarrow (4) + (1235);$$
$$L(x_1x_2) \rightarrow (12) + (345);$$
$$L(x_1x_3) \rightarrow (13) + (245);$$
$$L(x_1x_4) \rightarrow (14) + (235);$$
$$L(x_2x_3) \rightarrow (23) + (145);$$
$$L(x_2x_4) \rightarrow (24) + (135);$$
$$L(x_3x_4) \rightarrow (34) + (125);$$
$$L(x_1x_2x_3) \rightarrow (123) + (45);$$
$$L(x_1x_2x_4) \rightarrow (124) + (35);$$

$L(x_1x_3x_4) \rightarrow (134) + (25)$;

$L(x_2x_3x_4) \rightarrow (234) + (15)$;

$L(x_1x_2x_3x_4) \rightarrow (1234) + (5)$.

The design is of resolution V. The confounding patterns simplify if we assume

Run	x_1	x_2	x_3	$x_4 = x_1x_2$	x_1x_3	x_2x_3	$x_5 = x_1x_2x_3$
1	−	−	−	+	+	+	−
2	+	−	−	−	−	+	+
3	−	+	−	−	+	−	+
4	+	+	−	+	−	−	−
5	−	−	+	+	−	−	+
6	+	−	+	−	+	−	−
7	−	+	+	−	−	+	−
8	+	+	+	+	+	+	+

Defining relation:

$$I = x_1x_2x_4 = x_1x_2x_3x_5 = x_3x_4x_5.$$

Resolution III.

The eight linear combinations (corresponding to the column of all plus signs and the seven columns of the preceding table) estimate

$L_0 \rightarrow \mu + (124) + (1235) + (345)$;

$L_1 \rightarrow (1) + (24) + (235) + (1345)$;

$L_2 \rightarrow (2) + (14) + (135) + (2345)$;

$L_3 \rightarrow (3) + (1234) + (125) + (45)$;

$L_4 \rightarrow (12) + (4) + (35) + (12345)$;

$L_5 \rightarrow (13) + (234) + (25) + (145)$;

$L_6 \rightarrow (23) + (134) + (15) + (245)$;

$L_7 \rightarrow (123) + (34) + (5) + (1245)$.

Ignoring interactions of order 3 or higher, we have

$L_0 \rightarrow \mu; L_1 \rightarrow (1) + (24); L_2 \rightarrow (2) + (14)$;

$L_3 \rightarrow (3) + (45); L_4 \rightarrow (12) + (4)$;

$L_5 \rightarrow (13) + (25); L_6 \rightarrow (23) + (15)$;

$L_7 \rightarrow (5) + (34)$.

7.4-6 **(c)** The linear combinations from the fold-over estimate

$L_0' \rightarrow \mu$;

$L_1' \rightarrow (1) - (24) - (35) - (67)$;

that interactions of order 3 and higher are negligible.

7.4-3 Start with a 2^3 factorial, specify the four interaction columns, and associate x_4 with x_1x_2 and x_5 with $x_1x_2x_3$:

$L_2' \rightarrow (2) - (14) - (36) - (57)$;

$L_3' \rightarrow (3) - (15) - (26) - (47)$;

$L_4' \rightarrow (4) - (12) - (37) - (56)$;

$L_5' \rightarrow (5) - (13) - (27) - (46)$;

$L_6' \rightarrow (6) - (23) - (17) - (45)$;

$L_7' \rightarrow (7) - (34) - (25) - (16)$.

7.4-8 The design and its confounding patterns are described in Exercise 7.4-2.

Using a computer program (e.g., Minitab), we obtain the following estimates [the expressions to the right of the arrow indicate what these linear combinations estimate if we ignore interactions of order 4 and higher (3 and higher)]:

$2.8750 \rightarrow \mu$;

$L(x_1) = 0.8225 \rightarrow (1)$;

$L(x_2) = -1.2525 \rightarrow (2)$;

$L(x_3) = 0.3838 \rightarrow (3)$;

$L(x_4) = 2.7925 \rightarrow (4)$;

$L(x_1x_2) = 0.0550 \rightarrow (12)$;

$L(x_1x_3) = 0.0638 \rightarrow (13)$;

$L(x_1x_4) = -0.0950 \rightarrow (14)$;

$L(x_2x_3) = 0.0413 \rightarrow (23)$;

$L(x_2x_4) = -0.0450 \rightarrow (24)$;

$L(x_3x_4) = -0.3138 \rightarrow (34);$

$L(x_1x_2x_3) = 0.0013 \rightarrow (123) + (45)$

$= (45);$

$L(x_1x_2x_4) = -0.2875 \rightarrow (124) + (35)$

$= (35);$

$L(x_1x_3x_4) = 0.1863 \rightarrow (134) + (25)$

$= (25);$

$L(x_2x_3x_4) = -0.3063 \rightarrow (234) + (15)$

$= (15);$

$L(x_1x_2x_3x_4) = -0.8713 \rightarrow (5).$

The normal probability plot of the estimated effects shows that the main effects of factors 1, 2, 4, and 5 are large.

7.4-9 Using the Minitab computer software and the data in file Section7.4Exercise7.4-9 Viscosity, we obtain the following results: Significant main effect of $x_4 =$ mixing time (-1.481), and effects of $x_6 =$ spindle (-3.269) and $x_3 =$ mixing speed (2.344), which interact in their effects on viscosity.

Project 6 (c) Main effects A and G are significant; string of two-factor interactions $AC + BE + DG$ is borderline significant at the 10 % level.

Chapter 8

8.1-1 (b) $\hat{\beta}_0 = -7.877, \hat{\beta}_1 = 0.08676;$
$\hat{y}_1 = -7.877 + (0.08676)(117) = 2.27,$
$e_1 = 2.07 - 2.27 = -0.20;$
$\hat{y}_2 = -7.877 + (0.08676)(128) = 3.23,$
$e_2 = 2.80 - 3.23 = -0.43.$

8.1-2 (b) $\hat{\beta}_0 = 0.989, \hat{\beta}_1 = 1.99463;$
$\hat{y}_1 = 0.989 + (1.99463)(10) = 20.94,$
$e_1 = 21.2 - 20.94 = 0.26;$
$\hat{y}_3 = 0.989 + (1.99463)(11) = 22.93,$
$e_3 = 22.5 - 22.93 = -0.43.$

8.1-3 (b) $\hat{\beta}_0 = 54.95, \hat{\beta}_1 = 1.0641;$
$\hat{y}_1 = 54.95 + (1.0641)(50) = 108.16,$
$e_1 = 128 - 108.16 = 19.84;$
$\hat{y}_2 = 54.95 + (1.0641)(64) = 123.05,$
$e_2 = 159 - 123.05 = 35.95.$

8.1-8 (a) $\hat{\beta} = \sum x_i y_i / \sum x_i^2.$
(b) $\mathrm{var}(\hat{\beta}) = \sigma^2 / \sum x_i^2.$

8.2-1 $\hat{\beta}_1 = 0.35, s(\hat{\beta}_1) = (2.3/100)^{1/2} = 0.152.$
Test statistic $= (0.35 - 0.20)/0.152 = 0.99 <$
$t(0.05; 16 - 2 = 14) = 1.761;$
insufficient evidence to reject H_0: $\beta_1 = 0.20;$
we accept H_0;
confidence interval $(0.08 \le \beta_1 \le 0.62).$

8.2-2 **For Exercise 8.1-1:**

Source	SS	df	MS	F
Regression	2.306	1	2.306	28.9
Error	0.639	8	0.080	
Total	2.945	9		

$R^2 = 2.306/2.945 = 0.783.$
$\hat{\beta}_1 = 0.087, s(\hat{\beta}_1) = 0.016, t(\hat{\beta}_1) = \hat{\beta}_1 / s(\hat{\beta}_1) = 5.37;$ because $|t(\hat{\beta}_1)| = 5.37$ is larger than $t(0.025; 8) = 2.306$, we reject H_0: $\beta_1 = 0$ in favor of H_1: $\beta_1 \ne 0$; $(0.050 \le \beta_1 \le 0.124).$
$\hat{\beta}_0 = -7.877, s(\hat{\beta}_0) = 2.027, t(\hat{\beta}_0) = \hat{\beta}_0 / s(\hat{\beta}_0) = -3.89;$ because $|t(\hat{\beta}_0)| = 3.89$ is larger than $t(0.025; 8) = 2.306$, we reject H_0: $\beta_0 = 0$ in favor of H_1: $\beta_0 \ne 0.$

For Exercise 8.1-2:

Source	SS	df	MS	F
Regression	2306.4	1	2306.4	1949
Error	14.2	12	1.18	
Total	2320.6	13		

$R^2 = 2306.4/2320.6 = 0.994.$

$t(\hat{\beta}_1) = \hat{\beta}_1/s(\hat{\beta}_1) = 1.9946/0.0452 = 44.1 > t(0.025; 12) = 2.179$, reject $H_0: \beta_1 = 0$ in favor of $H_1: \beta_1 \neq 0$ at significance level $\alpha = 0.05$. The probability value is extremely small; very strong evidence that $\beta_1 \neq 0$. $t(\hat{\beta}_0) = \hat{\beta}_0/s(\hat{\beta}_0) = 0.989/0.859 = 1.15 < t(0.025; 12) = 2.179$; conclude that $\beta_0 = 0$.

For Exercise 8.1-3:

Source	SS	df	MS	F
Regression	1522.9	1	1522.9	4.09
Error	4834.9	13	371.9	
Total	6357.7	14		

$R^2 = 0.24$.
$\hat{\beta}_1 = 1.0641; s(\hat{\beta}_1) = 0.526; \hat{\beta}_1/s(\hat{\beta}_1) = 2.02$ is larger than $t(0.05; 13) = 1.771$; reject $H_0: \beta_1 = 0$ in favor of $H_1: \beta_1 \neq 0$ at significance level $\alpha = 0.10$.

8.2-7 $x_k = 60 : (108.04 \leq E(Y_k) \leq 129.56)$;
$x_k = 100 : (114.72 \leq E(Y_k) \leq 208.00)$.

8.3-5 (b) $\hat{\beta}_0 = 0.839; \hat{\beta}_1 = 0.489$.

Source	SS	df	MS	F
Regression	1830.7	1	1830.7	202
Error	63.3	7	9.04	
Total	1893.9	8		

$R^2 = 0.967$.
(c) Here, $t(\hat{\beta}_1) = \hat{\beta}_1/s(\hat{\beta}_1) = 0.489/0.034 = 14.2$ is much larger than the critical value $t(0.025; 7) = 2.365$; strong evidence that $\beta_1 > 0$.
(d) $t(\hat{\beta}_0) = \hat{\beta}_0/s(\hat{\beta}_0) = 0.839/2.15 = 0.39$ is smaller than the critical value 2.365; we conclude that $\beta_0 = 0$. Dispersion and age are proportional.
(e) $(49.59 \leq E(Y_k) \leq 59.67)$; extrapolation is dangerous because $x_k = 110$ is outside the experimental region.

8.3-8 (a) $\hat{\beta}_0 = 29.01, \hat{\beta}_1 = 1.06$.

Source	SS	df	MS	F
Regression	459.74	1	459.74	106
Error	65.17	15	4.34	
Total	524.92	16		

$R^2 = 0.876$.

(b) Residual plots show curvilinear patterns. Lag 1 autocorrelation $r_1 = 0.63$ is larger than $2/\sqrt{17} = 0.49$. The DW test is very small; positive autocorrelation among the residuals.

8.4-1 $\hat{\beta}_0 = 1, \hat{\beta}_1 = 0.5, \hat{\beta}_2 = 0.75$.

8.4-2 (a) $1.14 \pm (2.145)(0.16)$, or $(0.80 \leq \beta_1 \leq 1.48)$; $-0.83 \pm (2.145)(0.20)$, or $(-1.26 \leq \beta_2 \leq -0.40)$.
(b) $R^2 = 66/100 = 0.66$; $F = 13.59 > F(0.05; 2, 14) = 3.74$; reject $H_0: \beta_1 = \beta_2 = 0$ in favor of H_1 that at least one of the coefficients is different from zero.

8.4-3 (a)

Source	SS	df	MS	F
Regression	66	2	33.00	4.19
Error	134	17	7.88	
Total	200	19		

(b) $R^2 = 0.33$.
(c) $F = 4.19 > F(0.05; 2, 17) = 3.59$; we reject $H_0: \beta_1 = \beta_2 = 0$.
(d) $t(0.025; 17) = 2.11$. Estimate $\hat{\beta}_2$ is not significantly different from zero. We can simplify the regression model to $Y_i = \beta_0 + \beta_1 x_{i1} + \varepsilon_i$.

8.5-1 (a) $\hat{\beta}_0 = 86.436, \hat{\beta}_1 = -0.202$.

Source	SS	df	MS	F
Regression	24.5	1	24.5	0.29
Error	686.4	8	85.8	
Total	710.9	9		

$t(\hat{\beta}_1) = \hat{\beta}_1/s(\hat{\beta}_1) = -0.202/0.377 = -0.53$. $|t(\hat{\beta}_1)| < t(0.025; 8) = 2.306$, so we accept H_0. Residual plots reveal serious model inadequacies; quadratic term x^2 missing in the model.
(b) $\hat{\beta}_0 = 35.66, \hat{\beta}_1 = 5.26, \hat{\beta}_2 = -0.128$; $s(\hat{\beta}_0) = 5.62, s(\hat{\beta}_1) = 0.56, s(\hat{\beta}_2) = 0.013; |t(\hat{\beta}_2)| = 9.97 > t(0.025; 7) = 2.365; \beta_2 \neq 0$.

8.5-2 (b) $\hat{y} = 23.35 + 1.045x; R^2 = 0.955$. Residuals point to problems with this model; quadratic component x^2 is missing.

(c) $\hat{y} = 22.56 + 1.668x - 0.068x^2$; $R^2 = 0.988$; $\hat{\beta}_2/s(\hat{\beta}_2) = -6.59$ is significant; residuals look much better.

8.5-3 (a) $\hat{y} = 73.4 + 12.1x_1 + 3.12x_2 + 2.37x_1x_2$.

Source	SS	df	MS	F
Regression	1,299.37	3	433.12	17.6
Error	98.50	4	24.62	
Total	1,397.87	7		

$R^2 = 0.93$.

(b) $F = 17.6 > F(0.01; 3, 4) = 16.69$; reject H_0: $\beta_1 = \beta_2 = \beta_3 = 0$ at significance level $\alpha = 0.01$. The t-ratios for $\hat{\beta}_1, \hat{\beta}_2$, and $\hat{\beta}_3$ are given by 6.91, 1.78, and 1.35, respectively. Comparing these ratios with $t(0.025; 4) = 2.776$, we find that $\beta_1 \neq 0$. Only the main effect of temperature is significant.

8.5-5 (a) $\hat{y} = 56.25 + 1.79x_1 + 0.087x_2$.

(b) 17th soil sample is an outlier.

(c) After omitting this observation, we find that $\hat{y} = 66.47 + 1.29x_1 - 0.11x_2$; $R^2 = 0.525$.

(d) The simplified model is given by $\hat{y} = 62.57 + 1.23x_1$; $R^2 = 0.519$.

8.5-6 (a) $\hat{y} = 0.022 + 0.00131x_1 + 0.804x_2$; $R^2 = 0.347$.
$F = MSR/MSE = 2.39 < F(0.05; 2, 9) = 4.26$; very little reason to reject H_0: $\beta_1 = \beta_2 = 0$. For this particular data set, the GMAT scores have little explanatory power. The largest t-ratio fails to exceed $t(0.025; 9) = 2.262$.

8.5-13 (a) 0.932;

(b) $F = 110.35$;   reject H_0: $\beta_1 = \beta_2 = \beta_3 = 0$;

(c) $81.87 \pm (2.06)(47.82)$,   or $(-16.6$ to $180.4)$, can simplify the model by removing "bath."

8.5-14 (a) 0.924;

(b) $F = 152.30$; reject H_0: $\beta_1 = \beta_2 = 0$;

(c) no;

(d) explanatory variables are correlated.

8.5-15 Consider log transformations.

8.6-1 (a) $x_{1,0} = 1, x_{2,0} = 2$; maximum.

(b) $x_{1,0} = 1, x_{2,0} = 2$; minimum.

(c) $x_{1,0} = 1, x_{2,0} = 2$; saddle point.

(d) $x_{1,0} = 0.18, x_{2,0} = 0.77$; maximum.

8.6-2 $E(Y) = 11 - (x_1 + 2x_2 - 1)^2$.

8.6-3 Fitting the first-order model to the coded variables

$$x_1 = C - 13 \quad \text{and} \quad x_2 = \frac{T - 105}{5},$$

we obtain

$$\hat{y} = 80.5 + (2.25)x_1 + (5.25)x_2.$$

The design points on the path of steepest ascent are given by $(x_1 = 0.39, x_2 = 0.92)$ or $(C = 13.39, T = 109.6)$, $(x_1 = 0.78, x_2 = 1.84)$ or $(C = 13.78, T = 114.2)$, $(x_1 = 1.17, x_2 = 2.76)$ or $(C = 14.17, T = 118.8)$, and so on.

8.6-4 $\hat{y} = 34.36 + 0.7x_1 - 0.3x_2 - 2.83x_1^2 - 0.93x_2^2 + 0.98x_1x_2$. Stationary point $(x_{1,0} = 0.1054, x_{2,0} = -0.1058)$, or (temperature = 145.53, time = 88.94), corresponds to a maximum.

8.6-7 $\hat{y} = 82.17 - 1.01x_1 - 6.09x_2 + 1.02x_1^2 - 4.48x_2^2 - 3.60x_1x_2$. Stationary point $(x_{1,0} = -0.4121, x_{2,0} = -0.5141)$, or (time = 9.96, temperature = 239.72), corresponds to a saddle point.

8.6-9 The sample variance s^2 from the n_1 responses at the center point provides an estimate of σ^2. Then

$$\text{var(effect)} = s^2/2^k;$$

$$\text{var(difference of the two averages)} = (s^2/2^k) + (s^2/n_1).$$

Project 2 $\ln(\text{Price}) = -12.159 + 0.617\text{Temp} - 0.00387\text{Rain} + 0.00117\text{PRain} + 0.0239\text{Age}$; $R^2 = 0.828$.

Project 6 (a) $\hat{y} = 183.94 + 4.89x_1 + 6.48x_2 - 0.71x_3 - 4.42x_1^2 - 10.41x_2^2 - 1.92x_3^2 - 10.94x_1x_2 + 0.04x_1x_3 + 1.68x_2x_3$.

(c) $\hat{y} = 183.31 + 4.80x_1 + 6.39x_2 - 4.78x_1^2 - 10.95x_2^2 - 10.84x_1x_2$.

Stationary point $(x_{1,0} = 0.3903, x_{2,0} = 0.0986)$, or (GLDH = 138.7, pH = 7.47), corresponds to a maximum.

(e) Sketch the contours for $\hat{y} = (0.97)(184.56) = 179.02$; insensitive.

Appendix

C

STATISTICAL TABLES

Table C.1 Factors for Determining the 3σ Control Limits in $\bar{x}$-Charts and R-Charts

Number of Observations in Sample, n	Factors for $\bar{x}$-Charts		Factors for R-Chart	
	Using $\bar{s}$ A_3	Using $\bar{R}$ A_2	D_3	D_4
2	2.66	1.88	0	3.27
3	1.95	1.02	0	2.57
4	1.63	0.73	0	2.28
5	1.43	0.58	0	2.11
6	1.29	0.48	0	2.00
7	1.18	0.42	0.08	1.92
8	1.10	0.37	0.14	1.86
9	1.03	0.34	0.18	1.82
10	0.98	0.31	0.22	1.78
11	0.93	0.29	0.26	1.74
12	0.89	0.27	0.28	1.72
13	0.85	0.25	0.31	1.69
14	0.82	0.24	0.33	1.67
15	0.79	0.22	0.35	1.65
16	0.76	0.21	0.36	1.64
17	0.74	0.20	0.38	1.62
18	0.72	0.19	0.39	1.61
19	0.70	0.19	0.40	1.60
20	0.68	0.18	0.41	1.59

Source: Reproduced and adapted with permission from E. L. Grant, *Statistical Quality Control*, 2nd ed. (New York: McGraw-Hill, 1952), pp. 513 and 514.

Table C.2 Binomial Distribution Function (c.d.f.)

$$P(X \le x; p) = \sum_{k=0}^{x} \frac{n!}{k!(n-k)!} p^k (1-p)^{n-k}$$

$$P(X \le x; p) = P(X \ge n - x; 1 - p) = 1 - P(X \le n - x - 1; 1 - p)$$

n	x					p					
		0.05	0.10	0.15	0.20	0.25	0.30	0.35	0.40	0.45	0.50
2	0	0.9025	0.8100	0.7225	0.6400	0.5625	0.4900	0.4225	0.3600	0.3025	0.2500
	1	0.9975	0.9900	0.9775	0.9600	0.9375	0.9100	0.8775	0.8400	0.7975	0.7500
	2	1.0000	1.0000	1.0000	1.0000	1.0000	1.0000	1.0000	1.0000	1.0000	1.0000
3	0	0.8574	0.7290	0.6141	0.5120	0.4219	0.3430	0.2746	0.2160	0.1664	0.1250
	1	0.9928	0.9720	0.9392	0.8960	0.8438	0.7840	0.7182	0.6480	0.5748	0.5000
	2	0.9999	0.9990	0.9966	0.9920	0.9844	0.9730	0.9571	0.9360	0.9089	0.8750
	3	1.0000	1.0000	1.0000	1.0000	1.0000	1.0000	1.0000	1.0000	1.0000	1.0000
4	0	0.8145	0.6561	0.5220	0.4096	0.3164	0.2401	0.1785	0.1296	0.0915	0.0625
	1	0.9860	0.9477	0.8905	0.8192	0.7383	0.6517	0.5630	0.4752	0.3910	0.3125
	2	0.9995	0.9963	0.9880	0.9728	0.9492	0.9163	0.8735	0.8208	0.7585	0.6875
	3	1.0000	0.9999	0.9995	0.9984	0.9961	0.9919	0.9850	0.9744	0.9590	0.9375
	4	1.0000	1.0000	1.0000	1.0000	1.0000	1.0000	1.0000	1.0000	1.0000	1.0000
5	0	0.7738	0.5905	0.4437	0.3277	0.2373	0.1681	0.1160	0.0778	0.0503	0.0312
	1	0.9774	0.9185	0.8352	0.7373	0.6328	0.5282	0.4284	0.3370	0.2562	0.1875
	2	0.9988	0.9914	0.9734	0.9421	0.8965	0.8369	0.7648	0.6826	0.5931	0.5000
	3	1.0000	0.9995	0.9978	0.9933	0.9844	0.9692	0.9460	0.9130	0.8688	0.8125
	4	1.0000	1.0000	0.9999	0.9997	0.9990	0.9976	0.9947	0.9898	0.9815	0.9688
	5	1.0000	1.0000	1.0000	1.0000	1.0000	1.0000	1.0000	1.0000	1.0000	1.0000
6	0	0.7351	0.5314	0.3771	0.2621	0.1780	0.1176	0.0754	0.0467	0.0277	0.0156
	1	0.9672	0.8857	0.7765	0.6553	0.5339	0.4202	0.3191	0.2333	0.1636	0.1094
	2	0.9978	0.9842	0.9527	0.9011	0.8306	0.7443	0.6471	0.5443	0.4415	0.3438
	3	0.9999	0.9987	0.9941	0.9830	0.9624	0.9295	0.8826	0.8208	0.7447	0.6562
	4	1.0000	0.9999	0.9996	0.9984	0.9954	0.9891	0.9777	0.9590	0.9308	0.8906
	5	1.0000	1.0000	1.0000	0.9999	0.9998	0.9993	0.9982	0.9959	0.9917	0.9844
	6	1.0000	1.0000	1.0000	1.0000	1.0000	1.0000	1.0000	1.0000	1.0000	1.0000
7	0	0.6983	0.4783	0.3206	0.2097	0.1335	0.0824	0.0490	0.0280	0.0152	0.0078
	1	0.9556	0.8503	0.7166	0.5767	0.4449	0.3294	0.2338	0.1586	0.1024	0.0625
	2	0.9962	0.9743	0.9262	0.8520	0.7564	0.6471	0.5323	0.4199	0.3164	0.2266
	3	0.9998	0.9973	0.9879	0.9667	0.9294	0.8740	0.8002	0.7102	0.6083	0.5000
	4	1.0000	0.9998	0.9988	0.9953	0.9871	0.9712	0.9444	0.9037	0.8471	0.7734
	5	1.0000	1.0000	0.9999	0.9996	0.9987	0.9962	0.9910	0.9812	0.9643	0.9375
	6	1.0000	1.0000	1.0000	1.0000	0.9999	0.9998	0.9994	0.9984	0.9963	0.9922
	7	1.0000	1.0000	1.0000	1.0000	1.0000	1.0000	1.0000	1.0000	1.0000	1.0000

						p					
n	x	0.05	0.10	0.15	0.20	0.25	0.30	0.35	0.40	0.45	0.50
8	0	0.6634	0.4305	0.2725	0.1678	0.1001	0.0576	0.0319	0.0168	0.0084	0.0039
	1	0.9428	0.8131	0.6572	0.5033	0.3671	0.2553	0.1691	0.1064	0.0632	0.0352
	2	0.9942	0.9619	0.8948	0.7969	0.6785	0.5518	0.4278	0.3154	0.2201	0.1445
	3	0.9996	0.9950	0.9786	0.9437	0.8862	0.8059	0.7064	0.5941	0.4770	0.3633
	4	1.0000	0.9996	0.9971	0.9896	0.9727	0.9420	0.8939	0.8263	0.7396	0.6367
	5	1.0000	1.0000	0.9998	0.9988	0.9958	0.9887	0.9747	0.9502	0.9115	0.8555
	6	1.0000	1.0000	1.0000	0.9999	0.9996	0.9987	0.9964	0.9915	0.9819	0.9648
	7	1.0000	1.0000	1.0000	1.0000	1.0000	0.9999	0.9998	0.9993	0.9983	0.9961
	8	1.0000	1.0000	1.0000	1.0000	1.0000	1.0000	1.0000	1.0000	1.0000	1.0000
9	0	0.6302	0.3874	0.2316	0.1342	0.0751	0.0404	0.0207	0.0101	0.0046	0.0020
	1	0.9288	0.7748	0.5995	0.4362	0.3003	0.1960	0.1211	0.0705	0.0385	0.0195
	2	0.9916	0.9470	0.8591	0.7382	0.6007	0.4628	0.3373	0.2318	0.1495	0.0898
	3	0.9994	0.9917	0.9661	0.9144	0.8343	0.7297	0.6089	0.4826	0.3614	0.2539
	4	1.0000	0.9991	0.9944	0.9804	0.9511	0.9012	0.8283	0.7334	0.6214	0.5000
	5	1.0000	0.9999	0.9994	0.9969	0.9900	0.9747	0.9464	0.9006	0.8342	0.7461
	6	1.0000	1.0000	1.0000	0.9997	0.9987	0.9957	0.9888	0.9750	0.9502	0.9102
	7	1.0000	1.0000	1.0000	1.0000	0.9999	0.9996	0.9986	0.9962	0.9909	0.9805
	8	1.0000	1.0000	1.0000	1.0000	1.0000	1.0000	0.9999	0.9997	0.9992	0.9980
	9	1.0000	1.0000	1.0000	1.0000	1.0000	1.0000	1.0000	1.0000	1.0000	1.0000
10	0	0.5987	0.3487	0.1969	0.1074	0.0563	0.0282	0.0135	0.0060	0.0025	0.0010
	1	0.9139	0.7361	0.5443	0.3758	0.2440	0.1493	0.0860	0.0464	0.0233	0.0107
	2	0.9885	0.9298	0.8202	0.6778	0.5256	0.3828	0.2616	0.1673	0.0996	0.0547
	3	0.9990	0.9872	0.9500	0.8791	0.7759	0.6496	0.5138	0.3823	0.2660	0.1719
	4	0.9999	0.9984	0.9901	0.9672	0.9219	0.8497	0.7515	0.6331	0.5044	0.3770
	5	1.0000	0.9999	0.9986	0.9936	0.9803	0.9527	0.9051	0.8338	0.7384	0.6230
	6	1.0000	1.0000	0.9999	0.9991	0.9965	0.9894	0.9740	0.9452	0.8980	0.8281
	7	1.0000	1.0000	1.0000	0.9999	0.9996	0.9984	0.9952	0.9877	0.9726	0.9453
	8	1.0000	1.0000	1.0000	1.0000	1.0000	0.9999	0.9995	0.9983	0.9955	0.9893
	9	1.0000	1.0000	1.0000	1.0000	1.0000	1.0000	1.0000	0.9999	0.9997	0.9990
	10	1.0000	1.0000	1.0000	1.0000	1.0000	1.0000	1.0000	1.0000	1.0000	1.0000
11	0	0.5688	0.3138	0.1673	0.0859	0.0422	0.0198	0.0088	0.0036	0.0014	0.0005
	1	0.8981	0.6974	0.4922	0.3221	0.1971	0.1130	0.0606	0.0302	0.0139	0.0059
	2	0.9848	0.9104	0.7788	0.6174	0.4552	0.3127	0.2001	0.1189	0.0652	0.0327
	3	0.9984	0.9815	0.9306	0.8389	0.7133	0.5696	0.4256	0.2963	0.1911	0.1133
	4	0.9999	0.9972	0.9841	0.9496	0.8854	0.7897	0.6683	0.5328	0.3971	0.2744
	5	1.0000	0.9997	0.9973	0.9883	0.9657	0.9218	0.8513	0.7535	0.6331	0.5000
	6	1.0000	1.0000	0.9997	0.9980	0.9924	0.9784	0.9499	0.9006	0.8262	0.7256
	7	1.0000	1.0000	1.0000	0.9998	0.9988	0.9957	0.9878	0.9707	0.9390	0.8867
	8	1.0000	1.0000	1.0000	1.0000	0.9999	0.9994	0.9980	0.9941	0.9852	0.9673
	9	1.0000	1.0000	1.0000	1.0000	1.0000	1.0000	0.9998	0.9993	0.9978	0.9941
	10	1.0000	1.0000	1.0000	1.0000	1.0000	1.0000	1.0000	1.0000	0.9998	0.9995
	11	1.0000	1.0000	1.0000	1.0000	1.0000	1.0000	1.0000	1.0000	1.0000	1.0000

(continued)

Table C.2 Binomial Distribution Function (c.d.f.) (*Continued*)

n	x	0.05	0.10	0.15	0.20	0.25	0.30	0.35	0.40	0.45	0.50
						p					
12	0	0.5404	0.2824	0.1422	0.0687	0.0317	0.0138	0.0057	0.0022	0.0008	0.0002
	1	0.8816	0.6590	0.4435	0.2749	0.1584	0.0850	0.0424	0.0196	0.0083	0.0032
	2	0.9804	0.8891	0.7358	0.5583	0.3907	0.2528	0.1513	0.0834	0.0421	0.0193
	3	0.9978	0.9744	0.9078	0.7946	0.6488	0.4925	0.3467	0.2253	0.1345	0.0730
	4	0.9998	0.9957	0.9761	0.9274	0.8424	0.7237	0.5833	0.4382	0.3044	0.1938
	5	1.0000	0.9995	0.9954	0.9806	0.9456	0.8822	0.7873	0.6652	0.5269	0.3872
	6	1.0000	0.9999	0.9993	0.9961	0.9857	0.9614	0.9154	0.8418	0.7393	0.6128
	7	1.0000	1.0000	0.9999	0.9994	0.9972	0.9905	0.9745	0.9427	0.8883	0.8062
	8	1.0000	1.0000	1.0000	0.9999	0.9996	0.9983	0.9944	0.9847	0.9644	0.9270
	9	1.0000	1.0000	1.0000	1.0000	1.0000	0.9998	0.9992	0.9972	0.9921	0.9807
	10	1.0000	1.0000	1.0000	1.0000	1.0000	1.0000	0.9999	0.9997	0.9989	0.9968
	11	1.0000	1.0000	1.0000	1.0000	1.0000	1.0000	1.0000	1.0000	0.9999	0.9998
	12	1.0000	1.0000	1.0000	1.0000	1.0000	1.0000	1.0000	1.0000	1.0000	1.0000
13	0	0.5133	0.2542	0.1209	0.0550	0.0238	0.0097	0.0037	0.0013	0.0004	0.0001
	1	0.8646	0.6213	0.3983	0.2336	0.1267	0.0637	0.0296	0.0126	0.0049	0.0017
	2	0.9755	0.8661	0.6920	0.5017	0.3326	0.2025	0.1132	0.0579	0.0269	0.0112
	3	0.9969	0.9658	0.8820	0.7473	0.5843	0.4206	0.2783	0.1686	0.0929	0.0461
	4	0.9997	0.9935	0.9658	0.9009	0.7940	0.6543	0.5005	0.3530	0.2279	0.1334
	5	1.0000	0.9991	0.9924	0.9700	0.9198	0.8346	0.7159	0.5744	0.4268	0.2905
	6	1.0000	0.9999	0.9987	0.9930	0.9757	0.9376	0.8705	0.7712	0.6437	0.5000
	7	1.0000	1.0000	0.9998	0.9988	0.9944	0.9818	0.9538	0.9023	0.8212	0.7095
	8	1.0000	1.0000	1.0000	0.9998	0.9990	0.9960	0.9874	0.9679	0.9302	0.8666
	9	1.0000	1.0000	1.0000	1.0000	0.9999	0.9993	0.9975	0.9922	0.9797	0.9539
	10	1.0000	1.0000	1.0000	1.0000	1.0000	0.9999	0.9997	0.9987	0.9959	0.9888
	11	1.0000	1.0000	1.0000	1.0000	1.0000	1.0000	1.0000	0.9999	0.9995	0.9983
	12	1.0000	1.0000	1.0000	1.0000	1.0000	1.0000	1.0000	1.0000	1.0000	0.9999
	13	1.0000	1.0000	1.0000	1.0000	1.0000	1.0000	1.0000	1.0000	1.0000	1.0000
14	0	0.4877	0.2288	0.1028	0.0440	0.0178	0.0068	0.0024	0.0008	0.0002	0.0001
	1	0.8470	0.5846	0.3567	0.1979	0.1010	0.0475	0.0205	0.0081	0.0029	0.0009
	2	0.9699	0.8416	0.6479	0.4481	0.2811	0.1608	0.0839	0.0398	0.0170	0.0065
	3	0.9958	0.9559	0.8535	0.6982	0.5213	0.3552	0.2205	0.1243	0.0632	0.0287
	4	0.9996	0.9908	0.9533	0.8702	0.7415	0.5842	0.4227	0.2793	0.1672	0.0898
	5	1.0000	0.9985	0.9885	0.9561	0.8883	0.7805	0.6405	0.4859	0.3373	0.2120
	6	1.0000	0.9998	0.9978	0.9884	0.9617	0.9067	0.8164	0.6925	0.5461	0.3953
	7	1.0000	1.0000	0.9997	0.9976	0.9897	0.9685	0.9247	0.8499	0.7414	0.6047
	8	1.0000	1.0000	1.0000	0.9996	0.9978	0.9917	0.9757	0.9417	0.8811	0.7880
	9	1.0000	1.0000	1.0000	1.0000	0.9997	0.9983	0.9940	0.9825	0.9574	0.9102
	10	1.0000	1.0000	1.0000	1.0000	1.0000	0.9998	0.9989	0.9961	0.9886	0.9713
	11	1.0000	1.0000	1.0000	1.0000	1.0000	1.0000	0.9999	0.9994	0.9978	0.9935
	12	1.0000	1.0000	1.0000	1.0000	1.0000	1.0000	1.0000	0.9999	0.9997	0.9991
	13	1.0000	1.0000	1.0000	1.0000	1.0000	1.0000	1.0000	1.0000	1.0000	0.9999
	14	1.0000	1.0000	1.0000	1.0000	1.0000	1.0000	1.0000	1.0000	1.0000	1.0000

n	x	0.05	0.10	0.15	0.20	0.25	0.30	0.35	0.40	0.45	0.50
15	0	0.4633	0.2059	0.0874	0.0352	0.0134	0.0047	0.0016	0.0005	0.0001	0.0000
	1	0.8290	0.5490	0.3186	0.1671	0.0802	0.0353	0.0142	0.0052	0.0017	0.0005
	2	0.9638	0.8159	0.6042	0.3980	0.2361	0.1268	0.0617	0.0271	0.0107	0.0037
	3	0.9945	0.9444	0.8227	0.6482	0.4613	0.2969	0.1727	0.0905	0.0424	0.0176
	4	0.9994	0.9873	0.9383	0.8358	0.6865	0.5155	0.3519	0.2173	0.1204	0.0592
	5	0.9999	0.9978	0.9832	0.9389	0.8516	0.7216	0.5643	0.4032	0.2608	0.1509
	6	1.0000	0.9997	0.9964	0.9819	0.9434	0.8689	0.7548	0.6098	0.4522	0.3036
	7	1.0000	1.0000	0.9994	0.9958	0.9827	0.9500	0.8868	0.7869	0.6535	0.5000
	8	1.0000	1.0000	0.9999	0.9992	0.9958	0.9848	0.9578	0.9050	0.8182	0.6964
	9	1.0000	1.0000	1.0000	0.9999	0.9992	0.9963	0.9876	0.9662	0.9231	0.8491
	10	1.0000	1.0000	1.0000	1.0000	0.9999	0.9993	0.9972	0.9907	0.9745	0.9408
	11	1.0000	1.0000	1.0000	1.0000	1.0000	0.9999	0.9995	0.9981	0.9937	0.9824
	12	1.0000	1.0000	1.0000	1.0000	1.0000	1.0000	0.9999	0.9997	0.9989	0.9963
	13	1.0000	1.0000	1.0000	1.0000	1.0000	1.0000	1.0000	1.0000	0.9999	0.9995
	14	1.0000	1.0000	1.0000	1.0000	1.0000	1.0000	1.0000	1.0000	1.0000	1.0000
	15	1.0000	1.0000	1.0000	1.0000	1.0000	1.0000	1.0000	1.0000	1.0000	1.0000
16	0	0.4401	0.1853	0.0743	0.0281	0.0100	0.0033	0.0010	0.0003	0.0001	0.0000
	1	0.8108	0.5147	0.2839	0.1407	0.0635	0.0261	0.0098	0.0033	0.0010	0.0003
	2	0.9571	0.7892	0.5614	0.3518	0.1971	0.0994	0.0451	0.0183	0.0066	0.0021
	3	0.9930	0.9316	0.7899	0.5981	0.4050	0.2459	0.1339	0.0651	0.0281	0.0106
	4	0.9991	0.9830	0.9209	0.7982	0.6302	0.4499	0.2892	0.1666	0.0853	0.0384
	5	0.9999	0.9967	0.9765	0.9183	0.8103	0.6598	0.4900	0.3288	0.1976	0.1051
	6	1.0000	0.9995	0.9944	0.9733	0.9204	0.8247	0.6881	0.5272	0.3660	0.2272
	7	1.0000	0.9999	0.9989	0.9930	0.9729	0.9256	0.8406	0.7161	0.5629	0.4018
	8	1.0000	1.0000	0.9998	0.9985	0.9925	0.9743	0.9329	0.8577	0.7441	0.5982
	9	1.0000	1.0000	1.0000	0.9998	0.9984	0.9929	0.9771	0.9417	0.8759	0.7728
	10	1.0000	1.0000	1.0000	1.0000	0.9997	0.9984	0.9938	0.9809	0.9514	0.8949
	11	1.0000	1.0000	1.0000	1.0000	1.0000	0.9997	0.9987	0.9951	0.9851	0.9616
	12	1.0000	1.0000	1.0000	1.0000	1.0000	1.0000	0.9998	0.9991	0.9965	0.9894
	13	1.0000	1.0000	1.0000	1.0000	1.0000	1.0000	1.0000	0.9999	0.9994	0.9979
	14	1.0000	1.0000	1.0000	1.0000	1.0000	1.0000	1.0000	1.0000	1.0000	0.9997
	15	1.0000	1.0000	1.0000	1.0000	1.0000	1.0000	1.0000	1.0000	1.0000	1.0000
	16	1.0000	1.0000	1.0000	1.0000	1.0000	1.0000	1.0000	1.0000	1.0000	1.0000
17	0	0.4181	0.1668	0.0631	0.0225	0.0075	0.0023	0.0007	0.0002	0.0000	0.0000
	1	0.7922	0.4818	0.2525	0.1182	0.0501	0.0193	0.0067	0.0021	0.0006	0.0001
	2	0.9497	0.7618	0.5198	0.3096	0.1637	0.0774	0.0327	0.0123	0.0041	0.0012
	3	0.9912	0.9174	0.7556	0.5489	0.3530	0.2019	0.1028	0.0464	0.0184	0.0063
	4	0.9988	0.9779	0.9013	0.7582	0.5739	0.3887	0.2348	0.1260	0.0596	0.0245
	5	0.9999	0.9953	0.9681	0.8943	0.7653	0.5968	0.4197	0.2639	0.1471	0.0717
	6	1.0000	0.9992	0.9917	0.9623	0.8929	0.7752	0.6188	0.4478	0.2902	0.1662
	7	1.0000	0.9999	0.9983	0.9891	0.9598	0.8954	0.7872	0.6405	0.4743	0.3145
	8	1.0000	1.0000	0.9997	0.9974	0.9876	0.9597	0.9006	0.8011	0.6626	0.5000
	9	1.0000	1.0000	1.0000	0.9995	0.9969	0.9873	0.9617	0.9081	0.8166	0.6855

(continued)

Table C.2 Binomial Distribution Function (c.d.f.) (*Continued*)

n	x	0.05	0.10	0.15	0.20	0.25	0.30	0.35	0.40	0.45	0.50
	10	1.0000	1.0000	1.0000	0.9999	0.9994	0.9968	0.9880	0.9652	0.9174	0.8338
	11	1.0000	1.0000	1.0000	1.0000	0.9999	0.9993	0.9970	0.9894	0.9699	0.9283
	12	1.0000	1.0000	1.0000	1.0000	1.0000	0.9999	0.9994	0.9975	0.9914	0.9755
	13	1.0000	1.0000	1.0000	1.0000	1.0000	1.0000	0.9999	0.9995	0.9981	0.9936
	14	1.0000	1.0000	1.0000	1.0000	1.0000	1.0000	1.0000	0.9999	0.9997	0.9988
	15	1.0000	1.0000	1.0000	1.0000	1.0000	1.0000	1.0000	1.0000	1.0000	0.9999
	16	1.0000	1.0000	1.0000	1.0000	1.0000	1.0000	1.0000	1.0000	1.0000	1.0000
18	0	0.3972	0.1501	0.0536	0.0180	0.0056	0.0016	0.0004	0.0001	0.0000	0.0000
	1	0.7735	0.4503	0.2241	0.0991	0.0395	0.0142	0.0046	0.0013	0.0003	0.0001
	2	0.9419	0.7338	0.4797	0.2713	0.1353	0.0600	0.0236	0.0082	0.0025	0.0007
	3	0.9891	0.9018	0.7202	0.5010	0.3057	0.1646	0.0783	0.0328	0.0120	0.0038
	4	0.9985	0.9718	0.8794	0.7164	0.5187	0.3327	0.1886	0.0942	0.0411	0.0154
	5	0.9998	0.9936	0.9581	0.8671	0.7175	0.5344	0.3550	0.2088	0.1077	0.0481
	6	1.0000	0.9988	0.9882	0.9487	0.8610	0.7217	0.5491	0.3743	0.2258	0.1189
	7	1.0000	0.9998	0.9973	0.9837	0.9431	0.8593	0.7283	0.5634	0.3915	0.2403
	8	1.0000	1.0000	0.9995	0.9957	0.9807	0.9404	0.8609	0.7368	0.5778	0.4073
	9	1.0000	1.0000	0.9999	0.9991	0.9946	0.9790	0.9403	0.8653	0.7473	0.5927
	10	1.0000	1.0000	1.0000	0.9998	0.9988	0.9939	0.9788	0.9424	0.8720	0.7597
	11	1.0000	1.0000	1.0000	1.0000	0.9998	0.9986	0.9938	0.9797	0.9463	0.8811
	12	1.0000	1.0000	1.0000	1.0000	1.0000	0.9997	0.9986	0.9942	0.9817	0.9519
	13	1.0000	1.0000	1.0000	1.0000	1.0000	1.0000	0.9997	0.9987	0.9951	0.9846
	14	1.0000	1.0000	1.0000	1.0000	1.0000	1.0000	1.0000	0.9998	0.9990	0.9962
	15	1.0000	1.0000	1.0000	1.0000	1.0000	1.0000	1.0000	1.0000	0.9999	0.9993
	16	1.0000	1.0000	1.0000	1.0000	1.0000	1.0000	1.0000	1.0000	1.0000	0.9999
19	0	0.3774	0.1351	0.0456	0.0144	0.0042	0.0011	0.0003	0.0001	0.0000	0.0000
	1	0.7547	0.4203	0.1985	0.0829	0.0310	0.0104	0.0031	0.0008	0.0002	0.0000
	2	0.9335	0.7054	0.4413	0.2369	0.1113	0.0462	0.0170	0.0055	0.0015	0.0004
	3	0.9868	0.8850	0.6841	0.4551	0.2630	0.1332	0.0591	0.0230	0.0077	0.0022
	4	0.9980	0.9648	0.8556	0.6733	0.4654	0.2822	0.1500	0.0696	0.0280	0.0096
	5	0.9998	0.9914	0.9463	0.8369	0.6678	0.4739	0.2968	0.1629	0.0777	0.0318
	6	1.0000	0.9983	0.9837	0.9324	0.8251	0.6655	0.4812	0.3081	0.1727	0.0835
	7	1.0000	0.9997	0.9959	0.9767	0.9225	0.8180	0.6656	0.4878	0.3169	0.1796
	8	1.0000	1.0000	0.9992	0.9933	0.9713	0.9161	0.8145	0.6675	0.4940	0.3238
	9	1.0000	1.0000	0.9999	0.9984	0.9911	0.9674	0.9125	0.8139	0.6710	0.5000
	10	1.0000	1.0000	1.0000	0.9997	0.9977	0.9895	0.9653	0.9115	0.8159	0.6762
	11	1.0000	1.0000	1.0000	1.0000	0.9995	0.9972	0.9886	0.9648	0.9129	0.8204
	12	1.0000	1.0000	1.0000	1.0000	0.9999	0.9994	0.9969	0.9884	0.9658	0.9165
	13	1.0000	1.0000	1.0000	1.0000	1.0000	0.9999	0.9993	0.9969	0.9891	0.9682
	14	1.0000	1.0000	1.0000	1.0000	1.0000	1.0000	0.9999	0.9994	0.9972	0.9904
	15	1.0000	1.0000	1.0000	1.0000	1.0000	1.0000	1.0000	0.9999	0.9995	0.9978
	16	1.0000	1.0000	1.0000	1.0000	1.0000	1.0000	1.0000	1.0000	0.9999	0.9996
	17	1.0000	1.0000	1.0000	1.0000	1.0000	1.0000	1.0000	1.0000	1.0000	1.0000

n	x	\multicolumn{10}{c}{p}
		0.05	0.10	0.15	0.20	0.25	0.30	0.35	0.40	0.45	0.50
20	0	0.3585	0.1216	0.0388	0.0115	0.0032	0.0008	0.0002	0.0000	0.0000	0.0000
	1	0.7358	0.3917	0.1756	0.0692	0.0243	0.0076	0.0021	0.0005	0.0001	0.0000
	2	0.9245	0.6769	0.4049	0.2061	0.0913	0.0355	0.0121	0.0036	0.0009	0.0002
	3	0.9841	0.8670	0.6477	0.4114	0.2252	0.1071	0.0444	0.0160	0.0049	0.0013
	4	0.9974	0.9568	0.8298	0.6296	0.4148	0.2375	0.1182	0.0510	0.0189	0.0059
	5	0.9997	0.9887	0.9327	0.8042	0.6172	0.4164	0.2454	0.1256	0.0553	0.0207
	6	1.0000	0.9976	0.9781	0.9133	0.7858	0.6080	0.4166	0.2500	0.1299	0.0577
	7	1.0000	0.9996	0.9941	0.9679	0.8982	0.7723	0.6010	0.4159	0.2520	0.1316
	8	1.0000	0.9999	0.9987	0.9900	0.9591	0.8867	0.7624	0.5956	0.4143	0.2517
	9	1.0000	1.0000	0.9998	0.9974	0.9861	0.9520	0.8782	0.7553	0.5914	0.4119
	10	1.0000	1.0000	1.0000	0.9994	0.9961	0.9829	0.9468	0.8725	0.7507	0.5881
	11	1.0000	1.0000	1.0000	0.9999	0.9991	0.9949	0.9804	0.9435	0.8692	0.7483
	12	1.0000	1.0000	1.0000	1.0000	0.9998	0.9987	0.9940	0.9790	0.9420	0.8684
	13	1.0000	1.0000	1.0000	1.0000	1.0000	0.9997	0.9985	0.9935	0.9786	0.9423
	14	1.0000	1.0000	1.0000	1.0000	1.0000	1.0000	0.9997	0.9984	0.9936	0.9793
	15	1.0000	1.0000	1.0000	1.0000	1.0000	1.0000	1.0000	0.9997	0.9985	0.9941
	16	1.0000	1.0000	1.0000	1.0000	1.0000	1.0000	1.0000	1.0000	0.9997	0.9987
	17	1.0000	1.0000	1.0000	1.0000	1.0000	1.0000	1.0000	1.0000	1.0000	0.9998
	18	1.0000	1.0000	1.0000	1.0000	1.0000	1.0000	1.0000	1.0000	1.0000	1.0000

Table C.3 Poisson Distribution Function (c.d.f.)

$$P(X \le x) = \sum_{k=0}^{x} \frac{\lambda^k e^{-\lambda}}{k!}$$

$\lambda = E(X)$

x	0.1	0.2	0.3	0.4	0.5	0.6	0.7	0.8	0.9	1.0
0	0.905	0.819	0.741	0.670	0.607	0.549	0.497	0.449	0.407	0.368
1	0.995	0.982	0.963	0.938	0.910	0.878	0.844	0.809	0.772	0.736
2	1.000	0.999	0.996	0.992	0.986	0.977	0.966	0.953	0.937	0.920
3	1.000	1.000	1.000	0.999	0.998	0.997	0.994	0.991	0.987	0.981
4	1.000	1.000	1.000	1.000	1.000	1.000	0.999	0.999	0.998	0.996
5	1.000	1.000	1.000	1.000	1.000	1.000	1.000	1.000	1.000	0.999
6	1.000	1.000	1.000	1.000	1.000	1.000	1.000	1.000	1.000	1.000

x	1.1	1.2	1.3	1.4	1.5	1.6	1.7	1.8	1.9	2.0
0	0.333	0.301	0.273	0.247	0.223	0.202	0.183	0.165	0.150	0.135
1	0.699	0.663	0.627	0.592	0.558	0.525	0.493	0.463	0.434	0.406
2	0.900	0.879	0.857	0.833	0.809	0.783	0.757	0.731	0.704	0.677
3	0.974	0.966	0.957	0.946	0.934	0.921	0.907	0.891	0.875	0.857
4	0.995	0.992	0.989	0.986	0.981	0.976	0.970	0.964	0.956	0.947
5	0.999	0.998	0.998	0.997	0.996	0.994	0.992	0.990	0.987	0.983
6	1.000	1.000	1.000	0.999	0.999	0.999	0.998	0.997	0.997	0.995
7	1.000	1.000	1.000	1.000	1.000	1.000	1.000	0.999	0.999	0.999
8	1.000	1.000	1.000	1.000	1.000	1.000	1.000	1.000	1.000	1.000

x	2.2	2.4	2.6	2.8	3.0	3.2	3.4	3.6	3.8	4.0
0	0.111	0.091	0.074	0.061	0.050	0.041	0.033	0.027	0.022	0.018
1	0.355	0.308	0.267	0.231	0.199	0.171	0.147	0.126	0.107	0.092
2	0.623	0.570	0.518	0.469	0.423	0.380	0.340	0.303	0.269	0.238
3	0.819	0.779	0.736	0.692	0.647	0.603	0.558	0.515	0.473	0.433
4	0.928	0.904	0.877	0.848	0.815	0.781	0.744	0.706	0.668	0.629
5	0.975	0.964	0.951	0.935	0.916	0.895	0.871	0.844	0.816	0.785
6	0.993	0.988	0.983	0.976	0.966	0.955	0.942	0.927	0.909	0.889
7	0.998	0.997	0.995	0.992	0.988	0.983	0.977	0.969	0.960	0.949
8	1.000	0.999	0.999	0.998	0.996	0.994	0.992	0.988	0.984	0.979
9	1.000	1.000	1.000	0.999	0.999	0.998	0.997	0.996	0.994	0.992
10	1.000	1.000	1.000	1.000	1.000	1.000	0.999	0.999	0.998	0.997
11	1.000	1.000	1.000	1.000	1.000	1.000	1.000	1.000	0.999	0.999
12	1.000	1.000	1.000	1.000	1.000	1.000	1.000	1.000	1.000	1.000

					$\lambda = E(X)$					
x	4.2	4.4	4.6	4.8	5.0	5.2	5.4	5.6	5.8	6.0
0	0.015	0.012	0.010	0.008	0.007	0.006	0.005	0.004	0.003	0.002
1	0.078	0.066	0.056	0.048	0.040	0.034	0.029	0.024	0.021	0.017
2	0.210	0.185	0.163	0.143	0.125	0.109	0.095	0.082	0.072	0.062
3	0.395	0.359	0.326	0.294	0.265	0.238	0.213	0.191	0.170	0.151
4	0.590	0.551	0.513	0.476	0.440	0.406	0.373	0.342	0.313	0.285
5	0.753	0.720	0.686	0.651	0.616	0.581	0.546	0.512	0.478	0.446
6	0.867	0.844	0.818	0.791	0.762	0.732	0.702	0.670	0.638	0.606
7	0.936	0.921	0.905	0.887	0.867	0.845	0.822	0.797	0.771	0.744
8	0.972	0.964	0.955	0.944	0.932	0.918	0.903	0.886	0.867	0.847
9	0.989	0.985	0.980	0.975	0.968	0.960	0.951	0.941	0.929	0.916
10	0.996	0.994	0.992	0.990	0.986	0.982	0.977	0.972	0.965	0.957
11	0.999	0.998	0.997	0.996	0.995	0.993	0.990	0.988	0.984	0.980
12	1.000	0.999	0.999	0.999	0.998	0.997	0.996	0.995	0.993	0.991
13	1.000	1.000	1.000	1.000	0.999	0.999	0.999	0.998	0.997	0.996
14	1.000	1.000	1.000	1.000	1.000	1.000	0.999	0.999	0.999	0.999
15	1.000	1.000	1.000	1.000	1.000	1.000	1.000	1.000	1.000	0.999
16	1.000	1.000	1.000	1.000	1.000	1.000	1.000	1.000	1.000	1.000

x	6.5	7.0	7.5	8.0	8.5	9.0	9.5	10.0	11.0	12.0
0	0.002	0.001	0.001	0.000	0.000	0.000	0.000	0.000	0.000	0.000
1	0.011	0.007	0.005	0.003	0.002	0.001	0.001	0.000	0.000	0.000
2	0.043	0.030	0.020	0.014	0.009	0.006	0.004	0.003	0.001	0.001
3	0.112	0.082	0.059	0.042	0.030	0.021	0.015	0.010	0.005	0.002
4	0.224	0.173	0.132	0.100	0.074	0.055	0.040	0.029	0.015	0.008
5	0.369	0.301	0.241	0.191	0.150	0.116	0.089	0.067	0.038	0.020
6	0.527	0.450	0.378	0.313	0.256	0.207	0.165	0.130	0.079	0.046
7	0.673	0.599	0.525	0.453	0.386	0.324	0.269	0.220	0.143	0.090
8	0.792	0.729	0.662	0.593	0.523	0.456	0.392	0.333	0.232	0.155
9	0.877	0.830	0.776	0.717	0.653	0.587	0.522	0.458	0.341	0.242
10	0.933	0.901	0.862	0.816	0.763	0.706	0.645	0.583	0.460	0.347
11	0.966	0.947	0.921	0.888	0.849	0.803	0.752	0.697	0.579	0.462
12	0.984	0.973	0.957	0.936	0.909	0.876	0.836	0.792	0.689	0.576
13	0.993	0.987	0.978	0.966	0.949	0.926	0.898	0.864	0.781	0.682
14	0.997	0.994	0.990	0.983	0.973	0.959	0.940	0.917	0.854	0.772
15	0.999	0.998	0.995	0.992	0.986	0.978	0.967	0.951	0.907	0.844
16	1.000	0.999	0.998	0.996	0.993	0.989	0.982	0.973	0.944	0.899
17	1.000	1.000	0.999	0.998	0.997	0.995	0.991	0.986	0.968	0.937
18	1.000	1.000	1.000	0.999	0.999	0.998	0.996	0.993	0.982	0.963
19	1.000	1.000	1.000	1.000	0.999	0.999	0.998	0.997	0.991	0.979
20	1.000	1.000	1.000	1.000	1.000	1.000	0.999	0.998	0.995	0.988
21	1.000	1.000	1.000	1.000	1.000	1.000	1.000	0.999	0.998	0.994
22	1.000	1.000	1.000	1.000	1.000	1.000	1.000	1.000	0.999	0.997
23	1.000	1.000	1.000	1.000	1.000	1.000	1.000	1.000	1.000	0.999

Table C.4 Standard Normal Distribution Function (c.d.f.)

$$P(Z \le z) = \Phi(z) = \int_{-\infty}^{z} \frac{1}{\sqrt{2\pi}} e^{-w^2/2} dw$$

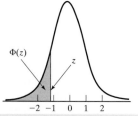

$\Phi(z)$

Normal Distribution Table (Each entry is the total area under the standard normal curve to the left of z, which is specified to two decimal places by joining the row value to the column value.)

z	0.00	0.01	0.02	0.03	0.04	0.05	0.06	0.07	0.08	0.09
−3.9	0.0000	0.0000	0.0000	0.0000	0.0000	0.0000	0.0000	0.0000	0.0000	0.0000
−3.8	0.0001	0.0001	0.0001	0.0001	0.0001	0.0001	0.0001	0.0001	0.0001	0.0001
−3.7	0.0001	0.0001	0.0001	0.0001	0.0001	0.0001	0.0001	0.0001	0.0001	0.0001
−3.6	0.0002	0.0002	0.0001	0.0001	0.0001	0.0001	0.0001	0.0001	0.0001	0.0001
−3.5	0.0002	0.0002	0.0002	0.0002	0.0002	0.0002	0.0002	0.0002	0.0002	0.0002
−3.4	0.0003	0.0003	0.0003	0.0003	0.0003	0.0003	0.0003	0.0003	0.0003	0.0002
−3.3	0.0005	0.0005	0.0005	0.0004	0.0004	0.0004	0.0004	0.0004	0.0004	0.0003
−3.2	0.0007	0.0007	0.0006	0.0006	0.0006	0.0006	0.0006	0.0005	0.0005	0.0005
−3.1	0.0010	0.0009	0.0009	0.0009	0.0008	0.0008	0.0008	0.0008	0.0007	0.0007
−3.0	0.0013	0.0013	0.0013	0.0012	0.0012	0.0011	0.0011	0.0011	0.0010	0.0010
−2.9	0.0019	0.0018	0.0018	0.0017	0.0016	0.0016	0.0015	0.0015	0.0014	0.0014
−2.8	0.0026	0.0025	0.0024	0.0023	0.0023	0.0022	0.0021	0.0021	0.0020	0.0019
−2.7	0.0035	0.0034	0.0033	0.0032	0.0031	0.0030	0.0029	0.0028	0.0027	0.0026
−2.6	0.0047	0.0045	0.0044	0.0043	0.0041	0.0040	0.0039	0.0038	0.0037	0.0036
−2.5	0.0062	0.0060	0.0059	0.0057	0.0055	0.0054	0.0052	0.0051	0.0049	0.0048
−2.4	0.0082	0.0080	0.0078	0.0075	0.0073	0.0071	0.0069	0.0068	0.0066	0.0064
−2.3	0.0107	0.0104	0.0102	0.0099	0.0096	0.0094	0.0091	0.0089	0.0087	0.0084
−2.2	0.0139	0.0136	0.0132	0.0129	0.0125	0.0122	0.0119	0.0116	0.0113	0.0110
−2.1	0.0179	0.0174	0.0170	0.0166	0.0162	0.0158	0.0154	0.0150	0.0146	0.0143
−2.0	0.0228	0.0222	0.0217	0.0212	0.0207	0.0202	0.0197	0.0192	0.0188	0.0183
−1.9	0.0287	0.0281	0.0274	0.0268	0.0262	0.0256	0.0250	0.0244	0.0239	0.0233
−1.8	0.0359	0.0351	0.0344	0.0336	0.0329	0.0322	0.0314	0.0307	0.0301	0.0294
−1.7	0.0446	0.0436	0.0427	0.0418	0.0409	0.0401	0.0392	0.0384	0.0375	0.0367
−1.6	0.0548	0.0537	0.0526	0.0516	0.0505	0.0495	0.0485	0.0475	0.0465	0.0455
−1.5	0.0668	0.0655	0.0643	0.0630	0.0618	0.0606	0.0594	0.0582	0.0571	0.0559
−1.4	0.0808	0.0793	0.0778	0.0764	0.0749	0.0735	0.0721	0.0708	0.0694	0.0681
−1.3	0.0968	0.0951	0.0934	0.0918	0.0901	0.0885	0.0869	0.0853	0.0838	0.0823
−1.2	0.1151	0.1131	0.1112	0.1093	0.1075	0.1056	0.1038	0.1020	0.1003	0.0985
−1.1	0.1357	0.1335	0.1314	0.1292	0.1271	0.1251	0.1230	0.1210	0.1190	0.1170
−1.0	0.1587	0.1562	0.1539	0.1515	0.1492	0.1469	0.1446	0.1423	0.1401	0.1379
−0.9	0.1841	0.1814	0.1788	0.1762	0.1736	0.1711	0.1685	0.1660	0.1635	0.1611
−0.8	0.2119	0.2090	0.2061	0.2033	0.2005	0.1977	0.1949	0.1921	0.1894	0.1867
−0.7	0.2420	0.2389	0.2358	0.2327	0.2296	0.2266	0.2236	0.2206	0.2177	0.2148
−0.6	0.2743	0.2709	0.2676	0.2643	0.2611	0.2578	0.2546	0.2514	0.2483	0.2451
−0.5	0.3085	0.3050	0.3015	0.2981	0.2946	0.2912	0.2877	0.2843	0.2810	0.2776
−0.4	0.3446	0.3409	0.3372	0.3336	0.3300	0.3264	0.3228	0.3192	0.3156	0.3121
−0.3	0.3821	0.3783	0.3745	0.3707	0.3669	0.3632	0.3594	0.3557	0.3520	0.3483
−0.2	0.4207	0.4168	0.4129	0.4090	0.4052	0.4013	0.3974	0.3936	0.3897	0.3859
−0.1	0.4602	0.4562	0.4522	0.4483	0.4443	0.4404	0.4364	0.4325	0.4286	0.4247
−0.0	0.5000	0.4960	0.4920	0.4880	0.4840	0.4801	0.4761	0.4721	0.4681	0.4641

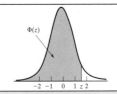
$\Phi(z)$

z	0.00	0.01	0.02	0.03	0.04	0.05	0.06	0.07	0.08	0.09
0.0	0.5000	0.5040	0.5080	0.5120	0.5160	0.5199	0.5239	0.5279	0.5319	0.5359
0.1	0.5398	0.5438	0.5478	0.5517	0.5557	0.5596	0.5636	0.5675	0.5714	0.5753
0.2	0.5793	0.5832	0.5871	0.5910	0.5948	0.5987	0.6026	0.6064	0.6103	0.6141
0.3	0.6179	0.6217	0.6255	0.6293	0.6331	0.6368	0.6406	0.6443	0.6480	0.6517
0.4	0.6554	0.6591	0.6628	0.6664	0.6700	0.6736	0.6772	0.6808	0.6844	0.6879
0.5	0.6915	0.6950	0.6985	0.7019	0.7054	0.7088	0.7123	0.7157	0.7190	0.7224
0.6	0.7257	0.7291	0.7324	0.7357	0.7389	0.7422	0.7454	0.7486	0.7517	0.7549
0.7	0.7580	0.7611	0.7642	0.7673	0.7704	0.7734	0.7764	0.7794	0.7823	0.7852
0.8	0.7881	0.7910	0.7939	0.7967	0.7995	0.8023	0.8051	0.8078	0.8106	0.8133
0.9	0.8159	0.8186	0.8212	0.8238	0.8264	0.8289	0.8315	0.8340	0.8365	0.8389
1.0	0.8413	0.8438	0.8461	0.8485	0.8508	0.8531	0.8554	0.8577	0.8599	0.8621
1.1	0.8643	0.8665	0.8686	0.8708	0.8729	0.8749	0.8770	0.8790	0.8810	0.8830
1.2	0.8849	0.8869	0.8888	0.8907	0.8925	0.8944	0.8962	0.8980	0.8997	0.9015
1.3	0.9032	0.9049	0.9066	0.9082	0.9099	0.9115	0.9131	0.9147	0.9162	0.9177
1.4	0.9192	0.9207	0.9222	0.9236	0.9251	0.9265	0.9279	0.9292	0.9306	0.9319
1.5	0.9332	0.9345	0.9357	0.9370	0.9382	0.9394	0.9406	0.9418	0.9429	0.9441
1.6	0.9452	0.9463	0.9474	0.9484	0.9495	0.9505	0.9515	0.9525	0.9535	0.9545
1.7	0.9554	0.9564	0.9573	0.9582	0.9591	0.9599	0.9608	0.9616	0.9625	0.9633
1.8	0.9641	0.9649	0.9656	0.9664	0.9671	0.9678	0.9686	0.9693	0.9699	0.9706
1.9	0.9713	0.9719	0.9726	0.9732	0.9738	0.9744	0.9750	0.9756	0.9761	0.9767
2.0	0.9772	0.9778	0.9783	0.9788	0.9793	0.9798	0.9803	0.9808	0.9812	0.9817
2.1	0.9821	0.9826	0.9830	0.9834	0.9838	0.9842	0.9846	0.9850	0.9854	0.9857
2.2	0.9861	0.9864	0.9868	0.9871	0.9875	0.9878	0.9881	0.9884	0.9887	0.9890
2.3	0.9893	0.9896	0.9898	0.9901	0.9904	0.9906	0.9909	0.9911	0.9913	0.9916
2.4	0.9918	0.9920	0.9922	0.9925	0.9927	0.9929	0.9931	0.9932	0.9934	0.9936
2.5	0.9938	0.9940	0.9941	0.9943	0.9945	0.9946	0.9948	0.9949	0.9951	0.9952
2.6	0.9953	0.9955	0.9956	0.9957	0.9959	0.9960	0.9961	0.9962	0.9963	0.9964
2.7	0.9965	0.9966	0.9967	0.9968	0.9969	0.9970	0.9971	0.9972	0.9973	0.9974
2.8	0.9974	0.9975	0.9976	0.9977	0.9977	0.9978	0.9979	0.9979	0.9980	0.9981
2.9	0.9981	0.9982	0.9982	0.9983	0.9984	0.9984	0.9985	0.9985	0.9986	0.9986
3.0	0.9987	0.9987	0.9987	0.9988	0.9988	0.9989	0.9989	0.9989	0.9990	0.9990
3.1	0.9990	0.9991	0.9991	0.9991	0.9992	0.9992	0.9992	0.9992	0.9993	0.9993
3.2	0.9993	0.9993	0.9994	0.9994	0.9994	0.9994	0.9994	0.9995	0.9995	0.9995
3.3	0.9995	0.9995	0.9995	0.9996	0.9996	0.9996	0.9996	0.9996	0.9996	0.9997
3.4	0.9997	0.9997	0.9997	0.9997	0.9997	0.9997	0.9997	0.9997	0.9997	0.9998
3.5	0.9998	0.9998	0.9998	0.9998	0.9998	0.9998	0.9998	0.9998	0.9998	0.9998
3.6	0.9998	0.9998	0.9999	0.9999	0.9999	0.9999	0.9999	0.9999	0.9999	0.9999
3.7	0.9999	0.9999	0.9999	0.9999	0.9999	0.9999	0.9999	0.9999	0.9999	0.9999
3.8	0.9999	0.9999	0.9999	0.9999	0.9999	0.9999	0.9999	0.9999	0.9999	0.9999
3.9	1.0000	1.0000	1.0000	1.0000	1.0000	1.0000	1.0000	1.0000	1.0000	1.0000

Selected Upper Percentage Points

Tail probability, α	0.100	0.050	0.025	0.010	0.005
Upper percentage point, $z(\alpha)$	1.282	1.645	1.960	2.326	2.576

Source: Reproduced in abridged form from Table 1 of E. S. Pearson and H. O. Hartley, *Biometrika Tables for Statisticians,* Vol. 1 (Cambridge: Cambridge University Press, 1954).

Table C.5 Upper Percentage Points of the Chi-Square Distribution: Values of $\chi^2(\alpha; r)$

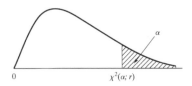

r	$\alpha = 0.995$	$\alpha = 0.99$	$\alpha = 0.975$	$\alpha = 0.95$	$\alpha = 0.05$	$\alpha = 0.025$	$\alpha = 0.01$	$\alpha = 0.005$	r
1	0.0^4393	0.0^3157	0.0^3982	0.00393	3.841	5.024	6.635	7.879	1
2	0.0100	0.0201	0.0506	0.103	5.991	7.378	9.210	10.597	2
3	0.0717	0.115	0.216	0.352	7.815	9.348	11.345	12.838	3
4	0.207	0.297	0.484	0.711	9.488	11.143	13.277	14.860	4
5	0.412	0.554	0.831	1.145	11.070	12.832	15.086	16.750	5
6	0.676	0.872	1.237	1.635	12.592	14.449	16.812	18.548	6
7	0.989	1.239	1.690	2.167	14.067	16.013	18.475	20.278	7
8	1.344	1.646	2.180	2.733	15.507	17.535	20.090	21.955	8
9	1.735	2.088	2.700	3.325	16.919	19.023	21.666	23.589	9
10	2.156	2.558	3.247	3.940	18.307	20.483	23.209	25.188	10
11	2.603	3.053	3.816	4.575	19.675	21.920	24.725	26.757	11
12	3.074	3.571	4.404	5.226	21.026	23.337	26.217	28.300	12
13	3.565	4.107	5.009	5.892	22.362	24.736	27.688	29.819	13
14	4.075	4.660	5.629	6.571	23.685	26.119	29.141	31.319	14
15	4.601	5.229	6.262	7.261	24.996	27.488	30.578	32.801	15
16	5.142	5.812	6.908	7.962	26.296	28.845	32.000	34.267	16
17	5.697	6.408	7.564	8.672	27.587	30.191	33.409	35.718	17
18	6.265	7.015	8.231	9.390	28.869	31.526	34.805	37.156	18
19	6.844	7.633	8.907	10.117	30.144	32.852	36.191	38.582	19
20	7.434	8.260	9.591	10.851	31.410	34.170	37.566	39.997	20
21	8.034	8.897	10.283	11.591	32.671	35.479	38.932	41.401	21
22	8.643	9.542	10.982	12.338	33.924	36.781	40.289	42.796	22
23	9.260	10.196	11.688	13.091	35.172	38.076	41.638	44.181	23
24	9.886	10.856	12.401	13.848	36.415	39.364	42.980	45.558	24
25	10.520	11.524	13.120	14.611	37.652	40.646	44.314	46.928	25
26	11.160	12.198	13.844	15.379	38.885	41.923	45.642	48.290	26
27	11.808	12.879	14.573	16.151	40.113	43.194	46.963	49.645	27
28	12.461	13.565	15.308	16.928	41.337	44.461	48.278	50.993	28
29	13.121	14.256	16.047	17.708	42.557	45.722	49.588	52.336	29
30	13.787	14.953	16.791	18.493	43.773	46.979	50.892	53.672	30

Source: Reproduced with permission from Table 8 of E. S. Pearson and H. O. Hartley, *Biometrika Tables for Statisticians,* Vol. 1 (Cambridge: Cambridge University Press, 1954).

Table C.6 Upper Percentage Points of the Student's t-Distribution: Values of $t(\alpha; r)$

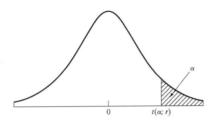

r	$\alpha = 0.10$	$\alpha = 0.05$	$\alpha = 0.025$	$\alpha = 0.01$	$\alpha = 0.005$
1	3.078	6.314	12.706	31.821	63.657
2	1.886	2.920	4.303	6.965	9.925
3	1.638	2.353	3.182	4.541	5.841
4	1.533	2.132	2.776	3.747	4.604
5	1.476	2.015	2.571	3.365	4.032
6	1.440	1.943	2.447	3.143	3.707
7	1.415	1.895	2.365	2.998	3.499
8	1.397	1.860	2.306	2.896	3.355
9	1.383	1.833	2.262	2.821	3.250
10	1.372	1.812	2.228	2.764	3.169
11	1.363	1.796	2.201	2.718	3.106
12	1.356	1.782	2.179	2.681	3.055
13	1.350	1.771	2.160	2.650	3.012
14	1.345	1.761	2.145	2.624	2.977
15	1.341	1.753	2.131	2.602	2.947
16	1.337	1.746	2.120	2.583	2.921
17	1.333	1.740	2.110	2.567	2.898
18	1.330	1.734	2.101	2.552	2.878
19	1.328	1.729	2.093	2.539	2.861
20	1.325	1.725	2.086	2.528	2.845
21	1.323	1.721	2.080	2.518	2.831
22	1.321	1.717	2.074	2.508	2.819
23	1.319	1.714	2.069	2.500	2.807
24	1.318	1.711	2.064	2.492	2.797
25	1.316	1.708	2.060	2.485	2.787
26	1.315	1.706	2.056	2.479	2.779
27	1.314	1.703	2.052	2.473	2.771
28	1.313	1.701	2.048	2.467	2.763
29	1.311	1.699	2.045	2.462	2.756
30	1.310	1.697	2.042	2.457	2.750
40	1.303	1.684	2.021	2.423	2.704
60	1.296	1.671	2.000	2.390	2.660
120	1.289	1.658	1.980	2.358	2.617
∞	1.282	1.645	1.960	2.326	2.576

Source: Reproduced with permission from Table 12 of E. S. Pearson and H. O. Hartley, *Biometrika Tables for Statisticians,* Vol. 1 (Cambridge: Cambridge University Press, 1954).

Table C.7 Upper Percentage Points of the F-Distribution: Values of $F(0.05; r_1, r_2)$

$\alpha = 0.05$

$F(0.05; r_1, r_2)$

r_2 = Degrees of freedom for denominator	r_1 = Degrees of freedom for numerator																		
	1	2	3	4	5	6	7	8	9	10	12	15	20	24	30	40	60	120	∞
1	161.4	199.5	215.7	224.6	230.2	234.0	236.8	238.9	240.5	241.9	243.9	245.9	248.0	249.1	250.1	251.1	252.2	253.3	254.3
2	18.51	19.00	19.16	19.25	19.30	19.33	19.35	19.37	19.38	19.40	19.41	19.43	19.45	19.45	19.46	19.47	19.48	19.49	19.50
3	10.13	9.55	9.28	9.12	9.01	8.94	8.89	8.85	8.81	8.79	8.74	8.70	8.66	8.64	8.62	8.59	8.57	8.55	8.53
4	7.71	6.94	6.59	6.39	6.26	6.16	6.09	6.04	6.00	5.96	5.91	5.86	5.80	5.77	5.75	5.72	5.69	5.66	5.63
5	6.61	5.79	5.41	5.19	5.05	4.95	4.88	4.82	4.77	4.74	4.68	4.62	4.56	4.53	4.50	4.46	4.43	4.40	4.36
6	5.99	5.14	4.76	4.53	4.39	4.28	4.21	4.15	4.10	4.06	4.00	3.94	3.87	3.84	3.81	3.77	3.74	3.70	3.67
7	5.59	4.74	4.35	4.12	3.97	3.87	3.79	3.73	3.68	3.64	3.57	3.51	3.44	3.41	3.38	3.34	3.30	3.27	3.23
8	5.32	4.46	4.07	3.84	3.69	3.58	3.50	3.44	3.39	3.35	3.28	3.22	3.15	3.12	3.08	3.04	3.01	2.97	2.93
9	5.12	4.26	3.86	3.63	3.48	3.37	3.29	3.23	3.18	3.14	3.07	3.01	2.94	2.90	2.86	2.83	2.79	2.75	2.71
10	4.96	4.10	3.71	3.48	3.33	3.22	3.14	3.07	3.02	2.98	2.91	2.85	2.77	2.74	2.70	2.66	2.62	2.58	2.54
11	4.84	3.98	3.59	3.36	3.20	3.09	3.01	2.95	2.90	2.85	2.79	2.72	2.65	2.61	2.57	2.53	2.49	2.45	2.40
12	4.75	3.89	3.49	3.26	3.11	3.00	2.91	2.85	2.80	2.75	2.69	2.62	2.54	2.51	2.47	2.43	2.38	2.34	2.30
13	4.67	3.81	3.41	3.18	3.03	2.92	2.83	2.77	2.71	2.67	2.60	2.53	2.46	2.42	2.38	2.34	2.30	2.25	2.21
14	4.60	3.74	3.34	3.11	2.96	2.85	2.76	2.70	2.65	2.60	2.53	2.46	2.39	2.35	2.31	2.27	2.22	2.18	2.13
15	4.54	3.68	3.29	3.06	2.90	2.79	2.71	2.64	2.59	2.54	2.48	2.40	2.33	2.29	2.25	2.20	2.16	2.11	2.07
16	4.49	3.63	3.24	3.01	2.85	2.74	2.66	2.59	2.54	2.49	2.42	2.35	2.28	2.24	2.19	2.15	2.11	2.06	2.01
17	4.45	3.59	3.20	2.96	2.81	2.70	2.61	2.55	2.49	2.45	2.38	2.31	2.23	2.19	2.15	2.10	2.06	2.01	1.96
18	4.41	3.55	3.16	2.93	2.77	2.66	2.58	2.51	2.46	2.41	2.34	2.27	2.19	2.15	2.11	2.06	2.02	1.97	1.92
19	4.38	3.52	3.13	2.90	2.74	2.63	2.54	2.48	2.42	2.38	2.31	2.23	2.16	2.11	2.07	2.03	1.98	1.93	1.88
20	4.35	3.49	3.10	2.87	2.71	2.60	2.51	2.45	2.39	2.35	2.28	2.20	2.12	2.08	2.04	1.99	1.95	1.90	1.84
21	4.32	3.47	3.07	2.84	2.68	2.57	2.49	2.42	2.37	2.32	2.25	2.18	2.10	2.05	2.01	1.96	1.92	1.87	1.81
22	4.30	3.44	3.05	2.82	2.66	2.55	2.46	2.40	2.34	2.30	2.23	2.15	2.07	2.03	1.98	1.94	1.89	1.84	1.78
23	4.28	3.42	3.03	2.80	2.64	2.53	2.44	2.37	2.32	2.27	2.20	2.13	2.05	2.01	1.96	1.91	1.86	1.81	1.76
24	4.26	3.40	3.01	2.78	2.62	2.51	2.42	2.36	2.30	2.25	2.18	2.11	2.03	1.98	1.94	1.89	1.84	1.79	1.73
25	4.24	3.39	2.99	2.76	2.60	2.49	2.40	2.34	2.28	2.24	2.16	2.09	2.01	1.96	1.92	1.87	1.82	1.77	1.71
26	4.23	3.37	2.98	2.74	2.59	2.47	2.39	2.32	2.27	2.22	2.15	2.07	1.99	1.95	1.90	1.85	1.80	1.75	1.69
27	4.21	3.35	2.96	2.73	2.57	2.46	2.37	2.31	2.25	2.20	2.13	2.06	1.97	1.93	1.88	1.84	1.79	1.73	1.67
28	4.20	3.34	2.95	2.71	2.56	2.45	2.36	2.29	2.24	2.19	2.12	2.04	1.96	1.91	1.87	1.82	1.77	1.71	1.65
29	4.18	3.33	2.93	2.70	2.55	2.43	2.35	2.28	2.22	2.18	2.10	2.03	1.94	1.90	1.85	1.81	1.75	1.70	1.64
30	4.17	3.32	2.92	2.69	2.53	2.42	2.33	2.27	2.21	2.16	2.09	2.01	1.93	1.89	1.84	1.79	1.74	1.68	1.62
40	4.08	3.23	2.84	2.61	2.45	2.34	2.25	2.18	2.12	2.08	2.00	1.92	1.84	1.79	1.74	1.69	1.64	1.58	1.51
60	4.00	3.15	2.76	2.53	2.37	2.25	2.17	2.10	2.04	1.99	1.92	1.84	1.75	1.70	1.65	1.59	1.53	1.47	1.39
120	3.92	3.07	2.68	2.45	2.29	2.17	2.09	2.02	1.96	1.91	1.83	1.75	1.66	1.61	1.55	1.50	1.43	1.35	1.25
∞	3.84	3.00	2.60	2.37	2.21	2.10	2.01	1.94	1.88	1.83	1.75	1.67	1.57	1.52	1.46	1.39	1.32	1.22	1.00

Source: Reproduced with permission from Table 18 of E. S. Pearson and H. O. Hartley, *Biometrika Tables for Statisticians*, Vol. 1 (Cambridge: Cambridge University Press, 1954).

Table C.7 Upper Percentage Points of the F-Distribution: Values of $F(0.01; r_1, r_2)$

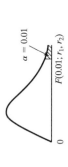

$\alpha = 0.01$

$F(0.01; r_1, r_2)$

r_2 = Degrees of freedom for denominator	r_1 = Degrees of freedom for numerator																		
	1	2	3	4	5	6	7	8	9	10	12	15	20	24	30	40	60	120	∞
1	4052	4999.5	5403	5625	5764	5859	5928	5982	6022	6056	6106	6157	6209	6235	6261	6287	6313	6339	6366
2	98.50	99.00	99.17	99.25	99.30	99.33	99.36	99.37	99.39	99.40	99.42	99.43	99.45	99.46	99.47	99.47	99.48	99.49	99.50
3	34.12	30.82	29.46	28.71	28.24	27.91	27.67	27.49	27.35	27.23	27.05	26.87	26.69	26.60	26.50	26.41	26.32	26.22	26.13
4	21.20	18.00	16.69	15.98	15.52	15.21	14.98	14.80	14.66	14.55	14.37	14.20	14.02	13.93	13.84	13.75	13.65	13.56	13.46
5	16.26	13.27	12.06	11.39	10.97	10.67	10.46	10.29	10.16	10.05	9.89	9.72	9.55	9.47	9.38	9.29	9.20	9.11	9.02
6	13.75	10.92	9.78	9.15	8.75	8.47	8.26	8.10	7.98	7.87	7.72	7.56	7.40	7.31	7.23	7.14	7.06	6.97	6.88
7	12.25	9.55	8.45	7.85	7.46	7.19	6.99	6.84	6.72	6.62	6.47	6.31	6.16	6.07	5.99	5.91	5.82	5.74	5.65
8	11.26	8.65	7.59	7.01	6.63	6.37	6.18	6.03	5.91	5.81	5.67	5.52	5.36	5.28	5.20	5.12	5.03	4.95	4.86
9	10.56	8.02	6.99	6.42	6.06	5.80	5.61	5.47	5.35	5.26	5.11	4.96	4.81	4.73	4.65	4.57	4.48	4.40	4.31
10	10.04	7.56	6.55	5.99	5.64	5.39	5.20	5.06	4.94	4.85	4.71	4.56	4.41	4.33	4.25	4.17	4.08	4.00	3.91
11	9.65	7.21	6.22	5.67	5.32	5.07	4.89	4.74	4.63	4.54	4.40	4.25	4.10	4.02	3.94	3.86	3.78	3.69	3.60
12	9.33	6.93	5.95	5.41	5.06	4.82	4.64	4.50	4.39	4.30	4.16	4.01	3.86	3.78	3.70	3.62	3.54	3.45	3.36
13	9.07	6.70	5.74	5.21	4.86	4.62	4.44	4.30	4.19	4.10	3.96	3.82	3.66	3.59	3.51	3.43	3.34	3.25	3.17
14	8.86	6.51	5.56	5.04	4.69	4.46	4.28	4.14	4.03	3.94	3.80	3.66	3.51	3.43	3.35	3.27	3.18	3.09	3.00
15	8.68	6.36	5.42	4.89	4.56	4.32	4.14	4.00	3.89	3.80	3.67	3.52	3.37	3.29	3.21	3.13	3.05	2.96	2.87
16	8.53	6.23	5.29	4.77	4.44	4.20	4.03	3.89	3.78	3.69	3.55	3.41	3.26	3.18	3.10	3.02	2.93	2.84	2.75
17	8.40	6.11	5.18	4.67	4.34	4.10	3.93	3.79	3.68	3.59	3.46	3.31	3.16	3.08	3.00	2.92	2.83	2.75	2.65
18	8.29	6.01	5.09	4.58	4.25	4.01	3.84	3.71	3.60	3.51	3.37	3.23	3.08	3.00	2.92	2.84	2.75	2.66	2.57
19	8.18	5.93	5.01	4.50	4.17	3.94	3.77	3.63	3.52	3.43	3.30	3.15	3.00	2.92	2.84	2.76	2.67	2.58	2.49
20	8.10	5.85	4.94	4.43	4.10	3.87	3.70	3.56	3.46	3.37	3.23	3.09	2.94	2.86	2.78	2.69	2.61	2.52	2.42
21	8.02	5.78	4.87	4.37	4.04	3.81	3.64	3.51	3.40	3.31	3.17	3.03	2.88	2.80	2.72	2.64	2.55	2.46	2.36
22	7.95	5.72	4.82	4.31	3.99	3.76	3.59	3.45	3.35	3.26	3.12	2.98	2.83	2.75	2.67	2.58	2.50	2.40	2.31
23	7.88	5.66	4.76	4.26	3.94	3.71	3.54	3.41	3.30	3.21	3.07	2.93	2.78	2.70	2.62	2.54	2.45	2.35	2.26
24	7.82	5.61	4.72	4.22	3.90	3.67	3.50	3.36	3.26	3.17	3.03	2.89	2.74	2.66	2.58	2.49	2.40	2.31	2.21
25	7.77	5.57	4.68	4.18	3.85	3.63	3.46	3.32	3.22	3.13	2.99	2.85	2.70	2.62	2.54	2.45	2.36	2.27	2.17
26	7.72	5.53	4.64	4.14	3.82	3.59	3.42	3.29	3.18	3.09	2.96	2.81	2.66	2.58	2.50	2.42	2.33	2.23	2.13
27	7.68	5.49	4.60	4.11	3.78	3.56	3.39	3.26	3.15	3.06	2.93	2.78	2.63	2.55	2.47	2.38	2.29	2.20	2.10
28	7.64	5.45	4.57	4.07	3.75	3.53	3.36	3.23	3.12	3.03	2.90	2.75	2.60	2.52	2.44	2.35	2.26	2.17	2.06
29	7.60	5.42	4.54	4.04	3.73	3.50	3.33	3.20	3.09	3.00	2.87	2.73	2.57	2.49	2.41	2.33	2.23	2.14	2.03
30	7.56	5.39	4.51	4.02	3.70	3.47	3.30	3.17	3.07	2.98	2.84	2.70	2.55	2.47	2.39	2.30	2.21	2.11	2.01
40	7.31	5.18	4.31	3.83	3.51	3.29	3.12	2.99	2.89	2.80	2.66	2.52	2.37	2.29	2.20	2.11	2.02	1.92	1.80
60	7.08	4.98	4.13	3.65	3.34	3.12	2.95	2.82	2.72	2.63	2.50	2.35	2.20	2.12	2.03	1.94	1.84	1.73	1.60
120	6.85	4.79	3.95	3.48	3.17	2.96	2.79	2.66	2.56	2.47	2.34	2.19	2.03	1.95	1.86	1.76	1.66	1.53	1.38
∞	6.63	4.61	3.78	3.32	3.02	2.80	2.64	2.51	2.41	2.32	2.18	2.04	1.88	1.79	1.70	1.59	1.47	1.32	1.00

Table C.8 Upper Percentage Points of the Studentized Range Distribution: Values of $q(0.05; k, \nu)$

Degrees of Freedom ν	Number of Treatments k								
	2	3	4	5	6	7	8	9	10
1	18.0	27.0	32.8	37.2	40.5	43.1	45.4	47.3	49.1
2	6.09	8.33	9.80	10.89	11.73	12.43	13.03	13.54	13.99
3	4.50	5.91	6.83	7.51	8.04	8.47	8.85	9.18	9.46
4	3.93	5.04	5.76	6.29	6.71	7.06	7.35	7.60	7.83
5	3.64	4.60	5.22	5.67	6.03	6.33	6.58	6.80	6.99
6	3.46	4.34	4.90	5.31	5.63	5.89	6.12	6.32	6.49
7	3.34	4.16	4.68	5.06	5.35	5.59	5.80	5.99	6.15
8	3.26	4.04	4.53	4.89	5.17	5.40	5.60	5.77	5.92
9	3.20	3.95	4.42	4.76	5.02	5.24	5.43	5.60	5.74
10	3.15	3.88	4.33	4.66	4.91	5.12	5.30	5.46	5.60
11	3.11	3.82	4.26	4.58	4.82	5.03	5.20	5.35	5.49
12	3.08	3.77	4.20	4.51	4.75	4.95	5.12	5.27	5.40
13	3.06	3.73	4.15	4.46	4.69	4.88	5.05	5.19	5.32
14	3.03	3.70	4.11	4.41	4.64	4.83	4.99	5.13	5.25
15	3.01	3.67	4.08	4.37	4.59	4.78	4.94	5.08	5.20
16	3.00	3.65	4.05	4.34	4.56	4.74	4.90	5.03	5.15
17	2.98	3.62	4.02	4.31	4.52	4.70	4.86	4.99	5.11
18	2.97	3.61	4.00	4.28	4.49	4.67	4.83	4.96	5.07
19	2.96	3.59	3.98	4.26	4.47	4.64	4.79	4.92	5.04
20	2.95	3.58	3.96	4.24	4.45	4.62	4.77	4.90	5.01
24	2.92	3.53	3.90	4.17	4.37	4.54	4.68	4.81	4.92
30	2.89	3.48	3.84	4.11	4.30	4.46	4.60	4.72	4.83
40	2.86	3.44	3.79	4.04	4.23	4.39	4.52	4.63	4.74
60	2.83	3.40	3.74	3.98	4.16	4.31	4.44	4.55	4.65
120	2.80	3.36	3.69	3.92	4.10	4.24	4.36	4.47	4.56
∞	2.77	3.32	3.63	3.86	4.03	4.17	4.29	4.39	4.47

Source: Reproduced with permission from Table 29 of E. S. Pearson and H. O. Hartley, *Biometrika Tables for Statisticians,* Vol. 1 (Cambridge: Cambridge University Press, 1954).

Table C.9 Factors k for Tolerance Limits

n	$1 - \alpha = 0.95$			$1 - \alpha = 0.99$		
	$p = 0.90$	$p = 0.95$	$p = 0.99$	$p = 0.90$	$p = 0.95$	$p = 0.99$
2	32.019	37.674	48.430	160.193	188.491	242.300
3	8.380	9.916	12.861	18.930	22.401	29.055
4	5.369	6.370	8.299	9.398	11.150	14.527
5	4.275	5.079	6.634	6.612	7.855	10.260
6	3.712	4.414	5.775	5.337	6.345	8.301
7	3.369	4.007	5.248	4.613	5.488	7.187
8	3.136	3.732	4.891	4.147	4.936	6.468
9	2.967	3.532	4.631	3.822	4.550	5.966
10	2.839	3.379	4.433	3.582	4.265	5.594
11	2.737	3.259	4.277	3.397	4.045	5.308
12	2.655	3.162	4.150	3.250	3.870	5.079
13	2.587	3.081	4.044	3.130	3.727	4.893
14	2.529	3.012	3.955	3.029	3.608	4.737
15	2.480	2.954	3.878	2.945	3.507	4.605
16	2.437	2.903	3.812	2.872	3.421	4.492
17	2.400	2.858	3.754	2.808	3.345	4.393
18	2.366	2.819	3.702	2.753	3.279	4.307
19	2.337	2.784	3.656	2.703	3.221	4.230
20	2.310	2.752	3.615	2.659	3.168	4.161
25	2.208	2.631	3.457	2.494	2.972	3.904
30	2.140	2.549	3.350	2.385	2.841	3.733
35	2.090	2.490	3.272	2.306	2.748	3.611
40	2.052	2.445	3.213	2.247	2.677	3.518
45	2.021	2.408	3.165	2.200	2.621	3.444
50	1.996	2.379	3.126	2.162	2.576	3.385
55	1.976	2.354	3.094	2.130	2.538	3.335
60	1.958	2.333	3.066	2.103	2.506	3.293
65	1.943	2.315	3.042	2.080	2.478	3.257
70	1.929	2.299	3.021	2.060	2.454	3.225
75	1.917	2.285	3.002	2.042	2.433	3.197
80	1.907	2.272	2.986	2.026	2.414	3.173
85	1.897	2.261	2.971	2.012	2.397	3.150
90	1.889	2.251	2.958	1.999	2.382	3.130
95	1.881	2.241	2.945	1.987	2.368	3.112
100	1.874	2.233	2.934	1.977	2.355	3.096
150	1.825	2.175	2.859	1.905	2.270	2.983
200	1.798	2.143	2.816	1.865	2.222	2.921
250	1.780	2.121	2.788	1.839	2.191	2.880
300	1.767	2.106	2.767	1.820	2.169	2.850
400	1.749	2.084	2.739	1.794	2.138	2.809
500	1.737	2.070	2.721	1.777	2.117	2.783
600	1.729	2.060	2.707	1.764	2.102	2.763
700	1.722	2.052	2.697	1.755	2.091	2.748
800	1.717	2.046	2.688	1.747	2.082	2.736
900	1.712	2.040	2.682	1.741	2.075	2.726
1000	1.709	2.036	2.676	1.736	2.068	2.718
∞	1.645	1.960	2.576	1.645	1.960	2.576

Source: Reprinted with permission from C. Eisenhart, M. W. Hastay, and W. A. Wallis, *Selected Techniques of Statistical Analysis* (New York: McGraw-Hill, 1947), p. 102.

Table C.10 2,000 Random Digits

98086	24826	45240	28404	44999	08896	39094	73407	35441	31880
33185	16232	41941	50949	89435	48581	88695	41994	37548	73043
80951	00406	96382	70774	20151	23387	25016	25298	94624	61171
79752	49140	71961	28296	69861	02591	74852	20539	00387	59579
18633	32537	98145	06571	31010	24674	05455	61427	77938	91936
74029	43902	77557	32270	97790	17119	52527	58021	80814	51748
54178	45611	80993	37143	05335	12969	56127	19255	36040	90324
11664	49883	52079	84827	59381	71539	09973	33440	88461	23356
48324	77928	31249	64710	02295	36870	32307	57546	15020	09994
69074	94138	87637	91976	35584	04401	10518	21615	01848	76938
09188	20097	32825	39527	04220	86304	83389	87374	64278	58044
90045	85497	51981	50654	94938	81997	91870	76150	68476	64659
73189	50207	47677	26269	62290	64464	27124	67018	41361	82760
75768	76490	20971	87749	90429	12272	95375	05871	93823	43178
54016	44056	66281	31003	00682	27398	20714	53295	07706	17813
08358	69910	78542	42785	13661	58873	04618	97553	31223	08420
28306	03264	81333	10591	40510	07893	32604	60475	94119	01840
53840	86233	81594	13628	51215	90290	28466	68795	77762	20791
91757	53741	61613	62269	50263	90212	55781	76514	83483	47055
89415	92694	00397	58391	12607	17646	48949	72306	94541	37408
77513	03820	86864	29901	68414	82774	51908	13980	72893	55507
19502	37174	69979	20288	55210	29773	74287	75251	65344	67415
21818	59313	93278	81757	05686	73156	07082	85046	31853	38452
51474	66499	68107	23621	94049	91345	42836	09191	08007	45449
99559	68331	62535	24170	69777	12830	74819	78142	43860	72834
33713	48007	93584	72869	51926	64721	58303	29822	93174	93972
85274	86893	11303	22970	28834	34137	73515	90400	71148	43643
84133	89640	44035	52166	73852	70091	61222	60561	62327	18423
56732	16234	17395	96131	10123	91622	85496	57560	81604	18880
65138	56806	87648	85261	34313	65861	45875	21069	85644	47277
38001	02176	81719	11711	71602	92937	74219	64049	65584	49698
37402	96397	01304	77586	56271	10086	47324	62605	40030	37438
97125	40348	87083	31417	21815	39250	75237	62047	15501	29578
21826	41134	47143	34072	64638	85902	49139	06441	03856	54552
73135	42742	95719	09035	85794	74296	08789	88156	64691	19202
07638	77929	03061	18072	96207	44156	23821	99538	04713	66994
60528	83441	07954	19814	59175	20695	05533	52139	61212	06455
83596	35655	06958	92983	05128	09719	77433	53783	92301	50498
10850	62746	99599	10507	13499	06319	53075	71839	06410	19362
39820	98952	43622	63147	64421	80814	43800	09351	31024	73167

Source: Reprinted with permission from pages 1–2 of *A Million Random Digits with 100,000 Normal Deviates,* by The Rand Corporation. New York: The Free Press, 1955. Used with permission.

D

LIST OF FILES (MINITAB AND TEXT FILES)

Chapter 1

Section1.2Exercise1.2-1Thermostat
Section1.2Temperatures
Section1.3Exercise1.3-2MeltingPoint
Section1.3Exercise1.3-3Lead1976
Section1.3Exercise1.3-8Thickness
Section1.3Exercise1.3-13Hurricane
Section1.3Exercise1.3-14BatchYield
Section1.3Exercise1.3-18Survey
Section1.3Table1.3-1Strength
Section1.3TestScoresN=58
Section1.3TestScoresN=44
Section1.4Exercise1.4-3Jaffe
Section1.4Exercise1.4-6TwoProcesses
Section1.4Table1.4-1MisfeedingLeads
Section1.4LakeNeusiedl
Section1.4Lead1976&1977
Section1.5Exercise1.5-4Fisher
Section1.5Exercise1.5-5Tukey
Section1.5Exercise1.5-6AirPollution
Section1.5Exercise1.5-7Time
Section1.5Exercise1.5-8ACT
Section1.5Exercise1.5-9Salary
Section1.5Table1.5-1Cars
Chapter1Project1MetalCutting
Chapter1Project2IowaFatalities
Chapter1Project2IowaVMT
Chapter1Project3Trucks
Chapter1Project9NHTemp

Chapter 3

Section3.1Exercise3.1-2Hours
Section3.1Exercise3.1-3Wind
Section3.6Exercise3.6-1Observations
Section3.6Exercise3.6-11Surgery

Chapter 4

Chapter4Project7ZieglerStudy1
Chapter4Project10Bootstrap
Chapter4Project11Permutation
Chapter4Project14Trucks

Chapter 5

Section5.1Exercise5.1-1Grant
Section5.1Exercise5.1-2Astro
Section5.1Exercise5.1-10Cartons
Section5.1Table5.1-1Chart
Section5.2Table5.2-1Capability
Chapter5Project1Breeding

Chapter 6

Section6.1Exercise6.1-2Cuckoo
Section6.1Exercise6.1-3Strength
Section6.1Exercise6.1-6Strength
Section6.1Exercise6.1-10GPA
Section6.1Exercise6.1-11Salary
Section6.1Table6.1-2Deflection
Section6.2Exercise6.2-4Bakery
Section6.2Exercise6.2-9Youden1
Section6.2Exercise6.2-10Youden2
Section6.3Exercise6.3-1WearTester
Section6.3Exercise6.3-2Resistors
Section6.3Exercise6.3-3Productivity
Section6.3Exercise6.3-4Reaction
Section6.3Exercise6.3-6Sodium
Section6.3Table6.3-1Strength
Section6.4Exercise6.4-2Marigold
Section6.4Exercise6.4-3Tire
Section6.4Exercise6.4-4Cheese
Section6.4Table6.4-5Yield

Chapter 7

Section7.1Exercise7.1-3Sales
Section7.1Exercise7.1-4Brightness
Section7.1Exercise7.1-5Break
Section7.1Exercise7.1-9StressTest
Section7.1Table7.1-5Popcorn
Section7.2Exercise7.2-1Pigment
Section7.2Exercise7.2-4SteelBeam
Section7.2Table7.2-1Rod
Section7.3Exercise7.3-4Conversion
Section7.3Exercise7.3-5Smoothness
Section7.3Exercise7.3-6Loss
Section7.3Exercise7.3-7Impurity
Section7.3Exercise7.3-9Yield
Section7.3Exercise7.3-10Conversion
Section7.3Exercise7.3-11Meredith
Section7.3Table7.3-5Fabric
Section7.4Exercise7.4-8Color
Section7.4Exercise7.4-9Viscosity
Section7.4Table7.4-2FractFact1
Section7.4Table7.4-3FractFact2
Chapter7Project6MotherJones

Chapter 8

Section8.1Exercise8.1-1Bets
Section8.1Exercise8.1-2Yield
Section8.1Exercise8.1-3Snedecor
Section8.1Table8.1-1Cars
Section8.3Exercise8.3-2Anscombe
Section8.3Exercise8.3-5Aerosol
Section8.3Exercise8.3-6Cars
Section8.3Exercise8.3-8Enrollment
Section8.3Exercise8.3-9Eggs
Section8.3Exercise8.3-11WeldStrength
Section8.3Table8.3-1Steam
Section8.4Exercise8.4-4CarsNew
Section8.4Exercise8.4-7TrainStoppingDistances
Section8.5Exercise8.5-1GrowthRate
Section8.5Exercise8.5-2CloudPoint
Section8.5Exercise8.5-5Soil
Section8.5Exercise8.5-6GPA
Section8.5Exercise8.5-7Traction
Section8.5Exercise8.5-8Oxygen
Section8.5Exercise8.5-10Yield
Section8.5Exercise8.5-11ToolLife

Section8.5Exercise8.5-15Trees
Section8.5Table8.5-6Durability
Section8.6Exercise8.6-3Yield
Section8.6Exercise8.6-4Reaction
Section8.6Exercise8.6-5PercYield
Section8.6Exercise8.6-7Synthesis
Chapter8Project2Wine
Chapter8Project4Neusiedl
Chapter8Project5Beans
Chapter8Project8Assay

INDEX

Design of experiments

Design	Model
Completely randomized one-factor experiment: Assign experimental units to the k treatment groups at random	$Y_{ij} = \mu + \tau_i + \varepsilon_{ij}$ $j = 1, 2, \ldots, n_i; i = 1, 2, \ldots, k;$ $\quad \sum n_i = N$ ε_{ij} independent $N(0, \sigma^2)$
Randomized complete block experiment: Group kb units into b homogeneous blocks of size k. Assign the units in each block to the k treatments at random	$Y_{ij} = \mu + \tau_i + \beta_j + \varepsilon_{ij}$ ε_{ij} independent $N(0, \sigma^2)$
Two-factor factorial experiment: Assign n units to each of ab combinations of factors A and B at random	$Y_{ijk} = \mu + \alpha_i + \beta_j + (\alpha\beta)_{ij} + \varepsilon_{ijk}$ ε_{ij} independent $N(0, \sigma^2)$
Nested factors: The b levels of factor B are nested within a levels of factor A. The a batches and b samples from each batch are chosen at random; n measurements are taken from each group	$Y_{ijk} = \mu + \alpha_i + \beta_{j(i)} + \varepsilon_{k(i,j)}$ α_i independent $N(0, \sigma_\alpha^2)$ $\beta_{j(i)}$ independent $N(0, \sigma_\beta^2)$ $\varepsilon_{k(i,j)}$ independent $N(0, \sigma^2)$

2^k factorial and 2^{k-p} fractional factorial experiments: k factors, each at a low $(-)$ and high $(+)$ level. For example, 2^3 and 2^{4-1}. In the latter, use 2^3 and take the fourth factor by $x_4 = x_1 x_2 x_3$. Various products, like $x_2 x_3$, give the signs that apply to the corresponding responses in determining effects, like (23).

Run	x_1	x_2	x_3	Response
1	$-$	$-$	$-$	Y_1
2	$+$	$-$	$-$	Y_2
3	$-$	$+$	$-$	Y_3
4	$+$	$+$	$-$	Y_4
5	$-$	$-$	$+$	Y_5
6	$+$	$-$	$+$	Y_6
7	$-$	$+$	$+$	Y_7
8	$+$	$+$	$+$	Y_8